4/9/91

BIOGEOGRAPHY OF THE WEST INDIES
PAST, PRESENT, AND FUTURE

BIOGEOGRAPHY OF THE WEST INDIES

PAST, PRESENT, AND FUTURE

Edited by CHARLES A. WOODS

Florida Museum of Natural History
Gainesville, Florida

WITH 50 CONTRIBUTORS

SANDHILL CRANE PRESS, INC.
Gainesville, Florida

1989

BIOGEOGRAPHY OF THE WEST INDIES: PAST, PRESENT, FUTURE

Acquisitions and Production Editor: Ross H. Arnett, Jr.

LIBRARY OF CONGRESS CATALOGING IN PUBLICATION DATA:

Biogeography of the West Indies : past, present, and future / edited by Charles A. Woods.
      p. cm.
   Includes bibliographical references.
   ISBN 1-877743-03-8 : $125.00
     1. Biogeography--West Indies. 2. Natural history--West Indies
3. Man--Influence on nature--West Indies. I. Woods, Charles A.
(Charles Arthur)
QH109.A1B56 1989
508.729--dc20   89-24172
                      CIP

Copyright 1989 by Sandhill Crane Press, Inc. All rights reserved. No part of this book may be reproduced in any form or by any means, without permission in writing from the publisher.

ISBN 1-877743-03-8

Manufactured in the United States of America

TABLE OF CONTENTS

Introduction vii
   Charles A. Woods

1. Old problems and new opportunities in West Indian biogeography 1
   Ernest E. Williams

2. Geological constraints and biological retrodictions in the evolution of the Caribbean Sea and its islands 47
   Michael R. Perfit and *Ernest E. Williams*

3. History of marine barriers and terrestrial connections: Caribbean paleogeographic inference from pelagic sediment analysis 103
   Thomas W. Donnelly

4. Peopling and repeopling of the West Indies 119
   Irving Rouse

5. Human exploitation of animal resources in the Caribbean 137
   Elizabeth S. Wing

6. Archaeological implications for Lesser Antilles biogeography: the small island perspective 153
   David R. Watters

7. Biogeography and evolution of the junipers of the West Indies 167
   Robert P. Adams

8. Recent vegetation changes in southern Haiti 191
   Antonia Higuera-Gundy

9. Quaternary biogeographical history of land snails in Jamaica 201
   Glenn A. Goodfriend

10. The potential use of amber fossils in the study of the biogeography of spiders in the Caribbean with the description of a new species of *Lyssomanes* from Dominican amber (Araneae: Salticidae) 217
   Jonathan Reiskind

11. The biogeography of West Indian butterflies (Lepidoptera: Papilionoidea, Hesperioidea): a vicariance model 229
    Lee D. Miller and Jacqueline Y. Miller

12. Zoogeography of the Antillean freshwater fish fauna 263
    George H. Burgess and Richard Franz

13. Evolution and biogeography of West Indian frogs of the genus *Eleutherodactylus*: slow-evolving loci and the major groups 305
    S. Blair Hedges

14. Phylogenetic relationships of the West Indian frogs of the genus *Eleutherodactylus*: a morphological analysis 371
    Rafeal L. Joglar

15. The relationships of Antillean *Typhlops* (Serpentes: Typhlopidae) and the description of three new Hispaniolan species 409
    Richard Thomas

16. A critique of Guyer and Savage (1986): cladistic relationships among anoles (Sauria: Iguanidae): are the data available to reclassify the anoles? 433
    Ernest E. Williams

17. Biogeographic patterns of predation in West Indian colubrid snakes 479
    Robert W. Henderson and Brian I. Crother

18. The shiny cowbird *Molothrus bonariensis* in the West Indian region - biogeographical and ecological implications 519
    Alexander Cruz, James W. Wiley, Tammie Nakamura, and William Post

19. The ecology of native and introduced granivorous birds in Puerto Rico 541
    Herbert A. Raffaele

20. Distribution, status, and biogeography of the West Indian manatee 567
    L. W. Lefebvre, T. J. O'Shea, G. B. Rathbum, and R. C. Best

21. Biogeography and population biology of the mongoose in the West Indies 611
    Donald B. Hoagland, G. Roy Horst, and C. William Kilpatrick

22. A review and analysis of the bats of the West Indies 635
    Karl F. Koopman

23. Distribution and systematics of bats in the Lesser Antilles 645
    J. Knox Jones, Jr.

24. Caribbean Island Zoography: A new approach using mitochondrial DNA to study Neotropical bats 661
Carleton J. Phillips, Dorothy E. Pumo, Hugh H. Genoways, and Phillip E. Ray

25. Fossil Chiroptera and Rodentia from the Bahamas, and the historical biogeography of the Bahamian mammal fauna 685
Gary S. Morgan

26. The biogeography of West Indian rodents 741
Charles A. Woods

27. The land mammals of Madagascar and the Greater Antilles: comparison and analysis 799
Charles A. Woods and John F. Eisenberg

28. Conservation trends and the threats to endemic birds in Jamaica 827
A. M. Haynes, R. K. Sutton, and Karen D. Harvey

29. A summary of conservation trends in the Bahamas 839
K. C. Jordan

30. A summary of conservation trends in the Dominican Republic 845
Jose Alberto Ottenwalder

31. A summary of conservation trends in Puerto Rico 851
Peter R. Ortiz

32. Conservation strategies and the preservation of biological diversity in Haiti 855
Paul Paryski, Charles A. Woods, and Florence Sergile

INTRODUCTION TO WEST INDIAN BIOGEOGRAPHY

Charles A. Woods[1]

Glover M. Allen (1911), Harold E. Anthony (1916), Thomas Barbour (1914), and W.D. Matthew (1915) began the systematic study of the biogeography of the West Indies. The early Twentieth Century was also the time that William L. Abbott made extensive collections of vertebrates in remote regions of Hispaniola (1916-1923), and Erik L. Ekman collected plants in Cuba and Hispaniola (See Ekman 1926, 1928). These collections led to the descriptions of many new species, and brought to light the realization of how diverse the flora and fauna of the West Indies is, with pockets of endemism occurring even on the same islands. The collections and published works of these biologists provided the first attempts to formulate theories on the biogeography of the West Indies.

The occupation of Haiti by the U.S. Marines (1915-1934), and the building of programs in agriculture modeled after the ones in place in the United States, provided an ideal opportunity for a number of American biologists to work in Haiti and other regions of the West Indies during the 1920s and 1930s. It was during this period that Philip Darlington (1935) journeyed to the top of Pic Macaya (October 1934), James Bond began a series of extensive field trips in the Massif de la Selle and other important regions of Haiti in 1927-28, Garrit S. Miller, Jr. (1930) and his associates from the Smithsonian collected fossil mammals throughout Hispaniola, Erik Ekman made his greatest botanical discoveries (Fig. 1), and Alexander Wetmore traveled to some of the most remote reaches of Hispaniola. There is a fine historical review of early biological work in the Greater Antilles (especially Hispaniola) in Wetmore and Swales (1931). The plant and animal collections made during this period were deposited in museums in Berlin, Hamburg, and Leiden, as well as the Philadelphia Academy of Natural Sciences, the Harvard Museum of Comparative Zoology, and the Smithsonian Institution.

These collections are the cornerstone of our present understanding of the flora and fauna of the Antilles. The analyses of these, and other collections, led to the first great syntheses of West Indian biogeography, such as the works of Darlington (1938, 1957) and Simpson (1940, 1943, 1956). The standard reference on the geologic history of the West Indies at the time was Schuchert (1935), who viewed the islands of the West Indies as having been relatively stable in geographic position throughout the Cenozoic.

A series of rapid advances in our understanding of the geologic history of the West Indies were made in the 1950s and 1960s. These revelations about the dynamic geologic history of the Antilles led to bold new syntheses. The work by Rosen (1976) ignited a firestorm of controversy as to how and when organisms dispersed from island to island,

---

[1] Charles Woods is Curator of Mammals at the Florida Museum of Natural History, University of Florida, Gainesville, FL 32611.

and what role plate tectonics played in biogeography. Biologists and geologists began to read one anothers papers with a keen interest as they searched for further evidence as to the history of the Antilles. The results of the geologic data are summarized in Pindell and Dewey (1982), while Pregill (1981a) and Hedges (1982) discuss the importance of the recent studies in plate tectonics to West Indian biogeography. Rosen (1985) made a second attempt to synthesize data from geology and biology into patterns that might explain West Indian biogeography. In addition, new collections of organisms from all regions of the Antilles began to bring to light new taxa of striking importance. The works of Storrs Olson (1976) on birds and Greg Pregill (1981b) on reptiles indicated that the history of some vertebrates in the Antilles may be more ancient than previously anticipated. The recent finds of fossil frogs, reptiles, and mammalian hair in amber from the Dominican Republic confirm these hypotheses (Poinar and Cannatella 1987). Most importantly, a series of new biochemical techniques, as well as the more widespread use of cladistics has made it possible for more rigorous testing of the various hypotheses on West Indian biogeography.

It is with these recent developments in mind, and the encouragement of a number of younger biologists with a significant body of new data on the flora and fauna of the West Indies, that Hugh Genoways and I decided to bring together as many investigators as possible for a symposium to explore all aspects of West Indian biogeography. We decided to encourage the broadest possible interpretations of the concepts of biogeography, and to make a place for historical factors such as anthropology and geology, as well as discussions of the conservation of endemic species. Our reasoning for this decision was that without taking into consideration the influence of changing landscapes due to past geologic events or human activities it would be impossible to develop an accurate picture of the significance to the modern distributions of organisms. Without considering ecology and conservation, it would be impossible to pass on the torch of West Indian biogeography to future generations of biologists with an interest in island biology. Therefore, we organized a symposium on the Biogeography of the West Indies: Past, Present, and Future. The symposium was held at the Florida State Museum (now Florida Museum of Natural History) on March 2-5, 1987. The proceedings of this symposium have been edited, and are presented here as a review of the status of biogeography in the West Indies in the 1980s. The book is designed to serve as a synthesis of our present knowledge of a broad range of topics of relevance to West Indian biogeography. It is also designed to be a beacon into the future to encourage further work in west Indian biogeography, and to encourage conservation of endemic species in the fragile but resilient West Indian Archipelago.

The papers are organized so that so that historical events are reviewed first. Where possible more than one authority has been selected to review a topic so that a variety of organisms and techniques can be covered. Ernest Williams, who has been studying the biogeography of the West Indies longer than any other active biologist, begins the book with a historical review that is intended to link earlier studies with suggestions for future directions. Professor Williams and Michael Perfit join together in the following chapter to analyze the joint significance of geologic events and biological data. Donnelly offers additional insight into geologic events based on new data from marine sediments. These chapters link geology and biology. The link between early human activities and biology are presented in chapters by Rouse on human history in the West Indies, Watters on the Lesser Antilles, and Wing on human exploitation of the fauna of the West Indies.

Biological aspects of West Indian biogeography are discussed in a series of papers that start with an analysis of West Indian junipers by Adams, and recent vegetational changes in Hispaniola by Gundy. The history of invertebrates in the Antilles is covered by Goodfriend (snails), Reiskind (fossil spiders), and Miller and Miller (butterflies). I feel fortunate to have been able to include these studies on botany and invertebrates in the book. They provide an excellent beginning for what I hope in the future will be a much larger body of data. Having reviewed much of the available data on West Indian biogeography, I have been struck by how few studies on plants and invertebrates there are in the overall body of data on West Indian biogeography compared with the information available on other groups.

Studies on the biology of vertebrates provide the largest body of knowledge in the study of West Indian biogeography, and therefore form the largest section of this book. The chapter by Burgess and Franz on the evolution and systematics of West Indian freshwater fishes provides important new data on a group that has also been under represented in the literature. Hedges and Joglar each present a separate chapter on the evolution and systematics of the amphibian genus *Eleutherodactylus*. These chapters provide an excellent opportunity to study and compare divergent data sets on the same taxon. Thomas closely analyzes the snake genus *Typhlops*. Williams reviews the recent paper by Guyer and Savage (1986) in which they summarize West Indian biogeography based on an analysis of anoles. Henderson and Crother present a novel link between ecology and biogeography in their analysis of colubrid snakes.

Birds have been widely studied in the West Indies, but few studies are available that discuss the biogeography of birds in the region. The chapter by Cruz et al. discusses the spread of the shiny cowbird throughout the West Indies. Raffaele discusses the niche utilization of small Puerto Rican passerines. I encourage other studies of West Indian biogeography that will provide additional data on the biogeography of Antillean birds. Many interesting problems that need to be investigated are outlined in the published works of James Bond, especially in his checklist and supplements on the birds of the West Indies.

Studies on mammals form the largest unit of this book. I accept the responsibility (and potential criticism) for this emphasis, since my background and professional ties are largely in mammalogy. However, it is also a reflection of the status of the literature of West Indian biogeography, since most available studies on Antillean biogeography have been on mammals. The section on mammals begins with analyses of the status of the West Indian manatee by Lefebvre et al., and the introduced mongoose by Hoagland, Horst, and Kilpatrick. It then turns to a series of papers on Antillean bats by Koopman (an overview), Jones (the Lesser Antilles), and Phillips et al. (new approaches using mitochondrial DNA). Morgan reviews the bats and rodents of the Bahamas. Woods analyzes the evolution and systematics of Antillean rodents. Woods and Eisenberg compare and contrast the recent and fossil mammals of the Greater Antilles with those of Madagascar.

The book concludes with a series of chapters on conservation. Haynes, Sutton, and Harvey discuss conservation trends in Jamaica, with an emphasis on the status of Jamaica's birds. Jordan (Bahamas), Ottenwalder (Dominican Republic), and Ortiz

(Puerto Rico) review conservation trends in the Greater Antilles. The book concludes with a chapter by Paryski, Woods, and Sergile on conservation strategies in Haiti. The rationale for this chapter is that if conservation programs can be successful in Haiti, they can work anywhere in the Antilles, and the mechanics of conservation activities in Haiti, therefore, are of widespread interest and importance.

In conclusion, I would like to thank all of the participants, and others who helped with the symposium. I also thank Rhoda Bryant for editorial advice and Diana Carver for assistance retyping sections of the book, (both from the Florida Museum of Natural History), and Bob O'Hara of the Museum of Comparative Zoology for his assistance converting computer programs. I appreciate the assistance of Hugh Genoways of the Nebraska State Museum in suggesting and helping organize the symposium, and Ross Arnett, Jr. of Sandhill Crane Press for his help in publishing this book.

There are various people to whom this book could (or should) be dedicated, such as the late Swedish botanist Erik Ekman for his long lonely years of fieldwork in the Greater Antilles, or the late Philip Darlington for his pioneering work on West Indian zoogeography. Each of these individuals has made extraordinary contributions to West Indian biogeography. This book could also be dedicated to the next generation of biologists who's task it will be to: 1) analyze the enormous body of knowledge that now exists on the flora and fauna of the Antilles; 2) synthesize the data between geology and biology; 3) look for and find important new fossils that will answer some of the lingering questions about the evolution and systematics of West Indian plants and animals; and 4) find ways to conserve the fragile flora and fauna of the widely scattered and geological divergent assemblage of islands known as the West Indies.

This book, however, is dedicated to two Cuban biologists who have done so much to document the radiation of West Indian mammals: Oscar Arredondo and Luis Varona. Without their dedication and hard work, it would not have been possible to discuss the biogeography of the mammals of the West Indies, which are now largely extinct. As this book goes to press Professor Arredondo is discribing a new primate from Cuba that will illuminate a much more diverse radiation of Cuban primates than previously known.

Unfortunatly, Luis Varona is no longer at work documenting the mammals of Cuba. Professor Varona died in Havana on July 15, 1987. He was born on September 3, 1923, and following his graduation from the University of Havana in 1947 he began a distinguished career as a mammalogist in Cuba. He worked at the Institute of Zoology of the Cuban Academy of Sciences from 1962 until his retirement in 1979. He had many friends around the world with whom he corresponded, and he was a consultant for a number of international organizations. His greatest contribution, however, was in his published works. Varona's 41 publications appeared in a variety of journals, both in Spanish and English. On most of these publications he was the sole author. However, Professors Varona and Arredondo co-authored three important papers together, which is an indication of their long association as friends and colleagues. Because of his fragile health Luis Varona was rarely able to work in the field. Intead, he worked diligently on specimens of recent and fossil vertebrates, and described many new species of *Capromys*. His monumnetal catalog of the living and extinct mammals of the Antilles, published in 1974, is the basis of many biogeographic analyses of West Indian mammals. His publications range over such diverse topics as whales, bats, ground sloths, primates, hutias, carnivores, insectivores, vultures, trogons, turtles, iguanas, anoles, and crocodiles. His now out of print book on the mammals of Cuba (1980) is a delight to all interested in the mammals of the West Indies. His last published work was a discussion of the mammals of the Isla de la Juventud (Isle of Pines), and included a description of another new species of hutia. Luis Varon was a major contributor in West Indian vertebrate biology. I will miss his interesting letters, and stimulating ideas. At the time of his death I owed him a letter, as did many of his colleagues who had a hard time keeping up with his voluminous correspondence. We will miss you, Luis!

List of Publications by Luis Varona

Varona, L.S. 1964. Un cráneo de *Ziphius cavirostris* del sur de Isla de Pinos. Poeyana 4.

———. 1965. *Balaenoptera borealis* Lesson (Mammalia: Cetacea) capturada en Cuba. Poeyana 7

———. 1966. Notas sobre los crocodílidos de Cuba y descripción de una nueva especie del Pleistocene. Poeyana 16.

———. 1967. *Capromys nana*, las más pequeña de las jutías de Cuba (Rodentia: Capromyidae). Trabajos de divulgación, Museo Filipe Poey 59.

———. 1969. Cuban Crocodile. Animal Kingdom (October Issue), New York.

———. 1970. Morphología externa y caracteres craneales de un macho adulto de *Mesoplodon europaeus* (Cetacea: Ziphiidae). Poeyana 69.

———. 1970. Nueva especie y nueva subgénero de *Capromys* (Rodentia: Caviomorpha) de Cuba. Poeyana 73.

———. 1970. Descripción de una nueva especie de *Capromys* del sur de Cuba (Rodentia: Caviomorpha). Poeyana 74.

———, and O. Garrido. 1970. Vertabrados de los cayos de San Felipe, Cuba, incluyendo una nueva especie de jutía. Poeyana 75.

———. 1972. Un dugóngido del Mioceno de Cuba (Mammalia: Sirenia). Memoria, Sociedad de Ciencias Naturales La Salle, Caracas, Venezuela 32.

———. 1974. Catálogo de los mamíferos vivientes y extinguidos de las Antilles. Instituto Zoología de la Academia Ciencias de Cuba 139.

———. 1974. *Capromys nana*, las más pequeña de las jutías de Cuba (Rodentia: Capromyidae). Torreia 34.

———. 1974. Nuevo reporte de *Lepidochelys olivacea* (Testudinata: Cheloniidae) de Cuba. Poeyana 137.

———. 1974. The Cuban Solenodon. Oryx 12(5).

———. 1975. Asociación de *Eidolon helvum* (Mammalia: Chiroptera) con aves marinas. Miscelánea Zoológica, Instituto de Zoología de Cuba 1.

———. 1976. Reemplazo de *Cubanocnus* por *Neocnus* (Mammalia: Edentata). Miscelánea Zoológica, Instituto de Zoología de Cuba 2.

———. 1976. Comportamiento agresivo en *Cathartes aura* (Aves: Cathartidae). Miscelánea Zoológica, Instituto de Zoología de Cuba 3.

———. 1976. *Caiman crocodilus* (Reptilia: Alligatoridae) en Cuba. Miscelánea Zoológica, Instituto de Zoología de Cuba 5.

———. 1977. Cuban Solenodon Surveyed. Oryx 14(1).

———. 1977. Report of Caribbean Monk Seals. Oryx 14(1).

_____. 1979. Subgénero y especie nuevos de *Capromys* (Rodentia: Caviomorpha) para Cuba. Poeyana 194.

_____, and O. Arredondo. 1979. Nuevos táxones fósiles de Capromyidae (Rodentia: Caviomorpha) Poeyana 195.

_____. 1980. The national bird of Cuba. Trinidad Naturalist 3(2).

_____. 1980. Protection in Cuba. Oryx 15(3).

_____. 1980. Mamíferos de Cuba. Editorial Gente Nueva, Havana.

_____. 1980. Sobre las jutías fósiles de Cuba (Rodentia: Capromyidae). Memoria, Sociedad Ciencias Naturales La Salle, Caracas, Venezuela 40(113).

_____. 1980. Una nueva subspecie de *Capromys pilorides* (Rodentia: Capromyidae). Memoria, Sociedad Ciencias Naturales La Salle, Caracas, Venezuela 40(114).

_____. 1981. Cuban crocodiles. The Rephiberary, Association of the Study of Retilia, Amphibia (Great Britain) 40.

_____, and O. Arredondo. 1981. Nuevos género y especie de mamífero (Carnivora, Canidae) del Holoceno de Cuba. Poeyana 218.

_____. 1983. Remark on the biology and zoogeography of *Solenodon (Atopogale) cubanus* Peters, 1861 (Mammalia: Insectivora). Bijdragen tot de Dierkunde 53(1).

_____. 1983. Nueva especie de jutía conga, *Capromys pilorides* (Rodentia: Capromyidae). Caribbean Journal of Science 19(3-4).

_____, and O. Arredondo. 1983. Sobre la validez de *Montaneia anthropomorpha* Ameghino, 1910 (Primates: Cebidae). Poeyana 255.

_____. 1984. Los crocodrilos fósiles de Cuba (Reptilia: Crocodyilidae). Caribbean Journal of Science 20(1-2).

_____. 1984. Nueva especie fósil de *Capromys* (Rodentia: Capromyidae) del Pleistoceno Superior de Cuba. Poeyana 285.

_____. 1984. Otra especie fósil de *Capromys* (Rodentia: Capromyidae). Poeyana 286.

_____. 1985. Modificaciones ontogénicas y dimorfismo sexual en *Mesoplodon gervaisi* (Cetacea: Ziphidae). Caribbean Jouranl of Science 21(1-2).

_____. 1985. Sistemática de iguanidae, sensu lato, y de anolinae en Cuba (Reptilia: Sauria). Doñana, Acta Vertebrata 12(1).

_____. 1985. The distribution of *Crocodylus acutus* in Cuba. Herpetological Review 16(4).

_____. 1986. Implicacion taxonómica de algunas caracteres externos de *Crocodylus acutus* (Reptilia: Crocodylidae). Poeyana 312(1-6).

_____. 1986. Algunas datos sobre la etología de *Crocodylus rhombifer* (Reptilia: Crocodylidae). Poeyana 313.

_____. 1986. Táxones del subgénero *Mysateles* en Isla la Juventud, Cuba. Descripción de una nueva especie (Rodentia: Capromyidae: *Capromys*). Poeyana 315.

LITERATURE CITED

Allen, G.M. 1911. Mammals of the West Indies. Bulletin of the Museum of Comparative Zoology 54(6):174-263.

Anthony, H.E. 1916. Preliminary report of fossil mammals from Porto Rico, with descriptions of a new genus of ground-sloth and two new genera of hystricomorph rodents. Annals of the New York Academy of Sciences 27:193-203.

Barbour, T, 1914. A contribution to the zoogeography of the West Indies, with special reference to the amphibians and reptiles. Memoirs of the Museum of Comparative Zoology 44:209-359.

Darlington, P.J., Jr. 1935. West Indian Carabidae ll.: Itinery of 1934; Forests of Haiti; new species; and a new key to Colpoides. Psyche 42(4):167-215

_____. 1938. The origin of the Greater Antilles, with discussion of dispersal of animals over water and through the air. Quarterly Review of Biology 13:274-300.

_____. 1957. Zoogeography: The Geographical Distribution of Animals. John Wiley and Sons, New York. 675 pp.

Ekman, E.L. 1926. Botanizing in Haiti. United States Naval Medical Bulletin 24(1):483-497.

_____. 1928. A botanical excursion in La Hotte, Haiti. Svensk Botanisk Tidskrift 22(1-2):200-219.

Guyer, C., and J.M. Savage. 1986. Cladistic relationships among anoles (Sauria: Iguanidae). Systematic Zoology 35:509-531.

Hedges, S.B. 1982. Caribbean biogeography: implications of recent plate tectonic studies. Systematic Zoology 31:518-522.

Matthew, W.D. 1915. Climate and evolution. Annals of the New York Academy of Science 24:171-213.

Miller, G.S., Jr. 1930. Three small collections of mammals from Hispaniola. Smithsonian Miscellaneous Collections 82(15):1-9.

Olson, S.L. 1981. Oligocene fossils bearing on the origins of the Todidae and Momotidae (Aves: Coraciiformes) Pp 111-119 in S.L. Olson (ed.). Collected papers in avian paleontology honoring the 90th birthday of Alexander Wetmore. Smithsonian Contributions in Paleontology 27.

Pindell, J., and J.F. Dewey. 1982. Permo-Triassic reconstruction of Western Pangea and the evolution of the Gulf of Mexico/Caribbean region. Tectonics 1:179-211.

Pregill, G.K. 1981a. An appraisal of the vicariance hypothesis of Caribbean biogeography and its application to West Indian terrestrial vertebrates. Systematic Zoology 30:147-155.

_____. 1981b. Late Pleistocene herpetofaunas from Puerto Rico. Miscellaneous Publications of the University of Kansas Museum of Natural History 71:1-72.

Poinar, G.O., Jr., and D.C. Cannatella. 1987. An Upper Eocene frog from the Dominican Republic and its implication for Caribbean biogeography. Science 237:1215-1216.

Rosen, D.E. 1976. A vicariance model of Caribbean biogeography. Systematic Zoology 24:431-464.

_____. 1985. Geological hierarchies and biological congruence in the Caribbean. Annals of the Missouri Botanical Garden 72:636-659.

Schuchert, C. 1935. Historical Geology of the Antillean-Caribbean region. John Wiley and Sons, New York. 811 pp.

Simpson, G.G. 1940. Mammals and land bridges. Journal of the Washington Academy of Science 30:137-163.

_____. 1943. Turtles and the origin of the fauna of Latin America. American Journal of Science 241:413-429.

_____. 1956. Zoogeography of West Indian land mammals. American Museum of Natural History Novitates 1759:1-28.

Wetmore, A., and B.H. Swales. 1931. The birds of Haiti and the Dominican Republic. Bulletin of the United States National Museum 155:1-483.

Figure 1. Photograph of Erik L. Ekman taken in December 1926 as he returned from having collected plants in the high ridges of the Massif de la Hotte.

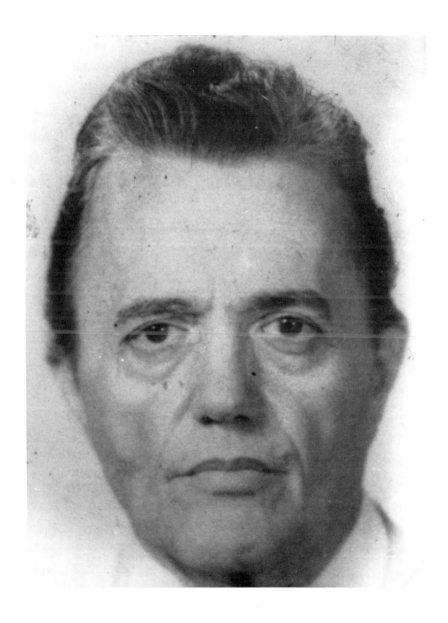

Figure 2. Photograph of Luis S. Varona, September 3, 1923-July 15, 1987 (compliments of the family).

OLD PROBLEMS AND NEW OPPORTUNITIES IN WEST INDIAN BIOGEOGRAPHY

Ernest E. Williams[1]

Abstract

The central problem in Caribbean biogeography is identified as the depauperate nature of the Antillean fauna. Brief historical sketches of the land bridge, cross-water dispersal, and vicariance hypotheses are given. Difficult and intermittent colonizations are required by the cross-water dispersal hypothesis, massive early extinctions by the land bridge and vicariance hypotheses. Necessary for firm decision is an old fossil record which might show times of arrival for the distinct groups, either nearly simultaneous (land bridge and vicariance scenarios) or at temporally spaced intervals (cross-water dispersal). Such a record is not now available. The few known old or possibly old fossils are enumerated - among them the rodent from Corozal, Puerto Rico, the Miocene fish from Haiti, the pelomedusid turtle from the San Sebastian Oligocene of Puerto Rico, and the arthropods and vertebrates from the Dominican amber. A plea is made for a more intensive search for old fossils on all the islands.

Introduction

The problems of island biogeography have always been a major concern of mine. I am therefore particularly happy in this introductory chapter for a Symposium on the Biogeography of the West Indies to have the function both of putting the subject of this Symposium in its more general context, and of relating that more general context to the specific case of the West Indies.

In particular the fauna of the West Indies has been for me a matter of lifelong interest, and I have been privileged to know several of the people who have been most involved in the exploration and study of these islands - most notably P.J. Darlington and G.G. Simpson. I was first introduced to the West Indies, while a student at Columbia University, when, at the instigation of Max K. Hecht, I began to examine the collections at the American Museum of Natural History made by G.K. Noble, Barnum Brown, and H.E. Anthony. My second paper (Williams 1950) was on West Indian tortoises collected by Barnum Brown, and for several of the early years I collaborated with Karl Koopman on some of the mammals collected by Anthony. For almost 40 years I have been at the Museum of Comparative Zoology, an institution that for most of its own history has had

---

[1] Dr. Williams is Curator Emeritus of Herpetology, Museum of Comparative Zoology, Harvard University, Cambridge, Massachusetts 02138.

a tradition of Caribbean involvement: the names, papers and books of Samuel Garman, Thomas Barbour, G.M. Allen, and Philip Darlington are classic. I have followed almost literally in their footsteps. I have visited the West Indies and the adjacent mainlands repeatedly. I have described, alone or with collaborators, new species or subspecies of reptiles and mammals, both Recent and fossil, from the Caribbean region. I have, in fact, been studying the evolution of the faunas of this area for all of my career.

Those who have carefully read P.J. Darlington's 1957 book Zoogeography will have noticed (pp. 476-477) that even then I was supposed to provide a summary of West Indian biogeography. That is a project that has long since lapsed. Here I will not be providing that summary; instead I will give a bit of history, comment on some recent studies, and sound what I hope will be a clarion call for further and more intense study of a puzzle still unsolved.

I shall begin by first stating what I believe to be the central problem that we confront.

Puzzles and solutions

The central problem

One problem is very characteristic of islands and is certainly manifest in the West Indies (Tables 1-2): island faunas are markedly impoverished as compared with adjacent mainlands. Two classic contrasting explanations have been given of this phenomenon:

1). Islands when they arise are totally without biota. Everything they eventually contain has been accumulated by colonization across a pre-existing barrier.

2). Island faunas are fragments of continental faunas. Their differences are primarily the results of extinction after a barrier has arisen.

These, in fact, reflect the extremes of a continuum along which islands belonging to different times and places fall in quite different positions. In all real islands presumably both colonization and extinction occur. For us, confronting any specific case, the issue is which has been more important, and how do we tell?

Under the first hypothesis the islands have never been connected to mainlands. Their fauna has had to reach them across water. Under the second hypothesis the islands have been connected with some mainland at some past time. The fauna has reached the islands across dry land.

In the simplest terms the problem of faunal exchange is one of two geographical points to be somehow connected: the continental source areas and the island or island group. Table 3 lists five modes that I see as possible for such connections. This list is immediately reminiscent of Simpson's (1940) biogeographic terms - "corridors," "filter bridges," and "sweepstakes routes." I have even adopted one of McKenna's (1973) additions - "Noah's arks" - to Simpson's package. I have not used any of Simpson's own terminology because of a difference in our emphases. Simpson was concerned primarily with continents and the connections between continents. Only his term "sweepstakes route" is applicable immediately and directly to islands. (I comment on this term below.) My own concern has long been primarily with islands, and my terminology is intended to provide an heuristic classification of connections between islands and continents and of islands with other islands. I use the term "heuristic" in the precise sense of requiring the reader to consider alternatives on his own.

For each of the phenomena in Table 3, I suggest the following specific expectations and predictions.

1). **Stable land bridges** ("filter bridges" in the sense of Simpson) - the expectation is one of free access of whole faunas in both directions. In our area the classic example is the Panamanian land bridge, stable since the Pliocene. The effect on the two sides of the land bridge will be mixtures of faunas that had been previously distinct. There will be some extinctions and some radiations, both time dependent. Ecology, including changing ecology, will strongly affect the timing and extent of mixture.

2). **Periodically interrupted land bridges** are like stable land bridges except: (1) that access in both directions is periodically interrupted by water gaps, and (2) faunas on one or both sides may be reduced by the extinction caused by the restriction of area during the periods of separation. Examples are such "continental islands" as Trinidad and Tobago, which are islands on the South American continental shelf, and other islands that are emergent peaks of island banks, such as the Virgin Islands, which are on the Puerto Rico bank and have been united with that island mainland several times during the Pleistocene, or Anguilla, St. Martin, and St. Barts, emergent islands of the Anguilla bank, which have again and again been united into a single island during the periods of lowered sea level in the Pleistocene. Mixture of faunas will again result. Even if the intervals between interruption and complete emergence of the land bridge are relatively short, differentiation may be minor, while the opportunities for competitive interaction may be substantial. Rates of extinction may be high, because of periodic restriction of island area.

3). **"Noah's Arks"** - whole lithospheric plates or fragments of plates carrying whole faunas from one source area to another or into the open sea - will be rare and characteristically unidirectional. The "Noah's arks" envisioned by McKenna (1973) were India and Australia, but much smaller arks are certainly part of the concept, and cases have since been mentioned by him (McKenna 1983). If the voyage of an ark is long, high extinction and marked differentiation might be expected in the island faunas. Further fragmentation of an ark should accentuate both processes.

These three categories are, in effect, land bridges and yield similar expectations and predictions. In the possibility of transmitting whole faunas, they differ very significantly from the two remaining categories. Noah's arks differ also in that the transport will ordinarily be one way.

4). **Stepping stone islands** are relatively permanent or temporary series of islands, variable in size, the water gaps between them moderate to small. These will provide "filter barriers" in the sense of Bussing (1985), relatively narrow barriers that nevertheless select out the more vagile and less vulnerable from the less vagile and more vulnerable components of faunas. More than for previous categories access will be expected to be partly stochastic and to show highly differential probabilities favoring quite selective transport. Whole faunas would not be expected to arrive over these filter barriers; instead individuals of individual species ("founder" individuals of Mayr 1942 or "propagules" of MacArthur and Wilson 1963) would be so transported. The example in Recent geography of our area that best fits this concept is the Lesser Antilles.

Direction need not be constrained in the case of stepping stone islands as it was in the case of "Noah's arks." However, more frequent transport from a larger to smaller fauna would be expected, i.e. the larger fauna and area should provide a larger number of "propagules" *sensu* MacArthur and Wilson (1963). Other factors may also enter, e.g. big islands will have big rivers, and on that account greater possibility of propagules. Stepping stone islands, however, will often not be big. In any case, the resulting island faunas should be "unbalanced," because of the filter effect inevitable with multiple water gaps. There should be relatively few arriving stocks, but these would be likely to proliferate and to radiate once they have reached some substantial land mass.

5). **Oceanic islands** are separated by extreme distances from mainlands, and they are the recipients of "waif dispersal." The difference between these and stepping stone islands is the much more extreme nature of the filter. The arriving stocks will be fewer and will have arrived at wider intervals. The aspect of imbalance must be expected to be greater for the resulting island faunas. Hawaii and the Galápagos are classic examples of "oceanic islands."

I have not employed Simpson's term "sweepstakes route" because, taken literally - Simpson did not mean it so - it would imply equal chances of dispersal for all taxa. In my view, it is of the essence of waif dispersal that, if it is a gamble, it is a dishonest one. Some taxa have, I believe, intrinsically much better chances than others. This is a point that has been demonstrated empirically many times. I have myself discussed this point at some length (Williams 1969); it is further elaborated in Perfit and Williams (this volume).

These two last categories are very different from the first three in precisely the impossibility of their transmitting entire faunas. Instead "founder" individuals must pass a pre-existing physiographic barrier. If the barrier is quite narrow, repeated founder events are to be expected. As gaps widen the founder events are expected to become increasingly rare.

I have not mentioned ecology except under the first of these geographic modes or models of colonization. It is, however, relevant to all five. The models consider only the absence, or presence and extent, of physical barriers. In reality ecology in all its aspects - physical and biotic - is equally an opportunity or a barrier. Savage (1982) has, for example, accounted for the isolation of his Nuclear Central American Fauna by two barriers: one geographic, an ocean gap to the south, and one ecological, a climatic zone to the north.

Ecology is subtler in its effects and not as readily categorized or, indeed, recognized as the geographic relationships I have listed. The simpler geographic phenomena have by themselves raised so much controversy that they must be dealt with first.

Bridge builders and rafters

The two extreme approaches to biohistory in the West Indies were already old when Matthew and Barbour disputed the issues in 1916. Thomas Barbour was vehemently, if politely, on the side of land bridges (Barbour 1916), denouncing in measured terms the views of William Diller Matthew, who declared for cross-water waif dispersal, stable continents, and permanent ocean basins (Mathew 1915).

The precise version of the land bridge hypothesis that Barbour espoused was losing ground even in 1916, succumbing to a new orthodoxy of which Matthew was the prophet and Simpson and Darlington the apostles. Until recently I had thought that views such as Barbour's had gone out of fashion. However Scharff, one of the major defenders of the old land bridge faith, is one of the two individuals to whom Nelson and Rosen's (1981) Vicariance Biogeography - A Critique is dedicated. I was also quite startled to find in the last issue of Systematic Zoology for 1986 an article by Luis Rivas that could have been written in 1916.

Figure 1 is taken from Scharff (1912). Note that one of his strange attenuate land bridges overlaps a small portion of the outline of present day South America. To choose this figure is in some regards unfair, since it is probably the most implausible of his figures. However, it certainly makes the point for which Rosen (in Nelson and Rosen 1981:4) commended him: that in Scharff's view there was "a biological necessity for an unstable geography."

Rivas' concern was a special group, the Antillean poeciliid fishes. Rivas was concerned to demonstrate that these freshwater fishes could not have dispersed to the Antilles except by a land bridge from Central America. Figure 2 is his map of present distances between Cuba and Jamaica and the adjacent mainlands. His mistake is to portray only emergent land, ignoring the shallower water shelves or banks that surround, sometimes widely, sometimes narrowly, both islands and continents. A fair picture of Recent geography must include the continental shelf and the island banks.

The concerns of the land bridge builders - Barbour, Scharff, Rivas, or many others - was always with biology and especially with disjunct distributions. For them it seemed the simplest (most parsimonious) explanation of any disjunction was to provide a direct dry land connection between the two related taxa at some time in the past. Done too casually for individual cases it always verged on the absurd. It provided strange attenuated links between continents (such as Fig. 1), and it appeared to make the sea floor go up and down at the whim of individual plants or animals. It was, in fact, surrendering geology to the biology of special groups. But the biology of the land bridge builders was faulty. The essence of land bridges is their "capability of transporting major segments of faunas." If you create a land bridge to transport any single animal or plant, there is no way to prevent much or all of the associated biota from coming with it. The classic land bridge builders damaged their case by insistent attention to special cases.

In 1935 a geologist Schuchert published a volume that is still valuable although badly outdated. It was a compendium of all that was then known about the geology of the Caribbean and its surrounding areas. He had a section on biogeography in which he came down on the side of land bridges. It is a rather fair and careful discussion, based, of course, on vastly more geological knowledge than that of most land bridge builders. Some of his remarks (p. 109) are interesting both geologically and biologically:

> "If [geologists and stratigraphers] know anything at all it is that the present Greater Antilles are not oceanic or volcanic islands and are now fragments of a once greater Central America; that this fragmentation began, in one place or another, shortly after mountain making in the late Cretaceous, and that the Greater Antilles were completely severed from the continent off and on; that the separation became permanent probably in Miocene and certainly in Pliocene time; and finally that the present island geography was initiated in the Pliocene and completed in the Pleistocene."

In another place (1935:52) he spells out the details of his geological history:

"Central America in lower Eocene time was widely connected with the Greater Antilles, making it possible for North American life to radiate into these lands. Nevertheless, no evidence is at hand to show that strong migrations of life spread from North America. This absence of evidence may mean that most of Central America and the Greater Antilles was forested highlands, which would prevent the plains life of North America from migrating south. During the later Eocene the Greater Antillean bridge was broken by submergence of Jamaica and Hispaniola. The bridge was again restored in the early Oligocene and for the last time in the late Miocene and early Pliocene, but there was no chance for intermigrations between North and South America until the Panama and Tehuantepec portals and Bolivar geosyncline of western Colombia were converted into land in the Pliocene. With the later Pliocene, interchange between the continents is in strong evidence, but then the Antillean bridge was broken. It is these unstable land conditions that cause the animal life of the Greater Antilles - lands of continental rock - to resemble that of oceanic islands."

Darlington (1957) was dismissive of Schuchert. He specially (p. 477) seized on Schuchert's phrase that if geologists and stratigraphers "knew anything at all" they knew that the Greater Antilles were not oceanic or volcanic islands and commented.

"This was unfortunate language. It now seems that the Greater Antilles are oceanic, volcanic islands! Obviously zoogeographers should use geological evidence with caution and should depend primarily on zoological evidence to decide the nature and history of island faunas. If the geological and zoological evidence agree, that is fine. If not, the reasons for the disagreement are worth looking for."

Simpson (1940, 1952, 1956, 1965) and Darlington (1938, 1957) were in the forefront of the counter-reaction to the land bridge point of view. They were concerned to refute its absurdities and firm in their belief in cross-water rafting of founder individuals, not of whole faunas. Success in these difficult voyages was, they admitted rare, but the few that were successful faced open ecological niches and therefore often radiated, often built up large populations.

Darlington, because of his close relationship with Barbour, remained sensitive to some of the criticisms that the land bridge builders brought against the rafters. He was able to dispose, rather convincingly, of the argument from the orderly, non-random aspect of the West Indian faunas by showing that dispersal could, indeed, produce orderly distributions, that it had its own regularities. He worried, however, about distances - preferring always to choose, like the land bridge builders, the shortest distance between two points. Darlington insisted on cross-water transport, not land connection. While Darlington favored origin from Central America for all of the Greater Antillean fauna (distance was a major reason for this preference), he was able to find in the proximity of Yucatan to Cuba and of the Nicaraguan Rise to Jamaica a way to reduce the distances that seemed so fearsome for fragile rafters (Fig. 3).

The number of dispersals required was another difficulty always pointed out by Barbour and other members of the land bridge school. Certainly this argument was influential with Darlington as well. A separate dispersal event for each Antillean species was, indeed, grossly improbable. Darlington preferred to believe in only a single dispersal for each genus, and Central American origin for all the Greater Antillean fauna made this story simple. He said (1957:517):

> "When the geological history of Central America and the fossil record of its fauna are better known, they may settle the question of the origin of the Greater Antillean fauna. But there are times, and this may be one of them, when the [present] distribution of animals tells certain things about the past better than geology or paleontology can. The history of Central America is not well known. North and South America were certainly separated during most of the Tertiary, and there is geological evidence of a water gap at the lower end of Central America, but geologists do not yet know when or whether Central America was separated from North America too. The composition and especially the distribution of the Greater Antillean fauna suggest that Central America was separated, that it was an island, not a peninsula of North America, through at least part of the Tertiary; that it had a mixed northern and southern fauna of mammals and other vertebrates; and that most Greater Antillean vertebrates were derived from it."

His confidence, like Schuchert's, was misplaced. Paleontological and geological evidence have destroyed both his predictions: 1) A gap at the isthmus of Tehuantepec is a discarded possibility: Central America was never a massive island (Durham, Arellano, and Peck 1952). 2) Miocene mammal faunas of wholly northern aspect have been discovered in Panama (Whitmore and Stewart 1965; see a summary of the known faunas in Rich and Rich 1983).

It is important that at this period Wegenerian continental drift still seemed to many of us only an historically interesting heresy. To most workers in geology and biology, stable continents seemed more than a hypothesis - nearly or quite established fact. Given this climate of opinion and its corollary that the West Indies had never been part of or attached to any continent, and aware also that waif dispersal was a sufficient though not a necessary explanation of depauperate faunas, the only conclusion that then seemed viable was that the West Indian fauna reached the islands across water. There remained as problems the factors of distance and the improbability, and especially the apparent unobservability, of rafting events.

There are, indeed, a number of anecdotal accounts of rafts seen or landings made (Barbour 1914, 1917; King 1962; Hardy 1982 and included older references; see also the accounts of pumice islands in Perfit and Williams, this volume), but these stories have never satisfied the skeptics. Hardy (1982), in fact, while recording that he himself had "rescued an anole from a tree floating near Tobago and had seen other reptiles drifting in marine waters," was quite incredulous that any frog could have made such a sea voyage. Papers by Heatwole and Levins (1972) and Levins and Heatwole (1963, 1973) are more substantive. They try to provide observational, experimental and theoretical approaches, but necessarily the data are limited. Again skeptics have not been convinced.

For me at the time, studying the relationships and distribution of anoline lizards, the required distances did not matter. Given stable islands and the relationships that I seemed to see in front of me, I was convinced that if cross-water voyages were improbable, they still must have occurred. I could not agree with Darlington that all the fauna of the Greater Antilles came from Central America: some taxa did, some did not. In part my problem was that I did not consider the anoline lizards a unit group. For me the relationship of one set of Antillean *Anolis*, species that occurred on the western islands of Jamaica and Cuba, was with the beta anoles of Etheridge (1960), and thus with species in Central America. The relationship of another set, all those of the more eastern islands, Hispaniola, Puerto Rico and the Lesser Antilles, and also some of those of Cuba, was with Etheridge's alpha anoles and with South American species for which there was no evidence that they had ever been in Central America. In 1969 I published my version of Antillean anoline lizard biogeography - one figure (here Fig. 4) with almost no text discussion.

My paper in 1969 was concerned to demonstrate some cases in which cross-water dispersal seemed to me inescapable. I dealt with small Recent islands and the biogeography of conspecifics or siblings that had rafted to quite distant islands. Figures 5 and 6 are from the 1969 paper and present the cases of two species of *Anolis* that have each made several cross-water voyages, some of them spectacular, and in these latter cases clearly as great as any that would be required for colonization of any West Indian islands at any time period.

I was, at that time, a less conservative (i.e., more extreme) dispersalist than Darlington, and I subscribed, of course, to all the corollaries of the dispersalist explanation of island faunas: 1) that dispersal was a rare and occasional event, 2) that certain organisms were more liable to dispersal than others, 3) that these when they reached their port of entry were often able to radiate in isolation, and 4) that the infrequency of dispersal, the selectivity of it, and the isolation provided by it were the necessary and sufficient factors in the evolution of the unique and depauperate faunas characteristic of "oceanic" islands.

Plate tectonics and vicariance biogeography

Darlington wrote his magnum opus Zoogeography in 1957. It was unfortunate timing. Within a few years the revolution in Earth Sciences destroyed the foundations of his theory. While many or most of the details of plate tectonics remain controversial, the basics clearly are not. Moving continents have become the basis of all our thinking. Geological stabilism remains in a special way a viable concept, but it is a new sort of stabilism that disputes only the amount and direction of movement, not its reality. The biogeographic consequences are serious. When stable continents and the West Indies as oceanic islands were conventional wisdom, waif dispersal to the Antilles was the only permissible hypothesis. With moving continents and the West Indies not necessarily "oceanic," the question of how the islands got their faunas became an open issue once again.

It was inevitable that the geological revolution, so conspicuously triumphant, would find its echoes in biology. For the West Indies the major name is, of course, Rosen, whose 1976 paper "A vicariance model of Caribbean biogeography" established a pattern for the whole new school of "vicariance biogeographers."

Rosen (1976: 447) quite clearly defines this school of biogeography: "The vicariance model...interprets modern patterns of biotic distribution as resulting from the subdivision of ancestral biotas in response to climatically, physiographically, or tectonically induced changes in geography. It admits the reality of dispersal, for without dispersal there would be no sympatry." Vicariance is thus primarily a fragmentation model, in contrast to dispersal, which, as applied to islands, is more strictly an accumulation model. The emphasis on "changes in geography" provides the relationship with mobilist models of geology.

It is unfortunate for geologists and biogeographers alike that the Caribbean area is tectonically not at all simple. The opening of the Atlantic by sea floor spreading, so dramatically documented for that ocean, is not so readily demonstrated for the Caribbean. There are basalt flows and complicated fault zones, rifts, volcanic arcs and subduction zones. So much evidence of complexity in geologic history has led to controversy that is opaque to the outsider. This by itself might be a minor problem, if on the biological side taxonomic studies were complete and firm for most, or even many, groups, and if on the paleontological side there were even an approximately adequate fossil record below the sub-Recent. Instead, the geology is in many respects uncertain, the phyletic analysis inadequate, and the fossil record wretched. We have if not the worst of all possible worlds, definitely a very bad one. The apparently plausible hope of vicariance biogeographers - the matching of taxonomic cladograms and area cladograms, i.e. the congruence of corroborated phylogenies and consensus geology - corresponds to no present reality in the case of the West Indies.

Rosen in 1976 endeavored to put together a complex mix of biological data, necessarily of very unequal value, and then in order to present any story at all he had to choose among competing and at least superficially incompatible geological models. Given the youth of modern geological study in the Caribbean, and given especially the complexity of the area, it is no surprise that there is no geological consensus. However, this puts upon the biogeographer a heavy burden. He is choosing his geology to fit his biological bias. Rosen in 1976 explicitly did this in postulating a proto-Antilles in the Panamanian position. Unfortunately, he gave a reference that appeared to accept Malfait and Dinkelman (1972) as his main authority (Patterson 1981; see, however, Pregill 1981, Briggs 1984, and compare Rosen's fig. 8 with Malfait and Dinkelman's 1972 fig. 1). He seems, in fact, to have relied more on other sources, including Tedford (1974), whom he referred to as incorporating Malfait and Dinkelman's plate tectonic model in a discussion of global biogeography. Tedford's figure 1 shows very clearly a proto-Antilles in the Panamanian position in the Jurassic.

Rosen's apparent inaccuracy in quotation is in any event trivial. His proto-Antilles is one of the geohistories seriously considered by geologists. There are, indeed, details in his geological reconstructions that are not quite right or not justified by any evidence that he provides, but these do not automatically invalidate the proto-Antillean aspect of his story.

Rosen, confronting the identical problem that Schuchert and Darlington had faced, was substituting a different mechanism. His biological evidence was not substantively different from the evidence that they had used. More taxa are cited, but the information content is not impressively greater. Schuchert had solved the biogeographical puzzle of the West Indies by postulating an eastward peninsula of Central America, ecologically somewhat peculiar (see above), but most significantly peculiar in that portions of it were emergent or submerged at different times. Darlington's solution was a Central American island with a mixed South American and North American fauna that transferred some substantial part of its fauna to the Greater Antilles by cross-water dispersal. Rosen's solution was his proto-Antilles, not Schuchert's submergent or emergent elements of a peninsula, but, in a vivid phrase used by MacFadden (1980), drifting islands rather than rafting animals.

Rosen's hypothesis was rebutted by Pregill (1981) on both biological and geologic grounds from a dispersalist point of view. In answer to one aspect of Pregill's criticism, Hedges (1982) was able to show that there was at least one model provided by geologists (Dickinson and Coney 1980) that showed a proto-Antilles in the Panamanian position, and that there was general agreement on some eastward motion of the pre-Antillean islands. He provided also a table listing "a variety of geologic scenarios available for interpretation by the biogeographer" and added the admonition that: "Thus it is obvious that caution must be used in using plate tectonic reconstructions of the Caribbean region to support biogeographic hypotheses."

From Rosen's own point of view his biohistory must be counted at least as important as his bow in the direction of modernist plate tectonic geology (see also the last sentence of his second (1985) paper on Caribbean biogeography). Scrutinized then as a vicariance model rather than a geological model, his scenario has an astonishing amount of cross-water dispersal in it, starting first with the colonization of his proto-Antilles. This is not especially apparent from his "simplified vicariance model" displayed in his fig. 20. It becomes blatant, however, in parts of his text and some of his detailed figures. Thus on p. 453 he says: "The combined vicariant and geophysical model...requires dispersal into new land of volcanic origin (e.g. from an ancestral Aves island arc across subsiding island stepping stones to the present Lesser Antilles, and by the same means...from mainland areas of North and South America onto ancestral connecting volcanic archipelagos." These were initial very early dispersals, presumably of high probability because of small water gaps. His figs. 13-17 make the dispersals from North or South America quite clear; he shows the proto-Antillean mini-islands black or partly black when they have received biota of North American affinities and left them white when they received only biota of South American affinities. His fig. 15 reports the situation after the proto-Antilles has moved eastward: "Further dispersal onto the proto-Antillean archipelago is postulated to have been greatly curtailed or to have ceased." His fig. 17 portrays a hypothetical "Late Tertiary phase of Caribbean history" in which "dispersal from the north and south once again populates a volcanic archipelago between Nuclear Central America and South America. The old inter-American archipelago [the former proto-Antilles] has now attained a modern aspect and shows a greatly diminished biota in the Lesser Antillean subregion - a consequence of the difficulty of biotic transferrals across subsiding stepping stones in a volcanic island arc."

There is in Rosen's work even in 1976 evident awareness of an important aspect of plate tectonic theory. The contact of major plates commonly involves subduction of one plate beneath another with attendant phenomena - volcanic island arcs - consistently present. Split-off continental fragments are less often encountered, and in the case of the West Indies they appear to be no part of the story - unless one western portion of Cuba is such a fragment. The Schuchertian story is no longer permissible even in a modernized and mobilist form. At least initially, by far the greatest part of the Greater Antilles must have been separated islands of a volcanic arc. Quite certainly these smaller islands united into the larger units that constitute the modern islands - and Rosen in 1976 pictures such accretions. This was not formally a part of vicariance theory as he defined it; it was itself an accretion to vicariance doctrine - required by geology.

Two of Rosen's postulates in 1976 have become axioms for advocates of vicariance theory; one of these is wholly false, the other is a partial truth. Currency for these two axioms is the responsibility of Colin Patterson (1981), who stated both far more clearly than did Rosen. I quote Patterson's formulation (p. 456):

> "Indeed, it is difficult to see how any new find of fossils could be incongruent with the vicariance method. Rosen emphasizes (p. 458) that one of the principal differences between dispersal and vicariance biogeography is in the age assigned to events. Dispersal biogeography sees present distributions as the result of relatively recent (especially Pleistocene) events, whereas the vicariance model implies that taxa were emplaced in remote times. Rosen stresses that fossils give only minimum ages, and because the vicariance requires that taxa be in place before a geological event, it could hardly be disturbed by fossils of any age."

The axiom that dispersal biogeography implies that present distributions must be relatively recent is a canard. Dispersal is an event that might take place in any time period; it occurs whenever any species expands its range, whether across a barrier or not. Darlington would have been incredulous of any other view; Simpson would have found the time limitation ludicrous.

Rosen based his restriction on the time of occurrence of dispersal on the claim (1976:445) that dispersal theories "incorporate a major, unexplained ingredient - namely, the coordinated movements via active migration and chance dispersal of countless organisms of vastly different biological properties." I would, to begin with, deny the postulated "coordination," but, even if the phenomenon were real and the objection is valid, I would need to inquire why such dispersal should have a higher probability in the Pleistocene than in any previous geological period.

The axiom that the age provided by any fossil is a minimum is, in a simplistic sense, quite true, but there are occasions when, in a geological sense, it is also near a maximum. More will be said on both these points in Perfit and Williams (this volume).

Apart from the novel concept of the eastward movement of the proto-Antilles as a unit, there is much in Rosen (1976) that Darlington, Mayr, or Simpson might have accepted with no hesitation. Stepping stone islands were neither strange nor unacceptable to them. Figure 7 from Mayr (1946) is unusual in picturing larger stepping stones, more like pieces of a mainland, than others have provided. His geology here is incorrect according to modern views, although based on the data of that period. Simpson (1943:416) says: "It is fairly certain that a few, random groups of land animals did enter South America while it was completely surrounded by sea; these may have had more than one origin, but some, at least, probably came from North America across the Central

American strait then existing, facilitated by islands in that flooded region, making it a sweepstakes route before it was a land bridge." Islands somewhere in the gap, during the long Tertiary separation of North and South America, have, indeed, been invoked repeatedly by almost everyone who has discussed the biogeography of South America. There were clearly groups - if only a few (and the fact that they were few is important) - that transferred between North America and South America during the interval of continental separation. Rosen's advance on the suggestions of Mayr or Simpson is the realization, prompted by the new geological evidence, that the island chain that may have provided stepping stones between North and South America in the later Cretaceous or early Tertiary is not the same as the later Panamanian island chain that fused into a dry-land connection in the Pliocene. Even Rosen's casual reference to the Aves Ridge as part of the proto-Antilles (and hence part of the chain of dispersal between North and South America) corresponds both to current geological opinion and to some paleontological opinion. (See Woods this volume.)

The biological novelty in Rosen's proposal was the suggestion that one island or several in the area of present Panama, in moving eastward to become the Antilles, brought with them from that area most of the elements that would become the present fauna. This claim is the only element that makes Rosen's a vicariance model. The essence of vicariance theory is that barriers have arisen secondarily. In one vivid metaphor (Croizat 1962:209)- especially applicable to islands - an original pane of glass is shattered repeatedly. It is quite unclear how or how well Rosen's model fulfills this classic vicariance expectation: his proto-Antilles explicitly begins (see again his fig. 13) as small islands, colonized cross-water from both the north and the south. These islands may later (in the case of the Greater Antilles, at least) amalgamate to produce the modern islands. Rosen 1985, confronted with further geological data, is explicit - it was only implicit in 1976 - that certain islands, e.g. Hispaniola and Cuba, did amalgamate. He then makes "accretionary" as well as "fragmentation events" expressly parts of his theory.

Rosen's model, thus, was never simple; it intimately involved dispersal and eventually added accretion. It was never a pure vicariance theory - despite the title of his 1976 paper, and its truly important claim was that the proto-Antilles was, in effect, a set of late Mesozoic or earliest Tertiary Noah's arks that carried most of the West Indian fauna, as we know it, eastward to their present distributions.

MacFadden (1980, 1981), partly, and McDowall (1978) and Pregill (1981), totally, reject this claim of Rosen. MacFadden (1981) makes his position quite clear: "I support the hypothesis that the historical biogeography of the Greater Antillean biota is exceedingly complex. The present-day biota could have resulted from an early vicariance event overprinted with numerous events of overwater dispersal to, and among, these Caribbean islands." McDowall (1978), pointing to Rosen's repeated invocation of dispersals, finds Rosen presenting only a model of dispersal from centers of origin, not a vicariance model at all. (I would have to subscribe to this point of view.) Pregill (1981) finds all the data consistent with "an insular biota, one evolving through time by organisms coming over water from different places and subsequently radiating from one or few colonizations."

For placental mammals the issue is very clear. (A useful summary is Savage 1974.) In the Late Mesozoic, when the proto-Antilles are postulated by Rosen to have received the North and South American elements now present in the West Indies, there were no placental mammals belonging to modern groups: most modern placental orders did not then exist. In the Paleocene there is the barest beginnings of modernization. By the Eocene - the Dawn of the Recent is the literal meaning of this age name - there may be modern suborders. Modern families date usually from the Miocene. The mammal fauna of the West Indies, except for the strange insectivores *Solenodon* and *Nesophontes*, which have relatives in the Eocene and late Paleocene of North America, are derived from forms that are Oligocene at the earliest. The insectivores are old enough - but only so by extension backward of their known time range - to have inhabited the proto-Antilles of Rosen's hypothesis. The ground sloths and all the rodents are significantly later. A rice rat (*Oryzomys*) is, indeed, known from only the Pleistocene and Recent of Jamaica; it is regarded as close to, probably a subspecies of, a Recent Central American species. The mammal fauna of the West Indies is clearly not a temporal unit, and quite certainly it was never as a whole on any proto-Antilles in the area of present Panama in the Late Mesozoic. Most of the known mammal fauna of the West Indies must have arrived cross-water after the Antilles had neared or reached their present positions. Some groups within the marsupials are old enough to have occurred on the proto-Antilles, but no marsupials of any kind are known within the Greater Antilles; a few taxa occur in the Lesser Antilles, but they are of Recent aspect - at best allospecies of a South American superspecies - as one critic of this paper has insisted to me, and again they quite surely have come cross-water.

The position of any of the islands of the Antilles in the Paleocene is, in fact, irrelevant to the mammals of the West Indies - unless, as MacFadden insists, the insectivores are an exception. In the case of even that exception, there is a strong counter-argument to MacFadden's vicariantist opinion: the absence in the Antilles of the other mammalian orders that should have entered the West Indies with the insectivores, if, in fact, the latter entered by a dry land connection.

Although for non-mammalian taxa the fossil record is not nearly so good as for mammals, the general phenomenon of different times of origin is consistent for all. The fauna of the West Indies is temporally diverse; there is the further implication that it has arrived not only at different times but, as Pregill (1981) has emphasized, from different places.

This is geologically not only plausible but required. Every geological model for the Caribbean has the major islands - both Greater and Lesser Antilles - in approximately their present position and with approximately their present size and shape during the last three to five million years. Whatever happened before this latest period, there is no question that dispersal must have gone on during it. There are island species in every major taxon that are conspecific with or sister to species of mainland taxa. These must have arrived cross-water by waif dispersal. With certainty they did not live on a Paleocene or even Eocene proto-Antilles. For these latest arrivals everyone accepts and has to accept waif dispersal - from many directions and many sources. Africa is not excluded: *Tarentola americana* in Cuba (and the Bahamas), with its relatives in southern Europe and northern Africa, is a conspicuous case, and Kluge (1969) has presented cogent arguments that at least some *Hemidactylus* must also be considered African colonizers not carried to the New World by man.

The Late Tertiary is impossible for vicariance; the islands were then nearly or quite in their present position. In the earliest Tertiary when vicariance was, perhaps, possible, the Paleocene and even the Eocene are almost too old for any placental mammalian

Antillean taxa and far too old for most. For certain elements of the herpetofauna, those known to be old, the question can remain open. But then the vicariating proto-Antilles - geologically permissible or not - is explaining, of all terrestrial vertebrates, only the herpetofauna (perhaps also some birds) and not all of these. The xantusiid lizard *Cricosaura*, relict on Cuba, might have been an inhabitant of a proto-Antilles, but the snakes *Nerodia fasciata compressicauda*, a subspecies with Floridian affinities, and *Tretanorhinus variabilis*, with conspecifics in Central America, were very surely not extant at so early a period.

However, whatever our reservations about Rosen's specific theory, it is very clear that we cannot return to Caribbean biogeography before Rosen. Although he was not the first biologist to make biological use of plate tectonic theory (to give just one example - Raven and Axelrod 1973), he was certainly the one who popularized the association of it with vicariance theory in a way that meant that only a very ignorant biologist could avoid confronting the problems and using plate tectonic data. Indeed, in spite of Rosen's conspicuous emphasis on vicariance as a new and revolutionary theory, it is his attempt - and it was only that - to harmonize biological and geological data, that will be the enduring inspiration to be derived from his 1976 paper.

Yet, very clearly, despite all Rosen's efforts, the West Indian problem is still with us. Much geological data has accumulated since Rosen (1976). Rosen (1985) was conspicuously unhappy with the diversity of new geological scenarios. He describes them correctly as "seemingly contradictory theories of Caribbean history," and his final phrases in 1985 show him less confident than in 1976 of the reassurance that was to have been provided by the matching - the reciprocal illumination - of geology and biology: "But if such corroboration is not forthcoming, as biologists we are bound by the message of biological data in describing a biotic history of those geographic areas regardless of any possible conflict with geologic theory."

For myself I have in this regard at once a more discouraged and a less discouraged view. On the side of discouragement, I am more than a little incredulous that we will ever have a precise knowledge of ancient biological or geological events. We will, of necessity, in the future, as we do now, have incomplete data on both sides - less incomplete, no doubt, but surely imperfect. Such imperfect data may well appear to support contradictory hypotheses, and all of these may be wholly or partially erroneous. Given this, it is important to realize that there is no less likelihood of error when partial biological and partial geological data are matched, than when hypotheses are drawn from one set alone. We do not, in fact, deal, in Rosen's phrase with "biological data" and "geologic theory"; we deal in both cases with data that is, not necessarily by any human error, partly defective, and interpretations that are necessarily faulty, since they are inadequately based. (More on this point will be said in Perfit and Williams (this volume)). On the side of encouragement, we are aware with complete certainty, as Rosen himself emphasized (e.g., Rosen 1978), that biology and geography - that is to say, geology - must share common histories. It is obvious that all forms of life are constrained by the spaces available for the way of life characteristic of each. We are entitled, therefore, to hope, as Rosen initially did, for reciprocal illumination. But this fortunate event will be a fact of the future, not of the present, and of approximation, not precision.

Savage is the next name that we must mention. Savage has published three papers that comment on West Indian biogeography, one in 1974 on mammals, one in 1982 on the herpetofauna, and one (1986) with his student (Craig Guyer) on anoline lizards. In only the last of these are his remarks on Caribbean biogeography more than an *obiter dictum* to a much more elaborate discussion of the biohistory of Central America.

Perhaps it was because Savage's major contributions have always been peripheral to the Caribbean, or out of irritation with Savage's 1974 paper, that Rosen (1985) totally ignored Savage's major 1982 paper, not even citing it in his bibliography. I cannot follow him in this regard. Savage's discussions are the most elaborate of the recent evaluations of West Indian biogeography (and its very relevant context the biogeography of Central America). I will need here to discuss Savage's earlier papers in order to put in context the more thoroughly developed West Indian hypothesis of Guyer and Savage (1986), especially well displayed in fig. 10 of the latter paper (here Fig. 8).

In 1974 Savage was, as regards West Indian terrestrial mammals, a convinced dispersalist - Simpsonian, by his own description - pointing out (p. 41) that: "The Antillean primate, bats, edentates and caviomorphs all appear to be advanced derivative groups rather than near the ancestral stocks that originally reached South America....For terrestrial mammals Simpson's three strata of overwater waifs...agree with the available evidence" and again (p. 48): "The mammal fauna of the West Indies is...a) ancient overwater waifs from North America, ancestors extinct - insectivores; b) Southern groups arriving by overwater transport from South America in Miocene-Pliocene times - endemic bats, edentates and caviomorph rodents; c) recent species by overwater transport from Central America - bats and *Oryzomys*, a rice rat."

At that time the issue of dispersal versus vicariance had not yet been raised. Savage's conversion to vicariance came only after Rosen's 1976 paper. However, he was already very much aware of an apparent conflict between the biogeography of mammals, Tertiary in its main aspect, and that of the herpetofauna, characterized by relatively ancient groups, with all orders and probably most families of living amphibians and reptiles already represented in the Mesozoic.

Savage's primary and insistent interest (starting with Savage 1966) has been in analysis of the herpetofauna of "Nuclear Middle America." He has argued for the distinctiveness of a "core Mesoamerican herpetofauna" and for the ancient relationship of its major elements with South American stocks. In explanation of the ancient relationship he invokes, as have many before him, an "Isthmian Link," a land connection of unspecified date, but not too early, perhaps the Cretaceous or earliest Tertiary. That link - call it Isthmian Link I - allowed dispersal of tropical biota from South America into proto-Nuclear Middle America. The later breakage of the link provided isolation by a water gap on the southern side; ecology - temperate climates to the north - completed the isolation that allowed the core Mesoamerican biota to differentiate. There was later mixture, first by invasion from the north, then significantly later from the south after the Pliocene reappearance of a land connection with South America - call it Isthmian Link II. None of the later events, according to Savage, compromised or made difficult the recognition of the core Mesoamerican herpetofauna.

There are assumptions here that Savage continues to be quite emphatic about: 1) explicitly, that he has correctly identified a distinctive tropical Mesoamerican fauna, 2) that it is South American in its affinities and quite distinct in this regard from a temperate fauna to the north of it, and 3) implicitly that he knows the approximate age of origin of the elements of this core Mesoamerican fauna, and, therefore, the approximate age of the connection with South America.

I would myself question his supposition that looking backward from the present, with no additional information - this is Savage's crucially important method - it is possible to retrodict past biological distributions and ecologies in detail. I would question as stringently his apparent hypothesis that, in the absence of relevant paleontological evidence, it is possible to infer, with any safety at all, the age of particular lineages. (On this point again more will be said in Perfit and Williams later in this volume.)

Already in 1974, and more conspicuously in his 1982 paper, Savage was evidencing some discomfort when he recognized that from the point of view of at least some geologists there was difficulty in accounting for his Isthmian Link I. Thus he says (1982:464): "...recent interpretations of the geology of the Isthmian region (Malfait and Dinkelman 1972; Marshall et al. 1979) raise doubts regarding the age of the pre-Eocene land connection between North and South America and place it so far back in time (100 myBP) as to antedate seemingly the origin of most extant Central American groups." By p. 535, however, he felt that he had rescued his crucial land connection: "...the emerging lines of evidence raise the possibility of a Late Cretaceous-Paleocene land connection between Nuclear Central and South America, lying to the eastward of the present Isthmus."

Savage (1982) is very positive about his Cretaceous-Paleocene land connection (p. 536): "Biogeography, if it is a science, must be able to predict pattern from pattern and estimate process from pattern. In the present case, there remains no recourse but to predict [his italics] that: there was a continuous land connection or series of proximate islands extending from northern South America to the area of Nicaragua in late Mesozoic and/or early Tertiary." At another place (p. 544) he is as emphatic:

> "Everything in the biotic history of Central America, except the too recently differentiated mammals, demands a land connection or its equivalent, a series of closely proximate islands between Central and South America in late Cretaceous-Paleocene. Geologic evidence for such a connection is absent or ambiguous....Still, it seems that if the tenets of scientific biogeography are sound, then biotic data can predict previously unrecognized geologic patterns. In essence, when in doubt, it is best to let the biota tell one what has occurred....The organisms speak for themselves. Their distributions require the presence of a late Cretaceous-Paleocene land connection to explain the relationships of the biotas of Central and South America and the Greater Antilles (Fig. 24). The biological evidence stands as a challenge to geologists and other biogeographers who doubtless will wish to invalidate the hypothesis. If they undertake the task, it is incumbent upon them to provide a better explanation than mine, based on a full evaluation of the evidence. I remain convinced that further studies will only enhance the explanatory power of the proposed model and will ultimately confirm the reality of the predicted early intercontinental connection. Hopefully, this challenge will stimulate a resurgence of interest in the biology and geology of the Central American region and that resurgence may lead to a concrete solution of the problem. Until then, I rest my case!"

He is here invoking a proto-Antilles, but, in contrast to Rosen, he is putting it, not in the Panamanian position, but explicitly eastward of this. If his fig. 24 is to be relied on, the major and earliest component, the future Greater Antilles, shown as a large unified block, is well to the north juxtaposed to Central America and initially without connection with South America. The Greater Antilles becomes only later associated with South America by the smaller separated islands of the Lesser Antilles, including initially the Aves Ridge.

If D, E, and F of Fig. 8 are a temporal sequence, as they surely must be, they are very poorly congruent with Savage's text and with his insistent plea for an isthmian explanation of the South American affinities of his "core Middle American herpetofauna." There is never an isthmus in this and no early connection with South America. Rosen's model, shown as A, B and C of the same figure, is better suited to support what he wants to explain.

It is possible, as Savage suggests on p. 535, that a more easterly connection with South America, such as the Aves Ridge, better corresponds to geological reality than any that could be postulated, as Rosen would have had it, in the position of present Panama. There are, however, some caveats to be taken into account in such a case. 1) The more eastern the hypothesized Cretaceous - earliest Tertiary connection, the more nearly the proto-Antillean archipelago must approach the position of the present West Indies. Donnelly (1985) - perhaps as a minority opinion - has already suggested that the West Indian islands have arisen nearly in place. In place of the thousand or so kilometers of drift that some have envisioned, he would reduce the drift, which he admits to be quite real, to hundreds of kilometers. 2) The biological consequence of this eastward positioning is to make any interchange between Central and South America via a proto-Antilles a longer colonization route with very probably a higher number of filter barriers to be passed. Such a proto-Antilles could not be an easy migration path, and any assessment of its biogeographic significance would require speculation about variables that there is not sufficient present evidence to constrain. (See Perfit and Williams this volume.)

Savage (1982) favorably mentions Coney's paper in the same volume as Savage's own. Coney argued for a Greater Antilles-Aves Arc island chain connecting to South America somewhere in the Venezuelan area. Savage felt that Coney's model corresponded very well to his own as portrayed in his fig. 24. Coney's own remarks (1982: 438) were tentative:

> "We do know that on several islands of the Greater Antilles rocks at least as old as Middle Cretaceous have the aspect of a submarine volcanic arc....This suggests that the proto-Greater Antilles formed as a submarine volcanic chain that presumably stood above a subduction zone. The most likely location of this subduction zone was off-shore of the Pacific-American margin far to the southwest of the present position of the Greater Antilles. It may have been part of the off-shore Cretaceous arc discussed earlier that stood southwest of Mexico. This arc apparently had accreted against North America in Early Tertiary time and may have simply extended southeastward across the growing gap between North and South America. It was subsequently swept north-

eastward through the widening gap that was to become the Caribbean Sea."

In 1986, returning (with Guyer) to the topic of the early land connection between North and South America, Savage has elected to present his Isthmian Link I in terms of his interpretation of the geological model provided by another pair of geologists, Pindell and Dewey (1982). Guyer and Savage's fig. 10 (here reproduced as Fig. 8) purports to be based on their work.

Guyer and Savage, in bringing items of new geological evidence before a wide biological audience, have performed a distinct service, and the clarity and simplicity of their figure ensures that it will often be copied. This, however, is not wholly fortunate. The first two panels of their figure have conspicuous errors.

Fortunately, the other panels of their figure, those dealing with periods from the Oligocene to the Recent, do, indeed, portray some of the more accepted elements of the New Geology of the West Indies (for more detail see Perfit and Williams this volume):

1) Jamaica rather well separated from the rest of the Greater Antilles, close to or in contact with Central America.

I would emphasize parenthetically (Perfit and Williams this volume will expand on this point) that "contact" to a geologist does not necessarily mean what contact implies to a terrestrial biologist: "contact" of lithospheric plates or platforms does not automatically imply contact of emergent land - "dry land connection:" there will be usually, although not inevitably, a salt water gap. Thus for the West Indian fish fauna Burgess and Franz this volume suggest that "Inoculation was most likely among nearby islands via contiguous brackish to marine shallows." Geologically such contiguity would be evaluated as "contact" although it conspicuously does not imply a dry land/isthmian connection.

I would add to Guyer and Savage's summary the point that Jamaica may have been totally submerged for part of the Eocene and all of the Oligocene and to have finally come above water only in the Early Miocene (Arden 1975; Buskirk 1985; see also Hedges this volume). Burgess and Franz mention, however, that Buskirk's phrase is "little, if any emergent land" and that Arden suggested the possibility of a few small islands. Burgess and Franz see in these admissions the opportunity for the hypothesis that "the elements of an archaic freshwater fish fauna could have survived after inoculation in the Paleocene." Hedges, Woods, and myself have elsewhere in this volume opted for the alternative hypothesis of total submergence as an explanation of the distinctiveness of the Jamaican fauna.

2) The southern part of Hispaniola was for a long time separated from, and well west of the rest of that island.

3) Eastern Cuba and part of Hispaniola are generally agreed to have been "united" during early Caribbean history.

4) Another part of Hispaniola was at one time "united" with Puerto Rico (geologically it still is).

5) The present island of Hispaniola is a relatively recent composite of at least three formerly discrete elements.

6) The Lesser Antilles, as we know it, is a relatively recent structure.

Here I would add two points. 1) the Aves Ridge is commonly regarded as a formerly emergent set of islands prior to and distinct from the modern Lesser Antilles. Rosen

(1976) implied that it was the original southern portion of his proto-Antilles. Savage (1982) was also aware of the importance of the Aves Ridge. Woods (this volume) makes it an important part of his scenario for the entrance of caviomorph rodents into the West Indies. 2) The northern Lesser Antilles might have been, from a geological viewpoint, part of the ancient Greater Antilles (see further in Perfit and Williams this volume).

The two panels in Fig. 8 for the Paleocene and Eocene are less congruent with accepted geology. The panel for the Paleocene in Fig. 8 shows a Cretaceous-Paleocene Isthmian Link, figured as very like the present Panamanian Isthmus but anchored not to Colombia, but eastward, according to the detail of the map, just a little west of the Lago de Maracaibo in Venezuela. In the figure this intercontinental link is a very substantial dry land connection - a true isthmus. In the panel for the Eocene this isthmus breaks off and a considerable land mass is carried to the west, leaving Jamaica and the southern island of Hispaniola behind.

These two panels are a very explicit revival of Savage's 1966 Isthmian link I (not there so named) ingeniously combined with Rosen's proto-Antilles, which is assumed to have resulted from the breakup of the isthmus. The effect of this reinterpretation of geology is to rescue Savage's long-held hypothesis about the origin of a "core Middle-American herpetofauna."

In fact, however, these two panels misrepresent both the timing and the details of the reconstructions proposed by their authorities, Pindell and Dewey (1982). Guyer and Savage's figure of the Paleocene most nearly corresponds to Pindell and Dewey's reconstruction of the Caribbean in the Late Cretaceous at 80 myBP and that of the Eocene to Pindell and Dewey's figure of 65 myBP (a date now conventionally considered to be the lower boundary of the Paleocene). Pindell and Dewey also nowhere suggest as solid a connection as Guyer and Savage suggest for the Paleocene nor as large a land mass as Guyer and Savage suggest for the residual Greater Antilles after Jamaica (and a southern portion of Hispaniola) separated from the remainder of the island mass.

On this issue the point must be made (further elaborated by Perfit and Williams this volume) that island arcs will first be series of submarine and later emergent active volcanos, initially without faunas and only accumulating these faunas by cross-water dispersal. Rosen (1976) was quite aware of this pattern, as his figs.13 to 17 cited above quite clearly show. Errors are made if the "platforms" and "masses" of geologists are automatically assumed to be emergent or equivalent to fragments of mainlands. Island arcs will not typically transmit full faunas, i.e. permit what Savage (1982) calls "concordant dispersal." If, in moving eastward, they become "Noah's arks," in McKenna's phrase, they will transmit not continental faunas - since they never had them - but more limited island faunas. Savage (1982) has called "a series of closely proximate islands" the "equivalent" of a land connection. I venture to contradict him - and by implication Rosen. A chain of islands is a set of sieves or filters that will let only portions of any fauna through - <u>filter barriers</u> more severe than Simpson's filter bridges, which, for him, were expressly the narrowed isthmian connections of continents - and the filtering agent primarily climate. (Hedges in this volume argues that the proto-Antilles was an old island arc and therefore may have been "relatively continuous.")

A "series of closely proximate islands" would be at best stepping stones. Coney (1982: 438) has written quite explicitly on the problem of faunal transfer: "The Cretaceous arc [= Rosen's proto-Antilles] is of interest, if it existed as portrayed, since it could have harbored faunas and floras upon its far flung volcanic islands, which were swept eastward as the arc migrated toward its present position. It could also have provided 'stepping stones' for dispersals between the two continents during Middle and Late Cretaceous time. The linkage was, however, no more than scattered islands and fringing reefs, probably much like the present day Lesser Antilles."

If Coney's comparison of the proto-Antilles with the modern-day Lesser Antilles is correct, it is devastating to any concept of the equivalence of "a series of closely proximate islands" to a dry land connection. Although the Lesser Antilles does have South American elements in its fauna, these attenuate severely in number and diversity to the north as a result, I infer, of the hazards of inter-island waif dispersal. If Savage were right, the Antilles would be even now as faunally rich as modern Central America.

The main point of Guyer and Savage is not, however, geological. It is vicariance biogeography after the model of Rosen, and in contrast to Rosen's broad based approach, Guyer and Savage tackle a specific case: anoline lizards. That happens to be in my area of special competence. I will not, however, address this special case here. I do so in a later chapter in this volume. Guyer and Savage's vicariance interpretation of it is unfortunately clouded by quite irrelevant errors. I here state only my appraisal of the essence of the case: the Scotch verdict "Not proven." In the classic language of proof (necessity and sufficiency) they do establish a possibility of some part of their suggested history - at least a partial answer to the criterion of sufficiency. The details, however, remain vague; proofs, or even strong indications, that the their postulated history must have, or even most probably, occurred, are not available. They fail the criterion of necessity.

I find that with Guyer and Savage (1986), as with Savage (1982) and Rosen (1976) some of the sounder parts of the presentations are geological. Whenever they have correctly reported their geological authorities, I see victories for the New Geology. It is less evident to me that these are victories for vicariance theory.

I point out that biogeographic theory has always had close ties with current viewpoints in geology. Classic land bridge concepts fell by the wayside when it became clear that ocean floors do not rise or fall with the frequency nor at the places that the more casual exponents of land bridge theory required. The reaction of the classic dispersalists was to believe all too firmly in the dogma of continental fixity and the permanence of ocean basins, and this has, in the end, defeated them. Vicariance biogeographers are in hardly better state. In the case of all the most recent biogeographers there is evident a very real unhappiness that plate tectonics has not shaken down to the degree of firmness in all details that the biological users demand of it. This has been the reason for the radical demands by Rosen and by Savage (quoted above) that if geology will not provide the geography desired by their theories, it is geology that must bow to the ineluctable logic of their theories - that biological necessities require specific unstable geographies.

I protest that biological science is at this time not in a position to make such demands of geology. Hypotheses in biology, no less than in geology or in the remainder of science, are somewhat below the level of theses. All that I know about the present state of biology tells me that most, in some regards all, of the data behind systematic and phylogenetic hypotheses in biology is defective as regards detail, sometimes also erroneous in part or entirety, inadequate in its sampling, irregular in its analysis, and insufficient to the claims made. I shall have occasion in the case of Guyer and Savage to make these statements all too vivid.

We must, I agree, try to make do with what we have. That is the only useful procedure. We have no choice but to begin with what is at hand at any moment. That is a necessary <u>starting point</u> - a means to frame a working hypothesis; that, however, must not be a <u>sticking point</u> - a resting hypothesis. We must look hard for additional confirmatory or contradictory evidence. That is the insistent theme of this introductory chapter. Only by searching can we look forward to the infrequent significant victories that will make it all worthwhile. In the meantime we had best make our biased statements with lowered voices.

Where are we now?

Rosen's paper is now more than ten years old. The most recent substantive papers published prior to this Symposium - Savage (1982), Rosen (1985), Guyer and Savage (1986) - are strongly on the vicariance side of the controversy. Does this this mean that there is a consensus for vicariance? The papers in this volume make it clear that there is not.

At this moment I myself see no compelling arguments for Caribbean vicariance, understood as the transfer of continental faunas as units from the mainlands to the islands. If, on the other hand, accretion is part of vicariance theory, then given the now accepted geological evidence for accretionary events within the Greater Antilles, island to island transfers of faunas must be considered at the least highly probable, and island to island vicariance fully acceptable.

The idea of continent to island transfer of faunas is, given a mobile earth, a possibility to be considered. But the actual occurrence of such an event is not established - nor is the hypothesis destroyed that the cross-relationships explained by vicariance are not equally well explained by dispersal - by assertion. Unhappily, this has been the method of argument of Rosen and, following him, Savage. Their rhetoric - much more than their actual models - appears to assume a dichotomy between dispersal and vicariance, an "either-or" proposition, for which I find no basis.

I would agree that cross-water dispersal from the American continents to the West Indian islands is not readily confirmable by empirical data. However, it is falsifiable by the discovery on the West Indian islands of floras and faunas - at whatever period - of continental type. Cross-water dispersal, because of its biased and selective properties, could not produce such faunas. Islands with such faunas are well known and have been classically designated "continental islands" and classically contrasted with "oceanic islands" which have faunas that are more than "depauperate." Oceanic islands are - this is the important point - selectively depauperate.

The East Indies - west of Wallace's line - are large "continental" islands. Darlington's 1957 discussion of these islands is quite adequate. They, in strong contradistinction to the West Indies, are only narrowly and shallowly separated from their source mainland - Asia. They were connected with Asia several times during the Pleistocene and are still separated by rather shallow seas from the sources of their biota, and the West Indies, in contrast, have never been connected with the American mainlands in the Pleistocene - or the Pliocene - and are, at least at present, sundered by both distance and deep water from any possible sources of their biota.

Under vicariance theory, great age of separation and massive extinctions during that long period of separation are the factors explaining the contrasting continental biota of the East Indies and the depauperate biota of the West Indies. Under dispersal theory the biota has never been continental; extinction has tended to increase an imbalance that already exists, new colonists have accumulated over some long period of time, but by the nature of the process, because of its inherently biased nature, the imbalance may be changed but it will not be cured.

Both dispersal and vicariance emphasize the importance of isolation as the factor permitting the distinctive radiations - island by island - characteristic of archipelagos. The West Indies offer prime examples of such unique - island by island - radiations. The two hypotheses differ, however, in the mechanism of isolation - whether by the primary or the secondary existence of a barrier. The choice of mechanism defines also the two explanations of depauperate faunas that were given at the beginning of this paper - either failure to arrive (dispersal) or extinction after initial presence (vicariance).

Between these two competing historical hypotheses I do not see any a priori grounds for choice. Drifting islands do not make vicariance per se more plausible or dispersal less so. In fact, in the whole history of the West Indies, as now understood, I do not see how dispersal on any present evidence can be ruled out. A moving Antillean complex alters distances between islands and mainlands and among the islands. It makes dispersal easier or more difficult. It says, however, nothing per se about faunas or how they got where they are. Nearness assists dispersal, distance makes dispersal difficult and very improbable. Nearness is not enough for vicariance, and distance makes it, by definition, impossible. To falsify dispersal in the Caribbean even in the earliest Tertiary will require evidence we do not now have. I shall plead for vigorous search for that relevant evidence below.

My judgment, thus, is that the honest statement to be made about the state of West Indian biogeography is that there are weaknesses in any present theory. At this moment I lean toward dispersal - not, however, on any direct evidence, but on the old and still uncontradicted argument that filter barriers provided by water gaps present from the beginning (although, we must now grant, not with the identical distances that exist today) better explain the extremely depauperate and highly unbalanced nature of the West Indian faunas. In my view if there were ever dry-land connections in the past, the fauna that came over those connections must have been both complex and diverse. It would seem to me more parsimonious to expect that on such substantial land masses as the Greater Antilles have been and are now some remnant or trace of that complexity and diversity would be retained into the present.

Vicariance theory requires Croizat's initial intact pane of glass that is later shattered - the initial large land mass that along with its fauna is subdivided by later gaps. Guyer and Savage (1986) are in their presentation closer to the classic vicariance model than was Rosen (1976): they propose a solid land connection - Savage's Isthmian Link I - uniting North and South America, break it off both to the north and the south, retain it briefly as a major land mass, and only later fragment it. Biogeographically this is critical. Such a solid land connection would have had a continental fauna. Savage's (1982) insistence on such a connection (repeated in Guyer and Savage 1986) derives precisely from this point: his belief that the age and relationships of the Central American herpetofauna require that the "concordant dispersal" made possible by continuous land connection be available at a particular time (for him the Late Cretaceous or the Paleocene). Such a large land mass, when it broke with the continents to the north and south and moved eastward as the progenitor of the Antilles, then must have had, at least initially, a continental fauna. The Antilles, according to this, were initially just as much continental islands as the East Indies are today.

There is here a first question whether the modern geological evidence is permissive at all of Guyer and Savage's hypothesis of a solid land connection, or of the proto-Antilles as a large subaerial land mass. Pindell and Dewey (1982) can be so interpreted, but I assert that the figure on which Guyer and Savage rely was not intended to have such an implication. Pindell and Dewey's figure shows an oval lying between South and Central America that encloses v's for emergent volcanos. The oval I interpret as only indicating the approximate area in which the vulcanism characteristic of a subduction zone was occurring. My own reading of the geological literature leads me to believe that only small recently emergent islands were components of a proto-Antilles; I reiterate my judgment that such islands would necessarily be filter barriers that could never transport whole faunas. This, however, is a problem that, since it involves a subjective evaluation of probabilities, will surely remain controversial.

The major issue, however, is not whether stepping stone islands, as a generalization, are filter barriers, but two explicit questions regarding real history: Did a proto-Antilles - a chain of islands somewhat west of the present islands - exist at an early period - perhaps the Paleocene? Did this proto-Antilles contain a rich and continental fauna that has persisted and is directly represented in the few relict elements left behind in the Recent and Pleistocene of the West Indies? I would at this moment answer the first question with an unqualified yes. I would answer the second question with a hesitant no.

These again are partly questions for which geology is pertinent. The West Indies are, indeed, surely an island arc, but an island arc appears to have two meanings: 1) the version which I interpret to be the common verdict of the geologists who have recently studied the Caribbean area - initially emergent chains of volcanic peaks formed at subduction zones; 2) rifted arcs that have broken off from continents. The first seems to me to be classically represented by the West Indies - faunally as well as geologically - highly depauperate faunas, temporally diverse and very poorly comparable to adjacent modern mainlands. The second seems to me to be represented by the Japanese Archipelago (see Uyeda's 1978 discussion of the origin of the Sea of Japan), which is geologically and also faunally a fragment of a continent. I see this as a clear-cut issue, on which, unless I seriously mistake the evidence, the verdict must be that the West Indies are not geologically the equivalent of Japan.

It should not be necessary to say that the association of geological phenomena with vicariance or dispersal is only a permissive rather than a necessary one. Mobilism in geology does not automatically imply vicariance. As Pregill (1981) and McKenna (1983) have insisted, plates may move under water, and no vicariance of terrestrial biota can possibly result. Vicariance does not by itself imply mobilism; faulting, mountain-building, climatic change - as Savage (1966, 1982) has insisted - can produce vicariance with no plate movement. Plate tectonics may revise much of the detail of biogeography but it does not, in and of itself, negate cross-water dispersal. It complicates rather than simplifies the story.

If it is once established that there was some land mass that transmitted a continental fauna to the proto-Antilles or some part of it - something not at this time demonstrated - a second question must inevitably be raised. If there once was such a fauna, where is it now? If there are relicts of that original continental fauna now, which are they? MacFadden (1980) endeavored to answer that question. In mammals he pointed to *Nesophontes* and *Solenodon*, the insectivores with their only known relatives in the early Tertiary of North America and, among birds, to the todies for which Storrs Olson (1976) has found a fossil relative in the Oligocene of Wyoming. But if these, why not much more? Did no other early Tertiary mammals arrive with the insectivores? Again we return to the vexing problem of this putative fauna of continental type. How do we explain its absence?

I can provide one hypothesis that will satisfy neither the fervent proponents of vicariance nor those emotionally committed to dispersal. I suggest a rather crudely Solomonic division of Caribbean history into two divisions - a Stage I and a Stage II.

There is on the geological record a mid-Tertiary phenomenon - not a plate tectonic phenomenon - that may imply such a division: To explain the possibly relictual nature of two anoline lizard genera, *Chamaeleolis* and *Chamaelinorops*, Etheridge in 1960 suggested the possibility of two stages in anoline biohistory. In 1969 I amplified the hypothesis by calling attention to Schuchert's 1935 mention of a mid-Tertiary inundation of the Greater Antilles, suggesting that, because of that period in which there was massive sinking of the entire Antilles and consequent disapppearance of some islands under the sea and great diminution in size for others, there were, indeed, two stages in the colonization of the Antilles, one prior to that extensive submergence, one after re-emergence. Geologically submergence is rather well documented (Perfit and Williams this volume), although it may not have been contemporaneous for all the islands.

Woods (this volume), noting that the vast majority of extinctions in the Antilles are associated with incursion of man, Amerindian or European, suggests that the sample that we have of at least the rodent fauna, late though the dates of presently known fossil rodents are, is a good sample, genuinely representing the major lineages that have existed. This may well be true for rodents, which are admitted to be one of the later arriving groups in the West Indies. Perhaps the rodents, the ground sloths, and the primates are Stage II invaders, while the insectivores and at least the relict genera of anoles and the lizard *Cricosaura*, relict on one peninsula in Cuba, belong to Stage I fauna, most of which has not survived.

This mid-Tertiary inundation, of course, does not, of and by itself, provide grounds for decision between the competing hypotheses of vicariance and dispersal. That decision must be made on other grounds. It does, however make vicariance a little more plausible by providing a mechanism for the extinction of the early Caribbean faunas that vicariance requires.

However, hypotheses are easy; the testing of them is not. In the argument between vicariance and dispersal, testability has often been argued, and especially the testability of dispersal has been denied. I argue that such a test of the two biogeographies is in principle quite possible. There are strong predictions to be made by both hypotheses. On the vicariance side the expectation for an original island fauna will be of a balanced continental type, one that will only decrease over time in major group diversity. On the dispersal side the expectation will be the opposite - of accumulation of major groups over time. The two diversity curves will be opposite in slope. A mid-Tertiary inundation would complicate, but would not defeat this test. What happened before and after the inundation would, however, be two different stories to be tested by the geological and biological evidence for two different time periods.

Neglected opportunities: Old fossils?

We have returned to the two questions with which we began. The fauna of the West Indies is strongly and selectively depauperate (see again Tables 1-2). Is it 1) because important elements of the fauna never arrived? Or is it 2) because significant elements of the fauna have gone extinct? In principle, a fossil record can answer these questions.

Colonization can explain the presence of taxa on islands. It is the explanation of the peculiarities of the West Indian biota preferred by those who believe that it has arrived across water. Extinction can be an explanation of absence. It is the explanation of the identical facts preferred by those who believe that most of the West Indian biota came from a single contemporaneous mainland source.

Savage (1982: 537), in rebutting Pregill's criticism of vicariance on the basis of the absence of marsupials, carnivores, and ungulates from the West Indian islands today, and the total lack of evidence for their fossil presence, commented:

> "The absence of groups from the fossil record of an area, especially a lowland tropical one, tells us very little about the history of its biota. There are no fossil records in Central America of marsupials, bats, primates, non-caviomorph rodents, most families of carnivores and almost all families of amphibians and reptiles that occur there today. Does this mean that none of these groups occurred there until very recently? Or tell us at what time they appeared in the region? There are hardly any records of fossil vertebrates from tropical South America, including most families present there today. Does this mean that the missing groups were absent from the region?"

Savage here at least appears to be inverting the logic of the fossil record. It is plausible to expect fossils in the same area as groups present there today. That is a logically permissible expectation. It is also true that absence in the present does not mean absence in the past. Perhaps Savage intended to say only this. Unfortunately, in his rhetorical emphasis of one truth, he has omitted to call attention to its inverse: that absence in the present does not imply presence in the past.

There can be valid evidence for an emphatic negative. 1) There is no evidence of native ungulates in the West Indies today. 2) There is plausible ground for inferring that they were never there: these are often large mammals. (A group old enough for a Paleocene vicariance event, as Ross MacPhee has reminded me, would be the pantodonts. These were usually large. Large fossils should by now have been found.) The two statements, however, are very different: No observers - native or foreign - have seen ungulates in the West Indies. There are, indeed, no historical accounts, no paintings, figures or artifacts that suggest their recent former presence; there are no remains in archeological deposits. There are no records at all. I would regard the evidence for the absence of ungulates in the West Indies in historic times as close to positive evidence of a negative as we can come. Their fossil absence may also be based on sound evidence. But it does not rest on the same observational ground. There are no known exposures in which some fragment of an ungulate jaw or limb has been seen protruding from a fossil bed. But there are not enough exposures in which anything has been seen, not enough exposures looked at or looked for. What we have now is not firm evidence of a negative. It is the absence of evidence.

The use of imagination as the alternative to fossil evidence is comparable to the old maps that filled unexplored blank spaces with the legend "Here be dragons." Savage is unfortunate in appearing to pretend that absence of a firm negative is compelling evidence of a positive.

Nothing, it is very clear, so exacerbates controversy as insufficient data. Biogeographers, no less than other scientists, must face up to the fragility or inadequacy of their data. We are trying to report biohistory. But in the West Indies we are doing something comparable to writing American history from a few files of recent newspapers - all scandal sheets, none of them the New York Times - and a few scattered undated volumes - none expressly historical. With this metaphor I do considerable injustice to the quantity of information that was presented in this Symposium and to the immense energy that many present and many not present have displayed in attempting to fill the blank spaces in the record with empirical data. I intend my hyperbole to highlight a very genuine concern. We do not have the kind of information we most need to have, i.e. a good fossil record.

Recently fossils have had a very poor press. They are denigrated as only partial evidence of the taxa they represent, as providing at best only minimal ages, as "plesions" not deserving of a place in formal taxonomy. I am aware of their deficiencies, of their ordinarily fragmentary nature, of their lack of many of the characters considered most important in the taxonomy of living forms, of their erratic occurrence, of their frequent difficulty of interpretation.

However, fossils are data. Parenti (1981: 492)) states: "The presence or absence of a taxon...constitutes the data of biogeographical analysis." Fossils are, indeed, essential, ineluctable historical data - not to be dismissed or discarded because incomplete or imperfect. (Any evidence that does not deceive is better than no evidence.) I admit the error that individual fossils may introduce, but the fossil record is more than the individual case. I applaud McKenna's (1983) observation that "...although it is generally recognized that the fossil record is imperfect, it is not likely that the fossil record is perverse...." Fossils can solve certain issues beyond a shadow of a doubt. If there were anywhere in the West Indies a fossil fauna clearly of continental type, we would forever be barred from invoking waif dispersal in the origin of that fauna in that place and time. Positive data are beyond remedy; they destroy hypotheses that depend upon its absence.

Which are the old fossils of the West Indies? Which are the presences we must take into account? There are some; even if not enough - some unfortunately only possibly old, but some also that are authentically so:

1) *Xenothrix mcgregori* Williams and Koopman 1952. This mysterious primate, sometimes put in its own family (Hershkovitz 1970), sometimes considered a giant marmoset, was described from a mandible collected by H.E. Anthony at Long Mile Cave,

Jamaica in 1920 (Williams and Koopman 1952). It has not been directly dated and might be young (MacPhee 1984). There is, however, a femoral fragment from a different Jamaican locality, Coco Ree Cave (Ford and Morgan 1986). Ford and Morgan regard it as a distinct lineage, but MacPhee (in litt.) regards it as identical with *Xenothrix mcgregori* on the basis of a very similar femur from Long Mile Cave that was not identified (not regarded as even primate) by Williams and Koopman. The Coco Ree fragment is dated at 38,000 yBP, very old by the standards of most West Indian vertebrate fossils. (This exact dating is contested by Goodfriend and Mitterer 1987; they estimate "an age probably between 30,000-50,000 yBP and almost certainly between 20,000-60,000 yBP.") MacPhee (in litt.) regards all published dates as inconclusive and has "no doubt that *Xenothrix* has had a long residence in Jamaica."

The Jamaican fossils are important in two regards: they certainly demonstrate that primates were native to the West Indies, and both the mandible and the femoral fragment document features in primate evolution not previously known. There are other fossil primates in the West Indies (see the summary discussions in Ford 1986a and b). These belong to quite distinct lineages and thus indicate several incursions into the West Indies. The Jamaican fossils afford the highest promise of true antiquity, primarily, however, because of their distinctive structure, not their presently known age.

2) "*Proechimys*" (=*Puertoricomys* Woods this volume) *corozalus* Williams and Koopman 1951. Again a mystery and a fossil without a date. Collected in 1930 by James Thorp and N.L. Britton "from a crevice in Corozal Limestone Quarry" in north central Puerto Rico. The specimen is well mineralized and might be old. On the advice of Albert Wood, Karl Koopman and I compared it with a Recent echimyine. Charles Woods finds it still puzzling but an echimyid and part of the heteropsomyine radiation. Again the peculiarities of the animal rather than any present information on its real age are the grounds for assigning this fossil importance and, perhaps, antiquity.

There is a fossil lizard femur associated with the mammal jaw that Karl Koopman and I did not identify. At my request Richard Estes and Gregory Pregill have re-examined it. It appears to be an iguanine comparable to the modern taxon *Cyclura* but not precisely identifiable to genus. Some turtle fragments are also associated.

The exact locality and stratigraphy of the "*Proechimys*" are still not known, but some unpublished notes by Clayton Ray during an expedition to Puerto Rico and Hispaniola in 1958 may possibly assist renewed investigation:

> "July 22....Turned off coast road at Toa Alta and proceeded to Corozal where we found only one active limestone quarry. This quarry lies at the west side of town just north of route 159. It is on the east side of the Corozal valley. An old unused quarry lies adjacent to this one, and another small unused one lies on the south side of town just beyond the Catholic college. One of these latter must be the source of *Proechimys corazalus*, as I understand neither has been used in the past 20 years and the active quarry began about 20 years ago. The active quarry is extremely interesting and looks like a good prospect for fissure material. We spent about three hours looking over the quarry wall. Rock types are quite varied -- crystalline limestone; limestone breccia with red clay matrix and some pebbles of conglomerate rock; deep rotten red

shale; fissures mainly with water worn walls and sometimes with drip stone deposits on walls filled with variable unconsolidated brown earth. A number of fissures terminated above, abruptly at the base of shale layers, suggesting that the fissure filling may be very old, but this is not certain as the erosion of Quaternary caverns may have been controlled in part by insoluble shale strata. This area deserves further study for fissure fossils."

Ross MacPhee and others (MS), using uranium series disequilibrium and electron spin resonance methods, have determined that the vertebrate faunule from Wallingford Roadside Cave in central Jamaica (MacPhee 1984) has a minimum age of 100,000 yBP, and may be as old as 180,000 yBP. This faunule contains the problematic rodent *Clidomys*, which thus becomes the oldest dated Antillean land mammal.

3) *Cichlasoma woodringi* Cockerell 1924. This, the only presently known fossil fish from the Antilles, comes from Tertiary beds in Haiti "in the arrondissement of Las Cahobas, along the road from Mirebalais to Las Cahobas, on the north side of the ravine at the foot of the mountains, on the north side of the gap." The same beds have terrestrial plant remains that were dated at the time of their description as Miocene. According to Rivas (in Burgess and Franz this volume) the fish fossil cannot be distinguished from a Recent species that was described later, *Cichlasoma haitiensis* Tee-Van 1935.

4) Four new species of fossil plants were described by Berry in 1923 from the fish locality, who in the same paper cited terrestrial plant fossils from 12 other localities in Haiti, spanning the period from the Cretaceous to the Pleistocene. Berry had previously (1922) described Tertiary plant fossils from five localities in the Dominican Republic and mentions a sixth locality from which determinable remains had been obtained. Hollick (1928), describing the "Paleobotany of Porto Rico" described in detail only the terrestrial plants from one deposit in northeastern Puerto Rico in "the ravine of the Collazo River from the falls above the Lares-San Sebastian Road bridge to a distance of about a kilometer below," but this is part of the same San Sebastian formation that has yielded two vertebrate fossils mentioned below.

These localities, as the fish fossil indicates, may be more important than the plant fossils that they contain. Berry (1923) mentions for the Haitian material that: "A large number of forms that might reasonably be expected from the Late Tertiary of this region are conspicuously absent, notably Leguminosae and members of the mangrove association. This appears to mean two things, namely, the absence of tidal mud flats and river estuaries, and suggests sandy or rocky shores. All of the leaves enumerated are of strand plants and all are maceration resisting forms, indicating that the other normal members of the Tertiary strand flora failed of preservation, at least at the discovered localities." For the Dominican Republic fossil plant flora, Berry (1922) comments: "The total number of forms identified is eleven, much too small a number to give a correct idea of the botanical facies or of the geological age beyond the obvious fact that they indicate a tropical habitat and a Tertiary age. There are no traces of ferns or palms, and the majority of the forms...are obviously strand types, as might well be true of the remainder. There are no traces of any of the typical plants of the Mangrove association, nor Lauraceae or Moraceae, all types normally present in Tertiary floras."

Hollick's Puerto Rican plant remains, although very restricted as to area, were much more abundant, better preserved, and more diverse. His comments are: "The flora of the Collazo shales represents one of a tropical environment; its habitat was in the vicinity and on the borders of lagoons or estuaries, in which brackish water was present; it is typically New World in its general facies; it is almost identical, generically, with the existing flora of Porto Rico and adjacent regions; it is Tertiary in age and referable to the lower-middle part of that period."

Graham and Jarzen (1969) studied pollen from three localities in the San Sebastian Formation, one of them a Hollick locality: "Collazo River, near or 'at' base of second falls below bridge." They report 165 morphological types of pollen spores of which 44 (about 25%) were identified to genus. They compare their identifications with the 56 genera recognized from the Collazo Shales, commenting that several of the latter were "tentative identifications and others need verification." They positively identified pollen for only six genera of Hollick's megafossils, but suggested that pollen of six other genera might be among the microfossils identified only to family.

Two of Graham and Jarzen's results are especially interesting. They did recover pollen of three genera (*Fagus, Liquidambar, Nyssa*) that they believed indicated temperate to cool-temperate climates, and, possibly, as they discuss, in Oligocene highlands that might have reached 3,000 meters. At the same they obtained no pollen indicative of the gymnosperms characteristic of very high elevations in the Greater Antilles today. They suggest that the time of introduction of high altitude gymnosperms was post-Oligocene.

5) "Pelomedusidae, *gen et sp. indet*." Wood (1972). A pelomedusid shell (AMNH 1836) from the same San Sebastian Formation as Hollick's plant material and dated as Oligocene, has been pieced together from small fragments that, when I saw them at the American Museum I considered unrecognizable except as turtle (Williams 1950). Although this totally unexpected record was published in the Breviora series of the Museum of Comparative Zoology in 1972, I have never seen another reference to it in the relatively voluminous literature of West Indian biogeography.

The characters of the reconstructed turtle are not merely pelomedusid but separate it quite clearly from the marine pelomedusids that are known fossil in North America. Wood's comments are: "The shape of the mesoplastron and the vertebral scute pattern might be considered sufficiently distinctive for recognition of either a new species of *Podocnemis* or perhaps even of a new genus....However, I prefer to refrain from giving it any kind of formal taxonomic designation, whether at the generic or specific level...."

Peter Meylan and I have re-examined this fossil at the American Museum of Natural History and we are in general agreement with Wood's assessment. It is a pelomedusid and quite probably not any known form. It would, we feel, be a major research project to go beyond Wood's conservative assignment and attempt any definitive description or discussion of relationships.

The collector of this extraordinary fossil, Narciso Rabell, is known. Wood reports: "According to the son of Señor Rabell Cabrero, who still lives in San Sebastian, the specimens that his father collected probably came from limestone exposures along the main road from San Sebastian to Lares (personal communication, Mr. and Mrs. Alan Patterson)." Additional material collected by Rabell was mentioned by Wood, but was assumed at the time of Wood's writing to be lost. More information is now available (Wyss and MacPhee in prep.). The Rabell collection has, in fact, remained in the care of the Rabell family since the death of the elder Rabell in 1927. By the intervention of Ross

MacPhee and Andre Wyss, who has endeavored to re-collect Narciso Rabell's localities, the collection has been donated by the family, the invertebrates to the Department of Biology of the University of Puerto Rico and the vertebrates to the Vertebrate Paleontology Department of the American Museum of Natural History. (The newly recovered collection at the American Museum contains, as Peter Meylan again showed me, additional material of Wood's pelomedusid that will undoubtedly help unravel the relationships of this San Sebastian turtle.)

The detailed records, including locality information, compiled by Narciso Rabell, have very unfortunately been destroyed by time and termites; however, Wyss and MacPhee (MS) have been able to establish, with high probability, that AMNH 1836 was collected along the gorge of the Collazo Creek, thus in the same general area that provided the most important Tertiary plant fossils reported by Hollick, many of the latter, in fact, collected by Narciso Rabell - among them eleven new plant species, including four palms.

A marine component of the same San Sebastian formation has produced the sirenian *Caribosiren turneri* (Reinhart 1959). The type is recorded as from "road between Sebastian and Lares...on side of hill in bluish to buff-colored arenaceous limestone; abundant foramifera and mollusca associated in the same member." Again some additional material was provided by the Rabell family.

Wood believed that "the close proximity, if not coincidence, of the localities" where *Caribosiren* and the pelomedusid were found implied that the pelomedusid was marine in habitat. Despite the fact that few fragments of matrix adhering to the shell are reported to have belonged to a "limestone facies," I hesitate to make the same judgment; no other members of this section of the pelomedusids occur in anything but freshwater, and the San Sebastian formation has a wide variety of lithofacies, as Wood, in fact, mentions, and, along the same road, there were many undoubted terrestrial plant fossils. Unquestionably in the case of this turtle we deal with near shore and immediately offshore deposits, but, I think, we must remain uncertain of the exact ecology and habitat of the turtle.

MacPhee and Wyss have discovered additional material of several sirenians and a one meter shell of "a pleurodire turtle" in the Miranda Sand member of the Cibao Formation - Miocene or at the Miocene-Oligocene boundary. The Miranda Sand Member is described as "a channel deposit composed apparently of fluvial sediments deposited near the coast." Again there are no convincingly terrestrial fossils of Oligocene-Miocene age yet found in Puerto Rico except Hollick's well-preserved plants. Wyss and MacPhee (in prep.) plan a summary of all the known pre-Quaternary vertebrates of Puerto Rico.

In Jamaica the yellow limestone of Eocene age has yielded one of the oldest dugongs, *Prorastomus sirenoides* Owen (R.J.G. Savage 1975). This is again a near shore deposit, and Versey (1962) reports that West of Kellitts, in Clarendon, fossil remains of palms have been collected in "the generally unfossiliferous Yellow Limestone clastics," which include clays, silts, sands and tuffs.

To me it is clear that strand and near shore strata such as these and the others I have mentioned appear to offer the best hope of old freshwater or terrestrial vertebrates.

6) The fossils of the Dominican amber: Two lizards - *Anolis dominicanus* Rieppel 1980, and *Sphaerodactylus dommeli* Böhme 1984- an *Eleutherodactylus* (Poinar and Cannatella 1987), mammalian hair, probably rodent (Poinar in press) and the very impor-

tant ant fauna (Wilson 1985a-e; Baroni-Urbani and Wilson 1987) have thus far been described from Dominican ambers, early Miocene or late Oligocene, or even Eocene in age. Other members of the faunas of Dominican amber continue to be discovered. Poinar in Poinar and Cannatella 1988 cites personal knowledge, including unpublished records, of six amber *Sphaerodactylus*, four *Anolis*, seven frogs, six amber pieces with mammalian hair, three bird feathers. A number of additional arthropod groups in addition to the ants are also known in substantial numbers. See Baroni-Urbani and Saunders 1982, Poinar 1982, Wunderlich 1986, Poinar and Cannatella 1988.

Both lizard genera have distinctive digital adaptations associated with arboreality that are readily seen in the amber fossils. There is the expected restriction in this type of fossil entrapment to small and arboreal taxa.

The *Anolis* was described by Rieppel as quite similar to the modern green anoles of Hispaniola, therefore to a relatively primitive ecomorph within the genus (Williams 1972, 1983). It is identified as an anole by the toe pads that are distinctive of the anoline group within iguanids. A fortunate glimpse of the caudal vertebrae shows them to be without transverse processes, thus the specimen is an alpha anole like the modern alpha anoles of the island. Rieppel thought he could go further and stated that *A. dominicanus* was closely related to Recent *A. coelestinus*. This, however, is an opinion that goes beyond the evidence. The reported characters, except for the caudal vertebrae, are all measurements, and are associated with ecology (i.e. are ecomorphic *sensu* Williams 1972, 1983). The species is, however, an anole of modern aspect and without evident specializations. According to Baroni-Urbani and Saunders (1982, text and plate III) high quality X-rays show "the skeleton to be almost completely preserved." No details have yet been published, and Rieppel (in litt.) tells me that the skeleton, while present, is poorly preserved.

The one *Sphaerodactylus*, thus far described, is more distinctive, especially so in a long neck that is unique in the genus (Böhme 1984). The toe pads again permit identification to genus.

The published identification of the amber *Eleutherodactylus* (Poinar and Cannatella 1987) is less illuminating: "Details of the osteology (shape of the vomer and its dentition, features of the vertebral column, pectoral and pelvic girdles) permit confident assignment to the genus *Eleutherodactylus* (family Leptodactylidae)...." The features mentioned are not further discussed but are of a sort that is ordinarily used at the subfamily, not generic level. The T-shaped terminal phalanges diagnostic of *Eleutherodactylus* (Hedges this volume) are not mentioned.

The ant fauna, much more adequately known than the two lizards or the frog, is richer in native (i.e, non-introduced) genera than the Recent fauna and contains the first army ants known from the West Indies, as well as the genus *Paraponera* and other forest genera no longer present in the islands, and most surprisingly (now apparently confirmed) *Leptomyrex*, a genus and subfamily now confined to the Indo-Australian region. Wilson's (1985d, 1988) summary supports the view that Hispaniola was "larger and closer to the mainland" than at present. He comments also that "moister tropical forests may have covered part of the island during the Tertiary Period."

Clearly there are old fossils, if not so far back as the Paleocene certainly as far back as the Eocene or Oligocene. Clearly they could have much to tell us. They tell us relatively little now, except that there may be surprises in store for us.

Surprises there are already, and it is not only old fossils that may provide them:

Olson and Kurochkin (1987) report two bones from widely separated Pleistocene cave deposits in Cuba (Isle of Pines off western Cuba and Camaguey Province in central Cuba) that are the only record of a bird superfamily (Furnarioidea) otherwise unknown in the West Indies. Olson and Kurochkin refer the bones to small (wren-sized) nearly flightless terrestrially adapted birds (genus *Scytalopus*) which are of South America affinities. How these birds got to Cuba is problematic. Olson and Kurochkin prefer cross-water transport, direct from South America on what some have called the "ever ready raft"; others will, without question, prefer the "convenient isthmus." In fact, at the moment, the discovery merely unveils a mystery.

Care must be taken not to over-interpret any discovery. Poinar and Cannatella (1987) have urged that the age of the amber *Eleutherodactylus* and the amber *Anolis* (from the same La Toca mine, now dated as Eocene, *fide* Lambert et al. 1985) tend to support vicariance rather than dispersal. This view has been contested by Mayer and Lazell (1988) on several grounds - the most important of which is that the ages of the amber fossils neither prove nor disprove vicariance. Eocene age is favorable to vicariance in the sense that it is permissive (i.e. possible according to current geological opinion). However, it is not definitive; dispersal might have occurred during the Eocene or earlier if island masses were available to be colonized cross-water - and when distances to relevant mainlands may have been less than now.

It is, in fact, from a vicariance point of view, uninformative when the old vertebrate fossils of the West Indies belong to groups, even genera, that are still there. Such fossils are not relevant to the point most critical in terms of vicariance theory - the demand for massive extinctions to explain the depauperate character of the present-day fauna. Support for the Rosen-Savage vicariance hypothesis will come from demonstration of the presence in Early Tertiary strata in a West Indian island of some taxon not now known in the Antilles, or preferably a fauna, that could have been present on some sundered portion of the mainlands in the Paleocene or Eocene. However, only a balanced continental fauna will be decisive. Among mammals a marsupial or an archaic ungulate, among lizards something perhaps crotaphytine-like would not be proof positive, but they would be stronger, more impressive relevant evidence, requiring dispersalists to re-examine their position.

Wilson's data, in this regard, is distinctly more favorable to a vicariance scenario. He has provided the only description of an old fauna from the West Indies, one with genera and one subfamily not previously known from the West Indies, even the apparent confirmed presence of one subfamily now known living only from the Indo-Australian region (although once almost world-wide). His own conclusions have been cautious. In keeping with his co-authorship of MacArthur and Wilson 1963, he has not mentioned the vicariance-dispersal controversy. Instead, he has emphasized the equilibrium aspect of the change between the Eocene/Miocene fauna and that now present. The number of ant genera has remained the same, despite a substantial change in generic identity.

I reiterate, that if we have only our present data, be it geological or biological and despite the old fossils at present known, there are no secure answers to be had. Each theorist-biogeographer will choose an answer to suit his own bias and his own background, and place upon his opponent the burden of proof. But it is plain beyond all argument that we all suffer under the burden of ignorance.

What the old fossils prove to me is that there are data to be found. I do not disparage the dedicated efforts that have been and are continuing to be made to gather data relevant to the West Indian puzzle. Much new information is reported in this symposium. But there is the clear fact in front of us that the one investigation that is most needed - an all-out paleontological search for old vertebrate fossils - has not been made. Paleontological discoveries have been made, but by whom? By mammalogists, ornithologists, herpetologists, botanists, and sometimes amateurs. Where have most of these people looked? In caves. Where have the oldest fossils thus far been found? Never in caves, but in dated beds with or adjacent to Tertiary plant fossils - or in amber.

A professional paleontological search needs to be made - by paleontologists, professionals both geologically trained and skilled in the latest paleontological methods. (Ross MacPhee and Andre Wyss have begun such a search in Puerto Rico.)

There is an obvious reason why the effort of the kind needed has not made. It is not at all evident that success would be immediate. If the West Indies dripped with large vertebrate fossils, discoveries would already have been made. If any locality, any conspicuous exposure in old rocks of any West Indian island, had demonstrated a rich fossil bed to a trained eye, that fossil bed would have already been harvested.

The old fossils have been hard to come by. There are several problems here. One is that on a luxuriantly vegetated tropical island exposures may not be readily seen. Reinhart, for example, in discussing the stratigraphy of *Caribosiren*, says: "Rank growth of vegetation plus a heavy soil mantle which covered and obliterated the strata of most of the surrounding area hindered determination of the precise stratigraphic position." MacPhee (in litt.) comments similarly: "...working in rain forest is very difficult." This difficulty is quite real, but there are arid areas on every - or almost every - West Indian island.

Another problem is clearly the fact that one of the two hypotheses about West Indian faunas says that the faunas were never rich in the continental sense, i.e. rich in diverse taxa. However, if the West Indian faunas show us anything, they show us remarkable and unexpected radiations of the few taxa that are in front of us. In any case, more important than any problem may be the simple remedy of trained eye and trained effort.

I have myself looked for fossils in caves, and found them. I have looked in more desultory fashion for the Limestone Quarry at Corozal and found nothing. I have been a describer of fossils, some of them important, but I have never in my life found, or, indeed, looked for, a significant old fossil. I was never trained to do so.

The search for old fossils in the West Indies cannot be desultory or untrained. It must be fully professional. What rewards can we offer the young professional, someone who has his name still to make? Two rewards: 1) the solution of a puzzle - this ought to be a reward in itself; 2) the decision of a controversy: fossil combined with geological evidence alone has the possibility of settling the dispersal-vicariance dispute. I despair of the solution of old problems unless new opportunities are utilized.

POSTSCRIPT: I am told by Malcolm McKenna and Ross MacPhee that the search for old fossils continues in Puerto Rico, and by Charles Woods and Gary Morgan that a similar search will soon begin in Hispaniola. There are still no new discoveries, but the hunt is on!

Acknowledgments

I am deeply grateful to Charles Woods for the opportunity to participate in this Symposium and for his encouragement at all times. I am very appreciative of the valuable comments made by him and by Michael Perfit, George Burgess, Gary Morgan, Karl Koopman, Malcolm McKenna, Richard Estes, Gregory Pregill, P.E. Vanzolini, Blair Hedges, Robert O'Hara, Glenn Flores, Robert Ross, and John Reiss.

Gregory Mayer has read multiple drafts, kept me from serious errors of quotation, and has helped in structuring the argument. Ross MacPhee has shared much of his unpublished information with me. Andre Wyss has shown me some of the newly recovered Rabell collection. Peter Meylan has been particularly helpful in the re-examination of the fossil pelomedusid from Puerto Rico. I am indebted also, more than I have ever said or can say, to the many who have over long years assisted me in West Indian exploration and study.

Literature cited

Arden, D.D., Jr. 1975. Geology of Jamaica and the Nicaragua Rise. Pp. 617-661 in A.E.M. Nairn and F.G.Stehli (eds.). The Ocean Basins and Margins 3: The Gulf of Mexico and the Caribbean. Plenum Press, New York.

Barbour, T. 1914. A contribution to the zoogeography of the West Indies, with special reference to amphibians and reptiles. Memoirs of the Museum of Comparative Zoology 44:209-359.

_____. 1916. Some remarks upon Matthew's "Climate and Evolution." Annals of the New York Academy of Science 27:1-15.

_____. 1917. Notes on the herpetology of the Virgin Islands. Proceedings of the Biological Society of Washington 30:97-104.

Baroni-Urbani, C. and J.B. Saunders. 1982. The fauna of the Dominican Republic amber: the present status of knowledge. Transactions of the 9th Caribbean Geological Conference, Santo Domingo, Dominican Republic 1980 [1982] 1:213-223.

_____, and E.O. Wilson. 1987. The fossil members of the ant tribe Leptomyrmecini (Hymenoptera: Formicidae). Psyche 94:1-8.

Berry, E.W. 1922. Tertiary fossil plants from the Dominican Republic. Proceedings of the United States National Museum 59:117-126.

_____. 1923. Tertiary fossil plants from the Republic of Haiti. Proceedings of the United States National Museum 14:1-10.

Böhme, W. 1984. Erstfund eines fossilen Kugelfingergecko (Sauria: Gekkonidae: Sphaerodactylinae) aus Dominikanischen Bernstein Oligozän von Hispaniola, Antillen. Salamandra 20:212-220.

Briggs, J.C. 1984. Freshwater fishes and biogeography of Central America and the Antilles. Systematic Zoology 24:428-435.

Buskirk, R.E. 1985. Zoogeographic patterns and tectonic history of Jamaica and the northern Caribbean. Journal of Biogeography 12:445-461.

Bussing, W.A. 1985. Patterns and distribution of the Central American ichthyofauna. Pp. 453-473 in F.G Stehli and D.Webb (eds.). The Great American Biotic Interchange. Plenum Press, New York.

Cockerell, T.B.A. 1924. A fossil cichlid fish from the Republic of Haiti. Proceedings of the United States National Museum 63(7):1-2.

Coney, P.J. 1982. Plate tectonic constraints on the biogeography of Middle America and the Caribbean region. Annals of the Missouri Botanical Garden 69:432-443.

Croizat, L. 1962 [1964]. Space, Time, Form: The Biological Synthesis. 881 pp. Published by author, Caracas.

Darlington, P.J. 1938. The origin of the fauna of the Greater Antilles, with discussion of dispersal of animals over water and through the air. Quarterly Review of Biology 13:274-300.

_____. 1957. Zoogeography: The Geographical Distribution of Animals. John Wiley and Sons, New York. 675 pp.

Dickinson, W.R. and P.J. Coney. 1980. Plate tectonic constraints on the origin of the Gulf of Mexico. Pp. 27-36 in R.H. Pilger (ed.). The Origin of the Gulf of Mexico and the Early Opening of the Central North Atlantic Ocean. Proceedings of a symposium at Louisiana State University, Baton Rouge, LA.

Donnelly, T.W. 1985. Mesozoic and Cenozoic plate evolution of the Caribbean region. Pp. 89-121 in F. G. Stehli and S. D. Webb (eds.). The Great American Biotic Interchange. Plenum Press, New York.

Durham, J.W., A.R.V. Arellano, and J.H. Peck. 1952. No Cenozoic Tehuantepec seaways. Bulletin of the Geological Society of America 63:1245.

Etheridge, R. 1960. The relationships of the anoles (Reptilia: Sauria: Iguanidae): An interpretation based on skeletal morphology. Unpublished Ph.D. Dissertation, University of Michigan, Ann Arbor, MI. 236pp.

Ford, S. M. 1986a. Subfossil platyrrhine tibia (Primates: Callitrichidae) from Hispaniola: A possible further example of island gigantism. American Journal of Physical Anthropology 70:47-62.

_____. 1986b. Systematics of the New World monkeys. Pp. 73-135 in D.R. Swindler and J. Erwin (eds.). Comparative Primate Biology 1: Systematics, Evolution and Anatomy. Alan R. Liss, Inc., New York.

Ford, S.M., and G.S. Morgan. 1986. A new ceboid femur from the Late Pleistocene of Jamaica. Journal of Vertebrate Paleontology 6:281-289.

Goodfriend, G.A., and R.M. Mitterer. 1987. Age of the ceboid femur from Coco Ree, Jamaica. Journal of Vertebrate Paleontology 7:344-345.

Graham, A., and D.M. Jarzen. 1969. Studies in Neotropical paleobotany. I. The Oligocene communities of Puerto Rico. Annals of the Missouri Botanical Garden 56:308-357.

Guyer, C., and J.M. Savage. 1986. Cladistic relationships among anoles (Sauria: Iguanidae). Systematic Zoology 35:509-531.

Hardy, J.D., Jr. 1982. Biogeography of Tobago, West Indies, with special reference to amphibians and reptiles: A review. Bulletin of the Maryland Herpetological Society 18:37-142.

Heatwole, H. and R. Levins. 1972. Biogeography of the Puerto Rican Bank: flotsam transport of terrestrial animals. Ecology 53:112-117.

Hedges, S.B. 1982. Caribbean biogeography: Implications of recent plate tectonic studies. Systematic Zoology 31:518-522.

Hershkovitz, P. 1970. Notes on Tertiary platyrhine monkeys and description of a new genus from the late Miocene of Colombia. Folia Primatologica 12:1-37.

Hollick, A. 1928. Paleobotany of Porto Rico. Scientific Survey of Porto Rico and the Virgin Islands 7:177-238.

King, W. 1962. The occurrence of rafts for dispersal of land animals into the West Indies. Quarterly Journal of the Florida Academy of Science 25:45-52.

Kluge, A.G. 1969. The evolution and geographical origin of the New World *Hemidactylus mabouia-brookii* complex (Gekkonidae, Sauria). Miscellaneous Publications of the Museum of Zoology, University of Michigan 138:1-78.

Lambert, J.B., J.S. Frye, and G.O. Poinar, Jr. 1985. Amber from the Dominican Republic: analysis by nuclear magnetic resonance spectroscopy. Archaeometry 27:43-51.

Levins, R., and H. Heatwole. 1963. On the distribution of organisms on islands. Caribbean Journal of Science 3:173-177.

_____. 1973. Biogeography of the Puerto Rican bank: introduction of species onto Palo minitos Island. Ecology 54:1056-1064.

MacArthur, R. H., and E.O. Wilson. 1963. An equilibrium theory of insular biogeography. Evolution 17:373-386.

MacFadden, B.J. 1980. Rafting mammals or drifting islands: biogeography of the Greater Antillean insectivores *Nesophontes* and *Solenodon*. Journal of Biogeography 7:11-22.

_____. 1981. Comments on Pregill's appraisal of historical biogeography of Caribbean vertebrates: Vicariance or dispersal or both. Systematic Zooloogy 30:370-372.

MacPhee, R.D.E. 1984. Quaternary mammal localities and heptaxodontid rodents of Jamaica. American Museum Novitates 2803:1-34.

Malfait, B.T., and M.G. Dinkelman. 1972. Circum-Caribbean tectonic and igneous activity and the evolution of the Caribbean Plate. Bulletin of the Geological Society of America 83:251-272.

Marshall, L.G., R.F. Butlar, R.E. Drake, G.N. Curtiss, and R.H. Tedford. 1979 Calibration of the Great American Interchange. Science 204:272-279.

Matthew, W.D. 1915. Climate and evolution. Annals of the New York Academy of Science 24:171-213.

Mayer, G.C., and J.D. Lazell, Jr. 1988. Significance of frog in amber. Science 239:1477-1478.

Mayr, E. 1942. Systematics and the Origin of Species Columbia University Press, New York. 334 pp.

_____. 1946. History of the North American bird fauna. The Wilson Bulletin 58:3-41.

McDowall, R.M. 1978. Generalized tracks and dispersal in biogeography. Systematic Zoology 27:88-104.

McKenna, M. C. 1973. Sweepstakes, filters, corridors, Noah's arks and beached Viking funeral ships in paleogeography. Pp. 295-308 in D.H. Tarling and S.K. Runcorn (eds.). Implications of Continental Drift to the Earth Sciences. Academic Press, New York.

_____. 1983. Holarctic landmass rearrangement, cosmic events and Cenozoic terrestrial organisms. Annals of the Missouri Botanical Garden 70:459-4.

Nelson, G., and D.E. Rosen (eds.). 1981. Vicariance Biogeography - A Critique. Columbia University Press, New York. 593pp.

Olson, S.L. 1976. Oligocene fossils bearing on the origins of the Todidae and Momotidae (Aves: Coraciiformes). Pp. 353-357 in S.L.Olson (ed.). Collected Papers in Avian Paleontology Honoring the 90th Birthday of Alexander Wetmore. Smithsonian Contributions to Paleobiology 27.

_____, and E.N. Kurochkin. 1987. Fossil evidence of a tapaculo in the Quaternary of Cuba (Aves: Passeriformes: Scytalopodidae). Proceedings of the Biological Society of Washington 100:353-357.

Parenti, L.R. 1981. Discussion [of Patterson (1981)]. Pp. 490-497 in G. Nelson and D.E. Rosen (eds.). Vicariance Biogeography - A Critique. Columbia University Press, New York.

Patterson, C. 1981. Methods of paleobiogeography. Pp. 446-489 in G. Nelson and D. E. Rosen (eds.). Vicariance Biogeography - A Critique. Columbia University Press, New York.

Pindell, J., and J.F. Dewey. 1982. Permo-Triassic reconstruction of Western Pangaea and the evolution of the Gulf of Mexico/Caribbean Region. Tectonics 1:179-211.

Poinar, G.O., Jr. 1982. Sealed in amber. Natural History 91(6):26-30.

_____, and D.C. Cannatella. 1987. An Upper Eocene frog from the Dominican Republic and its implication for Caribbean biogeography. Science 237:1215-1216.

_____. 1988. Response [to Mayer and Lazell 1988] . Science 239:1478.

Pregill, G.K. 1981. An appraisal of the vicariance hypothesis of Caribbean biogeography and its application to West Indian terrestrial vertebrates. Systematic Zoology 30:147-155.

Raven, P.N., and D.I. Axelrod. 1973. History of the flora and fauna of Latin America. American Scientist 63:420-429.

Reinhart, R.H. 1959. A review of the Sirenia and Desmostylia. Bulletin of the Department of Geological Sciences, University of California Publications 36:1-146.

Rich, P.V., and T.H. Rich. 1983. The Central American dispersal route: Biotic history and paleogeography. Pp. 12-34 in D.H. Janzen (ed.). Costa Rican Natural History. University of Chicago Press, Chicago, IL.

Rieppel, O. 1980. Green anole in Dominican amber. Nature 286:486-487.

Rivas, L.R. 1986. Comments on Briggs (1984): Freshwater fishes and biogeography of Central America and the Antilles. Systematic Zoology 35:633-639.

Rosen, D.E. 1976. A vicariance model of Caribbean biogeography. Systematic Zoology 24:431-464. [The date of this paper is inconsistently cited by its author. The abstract gives the date as 1976, but the cover and title page of the reprint give "December, 1975." Rosen (1978) in both text and bibliography cites the date of this paper as 1975. Rosen (1985) in both text and bibliography cites the identical paper as 1976.]

_____. 1978. Vicariant patterns and historical explanation in biogeography. Systematic Zoology 27:159-188.

_____. 1985. Geological hierarchies and biogeographic congruence in the Caribbean. Annals of the Missouri Botanical Garden 72:636-659.

Savage, J.M. 1966. The origins and history of the Central American herpetofauna. Copeia 1966:719-766.

———. 1974. The Isthmian link and the evolution of Neotropical mammals. Contributions in Science of the Los Angeles County Natural History Museum 260:1-51

———. 1982. The enigma of the Central American herpetofauna: dispersals or vicariance? Annals of the Missouri Botanical Garden 69:464-547.

Savage, R.J.G. 1977. Review of early Sirenia. Systematic Zoology 25:344-351.

Scharff, R.F. 1912. Distribution and origin of life in America. MacMillan, New York. 497 pp.

Schuchert, C. 1935. Historical geology of the Antillean-Caribbean region. John Wiley & Sons, New York. 811 pp.

Simpson, G.G. 1940. Mammals and land bridges. Journal of the Washington Academy of Science 30:137-163.

———. 1943. Turtles and the origin of the fauna of Latin America. American Journal of Science 241:413-429.

———. 1952. Probabilities of dispersal in geologic time. Bulletin of the American Museum of Natural History 99:163-176.

———. 1956. Zoogeography of West Indian land mammals. American Museum of Natural History Novitates 1759:1-28.

———. 1965. The Geography of Evolution. Chilton Books, Philadelphia, PA. 249 pp.

Tedford, R.H. 1974. Marsupials and the new paleogeography. Pp. 109-126 in C.A. Ross (ed.). Paleogeographic Provinces and Provinciality. Special Publication of the Society for Economic Paleontology and Mineralogy No. 21.

Tee-Van, J. 1935. Cichlid fishes in the West Indies with special reference to Haiti, including a description of a new species of Cichlasoma. Zoologica New York 10:281-300.

Uyeda, S. 1978. The new view of the earth. Moving continents and moving oceans. W.H. Freeman and Company, San Francisco. 217 pp.

Versey, H.R. 1962. Older Tertiary limestones Pp. 26-43. in V.A. Zans et al. Synopsis of the Geology of Jamaica.

Whitmore, F.C. and R.H. Stewart. 1965. Miocene mammals and Central American seaways. Science 148:180-185.

Williams, E.E. 1950. *Testudo cubensis* and the evolution of Western Hemisphere tortoises. Bulletin of the American Museum of Natural History 95:1-36.

———. 1969. The ecology of colonization as seen in the zoogeography of anoline lizards on small islands. Quarterly Review of Biology 44:345-389.

———. 1972. The origin of faunas. Evolution of lizard congeners in a complex island fauna: a trial analysis. Evolutionary Biology 6:47-89.

———. 1983. Ecomorphs, faunas, island size and diverse end points. Pp. 326-370, 481-483, 490 in R.B.Huey, E.R. Pianka, and T.W. Schoener (eds.). Lizard Ecology: Studies of a Model Organism. Harvard University Press, Cambridge, MA.

———, and K.F. Koopman. 1951. A new fossil rodent from Puerto Rico. American Museum Novitates 1515:1-9.

———, and K.F. Koopman. 1952. West Indian fossil monkeys. American Museum of Natural history Novitates 1546:1-16.

Wilson, E.O. 1985a. Invasion and extinction in the West Indian ant fauna: evidence from the Dominican amber. Science 229:265-267.

_____. 1985b. Ants of the Dominican amber (Hymenoptera: Formicidae). 1. Two new myrmicine genera and an aberrant *Pheidole*. Psyche 92:1-9.

_____. 1985c. Ants of the Dominican amber (Hymenoptera: Formicidae). 2. The first fossil army ants. Psyche 92:11-16.

_____. 1985d. Ants of the Dominican amber (Hymenoptera: Formicidae). 3. The subfamily Dolichoderinae. Psyche 92:17-37.

_____. 1985e. Ants of the Dominica amber (Hymenoptera: Formicidae). 4. A giant ponerine in the genus *Paraponera*. Israel Journal of Entomology 19:197-200.

_____. 1988. The biogeography of the West Indian ants (Hymenoptera: Formicidae). Pp. 214-230 in J K. Liebherr (ed.). Zoogeography of Caribbean Insects. Cornell University Press, Ithaca, NY.

Wood, R.C. 1972. A fossil pelomedusid turtle from Puerto Rico. Museum of Comparative Zoology, Breviora 392:1-13.

Wunderlich, J. 1986. Spinnenfauna gestern und heute. Fossile Spinnen in Bernstein und ihre heute lebenden Verwandten. Erich Bauer Verlag bei Quelle & Meyer, Wiesbaden. 283 pp.

Table 1. Flightless Land Mammal Families in the Greater Antilles.

PRESENT	ABSENT
Insectivores	Marsupials
Solenodontidae (CH)	Carnivores
Nesophontidae (CHP)	Ungulates
Monkeys	Elephants
Cebidae (H)	Rabbits
Callitricidae (JH)	Armadillos
Atelidae (C)	Anteaters
Ground Sloths	Most rodents
Megalonychidae (CHP)	All early Cenozoic orders,
Rodents	(e.g. Pantodonta,Taeniodonta,
Capromyidae (JCHP)	Dinocerata, Embrithopoda,
Heptaxodontidae (JHP)	Tillodontia, Condylarthra,
Echimyidae (CHP)	Multituberculata, etc.)
Cricetidae (J)	

J = Jamaica; C = Cuba; H = Hispaniola; P = Puerto Rico;

Table 2. Absences in the West Indian herpetofauna.

All salamanders	All caecilians
All pelobatids	All ranids
All microhylids	All dendrobatids
All sceloporines	All basiliscines
All gymnopthalmines	All elapids
All crotalines	Most turtles
	Most colubroids
	Polychrus

Table 3. Modes for faunal exchange between continents and islands.

FAUNAS/POPULATIONS
    1. Stable land bridges:
        (permanent/two way)
    2. Periodically interrupted land bridges:
        (periodic/two way)
    3. "Noah's Arks":
        (episodic/one way)
    4. Stepping stone islands:
        (stochastic/probability moderate)

SPECIES/FOUNDERS
    5. Oceanic islands:
        (stochastic/probability low)

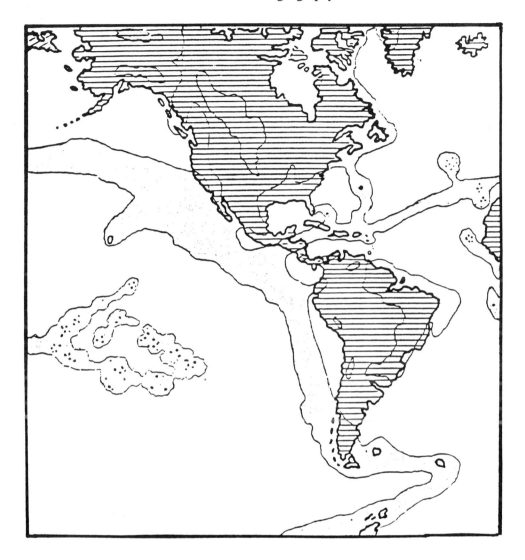

Fig. 1. North and South America at the beginning of the Tertiary. Land is indicated by a close stipple. From Scharff (1912).

Fig. 2. Present day water gaps between Central America and the Greater Antilles. From Rivas (1986).

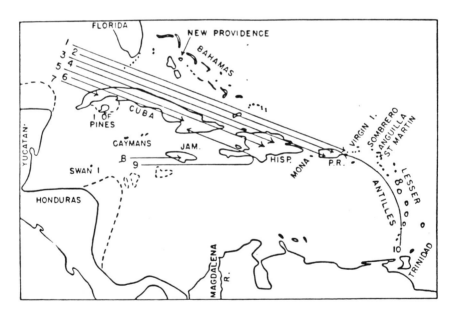

Fig. 3. West Indian distributions and distances. From Darlington (1957:510). The arrows were intended by Darlington to show distance and direction of dispersal of certain mammals, frogs and fish. He believed that this diagram demonstrated "the limited diversity of the Greater Antillean fauna, its orderliness and the fact that the most significant parts of it form a simple pattern of apparent immigration, mostly from Central America, with Cuba the most important port of entry."

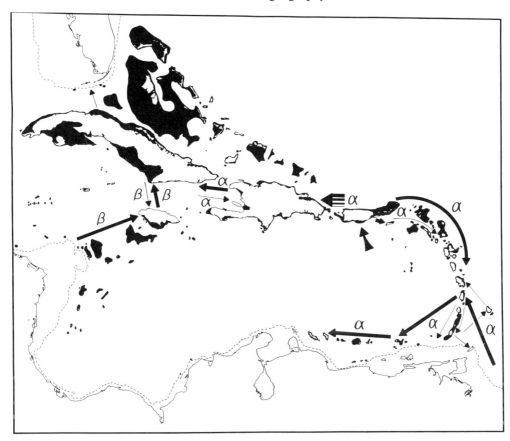

Fig. 4. Presumed paths of dispersal of the lizard genus *Anolis* to and within the Antilles. (Dispersal to the Bahamas and the small islands is not shown.) The beta anoles are shown colonizing Jamaica and Cuba only, with the alpha anoles colonizing the remaining islands and also invading Cuba. From Williams (1969). Since that paper I have (Williams 1983) chosen Hispaniola rather than Puerto Rico as the probable port of entry for alpha anoles into the Greater Antilles.

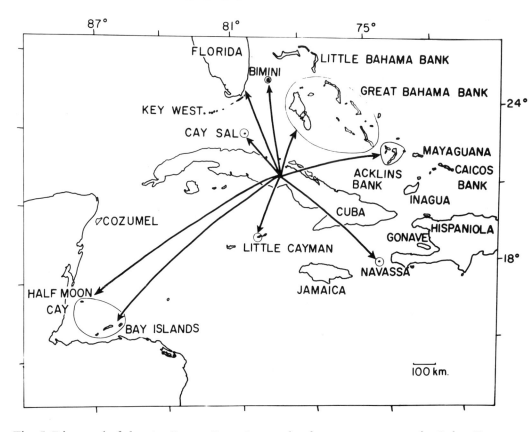

Fig. 5. Dispersal of the *Anolis carolinensis* complex from a source area in Cuba. From Williams (1969).

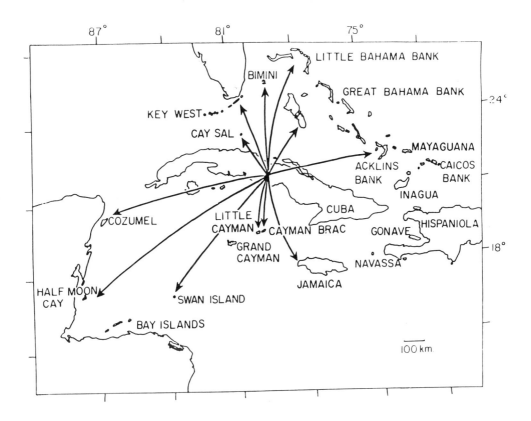

Fig. 6. Dispersal of *Anolis sagrei* from a source area in Cuba. From Williams (1969).

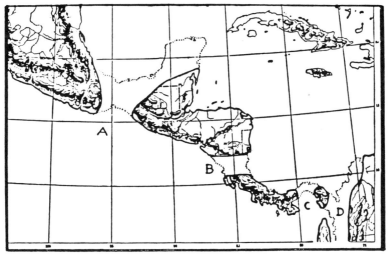

Fig. 7. Stepping stone islands between North and South America during the Tertiary as conceived by Mayr (1946). (Neither the land masses nor the water gaps correspond at all to modern ideas of the geology of the region.

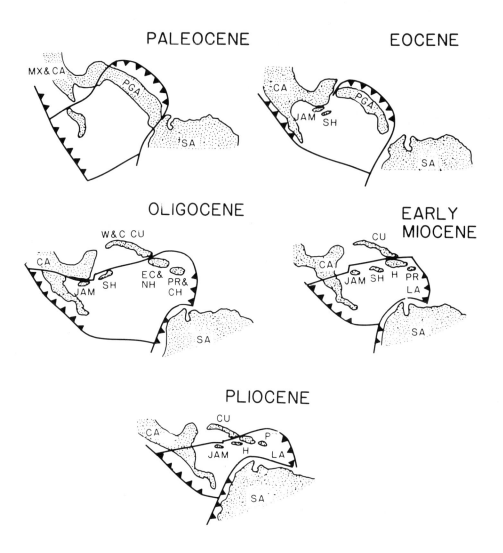

Fig. 8. Geological history of the West Indies *fide* Guyer and Savage 1986.

GEOLOGICAL CONTRAINTS AND BIOLOGICAL RETRODICTIONS IN THE EVOLUTION OF THE CARIBBEAN SEA AND ITS ISLANDS

Michael R. Perfit[1] and Ernest E. Williams[2]

Abstract

A geologist and biologist collaborate in the effort to integrate geological evidence and biological theory as they relate to Caribbean history. The methods used in obtaining geological evidence are described, and their limitations discussed. The present state of geological knowledge of the Caribbean region is summarized in terms of four periods stretching from the Late Cretaceous to the Present. Deficiencies in the present evidence are pointed out. The dispersal and vicariance theories of biogeography are contrasted, and their strengths and weaknesses estimated. A concluding appraisal of recent efforts at interpreting known geological history in terms of either of the competing biogeographic theories comes down on the side of caution and the need for new evidence.

Introduction

The historical biogeography of islands has to begin with geological data - the when, where, and how of the islands, their origins, emergence and submergence, changes in size, changes in position relative to mainlands and other islands. Biogeography, of necessity, is a discipline requiring the cooperation of biology and geology.

Unfortunately there is a real schism between geology and biology - a failure of communication. Fundamental ideas in both directions may be badly misunderstood, and, quite as importantly, what can and cannot be known is differently evaluated on the two sides.

The collaboration of a geologist and a biologist is here attempted with the intention of mitigating this problem, at least for the specific area of the West Indies.

It is our opinion that the deficiency in understanding is especially serious on the biological side where the new concepts of the still nascent area of global plate tectonics are concerned. The relative recency of plate tectonic theory is, perhaps, the reason. Plate tectonics - the modern concept of a dynamic rather than a stable earth - derives its foundations from the early theories of continental drift, but it did not become widely accepted until the revolution in the Earth Sciences in the mid-1960s. The major aspects of the instability of the earth's surface - spreading seafloors, lateral movement of conti-

---

[1] Michael R. Perfit is Associate Professor of Geology at the University of Florida, Gainesville, FL. 32611.
[2] Ernest E. Williams is Curator Emeritus at the Museum of Comparative Zoology, Cambridge, MA. 02138.

nents, formation of island arcs, earthquakes and volcanos, consumption of the sea floor - are now explicable, if still not completely understood, with the aid of the plate tectonic theory. However, much of the earth's geologic history has had to be reinterpreted as our understanding of plate tectonics has grown. An exciting and stimulating field of study encompassing all of the geological sciences, it is also conspicuously still a minefield of controversy. What started as a simple and elegant unifying theory consistent with most observed geological features has become increasingly complex as it tries to explain the wider ranges of observations and the more complex geologic histories.

Fortunately there are now some common perspectives on what are solid data and what still must be speculation. We will have to begin our attempts to decipher the geologic history of the Caribbean, as we would with any region - and this may be on any scale - by attempting to reconcile the observed or measured geologic and geophysical data with the basic precepts of plate tectonic theory. Then we will need to asess the adequacy of our data and the limitations of our methods. Only after these preliminaries can we go on to evaluate hypotheses about the history of the Caribbean sea and its islands, identifying the constraints on the evolution of so complex a region, keeping in mind always that these constraints - and all the evidence bearing on them - range over a full spectrum of higher to lower levels of certainty.

We will be concentrating our attention on the history of the Greater and Lesser Antilles, thus primarily on the northern and eastern boundaries of the Caribbean Plate. Necessarily we shall be mentioning the western boundary - Central America and the Panamanian isthmus - and the southern boundary - a complex region - but these are peripheral to the main concerns of this symposium. We shall cite them when they are important to our story.

Our purpose will be to describe what we know about the geological history of the islands - and how we have come to know what we know - as a framework for understanding the history of the biota. We have no doubt that the history of life and the history of land are strongly correlated (Croizat 1962).

We begin at a very elementary level by discussing the methods that must be used to gather geological evidence, and the special problems that have been the occasion of difficulty in the West Indies. After summarizing the evidence that has so far been collected and pointing out its limitations, we attempt a very provisional tectonic interpretation of this evidence, admitting alternative interpretations (with examples), and then point out the obvious: that more work is needed. We continue with a statement of the biological problems presented by islands, some suggested general principles for interpreting the relationship of geography to island biology - and then, with some diffidence, point out where, when, and to what extent the major suggested scenarios of Caribbean biogeography appear to be possible, <u>given the current state of our knowledge or lack of knowledge of Caribbean geology.</u>

The geological side of the story:
Constraints and interpretations

A geological preamble

A general remark first: A basic premise of plate tectonics is that a few large plates along with numerous smaller plates and blocks of lithosphere cover the surface of the earth and constantly move and interact with one another along plate boundaries: ridges, trenches, transform faults, and collision zones. The Caribbean is one of the smaller of these plates, yet it is divisible into more than 100 geologic provinces or terranes (Case et al. 1984). It is surrounded on all sides by major plates - the North American, South American, and Cocos and Nazca Plates. It has been at the mercy of these much larger, dominating plates, with resulting complexity and tectonic diversity all along the plate boundaries. In some ways the Caribbean Plate has acted as a buffer between the major plates. Our attempts to decipher the evolution of the Caribbean, therefore, must be concerned not only with the composition, age, and history of terranes, but also with the much grander plate interactions. The lesser phenomena cannot be understood except in terms of the greater.

As a consequence of the structure and physical properties of the outer layers of the earth many of the crucially important strata that record the geologic history of any region lie deep within the crust or are covered by oceans or seas beyond the reach of casual survey. The point must be emphasized that only a fraction of the crust that forms islands such as the Antilles is ever exposed at any one time, and the apparent separation of islands by bodies of water does not indicate that they are geologically distinct; they may be part of the same lithospheric plate or terrane. (For example, Hispaniola and Puerto Rico are now geologically on one plate.)

Similarly, the present physical integrity of a land mass need not imply that it was all one unit in the past. Transverse faulting and convergence along active plate boundaries can result in the dismemberment (or accretion) of seemingly rigid land masses. Cuba and Hispaniola are exemplary instances of islands that now are very recognizable units, but are composites of miniplates that once had quite different associations and geographic positions.

Getting the geological evidence

We may now go on to enumerate the several techniques that produce the evidence that, interpreted, must be the geological foundations for Caribbean biohistory.

Mapping.-- The first requirement in the study of the geological history of any region is geological mapping - a record of the geological strata determined from exposures, the nature of the exposed rocks, their structural relationships, and their ages, the latter initially at least, determined from the fossiliferous strata. Geological maps of good quality certainly exist for all the West Indies, but there are certain hazards and limitations special to these tropical islands. They are in large part forested, or the forests have been replaced by cultivation, and the exposures are in consequence fewer than might be ideal. The few outcrops are exposed to intense chemical and physical erosion, and crucially important strata typically extend beneath the surrounding seas. On land, faulting and complex structural deformation may make tracing strata difficult.

Subaerial exposures of bedrock, i.e. those that are above sea level, are only a minor component of the crust that actually forms continental or island masses. The crust that crops out on land is the tip of a lithospheric "iceberg" that floats on more dense but partially molten and plastic asthenospheric material - a portion of the upper mantle. The foundations of land masses extend to considerable depths (generally more than 20 km) below the subaerial surfaces, which are constantly being attacked and diminished by the several agents of erosion.

Drilling.-- In regions like the Antilles, where most of the crust is not exposed, the information from drilling must supplement routine mapping. Drilling has been done both on land and at sea by major oil companies but the poor results of this exploration and the economics of the world oil market in the 1980s has all but stopped exploratory drilling. Present well data are ordinarily in the archives of oil companies and have not been generally available (but see Arden 1975).

During 1969 and 1970-1971 eight holes were drilled in the Caribbean floor during Leg 4 and Leg 15 of the Deep Sea Drilling Project (Edgar et al. 1973). These drillings provided invaluable information about the type and ages of sediments in the Colombian and Venezuelan Basins as well as the nature of the uppermost underlying igneous crust/basement. The basalt that comprises the basement is similar in composition to the typical rocks of a spreading center. It cannot be dated directly (see discussion below), but the basaltic unit underlies or intrudes sediments that contain fossils of Late Cretaceous age (ca. 88-83 myBP). Unfortunately during the early years of the Ocean Drilling Program few holes could be drilled in any one area, and penetration into basement rocks ("hard rock") was limited to a few meters. Now completing Leg 120, the Ocean Drilling Vessel "Resolution" has the capability of re-entering holes and deepening "hard rock" drilling. Plans are being made to drill more and deeper holes in the Caribbean in the 1990s, but until these things are done, knowledge of the age and composition of the basins, rises, and ridges of the submarine Caribbean Plate will be very limited.

Dredging.-- A great deal of useful information about the geology of major submarine features has been obtained by dredging from oceanographic vessels. This rather imprecise technique involves dragging a weighted steel box or tube with a chain-link attachment along the ocean floor by means of a steel cable - generally 1000s of meters long - attached to the moving ship. Successful dredging requires locating bathymetric features with sufficient relief and rock outcrops such that the dredge can either scoop up pieces of rubble or break off rocks from the outcrops. This method further relies on knowing the exact position of the dredge relative to the ship - the position of which was itself not well known before the advent of satellite navigation - and the assumption that the recovered samples were in place or nearly so when the dredge picked them up. The technique is not unlike trying to make a geologic map from samples obtained by a bucket lowered from a helicopter 1000s of meters above the earth's surface.

A general picture of submarine surface geology emerges from repeated dredge hauls, but details inevitably remain obscure. Despite these limitations dredging is a relatively simple and inexpensive method for obtaining geologic information from large submerged areas - provided only a ship is available.

Most of the systematic dredging that has been done in the Caribbean was done by the late Bruce Heezen and his students at the Lamont-Doherty Geological Observatory of Columbia University. When coupled with information obtained by geophysical methods - seismic reflection and refraction, marine magnetics, gravity measurements, seismic data - dredging has resulted in a much clearer picture of the submarine geology of the Caribbean (Fox et al. 1971; Fox and Heezen 1975; Perfit and Heezen 1978, Perfit et al. 1980). Most of the dredging was done during the late 1960s and early 1970s. These results have been compiled by Nagle et al. (1978).

Coring.-- This is a method that involves dropping a length of pipe with a heavy weight attached to the top directly into sediments that cover the sea floor. In general, only the upper layers are recovered (10 meters or so). It is of limited utility, since these sediments will record only the most recent geological history (1 million years at 1 cm per 1000 years sedimentation rate).

Seismic reflection/refraction.-- Because the ocean floor cannot be observed directly - except for rare opportunities provided by submersibles - remote sensing techniques using surface ships must be employed. A brief overview of the geophysical methods commonly used to investigate the sea floor and underlying lithosphere is given below, along with some discussion of their limitations in the analysis of complex regions. More complete descriptions and discussion can be found in Kennett (1982) and Anderson (1986).

The geologic relief/bathymetry of the sea floor can be determined by seismic reflection in which sound waves are transmitted to and reflected from the ocean bottom. The time it takes for the echo or echos to return and the intensity of the returning echo allows the marine geologist to "see" features on and below the sea floor. Acoustic penetration may be as great as several hundred meters, and, when used in conjunction with piston coring and dredging, provides a "view" of the upper layers of the sea floor. Although modern narrow-beam echo sounders give very accurate information, most profiles of Caribbean sea floor were obtained in the 1960s and 1970s when it was difficult to interpret results from rough terranes, slopes, and geologically complex areas. Much of our knowledge of the northeastern Caribbean comes from seismic studies by the U.S. Navy, which has used the region as a testing ground since the 1930s.

Seismic waves generated from a surface ship propagate through and along different layers of the oceanic crust and are reflected or refracted at different lithological boundaries. Thus seismic refraction studies provide details about the thicknesses and distribution of different layers within the crust. The calculated velocities with which seismic waves propagate through the individual layers allows direct comparisons with velocities in typical rocks and sediments determined in geophysical laboratories.

Seismic refraction studies have been extremely useful for determining the crustal thicknesses and structures throughout the Caribbean, and, in concert with drilling, dredging, and coring, have allowed the division of the submarine portions of the Caribbean Plate into numerous smaller tectonic units or blocks. (See further below.)

Additionally these techniques allow determination of a crustal block as typically oceanic or continental or arc-like by providing information on the thicknesses of crustal layers and their respective densities. Some of the Caribbean blocks fall clearly into one of these categories, e.g. the Gulf of Mexico and the Yucatan Basin have oceanic crustal structure, the Nicaraguan Rise and Cayman Ridge have arc-like crust, and the seaward extension of the Florida-Bahaman platform is continental in type. Other pieces of the Caribbean are problematic in that they have transitional seismic crustal structures. Anomalously thick oceanic-type crust comprising much of the Venezuelan and Colombian Basins has led to the hypotheses (1) that the Caribbean Plate is an anomalous piece of thickened ocean crust (an oceanic plateau) from the eastern Pacific (Malfait and Dinkelman 1972; Pindell and Dewey 1982; Duncan and Hargraves 1984, Ghosh et al. 1984) or (2) that the plate formed as normal oceanic crust nearly in situ but experienced a major "flood basalt" event during the mid- to late Cretaceous (Donnelly 1973, 1975; Mattson 1979; Anderson and Schmidt 1983; Klitgord and Schouten 1986). It is obviously important to know which of these hypotheses is more nearly correct, but at present all we can state with confidence is that the foundations of the Caribbean Plate are oceanic (albeit anomalously thick and buoyant) and older than Tertiary. In similar fashion, it is difficult to ascertain if the thick and less dense crust that composes regions such as the Nicaragua Rise represents continental arcs with a continental basement (thus resembling the Mexican Volcanic Belt), or intra-oceanic island arcs built entirely on thin, dense oceanic basement (thus resembling the Lesser Antilles).

The evidence is clear that fragments of pre-Cretaceous crust exist throughout the Caribbean. Cuba and northern Central America have pre-Mesozoic continental basement, and a piece of Jurassic oceanic crust may be exposed on the island of La Desirade in the Lesser Antilles (Burke et al. 1984). In order to better constrain the development of the Caribbean as a whole and its individual blocks we must rely on results obtained from paleomagnetic and radiometric dating combined with detailed structural and petrologic studies of many areas.

Gravity and seismic studies.-- Regional gravity surveys (Bowen 1976) and earthquake distribution (seismic studies: Sykes et al. 1982; McCann and Sykes 1984) provide important information about recent plate interactions (neotectonics) in the Caribbean but are of limited value in determining events older than a few million years (Mann and Burke 1984).

Paleomagnetism.-- The study of the magnetism acquired by rocks during the time of their formation (paleomagnetism) has been very successful in the determination of both the ages of rocks and sediments and their relative position during formation or accumulation.

Paleomagnetic dating relies upon correlating a measured sequence of magnetic polarity reversals in rock successions or strata with the geomagnetic polarity time scale that has been developed by determining magnetic polarity sequences in terrestrial lavas that have been dated by radiometric methods. Since many rocks retain their original magnetic signatures and polarity reversals are synchronous world-wide, nearly continuous rock strata will have a magnetostratigraphy that corresponds to a specific time interval (Kennett 1982; Channell 1982). Magnetostratigraphy together with fossil data or radiometric dates provide geologists with good constraints on the age of strata on land and sediments on the sea floor. Paleogeographic positions can also be estimated by measuring the direction and inclination of the paleomagnetic vector.

Although paleomagnetic studies could potentially solve many tectonic problems in the Caribbean, few successful results have been obtained. To a large extent, this is a consequence of few and poor exposures of continuous rock sequences and of the extensive weathering which destroys the original magnetic signature. Some lithologies, particularly limestones, are relatively non-magnetic, so that only the newest most sensitive magnetometers are capable of resolving their magnetic properties. Local structural complexities in the strata (folding, faulting, rotation) that are not easily recognized make interpretation of inclination and declination measurements impossible. In addition, rocks that have formed near the equator have remnant inclinations (dips of magnetic vectors) that are very shallow. This makes it difficult to determine the paleolatitude of a sample and, without additional information, to determine if it formed in the northern or southern hemisphere. Even with very high quality data the uncertainties involved allow an estimation no better than plus or minus 2 degrees of latitude, which translates into uncertainties of 100s of kilometers and thus cannot place reasonable constraints on the paleolatitudes of Caribbean islands.

Changes in magnetic declination (direction to the N magnetic pole), used to determine polarity reversals, can also be used to determine whether a certain locality, presumably representative of a tectonic block, has been rotated and to what extent relative to larger plates. In the Caribbean large extents of both clockwise and counter-clockwise rotation relative to North America have been estimated for most of the Greater Antilles (Mann and Burke 1984; Van Fossen 1986). The results, however, apply only to the final position and do not tell us whether this rotation was the result of one major event in only one direction or if it represents a number of rotations. The lack of paleomagnetic data immediately north and south of the Caribbean also limits our ability to compare and contrast motions across plate boundaries.

Marine magnetic anomalies, measured by a magnetometer towed behind a ship, can also provide crucial information about the age and tectonic evolution of sea floor. Oceanic crust that is formed by spreading at mid-ocean ridges acquires a symmetric pattern of linear magnetic anomalies on each side of the ridge crest. The pattern of magnetic anomalies (positive and negative stripes) thus seen on the sea floor corresponds to the periods of normal and reversed magnetic polarity that are defined on the basis of terrestrial sequences. A detailed geomagnetic polarity time scale so constructed for the last 160 myBP has been further refined in the ocean basins using paleontological ages of basal sediments in drill-holes and estimates of crustal age based on spreading rates. As a consequence, the age of any given piece of the sea floor may be determined by comparison of the measured magnetic anomaly profiles with computer simulated anomaly profiles based on the magnetic polarity time scale. Offsets in the magnetic stripes are often an indication of transform faults between ridge segments, and missing stripes imply that some sea floor has been consumed, usually in a subduction zone.

Analysis of magnetic anomalies in the North Atlantic and South Atlantic Oceans has allowed geophysicists to reconstruct the opening of the Atlantic Ocean. The history of relative motion between the various plates - commonly known as plate kinematics - has been discussed by various authors (Pindell and Dewey 1982; Burke et al. 1984; Smith 1985; Duncan and Hargraves 1984; Pindell and Barrett 1988). This information together with pre-Triassic "fits" of the continents has been the basis of numerous reconstructions of the pre-rift configuration of North America, South America, and Africa (LePichon and Fox 1971; Pindell and Dewey 1982; Klitgord and Schouten 1986).

The position of the proto-Caribbean Plate is, in these reconstructions, controlled by the relative positions of the larger surrounding plates. At present there is still no consensus on the most nearly "correct" reconstruction, but it is generally believed that there was little room for the Caribbean Plate, as we know it today, between North and South America before approximately 210 myBP (see Fig. 3).

Ladd (1976) used sea floor magnetic anomalies to determine the relative positions of North America and of South America with respect to Africa for a series of time intervals. Using the method of finite difference rotations and assuming certain dynamic conditions during change, he was able to portray the the movement of South America relative to a fixed North American Plate from Triassic (ca. 180 myBP) to Miocene (9 myBP). As we mentioned above, the motions and tectonics of the Caribbean Plate have been largely dominated by the motions of the peripheral plates. More recent scenarios of North and South American plate motion are capable of explaining (constraining) many of the major tectonic events in the Caribbean (Klitgord and Schouten 1986; Pindell and Barrett 1988; Pindell et al. 1988).

Radiometric dates.-- In order to effectively use paleomagnetic data and geologic field data precise ages of the samples are required. Paleontological data can provide good relative age control and are invaluable for regional stratigraphic correlations. However, only radiometric dating, based on constant isotope decay rates, can provide the precision needed to establish an accurate geochronology for a region.

The common method used for dating relatively young igneous and metamorphic rocks (less than ca. 150 my) is the conventional K-Ar (Potassium-Argon) method. Most Caribbean igneous and metamorphic samples have been dated this way. Other less commonly used methods, at least historically, include 40Ar/39Ar, Rb-Sr, U-Th-Pb, and Sm-Nd. The principles and methodology of all of these systems are discussed by Faure (1986). Only the pitfalls and limitations of the methods will be discussed here.

The problems with radiometric dates do not stem from the analytical methods, which can give ages precise within a few million years or less, but instead from the rocks themselves. For example, the K-Ar method requires that crystalline rocks or constituent minerals retain the radiogenic 40Ar produced by the decay of 40K from the time of crystallization. Argon loss is a very common naturally occurring phenomenon; argon as a noble gas is not readily retained in the crystal structure of minerals even at low temperatures. Chemical weathering and alteration - so prevalent in the Caribbean - can not only enhance Ar loss but may change the K content of minerals, thereby disturbing the initial K-Ar systematics. Additionally, metamorphism or partial recrystallization of a rock caused by pressure or temperature will result in partial or complete loss of argon depending on both the temperature and the duration of the event.

In many rocks it is virtually impossible to tell if there has been argon loss unless a number of whole rocks and their constituent minerals yield identical K-Ar ages. In the past such extensive analysis was rarely done. K-Ar dating has primarily been done on volcanic and plutonic rocks from continents and island arcs because they have relatively high K contents, often contain K-rich minerals, and because secondary metamorphism and alteration should be discernible. Ocean floor basalts on the other hand have very low K contents, no K-bearing minerals, and typically have been altered by seawater to some extent. Studies have shown that ages determined for subaerial igneous rocks are commonly too low because of argon loss, whereas submarine basalts may trap excess Ar because of rapid cooling and high hydrostatic pressures, thereby yielding K-Ar ages that are too high (Faure 1986). Ages of metamorphic rocks presumably reflect the youngest tectonic event, but the minerals found are still subject to Ar loss by temperature alteration. In general, the most suitable minerals for K-Ar dating are biotite, muscovite and hornblende in plutonic igneous rocks and high grade metamorphic rocks and feldspars from volcanic rocks (Faure 1986).

Recently, the 40Ar/39Ar method of dating has been used with greater success and more confidence than the conventional K-Ar method. In this method a series of dates from a single sample is determined by releasing Ar in steps at increasing temperatures. If no loss or gain of Ar or K has occurred since initial cooling, the age calculated from 40Ar/39Ar should be constant. However, many minerals exhibit variable 40Ar/39Ar ratios during stepwise heating, yielding a variety of dates that must be carefully interpreted. The stepwise heating method allows geochemists to "see" events that may have affected the K-Ar systematics and therefore to make more nearly valid estimates of the initial age of igneous and metamorphic rocks. Clearly dates determined by classical K-Ar techniques in the past are open to great uncertainty and should be considered as "fact" only after careful evaluation.

Other methods of radiometric dating such as Rb-Sr, Sm-Nd, and U-Th-Pb have not been utilized to any great extent in the Caribbean. In part this is because the high-precision techiques required to measure small differences in isotopic ratios were not readily available, and ages determined for relatively young rocks (less than 100 myBP) were not very precise. Relative to the K-Ar method, the ages that have been determined by these methods offer a greater degree of certainty because the elements involved in the radiometric decay scheme are much less liable to secondary gain or loss. Although slight degrees of weathering or low-temperature alternation have little effect on the calculated dates, recrystallization by a later thermal event can "reset" the age of a rock. Hence, the ages of most metamorphosed rocks are likely to represent the latest thermal pulse which is, in most instances, a tectonic event.

Rock types.-- The most obvious pieces of information available to geologists are the rocks that are exposed on land or recovered from the seas. What information does their mineralogy (petrography) and geochemistry provide?

It is now well accepted among petrologists and geochemists that certain rock types and associations are characteristic of specific tectonic environments. Certainly there are rock types that can be found in more than one environment, and some environments are associated with a variety of rock types, but in general we can be reasonably certain of paleogeologic environment from the assemblage of rocks that are present.

As an example, magmas that form at mid-ocean spreading centers commonly erupt as pillow lavas and sheet flows and are fed from below by vertical conduits that solidify as sheeted dikes. Below the dikes is a magma chamber that slowly crystallizes to form a coarse grained gabbro and layers of ultramafic rock. This identical sequence or association of rocks, capped by a layer of deep-sea sediments, is often found as dismembered outcrops along plate margins and has been recognized as slivers of oceanic crust and mantle that were thrust onto land (obducted) rather than beneath it (subducted) as is the norm. In many instances the entire sequence of rocks, known as an ophiolite, is not present. Fortunately the chemical composition of mid-ocean ridge basalts (MORB) is so distinctive that they can be identified solely on the basis of their chemistry.

The presence of MORB on the seafloor suggests normal oceanic crust - and one should look for magnetic anomalies. However, when MORB and associated intrusive rocks outcrop on land, we can infer that they were obducted from an adjacent piece of seafloor. Massive outcrops of serpentinite (hydrated and metamorphosed ultramafic rocks) may also be an indication of sea floor obduction at some time in the past.

Geologists now recognize that ophiolites may represent fragments of back-arc basins - these are the crust beneath small seas on the landward side of island arcs - and as such do not imply the existence of a major ocean basin (e.g. Hawkins et al. 1984). The slightly different chemistry of back-arc basin basalts (BABB) compared to MORB and their association with arc-derived sediments and more shallow water fauna distinguish them from rocks formed at mid-ocean spreading centers.

Ophiolites and serpentinites are relatively common around the boundaries of the Caribbean Plate (Mattson 1979; Burke et al. 1984). It is obviously important to put some age constraints on these rocks because they may represent the initial foundations of Caribbean crust. Ages of metamorphic rocks associated with them may also provide the timing of major compressional events. Unfortunately, even if we determine the age of an ophiolite in the Caribbean, it may not be entirely clear whether it had its origin in the Pacific, the Atlantic, or the Caribbean Sea itself.

Igneous rocks formed along convergent plate boundaries - those associated with deep dipping seismic zones, submarine trenches, and active volcanoes - have petrographic and geochemical characteristics quite distinct from those of igneous rocks generated at ocean ridges or in old thick continental crust. In fact, the rock associations observed in island arcs, such as the Lesser Antilles, are easily identifiable even in ancient terranes. These rock types were at one time broadly classified as "andesites" after the rock type erupted from Andean volcanoes. However, we now recognize a spectrum of spatially associated rock types - pyroclastic deposits (e.g. tuffs, ash beds, volcanic ejecta), lava flows (basalts, andesites, dacites, rhyolites), and relatively small plutonic bodies that intrude the surrounding bedrock (plugs, stocks, dikes, plutons).

In modern active arcs these rock types are commonly erupted or intruded in relatively localized volcanic centers that may be 10s to 100s of square kilometers, and these centers may remain in one location for millions of years (Gill 1981). We also know that the location of the volcanic front - the zone at which a chain of volcanoes begins relative to the trench - is directly related to the presence and geometry of seafloor being subducted beneath the arc or continental margin. In general, when the subducted oceanic crust reaches a depth of 80 to 100 km, melting is initiated in the overlying mantle, and magmas ascend to feed volcanoes that, in turn, create new crust.

A key observation here is that arc volcanoes are nearly always associated with subduction zones. In regions where subduction has recently stopped, magmatism often terminates. By analogy then, where arc-type rocks and their erosional products are found, there most probably was active convergence of plates with a resulting trench and subduction zone. Questions remain: What was the direction or polarity of the subduction and did the arc migrate as the subduction zone changed angle or position?

In island arcs where the direction, or polarity, of subduction remains the same and no unusual tectonic events occur, the islands may develop in a systematic way over periods of a few to 10s of millions of years. The first stage presumably involves the eruption of submarine volcanoes upon which the primary volcanic front is built (Fig. 4). The reason for the uncertainty is that very little is known about the building of an island. Our understanding of arcs is based on studies of volcanic islands that are subaerially exposed, hence the foundations of the islands must necessarily be covered by 1000s of meters of younger rock. Alternatively, in old eroded arc terranes where the submarine foundations are now exposed, much of the overlying strata must have been removed by erosion and the arc may be deformed to such an extent as to make interpretations difficult.

In some island arcs the oldest lavas erupted have distinctive chemical characteristics and may be associated with marine sediments and features typical of submarine eruptions. These rock types comprise what is known as the "island arc tholeiitic series" (Gill 1981). In the Greater Antilles the oldest series of rocks, named the "primitive island arc" series by Donnelly and Rogers (1980), have chemical similarities to the island arc tholeiitic series, but there are some differences. Unfortunately it is not uncommon for primitive submarine eruptives in island arcs to have physical and chemical characteristics similar to those of MORB - lavas erupted at spreading centers. This has led to debate about the origin of ophiolites on some islands - La Desirade, for example. Are the igneous rocks of La Desirade related to the initial formation of an arc or are they pieces of oceanic crust?

A secondary, more extensive arc of subaerial volcanoes may develop on top of the primitive arc basement or slightly landward of the subduction zone. The various lavas erupted from the volcanoes and the plutonic rocks that crystallize in magma chambers at depth are generally known as the calc-alkaline series. The rocks that comprise this chemical group are the ones that are often, although incorrectly, called "andesites." In fact they show a wide range of chemical and physical characteristics including pyroclastic rock types. Because calc-alkaline igneous rocks are the most abundant rock type erupted along continental margins and island arcs (such as the Antilles), their presence has been accepted as indicative of the prior existence of a subduction zone and the resultant volcanic arc.

In some arcs - generally older and more mature - a third, less voluminous series, the alkalic or shoshonitic series, erupts behind the main arc over the deepest parts of the subducted plate or in parts of the main arc where rifting has occurred.

These generalized spatial and temporal changes in the position and chemical characteristics of arcs have led to the somewhat naive belief that the location and polarity of <u>ancient</u> subduction zones can be determined. Gill (1981) discusses the complexities of arc development and the generation of magmas along convergent plate boundaries. Recent studies of young arcs have clearly shown that the generalized model described above cannot be sustained for regions in which plate tectonic interactions have been complex. A number of modern arcs have been the locus of subduction polarity reversals ("flipping" subduction zones), terrane accretion (collision and suturing of blocks), or intra-arc extension (splitting of arcs to form back-arc basins) over periods less than a few million years (e.g., Johnson et al. 1978; Hawkins and Melchior 1985). At present there are no well defined subduction polarity reversals known in the Caribbean, but Mattson (1979) discusses evidence for flipped subduction zones during the Cretaceous. There is little that can be used as absolute evidence of such tectonic events, and we certainly cannot reconstruct them with great accuracy. Information obtained from the rock record, however, is considered reasonable circumstantial evidence.

Igneous rocks with petrological and geochemical characteristics of the island arc tholeiitic, calc-alkaline, and shoshonitic series can be used with confidence to infer the previous existence of a subduction zone. Whether it was an intraoceanic island arc (typically groups of small separate islands) or a continental margin arc (chains of volcanoes connected by thick, older crust) is difficult to resolve without additional geochemical and field data. Additionally, it is not possible to determine with any certainty the spatial and temporal relations between volcanism and tectonics without detailed structural, radiometric, and geophysical data.

A few other characteristics of convergent margins can provide some important constraints on deciphering the geologic history of a region. First, two rather distinct associations of metamorphic and sedimentary rocks commonly form in arcs because of the increased pressure created by convergence and subduction and the increased heat generated by volcanic processes. One association (a facies) is created in front of the volcanic arc, on the landward side of the trench by scraping off subducting material (mostly sediments) that accumulates on the trench and on the downgoing oceanic crust (Fig. 4). These sediments form an accretionary prism of complexly deformed and metamorphosed rocks in the forearc region. Barbados is an exposed portion of such a prism (Torrini et al. 1985). The width of this prism largely appears to be dependent on the availability of sediments through time (Karig and Sharman 1975). Due to the relatively high pressures and temperatures generated and the extensive thrusting caused by the subduction process, high-pressure, low temperature metamorphic rocks called blueschists, amphibolites, and eclogites are commonly found as exotic blocks in chaotic mixtures of oceanic and arc-derived sediments called melanges (or subduction zone complexes) exposed along some convergent margins.

The other rock association or facies is created within the crust of the arc as a result of increasing the temperature (at relatively low pressures) of the pre-existing igneous and sedimentary rocks during magma intrusion. These rocks typically have had subaerial rather than submarine histories and exhibit little deformation as a result of their metamorphism. The low-grade metamorphic rocks, or hornfels, as they are more specifically termed, exist landward of the subduction zone and accretionary prism and are spatially and temporally associated with shallow-level plutonic rocks (diorites, granodiorites, tonalites, granites).

When both rock associations exist in close proximity to one another, they are known as a paired metamorphic belt, and their relative positions can determine the direction of subduction (Miyashiro 1975). The positions of melanges alone, particularly when they include blueschists or eclogites, provide reliable information on the location of previous subduction zones - although not necessarily their polarity. In some instances fragments of ocean crust are caught up in the melanges and thrust up (obducted) to the surface. Ophiolites and high-pressure metamorphic rocks crop out in many places along the boundaries of the Caribbean plate and provide significant information with respect to the evolution of the Caribbean (Mattson 1979; Wadge et al. 1984; Burke et al. 1984).

A few other rock types that are worthy of mention here, because they provide information on paleoenvironments, are limestones, clastic redbeds, and tuffs. Limestones, generally speaking, reflect marine conditions but also can be used to specify whether they formed in deep open-ocean (pelagic or biogenic oozes) or on shallow, more restricted banks and reefs - coraline limestone, calcarenites, interbedded with intertidal sands and coarser clastic material deposited near shore by rivers.

Fringing reefs are very common around tropical islands but may not have great aerial extent. After major periods of volcanic activity (convergence) when islands begin to subside, reefs may continue to build at rates equivalent to subsidence rates, thus forming small coral islands and atolls (Darwin 1842 in Kennett 1982:316).

Macro- and micro-fossils in limestones provide evidence not only for relative ages of formation but also for environments of deposition. Inclusion of coarse clastic sediments in limestone is an indication of a terrestrial source that was probably nearby. However, the possibility cannot be dismissed that shallow water carbonates and subaerially derived sediments may have been transported over great distances by turbidity currents. In such a case, coarse turbidites will be interbedded with deep-sea pelagic deposits. Such mixtures of sediment types are not unexpected in arc environments where erosion may be rapid, currents strong, and submarine relief often extreme. Such environments are common in the Caribbean.

"Red beds" is a somewhat unspecific name given to clastic sediments deposited in terrestrial or shallow-water environments where oxidative conditions prevail, such as the seasonal rushing, rapidly eroding rivers that exist throughout the highlands of the Caribbean. The relatively coarse and unsorted nature of many redbeds (arkoses, lithic arenites, breccias) suggests rapid deposition and little transport from source areas. Typical environments of deposition include alluvial (rivers), deltaic, beach, and very shallow marine. In those regions where extremely thick deposits have been found in relatively narrow troughs or basins (e.g., the Triassic redbeds along the eastern United States, the Cenozoic Wagwater belt in Jamaica, the Enriquillo Valley in southern Hispaniola) it has been suggested that lithospheric rifting - sometimes accompanied by volcanism - may have created these special depositional environments (Burke et al. 1980; Mann et al. 1984, 1985).

Just as fossils are important in the interpretation of biogenic rocks such as limestones, individual rock and mineral fragments in coarse clastic rocks provide their own clues about the surrounding environments. In arcs, for example, an abundance of volcanic clasts and igneous minerals are sure signs of a nearby volcanic center. Unfortunately the age of deposits such as these is difficult to determine because of a lack of fossils (or the difficulty of finding them) and the inability to date them using radiometric techniques.

Tuffs are volcanic rocks composed of pyroclastic fragments and fine ash. Although they are formed by explosive volcanic eruptions, their mode of deposition is more like that of sedimentary rocks: they are wide-spread, time-stratigraphic units. Because they may be deposited over large areas in many different geologic environments and can be radiometrically dated, they can impose some important constraints on the position, type, and age of volcanism. In the Caribbean, however, pervasive chemical weathering rapidly turns tuffs and ash beds into barely recognizable and quite undatable layers of clay (bentonite).

Volcanic eruptions and their products have some special properties that should not be overlooked by biologists and biogeographers. The most explosive and devastating types of eruptions are those associated with composite cones, which are typical of island arcs and active continental margins. In fact the classification given to the most violent, gas-charged eruptions is peléan - after Mount Pelée on Martinique. It was the coastal town of St. Pierre that, in 1902, was destroyed by a glowing ash-cloud (nuée ardente) that rushed down the sides of the volcano with temperatures greater than 800°C. Obviously such catastrophic events, which are frequent, geologically speaking, can play a major role in extinctions on small islands.

Although frequent eruptions may inhibit the permanent colonization of an island, they may also aid in the dispersal of biota. One of the common products of a dominantly pyroclastic eruption is pumice - frothy blocks of lava that are so filled with holes where gas escaped that they can float for 1000s of kilometers on the ocean surface. There is plenty of evidence for the phenomenon. It is, for example, not uncommon to find the beaches of Australia's east coast littered with pumice from active volcanoes in the Tonga-Kermadec island arc (Perfit pers. obs. 1978). There are numerous accounts of pumice floating on the open ocean for over a year after the 1883 eruption of Krakatau, some pieces as far as 8170 km from the volcano (Simkin and Fiske 1983). Some large bays near Krakatau were choked with pumice for months, and there are many accounts of living animals and plants on the larger pieces. One such account by Captain Charles Reeves, who was travelling on the barque Umvoti from India to South Africa, is especially impressive (Simkin and Fiske 1983:50). There are other reports of pumice rafts capable of supporting sailors and carrying corpses across the Indian Ocean. In the Caribbean, Thomas Barbour (1917) tells of a report of pumice from the Mt. Pelée eruption that washed up on several of the Virgin Islands. Pumice from extensive Mexican eruptions in the Oligocene and early Miocene is abundant in deep-sea cores across the northern Caribbean (Donnelly pers. comm. 1988). It seems likely, therefore, that pumice rafts have been common in the Caribbean, and that, given the number of eruptions that undoubtedly occurred during active arc volcanism, some biota have been exchanged between the Caribbean islands on pumice rafts.

Interpreting the geological evidence

What can we say with confidence regarding the history of the Caribbean? The big picture of its tectonic evolution is emerging as more facts are obtained, and there is consensus developing with regard to the history of certain terranes or islands. However, with the data now available, subject as they are to the limitations we have outlined, it is not yet possible to complete the picture.

In particular, it is, at present, impossible to tell where each of the islands was at any given time in relation to the other islands or terranes. The best we can do is to approximate the relative locations of major tectonic blocks such as the Greater Antilles or Bahamas Platform. Relative plate motions determined from distant sea floor magnetic anomalies allow us to estimate what types of tectonic forces and movements may have occurred along the Caribbean plate boundaries. Accurately dated rocks associated with these tectonic events lend support to evolutionary models and provide better age constraints. Voluminous limestones that cover much of the Caribbean tell us where and approximately when seas were deep or shallow. Volcanic rocks, plutons, and terrestrial sediments mark the location of islands, but do not allow us to estimate the subaerial extent of these islands nor how long they were above sea level. The agents of erosion have been constantly at work, tearing the islands down and transporting them to the ocean floor. Hence, the lack of rocks of a certain age - particularly those that form islands - is insufficient proof that they did not exist in the past. Only a precise step-wise reconstruction of what did exist during each geologic epoch will allow us to determine the geologic history of the Caribbean even approximately.

Nevertheless, geologists, in spite of all these problems, have been attempting to reconstruct the paleogeography of the Caribbean region in a plate tectonic context since the early 1970's. The very great number of hypotheses and models that have been proposed in the last almost 20 years are themselves testimony to the incompleteness of our knowledge of Caribbean geology and to the paucity of the hard data that could allow us to choose a correct model.

At the present time each model of the the geologic/tectonic development of the Caribbean has its controversial points, and we can only insist that each is based on reasonable assumptions and permissible tectonics. A few of the most popular models include those presented by Malfait and Dinkelman (1972), Ladd (1976), Perfit and Heezen (1978), Pindell and Dewey (1982), Anderson and Schmidt (1983), Wadge and Burke (1984), Mattson (1984), Duncan and Hargraves (1984), Coney (1982), and Donnelly (1985). Even a cursory comparison of these models reveals the extraordinary diversity of opinion. Recent reviews of the available geologic and geophysical data have been presented by Burke et al. (1984), Mann and Burke (1984), Smith (1985), and Pindell and Barrett (1988).

It is not our purpose here to provide an exhaustive review of the geological literature on Caribbean geologic evolution, nor to present any new model. Instead we attempt to summarize the areas of agreement and point out the major areas of conflict. (A similar and complementary discussion is presented by Donnelly (1988).)

A few points are very clear: there is no escape from a mobilist view of Caribbean geological history. At the global level - the major plates and the major aspects of their movement - the story appears to be clear and irrefutable. At the level of the Caribbean area itself, on the other hand, it is equally clear that not only are many details obscure or disputed, but that some parts of the real story may never be known. Not only is there much that is now hypothetical, but there is also a residue that is likely to remain so.

We here summarize Caribbean history, as others have done, in terms of a series of vignettes, each corresponding to what we believe to be a significant period in the Caribbean story. We begin with the late Cretaceous and continue to the Recent. Geologically and biologically we see four Periods:

Period I (late Cretaceous to mid-Eocene):

Any discussion of the tectonic evolution and paleogeography of the Caribbean must begin with the origin of the Caribbean plate itself. Reconstructions of North America, South America, and Africa in the Early Mesozoic (see Smith 1985; Burke et al. 1984) leave little room for the Caribbean plate, and it is reasonably well established that the Caribbean developed as a consequence of the separation of the North and South American plates during the middle Mesozoic. Most of the early (pre-Late Cretaceous) evolution of the Caribbean can only be approximated from interpretations of the relative movements between the major continental plates and certain continental fragments, such as the Yucatan and Chortis blocks in Central America. These relative movements are fairly well constrained by paleomagnetic studies of the continental rocks and of magnetic anomaly patterns of the Atlantic ocean floor during the Mesozoic (see discussion of paleomagnetism above). It is also important to note that the Farallon plate in the Pacific must have played a major role in the tectonic development of the western Caribbean plate, but the earlier motions of this plate are poorly constrained. Burke et al. (1984) and Donnelly (1985) review the various interpretations of motion of South America relative to a fixed North America.

Two major assumptions that geologists make are 1) that the Caribbean plate, being a rather small piece of lithosphere, must react to changes in the relative positions of the larger surrounding plates and 2) that elastic deformation and tectonism (e.g. faulting, volcanism) primarily occur along plate boundaries. Theoretically, these assumptions are sound. However, the key to deciphering tectonic evolution in the Caribbean must come from actual data (radiometric, geochemical, paleomagnetic, paleontological) directly obtained from Caribbean crust or from crust that can be identified as existing prior to the formation of the Caribbean plate as we know it today (proto-Caribbean).

We can be fairly certain from a variety of geological evidence that Pangaea began to rift in the Triassic and that a proto-Atlantic Ocean was created between North America and western North Africa by the middle Jurassic, while South America still remained fixed to west southern Africa (Pindell 1985; Klitgord and Schouten 1986; Pindall et al. 1988). This led to a southeastward movement of South America-Africa (extension and left-lateral displacement) relative to North America between approximately 165 and 85 mBP (Pindell and Dewey 1982; Burke et al. 1984) and as a consequence the opening of the Caribbean. Unfortunately, the relatively well known plate motions only provide information about the size and shape of the region created between North and South America, not what land masses occupied this space nor how they got there.

In terms of plate tectonic theory it is reasonable to assume that oceanic crust first formed in the widening gap between the continents, and that this was probably generated at a ridge or ridges that were part of (offshoots from) the newly formed Mid-Atlantic Ridge. The southeastward movement of the Mexican and Chortis blocks along left-lateral shear zones apparently kept North and South America in contact during the Jurassic and did not allow extension of the ridge into the Pacific (Anderson and Schmidt 1983; Pindell 1985).

Although there is general agreement among geologists on this scenario, it still must be considered highly conjectural. Models of the origin of the Caribbean plate during the Jurassic and Early Cretaceous may, indeed, appear quite reasonable. However, they are not well constrained for three reasons: 1) the paucity or absence of rocks of pre-Late Cretaceous age; 2) the difficulties in finding and identifying Mesozoic marine magnetic lineations in Caribbean oceanic basins; and 3) inability to date accurately or determine the original provenance of the pieces of oceanic crust (ophiolites) that are exposed along the margins of the present Caribbean plate.

Without well-exposed Caribbean rocks of pre-Cretaceous age (and these in continuous stratigraphic section) we can say little about the earliest development of the area. It is not difficult to suggest an explanation for their absence. It is quite possible - we might even say probable - that if proto-Caribbean oceanic crust once existed, it has been entirely consumed by subduction that has occurred after the mid- to late Cretaceous. Scattered fragments that may be part of such a proto-Caribbean crust may exist (see below), but it is difficult to be positive about their identity, and impossible to ascertain their original geographic positions. Some workers include in their reconstructions pieces of Caribbean crust that no longer exist; such speculations are necessarily tenuous at best. It is possible that some such reconstructions are correct; many appear reasonable. However, the facts simply do not exist to confirm or deny them. Donnelly (1985) discusses many of these problems.

Burke et al. (1978) and Mattson (1979) have suggested that the thickened Caribbean crust represents an accumulation of lavas and sediments formed off-ridge, that is, it was too buoyant to subduct at trenches bordering the plate. Duncan and Hargraves (1984) suggest that it formed over the Galapagos hotspot as an oceanic plateau in the eastern Pacific. Drilling during Leg 15 of the Deep Sea Drilling Project (DSDP) in the Colombian and Venezuelan basins showed that the uppermost basaltic rocks comprising the uppermost Caribbean sea floor (so-called B" layer) are of Late Cretaceous age or older (Donnelly et al. 1973). Whether the crust was formed at a normal spreading center or represents some vast "flood basalt" episode, as Donnelly et al. (1973) and Donnelly (1985) suggest, is uncertain. We discuss the possible implications of the two interpretations below.

The oldest rocks and geologic features in the Caribbean region have been found along the boundaries of the plate. These are summarized by Burke et al. (1984). Jurassic to early Cretaceous reef and salt deposits exist in the Bahamas, on the North American plate off the north coast of Cuba, and probably on the southwest part of the Floridian shelf. Both south Florida and the Blake-Bahamas Plateau off east Florida are underlain by Triassic-Jurassic volcanic rocks and sediments related to the initial rifting of the Atlantic, but are well removed from the Caribbean-North American plate boundary. Early to middle Jurassic sedimentary rocks are well-documented in western Cuba, and some Jurassic and Late Cretaceous dates have been obtained from arc-related metamorphic rocks in central Cuba and the Isle of Pines respectively (Khudoley and Meyerhoff 1971). More recent data from oil company drill holes indicate high-grade metamorphic rocks that form the basement in central Cuba may be as old as Paleozoic (P. Mann pers. comm. 1987).

Many of the metamorphic rocks in western and central Cuba are unlike those in the Greater Antilles and have more similarities to older continental rocks in Central America. Results of dredging in the Yucatan Channel (Pyle et al. 1971; Perfit unpublished MS) suggest that western Cuba is part of the Yucatan block that probably existed as a continental fragment prior to the Mesozoic. Paleozoic radiometric ages have been obtained from schists recovered in basement drill holes in Yucatan (Pyle et al. 1971) and in DSDP holes 536 to 538 (Leg 77) between Yucatan and west Florida. It seems probable, although not proven, that much of the deep unexposed crust that bounds the Caribbean plate to the north is composed of continental crust and that this North American crust is at least as old as the early Paleozoic.

Metamorphic rocks of Paleozoic (possibly Precambrian) age also crop out in Honduras, which is part of the Chortis block - a piece of continental crust that is believed to have moved southeastward by left-lateral shearing to lie along southern Mexico. Exactly where the Chortis block started and how far it has moved are unresolved problems. Between the Chortis block and northern Colombia, no pre-Cretaceous rocks have been identified, and, if plate reconstructions are correct, little or no room could have existed between Mexico and northern South America.

In the Greater Antilles (excluding Cuba) the oldest rocks exposed are some primitive arc rocks that formed the foundations of the islands, and fragments of oceanic crust (ophiolites) that have been thrust onto the islands. Donnelly (1985) suggests that many of the ophiolite slivers in Guatemala, Cuba, Jamaica, southern Hispaniola, southern Puerto Rico, Tobago, Curacao, Aruba, northern Venezuela, eastern Panama, and Costa Rica may be pieces of Caribbean crust (B") with ages that range from around 125 myBP to 85 myBP - the date when the "basalt event" ended. However, Burke et al. (1984) are of the opinion that only a few of these ophiolites (Costa Rica, Panama, south Hispaniola) represent part of the proto-Caribbean floor (B") basalts and that the remainder are Atlantic or Pacific oceanic caught up in accretionary wedges at convergent boundaries (subduction zones) around the margin of the Caribbean plate. Without a great deal more field and chemical information it is impossible to determine from where these various oceanic fragments came.

If the Caribbean ocean floor was generated at a "normal" oceanic spreading center away from the magnetic equator, linear magnetic anomalies should be present in the major Caribbean basins and provide critical age information. Unfortunately, the few magnetic lineations recognized are of poor quality, making determinations of the age of the oceanic crust unreliable. For example, the latest Cretaceous age (76-67 myBP) east-west trending anomalies in the Colombian basin (Christofferson 1973) disagree with basement ages obtained by drilling (Donnelly et al. 1973). Ghosh et al. (1984) have identified northeast-southwest trending anomalies in the central Venezuelan basin that appear to have formed at very slow rates between 153 and 127 myBP. Furthermore they suggest that the Venezuelan Basin was created in the Pacific (west of Colombia) along a ridge system that extended from the central Atlantic into the Pacific. This mid-Jurassic to early Cretaceous crust migrated to the northeast as the Greater Antilles arc and subduction zone developed and cut off the ridge system from the Atlantic ridges. (See also Duncan and Hargraves 1984.)

In this context it is important to mention the ophiolitic rocks on La Desirade - the only old oceanic fragments exposed in the Lesser Antilles. These have isotopic ages of 149 to 85 myBP and Late Jurassic to Early Cretaceous fossils (Mattinson et al. 1980; Bouysse et al. 1983). Although often believed to represent pieces of Atlantic oceanic crust (Mattinson et al. 1980), it has also been suggested that they are fragments from the Venezuelan basin spreading center (Ghosh et al. 1984; Burke et al. 1984) or part of a primitive arc sequence (Donnelly 1985; Bouysse et al. 1983). The disputes about the origin of this one group of rocks emphasizes the difficulties that attend making definitive statements about the early evolution of the Caribbean.

In fact, whether the in situ or the Pacific origin of the Caribbean oceanic crust is eventually supported by critical evidence may have only indirect bearing on biogeographic problems. However, the locations of plate boundaries are important because it is along these zones that the initial emergent land massses formed that could provide avenues for biological colonization. It seems fairly certain that most of the Caribbean plate was oceanic during the Cretaceous, but that by at least 85 myBP compression along the circumference of the plate resulted in subduction, with related arc volcanism, and also obduction with the emplacement of ophiolites on newly emerging islands.

Early Cretaceous rocks with arc-like chemical characteristics have been recognized in Puerto Rico, the Virgin Islands, Bonaire, Venezuela, Hispaniola, and possibly in Jamaica, Cuba, Costa Rica, and La Desirade (Donnelly 1985, 1988; Westercamp et al. 1985; Pindell and Barrett 1988). These igneous rocks - many of which appear to have been erupted in shallow marine environments - represent Donnelly's primitive island arc, and most certainly mark the loci of subduction zones bordering the Caribbean oceanic crust discussed above. Cretaceous arc-related rocks have also been recovered from submarine escarpments along the Nicaraguan Rise, Cayman Ridge, Puerto Rico Trench, Aves Ridge, and nothern Lesser Antilles (Fox and Heezen 1975; Perfit and Heezen 1978; Bouysse et al. 1985).

It is at this time (ca. Late Cretaceous) that the precise location of the Caribbean oceanic crust with its fringing island arcs becomes important to both biogeographers and geologists. In theory the oldest arc rocks can provide data that constrain the extent and positions of islands in the Caribbean. In practice, however, geologists do not have the kind of information required to determine the paleogeographic positions of most of the islands nor the total area above sea level at any given time interval.

Radiometrically dated rocks from the Caribbean islands are typically plutonic rocks - the deep-seated intrusions beneath volcanic centers that are not exposed until most of the overlying volcanic and sedimentary rocks are stripped away by erosion. In such a case, while dating is possible, the evidence needed to determine the size of the island at that period may be largely missing. Similarly, dates from spatially associated plutons (assuming that their radiometric clocks have remained intact) permit us to postulate that a magmatic arc and associated trench existed for a given interval, but cannot tell us whether the volcanos were separate islands (as in the Aleutians or the present Lesser Antilles) or joined (as in the Alaskan Peninsula or Indonesia). Typically, however, oceanic arcs grow in size as volcanism continues, and islands may be united by tectonic processes, but initially the expectation is of small volcanic islands with regular and relatively large expanses of water between them.

The existence of Cretaceous shallow-water limestones and sediments or terrestrial "red bed" deposits associated with the volcanic plutons in the Caribbean also allows us to infer island arcs. The islands may be larger if sediments fill in basins or are uplifted by local or regional orogenic forces. Extensive shallow-water carbonates also are suggestive of uplift and of the potential for shallow marine "connections" between islands - connections in a geological sense - under water. But such phenomena do not necessarily indicate dry land connections - land bridges.

The earliest establishment of a proto-Antillean volcanic arc may, indeed, be marked by the 145 myBP siliceous igneous rocks on La Desirade, but it is clear that the major episode of arc-building occurred during the Mid-Late Cretaceous (ca. 110-65 myBP). The oldest rocks in the "primitive island arc series" in the Greater Antilles may be as old as the rocks on La Desirade (Donnelly pers. comm.). Recent compilations (Burke et al. 1984; Pindell and Barrett 1988) of radiometric and stratigraphic ages of Caribbean metamorphic and igneous rocks support a concentration of arc-related tectonic activity around 85 myBP, when compression was greatest along the northern plate boundary.

The presence of Late Cretaceous igneous rocks and associated terrestrial or shallow water sediments in most of the Greater Antilles, the Leeward or Dutch Antilles, Colombia, the Chortis block (Guatemala, Honduras), and along the Cayman Ridge-Nicaraguan Rise provides substantial proof that a broad emergent island arc existed by the end of the Cretaceous. Belts of dismembered and highly deformed metamorphic rocks - sometimes associated with ophiolites or melanges - are good indications of former subduction zones. However, the directions of subduction (subduction polarity), as based on the positions of the metamorphic belts, are difficult to determine, and the possibility is clear that reversals in subduction polarity may have occurred in relatively short time periods (less than a few million years.)

It is very clear that none of the data provide direct information on the paleogeographic positions of the islands during the Late Cretaceous/Paleocene, or on their positions relative to one another. We have no choice but to rely on inferences from plate tectonic reconstructions combined with paleomagnetic investigations of well-dated, relatively undeformed rock units. As we have already seen, the constraints imposed by reconstruction of the major plate motions are not precise enough to provide clear grounds for preferring one Caribbean model above another.

A crucial case is the dispute over the location of the Caribbean plate during the Late Cretaceous. The mobilist views (e.g., Pindell and Barrett 1988; Pindell 1988; Duncan and Hargraves 1984; Burke et al. 1984) have the Caribbean plate in the Pacific with the proto-Greater Antilles forming an arc slightly west of and between the Mexico-Chortis block and South America (Fig. 5c, d). The more stabilist (less mobilist) views (e.g., Donnelly 1985, 1988; Klitgord and Schouten 1986) have the Caribbean nearly in its present location and the proto-Antilles a broad arc extending from the Yucatan block to northeastern South America (Fig. 5a, b). Other models propose plate positions more or less intermediate between the two extreme models (e.g., Malfait and Dinkelman 1972; Perfit and Heezen 1978; Pindell and Dewey 1982). In most models the proto-Antilles are placed in the "gap" between North and South America. Whether they were somewhat west of North America, were directly between the continental blocks, or were further east, convex toward the Atlantic may possibly be of biogeographic importance, but this is also disputed. (See Donnelly 1988.)

What is distinctly more important than the exact position of the plate is the proximity of the islands of the proto-Antilles to one another and the continuity of the arc as a whole. We can be fairly certain that as volcanism continued through the Late Cretaceous and earliest Cenozoic the arc became more buoyant and subaerially extensive, thereby creating more chances for biotic interchange. In this regard we must discuss regions of the Caribbean that are presently submerged but were most assuredly emergent parts of the arc during the Late Cretaceous: the Cayman Ridge, the Nicaraguan Rise, and the Aves ridge.

The geology and tectonic history of the Cayman Ridge, the Nicaraguan Rise, and the intervening Cayman Trough (inappropriately called Cayman Trench in the past) have been discussed by Perfit and Heezen (1978), Holcombe and Sharman (1983), and Rosencrantz and Sclater (1986). The present North American-Caribbean plate boundary is marked by the Cayman Trough transform-ridge-transform zone that connects with the complex left-lateral fault zones in Guatemala (Motagua-Polochic faults) and Hispaniola (Enriquillo-Plantain Garden and Camus-Septentrional fault zones) and ultimately the Puerto Rico Trench fault zone (Mann and Burke 1984). Although there is still debate, the consensus is that the Cayman Trough began to form during the Eocene and by early Oligocene rifting resulted in the separation of the Cayman Ridge and Nicaragua Rise. Prior to this time (Late Cretaceous through Paleocene) the Cayman Ridge and Nicaraguan Rise may have been part of a northern Greater Antilles arc that extended as a chain of islands from Central America (Chortis block) to La Desirade or the Aves Ridge (see below).

Similarities between the ages and compositions of the arc-type rocks dredged from the Cayman Ridge and Nicaraguan Rise and those from central Central America (Chortis block) and the presently exposed parts of Jamaica, eastern Cuba, and Hispaniola led Perfit and Heezen (1978) to propose that a broad emergent arc formed above a southward dipping subduction zone during the Late Cretaceous. An abundance of Late Cretaceous to early Cenozoic shallow-water limestones and coarse clastic rocks from the Ridge and the Rise are like formations in the Greater Antilles and add substantive evidence that the northern boundary of the Caribbean plate was at relatively shallow depths and had numerous volcanically active islands.

The paleogeographic positions of these islands are unknown, but all the available evidence indicates that there was not a continuous connection among the islands or between the islands and the continent. Theoretically it is possible that some of the islands may have been in close proximity to the Chortis block where the arc could have had more continental affinities, i.e., it may have been more like a peninsula than separate islands. However, one must keep in mind that sea level at this time was at a global high (approximately 350 m higher than at present) (Vail et al. 1977; Haq et al. 1987) thus creating more difficulties for dry land connections. Again we emphasize that at present our picture of the paleogeography during the Cretaceous is inadequate.

We have very little information about the Yucatan basin between Cuba and the Cayman Ridge, except that it appears to be oceanic crust. When and how it formed has significant effects for our reconstructions, particularly if it is younger than the Cretaceous. Closure of this basin would bring western Cuba closer to the Cayman Ridge-Nicaraguan Rise, creating a very complex junction of arcs, trenches, faults, and crustal blocks (Fig. 6). Conversely, if the basin is pre-Cretaceous, more oceanic crust may have separated western Cuba and the Yucatan block from the proto-Antilles, and most of Cuba would have developed separately from the other islands (Fig. 5).

Geologic data from eastern Cuba, northern Hispaniola, and Puerto Rico suggest that there was a progressive northwest-to-southeast collision of the northeasterly migrating arc with the Bahamas Platform, beginning in the Late Cretaceous - Paleocene in the northwest and culminating in the early Oligocene in the southeast (near the Samana peninsula in northeastern Hispaniola). This model assumes an Atlantic subduction zone with southwestern polarity (Fig. 6). The locus of such a subduction zone may be marked by the present-day Puerto Rico Trench and high pressure metamorphic belts along the north coasts of Hispaniola and Cuba (Perfit et al. 1980; Pindell and Barrett 1988).

Assuming the Bahamas Platform has remained relatively stationary on the North American Plate, and given the collision history just mentioned, the proto-Greater Antilles could not have been significantly west of the Yucatan Block. It follows that the northeastward movement of the proto-Antilles was not simple, and individual blocks appear to have moved somewhat independently along major transcurrent faults.

Recent paleomagnetic studies of the Late Cretaceous to Eocene sedimentary rocks on Hispaniola and Puerto Rico indicate that at least parts of these islands have experienced significant northward movement and horizontal rotations (up to 45°) since middle Eocene (Van Fossen 1986). Paleomagnetic data from Haiti also suggest that the northern and southern portions were separate tectonic blocks during the early Cenozoic, a hypothesis supported by others on the basis of geological and geophysical evidence (Pindell and Dewey 1982; Mann et al. 1984; Sykes et al. 1982). Paleomagnetic studies in Jamaica suggest that relatively recent rotations have occurred, but little latitudinal displacement with reference to the North American plate has occurred since Late Cretaceous (Gose and Testamata 1983).

Such high quality paleomagnetic data are too scarce at present to allow detailed reconstructions of the paleogeography to be made; however, they do serve to illustrate that crustal blocks moved independently and in some cases collided to form larger composite islands. Estimates of left-lateral displacements along the northern part of the Greater Antilles are on the order of 100s of kilometers, but there is no single fault zone that can account for the estimated displacement of over 1000 km of Pindell and Barrett (1988).

The present Greater Antilles probably represent only the northern remnants of the proto-Antillean arc that extended between North and South America. It has been suggested that the Aves Ridge represents remnants of the southern parts of a Cretaceous proto-Antillean arc (Pindell and Dewey 1982). In the south, the ridge appears to be structurally linked to the Leeward Antilles. Magnetic, seismic, and gravimetric measurements indicate that the Aves Ridge has arc-like features. The composition and radiometric ages of dredged rocks (Fox et al. 1971) support the hypothesis that the Aves Ridge was part of an east-facing island arc during the Late Cretaceous.

Dredging along the middle and northern sections of the Aves ridge recovered only Eocene and younger limestones, marls, and cherts (Fox et al. 1971; Bouysse et al. 1985; Perfit unpublished data). Fossil evidence indicates shallow carbonate shelf environments from Eocene to Miocene, with more deep-water open-ocean conditions from mid-Miocene to the present. Sedimentological information from the Caribbean basin also indicates that the Aves Ridge was a structural prominence during the late Cretaceous to mid-Miocene (Holcombe and Moore 1977). The lack of volcanic rocks and abundance of deep-water sediments of Miocene and younger age conforms, in general, to the history of volcanic quiescence and regional subsidence deciphered from the rock record of the Nicaraguan Rise and Puerto Rico Trench (Perfit and Heezen 1978; Perfit et al. 1980).

Rare outcrops of Late Cretaceous sedimentary rocks interbedded with volcanic rocks have been reported in the southern Lesser Antilles (Grenadines) and the northeast submarine slope of the arc (Westercamp et al. 1985; Fox and Heezen 1975; Bouysse et al. 1985). Although strong evidence is lacking, it appears reasonable that the present northern Lesser Antilles formed on arc crust that was at least as old as Cretaceous and that included the Aves Ridge. Greatly diminished volcanic activity during the Paleocene, an eastward shift in the locus of volcanism and general submergence of the Caribbean basin has been associated with a change to more easterly motion of the Caribbean plate relative to the North American plate during the early Tertiary. Whether the northern Lesser Antilles were more volcanically active than the southern Lesser Antilles throughout this period is not yet answerable.

Late Cretaceous to Eocene igneous rocks exposed on some of the Leeward Antilles (Aruba, Bonaire) and the Caribbean Coast Ranges (Venezuela) imply the existence of an island arc system proximal to South America during the same period in which the Greater Antilles and Aves Ridge developed. The initial locus of this arc and its polarity and relationship to the more northern arc (or arcs) are unclear because of the intense deformation and strike slip dislocation that occurred along the southern boundary of the Caribbean plate. Pindell and Barrett (1988) have suggested that the Leeward Antilles magmatic arc was part of the Aves Ridge arc that was sutured onto South America as the Caribbean plate progressively pushed eastward during the Cenozoic.

It is, in fact, conceivable that all or parts of the Greater Antilles (including the Nicaraguan Rise-Cayman Ridge) plus the Aves Ridge and the proto-Leeward Antilles were part of a semi-continuous proto-Antillean island arc that first began to form between North and South America in the Early Cretaceous on the eastern edge of the proto-

Caribbean sea floor (Figs. 5c, d, 6a). Progressive eastward migration of the buoyant Caribbean plate (oceanic plateau) may have been accompanied by anomalous sea-floor spreading - the flood basalt event inferred by Donnelly (1985) - and continued arc development during the Cretaceous. As the expanding plate entered the Atlantic, the proto-Antillean plate may have broken into fragments (Greater Antilles, Aves Ridge, Leeward Antilles) and the boundaries sequentially deformed as they converged and slid past parts of the North and South American plates (Fig. 6a-d).

The progressively younger age of tectonic events toward the east along the northern and southern plate boundaries is very suggestive evidence for this generalized model (Pindell and Barrett 1988). If this is correct, it implies the progressive formation of islands from west to east via subduction and orogenesis, and also an increased relative movement between crustal blocks - a spreading out or dispersal of islands. Speed (1985) has suggested that the southern Lesser Antilles were not part of this migrating arc, but, instead, formed in the Eocene north and west of their present position. Geologic and geochemical evidence from northwest South America indicates that a volcanic arc existed to the west of Colombia during Cretaceous time and was separated from the active continental arc by oceanic crust, probably formed in a back-arc basin. During latest Cretaceous and earliest Cenozoic time this arc terrane was sutured onto the South American plate, as indicated by the regional metamorphic rocks of that age (Bourgois et al. 1987; Lebras et al. 1987).

How does the Central American arc (Panama-Costa Rica) fit into this model? If one assumes that since early Cretaceous the Caribbean plate has been approximately the same size that it is today and that the proto-Antilles were formed south of the Yucatan block, then the Central American block must have been forming several hundred kilometers to the west in the eastern Pacific Ocean. Even non-mobilist models suggest initial mid-Cretaceous development of a west-facing arc to the west of and unconnected to South America (e.g., Donnelly 1985; Klitgord and Schouten 1986) (Fig. 5a, b).

The first approach of this arc to South America and to northern Central America - via the Chortis and Yucatan blocks - did not occur until the very end of the Cretaceous or the beginning of the Tertiary (Donnelly 1985; Pindell and Barrett 1988). Deep sea sedimentological evidence, however, strongly argues against any continuous subaerial land mass in the Central American position during the early Tertiary (Holcombe and Moore 1977).

Period II (mid-Eocene to Miocene)

Major changes in relative plate motion during the Eocene resulted in diminished volcanic activity, rift formation (e.g. the Cayman Trough) and large scale subsidence of the Caribbean plate. As a consequence, the Caribbean Sea gradually encroached upon many islands that were simultaneously being worn away by incessant tropical erosion (Fig. 6d). The widespread deposition of limestone throughout the Greater Antilles and the transition from shallow-water to deep, open-ocean fossil assemblages reflect the increasing isolation of the individual land masses as subsidence continued from mid-Eocene to Miocene (Mattson 1984). Even for the Lesser Antilles, although this easternmost chain was being built at this time, no geological evidence exists to support connections between individual elements of this island chain or between the chain and South America.

To the west Donnelly (1985, this volume) has been able to show that there was a progressive restructuring of oceanic circulation between the Caribbean Sea and the Pacific Ocean during the late Eocene. He suggests that this is substantive evidence of an island chain (proto-Costa Rica and Panama) projecting south from Central America by mid-Miocene.

In the Greater Antilles the presence of some volcanic and clastic rocks of Eocene-Oligocene age and local unconformities (erosional breaks in the geologic record) indicate that certain elevated sections of the islands (Cuba, Hispaniola, Puerto Rico) may never have been completely submerged (Mattson 1984). Even in Jamaica, which is believed to have been largely submerged, there is evidence for locally emergent sections and certainly very shallow reefs in the Late Eocene and again in mid Oligocene to early Miocene (Arden 1975; Hendry 1987). The available, but scant, information suggests that islands that comprised regions such as the Nicaraguan Rise and the Aves Ridge began to subside by the Oligocene and did not become emergent again until the last glaciation. It should be noted, however, that this hypothesis is largely based on evidence from dredging (Fox and Heezen 1975; Perfit and Heezen 1978) that does not allow us to distinguish unconformities in the geologic record.

This local restriction of emergent land area and complete submergence of some islands has unquestionably significant biogeographic consequences. However, the extent of exposed land is not well-constrained by the geologic data. For example, thick deposits of Eocene-Oligocene limestone in western Jamaica support rapid subsidence and a marine transgression in this area, but the absence of this limestone to the east in the Blue Mountains could be interpreted either as evidence for persistent emergent land during the Eocene and Oligocene or as evidence for more recent uplift and erosion of the limestone.

Because elevated areas tend to shed sediments, not accumulate them, lack of sediments is often interpreted as evidence of subaerial exposure in the relevant time zone. In practice it is not possible to make reasonable estimates of the area or location of islands in the Caribbean during any given period without fairly complete stratigraphic sections; the record is in general less than complete, and extensive sampling of submerged parts of the Caribbean crust is mostly unavailable.

Any attempt to make sense of the geology of a Caribbean island, such as Jamaica, must keep in mind that even Jamaica, as we see it today, has, along with the Nicaraguan Rise, not only been transported 100s of kilometers eastward since the Eocene, but probably represents the accretion of more than one crustal block: western Jamaica was separated from eastern Jamaica (Draper 1986). Parallel fragmentations and accretions are known or are suspected on other islands. Pindell and Barrett (1988) present a model for the development of the Greater Antilles during the early Tertiary. According to this, southeastern Cuba, north-central Hispaniola, Puerto Rico, and the Virgin Islands formed a single magmatic arc in the Eocene but split by Early Miocene as a result of left lateral transcurrent motion along major faults, while at the same time the southern peninsula of Hispaniola, which had lain well to the west (Fig. 6c), moved eastward toward juncture with the central part of the island. The separation of the Aves Ridge from Puerto Rico is suggested to have occurred by mid-Miocene as the Anegada passage developed.

During this stage northern South America and the Leeward Antilles were uplifted and eroded. Deformation was largely limited to the southern Caribbean. Orogenesis and volcanism continued along the Pacific coast but not along the northern plate boundary, where motion was primarily lateral rather than convergent (Mattson 1984).

Period III (Miocene to mid-Pliocene)

A third major period in the geological history of the Caribbean began by mid-Miocene and continued into the Pliocene (ca. 5 myBP). During this period the Caribbean plate continued or renewed its eastward migration relative to the American plates. Slight northward relative motion of the South American plate caused some compressional features to develop along the southern plate boundary. As shown in Fig. 6e, Jamaica and southern Hispaniola moved further east as the Cayman Trough continued to grow. Strike-slip motion was accommodated along the Oriente Fault-Puerto Rico Trench fault zone, but the formation of numerous mountain ranges suggests that uplift accompanied this movement.

Episodic uplift and gentle compression during the Miocene are inferred from exposures of coarse clastic sedimentary rocks, local unconformities and broad folded mountains in Cuba, Jamaica, Hispaniola, central Puerto Rico, and St. Croix (Mattson 1984; Arden 1975; Lidz 1984). Neogene structures in Jamaica and Hispaniola are largely controlled by east-west transcurrent faults that have created both pull-apart basins and push-up elevations (Burke et al. 1980; Mann et al. 1984). Some of the extentional basins are associated with alkalic basaltic volcanism.

This was a period of emergence for most of the islands regardless of global sea level changes from low sea level stand at the end of the Miocene to a high stand at the beginning of the Pliocene. Rapid uplift of some areas is evident from Miocene and younger reef and coastal deposits and beach terraces that have been raised as much as 300-600m in eastern Cuba, northern Hispaniola, western Jamaica, and St. Croix (Mattson 1984; Lidz 1984; Hendry 1987). In other areas subsidence dominated during most of the Miocene. The submerged shelf off the north coast of Puerto Rico, for example, has subsided as much as 5000 m since Miocene-Pliocene time (Perfit et al. 1980; Moussa et al. 1987).

The Lesser Antilles arc-trench system now became well-developed as a result of continued westerly subduction of Atlantic ocean crust beneath the eastern edge of the Caribbean plate (Fig. 6e). Along the southern plate boundary right-lateral slip faulting continued to shear and rotate the Leeward islands, but compressional forces along the Venezuelan border resulted in renewed uplift of crustal blocks, thrusting, and rapid erosion into local basins along the Caribbean mountains (Mattson 1984; Pindell and Barrett 1988).

Wadge et al. (1984) suggest that the Central American arc has been more or less continuous with the Mexican continental arc since the Miocene and that it rotated in an anticlockwise sense during the Neogene. Such a reconstruction gives the Central American arc an east-west trend rather than a north-south strike as others (e.g., Pindell and Dewey 1982) have proposed. Their model also suggests that the development of the Central American arc is unrelated to the arc parallelling the northwest coast of South America. The geology of Panama is poorly known but the available data indicate a collision between the Central American arc and Colombia during the Late Miocene or earliest Pliocene (Wadge et al. 1984; Donnelly 1985). Recent geophysical data from the southern continental shelf of Panama has led Okaya and Ben-Avraham (1987) to propose development of the Panamanian arc (i.e. the extreme southeastern end of the Central American arc) in mid-Miocene, termination of subduction under western Panama during the Pliocene (3-5 myBP), and subsequent left-lateral transform motion. The latter had the result of collision between the Central American arc and northern South America (Colombia) and the terrestrial connection between North and South America.

Period IV (Pliocene to Present)

By this time the Caribbean was taking on a general appearance and configuration similar to the present, although, in detail, the individual tectonic blocks - and the contained islands - were continuing to respond to local tectonic stresses.

From the Pliocene to the present our tectonic models and paleogeographic reconstructions become tightly constrained. The rock record is better preserved, i.e., there has been less time for erosion, and multiple periods of tectonic activity simply did not occur and thus do not, as in the past, complicate or derail our understanding of geological relationship. The spatial relationships of the Greater and Lesser Antilles were from the beginning of this period essentially as they are today - most tectonism being controlled by left-lateral transform motion. More land was exposed, satellite islands were connected to their "mainlands," but many of the previously submerged islands (e.g. all of the Aves Ridge except Aves Island) remained below sea level.

The most significant geologic event was the complete closure of the Panamanian gap as the result of the collision of the Central American arc with the South American plate. The precise age of the connection is difficult to determine using terrestrial rock sequences. Recent investigations of Neogene pelagic sediments from DSDP cores from the Caribbean have suggested deep circulation in the western basins was terminated by 4 myBP and surface circulation was cut by 2 myBP (G. Keller pers. comm. 1987). Donnelly (this volume) supports the emergence of the Panamanian isthmus during the early Pliocene and discusses the biogeographic implications. Global drops in sea level between 5 and 2 myBP (Vail et al. 1977; Haq et al. 1987) obviously acccentuated the formation of dry land at this time.

During this period local tectonic adjustments seem to have predominated over regional movements but this may be an artifact of looking in great detail over a short period of time. A detailed account of the neotectonics of the Caribbean may be found in Mann and Burke (1984).

The biological side of the story:
Expectations and retrodictions

Geological and biological evidence compared

The preceding is a statement of what geology can plausibly tell us about the the paleogeography of the West Indies. The very diverse methods used in obtaining the evidence and the weaknesses of each have been described in some detail. We have then presented a four-period scenario of Caribbean tectonic history. There emerges a large canvas in which a sequence of temporal scenes shows patches of clarity that are separated by intractable obscurities. The obscurities tend to cluster in the older scenes, and only the near present contains more that is clear than is obscure.

A phrase commonly used by geologists is very revealing here: "strongly constrained."

The most recent period in the geological sequence, for example, has been said above to be "strongly constrained." The meaning, given the context, is perfectly clear. There is, for the near present and the present simply more evidence, and evidence sets limits to - "constrains" - speculation.

Geologists are very well aware of the gaps in the evidence that confronts them. They know that some strata have been eroded away and will never be available; indeed, this condition is so usual and familiar that there is a name for such gaps in the succession of strata: "unconformity." Geologists know that some strata are buried and that, while the evidence is there, it will not be accessible except under very special conditions or by special methods. Geologists know the limitations of certain techniques and are willing - as one of us (MRP) has done above - to compare dredging to "trying to make a geologic map from samples obtained by a bucket lowered from a helicopter 1000s of meters above the earth's surface." For geologists, weaknesses in the evidence are epitomized by the gentle but deadly phrase "not very well constrained."

It is very plain that, despite this, geologists are quite willing to speculate, and do speculate - extrapolate - well beyond their evidence. This is harmless - even heuristic - if they are conscious, as we believe they most often are, that only the absence of evidential boundaries allows their grander constructions.

Thus, while Caribbean geologists have made many very different reconstructions of the geological history of the region, these models have been made with an insider's awareness that other geologists will know reasonably well where the evidence is strong and where it is weak. Caribbean biologists who attempt to make use of these reconstructions - those who extrapolate beyond the extrapolations the geologists have already made - are, on the contrary, walking all too innocently and ignorantly into uncharted minefields.

In fact, despite all the complaints of Caribbean geologists about their own data - about the absence of sufficient constraints - the very diverse corroborative detail about many topics and time periods that the geologists have been able to gather is impressive. It is in strong contrast with the limitations - the factual poverty - of information that is the lot of Caribbean biologists.

The data of Caribbean biologists are: 1) poorly known modern distributions, much better known for some taxa than others, but always heavily disturbed and distorted by man's long occupation of the islands); 2) defective taxonomy, inadequate and very much in flux for even the best known groups); 3) a fossil record that even for the group paleontologically best known on continents, terrestrial mammals, has, on the islands, almost no temporal depth, no useful remains older than, at most, a little more than 100,000 years.

In grim fact this amounts to knowledge only of those patterns which, on imperfect and incomplete data, can be discerned in the Recent biota. Everything else must be derived from these easily misinterpreted patterns. Extrapolation has had to be the major method employed by all Caribbean biogeographers.

What Caribbean biologists have had to face is the treacherous problem of <u>retrodiction</u>, use of the possibly distorted information about the present-day as a means to extrapolate to the truth of the long past. In such a case, when evidence does not constrain speculation, theory fills the gaps. Land bridges, dispersal, and vicariance are all consequences of a felt need; they are all theoretical constructs that attempt to retrodict past events from present distributions.

Geologists' reconstructions are retrodictions also, but their freedom to speculate has been "constrained" by evidence that does bear directly on the issues interpreted. There is evidence that does pertain to relevant ages; there are real temporal sequences. For them it has not been necessary - nor permitted - to imagine everything. In particular, geologists use the rock record to infer processes and conditions in the past (as biologists might infer habitats). They do not speculate on earth history on the basis of Recent sediments.

Indeed, the Caribbean biogeographers, recognizing the fragility and insufficiency of their biological data, have invariably turned to the geologists to confirm their retrodictions. Insecure in their own base, they have sought more solid ground.

But geologists have their priorities and their own set of problems and questions - those problems and questions that they find both interesting and tractable within the limits of their techniques and methods. On the other hand, the questions that biologists most want answered are often - perhaps usually - not tractable for the geologists. The question, for example, "Did Jamaica or Cuba have a dry land connection with Central America at some specified time in the past?" may have no geologically verifiable answer. Certainly no such answer is at this time possible - and on strictly geological evidence an answer may never be possible.

What then can be done? First on both the biological and the geological side additional - and better - evidence should be sought - and ignorance admitted.

The present biota of the West Indies should be better known than it is - and the distribution of the taxa should be more accurately known. Despite the 100s of years of investigation the age of discovery is not yet over in the West Indies - even for vertebrates. Birds are sometimes believed to be all but completely known the world over, but a new species of warbler - *Dendroica angelae* - was only rather recently discovered in the montane forests of Puerto Rico (Kepler and Parkes 1972). The Pleistocene fossils of the Greater Antilles are relatively well known, but a new genus of rodent of phylogenetic significance is just now being described (Woods in press). More startling is the find (Olson and Kurochkin 1987) in cave deposits on the Isle of Pines and from somewhat older cave breccia in Camaguey, Cuba of a "practically" flightless wren-sized bird that belongs to an avian superfamily of South American affinities otherwise unknown in all of the West Indies. A remarkable new species of anoline lizard - *Anolis eugenegrahami* Schwartz 1978 - with semi-aquatic habits was first found in 1978 and is still known from one very limited brookside habitat in Haiti. New species of the frog genus *Eleutherodactylus* continue to be discovered almost routinely (see Hedges this volume). Thomas et al. (1985) report three new species of the rare burrowing thread snakes of the genus *Leptotyphlops*. New species of the larger burrowing blind snakes *Typhlops* are described by Thomas (this volume).

The systematics of no West Indian group is well worked out. One of us (EEW) has spent more than 30 years on the ecology and alpha taxonomy of the anoline group of lizards, all the while endeavoring to establish or encourage the establishment of a proper data base for the group as a whole. Much has been done, but that needed data base still does not exist. Worse, there are probably no groups for which there is better information.

The fossil record to this moment is only tantalizing. The richer record and the only one containing recognizable terrestrial mammals goes back at best to mid-Pleistocene. The intriguing and problematic rodent *Proechimys* [=*Puertoricomys*] *corozalus* (Williams and Koopman 1951), which might be older, is undated. Hair is known from early Tertiary Dominican amber, but it is not certain even to what mammalian order it belongs (Poinar 1988; Woods this volume). The known older fossils that have been found not in amber but in rock strata are either marine (sirenians) or suspected marine (a pelomedusid turtle). (See the list by Williams of all the old or possibly old Caribbean terrestrial fossils in the introductory chapter this volume.)

The amber fossils are at the moment the most important. Described so far are one frog and two taxa of lizards, all referred to genera still living on the islands, and a much richer fauna of invertebrates, especially ants. The misfortune of the amber fauna is the general inadequacy of the fossil record of these groups elsewhere and especially the absence of any sequence of fossil faunas containing them on the adjacent mainlands.

Biogeography: the biological correlate of geology

Preliminary remarks. Two theories compete for the biogeographic interpretation of the available biological and geological information for the Caribbean (Williams the introductory chapter this volume). One, the dispersal theory, is classically associated with the names of W.D. Mathew, Philip Darlington, and George Gaylord Simpson. The second, the vicariance theory, is, at least for the Caribbean, the creation of Donn E. Rosen. The "island biogeography" theory of MacArthur and Wilson (1967) assumes dispersal and can be subsumed under the dispersal theory, but elements of "island biogeography" relate extinction ("turnover") to island size and are very pertinent also for vicariance. Logically the two theories - dispersal and vicariance - are not mutually exclusive; the real question is one of the extent to which either is compatible with the available biological and geological evidence.

For the biologist confronting the biogeography of islands the issue that immediately arises is one of the amount of cross-water colonization. Here is the critical difference between the dispersal and vicariance theories. Pure disperalist theory for the Caribbean requires that all the biota came across water gaps. Pure vicariance theory assumes that none did. Given the stabilist interpretation of the Antillean archipelago, if the islands did arise in situ and experienced over their entire history only increases and diminutions in size, never any interisland connections, there is no way for these islands to have obtained their faunas or floras except cross-water. Given, on the other hand, a mobilist view of the Caribbean area, the fact and the importance of cross-water colonization becomes an open question.

What we believe we know about the geological history of the islands makes it clear, however, that the question is only partially open. For the last period of the geological sequence, because the islands were during all of this period substantially in their present positions, the question is not open at all. Apart from minor and local tectonic movement and the changes in area consequent on sea level changes during the Plio-Pleistocene the islands have been essentially as stabilists construed them to have been throughout their history. It follows that whatever fraction of the biota reached the islands in this last phase did come cross-water. This is not seriously controversial. Only those who refuse to admit any possibility of cross-water dispersal will be disturbed to make this admission. None of the serious vicariantists - neither Rosen nor Savage - have hesitated in this regard; they take this as given - a point not worthy of dispute. This part of the story, then, can be considered settled.

To say this, however, leaves our question truly open for most of Caribbean history - more than 80 million years. We here take the position that dispersal - by which we mean the dispersal of individual taxa - was a possibility at any time in this long period. We also, prior to the discussion below, assume that vicariance in some form might have been possible also during this entire early period. Vicariance in the case of islands, in contrast to continental vicariance, which might be ecological or climatic, has always been taken to imply discrete geological events. These might, however, be of many kinds. There might be actual rifting of land masses, either by breaking off of pieces of a continent or the break up of an isthmus, i.e., of a dry-land contact between two continents. The islands themselves might fragment. Or there might a gentler geologic change - subsidence or sea level change - that would result in water barriers between islands or between islands and continents. Alternatively, there might be the reverse kind of phenomenon - accretion of land masses, either of islands with each other, or of islands with continents.

Discrete geological events of this sort ought to be much more visible than dispersal on two grounds: 1) the geological event, which ought to have been at least moderately massive, has some substantive chance of leaving traces of its occurrence behind, and 2) fragmentation from a continent or dry land contact with a continent implies transmittal not of one taxon but of many, and, therefore, even with attrition over time, some substantial fraction of a <u>fauna</u> ought to have left a record that paleontology could <u>potentially</u> recover.

If this is correct, there are expectations - retrodictions, if you will - that can help us choose between dispersal and vicariance for the controversial at least 80 million years of Caribbean history. To make the eventual decision, to interpret any geological or paleontological evidence, we will need the background of some understanding of the basis of both dispersal and vicariance theory, and some detail about the retrodictions that appear to follow from them. We attempt below to provide some of this background; there is no effort to be exhaustive. Books have been and will continue to be written on these topics.

Sampling and fossils.-- Some first comments about 1) sampling and 2) fossils.

1) <u>Sampling</u>. Even Recent taxa and their distributions are never completely known. What are known are samples drawn from populations. These samples have been collected over the centuries with differing diligence and accuracy. Populations even in historical time may change in at least minor ways; their distributions may change much more. In the Caribbean very much was destroyed and much was altered before there was any thought of science as we know it. There are thus sure to be gaps and errors in the records, and then, inevitably, in the analysis. How serious are these gaps and errors for the most Recent time and the most assiduous collecting and rigorous analysis? And is the sample of the Recent period only - despite its known defects - sufficiently good by itself to allow reasonable projections backward in time - retrodictions? There are many who will say of the data at hand that this is all we have, that we must make do with what we have, that we cannot be blamed for mistakes made because evidence is lacking, and that, in any case, there is no ground for believing that the evidence we do have will mislead, if only our methods of analysis are themselves correct.

Our opinion is quite different. We are very conscious of the possible inadequacy of data. We are very aware that new evidence often destroys naive simplicities and annihilates the covert assumptions that seemed to follow with complete certainty from the incomplete data. When we try to look backward into the fog of time, there are always areas of complete obscurity as well as points of apparent clarity, and the obscurity tends - but not quite consistently - to increase with distance in time. Always there is reason for caution. And certainly every element that stands out in that fog - every fossil, every geological detail - must be seized and used - with proper care.

2) Fossils and temporal differentiation. A temporal record of some sort is essential to biogeography; it is, after all, one kind of history. Everywhere the temporal record of islands is more or less incomplete, both in terms of geology in the sense of the rock record and of biology in its paleontological aspect. In the Caribbean the geological data are conspicuously superior to the biological. We make as much use of the geological record as possible, despite known gaps. We need also to attempt to use the paleontological record - despite its greater deficiencies.

On the positive side, even a single valid datum point - any specimen, any fossil - is, like any rock type or radioactive date, a constraint - it sets a universe of discourse even though it does not bound it. Any future general statement that purports to be relevant to that datum point must reckon with it - must include it within its boundaries. At the same time every single datum point will need context before it becomes meaningful.

Within the biogeographic context that we seek, there are three aspects of a fossil that are significant: its taxonomic position, its locality, and its date. Discoveries are made when strange taxa are found in some strange locality at a date that is unexpected. "Strangeness" here consists in newness and especially in unexpectedness - in newness of information content.

It was definitely a discovery when the strange primate *Xenothrix mcgregori* (Williams and Koopman 1952) was found in Jamaica. This was a novelty both in taxonomy and locality. It was a hitherto unknown type of primate found on an island where no previous record of primates had been known. It was very much a discovery also when ground sloths and giant rodents were found fossil in the Greater Antilles. Ground sloths and giant rodents are known fossil on the mainlands; discovery on the islands was a novelty of locality. It was a discovery when the lizard genus *Anolis* was first found in Dominican amber (Rieppel 1980). This genus is well known and abundant on the islands today; the novelty of the fossil was the novelty of very early date. The important feature in all cases is, of course, is that these were - and are - datum points for biogeography - neither to be over nor undervalued.

The dating - contra some biogeographers - is quite as significant as the rest. The oldest known fossil of any lineage may still, as Rosen (1976) has insisted, give only a minimum age. This may be true of any individual fossil, but for well studied groups with well studied fossils it is important that there are evident limits beyond which lineages cannot be pushed backward in time.

That there were no cats in the Triassic, we take to be a statement that needs no defense. (It is, in fact, an implicit part of the recognition of the Triassic that there are no cats in strata named as Triassic - just as it is an explicit part of the definition that certain taxa do occur.) Similarly we have no sense of insecurity in saying that there were no cats in the Jurassic or the Cretaceous. Our hesitation saying that there were no cats in the Paleocene is very slightly greater; the record then shows that there were mammals that were at least remotely related to cats. If we travel forward to Miocene, there begin to be a wealth of felids, i.e., modernized, very recognizable cats.

The point here is that there are, indeed, groups that differentiated rather recently. Therefore, if your biogeographic hypothesis regarding the genus *Felis* depends upon association between a geologic event that occurred in the Late Cretaceous and a fossil on a West Indian island, any biogeographer will feel very much at liberty to doubt the dating of the geologic event, or the relevance of the event to the West Indian fossil, or the identification or date of the fossil called *Felis*, or all of these.

Cats (Felidae) may once have existed in the West Indies, as we know ground sloths (Megalonychidae) certainly did, but for both taxa their presence is only possible after they have evolved. For mammalian groups date of origin is determinable within a margin of error that eliminates certain possibilities.

There are, on the other hand, taxa for which there is no adequate relevant record. If your biogeographic hypothesis depends upon the association of a frog or lizard fossil on a West Indian island and some late Cretaceous event, the biogeographer will still want adequate documentation. He would still query the generic identity of the frog or lizard, inquire about the evidence for its date, verify the dating and reality of the geologic event and make sure of the frog's or lizard's relevance to the geologic event, but he would not feel automatically privileged to doubt the generic identity of the frog because of its age. We still do not have for frog or lizard genera or lineages the kind of record that can either ensure confidence or compel skepticism in regard to the dating of a fossil and its association with any well dated geologic event - unless the apparent discrepancy is very extreme.

The two competing biogeographic theories

Now as to the competing theories of dispersal and vicariance:

A problem in semantics, or at least nomenclature, confronts us at the outset. Platnick (1976) has distinguished between "dispersion" - the movement of individuals - and "dispersal" which he restricted to the spread of a taxon. We have not found that Platnick's proposed distinction has been widely followed or is even understood by any wide audience. We shall not ourselves employ Platnick's suggested language and shall treat his "dispersion" and "dispersal" as different aspects of one phenomenon: "dispersion" sensu Platnick is absolutely required for "dispersal" sensu Platnick. The real concern of Platnick, however, is to distinguish both phenomena from vicariance, which, as he defines it, is very strictly a population phenomenon. Platnick's concluding statement (p. 295) in 1976 is very clear:

> "It would seem then that the range of a taxon is the result of three different phenomena: vicariance, dispersion and dispersal. The original range of a taxon is established by vicariance at the time that the taxon splits from its sister group (Croizat, Nelson and Rosen 1974), and is maintained through time by dispersion; changes in that range can occur by dispersal. Of the three processes vicariance and dispersion may be deemed primary, being intrinsically necessary for the origin and continued existence of any taxon, while dispersal may be deemed secondary,

being only extrinsically necessary if changes in geography or ecology force (or allow) changes in the range of the taxon."

In this paragraph Platnick seems to me to state in an especially concise way the frame of reference in which "dispersal" and "vicariance" are used by Rosen, Nelson, Platnick, and their followers. To explain this frame of reference, which is the nexus of a very vigorous dispute, must be the object of our exposition.

(1) Dispersal:

All living species tend to disperse. This is partly a Malthusian phenomenon - a tendency to "natural increase" - but also partly a consequence of the fact that habitats change over time. The "living space" for any species - the space in which it can live and thrive - may expand, permitting and encouraging dispersal. Or the habitat upon which the species is dependent in any one place may be obliterated; its living space is then contracted, and if the species does not already have or has not found by dispersal some other place where it may survive, it will, of necessity, go extinct. Dispersal will always be by individuals, although, of course, an unspecifiable number - a group - may disperse simultaneously, given the opportunity.

Species ranges thus may expand or contract - or merely change position over time. They are in any case, at least at some microlevel, internally discontinuous. There is fluctuation in population size, and, as part of that fluctuation - very characteristically - there is local extinction - sometimes very local. The momentarily unoccupied space within the species may then be recolonized from neighboring populations.

Dispersal - understood as dispersal by individuals, i.e., dispersion sensu Platnick -is limited by barriers - environments in which the taxon is less viable than in its preferred habitat and in which it is unable to maintain a population. Water gaps are very obvious barriers to terrestrial species, but subtler habitat differences may also be effective barriers depending upon the physiology and habits of the animal. A barrier may be <u>permeable</u> in the sense that, while it does not offer living space, it does offer the possibility of transit of colonists - propagules sensu MacArthur and Wilson (1967) - to new living space.

The colonization of new living space - though always achieved by individuals - will result in a range extension if and only if the colonizations succeed. There must be many propagules that fail at the moment of entry or soon thereafter - for any number of reasons: the ecological conditions at the exact landing spot, the preemption of the available area by a taxon or taxa with nearly equivalent ecology that thus can deny access to - exclude - the invader. These failed colonizations would be dispersions without dispersal in Platnick's phraseology.

All of this is, so far, as applicable to continents as to islands. The significant barrier in the case of islands is, of course, water. Water, however, may be a very permeable barrier. It is possible to step across a creek, to swim across a river, to raft or sail across greater water barriers.

One of the factors in the permeability of water barriers - as with other barriers - is thus distance - the distance required for transit. This distance in the case of rafting, it should be noted, is not necessarily - perhaps not usually - straight line distance. Currents and winds will often make the voyage oblique or indirect.

Another variable very significant for a colonizing voyage is the propensities/potentialties of the taxon. A bird or a bat has the capacity to cross at least narrow water gaps, although it may not exercise the option available to it. A windblown or floating seed or a storm-driven bird or bat may cross water gaps without a choice in doing so. Lizards or rats may also cross on natural rafts or on man-built ships, and the different kinds of lizards or rodents are known to differ very much in the extent that their habits or habitats favor or disfavor the possibility of rafting - or human assistance. (See further below.)

Some of the same factors that are involved in the crossing of a water barrier may also feature in success after arrival. Certainly there are different probabilities from taxon to taxon in every aspect of cross-water colonization, from its beginning in the voluntary or passive inception of the voyage, through the hazards of the transit itself, to the difficulties of landing on a possibly hostile shore, and of establishing a continuing population thereafter. Some biota are <u>empirically</u> relatively "island-prone." One of us (Williams 1969) has discussed this phenomenon for anoline lizards in the Caribbean. Zoologists and botanists alike continually find examples on all islands.

There are, indeed, groups - the so-called "fugitive" or "tramp" species among animals and "weeds" among plants - whose preferred habitats may be very temporary - and whose life style and life story therefore depend always on colonization of a new, favorable, temporary environment. "Weediness" is a recognized character of a good colonizer: resistance to temporary hazards, resilience in the face of unfavorable conditions, even willingness to colonize. Size, physiology, and reproductive style are all factors in the making of a "good colonizer."

The other side of this story is, of course, that there are "poor colonizers" - biota that are <u>empirically</u> relatively "continent restricted." These include certain major taxa and also species even within groups that include good colonizers. It is part of the long-known lore of islands - particularly distant, "oceanic," islands - that the resident biota are always a small and peculiar fraction of the fauna of adjacent continents. The West Indies, in this regard, have the classic characteristics of "oceanic" islands, and this is, indeed, the fact that, along with the dogma of stable continents and oceans, convinced Simpson (1956), Darlington (1957), and Williams (1969) that cross-water dispersal was the only viable explanation of West Indian biogeography.

All historical discussion of cross-water dispersal is, as a matter of course, <u>a discussion of probabilities</u>. Cross-water colonizations unassisted by man are rarely documented even for historical time, although for birds stray individuals that are not the source of breeding populations are rather often recorded. (See the remarks on "wanderers" in Lack (1976, chapter 5).) Occasional successful cross-oceanic colonizations are well known. Some of the best information appears to be for New Zealand, where records have been kept since 1856. (The entry under "range changes" in Campbell and Lack 1985 is a useful summary.) For other groups accounts tend to be anecdotal (see Williams' introductory chapter this volume), but the undoubted cases lend credence to the anecdotes.

Long distance dispersals across significant barriers - water gaps in the case of islands - are for many critics of dispersal theory a major difficulty. Even in these cases - and for animals that must raft - e.g., lizards - there appear to be cases that must be true. The cases of *Anolis sagrei* and *A. allisoni* - species Cuban in origin that are found also on the fringes of the Central American mainland, but on offshore islands and keys or on the immediate shore only, are regarded by Williams (1969) as unquestionable cases of long distance cross-water voyages. Parallel cases exist among the lizards of the Pacific - *Anolis agassizi* of Malpelo Island, (Graham 1975), *Emoia arundeli* of Clipperton Island (Sachet 1963) and, of course, the *Tropidurus* and the iguanine lizards of the Galapagos (Wright 1984), as well as the even more distant and surprising iguanine *Brachylophus* of Fiji (Gibbons 1981, 1985)

Among vertebrates, birds, bats, lizards, rodents, snakes and frogs - in descending order - seem to show the greatest potentiality for cross-water colonization. Ungulates, carnivores, and salamanders - again in descending order - seem to show the least. This is, of course, a post facto judgment based on the known <u>characteristically</u> <u>biased</u> fauna found on distant islands. Darlington's 1957 chapter on "Island patterns" is still a useful summary. Carlquist's two books on islands, 1965 and 1974, document many of the idiosyncrasies characteristic of the biota of isolated islands.

Hawaii and the Galapagos are isolated archipelagos that there is almost indisputable geological reason for believing have not been connected with any continent. Hawaii is believed never to have had, among vertebrates, more than birds and bats. Mayr in 1943 estimated 14 bird invasions, which in terms of level of differentiation were very diverse - ranging from a radiation at the family level through endemic genera to endemic species to endemic subspecies and one species - a heron not differentiated at all). Mayr presumed a direct relationship between time since arrival and level of differentiation. The heron is on this ground believed to be a very recent arrival. The one bat is believed to be a similar case. The status of some Hawaiian lizards is uncertain, but none are verifiably native. The Galapagos have a larger surviving fauna - the colonizations required are still few - several bird invaders, including one that produced the classic radiation - Darwin's finches - a tortoise, a snake, some lizards, some rodents, and two bats.

These are cases that on current evidence must be dispersal, yet the biological record, it is clear, is not complete. Storrs Olson and Helen James (1982a and b), having discovered more than 40 extinct endemic bird species in Pleistocene or Holocene deposits on the Hawaiian islands, suggest that "the historically known avifauna represents only a third, or less, of the total number of endemic species of birds that were present in the Hawaiian islands when man first arrived there."

A question that is not fully answered, although the general problem has been addressed by MacArthur and Wilson (1967), is whether a sequential series of short cross water dispersals - stepping stone islands - is always more likely than a longer direct voyage. In an imaginable case the target island for a long voyage might be Hispaniola, a very large target, and the much smaller stepping stone islands might be the islands of the Aves Ridge or of the Lesser Antilles. If there were one low probability for the longer voyage, and a number of only somewhat higher probabilities for the sequence of inter-island stepping stone voyages, then it would seem plausible to multiply the several inter-island probabilities by one another, with the result then of a probability no higher than the low probability for the long voyage. In practice a number of factors might make this too-simple calculation invalid. All these probabilities are probabilities that must factor in not only time but also population size and conditions at the point of departure and the port of entry; if populations readily built up on the stepping stone islands, then the probabilities for the inter-island sequence might not be much lower than the individual probability between any two. Much would depend, as MacArthur and Wilson (1967) have made clear, on the size and number of the stepping stone islands, and their placement in relation to each other and the target island or mainland. Oceanic currents and storm tracks would also be important (Darlington 1938).

Size of island is clearly a factor in both the arrival and survival of any island biota. Many subsidiary factors contribute - the size of the island as target, the size of the island as related to habitat diversity, the size of the island understood as carrying capacity. MacArthur and Wilson (1967) have also postulated "saturation" - a situation in which the the success of a new invader must be correlated with the extinction of an old resident. In such a case "exclusion" or "replacement" are the alternatives.

Dispersal - it needs to be repeated - is always of individuals, although individuals may travel together, as in the dispersal of young individuals (spiders, frogs, or seeds from their parent source) or in land migrations of adults due to climatic or demographic pressures, or in cases of chance cross-water transport, whether by rafts or storms. Rafts - such as trees or pumice - may carry more than one element of the biota - lizards and also the insects they prey upon - as well as spores and seeds. Rafts, indeed, can be micro-vicariant events, much more frequent than the macrovicariant events that involve large land masses.

Dispersal may be active or passive - but species differ very much in the will to disperse, the physiological capacity to disperse, and in access to dispersal opportunity. As to the last point, rafts tend to form in special places - e.g., at the mouths of rivers - and at special times - times of flood; storms in having characteristic tracks have places where they preferentially pick up waifs as well as places at which they are more likely to leave them. Chance may be part of the story, but chances are very different according to occasion and taxon.

It should be remembered that chances occur to individuals and in the case of abundant species to many individuals. The chance of dispersal for any individual picked at random may appear to be infinitesimal, but the chance for a species may be quite otherwise. Multiply the miniscule opportunity of cross-water dispersal for an individual by large populations and long time periods, and the chance of success for a species may approach certainty - and in fact may easily imply successive multiple invasions - and in many cases has unquestionably done so.

Simpson's (1940) metaphor of "sweepstakes" was not totally unrealistic. Williams (in the introductory chapter to this volume) elected to discard it only in order to emphasize the unfairness of the biotic lottery. Simpson did distinguish "ticket holders" - those with some chance - from "non-ticket holders" - those with no chance, but the biotic lottery is much less fair than that implies. Not all ticket holders have the same chances. There are very real differential opportunities for members of some species to succeed where members of other species fail. Chance of cross-water dispersal varies, as we have indicated, with distance. With short distances many taxa may be winners. With increasing distance the number of taxa that can win falls off dramatically, and it is a very characteristic roster of taxa that reach the very distant islands. That roster is an empirical fact that dispersal theory does explain.

Summary for dispersal.-- Dispersal, it needs to be emphasized, must be regarded as a characteristic of life - the Malthusian tendency of life to increase without limit - unless controlled by extrinsic factors. MacArthur and Wilson's term "propagule" highlights the Malthusian aspect of all dispersal. But dispersal by itself does not automatically imply colonization. In cross-water dispersal what happens at the port of entry is critical. If in fact the propagule can seize upon an opportunity to disperse beyond the port of entry, the Malthusian factor again takes over: population size may reach numbers that make regression to zero extremely improbable. The propagule - the cross-water waif population - has then achieved innoculation: success in colonization.

That innoculation may not occur is part of the story of dispersal. Physical factors, perhaps confined to the chance port of entry, may be the relevant reason for failure, or, alternatively, biological factors such as competition, perhaps direct and obvious or, perhaps, sufficiently obscure to obtain the name "diffuse competition" (MacArthur 1972) may be the reality.

The island faunas resulting from dispersal are biased in the direction of those taxa that colonize easily; these are, not unexpectedly, often the same taxa that are readily introduced by man. Such faunas can be called "colonization faunas," and their degree of difference from continental faunas is highly correlated with distance from mainland source areas.

Dispersal, as a behavior that individuals exhibit, is not rare, not even occasional, not "confined to the Pleistocene" as Rosen (1976), Patterson (1981), and Poinar (1988) have suggested It is pervasive, continual, essentially ubiquitous - a consequence of the fact that life and reproduction are essentially synonymous. Dispersal may fail, and where barriers exist - water barriers in the case of islands - most attempted dispersals will fail. (This is, in fact, the essence of the definition of barrier.) The fact of attempt is, however, so nearly universal, that dispersal is not a concept to be discarded or denigrated, but a fact to be confronted. Colonization is the phenomenon that can result from dispersal. It is a rarer event, but non-random, strongly determined by the characters of the colonizing species.as well as by the environments at the point of departure and the port of entry.

The theoretical vulnerability of dispersal is its near ubiquity. It may happen almost anywhere at almost any time. For the biogeographer who is seeking to establish the possibility or probability of other explanations of distributions - particularly disjunct distributions - it is noise rather than signal. The frustration that such a biogeographer faces when his alternative is dismissed out of hand is certainly the phenomenon that has so inflamed the rhetoric of Croizat, Rosen, Nelson, and Savage. But - again we emphasize - the difficulty of determining whether dispersal for any given case is noise or signal is not a refutation of its frequent reality. The strength of the dispersal hypothesis is that it may be more often true than not.

2. Vicariance.

The presentday concept of vicariance derives from the allopatric model of speciation. It thus explicitly requires geographic isolation as the primary or only mode of species formation. The special frame of reference in which we have cited Rosen, Nelson and Platnick is "vicariance biogeography," and the key reference for the latter is Croizat, Nelson, and Rosen (1974), a paper later disavowed by its senior author (Croizat 1982). The major tenets of vicariance - at least those relevant to issues discussed in this paper - are to be found in the notes at the end (pp. 277-278) of the 1974 paper in which vicariance as a historical process is said to be "embodied" in the following premises and conclusions:

"a. Allopatric species (vicariants) arise after barriers separate parts of a formerly continuous population, and thereby prevent gene exchange between them.

b. The existence of races or subspecies of a species that are separated by barriers (vicariants) means that the population has subdivided, or, is subdividing, not that dispersal has occurred or is occurring across barriers.

c. The earliest stages (races and subspecies) of differentiation (vicariance), separated by complete or incipient barriers to gene exchange, are entirely allopatric.

d. Sympatry between species of a monophyletic group implies dispersal of one or more species into the range(s) of the other(s).

e. Allopatric speciation (vicariance) predominates over other forms of population differentiation; allopatry is the rule and sympatry the exception in the present-day distributions of a given monophyletic group.

f. Vicariance is, therefore, of primary importance in historical biogeography, and dispersal is a secondary phenomenon."

These are aphorisms - axioms employed to construct a theory. As axioms they have been treated as primary assumptions by the proponents of vicariance biogeography.

Yet their empirical base was, even in 1974, suspect. The possibility of allopatric speciation by the passage of a barrier was at that time not admitted nor discussed. This point, indeed, had to be emended by Platnick and Nelson (1978), but Mayr's founder concept, already published in 1942 and which has proved so important for tracing the radiation of the Drosophilidae in the Hawaiian series of hot spot islands (Carson 1984), was ignored. The requirement that the earliest stages of differentiation (vicariance) be entirely allopatric (a matter of separated populations) is an undocumented assertion, with which many workers on speciation would quarrel. Bush (1975) discussed other possibilities, e.g. "parapatric speciation" - species formation in contiguous populations. Brundin (1981) unequivocally rejected the Croizat et al. axioms e and f. (Whatever the incidence of sympatry at low taxonomic levels; it is, of course, essentially universal for high taxonomic categories.) Passed over without comment is the reinforcement concept (discussed by Mayr 1963:548-554) according to which the final stage in speciation is not realized until the secondary contact of allopatric populations tests species isolating mechanisms and species recognition modes. Speciation theory remains a very difficult and controversial topic. More than one mode and more than one variety of any mode seem likely to find real examples. It seems, on its face, unfortunate to make an exceptionally rigid and, it would appear, incompletely worked out version of one speciation mode the base of a major biogeographic theory.

Insistence that speciation be associated with macrogeological events was not an express part of the original set of axioms, cited above, but repeated reference in the 1974 paper to the Pliocene closure of the Panama gap, to the Galapagos as probably a detached part of South America, and to the splitting of Gondwanaland indicate a bias toward this sort of explanation. Plate tectonics assisted but was not the origin of this bias. Despite his later disavowal of Nelson-Rosen vicariance biogeography, Croizat had early advocated alternation of periods of mobilism and immobilism for biological taxa (see Croizat 1962 fig. 44 legend), the periods of mobilism (= migration in Croizat's usage) allowing the wide distribution of ancestral populations, while the periods of immobilism allowed "vicariant form-making." His thinking began with minor climatic/ecological separations, but with wide and repeated disjunct patterns he was very prepared to invoke macrogeographic phenomena. In his phrase (1964:91) "...geological considerations do forthwith prove necessary." Rosen (1976) elaborated and modernized Croizat's geology - inevitably since plate tectonics was in its youth or infancy when Croizat began, and Rosen's heavy emphasis remained, consistently in 1976 and thereafter, on macrogeological events.

For islands the initial macrogeological event in vicariance biogeography should logically be rifting or subsidence. Rosen's (1976) Caribbean biogeographic hypothesis began with a proto-Antillean island arc - thus with volcanos initially under water that later became emergent islands and still later aggregated into larger islands that moved eastward to become the Greater and Lesser Antilles. The curiosity of the Rosen 1976 model is that - although it was so named - it is not vicariance biogeography as defined in 1974 by Croizat et al., nor as modified by Platnick and Rosen in 1978. Guyer and Savage (1986), on the other hand, have proposed an authentic Caribbean vicariance model. They bypass the proto-Antilles as an island arc to put forward immediately a hypothesis of a Paleocene dry land connection - physically an isthmus very like the present day Costa Rica-Panama isthmus. Their Paleocene isthmus then secondarily shatters in classic vicariance fashion into the precursors of the modern islands.

That the Guyer and Savage isthmus is not really faithful to its supposed geological source (Pindell and Dewey 1982) may be irrelevant from the point of biogeographic theory. Theirs is an idealized and nearly pure version of the Caribbean vicariance biogeographic hypothesis, and as such provides an appropriate contrast to any idealized dispersal model.

The biotic exchange that should have been possible across the Late Cretaceous-Paleocene isthmus proposed by Guyer and Savage should not have differed significantly from the exchange that we know has actually occurred across the present Panamanian isthmus - except in terms of the biota then available to make the exchange. Savage, in fact, has stubbornly insisted on the neccessary existence of this earlier isthmus - Isthmian Link I of Williams (this volume) - because he believes his biological scenario for Central America requires it. Judging from the modern isthmus the exchange across the older one would be asymmetrical: not all groups even on the dry land of an isthmus get across (this is Simpson's filter bridge concept), and there would be deeper incursion by the fauna of one of the two continents into the foreign territory, and a lesser incursion by the fauna of the other continent. Still there would surely be a broad exchange, as there has been on the present one. The isthmus itself would be the area of initial mixture, and the expectation would be that this would be the area of maximal mixture also.

If this were true of the old isthmus - the isthmus of Guyer and Savage - then, when this land mass broke apart to become the Greater and Lesser Antilles, its initial fauna should show that maximal mixture. The one constraint that would be imposed upon the fauna of a vicariant proto-Antilles would be temporal: the initial fauna could only contain elements - lineages - that had differentiated - had had their inception - by the time of the vicariant event.

To spell out our retrodiction for Guyer and Savage's isthmus, it is that it should have had a very mixed fauna of both North and South American affinities, and that it should have contained only taxa temporally characteristic of the Late Cretaceous or the Paleocene.

This last retrodiction will be contested as meaningless. Fossils, it will be said, provide "only minimum age." This is an issue we have already dealt with above, when we discussed the meaning of the genus *Felis* for age determination. If you throw out relative age determination by fossils, you throw out a great deal of classical geology, confirmed now, quite independently, by radiometric dating.

To take the extreme case, species, and even more subspecies, common to an island and a mainland are reasonably suspected to have a very recent common origin. By this hypothesis either the geological event creating the island should be very recent or dispersal between island and mainland must be very seriously considered. A Late Cretaceous-Paleocene isthmus should not be the source of origin, nor vicariance the mode, for taxa that are conspecific or consubspecific now on mainland and islands.

Athough we think that this is a current majority view, Croizat et al. (1974) argue the alternative hypothesis of differential rates of evolution. They specifically raise the issue for bird species and subspecies found on both the classic Venezuelan island plateaus of "Pantepui" (Mayr and Phelps 1967) and their supposed montane centers of origin. Abstractly this is an alternative that may logically be considered, but, in practice, in each particular case, both the speed of evolution found from other evidence, including the paleontological, to be characteristic of the major group under consideration should plausibly be taken into account, and the validity of the current taxonomy sharply scrutinized. (See again the our discussion of the meaning of fossils above.)

As utilized by Rosen and by Savage, vicariance has been very dependent on specific plate tectonic models. This may not be the most useful approach. Two papers by Malcolm McKenna (1973, 1983) may be better. McKenna's endeavor was to describe realistic geological events - unspecific but clearly, given mobilist geology, within the realm of the possible - and to discuss the consequences for faunas involved in these events. He has provided for these events very vivid metaphorical names.

"Noah's arks" as described by McKenna (1973) - rifted fragments of land carrying with them their full biota - are classic vicariance agents for living faunas, rafting them to new associations. McKenna's postulated "Viking funeral ships" - named in the same paper - were again rifted fragments, each with its characteristic buried fossils - would be vicariance agents for fossils, carrying them to lands where they had never been alive. Each rifted fragment would as a Noah's ark carry the living fauna characteristic of the time at which it was launched. In the case of Viking funeral ships it would also carry a buried fauna set in place while the ship was being built. The fossil fauna of a Viking funeral ship could not change when the ship finally docked, but Noah's arks - if long at sea - might be expected to show evolutionary change, and also if arks collided with other arks, perhaps with substantially different faunas, each fauna might "board" the other ark with resultant mixtures. If a Noah's ark docked at a mainland other than its place of origin, again faunal mixture should result. These and others of McKenna's metaphors seem to us heuristic - especially useful because not tied to a specific time and place.

Another phenomenon that we would regard as a useful vicariance concept, cited by Rosen (1976) but not used as part of his own Caribbean story, is the make-break island sequence suggested by Ball (1971). Ball's statement of his hypothesis is a version of a land bridge scenario: "...a continuous land connection at any given point in time is not essential. It is only necessary that refugia should persist while various connections are made and broken." This is not unlike the hypothesis proposed by Schuchert (1935) of a peninsular connection between Central America and the Greater Antilles that was broken by intermittent submergence of part of the Antilles.

Ball was aware of Schuchert's specific suggestion but rejected it on the ground that the fresh-water planarians that were his special concern had their relatives in South America and therefore required a connection to the south not the east. Ball was, like Schuchert, relying on erroneous geology, in Ball's case citing Woodring (1954) to the effect that much of the Caribbean sea was land during the Cretaceous and possibly during the Eocene. Ball and Schuchert were both too committed to their own special scenarios. Their suggestions, however, of physical mechanisms for vicariance are, of course, themselves valid, whether or not the specific examples are themselves correct. Given mobilist geology we know that the situation was surely not as simple as Ball or Schuchert imagined. McKenna's arks better describe reality, but, in any case, the contact and separation of islands - and perhaps of islands and continents - is a very genuine part of the modern geological story for the Caribbean.

Summary for vicariance. - In contrast to dispersal, vicariance is not a matter of individuals actively or passively moving away from a source area but of two faunas separated by a secondary barrier. In the sense of the physical phenomena - fragmentation of faunas and the physical aggregation of faunas - it is, even in the Caribbean, a reality. If the islands have accreted and fragmented - and this seems clearly true in the Greater Antilles - vicariance has occurred. Vicariance biogeography, however, as method, explicitly expects to discover concordance of time and place in the separation of taxa and the separation of terranes - the matching of biological relationships and geological history.

In this - its hoped for merit - lies the present weakness of vicariance biogeography: only strongly corroborated biological hypotheses can be usefully matched with well constrained geological models. Ideally this should be valid. For the Caribbean it is at this moment impossible. The testable strength of the continent to island vicariance hypothesis lies in a specific prediction - that there was at the time of the vicariance event a separation not of individuals but of faunas. The test then is the discovery of fossil faunas. The alternative is to admit that the question is - at best - not decided.

Biogeography and the tectonic history of the Caribbean: Critique of attempts at synthesis

Against this very general background - part theory, part empirics - how do we deal with 80 million years of Caribbean history? Fig. 6 provides the opportunity for exploration of the possibilities. The four panels of that figure cover most of the total time period with which we are concerned, and each reflects in diagrammatic fashion what we consider to be the most probable geological configuration for four segments of that time. The most casual look will show that there are far fewer biological constraints than we could desire.

Ideally, we should first look at the geological configurations rather abstractly, endeavoring to determine just from the land masses available how much dispersal and how much vicariance might have occurred in each time period - and, importantly, where each biogeographic phenomenon might have been most likely.

If we so do, we will have to admit freely and frankly that we do not have decisive evidence on any point. Each of the panels represents a time period of several to tens of millions of years during which changes certainly occurred. We must therefore in our interpretation allow for alternative possibilities, nor may we suppose that we have thought of every alternative. The very best that we can hope for is that the alternatives that we suggest do, indeed, lie within the constraints of the known evidence.

Once we have looked at what seems geologically possible, we need to attempt to determine what may have been paleontologically possible - what groups may have entered where and by what means. To be realistic in this phase of our interpretation we would have to limit our hypotheses to one class of vertebrates only, the only group for which a terrestrial paleontological record is remotely satisfactory: mammals. We will leave to others the chore of going beyond the data to infer or to demonstrate the history of their favorite groups. We make a brief and cautious statement below.

In the earliest time for which useful data exists - Period I of our geological history above - we are dealing with the events and land masses well past the breakup of Pangaea and the origin of the Caribbean plate and of the Caribbean sea. Prior to this time - perhaps much prior to this - the existence of Pangaea assumes land connection in some more or less intimate sense between North and South America, and therefore presumably free access in both directions between the continents. From this very old land connection one might infer an association between the biota of North and South America greater than now exists, since the old connection is believed to have been broader than that now provided by the narrow Panamanian isthmus.

By the Late Cretaceous (Fig. 6a) subduction at the leading edges of a Caribbean plate has resulted in a chain of islands at the northeastern, eastern, and southeastern margins of the plate - an island arc that, in effect, provides stepping stones between North and South America. The arc was, of course, initially a series of submarine volcanos, only later becoming small emergent islands, perhaps rather evenly spaced, still later coalescing into the larger islands of our figure.

Even in this period the situation was complex, as our figure implies. Northeastward toward the North American plate there was the Cuban arc - CA of the figure - with some sort of proto-Hispaniolan-Puerto Rican complex eastward of it. Still further east, facing directly east toward the Atlantic, the figure implies that there were already in place the precursors of both the Aves Ridge and the Lesser Antillean arc. To the southeast these two abut on northwest South America.

To the west and north there is a Nicaraguan-Cayman arc, curving around the Chortis Block and continuing further to the north, fronting on the west coast of Mexico. To the southwest, inferred rather than known, our figure shows a southwest incipient arc - isolated in the Pacific - that is intended to represent the first indications of much later Costa Rica and Panama. Of these all except proto-Costa Rica-Panama might already be of biogeographic significance.

Indeed, at the stage figured cross-water dispersal onto these emergent islands from both North and South America should be occurring. Such dispersal would, of course, involve the selective/differential factors that have been mentioned above as favoring groups that show island-proneness. Some island-prone groups, whether from the north or from the south, might have had the possibility/likelihood of making it all the way between the two continents. If Fig. 6a is correct inter-continental dispersal should have been possible for some few groups - but only a few.

The existence of the island arc itself is uncontroversial. Exactly where the arc was and the extent to which the islands were enlarged and joined and where this occurred is not known (see Fig. 5). Any figure such as our Fig. 6a must choose rather arbitrarily from multiple possibilities. However, the position of the arc is not wholly a matter without biogeographic consequences. The length of the arc would presumably enter into any estimation of the difficulty of stepping stone colonization; the distances between islands and the size of islands would certainly enter into any estimation of cross-water dispersal.

The hypothesis of Guyer and Savage (1986) - and of Hedges (this volume) - adds the concept of a consolidation of the original arc of distinct small islands into a dry land isthmus between North and South America. Hedges, in particular, has emphasized that this connection might result, over the long period of time these arcs have existed, from accretion of the islands of the Cuban arc - and possibly also the Nicaraguan arc - with the islands of the Aves-Lesser Antilles arcs. This scenario cannot on geological grounds be definitively rejected, but equally it is not supported by any positive geological evidence. Hedges - and Guyer and Savage - predict - retrodict - the Cretaceous-Paleocene isthmus on the basis of biological hypotheses and the isthmus is made geologically plausible - if at all - by the age of the arc, coincident as its origin must have been with the very existence of a Caribbean plate.

The biogeographic expectations that would derive from such an isthmus are, as has been suggested above, exchange and mixture of full continental faunas. This, as we interpret the evidence, flies in face of the peculiarities of the known fauna of South America during both the Mesozoic and the early Tertiary. These peculiarities are abundantly documented in many references; we cite only two of the most recent - McKenna 1981 and Estes and Baez 1985. (See further below.) Our verdict is that none of the known biological data (neither the biological distributions in the present nor the paleontological data documenting the past) require such a massive exchange, and, in fact, are not congruent with it. We are compelled to regard the Cretaceous-Paleocene isthmus as an hypothesis that is not only not demonstrable geologically, but that goes against the weight of the paleontological evidence.

During this period - the Late Cretaceous - only the marsupials among mammals are at all likely to have been involved in a North American-South American interchange, and there is considerable dispute and ambiguity about what did occur and in what direction there was interchange. (Stehli and Webb's 1985 volume on "The Greatic Biotic Interchange" is an admirable sampling of these divergent views.) The early South American placentals - the edentates and the native South American ungulates - are especially puzzling. There is no consensus on how they got there (McKenna 1981). The Northern Hemisphere arctostylopids, so like the endemic South American ungulates that they were believed to be members of that South American radiation that had somehow migrated northward, are now considered to be Northern Hemisphere endemics that are merely convergent with the South American forms (Cifelli 1985, McKenna pers. comm.). Neither the marsupials nor any other mammals that might be relevant to this time period are at this time known from anywhere in the West Indies.

When in Fig. 6b - the Paleocene - we go beyond the first stages of Caribbean history, there is the beginning of approach to modern conditions. The northeast arc has collided with the Bahaman Platform, and vulcanism has diminished. There is still the opportunity for stepping stone interchange between the North American and South American continents, and to the north elements of the Greater Antilles have become more nearly recognizable as such, but barely so. Cuba is shown to be complexly fragmented, and northern Hispaniola and Puerto Rico are not recognizable except by general position. What will become southern Hispaniola is well to the west, along with Jamaica. In our figure the opportunities for anything but cross water dispersal are not at all obvious.

At some time approximating this, according to Guyer and Savage, the hypothetical proto-Antillean isthmus broke up. They postulate also contacts between the continent and some of the islands - for example Jamaica - and Central America at about this time. No direct geological evidence negates this suggestion, nor does any support it. For Jamaica contact resulting in early interchange of fauna might be irrelevant, since the island is known to have undergone submergence that may have been total. Part of Cuba is known to be very old and suspected to be a sliver of the North American plate. Certainly the situation in the Cuban-Yucatan area was particularly complex. If there was interchange, it is unclear what elements of that biota survived. Perhaps the xantusiid lizard *Cricosaura* dates from such an event.

However, it is clear that by the Paleocene larger land masses suggestive of a Greater Antilles were present to the north. If - and this is not indicated in our figure - one of these masses - perhaps part of Cuba - was closer to the northern mainland - perhaps Yucatan - there might have been an opportunity at the end of this period or the beginning of the next for the ancestors of *Nesophontes* and *Solenodon* to have entered the Antilles - whether by vicariance or dispersal. They could not have been on the isthmus - if that were real at all - in Late Cretaceous. The caveat must be put on record once again that if *Nesophontes* and *Solenodon* entered the Antilles by vicariance, they did not enter alone. They must, by the strong implication of vicariance, have been accompanied by other fauna - including other mammals. Placentals were then already highly differentiated and diverse. A paleontological record of mammals other than the insectivores but of similar antiquity should somewhere exist in the West Indies.

Examination of the remaining panels of our Fig. 6 provides even less probability of vicariance than those already examined. Distance from Central America increases, and any southern landmasses have been accreted to the South American mainland. Movement within the Antilles does occur, and there is both accretion and fragmentation. This is intra-Antillean vicariance; it cannot add anything to the fauna already present. Radiation may take place within the islands, adding species or even genera, but the major stocks must be already present.

There was a period of mid-Tertiary subsidence (Fig. 6d). Subsidence itself is well documented, but the extent of emergent land is not. The consequence of subsidence was surely extinction - quite possibly total obliteration of some stocks. But awareness that extinction ocurred does not tell us what elements of the biota went extinct, and certainly not what elements were present to go extinct. The significant evidence for vicariance must be found in the strata before the period of subsidence.

At least two - perhaps three - important components of the West Indian mammal fauna arrived well after the mid-Tertiary inundation. The ground sloths and the primates certainly did. The rodents perhaps - see Woods (this volume) - arrived earlier, but they are a diverse group, probably of several stocks. They most likely did not arrive - did not become established in the Greater Antilles - <u>before re-emergence.</u>

To summarize, our survey of 80 million years indicates a very early period in which vicariance - in the sense of continental faunas carried on fragments of a mainland or an isthmus - <u>might</u> have occurred. Our review of the evidence adds neither proof nor probability.

Aknowledgments

T.W. Donnelly and Malcolm McKenna provided very valuable criticism. An early draft was read by Karl F. Koopman. M.C. Lebron, P. Mann, D. Smith, R. Ross, B. Hedges, G. Mayer, R. O'Hara, and J. Reiss made helpful comments. M.R.P. thanks C. Woods for inviting him to participate in this Symposium and for introducing him to the realm of biogeographers. Laszlo Meszoly drew the figures.

Literature cited

Anderson, R.N. 1986. Marine Geology: A Planet Earth Perspective. John Wiley and Sons, New York.

Anderson, T.H., and V.A. Schmidt. 1983. The evolution of Middle America and the Gulf of Mexico-Caribbean Sea region during Mesozoic time. Bulletin of the Geological Society of America 94:941-966.

Arden, D.D., Jr. 1975. The geology of Jamaica and the Nicaraguan Rise. Pp. 617-661 in A.E.M. Nairn and F.G. Stehli (eds.). Ocean Basins and Margins. 3. Gulf Coast, Mexico and the Caribbean. Plenum Press, New York.

Ball, I.R. 1971. Systematic and biogeographical relationships of some *Dugesia* species (Tricladida, Paludicola) from Central and South America. American Museum of Natural History Novitates 2472:1-25.

Barbour, T. 1917. Notes on the herpetology of the Virgin Islands. Proceeedings of the Biological Society of Washington 30:97-104.

Bourgois, J., J-F. Toussaint, H. Gonzalez, J. Azema, A. Desmet, L.A. Murcia, A.P. Acevedo, E. Parra, and J. Tournon. 1987. Geological history of the Cretaceous ophiolitic complexes of northwestern South America (Colombian Andes). Tectonophysics 143:307-327.

Bouysse, P., R. Schmidt-Effing, and D. Westercamp. 1983. La Desirade Island (Lesser Antilles) revisited. Lower Cretaceous radiolarian cherts and arguments against an ophiolitic origin for the basal complex. Geology 11:244-247.

____, P. Andreieff, M. Richard, J.-C. Baubron, A. Mascle, R.-C. Maury, and D. Westercamp. 1985. Geology of Avis Swell and of the northern submarine shores of the Lesser Antilles (Rock dredging Cruise Arcante 3). Documents du Bureau de Recherches Géologiques, Géophysiques, et Minières 93:1-146.

Bowen, C. 1976. Caribbean gravity field and plate tectonics. Special Paper of the Geological Society of America 169:1-79.
Brundin, L.Z. 1981. Croizat's Panbiogeography versus phylogenetic biogeography. Pp. 94-138 in G. Nelson and D.E. Rosen (eds.). Vicariance Biogeography. A Critique. Columbia University Press, New York.
Burke, K., P.J. Fox, and A.M.C. Sengör. 1978. Buoyant ocean floor and the evolution of the Caribbean. Journal of Geophysical Research 83:3949-3954.
____, J. Grippi, and A.M.C. Sengör. 1980. Neogene structures in Jamaica and the tectonic style of the northern Caribbean plate boundary. Journal of Geology 88:375-386.
____, C. Cooper, J.F. Dewey, J.P. Mann, and J. Pindell. 1984. Caribbean tectonics and relative plate motions. In W.E. Bonini, R.B. Hargraves, and R. Shagam (eds.). The Caribbean-South American Plate Boundary and Regional Tectonics. Memoir of the Geological Society of America 162:31-64.
Bush, G.L. 1975. Modes of speciation. Annual Review of Ecology and Systematics 6:339-364.
Campbell, B., and E. Lack. 1985. A Dictionary of Birds. T. & A.D. Poyser, Calton. 670 pp.
Carlquist, S. 1965. Island Life. A Natural History of the Islands of the World. The Natural History Press, Garden City, New York. 451 pp.
____. 1974. Island Biology. Columbia University Press, New York. 660 pp.
Carson, H.L. 1984. Genetic revolutions in relation to speciation phenomena: the founding of new populations. Annual Review of Ecology and Systematics 15:97-131.
Case, J.E., T.L. Holcombe, and R.G. Martin. 1984. Map of geologic provinces in the Caribbean region. In W.E. Bonini, R.B. Hargraves, and R. Shagam (eds.). The Caribbean-South American Plate Boundary and Regional Tectonics. Memoir of the Geological Society of America 162:1-30.
Channell, J.E.T. 1982. Paleomagnetic stratigraphy as a correlation technique. Pp. 81-106 in G.S. Odin (ed.). Numerical Dating in Stratigraphy, Part I. John Wiley and Sons, U.K.
Cifelli, R.L. 1985. South American ungulate evolution and extinction. Pp. 249-266 in F.G. Stehli and S.D. Webb (eds.). The Great American Biotic Interchange. Plenum Press, New York.
Christofferson, E. 1973. Linear magnetic anomalies in the Colombian Basin, central Caribbean Sea. Bulletin of the Geological Society of America 84:3217-3230.
Coney, P.J. 1982. Plate tectonic constraints on the biogeography of Middle America and the Caribbean region. Annals of the Missouri Botanical Garden 69:432-443.
____. 1983. Un modelo de Mexico y sus relaciones con America del Norte, America del Sur, el Caribe. Revista del Instituto Mexicano Petroleo 15:6-15.
Croizat, L. 1962 (1964). Space, Time, Form: The Biological Synthesis. Salland Deventer, Netherlands. 879 pp.
____. 1982. Vicariance/vicariism, panbiogeography, "vicariance biogeography," etc.: A clarification. Systematic Zoology 31:291-304.
____, G. Nelson, and D.E. Rosen. 1974. Centers of origin and related concepts. Systematic Zoology 23:265-287.

Darlington, P.J. 1938. The origin of the fauna of the Greater Antilles, with discussion of dispersal of animals over water and through the air. Quarterly Review of Biology 13:274-300.

_____. 1957. Zoogeography: The Geographical Distribution of Animals. John Wiley & Sons, New York. 675 pp.

De Boer, J. 1979. The outer arc of the Costa Rican orogen (oceanic basement complexes of the Nicoya and Santa Elena peninsulas). Tectonophysics 56:221-259.

Donnelly, T.W. 1985. Mesozoic and Cenozoic plate evolution of the Caribbean region. Pp. 89-121 in F.G. Stehli and S.D. Webb (eds.). The Great American Biotic Interchange. Plenum Press, New York.

_____. 1988. Geologic constraints in Caribbean biogeography. In J. Liebherr (ed.). Zoogeography of Caribbean Insects. Cornell University Press, Ithaca, NY.

_____, W. Melson, R. Kay, and J.J.W. Rogers. 1973. Basalts and dolerites of Late Cretaceous age from the central Caribbean. In N.T. Edgar, A.G. Kaneps, and J.R. Herring (eds.). Initial Reports of the Deep Sea Drilling Project. National Science Foundation, Washington 15:989-1012.

_____, and J.J. Rogers. 1980. Igneous series in island arcs: the northeastern Caribbean compared with worldwide arc assemblages. Bulletin of Volcanology 43:347-382.

Draper, G. 1986. Blueschists and associated rocks in eastern Jamaica and their significance for Cretaceous plate-margin development in the northern Caribbean. Bulletin of the Geological Society of America 97:48-60.

Duncan, R.A., and R.B. Hargraves. 1984. Plate tectonic evolution in the mantle reference frame. In W.E. Bonini, R.B. Hargraves, and R. Shagam (eds.). The Caribbean-South American Plate Boundary and Regional Tectonics. Memoir of the Geological Society of America 162:81-84

Edgar, N.T., et al. 1973. Initial Reports of the Deep Sea Drilling Project 15. 1137 pp. U.S. Government Printing Office, Washington, D.C.

Estes, R., and A. Baez. 1985. Herpetofaunas of North and South America during the Late Cretaceous and Cenozoic: Evidence for interchange. Pp. 139-197 in F.G. Stehli and S.D. Webb (eds.). The Great American Biotic Interchange. Plenum Press, New York.

Faure, G. 1986. Principles of Isotope Geology. John Wiley and Sons, New York.

Fox, P.J., and B.C. Heezen. 1975. Geology of the Caribbean crust. Pp. 421-465 in A.E. Nairn and F.G. Stehli (eds.). The Ocean Basins and Margins. 3. The Gulf of Mexico and the Caribbean. Plenum Press, New York.

Fox, P.J., J. Schreiber, and B.C. Heezen. 1971. The geology of the Caribbean crust: Tertiary sediments, granitic and basic rocks from the Aves Ridge. Tectonophysics 12:89-109.

Ghosh, N., S.A. Hall, and J.F. Casey. 1984. Seafloor spreading magnetic anomalies in the Venezuelan Basin in W.E. Bonini, R.B. Hargraves, and R. Shagam (eds.). The Caribbean-South American Plate Boundary and Regional Tectonics. Memoir of the Geological Society of America 162:65-80.

Gibbons, J.R.H. 1981. The biogeography of *Brachylophus*, including description of a new species, *B. vitiensis* from Fiji. Journal of Herpetology 15:255-273.

_____. 1985. The biogeography and evolution of Pacific island reptiles and amphibians. Pp. 125-142 in G. Grigg, R. Shine, and H. Ehmann (eds.). Biology of Australasian Frogs and Reptiles. Royal Zoological Society of New South Wales.

Gill, J.B. 1981. Orogenic Andesites and Plate Tectonics. Springer-Verlag, Berlin. 390 pp.

Gose, N.A., and M.M. Testamata. 1983. Paleomagnetic results from sedimentary rocks in Jamaica: Initial results. Journal of the Geological Society of Jamaica 22:16-24.

Graham, J.B. 1975. The biological investigation of Malpelo Island, Colombia. Smithsonian Contributions to Zoology 176:1-98.

Guyer, C., and J.M. Savage. 1986. Cladistic relationships among anoles (Sauria:Iguanidae). Systematic Zoology 35:509-531.

Haq, B.U., J. Hardenbol, and P.R. Vail. 1987. Chronology of fluctuating sea levels since the Triassic. Science 235:1156-1167.

Hawkins, J.W., and J.T. Melchior. 1985. Petrology of the Mariana Trough and Lau Basin basalts. Journal of Geophysical Research 90:11431-11468.

_____, S.H. Bloomer, C.A. Evans, and J.T. Melchior. 1984. Evolution of intra-oceanic arc-trench systems. Tectonophysics 102:174-205.

Hendry, M.D. 1987. Tectonic and eustatic control on late Cenozoic sedimentation within an active plate boundary zone, west coast margin, Jamaica. Bulletin of the Geological Society of America 99:718-728.

Holcombe, T.L., and W.S. Moore. 1977. Paleocurrents in the eastern Caribbean: geological evidence and implications. Marine Geology 23:35-56.

_____, and G.F. Sharman. 1983. Post-Miocene Cayman Trough evolution: a speculative model. Geology 11:714-717.

Johnson, R.W., D.E. MacKenzie, and I.E.M. Smith. 1978. Volcanic rock associations at convergent plate boundaries; reappraisal of the concept using histories from Papua New Guinea. Bulletin of the Geological Society of America 89:96-106.

Karig, D.E., and G.F. Sharman. 1975. Subduction and accretion in trenches. Bulletin of the Geological Society of America 86:377-389.

Kennett, J.P. 1982. Marine Geology. Prentice Hall, Englewood Cliffs, NJ. 813 pp.

Kepler, C., and K.C. Parkes. 1972. A new species of warbler (Parulidae) from Puerto Rico. Auk 89:1-18.

Khudoley, K.M., and A.A. Meyerhoff. 1971. Paleogeographic and geologic history of the Greater Antilles. Memoir of the Geological Society of America 87:969-976.

Klitgord, K.D., and H. Schouten. 1986. Plate kinematics of the central Atlantic. Pp. 351-378 in P.R. Vogt and B.E. Tucholke (eds.). The Geology of North America. M: The Western North Atlantic Region. Geological Society of America.

Lack, D. 1976. Island Biology Illustrated by the Land Birds of Jamaica. University of California Press, Berkeley and Los Angeles. 445 pp.

Ladd, J.W. 1976. Relative motion of South America with respect to North America and Caribbean tectonics. Bulletin of the Geological Society of America 87:969-976.

Lebras, M., F. Megard, C. Dupuy, and J. Dostal. 1987. Geochemistry and tectonic setting of pre-collision Cretaceous and Paleogene volcanic rocks of Ecuador. Bulletin of the Geological Society of America 99:569-578.

LePichon, X., and P.J. Fox. 1971. Marginal offsets, fracture zones, and the early opening of the North Atlantic. Journal of Geophysical Research 76:6294-6308.

Lidz, B.H. 1984. Neogene sea level change and emergence, St. Croix, Virgin Islands: evidence from basinal carbonate accumulations. Bulletin of the Geological Society of America 95:1268-1279.

MacArthur, R.H. 1972. Geographical Ecology. Patterns in the Distrubution of Species. Harper and Row, New York. 269 pp.

____, and E.O. Wilson. 1967. The Theory of Island Biogeography. Princeton University Press, Princeton, NJ. 203 pp.

Malfait, B.T., and M.G. Dinkelman. 1972. Circum-Caribbean tectonic and igneous activity and the evolution of the Caribbean plate. Bulletin of the Geological Society of America 83:251-272.

Mann, P., and K. Burke. 1984. Neotectonics of the Caribbean. Review of Geophysics and Space Physics 22:309-362.

____, K. Burke, and T. Matsumoto. 1984. Neotectonics of Hispaniola: plate motion, sedimentation and seismicity at a restraining bend. Earth and Planetary Science Letters 70:311-324.

____, G. Draper, and K. Burke. 1985. Neotectonics of a strike-slip restraining bend system, Jamaica. In K.T. Biddle and N. Christie-Blick (eds.). Strike-slip Deformation, Basin Formation and Sedimentation. Special Publication of the Society for Economic Paleontology and Mineralogy 37:211-226.

Marsh, B.D. 1979. Island-arc volcanism. American Journal of Science 67:161-172.

Mattinson, J.M., L.K. Fink, Jr., and C.A. Hopson. 1980. Geochronologic and isotopic study of the La Desirade basement complex: Jurassic oceanic crust in the Lesser Antilles. Contributions to Mineralogy and Petrology 71:237-245.

Mattson, P.H. 1979. Subduction, buoyant braking, flipping, and strike-slip faulting in the northern Caribbean. Journal of Geology 87:293-304.

____. 1984. Caribbean structural breaks and plate movements. In W.E. Bonini, R.B. Hargraves, and R. Shagam (eds.). The Caribbean-South American Plate Boundary and Regional Tectonics. Memoir of the Geological Society of America 162:131-152.

Mayr, E. 1943. The zoogeographic position of the Hawaiian Islands. Condor 45:45-48.

____. 1963. Animal Species and Evolution. Harvard University Press, Cambridge, MA. 797 pp.

____, and W.H. Phelps, Jr. 1967. The origin of the bird fauna of the south Venezuelan highlands. Bulletin of the American Museum of Natural History 136:269-328.

McCann, W., and L. Sykes. 1984. Subduction of aseismic ridges beneath the Caribbean plate: implications for the tectonics and seismic potential of the northeastern Caribbean. Journal of Geophysical Research 89:4493-4519.

McKenna, M. 1973. Sweepstakes, filters, corridors, Noah's arks and beached Viking funeral ships in paleogeography. Pp. 295-308 in D.H. Tarling and S.K. Runcorn (eds.). Implications of Contintental Drift to the Earth Sciences.

____. 1981. Early history and biogeography of South America's extinct land mammals. Pp. 43-77 in R.L. Ciochon and A.B. Chiarelli (eds.). Evolutionary Biology of the New World Monkeys and Continental Drift. Plenum Press, New York.

____. 1983. Holarctic landmass rearrangement, cosmic events and Cenozoic terrestrial organisms. Annals of the Missouri Botanical Garden 70:459-489.

Miyashiro, A. 1961. Evolution of metamorphic belts. Journal of Petrology 2:277-311.
———. 1975. Volcanic rock series and tectonic setting. Annual Review of Earth and Planetary Science 3:251-269.
Moussa, M.T., G.A. Seigle, A.A. Meyerhoff, and I. Taner. 1987. The Quebrdillos limestone (Miocene-Pliocene), northern Puerto Rico, and tectonics of the northeastern Caribbean margin. Bulletin of the Geological Society of America 99:427-439.
Nagle, F., R.N. Erlich, and C. Canovi. 1978. Caribbean dredge haul compilation: summary and implications. Geologie en Minjbouw 57(2):267-270.
Okaya, D.A., and Z. Ben-Avraham. 1987. Structure of the continental margin of southwestern Panama. Bulletin of the Geological Society of America 99:792-802.
Olson, S.L., and H.F. James. 1982a. Fossil birds from the Hawaiian Islands: Evidence for wholesale extinction by man before Western contact. Science 217:633-635.
———. 1982b. Prodromus of the fossil avifauna of the Hawaiian Islands. Smithsonian Contributions to Zoology 365:1-59.
———, and E.N. Kurochkin. 1987. Fossil evidence of a tapaculo in the Quaternary of Cuba (Aves: Passeriformes: Scytalopodidae). Proceedings of the Biological Society of Washington 100:353-357.
Patterson, C. 1981. Methods of paleobiogeography. Pp. 446-489.
Perfit, M.R., and B.C. Heezen. 1978. The geology and evolution of the Cayman Trench. Bulletin of the Geological Society of America 89:1155-1174.
———, B.C. Heezen, M. Rawson, and T.W. Donnelly. 1980. Chemistry, origin and tectonic significance of metamorphic rocks from the Puerto Rico Trench. Marine Geology 34:125-156.
Pindell, J.L. 1985. Alleghenian reconstruction and the subsequent evolution of the Gulf of Mexico, Bahamas, and Proto-Caribbean Sea. Tectonics 4:1-39.
———, and S.F. Barrett. 1988. Geological evolution of the Caribbean: a plate tectonic perspective. The Geology of North America: The Caribbean Region. In press. Geological Society of America.
———, and J.F. Dewey. 1982. Permo-Triassic reconstruction of western Pangea and the evolution of the Gulf of Mexico/Caribbean region. Tectonics 1:179-212.
———, J.F. Dewey, S.C. Cande, W.C. Pitman, III, D.C. Rowley, J. LaBrecque, and W. Hexby. 1988. Plate-kinematic framework for models of Caribbean evolution. Tectonophysics 155: 121-138.
Platnick, N. 1976. Concepts of dispersal in historical biogeography. Systematic Zoology 25:294-295.
———, and G. Nelson. 1978. A method of analysis for historical biogeography. Systematic Zoology 27:1-16.
Poinar, G.O., Jr. 1988. Hair in Dominican amber: Evidence for tertiary land mammals in the Antilles. Experientia 44:88-89.
Pyle, T.E., A.A. Meyerhoff, D.A. Fahlquist, J.W. Antoine, J.A. McCrevey, and P.C. Jones. 1971. Metamorphic rocks from northwestern Caribbean Sea. Earth and Planetary Science Letters 18:339-344.
Rieppel, O. 1980. Green anole in Dominican amber. Nature 286:486-487.
Rosen, D. 1976. A vicariance model of Caribbean biogeography. Systematic Zoology 24:431-464.

Rosencrantz, E., and J.B. Sclater. 1986. Depth and age in the Cayman Trough. Earth and Planetary Science Letters 79:133-144.

Savage, J. 1982. The enigma of the Central American herpetofauna: Dispersals or vicariance? Annals of the Missouri Botanical Garden 69:464-547.

Sachet, M.-H. 1963. History of change in the biota of Clipperton Island. Pp. 525-534 in J. L. Gressitt (ed.). Pacific Basin Biogeography. Bishop Museum Press, Honolulu, HI.

Schuchert, C. 1935. Historical Geology of the Antillean-Caribbean Region. John Wiley & Sons, London.

Schwartz, A. 1978. A new species of aquatic *Anolis* (Sauria, Iguanidae) from Hispaniola. Annals of Carnegie Museum 47:261-279.

Simkin, T., and R.S. Fiske. 1983. Krakatoa 1883. The Volcanic Eruption and its effects. Smithsonian Institution Press, Washington, D.C. 464 pp.

Simpson, G.G. 1940. Mammals and land bridges. Journal of the Washington Academy of Science 30:137-163.

_____. 1956. Zoogeoography of West Indian land mammals. American Museum of Natural History Novitates 1759:1-28.

Smith, D.L. 1985. Caribbean relative plate motions. Pp. 17-48 in F.G. Stehli and S.D. Webb (eds.). The Great American Biotic Interchange. Plenum Press, New York.

Speed, R.C. 1985. Cenozoic collision of the Lesser Antilles Arc and continental South America and the origin of the El Pilar fault. Tectonics 4:41-69.

Stehli, F.G., and S.D. Webb (eds.). 1985. The Great American Biotic Interchange. Plenum Press, New York. 532 pp.

Sykes, L.R., W.R. McCann, and A.L. Kafka. 1982. Motion of the Caribbean plate during the last 7 million years and implications for earlier movements. Journal of Geophysical Research 87:10656-10676.

Thomas, R., R.W. McDiarmid, and F.G. Thompson. 1985. Three new species of thread snakes (Serpentes:Leptotyphlopidae) from Hispaniola. Proceedings of the Biological Society of Washington 98:204-220.

Torrini, R.J., Jr., R.C. Speed, and G.S. Mattioli. 1985. Tectonic relationships between forearc-basin strata and the accretionary complex at Bath, Barbados. Bulletin of the Geological Society of America 96:861-874.

Vail, P.R., R.M. Mitchum, and S. Thompson. 1977.Seismic stratigraphy and global changes of sea level. Memoir of the American Association of Petroleum Geologists 26 49-212.

Van Fossen, M.C. 1986. The Paleomagnetism of Late Cretaceous, Paleocene and Eocene Rocks from Haiti and Southwest Puerto Rico: Tectonic implications and geomagnetic polarity stratigraphy. Unpublished M.S. thesis, University of Forida, Gainesville, FL. 189 pp.

Wadge, G., and K. Burke. 1983. Neogene Caribbean plate rotation and associated Central American tectonic evolution. Tectonics 2:633-643.

_____, G. Draper, and J.F. Lewis. 1984. Ophiolites of the northern Caribbean: a reappraisal of their roles in the evolution of the Caribbean plate boundary. In I.G. Grass, S.J. Lippard, and J.W. Shelton (eds.). Ophiolites and Oceanic Lithosphere. Special Publication of the Geological Society of London 13:367-380.

Wadge, G., T.A. Jackson, M.C. Isaacs, and T.E. Smith. 1982. The ophiolitic Bath-Dunrobin Formation, Jamaica: significance for Cretaceous plate margin evolution in the northwestern Caribbean. Journal of the Geological Society of London 139:321-333.

Westercamp, D., P. Andreieff, P. Bouysse, A. Mascle, and J.C. Baubron. 1985. Geologie de l'archpiel des Grenadines. Documents du Bureau de Recherches Geologiques et Minieres 92: 198 pp.

Williams, E.E. 1969. The ecology of colonization as seen in the zoogeography of anoline lizards on small islands. Quarterly Review of Biology 44:345-389.

_____, and K.F. Koopman. 1951. A new fossil rodent from Puerto Rico. American Museum of Natural History Novitates 1515:1-9.

_____. 1952. West indian fossil monkeys. American Museum of Natural History Novitates 1546:1-16.

Williamson, M. 1981. Island Populations. Oxford University Press, Oxford. 286 pp.

Woodring, W.P. 1954. Caribbean land and sea through the ages. Bulletin of the Geological Society of America 65:719-732.

Woods, C.A. In press. A new capromyid rodent from Haiti: the origin, evolution and extinction of West Indian rodents and their bearing on the origin of New World hystricognaths. Contributions in Science of the Natural History Museum of Los Angeles County.

Wright, J.W. 1984. The origin and evolution of the lizards of the Galapagos Islands. TERRA March/April 1984:21-27.

Evolution of the Caribbean

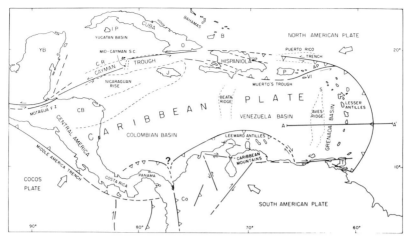

Figure 1. Location map of places and major tectonic features in the Caribbean region. Plate boundaries and major fault zones marked by heavy lines (broken when inferred), dashed light lines indicate mostly submarine features. General plate motions with respect to the Caribbean plate are indicated by large open arrows (after Smith 1985). Active subduction zones are indicated by tooth-patterns pointing in the direction of subduction. Cross-section A-A' is represented in Fig. 3. Abbreviations are as follows: Co, Colombia; CB, Chortis Block; YB, Yucatan Block; IP, Isla de Pinos; O, Oriente Peninsula; B, Bahamas Block; J, Jamaica; P, Puerto Rico; AP, Anegada Passage; VI, Virgin Islands; S, Saba Bank; T, Trinidad.

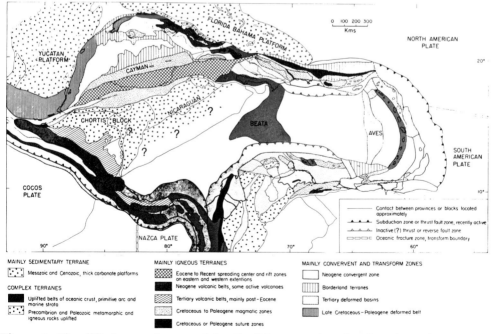

Figure 2. Simplified tectonic map of the Caribbean Region showing the major tectonic terranes, crustal blocks, and fault zones (modified after Case et al. 1984).

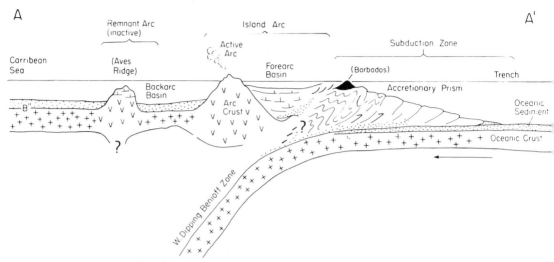

Figure 3. Generalized cross-section across the Lesser Antilles subduction zone and island arc (section A-A' of Fig. 1). Features shown are modified from Torini et al. (1985), and terminology is largely from Fox et al. (1971) and Karig and Sharman (1975). Layer B" in the Caribbean Sea is the oceanic crustal layer that gives prominent seismic reflections (see text for discussion).

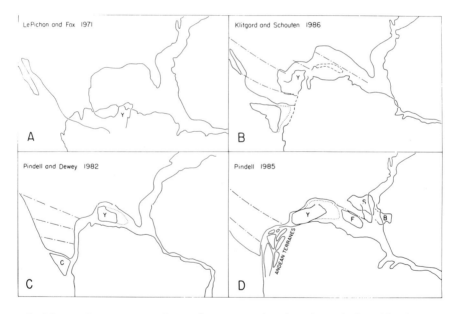

Figure 4. Mesozoic reconstructions of Pangaea showing the relationships between the major continental plates and the proto-Caribbean. Possible extensions of continental blocks are dotted, and major fault zones are represented as dot-dash lines. Y, Yucatan Block; C, Chortis Block; B, Bahamas Block; F, Florida Straits Block; S, Sabine Uplift. Note significant continental overlap in model A in which major fault zones in Mexico are not reconstructed.

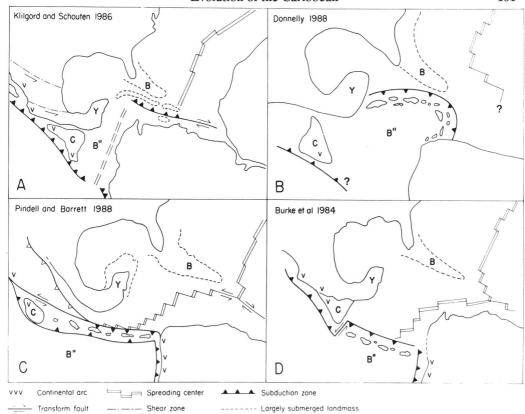

Figure 5. Four different Early Cretaceous plate reconstructions of the Caribbean region, emphasizing the various possible models and interpretations. Spreading ridges are shown as thin double lines in ocean basins. Subduction zones have tooth-patterns on the overriding plate. Continental margin arcs are indicated by small v's, whereas island arcs are denoted as small islands. Note that the size and distribution of the islands are diagrammatic and not to be interpreted as their exact sizes or positions. Dashed lines represent extensions of continental blocks or partly submerged islands or peninsulas. B" represents the sea floor which forms the uppermost layer of the present Caribbean oceanic crust.

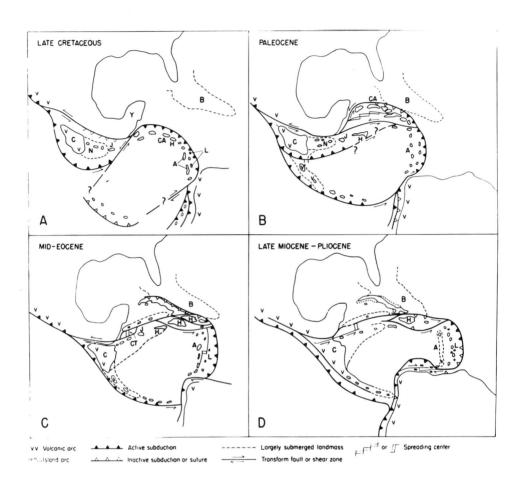

Figure 6. Summary of the geologic/tectonic history of the Caribbean from the Late Cretaceous through Miocene time (modified from Perfit and Heezen 1978; Burke et al. 1984; Wadge et al. 1984; and Pindell and Barrett 1988). See text for discussion of the periods of Caribbean evolution and details of the models presented.

HISTORY OF MARINE BARRIERS AND TERRESTRIAL CONNECTIONS: CARIBBEAN PALEOGEOGRAPHIC INFERENCE FROM PELAGIC SEDIMENT ANALYSIS

Thomas W. Donnelly[1]

Abstract

Pelagic oceanic sedimentation records several important chemical characteristics of the overlying water masses. Caribbean, Atlantic, and Pacific sediments recovered during the Deep Sea Drilling Project record at different times three phenomena which are significant for understanding the formation of inter-oceanic barriers, which in turn are the equivalent of semi-emergent corridors for dispersal of terrestrial organisms. During the latest Cretaceous and early Cenozoic all of the peri-American oceanic waters carried a high silica content and produced a characteristically highly siliceous sediment. Segregation first of Atlantic waters during the late Eocene and second of Caribbean waters during the early Miocene was caused by the formation of Antillean (first) and Central American (second) barriers to deep water circulation; this segregation is recorded by decreasingly siliceous sediment. These barriers, which were continuously shallower than 1,500 m and probably surmounted by an island chain, could have provided filter bridges for island-hopping organisms. Inter-American connections in the form of continuous inter-continental island chains are ruled out in the early Cenozoic.

During the late Miocene, a deep water connection (deeper than 4,000 m) between the Atlantic and Caribbean was again in existence. This breach in the Antillean arc probably persisted until some time in the Pliocene or Pleistocene. In the early Pliocene, surface water connections between the Caribbean and Pacific were severed with the emergence of the Panamanian isthmus, as shown by the record of oxygen isotopes in foraminifera living in surface waters and by development of provincialism in foraminiferal faunas.

Pelagic sedimentation records in the two earlier cases the timing of barrier formation; however, in neither case should the barrier be considered to be a complete terrestrial connection.

---

[1] Dr. Donnelly is Professor of Geology. Department of Geological Sciences, State University of New York at Binghamton, NY, 13901

Introduction

One of the enduring problems of biogeography of Middle America is the history of terrestrial connections between North and South America along either an Antillean or Central American emergent corridor. A recent account of the faunal imperatives for historical inter-American connections has been provided by several authors in Stehli and Webb (1985). The consensus is that, subsequent to an early Jurassic separation of the Americas, there might have been limited Cretaceous connections between the Americas, no connections during the early Paleocene and Eocene, limited connections during the Oligocene and Miocene, and extensive connections following the Pliocene emergence of the Panamanian isthmus (the "great American biotic interchange"). In general the authors do not require that connections be terrestrial, and indeed the overall conclusion drawn from a perusal of these chapters is that the limited trans-American faunal similarities would best be explained by limited over-water dispersals.

Geological evidence (summarized in Donnelly 1985) provides a framework for understanding the sequence of geological events that lead to formation of the proto-Antillean island arcs and the Central American volcanic arc. This evidence also provides constraints on the timing of collisional (suturing) events and submergences and emergences. What geological evidence generally fails to provide is evidence that an island arc may or may not have spanned the entire distance between North and South America, and thus provided a chain of islands suitable for limited dispersal of terrestrial organisms. For instance, the origin of the Greater Antilles is a keenly debated topic in contemporary Caribbean geology, with views ranging from an original position in the present Panamanian isthmus to a location approximating its present position. However, evidence that either end was anchored on continental crust or that it was continuous along its length is lacking.

The purpose of this brief contribution is to illustrate the application of oceanic sediment studies to these problems. Pelagic oceanic sediments have two characteristics appropriate for this application: 1) they are commonly dateable with a precision unmatched in other stratified geological materials; and 2) in many cases their Mineralogical, paleontological, or chemical character reflects characteristics of the overlying water masses. It is this second characteristic that I propose to exploit.

The sediments examined in the present study were recovered by the Deep Sea Drilling Project: a project funded by the National Science Foundation and managed by Scripps Institution of Oceanography from 1968 to 1983. The Atlantic and Caribbean sites considered here are shown in Fig. 1; about two dozen Pacific sites that were investigated were not shown.

Three approaches are considered here. The first is the relationship between dissolved silica in intermediate and deep ocean water and excessive (=biological) silica in sediments. The second is the record of the paleodepth at which calcareous microfossil debris dissolves in deep water (the so-called CCD, or calcite compensation depth). The third, which has been extensively developed by Keigwin (1982a), is the observation of divergent oxygen isotope values for pelagic formaminifera from populations which were separated by the emerging Panamanian isthmus.

Early Cenozoic.-- Excessive silica in sediments and the origin of Eocene and Oligo-Miocene inter-oceanic barriers.

The geological history of excessive silica in sediments is important because it reflects the geological history of isolation of Pacific and Atlantic water masses resulting from the closure of inter-oceanic barriers, which span the gap between the Americas. Pelagic sediments record this history of oceanic water isolation.

Figure 2 shows the vertical distribution of silica in representative tropical Atlantic and east Pacific water columns. At both sites (and everywhere in the ocean) silica is almost absent from surface waters and increases with depth. At the Atlantic site, however, the increase is very small, while at the Pacific site silica is relatively abundant below about 1,500 m. In the present world ocean the Atlantic is a relatively low-silica water mass and the Pacific a high-silica water mass, with the southern Atlantic and Indian Oceans intermediate between the two extremes. The explanation for this difference is that the north and south ends of the Atlantic are the sites of downwelling of surface water and the central and northern Pacific the sites of general upwelling of deep water. Silica is removed almost quantitatively from shallow water by microorganisms whose skeletal remains carry silica to deep water, where it redissolves and accumulates. When this deeper water upwells, the silica is immediately removed by organisms and returned to depth after their rather brief life. Because the water that passes from the Pacific to the Atlantic (almost entirely via the southern ocean) is shallow, it is also silica free.

But the intermediate and deep water passing from the Atlantic to the Pacific (also via the southern ocean) is siliceous, and the final result is that silica accumulates in the Pacific intermediate and deep water. Because a fraction of silica brought down in microorganism skeletal debris is preserved on the sea floor in an undissolved state, Pacific sediments (with the minor exceptions of the far western Pacific and the centers of the major gyres in the north and south Pacific) are almost invariably excessively siliceous. This subject has been discussed with great clarity by Broecker (1974).

The chemical composition of pelagic sediments enables us to quantify the record of silica sedimentation. The chemical method is preferred over microscopical techniques for the examination of sediments older than Recent not only because it is arguably far more quantitative than any microscopical method, but also because in older sediment siliceous skeletal material reacts with pore water to form a variety of authigenic materials, mainly but not only chert. The silica content of pelagic sediments has two main sources: siliciclastic debris (mainly clay minerals) which are derived from emergent continental areas and carried to the ocean by rivers or by wind, and biological silica, which is the opaline tests of microorganisms. Siliciclastic debris also contains abundant aluminum: and,the Si/Al ratio of siliciclastic debris varies between narrow limits for pelagic sediments. Thus the measurement of Si and Al in sediment enables us to separate two silica fractions: that which belongs to the siliciclastic fraction, and that which is biological, for which I also use the term "excessive".

The relationships between excessive sedimentary silica and time are shown in Fig. 3. Only Caribbean and Atlantic sites are shown; numerous sites examined in the Pacific all show excessive silica throughout this time interval. In Fig. 3, the base value of Si/Al which represent pure siliciclastic debris is shown as a pair of lines which encompass the limits of variation for this debris. Tropical debris tends to have slightly lower Si/Al ratios (more kaolin), and higher latitude debris higher ratios (more illite). This value range is taken from an extensive study of surface sediments of the tropical, subtropical, and temperate North Atlantic (Donnelly, unpub. data) and supported from examination of numerous sediments from other oceans. The ages are taken from micropaleontological determinations recorded in the appropriate volumes of the series "Initial Reports of the Deep Sea Drilling Project".

The history of excessive silica sedimentation in the Atlantic is assembled from recovered sediment at seven sites (Fig. 3). Although these sites are broadly distributed (Fig. 1), they agree among each other in showing abundant excessive silica prior to 45 million years (Eocene), and diminished excessive silica until about 35 myBP. Because Pacific sites (not shown) show excessive silica throughout the entire time interval, we can conclude that the gradual upward disappearance of excessive silica at the Atlantic site represents the formation of a barrier between the oceans which constricts and then shuts off the inter-oceanic circulation of silica-bearing intermediate and deep waters. This barrier, if we can extrapolate present observations back in time, must be at about 1,500 m depth, compared with a normal oceanic depth closer to 4,000 meters.

In the Venezuelan Basin of the Caribbean (Site 149, Fig. 3) excessive silica accumulation has diminished considerably by 35 myBP but continued until about 15 myBP (late early Miocene). This history differs from that of the Atlantic and shows that the Caribbean was more broadly connected with the pacific than the Atlantic during the period 35 to 15 myBP, a second barrier is required; the western will be termed the "proto-Central American". The barriers on the east (proto-Antillean) and west (proto-Central American) sides of the Caribbean differ in age by about 20 million years.

The proto-Antillean barrier is represented today by the Greater Antilles (which might have had a rather different configuration of islands than is at present; and by fragments of this antiquity underlying the northern, central, and southern Lesser Antilles, as well as the Aves Ridge. Although the spatial reconstruction of these fragments into an island arc is unclear, the rocks exposed in these fragments confirm that the barrier was a chain surmounted by emergent volcanic islands. The barrier must have spanned the gap between the Americas. On the other hand, the barrier could not have been completely emergent; expected microfossal provinciality that would have resulted from inter-oceanic isolation is absent for Eocene microfossil populations from the tropical Atlantic, Caribbean, and Pacific. We are drawn irresistibly to an Eocene proto-Antillean arc which has the characteristics of the modern Greater and Lesser Antilles with, presumably, the same opportunities for terrestrial organisms to disperse across short oceanic gaps. This arc would have most likely occupied the site of the present Aves Ridge rather than the present Lesser Antilles. In spite of containing diverse fragments of an older island-arc, the Lesser Antilles does not seem to have the roots of an entire arc in its substructure. The Aves Ridge, on the other had, appears to be the remnants of a continuous early Tertiary arc.

There are alternative arguments against the concept of one and then two barriers spanning a gap, as outlined above. For example, one might suggest that a barrier existed during the early Cenozoic, but that oceanic deep water only began to be driven from Atlantic to Pacific during the Eocene. This is not an easy argument to dismiss, but it appears to fly in the face of Atlantic Ocean reconstructions. The present formation of oceanic deep water is the result of formation of more saline (denser) and cooler (denser) waters in the Atlantic. The availability of large quantity of cold water is probably a relatively young characteristic of the Atlantic, both because of late Cenozoic cooling and because of the Arctic opening of the Atlantic in the middle Cenozoic. However, the earlier Atlantic was considerably narrower than the present ocean, and this earlier ocean must have been more saline than the present because of aridity. Further, we observe that the Pacific throughout this interval is documented as being a persistent recipient of silica, showing that the circulation observed today must have persisted back through the Cenozoic.

A second argument against the usage of the Si/Al ratio as providing a record of silica content of water might be that oceanic productivity was so low or upwelling so limited that siliceous organisms were never produced. The weakness of this argument is seen in two observations: 1) only a very minor part of the north Pacific Ocean does not presently exhibit excessively siliceous sediment, even though the zones of high productivity are sharply limited; and 2) at site 149 the production of calcareous sediment continued at a high rate during the gradually extinction of the siliceous biota.

Prior to the formation of the barrier there was a broad water gap between the Pacific and Atlantic for at least the first 20 million years of the Cenozoic. After 45 myBP, the Antilles is at least a possible island arc with the capability of providing a filter bridge. And at 15 myBP, the Central America Isthmus might have provided the same capability. Of course, once the Central American barrier closed, there will be no way of ascertaining whether an Antillean barrier persisted, because both the Caribbean and Atlantic were then isolated from the Pacific and thus were similarly depleted in silica. We can set at rest, however, any ideas of a Paleocene island arc connection between the Americas such that might explain similarities, for example, between edentate, notoungulate, and dinocerate faunas of North and South America (Gingerich 1985). Either these early mammals dispersed across a substantial width of ocean or dispersed by some more complicated pathway through Asia.

Late miocene.-- The CCD passes through a breached Antillean barrier.

Throughout the Atlantic Ocean in the late Miocene the depth at which there is massive solution of calcite skeletal debris in deep water (the CCD) abruptly deepened. The reasons for this deepening are not completely understood, but probably largely result from the gradual exclusion of Pacific deep water from the Atlantic as a result of the formation of the Central American barrier (see above). Regardless of explanation, the consequences of the deepening of the CCD can be observed at nearly every Atlantic drilling site where pelagic sediment of this age has been recovered: Late Miocene sediment which is nearly barren of calcareous fossils is overlain by increasingly calcareous marls and oozes. Figure 4 shows the record of passage of the CCD at two sites on either side of the Antillean arc: sites 29, in the Venezuelan Basin, and 543 northeast of Barbados (Fig. 1). There is little reason to believe that either paleodepth was significantly different than the present water depth. Unfortunately the earlier drilling at Site 29 produced an inferior core, whose fragmentary recovery and lower accumulation rate result in its having a far lower stratigraphic precision than that of site 543. However, both sites show that at about the Pliocene- Miocene boundary there was a passage through the CCD. Because the CCD must have been in excess of 4,000 m at that time, I conclude that the Venezuelan Basin had a water connection with the Atlantic of at least this depth. At present the greatest depth is 1,900 m (Fig. 5). The location of this passage within the Antillean chain cannot he identified, but the indication of a CCD movement within the sedimentary section suggests that the chain was substantially broken in at least one place. The conclusion is that, following the formation of the Central American barrier, the Antillean barrier was breached and has healed again only later in the Pliocene or in the Pleistocene.

Pliocene.-- The emergence of the Panamanian isthmus.

The Pliocene emergence of the Panamanian isthmus has captured by far the majority of the attention directed to Caribbean biogeography. First anticipated by Darwin during the voyage of the Beagle, the rise of the isthmus has been the central cause not only of the "great American biotic interchange" (Stehli and Webb 1985), but also for the isolation of marine faunas on either side of the isthmus. Woodring (1966) pointed out the origin and divergence of "Caribphile" and "Paciphile" elements in the molluscan faunas, showing that since the middle Miocene the formerly uniform faunal province began to split into two provinces. More recently Keigwin (1978, 1982a, 1982b) provided the means towards a precise documentation of the isolation of Caribbean and Pacific surface waters by means of an analysis of microfossil assemblages and oxygen isotopes in calcareous microfossil skeletal debris from Caribbean and Atlantic sediment cores.

The microfossil studies deserve especially close scrutiny because they can provide dates for the emergence of the Panamanian isthmus far more precise than, but still remarkably similar to, results drawn from the stratigraphy of terrestrial fossiliferous sites. Jones and Hasson (1985) provide an summary of these results (as well as for macrofossils). The microfossil results are not easily interpretable; evidently populations of microorganisms remained in a state of relative stasis for variable delay times following inter-oceanic isolation. A more compelling analysis comes from differences in oxygen isotopes of skeletal debris of a planktonic foraminifera (Keigwin 1982a), which reflect variations in the fine balance of precipitation to, or evaporation from, surface ocean water. Keigwin shows convincingly that at 4.1 myBP Pacific and Caribbean surface waters first display differing oxygen isotopes and continue to the present day with about the Same amount of difference: reflecting the higher salinity (enhanced evaporation) of Atlantic and Caribbean surface waters.

Neither the date of isotopic divergence nor the slightly younger dates for microfossil provinciality directly indicate a completely dry (emergent) connection. However, the jump of isotopic differences to present-day values is an instantaneous response to a virtual cessation of surface water transport across the isthmus. Thus, the 4.1 mybp date should be considered the most important for biogeographic studies.

Recent.-- The Caribbean as a newly established independent ocean basin.

The question might well be asked, what is the present oceanographic state of the Caribbean Sea? The oceanographer thinks of the Atlantic as having oxygen-rich bottom water in which solution of calcareous debris takes place only in very deep (deeper than 4,000 m) water; this ocean contains little silica in its intermediate and deep water, and passes what it has to the Pacific. The Pacific Ocean, on the other hand, has much less oxygen in its deep water, and calcareous debris begins to dissolve in variably shallow water, commonly shallower than 2,000 m. The intermediate and deep waters have substantial quantities of dissolved silica. The Caribbean (Fig. 5) is presently enclosed behind barriers which allow essentially no contact with intermediate or deep waters. Thus silica which enters the Caribbean (probably dominantly from the Magdalena River) cannot escape and is presently substantially greater than that of the adjacent Atlantic (Fig. 2, data from Atwood et al. 1979).

Siliceous microfossil remains are found in the very youngest Quaternary sediments of the Caribbean (Riedel and Sanfilippo 1973; Riedel and Westberg 1982), suggesting that the barriers which are now allowing silica to accumulate in the Caribbean must have been relatively recently raised. The site of the last closure of the Caribbean is not known but most likely to lie within the Lesser Antilles arc, possibly at its southern end. The implication of this closure, wherever it occurred, is that biogeographers must consider that the passage of terrestrial organisms from South America to the Antilles is almost undoubtedly over a total water path that is presently shorter than was the case in the relatively recent past.

Summary and conclusions

The history of inter-oceanic barriers, which are also inter-American island chains is shown in Fig. 6, which depicts by a series of cartoon sketches the development of these features. For the purposes of this figure I omit an eastward motion of the Caribbean plate because I believe such motion is not germane to paleobiogeographical analysis.

Figure 6a begins at an unspecified time in the late Cretaceous. Although Cretaceous conditions have not been discussed in this paper, and although paleoceanographic conditions cannot be inferred in the Pacific because of a paucity of recovered pelagic sediment in this age range, nevertheless there are two reasons for believing that an island chain spanned the gap between the Americas. The first is the observation that some North and South American Cretaceous reptiles bore some resemblance to each other (Estes and Baez 1985); there is also some resemblance in marsupial fauna (Webb 1985). A geological reason for postulating this gap is found in the remnants of middle Cretaceous island arc fragments on Bonaire, Tobago, La Desirade (part of Guadeloupe), and the Greater Antilles. Evidently this island chain was disrupted sometime in the late Cretaceous by eastward movement of the Caribbean plate (Donnelly 1985).

Figure 6b shows the configuration of North America, South America, and the Caribbean at the end of the Cretaceous. A possibly once continuous late Cretaceous island chain (circles) had been broken, perhaps by eastward movement of the Caribbean plate. Ocean waters flowed without impediment through the Caribbean and there was no distinction among the waters of, or the sediments beneath, the west tropical Atlantic, Caribbean, and east tropical Pacific.

Figure 6c shows the condition at 35 myBP, or approximately the end of the Eocene. A proto-Antillean barrier, shown as an island chain by a row of circles, now spanned the gap between the Americas and impeded the flow of Pacific deep water back into the tropical Atlantic. At this time siliceous sedimentation had ceased in the Atlantic but still continued in the Caribbean.

Figure 6d shows the configuration at about 15 myBP (late middle Miocene), when the proto-Central America barrier had been closed. At this time there was no more Pacific siliceous intermediate and deep water penetrating back into the Caribbean, and siliceous sedimentation had ceased in the Caribbean.

Figure 6e shows the configuration at a time prior to about 5 myBP, when the proto-

Antillean barrier had been breached to a depth sufficient to allow deep waters at the contemporaneous CCD to circulate between the Caribbean and Atlantic basins.

Figure 6f shows the situation at the present time. The Caribbean is now encircled by barriers to deep and intermediate water circulation, perhaps for the first time in geologic history. The Recent Caribbean has taken on oceanographic characteristics of its own and is already more siliceous than the adjacent Atlantic.

Caribbean biogeographers have recently "discovered" Caribbean geology. But whose geology have they discovered? Few of the recent biological-geological syntheses that have been published recently reflect the flavor of the controversy that dominates almost all aspects of the historical geological synthesis that is slowly and painfully emerging.

A large part of the problem is that geologists do not address some of the questions that biologists need for their analyses. To a geologist it is relatively unimportant if an island arc is broadly emergent or mainly submergent, whereas the biologist is vitally interested in that information. On the other hand, the movement of the "Caribbean plate" eastward from putative Pacific location is a question close to the hearts of geologists, but probably of relatively little interest to biologists. A "proto-Antillean arc" may be of value for biological purposes regardless of whether it occupies the position of present Central America or the present Antilles, as long as connections between North and South America are the major issue.

What oceanography, and more specifically, the study of pelagic sediments, contributes is a history of the formation of barriers that might serve also as island chains which are available for dispersal of terrestrial organisms equipped to overcome relatively short over-water transits. But we must realistically consider the scale of oceanic circulation involved; a Paleocene - Eocene opening between the Americas that allows the passage of water deeper than 1,500 m in sufficient quantities to allow siliceous sedimentation in both the Atlantic and Pacific must be measured in hundreds and not tens of kilometers.

Literature cited

Atwood. D.K., P.N. Froelich, M.E.Q. Pilson, M.J. Barcelona, and J.L. Vilen. 1979. Deep silicate content as evidence of renewal processes in the Venezuela Basin. Caribbean Sea. Deep Sea Research 26A:1179-1184.

Bainbridge, A. 1981. GEOSECS. Atlantic expedition. Vol. 1. Hydrographic data, 1972-1973. Superintendent of Documents. Washington, 121 pp.

Broecker. W.S. 1974. Chemical Oceanography. Harcourt Brace Jovanovich, New York, 214 pp.

Broecker, W.S.. D.W. Spencer. and H. Craig 1982. GEOSECS. Pacific Expedition. Vol. 3. Hydrographic data 1973-1974. Superintendent of Documents, Washington, 137 pp.

Donnelly, T.W. 1985. Mesozoic and Cenozoic plate evolution of the Caribbean region. Pp. 89-121 in F.G. Stehli and S.D. Webb (eds.). The Great American Biotic Interchange. Plenum Press, New York.

Estes. R. and A. Baez. 1985. Herpetofaunas of North and South America during the late Cretaceous and Cenozoic: evidence for interchange? Pp.140-197 in F.G. Stehli and S.D. Webb (eds.). The Great American Biotic Interchange. Plenum Press, New York.

Gingerich. P.D. 1985. South American mammals in the Paleocene of North America. Pp.123-137 in F.G. Stehli and S.D. Webb (eds.). The Great American Biotic Interchange. Plenum Press, New York.

Jones. D.S. and P.F. Hasson. 1985. History and development of the marine invertebrates faunas separated by the central American isthmus. Pp. 325-355 in F.G. Stehli and S.D. Webb (eds.). The Great American Biotic Interchange. Plenum Press, New York.

Keigwin. L.D. Jr. 1978. Pliocene closing of the isthmus of Panama, based on biostratigraphic evidence from nearby Pacific Ocean and Caribbean Sea cores. Geology 6:630-634.

———. 1982a. Isotopic paleoceanography of the Caribbean and east Pacific: role of Panama uplift in late Neogene time. Science 217:350-353.

———. 1982b. Neogene planktonic foraminifera from Deep Sea Drilling Project sites 502 and 503. Pp. 269-288 in W.L. Prell and J.V. Gardner (eds.). initial Reports of the Deep Sea Drilling Project, Volume 68. U.S. Government Printing Office, Washington, DC.

Riedel. W.R. and A. Sanfilippo. 1973. Cenozoic radiolaria from the Caribbean. Deep Sea Drilling Project. Leg 15. Pp.705-751 in N.T. Edgar and J.B. Saunders (eds.). Initial Reports of the Deep Sea Drilling Project. Volume 15. U.S. Government Printing Office, Washington, DC

———, and M.J. Westberg. 1982. Neogene radiolarians from the eastern tropical Pacific and Caribbean. Deep Sea Drilling Project Leg 68. Pp. 289-300 in W.L. Prell and J.V. Gardner (eds.). Initial Reports of the Deep Sea Drilling Project, Volume 68. U.S. Government Printing Office, Washington, DC

Stehli, F.G. and S.D. Webb (eds.). 1985. The Great American Biotic Interchange. Plenum Press, New York, 532 pp.

Webb. S.D. 1985. Main pathways of mammalian diversification in North America. Pp. 201-217 in F.G. Stehli and S.D. Webb (eds.). The Great American Biotic Interchange. Plenum Press, New York.

Woodring. W.P. 1966. The Panama land bridge as a sea barrier. Proceedings of the American Philosophical Society 110:425-433

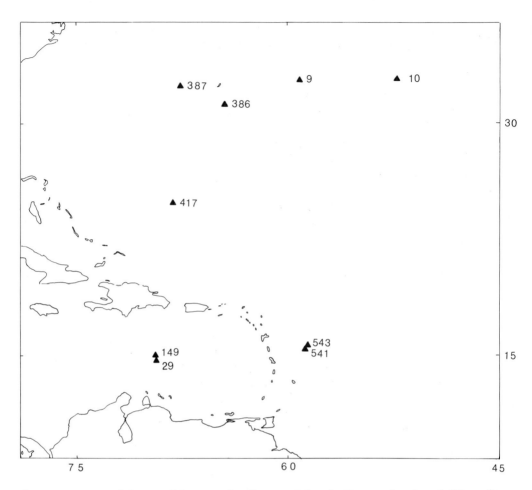

Figure 1. Map of the Caribbean and adjacent Atlantic Ocean showing drilling sites (Deep Sea Drilling Project) referred to in text.

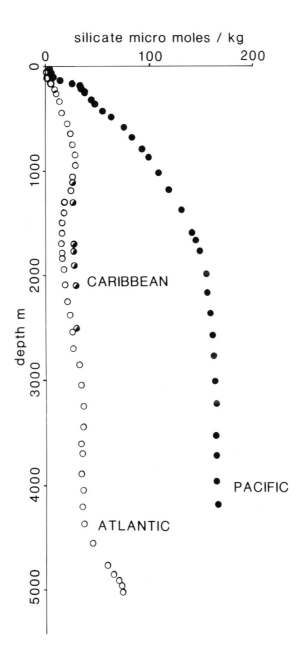

Figure 2. Vertical distribution of dissolved silica at stations in the Atlantic Ocean (Open circles; site 37, GEOSECS. lat. 12°1' N; long. 50°59' W; Bainbridge. 1981); Pacific Ocean (Filled circles; site 343. GEOSECS. lat 16°31' N, long 123°1' W; Broecker et al., 1982), with a few Caribbean values taken from Venezuelan Basin stations (Half-filled circles; lat. between 14° and 15° N, long. 67° W; Atwood et al. 1979).

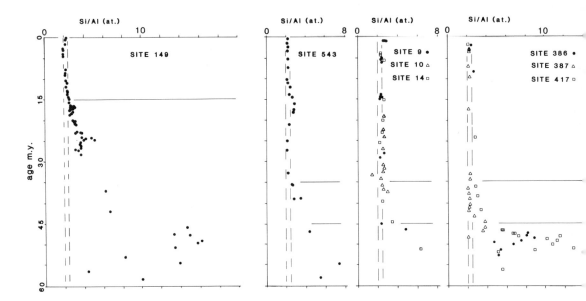

Figure 3. Diagrams showing variation of Si/Al (atomic) vs. age for pelagic sediment. for one Caribbean site (149) and 7 western Atlantic sites (543. 9, 10, 14. 386, 387. 417) from the Deep Sea Drilling Project (Fig. 1). The two dashed vertical lines for each diagram show the Si/Al (atomic) range for terrigenous sedimentary debris. Values lying to the right of these lines are considered to have excessive (=biological) silica. Horizontal lines for the Atlantic sites show ages of 45 and 35 million years, at which the excess silica in the sediment diminished sharply, and essentially vanished. The horizontal line for site 149 shows that the excessive silica vanished at about 15 million years. The scattering of values in the lower parts of these diagrams reflects the heterogeneous secondary redistribution of biological silica during subsequent diagenesis, which produces chert layers.

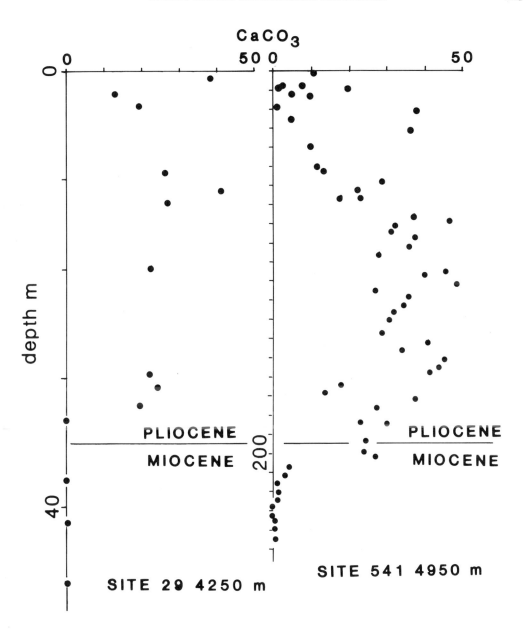

Figure 4. Diagram showing the calcium carbonate content (percentage) vs. sub-bottom depth for pelagic sediment at two drilling sites: 29 (Caribbean) and 541 (west Atlantic), located in Figure 1. The vertical scales have been adjusted for the two sites to bring the Pliocene - Miocene horizon into coincidence.

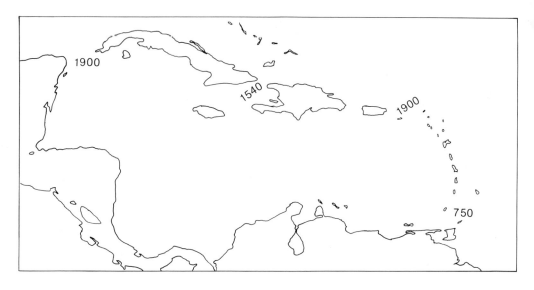

Figure 5. Map showing approximate sill depths (meters) around the Caribbean.

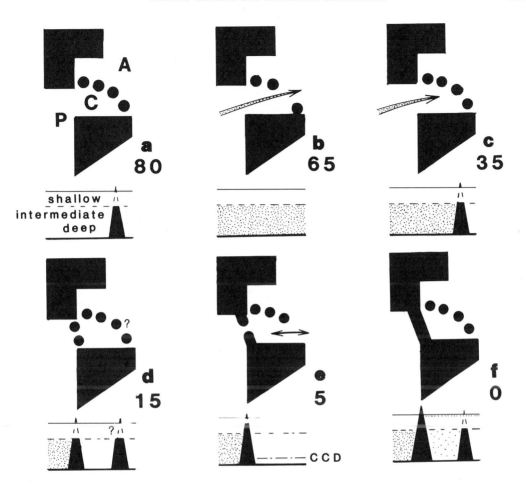

Figure 6. Sketches showing the approximate configuration of Caribbean island chains or land bridges (upper) with corresponding cross-sections of the ocean (lower) at 6 times. Figure 6a is late Cretaceous, approximately 80 million years. Oceanographic conditions are not clear, because of lack of sediment data, especially from the Pacific. A= Atlantic; C= Caribbean; P= Pacific. The circles show a putative island chain in existence sometime during the late Cretaceous. The chain is shown as a partial barrier in the section. Figure 6b shows the island-chain barrier breached at 65 myBP (end Cretaceous). The stippled arrow shows circulation of silica-bearing pacific deep and intermediate water into the Atlantic. (also shown in section). Figure 6c shows a barrier again in existence at 35 MyBP (end Eocene). Figure 6d shows a Central American barrier in existence at 15 myBP (end early Miocene). The status of the Antillean barrier is unknown at this time. Figure 6d shows at 5 myBP (end Miocene) the Central American barrier nearly closed and the Antillean barrier breached to a depth allowing a connection with the Atlantic at the level of the CCD. Figure 6e is the present configuration, with the Caribbean deep and intermediate water isolated and taking on characters intermediate between the Atlantic and the Pacific (light stipples. showing an elevated silica content. in the section). Also shown is isotopically contrasting surface water (line of dots).

PEOPLING AND REPEOPLING OF THE WEST INDIES

Irving Rouse[1]

Abstract

Three ethnic groups, each with its own language and culture, inhabited the West Indies in the time of Columbus: Guanahatabeys in western Cuba, Tainos in the Bahamas and rest of the Greater Antilles, and Island-Caribs in the Lesser Antilles. The Guanahatabeys were a relict of the original peopling of the Indies by preceramic food-gatherers, who appear to have arrived from Middle America about 5,000 B.C. and from South America about 3,000 B.C. The ancestors of the Tainos repeopled the islands from South America about the time of Christ, introducing pottery, agriculture, the worship of zemis, and the Arawakan family of languages. They pushed the ancestors of the Guanahatabeys back through a succession of frontiers to the one in existence at the beginning of historic time. They were followed by the ancestors of the Island-Caribs shortly before the arrival of Columbus. Most authorities assume that these Indians, too, repeopled the islands they occupied. The linguistic evidence indicates, however, that they came as relatively small war parties, which conquered the local population but were assimilated into it.

Introduction

While wildlife is classified only into biologically defined population groups, members of the human species are assigned to three overlapping kinds of groups: biologically defined races; speech communities, each using a different language; and peoples, each with its own culture. In the study of mankind, as a result, the Linnaean hierarchy has to be supplemented by linguistic and cultural hierarchies.

The biological distinction between Linnaean and popular categories also carries over into the study of mankind. Anthropologists differentiate their systematically defined units--races, speech communities, and peoples--from ethnic groups, which are the units of ordinary discourse (Rouse 1986:110-11). Ethnic groups are defined by a mixture of racial, linguistic, and cultural criteria.

The four kinds of population groups will be discussed here in reverse order, beginning with ethnic groups and ending with races. Ethnic groups are given priority because experience has shown that the best results are obtained by working back from them, that is, from history into prehistory. Peoples and their cultures are considered next because this paper is about the peopling and repeopling of the West Indies. Research on speech

---

[1] Dr. Rouse is Charles J. MacCurdy Professor Emeritus of Anthropology, Yale University, New Haven, CT 06520.

communities and their languages is then summarized in order the check the conclusions reached through the study of peoples and cultures. Races are left until last, despite their central position in this conference, because little is known about them in the West Indies.

Ethnic groups

Our information about the ethnic groups comes primarily from documentary research by ethnohistorians. The groups were assimilated into the modern population too soon to be observed in detail by ethnologists. Three major groups are recognized: Guanahatabeys, who lived in western Cuba; Tainos, who occupied the Bahamas and the rest of the Greater Antilles; and Island-Caribs, in the central and southern parts of the Lesser Antilles (Fig. 1). The nature of the Indians who inhabited the northern part of the Lesser Antilles is not known ethnohistorically; the area may have been largely depopulated in the time of Columbus.

The Guanahatabeys have been erroneously termed Ciboneys. (That name actually refers to a Taino subgroup; see e.g. Alegría 1981:5-6.) The documents inform us that they dwelled in caves, lived by hunting and gathering rather than agriculture, and were organized into bands instead of villages. They must have had a separate language because Columbus' Taino interpreter was unable to speak with them. Ethnohistorians have inferred from the Guanahatabeys' remote position and the primitiveness of their culture that they were survivors of the original population of the West Indies, pushed back into western Cuba by the ancestors of the Tainos (Lovén 1935:3-6).

The Tainos are also known as Arawaks, but this is another misnomer; the documents indicate that the people who called themselves Arawaks were restricted to the land around the mouth of the Orinoco River in South America (Boomert 1984). The Tainos had a different culture and a different language than either the Arawaks or the Island-Caribs. They were agriculturally more advanced, lived in larger villages, and they alone had a hierarchy of chiefs, who inherited their offices and derived their power largely from personal deities known as zemis. The Tainos carved statues of these zemis, portrayed them on their utensils and ornaments, and worshiped them publicly as well as in their homes (Arrom 1975). Ethnohistorians have concluded that the ancestors of the Tainos came from South America because they share so many cultural and linguistic traits with the Arawaks and their neighbors on the mainland (Lovén 1935).

The Island-Caribs--so-called to distinguish them from another group of Caribs that lived on the mainland--emphasized warfare rather than religion. They raided Taino villages, ate the flesh of their male captives in order to acquire their power, and stole the Taino women, whom they set up in single-family houses apart from the men's houses in which they themselves lived. They had only elected war chiefs, and spoke a special language in the men's houses. According to their traditions, they came from South America shortly before the time of Columbus and conquered the previous Igneri population. Most authorities (e.g. Allaire 1980) have assumed that they replaced their predecessors as the ancestors of the Tainos had done before them, but Gullick (1980:479) has hypothesized that they married into the Igneri population. This is consistent with their treatment of Taino women.

Peoples and their cultures

Each of the foregoing ethnic groups can also be regarded as a people, since each had a different culture. Archeologists have extended the range of the Taino people into the ethnically unknown northern part of the Lesser Antilles by demonstrating that the latest aboriginal remains in that area resemble the contemporary finds in the Greater Antilles (Allaire 1985). They have also drawn a distinction between the Classic Taino people, who centered in Hispaniola and Puerto Rico, and the surrounding Sub-Tainos, who had less highly developed chiefdoms and zemiism (Fig. 2). The Classic Tainos are further distinguished by earth- and stone-lined plazas, used for public ceremonies and for playing ball (Alegría 1983).

The chronology shown in Fig. 3a has been constructed to assist in tracing the ancestry of the peoples and cultures just defined. Most of the areas across its top are named after the water passages between the islands, rather than the islands themselves, because the Taino remains on either side of the passages resemble each other more closely than they do the finds elsewhere on the same islands (Rouse 1982:52, fig. 2). The Bahama Channel area thus includes central Cuba as well as the Bahamas; and the Jamaica Channel area, southwestern Hispaniola as well as Jamaica. The Windward passage area extends from eastern Cuba into northwestern Hispaniola, the Mona Passage area from eastern Hispaniola into western Puerto Rico, and the Vieques Sound area from eastern Puerto Rico into the Virgin Islands. These divisions reflect the fact that the Tainos, being expert canoeists, interacted more closely across the passages than overland, contrary to the present inhabitants of the islands. The periods along the side of the chart are based upon studies of stratigraphy and seriation and the dates, upon radiocarbon analysis.

The temporal and geographical extent of the historic peoples and cultures is depicted beneath the areas at the top of Fig. 3a. Archeologists have traced the Guanahatabeys back to the prehistoric units under them by their stonework, which constitutes 95 percent of their artifactual remains. They have traced the ancestry of the Sub-Tainos and Classic Tainos primarily in terms of their pottery, for the same reason. They have also attempted to identify the ancestors of the Island-Caribs through their pottery, but so far have been unable to reach agreement about the nature of the ancestral pottery (Allaire 1980). Consequently, the following discussion will be limited to the Guanahatabeys and the Tainos.

Guanahatabeys.--Archeologists have discovered a number of highly localized, preceramic peoples and cultures, each with its own distinctive complex of stone artifacts. They have classified them into subseries, whose names end in the suffix -an, and series, whose names end in the suffix -oid. The terminology varies from author to author. They trace the Guanahatabeys back through the Redondan and Courian subseries of the Casimiroid series to an original Casimiran subseries (Fig. 3a).

The Casimiroid series is characterized by the development of points, knives, and scrapers made from macroblades that were struck off prismatic flint cores (Cruxent and Rouse 1969). MacNeish (1982:38-42) has carried this Casimiroid tradition of flint chipping back to 7500 B.C. in Belize on the Yucatan Peninsula. Since his finds antedate the arrival of the tradition in the West Indies, we may hypothesize a population movement from Yucatan into Cuba and Hispaniola. So far, there is no evidence that it continued onto the other islands.

When the Casimiroid peoples arrived in the Antilles, they seem to have been in the Lithic age, that is, they made only chipped stone artifacts (Fig. 3a). They advanced into the Archaic age by learning to grind stone axes and vessels, some of which they engraved with rectilinear parallel-line designs (Rouse 1982b, figs. 1,2, pls. 1,2).

The three Casimiroid subseries have an irregular distribution in Figure 3, presumably because their peoples were oriented towards the land rather than the sea and hence did not conform to the Taino passage areas. Kozlowski (1980:64-70) has suggested that they moved seasonally up and down the major streams. This parallels the situation hypothesized by MacNeish (1982:38-9) for their Sand Hill ancestors in Central America.

The original Casimiroid peopling of Cuba and Hispaniola took place about 5,000 B.C. There is less satisfactory evidence for the arrival of a second, Ortoiroid series of peoples from South America two millennia later (Figs. 3a,b). These peoples were also in the Archaic age, but are characterized by simple flint flakes and edge grinders (Rouse 1960, Figs. 4,7). They appear to have come into contact with the Casimiroid peoples when they reached Puerto Rico and to have been pushed back with them into western Cuba. If so, they, too, contributed to the composition of the Guanahatabeys.

Tainos.--All of the Tainos made pottery belonging to a single Ostionoid series of local styles. The western Sub-Tainos can be traced back through its Meillacan and Ostionan subseries, the Classic Tainos through its Chican and Ostionan subseries, and the eastern Sub-Tainos through its Elenan subseries (Fig. 3a). From these units the trail leads via the Cedrosan Saladoid pottery of Puerto Rico, the Lesser Antilles, and the Guianan and east Venezuelan coasts to Ronquinan Saladoid pottery in the Orinoco Valley (Fig. 3b).

The progress of the ancestors of the Tainos as they moved along this trail can be seen in Fig. 3 by following the dotted line that marks the beginning of the Ceramic age. The vertical jags in the line indicate frontiers, at which first the Saladoid and then the Ostionoid peoples halted for appreciable periods of time (Fig. 4). The first four frontiers are prehistoric, while the fifth is the one that separated the Tainos from the Guanahatabeys in the time of Columbus.

The Ronquinan Saladoid peoples lived behind Frontier 1, which was at the head of the Orinoco Delta (Fig. 4). They made modeled-incised and white-on-red painted pottery and used clay griddles to bake bread from cassava flour (Rouse 1986, figs. 25, 26d). Their pottery at the head of the delta was simpler than that upstream, a condition that can be attributed to the founders' effect (Rouse 1986:136).

During the first millennium B.C. the Ronquinan Saladoid peoples broke through the delta into the Guianas, where they established a second frontier in the present country of Suriname (Fig. 4). Behind this frontier they are hypothesized to have developed Cedrosan Saladoid pottery, which is distinguished by zoned-incised crosshatching (Rouse 1986, fig. 26).

Shortly before the time of Christ, the Cedrosan Saladoid potters radiated westward as far as Margarita Island in eastern Venezuela and northward through the Lesser Antilles and Puerto Rico to the eastern tip of Hispaniola, where they established Frontier 3 (Fig. 4). The movement westward was slow, probably because the migrants faced resistance by the previous Ortoiroid peoples and had to adapt from a riverine to a maritime environment, there being no large rivers in their new territory (Fig. 3b). The movement north was surprisingly rapid (Fig. 3a). There is reason to believe that the migrants in this direction were opposed by few, if any, Archaic-age Indians and that they at first settled only the larger islands which had sizable rivers, comparable to those in the Guianas (Watters 1980:297-308; Keegan 1985:51-3). They concentrated on riverine resources at the expense of the maritime resources that their relatives on the mainland were beginning to exploit (Barrau and Montbrun 1978). Only the northward migrants, therefore, retained their original ecological niche (Keegan and Diamond 1987).

If our knowledge were limited to pottery, we could not be sure whether the Cedrosan Saladoids actually entered the Antilles or merely passed their ceramics on to its Archaic-age inhabitants. We have found, however, that the pottery was accompanied in its spread into the Antilles not only by the original settlement pattern but also by the first evidences of agriculture and of zemiism, the latter in the form of small, three-pointed objects of stone, shell, coral, or pottery (Rouse 1986, fig. 27b). We may conclude, therefore, that the Cedrosan Saladoids conquered the islands as far north as Puerto Rico and, in effect, repeopled them.

It is assumed that the invaders halted at Frontier 3, on the eastern end of Hispaniola, because their forward progress was barred at that point by the first large population in their path, the Courian Casimiroids, and because they needed time to fully populate the islands they had already occupied. As they learned to exploit new resources they expanded from their riverine settlements to the coast and the mountainous interior (Rouse 1952:567; Goodwin 1979:379-474). Immediately behind the frontier, their pottery devolved by a process of simplification from the Saladoid into the Ostionoid series. The initial Ostionan pottery on the frontier was almost completely plain, retaining only the red slip and the simplest modeling of the previous Saladoid series (Rouse 1986, Figs. 28b,c).

About 600 A.D., the Ostionoid peoples resumed the previously Saladoid advance in two directions, along the southern coast of Hispaniola into Jamaica and through the northern valleys of that island onto the eastern tip of Cuba, where they established Frontier 4 (Figs. 3a, 4). Behind this frontier, they developed a new Meillacan subseries, which is characterized by rectilinear parallel-line incision borrowed from the previous Courian Casimiroid people (Rouse, 1986, figs. 28d,e).

Between 800 and 1,200 A.D., they completed their expansion by occupying central Cuba and the Bahamas. This brought them to the frontier behind which Columbus found them (Figs. 3a, 4). In the Bahamas, their Meillacan Ostionoid pottery degenerated into a crude Palmetto ware, which is almost completely undecorated (Rouse 1986, fig. 28f).

Meanwhile, the Ostionan Ostionoid peoples in the Mona Passage area, back from the frontiers, were developing a new Chican Ostionoid form of pottery, marked by a revival of the Cedrosan Saladoid modeling-incision but without its painting (Rouse 1986, fig. 29). They also elaborated the worship of zemis. The small, plain three-pointers of Cedrosan-Saladoid time now became large objects carved into animal and human figures (Rouse 1986, fig. 27, cf. b and c). In addition, the central peoples began to build ceremonial plazas and ball courts, thereby advancing from the Ceramic into the Formative age, which is defined by public monuments and hierarchical chiefdoms (Fig. 3a). Thus they transformed themselves into the Classic Tainos.

Speech communities and their languages

Each of the ethnic groups functioned as a separate speech community, with its own language, as well as a separate people, with its own culture (Fig. 1). Each had a different ancestry as a speech group than it had as a people, owing to the fact that languages and cultures can and normally do develop independently.

Granberry (1986:53-5) has assigned the Guanahatabey language to the Chibchan family, which was widespread in Colombia and Central America. The linguistic evidence is not sufficient to warrant this conclusion (Floyd G. Lounsbury pers. comm.), but it is consistent with the archeological evidence that the West Indies were originally peopled from Central America.

The Taino language is a member of the Arawakan family. While Island-Carib was originally assigned to the Cariban family because of its name (e.g., Steward and Faron 1959:23), it has also proved to be Arawakan. Apparently the Island-Carib warriors who conquered the Igneris of the Lesser Antilles adopted the Igneris' language, just as the Norman invaders of England gave up French in favor of English. The warriors retained only a secondary, pidgin language belonging to the Cariban family, which they spoke in their men's houses (Taylor and Hoff 1980).

Working back from the Taino, Island-Carib, and Arawak languages, linguists have constructed the phylogeny of the Arawakan family that is shown in Fig. 5. They find that the original, Proto-Arawakan language most probably developed in the middle of the Amazon Basin. The speakers of that language originally expanded upstream to the headwaters of the Amazon, as shown on the right side of the diagram. Other Proto-Arawakan speakers moved up the Negro River, a northern tributary of the Amazon, passed through the Casiquiare Canal, and entered the Orinoco Valley (Fig. 6). Somewhere along that route they developed a new, Proto-Maipuran language, which evolved into Proto-Northern after their arrival in the Orinoco Valley.

The speakers of the Proto-Northern language subsequently spread into the Guianas and the West Indies. The Proto-Northerners who remained in the Guianas intercommunicated primarily among themselves and as a result developed their own Arawak language, which later became Lokono (Fig. 5). The Proto-Northerners who settled in the Lesser Antilles similarly produced their own Igneri language, which they transmitted to their Island-Carib conquerors. The Proto-Northern speakers who continued into the Greater Antilles and intercommunicated among themselves there, developed the Taino language and carried it into the Bahamas.

The dates along the side of Fig. 5 have been obtained by glottochronology, a technique that estimates the length of time since two languages began to diverge by counting the number of differences in their basic vocabularies and dividing that figure by the rate of change known for historic languages. Glottochronology indicates that the Proto-Arawakan language arose about 3,500 B.C.; Proto-Maipuran about 1,500 B.C.; Proto-Northern during the first millennium B.C.; and Arawak, Island-Carib, and Taino within the Christian era. We may therefore conclude that the ancestors of the Taino speakers entered the West Indies about the time of Christ (Rouse 1986:120-6).

Races

Imbelloni (1938) has classified the natives of the West Indies in an Amazonid race, which extended northward from the Amazon Basin to the Florida Peninsula and was paralleled on the west side of the Caribbean Sea by an Isthmid race. He has been criticized by Newman (1958:72-8) for utilizing cultural as well as biological criteria, for failing to make detailed biological comparisons, and for not plotting his units in time as well as in space, as archeologists and linguists do. Nevertheless, his classification suggests a pair of hypotheses that would be worth investigating.

One is that the Guanahatabeys belonged to the Isthmid race (Fig. 7). This could be tested by comparing the human skeletal remains from preceramic sites in Central America with those from Cuba and Hispaniola in terms of their biological traits, especially their dental morphology since it has proved useful in tracing population movements in other parts of the world (Rouse 1986:51, 83).

The other is that both the Tainos and the Island Caribs belonged to the Amazonid race (Fig. 7). Testing of this hypothesis would require redefinition of the race and comparison of the Ceramic-age skeletal material of the Amazon and Orinoco Basins with that of the West Indies in terms of the revised criteria. The comparison should be done period by period, beginning with Period II when the Ceramic-age Indians began to displace the preceramic peoples (Fig. 3, left and right sides).

Conclusions

The ethnohistorical, archeological, and linguistic research may be synthesized in terms of the periods shown in Fig. 3. Period I, during which the islands were first peopled, is currently known only through archeological research, which indicates that Casimiroid peoples moved from Central America into Cuba and Hispaniola about 5,000 B.C. and that Ortoiroid peoples expanded from South America into the Lesser Antilles and Puerto Rico about 3,000 B.C. As yet, no remains dating from Period I have been found in either Jamaica or the Bahamas (Fig. 3).

Archeologists have discovered that Saladoid peoples from the Guianas repeopled the Lesser Antilles and Puerto Rico during Period II, beginning about the time of Christ. This is confirmed by the linguists' conclusion that the Proto-Northern speech community arrived about the same time. The Saladoid peoples evidently brought the Proto-Northern language with them.

Period III, from 600 to 1,000 A.D., appears to have been a time of divergence in both culture and language. The Saladoid peoples split into two groups, Troumassoids in the Windward Islands and Ostionoids farther north, and the latter group expanded through the territory of the Casimiroid peoples in the Greater Antilles to the frontier encountered by Columbus in western Cuba. Simultaneously, the Proto-Northern language diverged in the south into Igneri--later to be known as Island-Carib--and in the north into Taino. Presumably, the Ostionoids carried the Taino language with them as they expanded through the Greater Antilles during Period III and into the Bahamas during Period IV.

Both the ethnohistorical and the linguistic evidence indicate that Carib warriors invaded the Windward Islands from the Guianas during Period IV, that is after 1000 A.D. Linguists have concluded that invaders were absorbed into the local, Igneri population, adopting its Arawakan language in place of their own Cariban language. The Carib men did, however, retain a secondary, pidgin language for use in speaking among themselves (Taylor and Hoff 1980). If so, the Carib invaders must be regarded as immigrants rather than repeoplers, despite the fact that they imposed their own name on the Igneri population and on its language. Archeologists, however, still seek ceramic evidence that the Caribs repeopled the Windward Islands (e.g. Allaire 1980).

Literature cited

Alegría, R. E. 1981. El uso de la terminología etno-histórica para designar las culturas aborígenes de las Antillas. Seminario de Historia de América, Universidad de Valladolid, Valladolid. 30 pp.

_____. 1983. Ball courts and ceremonial plazas in the West Indies. Yale University Publications in Anthropology 79:1-185.

Allaire, L. 1980. On the historicity of Carib migrations in the Lesser Antilles. American Antiquity 45(2):238-245.

_____. 1985. The archaeology of the Caribbean. Pp. 370-371 in Christine Flon (ed.). The World Atlas of Archaeology. G.K. Hall and Co., Boston, MA.

Arrom, J. J. 1975. Mitología y artes prehispánicas de las Antillas. Siglo Veintiuno Editores, Mexico City. 191 pp.

Barrau, J., and C. Montbrun. 1978. La mangrove et l'insertion humaine dans les écosystèmes insulaires des Petits Antilles: le cas de la Martinique et de la Guadeloupe. Social Science Information 17(6):897-919.

Boomert, A. 1984. The Arawak Indians of Trinidad and coastal Guiana, ca. 1500-1650. Journal of Caribbean History 19:123-188.

Cruxent, J. M., and I. Rouse. 1969. Early man in the West Indies. Scientific American 221(5):42-52.

Goodwin, R. C. 1979. The prehistoric cultural ecology of St. Kitts, West Indies: a case study in island archeology. Unpublished Ph.D. dissertation, Arizona State University, Tempe, AZ. 514 pp.

Granberry, J. 1986. West Indian languages: a review and commentary. Journal of the Virgin Islands Archaeological Society 10:51-56.

Gullick, C. J. R. M. 1980. Island Carib traditions about their arrival in the Lesser Antilles. Pp. 464-472 in M. Lowenstein (ed.). Proceedings of the Eighth International Congress for the Study of the Pre-Columbian Cultures of the Lesser Antilles. Anthropological Research Papers 22, Arizona State University, Tempe, AZ.

Imbelloni, J. 1938. Tabla clasificatoria de los Indios: regiones biológicas y grupos raciales humanos de América. Physis: Revistade la Sociedad Argentina de Ciencias Naturales 12:229-249.

Keegan, W. 1985. Dynamic horticulturalists: population expansion in the prehistoric Bahamas. Unpublished Ph.D. dissertation, University of California, Los Angeles, CA, 358 pp.

_____. 1987. Colonization of islands by humans: a biogeographical perspective. Pp. 49-92 in M. B. Schiffer (ed.). Advances in Archaeological Method and Theory 10. Academic Press, New York.

Kozlowski, J. K. 1980. In search of the evolutionary patterns of the preceramic cultures of the Caribbean. Bolétin del Museo del Hombre Dominicano 13:61-79.

Lovén, Sven. Origins of the Tainan Culture, West Indies. Elanders Bokfryckeri Äkfiebolag, Göteborg. 697 pp.

MacNeish, R. S. 1982. Third annual report of the Belize Archaic archaeological reconnaissance. Phillips Academy, Andover, MA. 91 pp.

Newman, M. T. 1951. The sequence of Indian physical types in South America. Pp. 69-97 in W. S. Laughlin (ed.). Papers on the Physical Anthropology of the American Indian Delivered at the Fourth Viking Fund Summer Seminar in Physical Anthropology Held at the Viking Fund, September 1949. The Viking Fund, Inc., New York. 202 pp.

Rouse, I. 1952. Porto Rican prehistory. New York Academy of Sciences, Scientific Survey of Porto Rico and the Virgin Islands, 18(3-4):307-577.

_____. 1960. The entry of man into the West Indies. Yale University Publications in Anthropology 61. 26 pp.

_____. 1982a. Ceramic and religious development in the Greater Antilles. Journal of New World Archaeology 5(2):45-55.

_____. 1982b. The Olsen collection from Ile à Vache, Haiti. Florida Anthropologist 35(4):169-185.

_____. 1986. Migrations in Prehistory: Inferring Population Movement from Cultural Remains. Yale University Press, New Haven, CT. 202 pp.

Steward, J. H., and L. C. Faron. 1959. Native Peoples of South America. McGraw-Hill, New York. 481 pp.

Taylor, D. R., and B. J. Hoff. 1980. The linguistic repertory of the Island-Carib in the seventeenth century: the men's language--a Carib pidgin. International Journal of American Linguistics 46:301-112.

Watters, D. R. 1980. Transect surveying and prehistoric site locations on Barbuda and Montserrat, Leeward Islands, West Indies. Unpublished Ph.D. dissertation, University of Pittsburgh, PA. 416 pp.

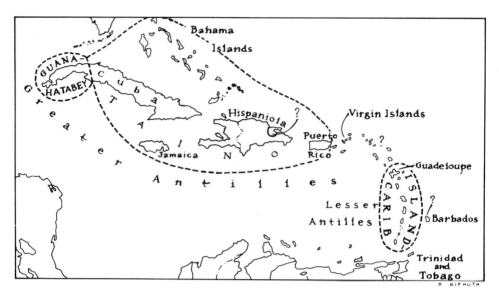

Figure 1. Ethnic groups in the West Indies at the time of European discovery.

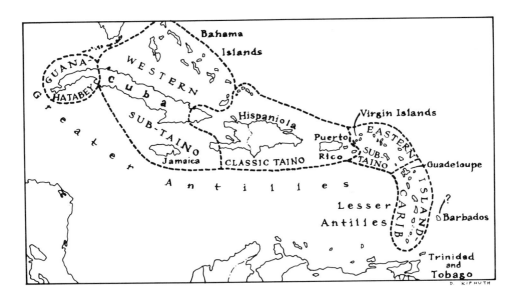

Figure 2. Peoples and cultures in the West Indies at the time of European discovery.

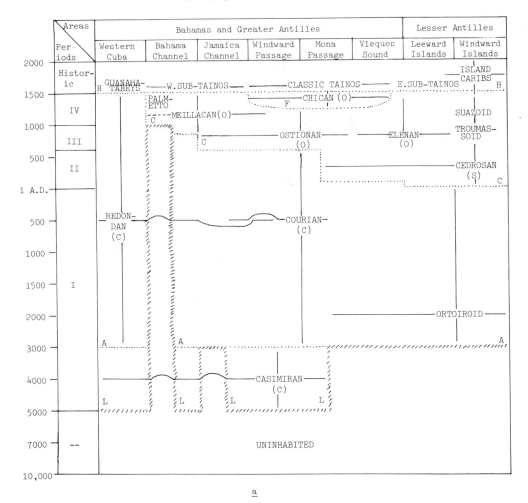

Figure 3. Chronology of the peoples and cultures of the Caribbean area: (a) the West Indies; (b) the adjacent coast and the Orinoco Valley. Ages: L = Lithic, A = Archaic, C = Ceramic, F = Formative, H = Historic. Series: (C) = Casimiroid, (O) = Ortoiroid, (S) = Saladoid, (B) = Barrancoid, (D) = Dabajuroid.

Peopling and repeopling of the West Indies

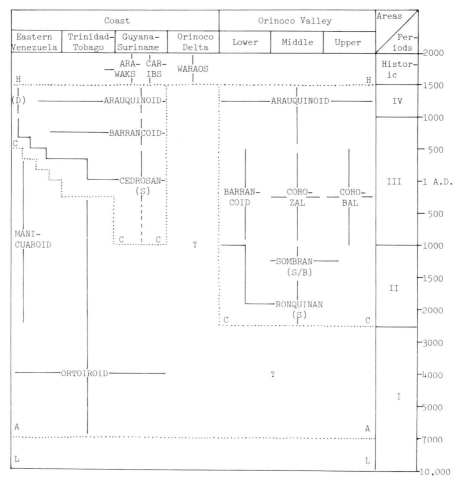

b

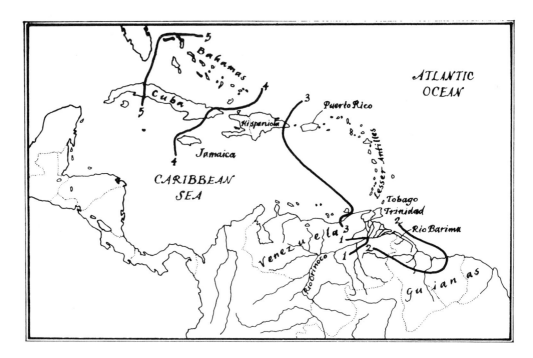

Figure 4. Advance of the Ceramic-Archaic age frontier from South America into the West Indies.

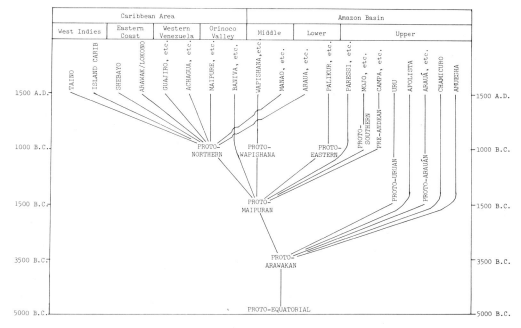

Figure 5. Phylogeny of the Arawakan family.

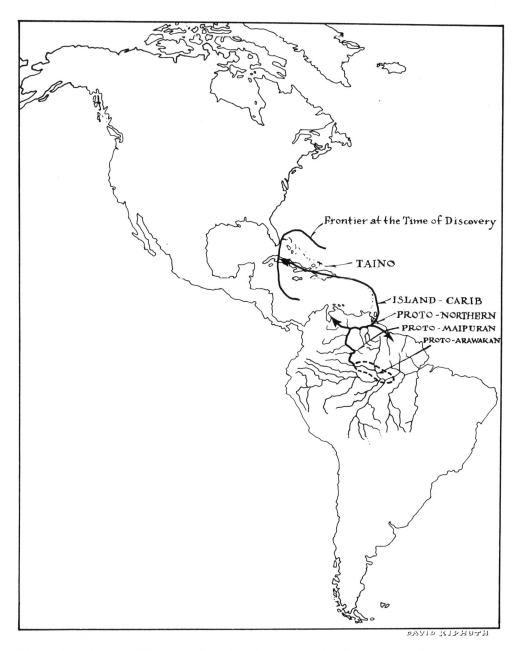

Figure 6. Advance of the Arawakan speech communities from Amazonia into the West Indies.

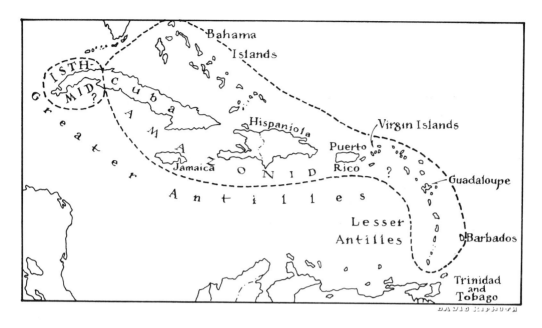

Figure 7. Hypotheses about the racial composition of the West Indies at the time of European discovery.

Table 1. Composition of the Volatile Leaf Oils of *Juniperus barbadensis* from St. Lucia, BWI along with the volatile leaf oils of other Caribbean junipers previously reported (Adams 1983; Adams and Hogge 1983). Compounds are listed in order of their elution from a DB1 column. BA = *J. barbadensis*, St. Lucia, BWI; LJ = *J. lucayana*, Jamaica; LB = *J. lucayana*, Bahama Islands; BM = *J. bermudiana*, Bermuda; EK = *J. ekmanii*, Haiti; GR = *J. gracilior*, Dominican Republic; VS = *J. virginiana* var. *silicicola*, Florida, USA; and VG = *J. virginiana* var. *virginiana*, Washington, D.C. USA. Those 70 components denoted with an asterisk (*) were utilized in principal coordinate analysis. Compositional values in parenthesis indicate that a compound runs at that retention time but no mass spectrum was obtained. Compound names in parenthesis are tentatively identified. T indicates the compound was present in trace amounts (less than 0.5% of the total oil).

Compound	% total oil							
	BA	LJ	LB	BM	EK	GR	VS	VG
Percent Yield*	0.6	0.6	0.2	0.3	1.4	0.8	0.4	0.2
Unknown 1*, RRT=0.143	-	-	T	-	(T)	0.8	-	(T)
Unknown 2*, RRT=0.151	-	-	T	-	(T)	0.8	-	T
Tricyclene + α-Thujene*	0.9	0.6	0.5	T	1.9	1.4	T	T
α-Pinene*	7.4	49.1	33.0	22.3	1.3	1.8	2.4	1.4
Camphene*	T	T	T	0.7	1.9	1.2	T	T
Sabinene*	31.0	9.7	8.3	2.8	5.0	10.1	T	6.7
β-Pinene*	T	1.1	1.2	0.6	T	T	T	T
1-Octen-3-ol*	-	T	T	1.0	T	T	0.9	-
Myrcene*	3.8	3.2	4.0	2.9	2.5	1.9	0.9	0.9
2-Carene	-	T	T	-	T	-	T	T
α-Phellandrene	T	-	-	T	-	T	-	-
3-Carene	-	-	-	T	-	-	T	(T)
α-Terpinene*	1.7	T	T	T	0.9	1.7	T	T
p-Cymene*	T	T	T	0.5	0.5	1.4	T	-
β-Phellandrene	T	-	-	-	-	-	-	T
Limonene*	34.2	25.9	18.0	35.3	9.6	7.3	33.3	18.9
trans-Ocimene*	0.7	T	-	T	-	-	-	(T)
Γ-Terpinene*	2.7	0.8	0.7	0.7	1.7	3.5	T	T
(p-menth-1(7),3-diene)*	T	-	T	-	-	-	T	
(cis-p-menth-2-ene-ol)*	0.9	-	T	-	0.9	1.1	-	T
Terpinolene*	1.2	1.0	0.8	0.8	0.6	0.9	(T)	0.5
(trans-p-menth-2-ene-ol)*	0.7	-	-	-	-	T	-	-
4-Terpinenyl acetate	T	T	-	-	-	-	-	T
Linalool*	-	-	T	1.1	0.6	2.6	1.5	4.4

HUMAN EXPLOITATION OF ANIMAL RESOURCES IN THE CARIBBEAN

Elizabeth S. Wing[1]

Abstract

Conclusions about the nature of prehistoric and early historic exploitation of animal resources in the Caribbean are based on the identification and analysis of samples of animal remains excavated from archeological sites. Such information has innate biases which include differences in prehistoric disposal patterns and in the preservation of faunal remains as well as in their recovery and analysis. Nevertheless some patterns of exploitation are emerging which warrant further investigation. The faunal samples from the earliest occupations have a greater relative abundance of terrestrial resources than the samples from the later occupations. A number of animals were introduced both in prehistoric and historic times. Both of these characteristics suggest that the first people to move into the Antilles were attempting to adhere to a tradition of exploitation developed on the continental mainland with its abundant terrestrial fauna. In the samples from the later occupations one can see the results of an economy developed on the Caribbean Islands with diverse fishing and shellfish gathering and some evidence for partitioning of resources. Further analysis of faunal samples from archeological sites is necessary to substantiate and refine these findings.

Introduction

Animal remains excavated from archeological sites provide an understanding of the nature of the pre-Columbian and early historic exploitation of animal resources in the Caribbean. The use of animals by people living in preindustrial conditions is dictated by the animals that are available and the human motivation and technology to procure them. Species most frequently used are those which live close to the human occupation as well as those species which are captive, tame, or domestic. When we look across cultures at the array of organisms that are consumed or rejected, we see a great variety of customs vigorously adhered to by the people of each culture. In the Caribbean, several opportunities exist for examining the choices different people have made in establishing themselves in the islands and adapting to use of the island resources.

Human modification of the organic environment in the West Indies began early in prehistoric times and continues to this day. The history of the changes wrought by

---

[1] Dr. Wing is Curator of Zooarchaeology at the Florida Museum of Natural History, University of Florida, Gainesville, FL 32611.

humans has been documented for the islands of Antigua, Barbuda, and Anguilla by Harris (1965). According to his calculations one fifth of the flowering plant species from Antigua and Barbuda are alien species deliberately or accidentally introduced (Harris 1965:53). Major portions of these islands are under cultivation today. Clearing of the land began with the Amerindian cultivation of crops such as manioc, corn, sweet potato, tania, peanuts, common beans, peppers, as well as other useful plants (ibid.:74). This paper does not discuss the history of human uses of plants in the Caribbean but it musbe understood that plant harvesting, clearing, and cultivation modified the environment which in turn affected the terrestrial animal populations. Both the exploitation and the introduction of plants and animals has had a profound effect on the native fauna and flora of the islands (Olson 1978 and elsewhere).

The intent of most animal exploitation is their use as a source of food, although other uses of animals and animal products exist. In the Bahamas, where the native rock is relatively soft limestone, shell was used as a raw material for tools. The manufacture of tools would not necessarily be the exclusive use of the animal as the soft parts of the mollusc could still be consumed. Evidence from accounts and from archeological remains document the development of an early Spanish hide industry in Haiti based on cattle introduced in the early 16th Century (Reitz 1986). Tallow, glue, and gelatin in addition to hides were probably extracted from the beef carcass and the meat was jerked or dried for preservation (Reitz 1986). Domestic animals had many roles which contributed to the pattern of life in the Caribbean. Wild animals may also have had multiple uses that we can not as easily detect.

The sources of traditions of foodways and patterns of animal use of the people of the Caribbean were derived, as were the people themselves, from continental mainlands. These mainland traditions developed where land animals were abundant and varied. These traditions were then modified as a result of experiences gained during life on the island chain. The continents from which people moved into the islands were initially from Middle America, secondly from northeastern South America, and then finally from the Iberian Peninsula (Rouse this symposium). In this paper, only the South American and Spanish adaptation to Caribbean life will be discussed simply because too little archeological evidence of animal use exists for the earliest Casimiroid people.

Materials and methods

The data upon which this paper is based are primarily from a series of faunal samples excavated from archeological sites (Table 1, Map 1) in the Caribbean (Wing and Reitz 1982). These are chosen to represent sites from different time periods as well as different regions of the Caribbean. Clearly more samples are necessary to document the full range of human adaptation to life in the Caribbean.

The samples differ in a number of respects which can introduce bias. Innate differences in the samples may be the result of differences in the prehistoric disposal patterns or in the preservation of the remains. Today in the Turks and Caicos islands conch meat is a commercial item, dried to be sold in the northern Bahamas. Fishermen gather conch, extract the meat on the spot, and toss the shell overboard thereby bringing no hard evidence of this exploitation to the home site. The scarcity of conch shell in the Caicos sites may not be a true reflection of its past use.

Another source of bias are the conditions of preservation. Bone preservation is closely correlated with the acidity of the soil. The more acidic the soil the greater the loss of bone. Consequently, bone is usually well preserved in coastal shell middens and less well preserved in inland sites in the absence of shell.

Biases have also resulted from the method of excavation and recovery. Early archeological procedures employed screening of archeological material with a coarse gauge screen (6 mm or larger). Archeologists have come to realize that much of the plant and animal remains are lost by this method. Improved recovery of biological remains is now achieved with the use of finer gauge screens (1 mm). Another improvement in the archeological procedure is to maintain the integrity of the entire faunal sample so that the analyst can evaluate the relative abundances of different classes and phyla in the faunal assemblage. Strides have been made in zooarcheological methods and one of these is the awareness of the biases introduced by various factors of deposition and recovery.

Standard zooarcheological techniques are used in quantifying the faunal samples discussed in this paper. Most of the data are presented in percentage of the calculated minimum numbers of individuals. This method, despite its problems, is the most satisfactory when the relative abundance of different phyla and classes, each with different numbers of preservable parts, are compared.

Results

Some of the promenant features of the faunal samples are most easily discussed in two segments which describe the terrestrial component and the aquatic component. These two components are, of course, interrelated. In fact, the relative abundance of aquatic and terrestrial organisms in the samples is significant for an understanding of the degree of adaptation to the use of Caribbean Island resources.

Native terrestrial species

The native species that are repeatedly found or abundantly represented in West Indian faunal samples are oryzomyine and capromyid rodents, a number of larger ground dwelling or nesting and flightless birds, iguanid lizards, and land crabs. Other rodents, insectivores, and shore birds have been identified from archeological samples but they are neither abundant in any one site nor frequently encountered in a number of sites studied thus far (Morgan and Woods 1986).

The oryzomyine and capromyid rodents, the two most important mammals, were widely and intensively exploited. A number of described and undescribed oryzomyine rodents are represented in Lesser Antillian (on Antigua, St. Eustatius, Marie Galante, St. Lucia, and Martinique) and Jamaican sites. The capromyids represented in sites are primarily *Geocapromys* in Jamaican sites (White Marl, Rio Bueno, Rio Nuevo, Bellevue, Cinnamon Hill) and Bahaman sites (Palmetto Grove and MC 6), and *Isolobodon* in a Haitian site (En Bas Saline) and several Puerto Rican sites (on Vieques Island [Narganes 1982], El Bronce [Reitz MS.], and Hacienda Grande). Rodents predominate in the faunal sample of only one site, the Bellevue site in Jamaica. This site is located on the inland side of the city of Kingston, over 6 km from the shore. At this location, some distance from the shore, the hutia (*Geocapromys brownii*) constitutes the major portion,

89 percent, of this sample (Table 2). This degree of intensive exploitation of one species is usually seen in sites where animals were domestic.

Birds that predominate in archeological sites of the West Indies are all moderately large and tend to be ground dwelling or nesting. Such birds are Audubon's shearwater (*Puffinus lhermanieri*), which is abundant at one site on Antigua (Mill Reef site), and a rail, probably the extinct flightless rail *Nasotrochis* which occurs in northern Haiti (En Bas Saline). Olson (1982) has reported the flightless rail from midden sites in Puerto Rico and the Virgin Islands and believes they were reared in captivity. A group of birds frequently encountered in sites is the pigeons (Columbidae), particularly the scaled pigeon *Columba squamosa*. Scaled pigeons have been reported from sites on Vieques Island (Narganes 1982), in St. Eustatius (Golden Rock [van der Klift 1985]), Montserrat (Trant's [Steadman et al. 1984]), Antigua (Mill Reef), Marie Galante (Folle Anse), and possibly St. Kitts (Sugar Factory Pier site).

The iguanid lizards of the genera *Cyclura* and *Iguana* are terrestrial reptiles that were clearly important to the Amerindian subsistence. *Cyclura* has been identified from the following sites: Jamaica (Cinnamon Hill, Bellevue, White Marl), the Turks and Caicos (MC 6, MC 12, Pine Island), and the Bahamas (Palmetto Grove). *Iguana* have been identified from sites on the Virgin Islands, St. Eustatius (Golden Rock [van der Klift 1985]), St. Kitts (Sugar Factory Pier), Antigua (Mill Reef), Marie Galante (Folle Anse and Taliseronde), St. Lucia (Grand Anse), and Grenada. They were clearly as much esteemed in the past as they are today.

The final group of terrestrial animals that was frequently used but for which we do not have a full record of their occurrence in faunal samples, is the land crabs. The two genera of large land crabs that are found are *Gecarcinus* and *Cardisoma*. They are the basis of the descriptive term "crab culture" coined by Rainey (1940) to describe the earlier West Indian cultures. Abundant remains of these land crabs are recorded for sites on Puerto Rico (Hacienda Grande), St. Kitts (Cayon and Sugar Factory Pier), St. Eustatius (Golden Rock [van der Klift 1985]), and Antigua (Marmora Bay).

Introduced terrestrial species

The introduction of domestic or tamed land animals is a means of enriching an otherwise meager land fauna and maintaining contact with familiar animals. Domestic animals are the result of controlled selection and breeding by man and thus may be thought of as man-made animals (Clutton-Brock 1981). Exploited captive animals are those whose breeding remains more under natural influences than under human selection. The exploitation of captive animals is normally hard to detect in the fragmentary archeological remains of these animals. The captivity and transport of animals can be documented throughout the Caribbean by the occurrence of species in archeological contexts and where there is no history of natural occurrence in the fossil record on the same island (Olson 1982; Morgan and Woods 1986).

The one fully domestic animal which has accompanied man around the world is the dog. Dog remains have been recorded from sites on Grenada, Barbados, St. Lucia (Grand Anse), St. Kitts (Sugar Factory Pier), Montserrat (Trant's), Puerto Rico (El Bronce, Hacienda Grande, and Vieques Island), Dominican Republic (Lawrence 1977), Jamaica (White Marl and Bellevue), Cuba (Miller 1916), and the Turks and Caicos (MC 12). Many of these dog remains are associated with human burials suggesting that they were valued as companions (Lawrence 1977). Since many were afforded a special place in the prehistoric burial practices, their remains are not frequently disposed of in midden deposits. Therefore, as most of the deposits examined thus far have been from middens, dogs may be underrepresented in the faunal samples studied.

Guinea pigs, *Cavia porcellus*, have a long history of domestication in the Andean area and are also recorded from sites in the Caribbean. Their remains were found at the Mill Reef site in Antigua (Wing et al. 1968), the Hacienda Grande site (in mixed levels 20 to 40 cm. deep) in Puerto Rico, and at the Anadel site in the Dominican Republic (Miller 1929). These finds are questionably prehistoric by virtue of their being in superficial levels of the sites or in association with the European rat *Rattus*. On the other hand, guinea pigs were available in northern South America based on the finds of their remains in the prehistoric site of Turen located on the Apuré River tributary of the Orinoco River in the western llanos of Venezuela (Garson pers. comm.).

The other animals that must at this time be thought of as exploited captive animals are the opossum (*Didelphis*) and the agouti (*Dasyprocta*) (Wing et al. 1968). The opossum is recorded only from the southern Lesser Antilles (Grenada and St. Lucia). The agouti was much more widely distributed. Their remains have been recorded from sites in Grenada, St. Lucia (Grand Anse), Martinique (Macabou and Paquemar), Marie Galante (Taliseronde and Folle Anse), Antigua (Mill Reef), St. Kitts (Sugar Factory Pier), and St. Eustatius (Golden Rock [van der Klift 1985]). They existed in the wild on these islands well into modern times and, in fact, some populations may still be extant. They were presumably killed off by introduced preditors such as the fer-de-lance and mongoose as well as displaced by the increase of cleared land (Westerman 1953).

In addition to the introduction of animals from the mainland of South America, native West Indian animals were moved from one island to another by the Amerindians. The large rodent, *Isolobodon portoricensis*, is believed to be native only in Hispaniola and introduced to Puerto Rico and the Virgin Islands by Amerindians (Miller 1929; Morgan and Woods 1986). Olson (1982a and b) has reviewed the cases that can be made for the trade and transportation of a number of West Indian birds and mammals. Among those animals thought to have been transported are the Cuban *Capromys pilorides* to Hispaniola and *Geocapromys ingrahami* to some of the Bahaman Islands. The flightless rail *Nesotrochis debooyi* is believed to have been reared in captivity and transported to the Virgin Islands, and the macaw *Ara autocthones* may have been traded to St. Croix.

European introductions

Although the impact of the Amerindian introductions of animals in the West Indies was great in modifying the land faunas of the islands, the historic introductions of plants, plantation cultivation, and animals profoundly effected the biota of the islands. These introductions coincided with the first Spanish encounter of the Caribbean Islands. In fact, it would appear that as the Santa Maria, Columbus' flagship in his historic trip in 1492, floundered on a rock during the fateful Christmas night, off of what is now Haiti, rats (*Rattus rattus*) were fleeing the ship for safer ground. Their remains, along with the remains of pigs (*Sus scrofa*), are identified in a faunal sample from the site of En Bas Saline which is thought to be the location where Columbus' sailors erected a fortified settlement called Navidad (Deagan pers. comm.).

Other European domestic animals soon followed the rat and the pig and prospered in the New World setting (Table 3). Cattle were introduced at Puerto Real, a town close to En Bas Saline founded in 1502 and officially closed in 1578 (Reitz 1986). In that short span of time, cattle and pigs became feral and multiplied to such an extent that cow hides and other carcass products formed the basis of a commercial enterprise while pigs became a major food item in the town (McEwan 1983). Other European animals that were introduced to the Caribbean early in the historic period are Old World dogs, cats, sheep and/or goat, horse, and chickens. Their remains are not as abundant in early 16th century sites as cattle and pigs, however, in time they played their part in exterminating much of the native West Indian land fauna.

Aquatic component of the faunal samples

The aquatic resources that were used in prehistoric times are varied and include species typically found in different habitats. This exploitation is most easily discussed in terms of the aquatic habitat of the represented species in the faunal assemblage. Clearly some species may be found in more than one habitat throughout their lives in which case we have grouped them in the habitat in which they are most frequently found or in a catagory of mixed habitats. The three major habitats recognized as most frequently represented by the species encountered in the samples are inshore estuarine and tidal flats habitat, reefs and banks habitat, and offshore pelagic. The following is a discussion of each of these and the evidence that exists for an understanding of human uses of the resources in each habitat.

Coral reefs and rocky banks

This habitat is renowned for its diverse vertebrate and invertebrate fauna. The families of fishes that we associate with this habitat are squirrelfishes (Holocentridae), groupers (Serranidae), some jacks (Carangidae), snappers (Lutjanidae), grunts (Haemulidae), parrotfishes (Scaridae), wrasses (Labridae), and surgeonfishes (Acanthuridae). When these fishes are encountered in archeological sites we assure that this habitat was exploited. This reef assemblage predominates in the faunas of a majority of Caribbean sites and is particularly well represented at the Palmetto Grove site, MC 12, and Mill Reef. The fishes that are particularly abundantly represented are the parrotfishes, especially *Sparisoma viride*, and the surgeonfishes in the genus *Acanthurus*.

The composition of populations of reef fishes are different in extended barrier reefs and isolated patch reefs (Bardach 1959). Extended reefs have relatively more herbivorous and omnivorous fish species while isolated reefs have relatively more carnivorous species (Table 4). The faunal assemblages in the sites from Middle Caicos reflect the composition of the faunas of the type of reef adjacent to the site. There is an extended reef off shore from MC 12 and a lagoon with isolated reefs close to MC 6 and Pine Cay.

The two sites, MC 6 and MC 12, are located just 5 km apart yet their occupants clearly fished in the closest waters as reflected by the differences in the reef fish composition. Even more distinctive is the dependence on tidal flat and lagoon animals at the MC 6 site located close to the lagoon on the south side of the island (Table 5, Map 2). This must have meant the use of different fishing techniques at the two neighboring sites which resulted in the predominant capture of different species.

Inshore estuarine and tidal flats

The shallow inshore waters harbor a great number of the invertebrates and fishes that were used. The major families of fishes that are associated with this habitat are bone fish (Albulidae), snook (Centropomidae), some jacks (Carangidae), mojarra (Gerreidae), sheepshead (Sparidae), and sleepers (Eleotridae). By the measure of the occurrence of these animals, the use of this habitat was especially important to people dwelling along the coast of the Greater Antilles. An inshore estuarine assemblage is prominent in the sample from White Marl on the south coast of Jamaica, En Bas Saline on the north coast of Haiti, and Hacienda Grande in Puerto Rico. Bonefish (*Albula vulpes*) and other inshore fishes are also abundant at MC 6 which is located on the edge of the large lagoon on the south side of Middle Caicos. Needless to say, the use of this inshore assemblage occurs where this habitat is most extensive and accessible.

Off shore pelagic

What is called off shore pelagic here may not, in fact, be far off shore but in areas where the shelf is narrow and deep water is close to shore. The tuna fishes (Scombridae) are representative of off shore pelagic fauna. A small sample of finely sceened material from an archeological survey on Barbados has added the flying fish to this list. Off shore pelagic fishes constitute a major portion of the fauna from sites on St. Kitts (Cayon and Sugar Factory Pier Table 6), Montserrat (Trant's), and Marie Galante (Folle Anse). These islands are in the volcanic arc and have relatively narrow shelves.

Relative importance of terrestrial species

The calculation of the relative importance of terrestrial animals in a fauna is profoundly affected by the biases of recovery and analysis. If the faunal sample was recovered with coarse gauge screen, thereby possibly losing a segment of the aquatic fauna, especially small fishes, the relative importance of terrestial animals would be inaccurate. Likewise, the relative abundance of terrestrial animals will differ when all or only part of the fauna is taken into consideration. For the purpose of this analysis, only the vertebrates are being considered. Keeping these sources of bias in mind, some trends in the relative abundance of terrestrial animals are apparent.

When the sites are grouped according to location and period of occupation, the samples from the sites in the Greater Antilles and from the early occupation in the Lesser Antilles show, on the average, a greater relative abundance of terrestrial animals than do the samples from the later occupations in the Lesser Antilles and the samples from the Bahamas (Table 7). The variation between the samples is, however, very great. This may reflect local differences in the availability of animals as well as archeological biases. Clearly, this is an aspect of Caribbean zooarcheology that needs more study.

Diversity of species represented

An analysis of species diversity suffers from the same biases as does the analysis of terrestrial species representation. Nevertheless some extremes in the diversity of the animals represented in the site can be discussed. The site of En Bas Saline from the north coast of Haiti is an example of a site with a very diverse fauna. Represented in the sample are 68 vertebrates and 64 invertebrates. We believe little was lost in the recovery process because the flotation samples from the site added no new information. This sample is in sharp contrast to the sample from the Bellevue site, in which 16 species were identified and *Geocapromys* constitutes 89 percent of the fauna, and with the sample from Locus 39 at Puerto Real, in which 10 species were identified and, of these, cattle and pig make up 80 percent of the fauna. These extremes in the diversity of the fauna are related to both the availability of resources in the vicinity of the site and to the economy of the occupants of the site.

Conclusions

It is hard to escape the conclusion that more studies of faunal samples from Caribbean sites using the latest research techniques are needed to fully understand the relative importance of vertebrates and invertebrates in prehistoric and early historic economies. Equally important is a consideration of the uses of plants for a more complete understanding of the subsistence systems and the extent and history of human manipulation of the island environments.

We know from archeological investigations (Rouse this volume) and history that the people who moved into the Caribbean, whether they came from South America or from Europe, came from large continental land masses with diverse terrestrial faunas. Faced with life in a new environment, the immigrants attempted to duplicate their traditional customs and foodways as closely as possible. One aspect of this was to retain access to the familiar land animals of their native land. To insure such access, we know from the Puerto Real and En Bas Saline data that a variety of domestic animals, such as pigs, cattle and chickens, accompanied the European explorers to the New World. Evidence from the faunal samples indicates that the Amerindians also carried along captive animals when embarking on voyages of colonization in the Antilles from the South American mainland. Captive and domestic animals were carried into the Antilles from South America as well as between the Islands of the Caribbean.

Another means of replicating traditional exploitation patterns is to engage in hunting, fishing, and gathering activities in accustomed levels. The data (Table 7) on the terrestrial component of the faunal samples from the early levels from the Lesser Antilles and the Greater Antilles suggest the initial colonists in the Caribbean relied on land species for about a third of their total catch of animals. Through subsequent adaptation to greater dependence upon fishing, this reliance on land animals was diminished to slightly less than 20 percent of the total catch.

Many of the faunal samples document the use of faunal assemblages closest to the habitation site, implying flexibility in procurement techniques and an acceptance of diverse animals for food. The samples that illustrate this most clearly are the two samples from Middle Caicos Island. Though these two site are within easy walking distance from each other, each has a distinctive faunal assemblage characteristic of the aquatic habitat closest to the site.

This research sets the stage for a number of further investigations. As already mentioned, study of the plant component of the economies of the different Caribbean colonists is of great importance. Generally poor preservation of plant remains in archeological contexts may require augmentation of the studies of the plant remains by indirect analyses. Some of these indirect methods rely on correlations between diet and the trace mineral and isotope characteristics of bone. When applied to human bone they may reveal the relative amounts of plant and animal foods in the diet. Zooarcheolgical studies present the unique possibility of learning more about the nature of human manipulation of animals through the captivity and transportation of different mammals and birds. This aspect of animal use is important on its own merits and also as a preliminary step towards animal domestication. Captivity of animals is not easily detected in archeological materials. In this respect the West Indian faunal samples, as well as samples from other island chains, offer special opportunities for examining the question of animal captivity as introduced captive animals can be detected among isolated endemic faunas of the islands.

Literature cited

Bardach, J.E. 1959. The Summer Standing Crop of Fish on a Shallow Bermuda Reef. Limnology and Oceanography 4:77-85.

Clutton-Brock, J. 1981. Domesticated Animals from Early Times. University of Texas Press. Austin, TX 208 pp.

Crosby, A.W. 1986. Ecological Imperialism: The Biological Expansion of Europe 900-1900. Cambridge University Press Cambridge, UK 368 pp.

Ewen, C.R. 1987. From Spaniard to Creole: The Archaeology of Hispanic American Cultural Formation at Puerto Real, Haiti. Unpublished Ph.D. dissertation, University of Florida, Gainesville, FL 259 pp.

Harris, D.R. 1965. Plants, Animals, and Man in the Outer Leeward Islands, West Indies. University of California Publications in Geography Vol 18, 164 pp.

Klift, H.M. van der. 1985. Animal and Plant Remains from the Golden Rock Site on St. Eustatius. Pp. 12-23 in Archaeological Investigations on St. Eustatius (Netherlands Antilles). Interim Report. R. U. Archeologisch Centrum Leiden, Netherlands.

Lawrence, B. 1977. Dogs from the Dominican Republic. Cuadernos del Cendia 8:3-19.

McEwan, B.G. 1983. Spanish Colonial Adaptation on Hispaniola: The Archaeology of Area 35, Puerto Real, Haiti. Unpublished Masters Thesis, University of Florida, Gainesville, FL 175 pp.

Miller, G.S. jr. 1929. Mammals eaten by Indians, Owls, and Spaniards in the Coast Region of the Dominican Republic. Smithsonian Miscellaneous Collections 82(5):1-16.

Morgan, G.S. and C.A. Woods. 1986. Extinction and the Zoogeography of West Indian Land Mammals. Biological Journal of the Linnean Society 28:167-203.

Narganes S., Y.M. 1982. Vertebrate Faunal Remains from Sorcé, Vieques, Puerto Rico. Unpublished Masters Thesis, University of Georgia, Athens, GA 110 pp.

Olson, S.L. 1978. A Paleontlogical Perspective of West Indian Birds and Mammals. Pp. 99-117 in F. B Gill (ed.). Zoogeography in the Caribbean. The Leidy Medal Symposium, Academy of Natural Sciences of Philadelphia Special Publication No. 13.

_____. 1982a. Biological Archeology in the West Indies. The Florida Anthropologist 35(4):162-168.

_____. 1982b. Fossil Vertebrates from the Bahamas. Smithsonian Contribution to Paleobiology 4:1-60.

Rainey, F.G. 1940. Porto Rican Archaeology: Scientific Survey of Porto Rico and the Virgin Islands. New York Academy of Sciences 18(1):1-208

Reitz, E.J. 1986. Cattle at Area 19, Puerto Real, Haiti. Journal of Field Archaeology 13:317-328.

———. 1985. Vertebrate Fauna from El Bronce, Puerto Rico. MS.

Steadman, D.W., D.R. Watters, E.J. Reitz, G.K. Pregill. 1984. Vertebrates from Archaeological Sites on Montserrat, West Indies. Annals of Carnegie Museum 53(1):1-29.

Watters, D.R., E.J. Reitz, D.W. Steadman, G.K. Pregill. 1984. Vertebrates from Archaeological Sites on Barbuda, West Indies. Annals of Carnegie Museum 53(13):383-412.

Westermann, J.H. 1953. Nature Preservation in the Caribbean. Foundation for Scientific Research in Surinam and the Netherlands Antilles, Utrecht 9:1-106.

Wing, E.S., C.E. Ray and C.A. Hoffman Jr. 1968. Vertebrate Remains from Indian Sites on Antigua, West Indies. Caribbean Journal of Science 8(3-4):123-129.

———. 1969. Vertebrate Remains Excavated from San Salvador Island, Bahamas. Caribbean Journal of Science 9(1-2):25-29.

———. and S.J. Scudder. 1980. Use of Animals by the Prehistoric Inhabitants on St. Kitts, West Indies. Proceedings of the 8th International Congress for the Study of the Pre-Columbian Cultures of the Lesser Antilles. Arizona State University Anthropological Research Papers 22:237-245.

———. and E.J. Reitz. 1982. Prehistoric Fishing Economies of the Caribbean. Journal of New World Archaeology 5(2):13-32.

———. and S.J. Scudder. 1983. The Tropical Marine Edge Animal Exploitation by Prehistoric People. Pp. 197-210 in J. Clutton-Brock and C. Grigson (eds.). Animals and Archaeology: Vol. 2, Shell Middens, Fishes and Birds. BAR International Series 183:197-210.

Table 1. Faunal samples excavated from archeological sites and referred to in this paper.

1. St. Kitts (Wing and Scudder 1980)
       a. Cayon AD 1-300
       b. Sugar Factory Pier AD 700-1000
2. Antigua (Wing et al. 1968)
       a. Mill Reef AD 500-1150
3. Middle Caicos (Wing and Scudder 1983)
       a. MC 6 AD 750-1500
       b. MC 12 AD 750-1500
4. San Salvador (Wing 1969)
       a. Palmetto Grove AD 1000-1500
5. Jamaica (Wing 1972)
       a. White Marl AD 900-1500
       b. Bellevue AD 900-1000
6. Haiti
       a. En Bas Saline AD $\pm$ 1492
       b. Puerto Real AD 1503-1578 (McEwan 1983, Reitz 1986, Ewen 1987)

Table 2. The exploitation of different habitats at two sites in Jamaica.

HABITAT EXPLOITED	WHITE MARL % MNI	BELLEVUE % MNI
LAND	64	89
FRESHWATER	0	0
BEACH	3	0
ESTUARINE	22	8
BANKS AND REEFS	10	2
PELAGIC	TR	1
TOTAL NUMBER	690	102

Table 3. Exploitation of vertebrates summarized for the pre and post contact components at En Bas Saline and the early historic sample from Puerto Real (McEwan 1983).

VERTEBRATE CLASS	EN BAS SALINE PRE CONTACT % FRAGMENT	EN BAS SALINE POST CONTACT % FRAGMENT	PUERTO REAL % FRAGMENT
MAMMAL	9	9	57
BIRD	3	2	TR
AMPHIBIAN	1	TR	0
REPTILE	10	4	42
FISHES			
ESTUARINE	13	18	0
REEF	62	59	1
PELAGIC	2	7	0
TOTAL NUMBER	2745	2377	28876

Table 4. Comparison between the relative abundance of reef omnivores and carnivores using A. reef census data (Bardach 1959) and B. data on the reef fishes from the sites on Middle Caicos.

A.

Fishes	Extended Barrier Reef		Isolated Patch Reef	
	number of individuals	%	number of individuals*	%
Omnivores	109	87	581	29
Carnivores	16	13	1400	71

*This number excludes fishes that weigh less than 20 gms in order to more closely compare with the archeological faunas which had no fishes in that small size range.

B.

Fishes	MC 12		MC 6		Pine Cay	
	MNI	%	MNI	%	MNI	%
Omnivores	44	72	13	32	25	31
Carnivores	17	28	28	68	55	69

Table 5. Exploitation of vertebrates at two sites on Middle Caicos.

HABITATS EXPLOITED	MC 6 % MNI	MC 12 % MNI
LAND	18	32
TIDAL FLATS AND LAGOON	41	7
REEF OMNIVORES	13	44
REEF CARNIVORES	28	17
TOTAL NUMBER	140	86

Table 6. Exploitation at two sites on St. Kitts occupied at different time periods.

HABITAT EXPLOITED	CAYON % MNI	SUGAR FACTORY PIER SITE	
		EARLY % MNI	LATE % MNI
LAND	57	31	19
BEACH	1	0	1
SHALLOW REEF	7	11	5
DEEP REEF	17	32	28
PELAGIC	17	31	49
TOTAL NUMBER	109	184	126

Table 7. Relative abundance of terrestrial animals from Caribbean faunal samples.

REGION	NUMBER OF SAMPLES	AVERAGE	RANGE
Greater Antilles	12	34%	11%-89%
Lesser Antilles			
early	5	38%	3%-66%
late	8	19%	10%-32%
Bahamas	3	17%	0.3%-32%

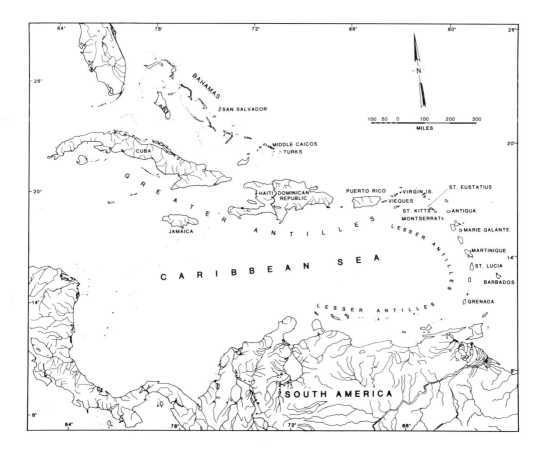

Map 1. Map of the Caribbean Basin.

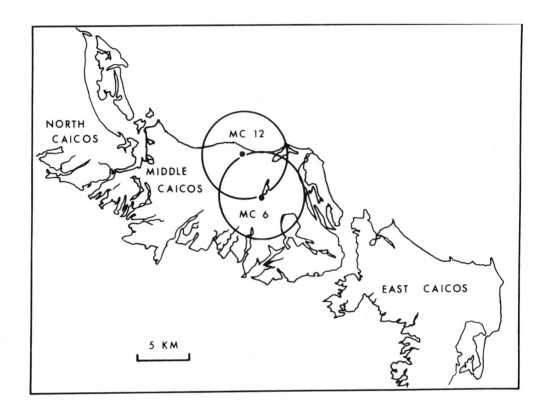

Map 2. Map of the Caicos Islands showing collecting locations.

ARCHAEOLOGICAL IMPLICATIONS FOR LESSER ANTILLES BIOGEOGRAPHY: THE SMALL ISLAND PERSPECTIVE

David R. Watters[1]

Abstract

Archaeological faunal remains have important implications for certain biogeographic issues in the Lesser Antilles. Faunal remains, which occur in concentrations in archaeological sites, can provide data of a diachronic nature that are crucial to issues of extinction and introduction of species. Determination of the context of archaeological faunal remains in excavated sites is an essential prerequisite to the use of those remains for biogeographic studies. The relevance of archaeology for biogeography depends on such factors as the intensity of research on various islands, the recovery techniques used in the field, and the phases of human occupation under investigation. For the period after human colonization, which is the only period archaeologists can address directly, the archaeology of the prehistoric (=Amerindian) occupation is better known than the historic (=post-Columbian). It is in combination with paleontological materials, which pertain to the periods before and after human colonization, that archaeological faunal remains have their greatest utility for biogeographic studies.

Introduction

Recent research in the Pacific islands has shown that the archaeological record contains important data for persons interested in biogeographic analysis and interpretation. While this research has focused mainly on the issue of extinction of various animal forms on islands (Olson and James 1982; Steadman and Olson 1985), it is becoming evident that materials from archaeological sites have implications for other biogeographic concerns, such as endemism or range studies (Steadman 1988).

The research in the Pacific comes at an opportune time for those interested in Caribbean biogeography (Watters 1982). The Pacific work has shown the merit of increased interaction and information flow among disciplines. The presence of archaeologists in a symposium on Caribbean biogeography organized by natural scientists indicates there also is an awareness of the need for this kind of information exchange among Caribbeanists.

At the same time, archaeologists have become increasingly interested in colonization

---

[1] Dr. Watters is Associate Curator of Anthropology, The Carnegie Museum of Natural History, Pittsburgh, PA, 15213.

of islands by human populations on a worldwide basis, as is shown by the session on "Processual Studies in Island Archaeology" held at the 1986 Society for American Archaeology annual meeting, as well as colonizations within the Caribbean (e.g. Keegan 1985). One aspect of this increased concern with human colonization is the impact of human populations on insular floras and faunas.

Purposes

Elsewhere in this symposium, Rouse presents an overview of human occupation of the entire Caribbean region and Wing discusses exploitation of animal resources by humans. In an effort to avoid duplication, I have chosen a different approach. This paper is one archaeologist's perspective on the potential usefulness of archaeological research to other disciplines interested in biogeographic questions, and it addresses both the benefits and constraints of such archaeological research. The paper is not specifically concerned with establishing the utility of research by other disciplines to archaeology.

The purposes of this paper are twofold. First, it explores some implications of archaeological research for biogeographic studies of the Lesser Antilles. Second, it presents a general chronological framework for ordering archaeological data that are relevant to biogeography. The importance of time depth for biogeographic studies is a corollary of this chronological ordering.

It is not the intent of this paper to present historical summaries or literature reviews of either archaeological or biogeographic studies of the Lesser Antilles in particular or the Caribbean in general. Persons interested in such studies are referred to Olson (1978), Allaire (1973), Rouse and Allaire (1978), or Myers (1981).

Definitions

Discussions among disciplines can open a Pandora's box of ill-defined or misinterpreted terms because the implications or connotations of any term are not necessarily equivalent. In an effort to avoid such misinterpretations, a number of terms used in the paper are defined below.

The Lesser Antilles, the geographic scope of this paper, range from Sombrero in the north to Grenada in the south (Fig. 1). This chain of islands in the eastern Caribbean includes the volcanic and limestone arcs, both of which occur on the Lesser Antilles Ridge (Uchupi 1975:26-28). As defined here, the Lesser Antilles does not include Barbados, Trinidad, or Tobago. In comparison to the Greater Antilles, the Lesser Antilles are decidedly small islands.

The term "archaeological" with regard to faunal materials, refers to animal remains found in legitimate and verifiable association ("good context") with cultural materials. Archaeological faunal remains have been deposited culturally, as a result of human activity.

Archaeological faunal remains in the Lesser Antilles may have been deposited by "prehistoric" (Amerindian) or "historic" (European, African, Asian) populations during the "pre-Columbian" or "post-Columbian" periods respectively. For now, we will ignore the complications of the "contact period" when cultures from the Old and New World met in the Caribbean.

"Paleontological" as used herein, refers to faunal remains that were "naturally" rather than culturally deposited. Such remains are not found in association with cultural materials. Paleontological and archaeological are terms primarily referring to the modes of deposition, either by non-human or human agencies.

The term "biogeographic" is used in a very loose sense in this paper to include everything from systematics to morphology to distribution studies to extinctions because, to some extent, faunal remains from archaeological sites can be pertinent to all those studies. Moreover, we do not attempt to make a distinction between paleontological and zoological (i.e., modern living species) studies because they are aspects of a continuum, rather than discrete units, insofar as the archaeologist viewing it from the outside is concerned. We do recognize that such a distinction is regularly made among our natural science colleagues, a fact that can frustrate the time depth perspective mentioned earlier.

One set of related terms that we are not about to try to define is fossil, subfossil, faunal remains, ecofacts, and archaeofaunas, all of which are terms that appear in the literature. We could probably devote an entire symposium to defining these terms and identifying their connotations and usages. For this symposium, we have chosen "faunal remains" as the generic term that is broadest in scope and least "loaded" with connotation.

Status of archaeological research

It is not my intention to dwell in any detail on the relationship of the Lesser Antilles to other areas of the Caribbean. It is necessary to note that the Lesser Antilles have received less attention from archaeologists than have the Greater Antilles islands, and study of the Lesser Antilles began more recently.

"Variable" -- this is the term that most appropriately characterizes our current archaeological knowledge of the Lesser Antilles islands. Sustained archaeological study has occurred on some islands such as Martinique or Antigua whereas others, such as Barbuda or Anguilla, have only recently been investigated, and still others (e.g. St. Barts) are largely unknown. In no case can a Lesser Antilles island be deemed "thoroughly studied" from an archaeological perspective. Also, while interest in prehistoric archaeology is long-standing, interest in historic archaeology developed in a major way only in the past decade.

Of more direct interest for biogeographic matters is the variety of archaeological research projects undertaken in the Lesser Antilles. Most islands have had some kind of general surface survey to locate prehistoric sites, although the intensity of these surveys is widely variable. Surveys have been conducted by professional archaeologists, by avocational archaeologists (usually members of a local archaeology society), or by a combination of the two.

Excavations at prehistoric sites in the Lesser Antilles have been limited for the most part. While a few sites, such as the Sugar Factory Pier site on St. Kitts or the Indian Creek site on Antigua, have had multiple test pits dug, many more sites have not been dug or have had but a few test pits excavated. Thus, the excavated area within a site may be very restricted, and the number of tested sites on an island may be a small proportion of the total known. For example the estimated areal extent of the Trant's site, Montserrat, is 42,500 m^2 based on artifacts observed on the surface (Watters 1980: Table 8). A total of 4 m^2 has been excavated at this site.

Much of the previous archaeological research was mainly concerned with artifacts, both to define categories of artifacts and to establish their chronology. This work has provided a firm grounding in basic chronology (which is still being refined), but this research has limited value for biogeographic studies. More recently there has been an increase in interest among archaeologists working in the Lesser Antilles to make better use of faunal remains recovered from archaeological sites. This interest is usually expressed in terms of cultural ecology, human adaptation, paleoecology, or some similar concept relating past human populations to their environments. For at least some Lesser Antilles archaeologists, the stage is set for increased interaction and information exchange among disciplines.

Limits of archaeological data

Faunal remains from archaeological sites are important to archaeologists and biogeographers alike. However, there are definite limits on the suitability or appropriateness of archaeological faunal remains for biogeographic issues.

Archaeological sites do not contain a complete record of animal forms that were present on an island during human occupation. Thus, the taxa that are recovered from archaeological sites already form a biased and restricted sample of an island's total fauna. The species diversity observed for archaeological faunal remains is not equivalent to the original diversity.

The absence of some taxa in archaeological sites can be attributed to several factors. First, not all taxa were deposited in archaeological sites. If humans did not make use of a particular animal, then its remains will not show up in sites (unless it was inadvertently deposited). Second, not all taxa that were deposited are necessarily preserved. Species lacking hard body parts would be preserved only in unusual circumstances. Third, cultural selectivity plays an important role in determining what taxa were or were not deposited. Thus, we should not look to the archaeological record for a complete inventory of island fauna, either native or introduced.

Amerindian populations of the Lesser Antilles relied heavily on marine organisms as sources of protein. This includes finfish, shellfish, and marine mammals. Reliance on marine species is particularly evident in sites on islands that are agriculturally marginal, such as Barbuda and Anguilla, where fish bones and shell remains are abundant. We have already documented the manatee (*Trichechus manatus*) in prehistoric context on Barbuda (Watters et al. 1984) and suspect it is present in a site recently excavated on Anguilla. Marine organisms, because they are not constrained by circumscribed island land masses, may have little direct bearing on biogeographic issues for land animals.

In general, faunal materials collected from surfaces of archaeological sites have limited utility for biogeographic studies. Materials exposed to surface elements can decompose relatively rapidly but more importantly, these remains cannot be verified as being in good archaeological context. The time of their deposition cannot be accurately determined. Moreover, most prehistoric sites have been disturbed by cultivation activities (especially hoeing) during the historic period. A rule of thumb is that the uppermost

20-30 cm in a site is usually disturbed and the context of the materials therefore is questionable.

Faunal remains from excavated sites are preferable, both because of enhanced preservation and good context. As was already noted, only a few Lesser Antilles sites have had large scale excavations. In most instances where excavations have occurred, the number of test pits is limited. This in turn limits our interpretation, both from an archaeological and biogeographic perspective.

Probably more important than site area excavated is the field technique employed by an archaeologist. In some earlier excavations, faunal remains found in sites were discarded entirely or a "sample" was saved. In some cases, the deposits were not screened, so whatever faunal remains were recovered were subjectively picked out by the archaeologist. Such collecting strategies tend to favor larger and more intact remains at the expense of smaller ones. Also, there has been a bias toward retention of bone material over shell.

Even when all deposits have been screened, there is considerable variation in the faunal remains recovered based on the mesh size of the screen. Archaeologists traditionally have used 1/4-inch mesh (occasionally 1/2-inch), which means that small faunal remains fall through the screen and are lost.

Smaller mesh (1/8-inch or rarely 1/16-inch) screens retain many more faunal remains (Thomas 1969) but such screens have been used only recently. When evaluating the faunal remains from excavations, it is important to know the field techniques employed by the archaeologist because a comparison of faunal remains from different test pits is going to be of limited value if excavation techniques differed. This applies to basic counts as well as any more sophisticated analysis.

Benefits of archaeological data

Given the limits noted above, it might seem that archaeological faunal remains have limited usefulness for biogeographic studies. Such is definitely not the case.

Archaeological sites are an excellent source of faunal remains because of the concentration of such remains in a restricted area. Regardless of whether the animal remains in sites are deposited as a result of food procurement or some other function (e.g. materials for tool manufacturing), the significant point is that they occur in concentrations that will not be duplicated by natural deposition. Of 427 bones recovered from undisturbed levels at the Trant's site test pit on Montserrat, the minimum number of individuals present was 56 (Steadman et al. 1984b). At the test pit at the Indiantown Trail site on Barbuda, 2405 bones representing 136 individuals were recovered (Watters et al. 1984).

Certain systematic studies such as comparative morphology can benefit greatly from the abundance of archaeologically derived samples. For example, the undisturbed levels of the Trant's site yielded 29 bones (MNI=8) of the tribe Oryzomyini (rice rat), which represent at least two species and possibly four. The Indiantown Trail site on Barbuda yielded 34 bones (MNI=6) of one species of Oryzomyini. These materials were recovered from 2 x 2 m test pits with the sediments screened through 1/8-inch mesh.

For studies of extinction, introduction of non-native species, or for diachronic studies in general, it is especially important that the faunal remains be recovered in good context; that is, in verifiable association with cultural materials. This permits us to establish the contemporaneity of a species with human occupation of that site. In stratified sites showing evidence for lengthy human occupation, context is especially critical because of possible variation (presence or absence; frequency of occurrence) between older and younger deposits. Steadman's (1988) work on birds in Polynesia clearly shows that extinct forms are concentrated in the lower (=older) strata of the archaeological sites.

There are other benefits to recovering faunal remains in good context in archaeological sites. Of major importance is the possibility of securing absolute dates from carbon or other materials suitable for radiometric dating that are found in the same stratum. This allows us to establish the existence of a species at a particular place and time.

In the absence of suitable material for radiometric dating, we still have the possibility of relative dates of faunal materials, derived from the fact that artifact assemblages change through time. This allows us to say that a particular species is found in association with a specific artifact assemblage, which is older (or younger) than another assemblage that does not yield that species.

There is one caveat regarding faunal remains from archaeological sites. So far, we have assumed that all faunal remains within a site were deposited culturally. This may be an acceptable working assumption for an archaeologist, but it is necessary to point out that some faunal remains may be introduced into a site by natural rather than cultural processes.

For example, scavenging animals may be attracted to a site (either occupied or abandoned) because of the refuse concentrated there. Their remains could be incorporated into the sediments on an "accidental" basis that had nothing to do with human actions. Land snails in particular are notorious for accidental intrusion. However, the same situation may apply to other species found in sites. It can be quite difficult to distinguish accidental from cultural deposition or natural from cultural bone (Thomas 1971).

Chronological framework

The chronological framework discussed below is presented in an effort to order data in a time-based sequence. The data on faunal remains may be derived from either paleontological or archaeological research.

In terms of the biogeography of the West Indies, it seems logical to distinguish between two basic periods, the first being prior to human colonization and the second after human colonization. By making the appearance of humans the breakpoint in this framework, we separate those biogeographic issues in which humans played no part from those in which humans may have played a role (Fig. 2).

Prior to human colonization, a period that stretches back to when the Lesser Antilles first emerged from the ocean as landmasses, biogeographic data are derived only from paleontological sources. Archaeologists cannot contribute directly to biogeographic issues in this pre-human period. Thus, certain concepts of great interest to biogeographers in the Caribbean, such as the vicariance hypothesis, are outside of the domain of archaeology.

In the period after initial human colonization of the West Indies, archaeological data become relevant. However, data from paleontological sources also are pertinent in the period after human colonization because faunal remains continue to be deposited naturally. In brief, natural deposition occurred in both periods but cultural deposition occurred only in the latter period.

The period after human colonization may be viewed as a single period dating from that initial colonization to the present, or as two subperiods (prehistoric=Amerindian or pre-Columbian; historic=European, African, Asian, or post-Columbian).

The two subperiods allow us to distinguish between different human populations on a gross level. More importantly, this distinction relates to the different kinds (not just degrees) of impacts these populations had on the flora and fauna of the Lesser Antilles.

Finally, when warranted, these subperiods may be even further divided on the basis of different artifact assemblages or radiometric dates. For example, the prehistoric subperiod might be divided into Archaic and Ceramic components.

At this time, however, I prefer to discuss the ordering of data at the periods and subperiods levels. It seems to me that it is important to make distinctions between faunal remains found in the pre-human and human periods, as well as between the prehistoric and historic subperiods. Two examples will illustrate the rationale for ordering such data in a time-based sequence.

Geochelone carbonaria is the tortoise native to northern South America that is found on a number of Lesser Antilles islands (Williams 1960). The presumption has been that this species was introduced into the Lesser Antilles by Amerindian groups migrating northward from South America. However, to the best of my knowledge, this species has never been documented in a prehistoric context from an archaeological site. Thus, although the assumption of introduction by Amerindians might be a logical one, we do not have archaeological verification of such an introduction. I would suggest that an equally plausible explanation was that *Geochelone carbonaria* was introduced into the Lesser Antilles by early European explorers in the historic subperiod. In the absence of verified archaeological association, either prehistoric or historic, any attribution of which human population introduced this species into the Lesser Antilles remains questionable.

The extinct lizard *Leiocephalus cuneus* provides a good example of the implications of ordering faunal remains in a time-based sequence. Etheridge (1964) identified this species from cave deposits near Two Foot Bay on Barbuda. The materials were found in paleontological deposits presumed to be of latest Pleistocene age. Whether those paleontological deposits were laid down during the pre-human or human period is uncertain.

Subsequent archaeological research on Barbuda, at the Indiantown Trail site, recovered *L. cuneus* bones in association with Amerindian cultural materials, and thus verified the occurrence of the curly-tailed lizard in prehistoric context (Watters et al. 1984). The ceramic assemblage is post-Saladoid (=Elenoid in Rouse and Allaire's 1978 terminology) and dates to the second half of the first millennium A.D., which indicates this now extinct lizard survived on Barbuda until at least 500 A.D. and probably more recently. Later research (Steadman et al. 1984a) on nearby Antigua found *L. cuneus* bones in a limestone fissure at Burma Quarry. Artifacts, presumably Archaic, also were found in this quarry but their context, in relationship to the sediments and the *L. cuneus* bones, needs further explication. (Note in proof: see Pergill et al. 1988)

For *L. cuneus*, we have a paleontological record (whether of the pre-human or human period is uncertain) and a prehistoric archaeological record (Ceramic) from Barbuda as well as a possible prehistoric record (Archaic) from Antigua. Of particular interest to me would be the question of whether *L. cuneus* survived on Barbuda into the historic period (settlement began in the 1600s). The curly-tailed lizard has not been found in historic context, but very little historic archaeology has been done on Barbuda.

This raises what to me is a very important issue that largely has been neglected in Lesser Antilles archaeology and biogeography. I refer to the impact of occupation of the islands by European, African, and, later, Asian populations. Occupation of these islands during the historic subperiod led to massive habitat modification through deforestation and plantation (especially sugarcane) agricultural practices. It also brought about the introduction of numerous alien floral and faunal species that competed with native species, often to the detriment of the latter. This is not meant to imply that habitat modification and introduction of new species was unknown in the prehistoric subperiod. Such processes certainly occurred but seemingly not to the extent they did in the post-Columbian years.

I would submit that the loss of many indigenous faunas in the Lesser Antilles may have occurred in the historic subperiod rather than during Amerindian occupation. As more historic archaeological excavations take place in the Lesser Antilles, and as analysis of faunal remains becomes an integrated part of such work, I suspect some species thought to have become extinct because of Amerindian exploitation will in fact be found to have persisted into the historic subperiod. Of particular importance in this regard are the "contact" sites displaying evidence of interaction between Amerindian and European peoples, as well as the earliest European settlements in the Lesser Antilles.

From the archaeologist's perspective, the most dramatic of the extinctions in the historic subperiod was the almost complete annihilation of the Amerindian peoples of the Lesser Antilles.

The context issue

It is evident that any attempt to establish accurately the contemporaneity of specific taxa with human occupations in the Lesser Antilles and, by extension, to order such data chronologically rests at the base level on legitimate and verifiable contexts for the archaeological faunal remains. Questionable contexts and dubious associations of faunal remains and artifacts tend to muddle, not clarify, the situation.

A classic example of the context issue in the Lesser Antilles involves the extinct giant rodent, *Amblyrhiza inundata*, first mentioned by E.D. Cope in 1868, described in 1869, and further discussed in 1883. The circumstances of the discovery of these specimens as described by Cope (1869:183-184; 1883:1, 9-14) are pertinent. Cope first observed the bones in a shipment of "...cave earth, limestone fragments, and bone breccia..." (1883:1) that arrived at a firm in Philadelphia for the purpose of ascertaining its value as a fertilizer. The material had been obtained from a cave on the island of Anguilla. Cope then got in contact with H. E. van Rijgersma, a Dutch doctor on the nearly island of St. Maarten (Holthuis 1959), who went to Anguilla, examined the debris from the mined cave, and sent additional materials he found to Cope.

Cope did not visit Anguilla to observe the cave site. Moreover, it is clear that van Rijgersma sent additional bones of *A. inundata* but he did not provide any important information on the context of the cave. Cope writes "unfortunately no notes were taken as to the relations of the parts of the cave deposit, or whether any stratification was observed" (1883:1). The occurrence of *A. inundata* would be of no direct concern for archaeologists had not an artifact of undoubted human manufacture also been found in the deposits.

The artifact, which Cope (1869:187-188; 1883:26) terms a chisel, was provided by van Rijgersma, so it could not have been in the original shipment to Philadelphia but instead was with a later shipment of materials collected by the Dutch doctor. Cope indicates the artifact had been found with bones and teeth that "...occurred loose in a red earth in cavities of the breccia" (1883:26). All of these items had similar reddish coloration. The artifact in question is illustrated (Cope 1883:Plate I, figs. 12 and 12a) and clearly is an example of the shell celt or adze that is widespread in the West Indies. It was manufactured from the outer whorl (flaring lip) of the marine gastropod *Strombus gigas*.

Cope (1869:188) indicates the evidence is not conclusive but suggests there is a very strong inference that *A. inundata* and humans were contemporaries. In the subsequent article (Cope 1883:26-27), he displays greater reserve and indicates the contemporaneity issue must be left for future investigators. From the archaeologist's perspective, there is no legitimate and verifiable contextual relationship between the *A. inundata* finds and the shell "chisel". The lack of good context precludes any association between these materials which in turn means they cannot be used justifiably to establish contemporaneity between the extinct rodent and prehistoric human populations. This does not mean we are denying the possibility that they are contemporaneous, but instead means these particular specimens cannot be used to resolve the issue. In fact, given the large size of *Amblyrhiza inundata*, it might have been a prime quarry for Amerindian peoples on Anguilla if it existed during human occupation.

Conclusions

Archaeological faunal remains from Lesser Antilles sites can yield important data about the contemporaneity of specific taxa with human populations. Remains found in good context within prehistoric or historic sites can be used to address numerous biogeographic issues ranging from extinctions and introduction to morphology and systematics in general. Archaeological faunal remains, when they can be dated radiometrically, provide confirmation of the existence of a particular taxon at one place and time.

Archaeological sites can be viewed as providing another data set relevant to biogeographic issues. The archaeological data are most useful in combination with data derived from paleontological and modern zoological studies. There are definite limitations on the usefulness of archaeological faunal remains for biogeographic studies. These limits reflect both cultural selectivity in terms of species deposited in sites and variation in preservation of the deposited materials. Nonetheless, biogeographers, especially those interested in diachronic studies, would do well to familiarize themselves with the range of archaeological research undertaken in the Lesser Antilles. This is especially important in distinguishing biogeographic issues that archaeology can address from those it cannot. A recent study of extinction and zoogeography of West Indian land mammals by Morgan and Woods (1986) demonstrates the importance of integrating data derived from archaeological sites with other data sets.

Archaeological faunal remains obviously are useful only in the period after human colonization of the Lesser Antilles. However, in the framework for chronological ordering of data presented in this paper, archaeological faunal remains are a crucial resource for many biogeographic questions. Yet, to be useful, those archaeological faunal remains must be found in verifiable association with cultural materials. Without the certainty of good context, the usefulness of archaeological faunal remains for biogeographic studies is severely restricted.

Literature cited

Allaire, L. 1973. Vers une Préhistoire des Petites Antilles. Centre de Recherches Caraïbes, Martinique, 53pp.

Cope, E.D. 1868. [discussion of exhibited Anguilla rodent remains]. Proceedings of the Academy of Natural Sciences of Philadelphia 20:313.

_____. 1869. Synopsis of the Extinct Mammalia of the Cave formations in the United States, with observations on some Myriapoda found in and near the same, and on some Extinct Mammals of the Caves of Anguilla, W.I., and of other localities. Proceedings of the American Philosophical Society 11:171-192.

_____. 1883. On the Contents of a Bone Cave in the Island of Anguilla (West Indies). Smithsonian Contributions to Knowledge 25(2):1-30.

Etheridge, R. 1964. Late Pleistocene lizards from Barbuda, British West Indies. Bulletin of the Florida State Museum, Biological Sciences 9:43-75.

Holthuis, L.B. 1959. H. E. van Rijgersma--a little-known naturalist of St. Martin (Netherlands Antilles). Studies on the Fauna of Curaçao and other Caribbean Islands 9(39):69-78.

Keegan, W.F. 1985. Dynamic Horticulturalists: Population Expansion in the Prehistoric Bahamas. Unpublished Ph.D. dissertation, University of California, Los Angeles, 358 pp. (University Microfilms International, Ann Arbor, #8519111).

Morgan, G.S. and C.A. Woods. 1986. Extinction and the zoogeography of West Indian land mammals. Biological Journal of the Linnean Society 28:167-203.

Myers, R.A. 1981. Amerindians of the Lesser Antilles: A Bibliography. HRAFlex Books, New Haven, CT, 158 pp.

Olson, S.L. 1978. A Paleontological Perspective of West Indian Birds and Mammals. Pp 99-117 in F.B. Gill (ed.). Zoogeography in the Caribbean, Special Publication 13, Academy of Natural Sciences, Philadelphia, 128 pp.

_____, and H.F. James. 1982. Fossil Birds from the Hawaiian Islands: Evidence for Wholesale Extinction by Man Before Western Contact. Science 217:633-635.

Pregill, G.K., D.W, Steadman, S.L. Olson, and F.V. Grady. 1988. Late Holocene Fossil Vertebrates from Burma Quarry, Antigua, Lesser Antilles. Smithsonian Contributions to Zoology 463, 27pp.

Rouse, I. and L. Allaire. 1978. Caribbean. Pp 431-481 in R.E. Taylor and C.W. Meighan (eds.). Chronologies in New World Archaeology, Academic Press, New York, 587 pp.

Steadman, D.W. 1988. Fossil Birds and Biogeography in Polynesia. Pp. 1526-1534 in H. Ouellet (ed.). Acta XIX Congressus Internationalis Ornithologici, National Museum of Natural Sciences, Ottawa, 2815pp.

_____, and S.L. Olson 1982. Bird remains from an Archaeological site on Henderson Island, South Pacific: Man-caused extinctions on an "uninhabited" island. Proceedings of the National Academy of Sciences 82:6191-6195.

_____, G.K. Pregill and S.L. Olson. 1984a. Fossil vertebrates from Antigua, Lesser Antilles: Evidence for late Holocene human-caused extinctions in the West Indies. Proceedings of the National Academy of Sciences 81:4448-4451.

_____, D.R. Watters, E.J. Reitz, and G.K. Pregill. 1984b. Vertebrates from Archaeological Sites on Montserrat, West Indies. Annals of Carnegie Museum 53(1):1-29.

Thomas, D.H. 1969. Great Basin Hunting Patterns: A Quantitative Method for Treating Faunal Remains. American Antiquity 34(4):392-401.

_____. 1971. On Distinguishing Natural from Cultural Bone in Archaeological Sites. American Antiquity 36(3):366-371.

Uchupi, E. 1975. Physiography of the Gulf of Mexico and Caribbean Sea. Pp 1-64 in A.E.M. Nairn and F. G. Stehli (eds.). The Ocean Basins and Margins, Volume 3, The Gulf of Mexico and the Caribbean, Plenum Press, New York.

Watters, D.R. 1980. Transect Surveying and Prehistoric Site Locations on Barbuda and Montserrat, Leeward Islands, West Indies. Unpublished Ph.D. dissertation, University of Pittsburgh, 416 pp. (University Microfilms International, Ann Arbor, #8112643).

_____. 1982. Relating Oceanography to Antillean Archaeology: Implications from Oceania. Journal of New World Archaeology 5(2):3-12.

_____, E.J. Reitz, D.W. Steadman, and G.K. Pregill. 1984. Vertebrates from Archaeological Sites on Barbuda, West Indies. Annals of Carnegie Museum 53(13):383-412.

Williams, E.E. 1960. Two species of tortoise in northern South America. Brevoria 120:1-13.

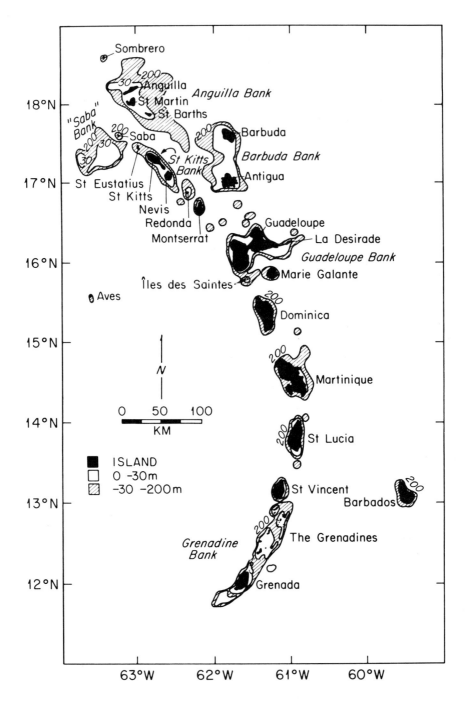

1. The current configuration of the Lesser Antilles islands is significantly different from that of the Holocene, when sea level was lower and islands expanded in size. This has important implications for biogeography and may have for archaeology as the initial date of Amerindian colonization is pushed further back in time.

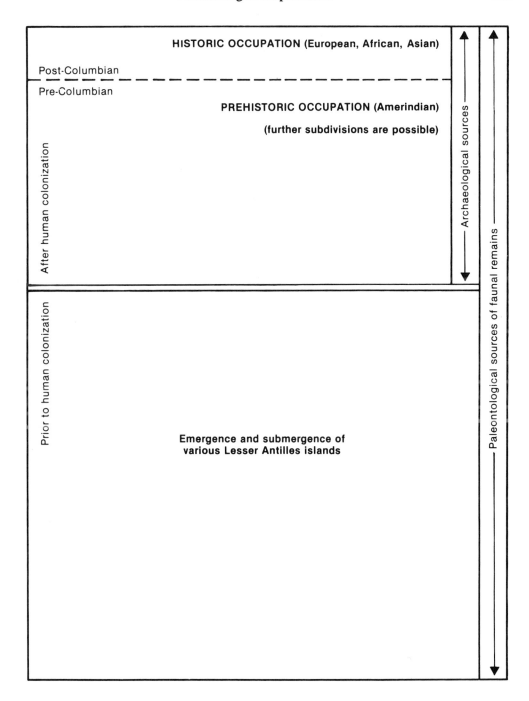

2. A major distinction is made between the periods prior to human colonization and after human colonization. Archaeological research is only relevant to biogeographic studies in the period after human colonization (including the prehistoric and historic subperiods).

BIOGEOGRAPHY AND EVOLUTION OF THE JUNIPERS OF THE WEST INDIES

Robert P. Adams[1]

Abstract

The volatile leaf oils of the junipers of the West Indies were examined and comparisons made using principal coordinate analysis and minimum spanning networks. The taxa studied included: *J. barbadensis, J. ekmanii, J. gracilior, J. lucayana,* and *J. saxicola* of the Caribbean as well as *J. bermudiana* from Bermuda and *J. virginiana* var. *silicicola* and var. *virginiana* of the southeastern United States. This is the first report on the composition of the volatile leaf oil of *Juniperus barbadensis* since its recent rediscovery on St. Lucia, BWI. The oil is dominated by limonene (34%) and sabinene (31%) with moderate amounts of α-pinene, 4-terpineol, myrcene, r-terpinene, α-terpinene and terpinolene. Several components normally found only in the heartwood were found in the leaves of this species. *Juniperus barbadensis* has been extinct on Barbados (the type locality) since before 1700 and is now known from only one small population on Petit Piton in St. Lucia, BWI. All of the junipers of the West Indies are in section *Sabina*, series entire, and as such are in the North American group related to *J. virginiana*. Examination of the minimum spanning network, based on 70 leaf terpenoids, revealed that the West Indian junipers appear to have arisen from the ancient Appalachian region (*J. virginiana* or its ancestor) and not from the junipers of southern Mexico and Guatemala, which are in series denticulate of section *Sabina*. The biogeography of the West Indian junipers supports floristic affinities with the eastern United States rather than affinities with Central America.

Historical treatments of *Juniperus* of the West Indies

Linnaeus (1753) described only three junipers from the New World (*J. virginiana* L., "Virginia and Carolina"; *J. barbadensis*, "America"; and *J. bermudiana*, "America"). However, Hemsley (1883) equated *J. barbadensis* with *J. bermudiana*, adopting *J. bermudiana* as the name for all the junipers of the Caribbean. Sargent (1902) recognized *J. barbadensis* and said it occurred along the Atlantic coast of Georgia and Florida as well as "on the Bahamas, San Domingo (Dominican Republic), mountains of Jamaica and on Antigua." Britton (1908) recognized *J. lucayana* in the Bahamas and reserved *J. barbadensis* for the plants of southern Georgia, Florida and the rest of the Caribbean. Pilger (1913) equated *J. bermudiana* and *J. barbadensis*, but used *J. barbadensis* for the name of

---

[1] Dr. Adams is Director of the Center for the Study of Famine and Agricultural Alternatives, Box 7372, Baylor University, Waco, TX. 76798.

the common juniper of the Caribbean on the grounds that it was listed first by Linnaeus (1753). Florin (1933) reviewed the junipers of the Caribbean and recognized 5 species: *J. saxicola* Britton and Wilson from Cuba; *J. lucayana* Britton from Cuba, Haiti, Jamaica and the Bahamas; *J. gracilior* Pilger from Haiti and Dominican Republic; *J. ekmanii* Florin from Haiti; and *J. urbaniana* Pilger and Ekman from Haiti. Carabia (1941) recognized *J. barbadensis*, throughout the Caribbean, *J. bermudiana* on Bermuda and *J. virginiana* in the United States. Gillis (1974) treated the Bahamian junipers as *J. bermudiana*. Correll and Correll (1982) recognized the juniper of the Bahamas as *J. barbadensis*. Just from a review of the nomenclatural literature alone, one can sense that the taxa are not very distinct, morphologically.

The recent rediscovery of a small population of *J. barbadensis* at the summit of Petit Piton on St. Lucia, BWI (Adams et al. 1987a) afforded an opportunity to collect fresh foliage and analyze the volatile leaf oils. Although the original populations of *J. barbadensis* on Barbados were apparently cut out before 1700 (Adams et al. 1987a), the rediscovery of the small, relictual population on St. Lucia, only a 150 kilometers from the type locality (Barbados), was very fortuitous. No other natural populations of juniper are known from the lesser Antilles, with the closest natural populations of juniper being *J. ekmanii* in Haiti, *J. gracilior* in the Dominican Republic and *J. lucayana* in Jamaica.

Morphologically all the Caribbean juniper species except *J. bermudiana* and *J. saxicola* are very similar and difficult to distinguish. The variable nature of leaves, even on a single branch, has resulted in confusion in the taxonomy of the junipers of the Caribbean. In fact, it is unlikely that the systematic relationships could ever be determined based solely on morphology. However, the leaf volatile oils have evolved into quite different patterns and the taxa are easy to separate using these chemical data (Adams 1983a; Adams and Hogge 1983; Adams et al. 1987b). In previous studies of the Caribbean junipers, the volatile oil compositions of *Juniperus bermudiana* L., *J. ekmanii* Florin, *J. gracilior* Pilger, *J. lucayana* Britton, *J. saxicola* Britt. & Wils., *J. silicicola* (Small) Bailey and *J. virginiana* L. have been reported and the systematic relationships examined among the taxa (Adams 1983; Adams and Hogge 1983; Adams et al. 1987b). Examination (Adams 1986) of both the volatile leaf oils and morphology of natural populations of *J. virginiana* and *J. silicicola* indicated that these taxa are conspecific and the juniper of the coastal foredunes of the southeastern United States (*J. silicicola*) was therefore treated as a variety of *J. virginiana* L. (i.e., *J. virginiana* var. *silicicola* (Small) E. Murray).

The purposes of this paper are to present the first report on the composition of the volatile leaf oil of *J. barbadensis*, compare its oil with the oils of the other junipers of the West Indies, Bermuda and the southeastern United States to determine if *J. barbadensis* on St. Lucia is conspecific with *J. lucayana* from the Bahamas and Jamaica, and discuss the biogeography and evolution of the junipers of the West Indies, Bermuda and southeastern United States.

Juniper populations examined and status

Figure 1 shows the populations sampled over the past several years. Samples of *Juniperus barbadensis* (BA, Fig. 1) were collected (**Adams 5367-5371**; Lat. 14° 10" N, Long. 61° 03" W, elev. 730 m, Petit Piton, St. Lucia, BWI). The population consists of approximately 25 trees, all within 30 m of the summit of Petit Piton. The species is bearing seed and two young trees (ca. 3-6 yrs. old) were found near the top on the west side. No other populations of *J. barbadensis* are known and it must be considered threatened on St. Lucia due to having only one very small population. Nearby islands such as Martinique or Dominica should be reexamined for relictual populations. A visit to Barbados revealed that the habitat has been eliminated (converted to sugar cane fields) over 280 years ago.

Juniperus bermudiana L. (BM, Fig. 1) is endemic to Bermuda. Sometime prior to 1942, two scale insects, *Lepidosaphes newsteadi* and *Carulaspis minima*, were introduced into Bermuda (Bennett and Hughes 1959) from the United States mainland (Groves 1955). These insects infested *J. bermudiana* causing defoliation and death. By 1954, most of the trees of *J. bermudiana* were infected. Biological control was tried with no success. Groves (1955) estimated that 90 percent of the trees were dead by 1955. In 1978, Dr. W. E. Sterner, Director of the Bermuda Biological Station (pers. comm.) estimated that perhaps 99 percent of the original trees had died. The younger trees appear to have some resistance (or perhaps just youthful vigor). The younger trees are now of reproductive age, so perhaps some resistance has developed. The introduction of the resistant *J. virginiana* from the eastern United States into Bermuda should be avoided because the unique germplasm on Bermuda (see below) might be diluted if hybridization occurred with *J. virginiana*.

Juniperus ekmanii Florin (EK, Fig. 1) is endemic to the Morne de la Selle in Haiti. In 1981, only two trees of *J. ekmanii* were found in the Mare Rouge area (**Adams 3106, 3107**; Mare Rouge, Morne de la Selle, 17 km NE of Seguin, Haiti; Lat. 18° 20"N, Long. 72° 04" W, elev. 1,770 m). Most of this region was logged about 1965. Old stumps of *J. ekmanii* with DBH of up to 2 m are still in the area. Young trees of the species are reported to occur in the Morne la Visite area (Paul Paryski pers. comm.) but their status is threatened by local cutting. *Juniperus ekmanii* must be considered endangered and is most likely headed for extinction before the end of this century.

Juniperus gracilior Pilger (GR, Fig. 1) is endemic to the Dominican Republic and probably Haiti. Collections were made west of Constanza in 1980 (**Adams 2785-2794**; 14.6 km W of Constanza, Dom. Rep., Lat. 18° 55"N, Long. 70° 50" W, elev. 1,400 m). The species appears to occupy a considerable area west and south of Constanza, Dominican Republic. However the area is heavily utilized and many trees were being cut west of Constanza in 1980. It appears to be the least endangered juniper of Dominican Republic/Haiti at present.

Juniperus lucayana Britton was collected in the Bahama Islands, Cuba and Jamaica. In the Bahamas, *J. lucayana* (LG, Fig. 1) was collected on Great Abaco Island (**Adams 2686 -2695**) where two small populations were found: 5 to 10 trees (55 km south of Marsh Harbour, Lat. 26° 9" N, Long. 77° 12" W, elev. ca. 9 m) on coppice in thick underbrush; and a second site with 15 to 20 trees (5 km N of Hole-in-the-Wall, Lat. 25° 55" N, Long. 77° 14" W, elev. ca. 30 m) on the coppice. On Andros Island (**Adams 2696-2705**), 15 to 20 trees of *J. lucayana* (LA, Fig. 1) were found scattered in the coppice with pines (5 km S of Andros Town/ Fresh Creek at the junction of the main road and airport road, Lat. 24° 39" N, Long. 77° 48" W, elev. ca. 30 m). On Grand Bahama Island (**Adams 2706-2715**), 20 to 30 trees of *J. lucayana* (LB, Fig. 1) were found on coppice (23 km W of Freeport and thence 3 km N, Lat. 26° 38" N, Long. 78° 55" W, elev. ca. 3 m). These trees have the crowns flattened (from wind damage?) and a considerable number of dead branchlet tips. Many dead and partially dead trees were in the area.

With the current travel restrictions in Cuba, it is difficult to assess the status of juniper populations. On a visit in 1985, the author talked with Dr. Antonio Lopez Almirall, who felt that *Juniperus lucayana* has now been largely cut out except in the Sierra de Nipe region, Holquin Province (this region was previously a part of the Oriente Province). All the recent specimens of *J. lucayana* from the mainland of Cuba were examined at the Jardine Botanico Nacional de Cuba (HAJB!). These were found to be from the mountainous region in eastern Cuba (**J. Bisse 15775**, April 1980, 700-1,000 m, Prov. Holquin, Lat. 20° 29" N, Long. 79° 48" W, HAJB!; **J. Bisse, Mayer, Bassler, Alvarez & Gutierrez 35818**, Oct. 1972, Loma de Mensura, Prov. Holquin, HAJB! and **J. Bisse** and **Lippold 10222**, Oct. 1968, 500-600 m, Pinar de Monte Cristi Sobre, Prov. Guantanamo, Lat. 20° 20" N, Long. Guantanamo, Lat. 20° 20" N, Long. 75° 10" W, HAJB!). Fortunately, samples (**Adams 5259-5280**) could be obtained two trees of *J. lucayana* are still growing at the Havana Botanical Garden. These trees came from seed collected by J. T. Roig sometime before 1960 from the Cuban mainland (possibly Sierra de Nipe?).

Juniperus lucayana is also found in the swamps in the south-central portion of the Isle of Pines (= Isla de Pinos, = Isla de la Juventud). Dr. Lopez said these populations appear to be stable and not threatened by wood harvesting, etc. A number of trees (15 to 20) of *J. lucayana* have been raised from seed collected from the Isle of Pines and these have been growing at the National Botanical Garden (Jardine Botanico Nacional de Cuba) near Habana since 1977. The female trees are producing female cones that are blue-black, and noticeably flattened. Samples from 2 trees were collected (**Adams 5281-5282**).

Juniperus lucayana (LJ, Fig. 1) was also collected from Jamaica (**Adams 2875-2884**), 15 to 20 trees were seen (ridge at the Forestry nursery at Clydesdale and along the road to Hardware Gap, Blue Mtns., Lat. 18° 14" N, Long. 76° 41" W, elev. 1,100 to 1,200 m, St. Andrews Parrish, Jamaica). These trees were very scattered in the forest and their occurrence at high elevations differs greatly from the situation in the Bahamas. Their habitat would seem to be more similar to that of the Sierra de Nipe in eastern Cuba (see above). The plants appeared to be reproducing themselves but the selection of juniper wood to produce tourist souvenirs could easily eliminate *J. lucayana* from Jamaica.

Juniperus saxicola Britt. & Wils. (SX, Fig. 1) is endemic to Cuba where it is only known from the Pico Turquino region at 1,200 to 1,700 m. The species has apparently been fixed by neoteny in the juvenile-leafed growth stage, because even the adult, reproductive individuals have only juvenile leaves (an awn-shaped blade which diverges from the stem at 45 to 60 degrees and a sheath portion that clasps the stem). Normally, in section *Sabina*, only young (up to 4 or 5 years old) junipers have the awn-shaped leaves. The scale (adult) leaves are then produced throughout the life span, except juvenile (awn like) leaves are produced at the tips of branches during a rapid growth period. During the past 20 years the author has examined thousands of junipers in the field and 3 or 4 plants have been found that have juvenile foliage on otherwise mature trees. All the specimens examined of *J. saxicola* have awn-like leaves and no other juniper species have been found from the Pico Turquino region. Thus, it appears that this small region has a reproductive population in which the juvenile leaf gene(s) have become fixed. Pico Turquino is quite isolated from major urban areas and is also somewhat protected by the government as a memorial because this is the region that Fidel Castro used as a sanctuary in the revolution. However, branches and berries (female cones) are currently being harvested for medicinal use so the endemic population should be considered as possibly threatened. Due to the short amount of time for field work only two trees of *J. saxicola* were sampled (**Adams 5284,5285**; west slope of Pico Turquino, Lat. 20° 1" N, Long. 76° 51" W, ca. 1,200 m & ca. 1,550 m, Prov. Granma/ Santiago de Cuba boundary, Cuba). No estimate of the population size could be made due to the dense forest growth.

Juniperus virginiana and var. *silicicola* were sampled near Washington, D.C. (**Adams 2409-2423**; 16 km E of Dulles Airport on I495, Fairfax Co., VA) and Oak Hill, Florida (**Adams 2775-2784**; 1.6 km E of the jct. of US highway 1 and E. Halifax St, Oak Hill, Volusia Co., FL). *Juniperus virginiana* L., the eastern red cedar, is rapidly increasing the density of plants within its range in the eastern United States by the invasion of old fields. *Juniperus virginiana* var. *silicicola* (Small) E. Murray is confined to coastal fore-dunes and sandy areas near the coast in the southeastern United States. It appears to be stable in population sizes but the populations often consist of only a few hundred trees (or less).

Two additional populations of *Juniperus* were sampled. Examination of Ekman's *J. lucayana* specimens at the Agricultural College, Damien, Haiti (**Ekman 3258,3643**) revealed that those specimens are similar to *J. lucayana* from the Bahamas. Field trips to St. Michael de l'Attalaye and Bassin Blue have been unsuccessful in locating any naturally occurring, extant trees. The area has been thoroughly cut-over since 1965. Two collections were made near St. Michael. Both are cultivated plants transplanted from the surrounding region about 1965. The first site (NH) is at an abandoned monastery and the second site was in a church courtyard at Ennery. The plants at the church at Ennery were reported to have been obtained about 4 km east of Ennery. Since these plants appeared to have affinities to *J. gracilior* (Adams 1983), this population (NH) is included in this analysis (**Adams 2676-2685**; 12 km N of Ennery on route 100, at the summit, Lat. 19° 33" N, Long. 72° 28" W, elev. 950 m, Dept. de L'Artibonite, Haiti). As far as is known these junipers exist in only two cultivated locations at present.

A second population of junipers, referred to me by Dr. Tom Zanoni of the Jardine Botanico Nacional of Santo Domingo, is in the Isla de Pelempito region of the Dominican Republic. Samples were collected (**Adams 3097-3106**; 4 km N of northernmost Alcoa bauxite mine, Lat. 18° 10" N, Long. 71° 36" W, elev. 1,000 to 1,200 m, Pedernales Prov., Dominican Republic). Only 9 live trees and 12 dead trees were found. The area is heavily burned by man-made fires which presents considerable danger to this population. The junipers are found on coppice with eroded limestone at the edge of the pine forests. This population, with affinities to *J. gracilior* and *J. lucayana* (Adams 1983), is referred to in this analysis as PL.

Unfortunately, several unsuccessful attempts have been made to collect *J. urbaniana* Pilger & Ekman, a spreading shrub from near the summit of Pic la Selle, Haiti. The population is apparently restricted to a very small area but still viable in 1984 (Walter Judd pers. comm.). Due to the close similarity in the morphology of *J. ekmanii* and *J. urbaniana*, it would seem likely that they are sibling species. Surprisingly, the very small population of *J. urbaniana* may be the least endangered of any of the junipers in Haiti or Dominican Republic. The site is very inaccessible and above sources of water. The plants are not trees and therefore of no importance for timber or fence posts.

Herbarium vouchers for all of the aforementioned collections are deposited at BAYLU!

Extraction of volatile leaf oils

The volatile leaf oils were extracted by steam distillation of approximately 200 g of foliage for 2 h (Adams 1975a). The oils were concentrated with nitrogen, tightly sealed in glass vials with foil-lined caps and stored at -20° C until analyzed.

Identification of oil components

Mass spectra were recorded with a Finnigan Ion Trap (ITD) mass spectrometer model 700 directly coupled to a Varian 6500 gas chromatograph, using J & W DB5, 0.25 mm x 60 m, 0.25 micron coating thickness, and J & W DB1, 0.25 mm x 30 m, 0.25 micron coating thickness, fused quartz capillary columns. The GC/ITD was operated under the following conditions: Injector temperature: 180°C; temperature programmed for DB1: 60 - 96°C @ 2°/min; 96 - 156°C @ 3°/min; 156 - 230 C @ 6°/min; for DB5: 60 - 240 C @ 3°/min. Carrier gas using DB5 column: He @ 22.9 cm/sec (60°C), 19.1 cm/sec (220°C); using DB1 column: He @ 53.6 cm/sec (60°C), 44.7 cm/sec (220°C), 0.1 ul (20% soln), split 1:50. Tuning values for the ITD were 40, 48, 85, 95 using cedrol as a tuning standard. Cedrol is well-suited for tuning the ITD because its mass spectrum is very sensitive to changes in tuning values. N-octane, n-eicosane and hexadecyl acetate were added as internal standards.

Quantification was made by FID using a DB1 column (see above) in a Varian 6500 gas chromatograph with He as the carrier gas with an average linear velocity of 30 cm/sec (60°C); 25 cm/sec (220°C), 0.1 ul (20% soln.), split 1:30, temperatures as above, except the FID @ 240°C. Identifications were made by comparisons of the mass spectrum of each component in the oils with those of the known terpenes and by searches of spectra from the Finnigan library based on the National Bureau of Standards (NBS) data. Relative retention times (RRT hexadecyl acetate = 1.00) were also compared with the RRT of known terpenoids run under the same conditions. Peak areas were quantitated using a Columbia Scientific Industries Supergrator-2 electronic digital integrator.

Numerical analyses

The chemical data were coded and analyzed by one-way analysis of variance (ANOVA) with 15 treatments (14 df). Compounds that were never larger than a trace (0.5% of the total oil) in the average of any taxon and those with F ratios less than 1.0 were eliminated from use in computing similarity measures. Thus, the initial set of 150 chemical characters was pared to 70 compounds that had F ratios greater than 1.0 and occurred with an average amount greater than 0.5 percent of the total oil in at least one taxon. These 70 compounds (denoted by an asterisk in Table 1) were then used for canonical variate analysis (CVA) and to compute similarity measures among the 15 taxa. The similarity measure used was the Manhattan metric, scaled by the range (=Gower metric, Gower 1971) and weighted by F-1 (from ANOVA) as formulated by Adams (1975b; 1982). Principal coordinate analysis followed the formulation of Gower (1966). Canonical variate analysis followed the formulation of Blackrith and Reyment (1971), Cooley and Lohnes (1971) and Pimentel (1979).

Volatile oil composition of *Juniperus barbadensis*

The compositions of the steam volatile leaf oils from *J. barbadensis* and related species are shown in Table 1. The oil of *J. barbadensis* is pale yellow and yields were 0.6 percent dry wt. (2 hr.) and 2.0 percent (24 hr). *Juniperus barbadensis* leaf oil is dominated by limonene (34%) and sabinene (31%), with moderate amounts of α-pinene, myrcene, α-terpinene, r-terpinene, terpinolene and 4-terpineol. Particularly surprising is the presence of several sesquiterpenoids that have previously been found almost exclusively in the juniper wood oils (cuparene, widdrol, cedrol). The leaf oil of *J. barbadensis* does share a profile somewhat similar to that of *J. lucayana* (LB, LJ, Table 1) in that these taxa contain large amounts of limonene with generally small amounts of sesquiterpenoids. One new unknown (#11), greater than a trace, was found in the oil. Unknown 11, RRT = 1.099, (m/z [%]) MW 272, 131(100), 187(73), 243(40), 43(30), 145(19), 253(13), 188(13), 91(10), a diterpene. All of the other unknown compounds in Table 1 have been previously discussed (Adams and Hogge 1983, Adams et al. 1987b).

Patterns among the junipers of the West Indies

In order to assess the relationships among the junipers of the West Indies, weighted Gower metric similarities were computed among the 15 OTUs using F-1 weighing (F ratio from ANOVA). The resulting similarity matrix was then factored using principal coordinate analysis (PCOOR). First 10 eigenroots (coordinates) accounted for 22.7, 16.3, 9.0, 8.3, 7.8, 6.8, 6.5, 5.1, 4.7 and 4.0 percent (total of 91.4%) of the variation among the 15 OTUs. The first coordinate (22.7%) separates the junipers of Hispaniola (EK, GR, NH, PL) from all the other taxa in the study (Fig. 2). The second coordinate (16%) separates the two varieties of *J. virginiana* (VG, VS) from all the other taxa (Fig. 2). The third coordinate (9%) separates *J. bermudiana* (BM) from the other taxa and also separates the two populations, NH and PL, from *J. ekmanii* (EK) and *J. gracilior* (GR), see Fig. 2. The minimum spanning network that is superimposed (Fig. 2) is based on the distances using the first 10 eigenroots and thus is useful in sensing the distortion of viewing these 15 OTUs in a 3-dimensional projection (see discussion below).

Figure 3 depicts the effects from coordinates 4 and 5. Coordinate 4 (8%) clearly separates *J. saxicola* (SX) from the other taxa (Fig. 3). Coordinate 5 (8%) serves to further separate *J. saxicola* (SX) but it also separates *J. ekmanii* (EK) and *J. gracilior* (GR) from the northern Haiti (NH) and the Pelempito (PL) populations (Fig. 3). Coordinate 6 (7%) appears to further separate the Hispaniola populations, but particularly resolves *J. barbadensis* (BA) from the other taxa (Fig. 4). *Juniperus barbadensis* is further resolved (Fig. 4) on coordinate axis 7 (7%). Coordinates 8, 9, and 10 only accounted for 5.1, 4.7, and 4.0 percent of the variation and only minor separations between the taxa were present.

Variation within *Juniperus lucayana*

Due to the close similarities among the populations of *J. lucayana* (Figs. 2-4), additional analyses were made using the 6 populations of *J. lucayana* (LA, LB, LC, LG, LJ, LP) in ANOVA. Forty compounds were found with F ratios greater than 1.0 and a population average greater than 0.5 percent of the total oil. Similarities were calculated as above and principal coordinate analysis performed on the similarity matrix. Principal coordinate analysis resulted in 5 eigenroots that accounted for 43.4, 23.8, 14.1, 12.7, and 5.9 percent of the variation (total of 100%) among the 6 taxa.

The first coordinate accounted for the major trend among the populations (43%) by separating the Cuban junipers (LC, LP) from the other populations (Fig. 5). The second coordinate (24%) was due to the differentiation of the Jamaica population (LJ) from the other population (Fig. 5). The third coordinate (14%) revealed minor differences between the Cuban populations (LC, LP; Fig. 5) as well as small differences among the populations from the Bahama Islands (LA, LB, LG; Fig. 5). Coordinates 4 (12.7%) revealed additional differentiation of the Isle of Pines population (LP) from the other populations. Coordinate 5 (5.9%) was due to a minor separation between the Grand Bahama (LB) and Great Abaco (LG) populations. The overall minimum spanning network (based on the combined distance along all 5 coordinate axes) is shown in Fig. 5. This analysis revealed only a minor shift of LA (Andros Island) to link with LG (Great Abaco) instead of with LB (Grand Bahama Island) [cf. Figs. 5 and 2].

It might be noted that canonical variate analysis (CVA) was also performed using the same 70 compounds (by 15 taxa) as analyzed by principal coordinates (see above). The first 10 canonical variates accounted for 31.39, 25.36, 9.76, 8.91, 7.56, 5.28, 4.14, 3.36, 2.22, and 0.71 percent (total of 96.69%) of the variation among the 15 OTUs. All 10 canonical variates were highly significant from zero (Bartlett's test of sphericity). The results from CVA were similar to PCOOR except the taxa were not as well resolved on individual axes. The minimum spanning network derived from distances on canonical axes was not as amenable to interpretation in view of the geological facts (see discussion below). Pimentel, in his discussion of canonical variate analysis (1979:222), states "...the best criterion for any decision is that it makes good biological sense." Thus, in the case of the junipers of the Caribbean, it appears that canonical variate analysis at the species level was not as useful as F-weighted, principal coordinate analysis. The removing of correlation between variables in CVA appears to obscure some of the relationships among the taxa. It appears that the patterns of correlation among the original variables (terpenoids in this case) are very important in defining relationships between taxa (see examples in *Juniperus* and sunfish, Adams 1982). A second problem is that CVA's assumption of equal sample sizes and equal variances are not met and may seldom be met when sampling natural populations. For example, only two mature trees of *J. ekmanii* (endemic to Haiti) are known to exist; therefore, obtaining an adequate sample set was not possible.

Minimum spanning network analyses

The minimum spanning network based on the distances on the first 10 principal coordinates using 70 terpenoids, is shown in Fig. 6. Note the central nature of *J. lucayana* populations (LA, LB, LC, LG, LJ, LP) in the network. The junipers appear to be divided into 6 groups: *J. barbadensis*, BA; *J. bermudiana*, BM; the Hispaniolan junipers, EK, GR, NH, PL; the *J. lucayana* junipers, LA, LB, LC, LB, LJ, LP; *J. saxicola*, SX; and the eastern United States junipers, VG, VS. *Juniperus virginiana* var. *virginiana* (VG) is linked to the West Indian junipers through *J. lucayana* from Andros Island (LA, 82.9). However, notice that var. *silicicola* (VS, dashed line in Fig. 6) is almost equally linked to *J. lucayana* from Grand Bahama Island (83.4, Fig. 6). *Juniperus saxicola* is linked through *J. lucayana* (LP, 74.4) from the Isle of Pines but the next shortest link is with *J. lucayana* from Grand Bahama Island (LB, 75.1, link not shown). The linkage of the Hispaniola junipers is through the Pelempito population (PL) to either the Isle of Pines (LP, 72.3, Fig. 6) or the Cuba mainland (LC, 72.5, Fig. 6, dashed line).

Superimposition of the minimum spanning network onto the geographical map of the populations sampled is shown in Fig. 7 (one should bear in mind that the origin of LC [*J. lucayana* from the Havana Botanical Garden] cannot be precisely located and may well have come from eastern Cuba). One is immediately impressed with the great distances that separate *J. barbadensis* from the nearest extant junipers (over 1,200 km) and *J. bermudiana* from the Bahamas (1400 km). A second important feature is the degree of divergence among the populations of juniper in Hispaniola (note distances of 61 [EK-NH] and [EK-GR]). There is a definite north-south tendency in the network which may reflect bird migration patterns. Juniper seeds are very effectively transported by birds (Livingston 1972; McAtee 1947; Phillips 1910; Poddar and Lederer 1982). Several birds commonly feed on juniper berries (i.e. female cones) such as cedar waxwings, crows, bluebirds, robins, and starlings (Livingston 1972; McAtee 1947; Phillips 1910; Poddar and Lederer 1982). These and other birds' migration pathways need to be further examined.

Speciation of *Juniperus* into the West Indies

The genus *Juniperus* is divided into 3 sections: *Caryocedrus*, *Oxycedrus* and *Sabina* (Gaussen 1968). The junipers of the Western Hemisphere are found in two sections: section *Oxycedrus* with only one circumboreal species, *J. communis* in North America; and section *Sabina* with 38 taxa in North America (Zanoni 1978). The section *Sabina* has been divided into 2 informal series (Gaussen 1968) based on the leaf margins: entire and denticulate (at 40x magnification). The series denticulate likely originated in the highlands of Mexico as a part of the Madro-Tertiary geoflora and radiated out into the arid and semi-arid regions of the southwestern United States and throughout the highlands of Mexico as far south as northern Guatemala (Fig. 8). The southernmost species in the continental North America (*J. comitana* Martinez, *J. gamboana* Martinez, *J. standleyi* Steyermark) are now found in the highlands of Chiapas, Mexico and Guatemala (Fig. 8) at elevations ranging from 1,300 to over 3,000 m (Zanoni and Adams 1979, Adams et al. 1985). These species are all in the denticulate group (or series) and appear to have been part of the radiation and adaptation from the ancestral denticulate junipers (Fig. 8). The series entire is composed of *J. blancoi* Martinez, *J. horizontalis* Moench, *J. scopulorum* and *J. virginiana* (and var. *silicicola*) on the continental North America as well as all the taxa in the West Indies and Bermuda (Fig. 8). The series entire on continental North America appears to have arisen from *J. virginiana* (or its ancestor) from the ancient land mass of Appalachia (Anderson 1953, Flake et al. 1969). Elements of the eastern North American flora are thought to be closely related to the Old World (Fernald 1931) and this is certainly true between *J. virginiana* of the eastern United States and *J. sabina* of southern Europe.

Juniperus scopulorum, a sibling species of *J. virginiana*, has been treated as a variety (*J. virginiana* var. *scopulorum* [Sarg.] Lemmon, Handb. West-Amer. Cone Bearers ed. 4, 114, 1900; *J. virginiana* var. *montana* Vassey, Rep. U. S. Commiss. Agric. 1875, 184, 1876; Zanoni 1978). Several studies have confirmed hybridization between these taxa (Fassett 1944; Van Haverbeke 1968; Schurtz 1971; von Rudloff 1975; Flake, von Rudloff, and Turner 1978; Adams 1983b; Comer, Adams and Van Haverbeke 1982). *Juniperus scopu-*

lorum is postulated (Adams 1983b) to have arisen from *J. virginiana* (Fig. 8). *Juniperus horizontalis* is closely related to both *J. scopulorum* and *J. virginiana* and has been treated as a variety, *J. virginiana* var. *prostrata* (Persoon) Torrey (Fl. New York 2, 235, 1843). *Juniperus horizontalis* hybridizes in several areas with *J. virginiana* (Fassett 1945a,b; Schurtz 1971; Palma-Otal et al. 1983). *Juniperus horizontalis* also hybridizes with *J. scopulorum* (Fassett 1945c; Schurtz 1971; von Rudloff 1975; Adams 1983b). *Juniperus horizontalis* is postulated to have been derived from *J. virginiana* or its ancestor (Fig. 8). *Juniperus blancoi* is very closely related to *J. scopulorum* (Adams 1983b) and is confined to a few locations in central and northern Mexico along flowing streams. It is postulated to have been derived from *J. scopulorum* (in fact it may be conspecific, see Adams 1983b). The only other member of series entire is *J. virginiana* var. *silicicola* which is barely distinct from *J. virginiana* (Adams 1986) and may have been derived as recently as the Pleistocene (Fig. 8).

Speciation of *Juniperus* into the West Indies is postulated to have occurred by long distance dispersal of *J. virginiana* (or its ancestor) by birds to the Bahama Islands and then to Bermuda, Cuba, Jamaica, and Hispaniola. *Juniperus saxicola* most likely evolved from a *J. lucayana* ancestor in eastern Cuba from seeds carried into the Pico Turquino region. Either by a chance founder's effect or by genetic drift, the gene(s) for controlling the conversion from juvenile (awn-like) to adult (scale-like) leaves become fixed such that all adults now have only juvenile leaves. *Juniperus barbadensis* appears to have arisen from *J. lucayana,* possibly from Cuba. The large distance from Cuba to St. Lucia and the lesser Antilles render this postulate somewhat tentative. The alternative mode, island hopping from Hispaniola is less attractive because suitable habitat would seem unlikely on many of the intervening islands.

The situation in Hispaniola is still far from resolved, except to say that *J. lucayana* appears to have been involved in colonization. *Juniperus ekmanii* and *J. gracilior* probably were derived from a *J. lucayana*-like ancestor. *Juniperus urbaniana,* although not sampled in this study, would appear to be very closely related to *J. ekmanii* and probably arose from *J. ekmanii* or its ancestor. Additional studies (in progress) may resolve the apparent diversity in Hispaniola.

The introduction of *J. bermudiana* to Bermuda must have been relatively recent as Bermuda's soil was only formed during the first interglacial period of the Pleistocene (1 million yBP; Sayles 1931; Bryan and Cady 1934; Cox 1959). Considering the genetic bottleneck that the Bermuda junipers have gone through in their current fight for survival, we cannot be sure that extant trees fairly represent the gene pool that evolved on Bermuda. This may account in part, for the divergence of *J. bermudiana* from the Bahama junipers.

Significance to floristic affinities of the West Indies

This study of the Caribbean junipers may shed some light on the origin of the vegetation of the Caribbean Islands. Howard (1973) suggests that the major vegetation affinities of the Antilles lie with Central America and then to the northern part of South America. Rosen (1975) came to a similar conclusion based primarily on fresh water fish distributions. He even postulated that continental drift accounts for the affinities by proposing that the Caribbean islands are part of the Pacific plate, which pushed its way between North and South America, northeasterly to its present position. This would have allowed contact between the proto-Antilles land mass and the southern Mexico-Guatemala (to the north) and the South American land mass (to the south). Thus, according to Rosen, the flora and fauna from Central America and South America could have easily transferred to the proto-Antilles land mass and rode the drift to the present position in the Caribbean.

However, Khudoley and Meyerhoff (1973) were firmly opposed to the idea of continental drift in explaining the origin of the Antilles. Interestingly though, they (Khudoley and Meyerhoff 1973) state "Faunal and floral similarities between Central America and the Greater Antilles suggest that direct land connections existed, perhaps via western Cuba or via the Cayman Ridge and Nicaragua Rise." It should be also be noted however, that Khudoley and Meyerhoff (1973) are diametrically opposed on the origin of the Antilles with Khudoley believing that "the area of the present Caribbean Seas was a land mass until Cretaceous time", and Meyerhoff believing that "the whole of the Greater Antilles was a late Jurassic through early Tertiary island arc...". If Rosen (1975) or Khudoley (Khudoley and Meyerhoff 1973) are correct then the land mass has been large and contiguous for a long time whereas if Meyerhoff (Khudoley and Meyerhoff 1973) is correct, one would expect to find more endemic species on the various islands.

This study favors the Meyerhoff theory in that each island has seemed to spawn a new taxon of *Juniperus*. All of the junipers of the Caribbean have smooth leaf margins (series entire) and no junipers from the denticulate (serrate) leaf margined junipers (series denticulate). Only the denticulate leaf margined junipers are found in southern Mexico and Guatemala (the southern-most range of *Juniperus* in the continental western hemisphere). There are therefore no affinities between the Caribbean junipers and those of Central America. The spread of the junipers across the Caribbean islands has most likely been by birds from eastern North America. The differentiation of these island populations has been affected both by selection and founder's effects. Genetic drift may have also played a part in the diversity of the island junipers because of the expansion and contraction of their ranges during the Tertiary and Pleistocene. According to Curray (1965), the Caribbean sea level dropped approximately 122 m, about 19,000 yBP as well as another drop in sea level of 146 m at 40,000 yBP. Rosen (1978) shows that these drops in sea level would unite several of the Bahamian Islands. Conversely, a rise in the ocean level of only a few meters would inundate many juniper sites in the Bahamas where *J. lucayana* often occurs at 1 to 2 m above sea level. Broecker (1965) reported evidence for higher levels about 80,000 yBP in the Bahamas. Thus, there is ample evidence of changes in available juniper habitat, which in turn has probably led to local extinctions as well as range expansions. This, coupled with limited gene flow between the islands, has led to the considerable amount of diversity and differentiation in the leaf oils of the Caribbean junipers.

The results from this study indicate that *Juniperus barbadensis* is quite distinct from the other junipers in its leaf oil. It is most similar to *J. bermudiana* and *J. lucayana,* but does not appear to be conspecific. The West Indies junipers appear to have arisen from *J. virginiana* (or its ancestor) in eastern North America, with the West Indies having been populated by long distance dispersal of seeds by birds.

Acknowledgements

I would like to thank Larent Jean-Pierre and Verna Slane for their assistance on our expedition to the top of the Petit Piton Mountain in St. Lucia to help collect the specimens of *J. barbadensis* and Mark Donsky for help in collecting *J. saxicola* from Pico Turquino in Cuba. A portion of this research (work in Bermuda, the Bahamas, Hispaniola, and Jamaica) was supported by funds from NSF grant DEB79-21757.

Literature cited

Adams, R.P. 1975a. Gene flow versus selection pressure and ancestral differentiation in the composition of species: Analysis of populational variation in *Juniperus ashei* Buch, using terpenoid data. Journal of Molecular Evolution 5:177-185.

_____. 1975b. Statistical character weighting and similarity stability. Brittonia 27:305-316.

_____. 1982. A comparison of multivariate methods for the detection of hybridization. Taxon 31:646-661.

_____. 1983a. The junipers (*Juniperus*: Cupressaceae) of Hispaniola: Comparison with other Caribbean species and among collections from Hispaniola. Moscosa 2:77-89.

_____. 1983b. Infraspecific terpenoid variation in *Juniperus scopulorum*: evidence for Pleistocene refugia and recolonization in Western North America. Taxon 32:30-46.

_____. 1986. Geographic variation in *Juniperus silicicola* and *J. virginiana* of the southeastern United States: Multivariate analyses of morphology and terpenoids. Taxon 35:61-75.

_____ and L. Hogge. 1983. Chemosystematic studies of the Caribbean junipers based on their volatile oils. Biochemical Systematics and Ecology 11:85-89.

_____, C.E. Jarvis, V. Slane and T.A. Zanoni. 1987a. Typification of *Juniperus barbadensis* L. and *J. bermudiana* L. and the rediscovery of *J. barbadensis* from St. Lucia, BWI. Taxon 36(2):441-445.

_____, A.L. Almirall and L. Hogge. 1987b. Chemosystematics of the junipers of Cuba, *J. lucayana* and *J. saxicola* using volatile leaf oils. Flavour and Fragrance Journal 2:33-36.

_____, T.A. Zanoni and L. Hogge. 1985. The volatile leaf oils of the Junipers of Guatemala and Chiapas, Mexico: *Juniperus comitana*, *J. gamboana* and *J. standleyi*. Journal of Natural Products 48:678-680.

Anderson, E. 1953. Introgressive hybridization. The Biological Review 28:280-307.

Bennett, F.D. and I.W. Hughes. 1959. Biological control of insect pests in Bermuda. Bulletin Entomological Research 50:423-436.

Blackrith, R.E. and R.A. Reyment. 1971. Multivariate morphometrics. Academic Press, London, 412 pp.

Britton, N.L. 1908. North American trees. Henry Holt & Co., New York, 894 pp.

Broecker, W.S. 1965. Isotype geochemistry and the Pleistocene climatic record. Pp 737-754 in The Quaternary of the United States. H.E. Wright and D.G. Frey. (eds.). Princeton University Press, Princeton, NJ. 922 pp.

Bryan, K. and R.C. Cady. 1934. The Pleistocene climate of Bermuda. American Journal of Science 27:241-264.

Carabia, J.P. 1941. Contribuciones al estudio del flora Cubana. The Caribbean Forester 2:83-92.

Cooley, W.W. and R.P. Lohnes. 1971. Multivariate data analysis. John Wiley & Sons, Inc., New York, 363 pp.

Comer, C. W., R. P. Adams, and D. R. Van Haverbeke. 1982. Intra-and inter-specific variation of *Juniperus virginiana* L. and *J. scopulorum* Sarg. seedlings based on volatile oil composition. Biochemical Systematics and Ecology 10:297-306.

Correll, D.S. and H.B. Correll. 1982. Flora of the Bahama Archipelago. Cramer, Hirschberg, West Germany, 1692 pp.

Cox, W.M. 1959. Bermuda's beginning. C. Tinling & Co., Ltd., Liverpool, UK. 24 pp.

Curray, J.R. 1965. Late Quaternary history, continental shelves of the United States. Pp 723-736 in The Quaternary of the United States. H.E. Wright and D.G. Frey, (eds.). Princeton University Press, Princeton, NJ. 922 pp.

Fassett, N.C. 1944. *Juniperus virginiana*, *J. horizontalis* and *J. scopulorum*-II. Hybrid swarms of *J. virginiana* and *J. scopulorum*. Bulletin of the Torrey Botanical Club 71:475-483.

_____. 1945a. *Juniperus virginiana* , *J. horizontalis*, and *J. scopulorum*-IV. Hybrid swarms of *J. virginiana* and *J. horizontalis*. Bulletin of the Torrey Botanical Club 72:379-384.

_____. 1945b. *Juniperus virginiana*, *J. horizontalis* and *J. scopulorum*-V. Taxonomic treatment. Bulletin of the Torrey Botanical Club 72:480-482.

_____. 1945c. *Juniperus virginiana*, *J. horizontalis*, and *J. scopulorum*-III. Possible hybridization of *J. horizontalis* and *J. scopulorum*. Bulletin of the Torrey Botanical Club 72:42-46.

Fernald, M.L. 1931. Specific segregations and identities in some floras of eastern North America and the Old World. Rhodora 33:25-62.

Flake, R.H., E. von Rudloff, and B.L. Turner. 1969. Quantitative study of clinal variation in *Juniperus virginiana* using terpenoid data. Proceedings of the National Academy of Sciences of the United States of America 62:487-494.

Florin, R. 1933. Die von E.L. Ekman in Westindien gesammelten Koniferen. Arkiv foer Botanick 25A(5):1-22.

Gaussen, H. 1968. Les gymnospermes actuelles et fossiles. Les cupressacees. Trav. Lab. Forest. Toulouse. Tome II, Sect. I, Vol 1, Partie II, Fasc. 10.

Gillis, W.T. 1974. Name changes for the seed plants in the Bahama flora. Rhodora 76:67-138.

Gower, J. C. 1966. Some distance properties of latent root and vector methods used in multivariate analysis. Biometrika 53:315-328.

_____. 1971. A general coefficient of similarity and some of its properties. Biometrics 27:857-874.

Groves, G.R. 1955. The Bermuda cedar. World Crops 7:1-5.

Howard, R.A. 1973. The vegetation of the Antilles. Pp. 1-38 in Vegetation and vegetational history of northern Latin America. Alan Graham, (ed.). Elsevier Scientific Publishing Co., New York, 393 pp.

Hemsley, W.B. 1883. The Bermuda cedar. Garden Chronicle 19 (n.s.) May 26, 656-567.

Khudoley, K.M. and Meyerhoff, A.A. 1973. Paleogeography and geological history of the Greater Antilles. Memoir 129, Geological Society of America, Boulder, CO. 199 pp.

Linnaeus, C. 1753. Species plantarum. Stockholm, 1200 pp.

Livingston, R. B. 1972. Influence of birds, stones and soil on the establishment of pasture juniper, *Juniperus communis* and red cedar, *J. virginiana* in New England pastures. Ecology 53:1141-1147.

McAtee, W.L. 1947. Distribution of seeds by birds. American Midland Naturalist 38:214-223.

Palma-Otal, M., W.S. Moore, R.P. Adams and G.R. Joswiak. 1983. Genetic and biogeographical analyses of natural hybridization between *Juniperus virginiana* and *J. horizontalis* Moench. Canadian Journal of Botany 61:2733-2746.

Phillips, F.J. 1910. The dissemination of junipers by birds. Forestry Quarterly 8:11-16.

Pilger, R. 1913. IX. Juniperi species antillanae. Symbolae Antillanae 7:478-481.

Pimentel, R.A. 1979. Morphometrics. Kendall/Hunt Publ. Co., Dubuque, IA., 276 pp.

Poddar, S. and R.J. Lederer. 1982. Juniper berries as an exclusive winter forage for Townsend's Solitaires. American Midland Naturalist 108:34-40.

Rosen, D.E. 1975. A vicariance model of Caribbean biogeography. Systematic Zoology 4:431-464.

_____. 1978. Vicariant patterns of historical explanation in biogeography. Systematic Zoology 27:159-188.

Sargent, C.S. 1902. Silva of North America. Vol. 14 (1947 reprint) Peter Smith, New York, 433 pp.

Sayles, R.W. 1931. Bermuda during the Ice Age. American Academy of Arts and Sciences 65:279-468.

Schurtz, R.H. 1971. A taxonomic analysis of a triparental hybrid swarm in *Juniperus* L. Unpublished Ph.D. dissertation, University of Nebraska, Lincoln, NE. 90 pp.

Van Haverbeke, D. F. 1968. A population analysis of *Juniperus* in the Missouri river basin. University of Nebraska Studies, n.s. 38. 82 pp.

von Rudloff, E. 1975. Volatile oil analysis in chemosytematic studies of North American conifers. Biochemical Systematics and Ecology 2:131-167.

Zanoni, T.A. 1978. The American junipers of the section *Sabina* (*Juniperus*, Cupressaceae)-a century later. Phytologia 38:433-454.

_____, and R.P. Adams. 1979. The genus *Juniperus* (Cupressaceae) in Mexico and Guatemala. Synonymy, key and distributions of the taxa. Boletin de la Sociedad Botanica de Mexico 38:83-121.

Table 1. Composition of the Volatile Leaf Oils of *Juniperus barbadensis* from St. Lucia, BWI along with the volatile leaf oils of other Caribbean junipers previously reported (Adams 1983; Adams and Hogge 1983). Compounds are listed in order of their elution from a DB1 column. BA = *J. barbadensis*, St. Lucia, BWI; LJ = *J. lucayana*, Jamaica; LB = *J. lucayana*, Bahama Islands; BM = *J. bermudiana*, Bermuda; EK = *J. ekmanii*, Haiti; GR = *J. gracilior*, Dominican Republic; VS = *J. virginiana* var. *silicicola*, Florida, USA; and VG = *J. virginiana* var. *virginiana*, Washington, D.C. USA. Those 70 components denoted with an asterisk (*) were utilized in principal coordinate analysis. Compositional values in parenthesis indicate that a compound runs at that retention time but no mass spectrum was obtained. Compound names in parenthesis are tentatively identified. T indicates the compound was present in trace amounts (less than 0.5% of the total oil).

Compound	% total oil							
	BA	LJ	LB	BM	EK	GR	VS	VG
Percent Yield*	0.6	0.6	0.2	0.3	1.4	0.8	0.4	0.2
Unknown 1*, RRT=0.143	-	-	T	-	(T)	0.8	-	(T)
Unknown 2*, RRT=0.151	-	-	T	-	(T)	0.8	-	T
Tricyclene+p-Thujene*	0.9	0.6	0.5	T	1.9	1.4	T	T
α-Pinene*	7.4	49.1	33.0	22.3	1.3	1.8	2.4	1.4
Camphene*	T	T	T	0.7	1.9	1.2	T	T
Sabinene*	31.0	9.7	8.3	2.8	5.0	10.1	T	6.7
β-Pinene*	T	1.1	1.2	0.6	T	T	T	T
1-Octen-3-ol*	-	T	T	1.0	T	T	0.9	-
Myrcene*	3.8	3.2	4.0	2.9	2.5	1.9	0.9	0.9
2-Carene	-	T	T	-	T	-	T	T
α-Phellandrene	T	-	-	T	-	T	-	-
3-Carene	-	-	-	T	-	-	T	(T)
α-Terpinene*	1.7	T	T	T	0.9	1.7	T	T
p-Cymene*	T	T	T	0.5	0.5	1.4	T	-
β-Phellandrene	T	-	-	-	-	-	-	T
Limonene*	34.2	25.9	18.0	35.3	9.6	7.3	33.3	18.9
trans-Ocimene*	0.7	T	-	T	-	-	-	(T)
γ-Terpinene*	2.7	0.8	0.7	0.7	1.7	3.5	T	T
(p-menth-1(7),3-diene)*-	T	-	T	-	-	-	T	
(cis-p-menth-2-ene-ol)*0.9	-	T	-	0.9	1.1	-	T	
Terpinolene*	1.2	1.0	0.8	0.8	0.6	0.9	(T)	0.5
(trans-p-menth-2-ene-ol)*	0.7	-	-	-	-	T	-	-
4-Terpinenyl acetate	T	T	-	-	-	-	-	T
Linalool*	-	-	T	1.1	0.6	2.6	1.5	4.4

Table 1 Continued

Unknown 3, RRT=0.337*	-	-	-	(T)	1.6	2.0	-	-
(3-Cyclopentene-1-acetaldehyde, 2,2,3-tri-methyl)	(T)	T	T	T	-	-	-	-
cis-Sabinene hydrate	-	T	T	T	-	-	-	T
(cis-Dihydrocarveol)*	T	-	-	-	0.5	0.8	-	-
Camphor*	(T)	T	T	6.5	5.8	1.1	T	3.7
trans-Pinocarveol*	-	-	-	1.1	-	-	-	-
(trans-Dihydrocarveol)*	T	T	(T)	-	(T)	0.7	-	-
trans-Sabinene hydrate	-	-	-	-	-	-	-	T
Camphene hydrate*	-	-	(T)	1.4	2.2	1.4	T	T
Borneol*	T	T	T	2.1	5.1	2.0	-	0.8
4-Terpineol*	6.5	1.6	2.5	1.4	6.3	11.6	T	1.5
Myrtenal*	-	-	(T)	0.7	-	-	T	T
α-Terpineol*	T	T	T	T	0.8	0.9	-	T
Estragol*	-	-	-	-	-	-	0.5	T
(p-Cymen-9-ol)	-	T	T	T	-	-	T	T
cis-Piperitol*	T	-	-	-	T	T	-	-
trans-Piperitol	T	-	-	-	-	-	-	-
Unknown 4, RRT=0.426*	-	-	-	-	(T)	1.2	-	-
Carvone*	T	T	(T)	1.0	(T)	T	T	T
Citronellol*	T	T	0.9	T	(0.6)	0.6	T	2.3
Piperitone*	-	-	0.6	-	-	T	T	(T)
Isosafrole*	(T)	-	(T)	(T)	-	-	3.6	6.7
Bornyl acetate*	T	0.6	4.1	4.2	43.9	35.7	T	2.1
Safrole*	(T)	T	(T)	-	-	(T)	13.7	10.9
Sabinyl acetate*	(T)	-	T	0.8	(T)	(T)	-	-
Methyl eugenol*	-	T	(T)	(T)	-	T	8.2	2.9
Caryophyllene	T	T	T	T	T	T	T	T
Thujopsene*	T	-	(T)	2.1	(T)	T	(T)	T
α-Cadinene*	-	-	T	-	T	T	-	T
Germacrene isomer 2*	(T)	-	T	-	T	T	T	T
Germacrene D*	T	0.6	T	T	-	-	(T)	T
Cuparene	T	-	-	-	-	-	-	-
β-Cubebene*	-	-	0.8	-	T	T	-	-
Unknown 5, RRT=0.658*	-	-	0.9	-	-	-	-	-
α-Muurolene	(T)	T	-	T	T	T	T	T
ϒ-Cadinene*	T	-	2.8	-	T	-	T	T
δ-Cadinene*	T	T	0.7	T	T	T	0.7	0.8
Unknown 6, RRT=0.692*	-	-	0.6	-	-	-	T	T
Elemicin*	-	-	-	-	T	T	-	T
Elemol*	(T)	T	T	T	-	-	12.1	8.2
Cadinol isomer 1*	T	T	-	-	-	-	-	-

Table 1 Continued

Compound								
Unknown 7, RRT=0.715*	(T)	-	1.9	-	-	-	0.9	T
Cadinol isomer 2*	-	T	0.9	-	(T)	(T)	(0.8)	0.7
Unknown 8, RRT=0.732*	T	T	2.2	T	-	-	0.7	2.0
Widdrol	T	-	-	-	-	-	-	-
Cedrol	T	-	-	-	-	-	-	-
(Cubenol)*	-	-	1.2	T	T	T	(0.7)	0.9
T-Eudesmol*	-	-	-	-	-	-	3.4	2.8
r-Cadinol*	-	T	-	-	T	T	2.0	T
r-Muurolol*	-	T	1.6	T	T	-	-	2.4
β-Eudesmol*	-	-	-	-	-	-	2.8	1.7
Cadinol isomer 4*	-	0.5	-	-	-	-	-	-
α-Cadinol*	T	-	-	-	-	-	-	-
α-Eudesmol*	-	-	-	-	-	-	3.4	3.1
Unknown 9, RRT=0.769*	T	-	2.3	T	(T)	T	-	-
Unknown 10, RRT=0.791*	T	T	0.9	T	T	T	T	-
Acetate II*	-	-	-	-	-	-	1.8	3.5
Abietatriene	T	-	T	T	-	T	T	-
Manool*	(T)	-	T	-	-	-	T	(T)
(Kaur-16-ene)*	T	-	T	0.7	-	T	-	-
Unknown 11, RRT=1.099*	0.6	-	-	-	-	-	-	-

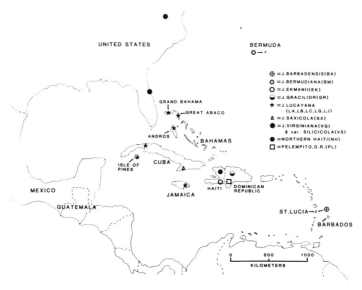

Figure 1. Population vicinity map of the study area with population locations. The endemic juniper, *J. urbaniana* occurs near *J. ekmanii* in Haiti but was not sampled in this study. Additional populations of *J. lucayana* (not sampled) occur on a few of the islands of the Bahamas and in eastern Cuba.

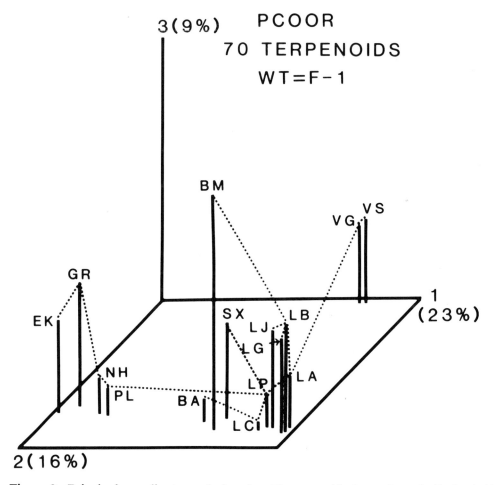

Figure 2. Principal coordinate analysis using 70 terpenoids from the volatile leaf oils. The individual components were weighted by F-1 (F from ANOVA of the 15 OTUs) in the similarity measure. The minimum spanning network (dashed line) distances are based on the first 10 principal coordinates. The OTU codes are: BA = *J. barbadensis*, St. Lucia; BM = *J. bermudiana*, Bermuda; EK = *J. ekmanii*, Haiti; GR = *J. gracilior*, Dominican Republic; LA, LB, LC, LG, LJ, and LP = *J. lucayana*, from, respectively, Andros Island, Grand Bahama Island, Cuba (mainland), Jamaica, Great Abaco Island, and Isle of Pines; NH = northern Haiti; PL = Pelempito region of Dominican Republic; SX = *J. saxicola*, Cuba; VG = *J. virginiana*, Washington, D.C.; and VS = *J. virginiana* var. *silicicola*, Florida, USA. The percentage number on each axis is the percent of the total variation among the OTUs accounted for by that axis.

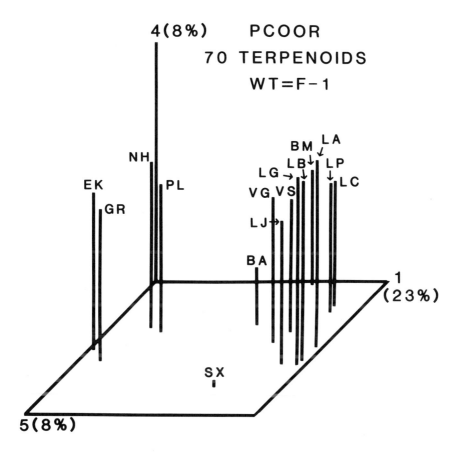

Figure 3. Principal coordinate analysis based on terpenoids. See Fig. 2 for OTU codes. Note the separation of SX (*J. saxicola*) and BA (*J. barbadensis*) on axis 4. Axis 5 primarily separates EK, GR from NH, PL and *J. saxicola* (SX) from the other junipers.

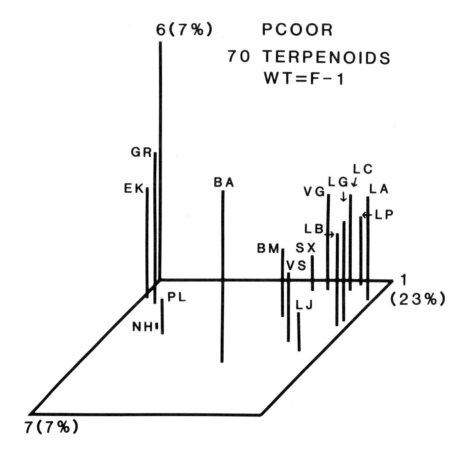

Figure 4. Principal coordinate analysis for coordinates 1, 6, and 7. Coordinate 6 further separates the populations from Hispaniola into EK, GR and NH, PL as well as separating the two varieties of *J. virginiana*, VG and VS.

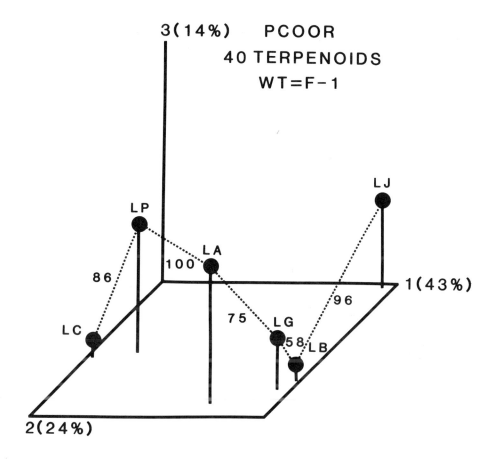

Figure 5. Principal coordinate analysis of the six *J. lucayana* populations (see Fig. 2 caption for codes) using 40 terpenoids. The first coordinate primarily separates the two Cuban populations (LC, LP) from the other populations. The second axis separates the Jamaica population (LJ). The third axis separates the two Cuban populations (LC, LP) from each other and separates out two of the Bahamian populations (LG, LB). The dashed line is the minimum spanning network based on distances on the first 5 principal coordinates.

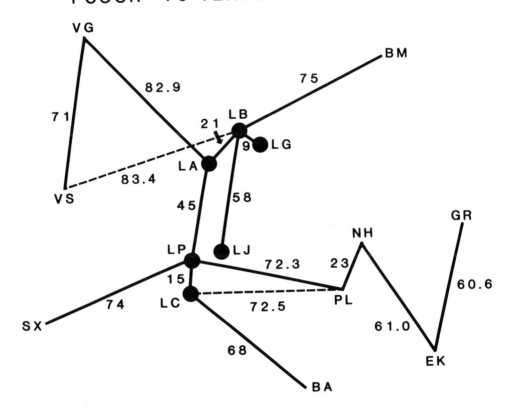

Figure 6. Minimum spanning network based on the first 10 coordinates from PCOOR using 70 terpenoids and 15 OTUs. All the distances have been scaled by 100 to eliminate most of the decimals. The dashed line from VS to LB is the smallest distance from VS (*silicicola*, Florida) to any OTU off the mainland. The dashed line from PL to LC is the second shortest distance to an OTU not on Hispaniola. OTU codes are as defined in Fig. 2.

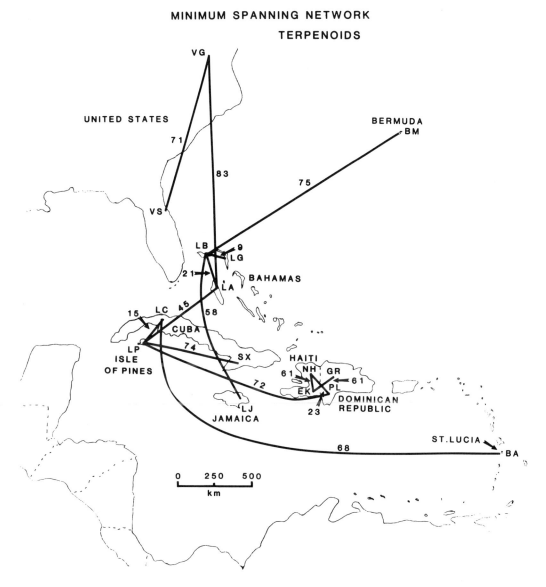

Figure 7. Minimum spanning network (from Fig. 6) mapped onto the geographical map of the West Indies and surrounding region. OTU codes are as defined in Fig. 2. See text for discussion.

190   Biogeography of the West Indies

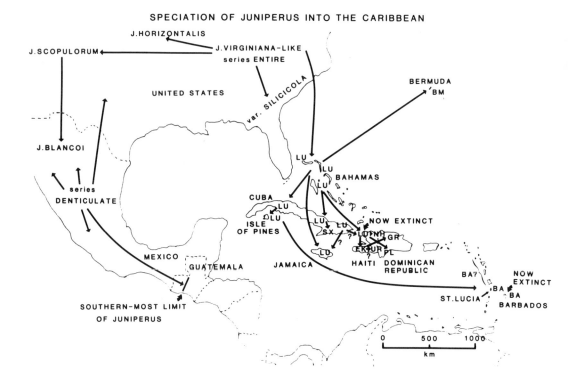

Figure 8. Proposed Speciation of *Juniperus* into the Caribbean. Note especially that the junipers from the southern Mexico/ northern Guatemala area are in series denticulate, whereas all the junipers of the Caribbean, Bermuda and eastern United States are in series entire. See text for additional discussion.

RECENT VEGETATION CHANGES IN SOUTHERN HAITI

Antonia Higuera-Gundy[1]

Abstract

A 58-cm sediment core from Lake Miragoane, Haiti preserves the last millennium of local environmental history. The bottom half of the core (58-30 cm) contains deposits of pre-Columbian age (1,000-500 yBP), and is rich in organic matter and pollen grains. Prior to European arrival, riparian lowland vegetation was characterized by open, dry and mesic plant communities. Upland moist forests covered nearby mountainsides. Pollen and chemical evidence from pre-Columbian deposits indicates minimal disturbance of local vegetation and soils. However, some impact is recorded by the presence of carbonized fragments and pollen grains of successional trees (*Trema, Cecropia, Celtis*). Interference with local vegetation is attributable to early Arawak settlers (1,400-500 yBP), and/or natural events, such as spontaneous fires. Deposits from the upper half of the section (30-0 cm) reflect riparian disturbance of local vegetation and soils since European arrival (500 yBP to present). Deforestation is recorded by increasing amounts of weed relative to tree pollen. Vegetation removal was followed by increased erosion, detected in higher concentrations of sedimentary carbonates, presumably of detrital origin. Surface deposits are rich in carbonates and dominated by weed pollen, and clearly reflect present environmental degradation of the Miragoane watershed.

INTRODUCTION

The historical ecology of Haiti is being investigated by stratigraphic study of physical, chemical, and biological characteristics of sediment cores from a deep, freshwater lake on the southern peninsula. The research will allow biogeographers to draw on the paleoecological record to interpret contemporary distributions and abundances of Haitian flora and fauna.

Following reconnaissance and preliminary coring work in 1983, Lake Miragoane was selected for intensive study. In 1985, a long core, and two short, mud-water interface sections were collected from deep water near the center of the lake. The long core has a basal date > 10,000 YBP, and therefore contains a complete record of Holocene sedimentation. Pollen analysis of the section is now under way and will be invaluable for reconstructing the long-term vegetation history of Haiti.

---

[1] Antonia Higuera-Gundy is a doctoral student at the Florida Museum of Natural History, University of Florida, Gainesville, FL, 32611

The short cores span the last millennium, and thus contain about five centuries of pre-Columbian deposits as well as a complete environmental record since European settlement. This paper reports pollen and chemical data from one of the short cores. The sedimentary record shows that widespread land clearance began in the region about five hundred years ago.

Study site

Lake Miragoane lies in limestone terrain at 18° 24' N Lat, 73° 05' W Long. The basin is set in a tectonic rift system on the northern edge of Haiti's southern peninsula (Fig. 1). The lake surface covers 7.06 km^2 and the lakewater conductivity is 350 uS cm^{-1}, making it one of the largest, if not the largest, truly freshwater lakes in the Caribbean (Binford et al. 1986). Miragoane is nearly circular, with the exception of a bay in the southwest area of the lake, making shoreline development 1.19 (Fig. 1). The lake surface lies at 20 m above sea level, and with a maximium depth of 41 m, the basin has a cryptodepression of 21 m. Site visits in August 1983 and July 1985 showed the lake to be thermally stratified. Bottom waters below 25 m contained less than 1 mg/L of dissolved oxygen. As limnological investigations have been restricted to summer months, the mixing pattern of the lake waters is unknown. Based on geographic and morphometric considerations, the lake is presumed to be a warm, monomictic system (Lewis 1983).

Another, unexplored lake lies to the east of Miragoane. The two basins are separated by an extensive marsh that may receive outflow from Lake Miragoane. Common plant taxa of the marsh are *Cladium* (sawgrass), *Typha* (cat-tail), *Nymphaea* (water lily), and others. Within Lake Miragoane, *Potamogeton* (pondweed) appears to be the most abundant submerged macrophyte. *Nelumbo lutea* (American lotus) forms large dense patches in several littoral areas, generally at some distance from the marsh.

Haiti is typified by warm temperatures throughout the year. Mean annual temperatures of 26-27° C are recorded at lowland coastal sites near Lake Miragoane. At higher altitudes, mean annual figures are somewhat lower and temperature decreases at a rate of about 1° C per 150 m increase in elevation (Woodring et al. 1924). Lowland sites near the lake receive between 1,000 and 2,000 mm of precipitation annually. Higher rainfall is reported for mountainous areas. With little temperature fluctuation throughout the year, seasonality in Haiti is defined by dry and wet seasons. There are two wet periods, one in spring and one in fall. Spring rainfall attains a maximum in May, while autumn rainfall is greatest in October or November. July is the driest and usually the warmest summer month (Woodring et al. 1924).

Topographic variation in Hispaniola is responsible for the irregular distribution of rainfall over the island. Moisture-laden trade winds, blowing from the east and northeast, sweep through the high mountains of the Dominican Republic, where they deliver much of their rainfall before continuing westward to Haiti. Much of Haiti lies in rain shadow, and is considerably drier than the Dominican Republic. The southern peninsula of Haiti receives more precipitation, as it projects beyond the rain shadow. The north shore of the peninsula, where Miragoane is located, is wetter than the south shore (Woodring et al. 1924).

Plant assemblages found along the Haitian coast range from tropical lowland dry forest associations to more xeric, cactus-dominated communities (Holdridge 1945). The types and distributions of these plant associations are determined to some extent by natural environmental factors (e.g. rainfall), but are also controlled by human activities. While the Miragoane watershed and surrounding areas supported tropical lowland dry forest in the past, the region is largely deforested today.

Moist forests near Lake Miragoane are associated with two mountainous areas, the Massif de la Selle and the Massif de la Hotte (Fig. 1). Moist forest vegetation extends in an east-west direction over nearly the full length of the peninsula. The continuity of the moist forest is interrupted by a 30-km-wide band of dry forest vegetation that lies just east of Miragoane. Low- and high-elevation mesic forests are rich in both arboreal and non-arboreal taxa. The summits of the massifs support conifer forests comprised of the only pine species in Hispaniola, *Pinus occidentalis*. Pines also grow in suitable habitats at lower elevations, but are present in lower densities than on the summits. Pure pine stands exist in areas that dry out and burn during the winter dry season. In areas with infrequent fires, hardwoods invade the pine forests (Holdridge 1945).

Methods

In July, 1985, two mud-water interface cores (< 1m long) were obtained from the deep, central part of Lake Miragoane (Fig. 1) using a 4-cm-diameter piston corer. A long section (7.67 m) was obtained using a square-rod piston corer (Wright et al. 1984). This paper presents chemical and palynological results from a 72-cm, mud-water interface core, designated 17-VII-85-1. The core was extruded in the field at 1-cm intervals down to 10 cm, 2-cm intervals from 10 cm to 30 cm and at 4-cm intervals from 30 cm to the bottom. Samples were stored in whirl-pak bags and shipped to the Florida Museum of Natural History, Gainesville, where they have been kept under refrigeration. Samples below 58 cm were set aside for radiocarbon dating.

In the laboratory, several sets of volumetric (1 cm^3) subsamples were removed for physical (e.g. dry weight per wet volume, percent dry mass), chemical (organic matter, carbonate, cation, nutrient) and biological (pollen) analyses. Water content was evaluated by weight loss on drying for 24 hours at 110° C. Dried sediments were ground to a fine powder in a mortar. Organic matter content was measured by weight loss on ignition at 550° C and carbonate content was measured by weight loss, as CO_2, between 600° and 990° C (Dean 1974).

Samples for pollen analysis were pretreated with 10 percent HCl to remove carbonates. The sediments were then deflocculated by boiling in 5 percent KOH. Dissolution of silicates with 40 percent HF was followed by acetolysis and dehydration. The residue, consisting of pollen and organic matter resistant to oxidation, was brought to a volume of 5 ml with tertiary butyl alcohol (TBA). The TBA-pollen suspension was homogenized, and one or more aliquots (0.02 ml) were removed to prepare each slide. Pollen counting was done at 200X, though higher magnification was used for some identifications. All pollen types, both known and unknown, were included in the pollen sum. Proportions of each taxon are expressed as a percentage of the pollen sum. Carbonized fragments were counted, but were not distinguished as having come from monocots or dicots.

The chronology of the core was established by ^{210}Pb dating. The limits of the technique were reached at a depth of 8 cm, where the age is 133 years (before 1985). The mean rate of net sediment accumulation in the upper 8 cm of the core (g/cm^2/yr) was calculated from the total dry mass sedimented per cm^2 above 8 cm, divided by the number of years it took to accumulate. Next, the sedimentation rate, on a mass basis, was assumed to have remained constant over the time span covered by the core. Dates were assigned to levels below 8 cm in the core by measuring the cumulative mass over all contiguous sections of the section, and are therefore extrapolations of the ^{210}Pb dates.

Results

Figure 2 summarizes the pollen, organic matter and carbonate stratigraphy of Lake Miragoane core 17-VII-85-1. The "trees" and "herb" categories in Fig. 2 represent several taxa. The major components of the arboreal assemblage are Moraceae, *Celtis*, and *Trema*. At each level they accounted for 5-10 percent of the pollen sum. *Bursera*, though uncommon (2-5%), was found throughout the core. Taxa present in low proportions include *Alchornea*, *Phyllostylon*, *Sapindus*, *Caesalpinia* (a legume), Sapotaceae, Myrtaceae, and Meliaceae (probably *Swietenia*). These taxa were all included in the arboreal pollen percentage diagram. However, *Pinus* and *Cecropia* were excluded. Pine pollen accounts for 4-5 percent of the total pollen sum throughout the core, except in the surface sediments where it reaches a maximum of 10 percent. *Cecropia* is the most abundant pollen type in the sediments, averaging about 35 percent of the total pollen sum, and about 50 percent of the arboreal pollen sum.

Grasses are the major contributor to the "weeds" category, representing 10-20 percent of the pollen sum. The Cyperaceae and Chenopodiaceae/Amaranthaceae (Cheno/Am group) each average less than 10 percent. Less abundant components of the "weeds" group include *Ambrosia* and other composites, Solanaceae, *Pilea*, and additional herbs. A few grains of *Zea mays* were counted in the uppermost 6 cm of the core.

The mud-water interface core can be divided into an upper and lower section based on the organic matter/carbonate content of the dried sediments. Lake sediments below 30 cm average about 30 percent organic matter and about 15 percent carbonates (as CO_2). In the topmost 30 cm of the core, the organic matter content averages about 15 percent, while carbonates account for some 27 percent of the dry weight.

Figure 3 shows the concentration of pollen and carbonized plant fragments throughout the core. Weed and arboreal taxa include the taxa as in Fig. 2, except that *Pinus* has been added. *Cecropia* remains excluded from the diagram. The "shrubs" category represents pollen of Melastomataceae and Euphorbiaceae. The "aquatics" include *Cladium*, *Typha*, *Potamogeton*, *Nymphaea*, *Nelumbo*, and less common taxa.

Discussion

Two stratigraphic zones can be distinguished in the Miragoane mud-water interface core (Fig. 2). Below 30 cm, sediments are brown in color and rich in organic matter. Above 30 cm, the sediments are gray and contain high levels of carbonate. The lithologic shift at 30 cm is coincident with a change in the pollen spectrum. Below the 30-cm horizon, arboreal pollen is plentiful, but above this level, weed pollen is more abundant.

The bottom half of the core preserves five centuries of the pre-Hispanic ecological record (1,000-500 YBP). The pollen and lithology of the section suggest a prolonged period of vegetation and soil stability in the region. Chemical and palynological changes in the top half of the core reflect nearshore human disturbances, including vegetation clearance and increased erosion. According to the extrapolated, ^{210}Pb-based chronology, major human impact began about 500 years ago, and continues today. The environmental deterioration coincides with the arrival of Europeans on Hispaniola in 1492.

Palynological evidence suggests that dry and mesic vegetation prevailed during pre-Columbian times. Evidence of dry forest is found in the pollen of Sapotaceae, *Phyllostylon*, Meliaceae (probably *Swietenia*), *Bursera*, and *Caesalpinia*. These plants were probably common in the riparian vegetation. *Trema* and *Celtis* were also abundant in the local flora, based on the large numbers of their pollen in the sediments. *Trema* and *Bursera*, and perhaps *Celtis* too, have broad ecological ranges, but are considered to be successional genera (Leyden 1985). Presence of these secondary trees as well as abundance of grass and shrub pollen in the bottom half of the core, indicate that the dry forest was rather open.

The moist forests of Haiti are currently species-rich. In the sediments, evidence of pre-Columbian moist forests is found in the pollen of *Cecropia* and several genera of Moraceae (Fig. 3). Presently, *Cecropia* is a very common component of lowland mesic forests on Hispaniola. High concentrations of *Cecropia* pollen in the bottom half of the core indicate that the genus was an important forest component prior to European settlement. The Moraceae pollen grains in the lower section of the core were probably derived from moist forests growing at higher elevations, some distance from the lake.

While the Miragoane region appears to have been forested during the period between 1,000 and 500 YBP, the area had been settled even earlier by native inhabitants. Ceramic remains indicate that earliest settlement of the region began around 600 A.D. More recent pottery styles are indicative of two additional settlement episodes dating to between 900 and 1,500 A.D. (Rouse and Moore 1984). The early Arawak settlers practiced fishing and swidden cultivation, planting *Manihot* (cassava) as a staple, as well as corn and other crops. Archaeological data for the region are limited and shed little light on prehistoric population densities or proximity of Arawak settlements to the lake shore. It remains uncertain as to whether the area was continuously occupied between 600 and 1,500 A.D.

Pollen of successional trees (*Cecropia*, *Trema*, and *Celtis*) as well as abundant carbonized fragments in the lower section of the core suggest pre-Columbian disturbances in the local forests. The presence of *Cecropia* pollen and carbonized fragments has been used to document the practice of shifting cultivation by the Maya at both lowland and highland sites in Guatemala (Tsukada and Deevey 1967, Vaughan 1979, Higuera-Diaz 1983). Prehistoric agriculturalists near Lake Miragoane undoubtedly contributed to local deforestation, but the occurrence of these microfossils in the lake sediments does not necessarily reflect riparian human activities. Preliminary study of deep sediment samples from the long (7.67 m) core revealed carbon fragments and *Cecropia* grains, together with *Trema* and *Celtis* pollen. The deep stratigraphic samples presumably predate human occupation of the area, and suggest that fires and consequent vegetation succession were natural processes in the Haitian forest. Until baseline (predisturbance) vegetation conditions are fully reconstructed, and archaeological settlement data are available, it will be difficult to discern natural from anthropogenic disturbances of the pre-Columbian forests. However, the pollen data clearly show that modern vegetation disruption is far more severe than at any time prior to the arrival of Europeans.

Pre-Columbian sediments, deposited between 1,000 and 500 YBP, are rich in organic matter. Below 30 cm, the sediments vary between 24.3 percent and 35.6 percent L.O.I. 550° C (Fig.2). The deposits are very rich in pollen. High organic matter and pollen concentrations suggest high input rates of pollen from local forests and slow bulk sediment accumulation rates, with little contribution from nearshore mineral soils. Prior to forest clearance, riparian vegetation stabilized basin soils, thereby preventing erosion.

Pollen and chemical data from the top 30 cm of the core provide evidence of a long-term period of deforestation and consequent soil erosion. The episode of human disturbance dates to the time of European contact, and continues at present in the Miragoane watershed. Three lines of evidence support the hypothesis of forest removal and downwasting of riparian soils: 1) total pollen concentrations in the sediment drop sharply above 30 cm 2) there is a shift in the relative abundance of pollen types above 30 cm, with weeds replacing trees 3) carbonates in the sediment, presumably of detrital origin, increase relative to organic matter above 30 cm (see Figs. 2 and 3).

At the transition point, where human disturbance becomes quite apparent, the total pollen concentration in the sediments drops from 6.5×10^5 to 0.13×10^5 grains/g (Binford et al. 1987). Forest removal eliminated much of the vegetation responsible for producing the pollen rain, thereby lowering the pollen numbers in the sediment. Additionally, as watershed vegetation was removed, riparian soils eroded at greater rates. The higher erosion rates are indicated by the elevated carbonate content of the sediments. The resulting increased sedimentation rates further diluted sedimenting pollen grains. Higher sedimentation rates during this period are corroborated by the low concentrations of carbonized fragments. Production rates of carbonized fragments are assumed to have been high during the phase of massive land clearance, but these fossils too, were diluted by high rates of allochthonous input to the lake. If sedimentation rates during the early land clearance period were substantially higher than those measured for the last century, downcore extrapolation of the ^{210}Pb chronology results in erroneously old ages being assigned to deeper levels in the section. On the other hand, if predisturbance sedimentation rates were much lower than rates measured in the top 8 cm, the true basal date for the core may be much older. Lacking radiocarbon dates from the section, and anticipating difficulties with hard-water-lake errors (Deevey and Stuiver 1964), the preliminary calculation of the age/depth relationship will have to suffice.

Sediments between 30 and 10 cm were deposited over three centuries and correspond to the periods of Spanish (1,500-1,700 A.D.) and French (1,700-1,800 A.D.) occupation of Haiti. Pollen and lithologic changes in this sediment interval indicate disturbance of regional and local vegetation and soils. Forest removal is clearly recorded by the progressive decline in pollen of trees relative to herbs (Fig. 2). During this time, successional trees such as *Cecropia* and *Trema* grew in some areas. *Cecropia* seems to have been more abundant than *Trema* (see Fig. 3 in which *Trema* is included in the "arboreal" category), perhaps because *Cecropia* can grow very rapidly in freshly cleared areas. The persistence of these trees suggests that while much of the landscape was under cultivation, some areas were recolonized by successional vegetation.

According to the ^{210}Pb chronology, sediments from 8 to 6 cm were deposited between 1852 and 1944 A.D. This horizon contains high concentrations of pollen grains, comparable to the amounts found in pre-Columbian sections of the core, below 30 cm. Between 8 and 6 cm there is an increase in the abundance of tree pollen relative to weeds, and the organic content of the mud increases relative to carbonates. The evidence suggests some recovery of the local forest and stabilization of the watershed soils. *Cecropia*, *Trema*, *Bursera* and *Celtis* show higher percentages in this horizon, indicating temporary recovery of successional forest.

Because the two centimeter section (8-6 cm) corresponds to nearly a century of sedimentation, it is impossible to define accurately the timing of the recovery event. Nevertheless, the time frame during which these sediments were deposited corresponds to a period in Haitian history when social and political changes may well have encouraged reforestation. Following Haiti's independence from France in 1804, large, French-run plantations were replaced by small plots used for subsistence agriculture. The land distribution program created numerous small farms throughout the country, and higher elevation sites were exploited for the first time (Woods 1987, Holdridge 1945). Thus, reforestation may have been promoted by the elimination of the plantation system, coupled with a return to subsistence practices, and the possible emigration of people from the Miragoane watershed to newly available farm sites. The degree and areal extent of forest regrowth during the latter part of the 19th and early part of the 20th century is an important question requiring additional study.

The uppermost 6 cm of the Miragoane short core record the last four decades (1944-1985 A.D.) of environmental history. These deposits are characterized by carbonate-rich sediments with low organic content, and very low concentrations of all pollen types (Figs. 2 and 3). Once again, the sediments preserve a record of acute local deforestation and consequent soil erosion. The watershed is nearly totally deforested today. Few trees remain, and those that do are virtually all domestic fruit trees. Elsewhere, the drainage is characterized by agricultural fields, both in use and abandoned, and some isolated patches of secondary growth. Steep slopes on the south side of the lake are characterized by badly eroded soils with exposed bedrock.

Sediments from Lake Miragoane contain a well-preserved record of long-term human disturbance of riparian vegetation. The Miragoane watershed, with its abundant supply of fresh water, presented an ideal location for both early Arawak and later European settlement. As human population densities increased, other lowland sites were undoubtedly exploited. Today, lowland forests throughout Haiti have been decimated, a process that probably began centuries ago at many sites. Interpretations of contemporary Hispaniolan biogeography must consider the long history of human impact on Haitian landscapes.

Acknowledgements

This work was supported by NSF grant BSR 85-00548 awarded to M.W. Binford and E.S. Deevey. I thank M.W. Binford and M. Brenner for obtaining the core and collecting limnological data. J.D. Skean collected and identified aquatic and terrestrial plants. Chemical analysis of the core was done by M. Brenner, and M.W. Binford ^{210}Pb-dated the section. I thank M. Brenner, M.W. Binford and E.S. Deevey for help with preparation of the manuscript.

Literature cited

Binford, M.W., M. Brenner, and A. Higuera-Gundy. 1986. Limnology and sedimentology of Lake Miragoane, Haiti. Bulletin of the American Society of Limnology and Oceanography (Abstract).

_____, M. Brenner, T.J. Whitmore, A. Higuera-Gundy, E.S. Deevey, and B. Leyden. 1987. Ecosystems, paleoecology and human disturbance in subtropical and tropical America. Quaternary Science Reviews 6:115-128.

Dean, W.E., Jr. 1974. Determination of carbonate and organic matter in calcareous sediments and sedimentary rocks by loss on ignition: comparison with other methods. Journal of Sedimentary Petrology 44:242-248.

Deevey, E.S. and M. Stuiver. 1964. Distribution of natural isotopes of carbon in Linsley Pond and other New England lakes. Limnology and Oceanography 9:1-11.

Higuera-Diaz, A. 1983. A paleolimnological record of human disturbance in Lakes Atitlan and Ayarza, Guatemala. Unpublished M.S. thesis, University of Florida, Gainesville, FL. 66 pp.

Holdridge, L.R. 1945. A brief sketch of the flora of Hispaniola. Pp. 76-78 in F. Verdoorn (ed.). Plants and Plant Science in Latin America. Chronica Botanica Co., Waltham, MA. 381 pp.

Lewis, W.M., Jr. 1983. A revised classification of lakes based on mixing. Canadian Journal of Fisheries and Aquatic Sciences 40:1779-1787.

Leyden, B.W. 1985. Late Quaternary and Holocene moisture fluctuations in the Lake Valencia basin, Venezuela. Ecology 66:1279-1295.

Rouse, I. and C. Moore. 1984. Cultural sequence in southwestern Haiti. Bulletin de Bureau National d'Ethnologie 1:25-38.

Tsukada, M. and E.S. Deevey. 1967. Pollen analyses from four lakes in the southern Maya area of Guatemala and El Salvador. Pp. 303-331 in E.J. Cushing and H.E. Wright, Jr. (eds.). Quaternary Paleoecology. Yale University Press, New Haven, CT. 433 pp.

Vaughan, H.H. 1979. Prehistoric disturbance of vegetation in the area of Lake Yaxha, Peten, Guatemala. Unpublished Ph.D. dissertation, University of Florida, Gainesville, FL. 176 pp.

Woodring, W.P., J.S. Brown, and W.S. Burbank. 1924. Geology of the Republic of Haiti. Republic of Haiti Department of Public Works, Geologic Survey of the Republic of Haiti, Port-Au-Prince. 631 pp.

Woods, C.A. 1987. The threatened and endangered birds of Haiti: lost horizons and new hopes. Proceedings Delacour/IFCB Symposium 2:385-430.

Wright, H.E., D.H. Mann, and P.H. Glaser. 1984. Piston corers for peat and lake sediments. Ecology 65:657-659.

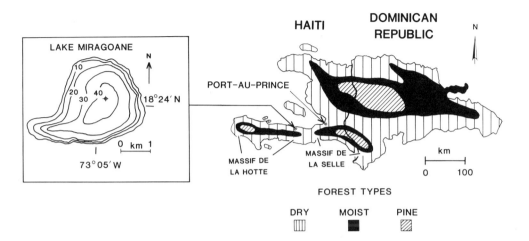

Figure 1. Vegetation map of Hispaniola showing the distribution of dry, moist and pine forests (modified from Holdridge 1945). Lake Miragoane is located on the north shore of Haiti's southern peninsula. The inset map shows the lake bathymetry at 10-m contour intervals. Mud-water interface core 17-VII-85-1 was taken in deep water at the center of the lake.

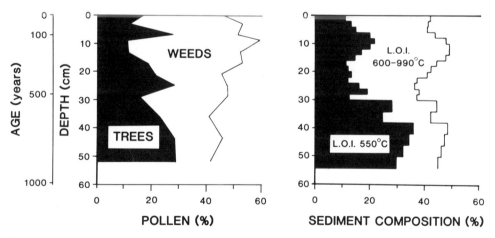

Figure 2. Pollen, organic matter and carbonate stratigraphy of Lake Miragoane core 17-VII-85-1 (modified from Binford et al. 1987). The "trees" category includes pollen of Moraceae, *Celtis*, *Trema*, *Bursera*, *Alchornea*, *Phyllostylon*, *Sapindus*, *Caesalpinia*, Sapotaceae, Myrtaceae, and Meliaceae. *Pinus* and *Cecropia* are excluded. The "weeds" include grasses, Cyperaceae, Chenopodiaceae/Amaranthaceae, *Ambrosia* and other composites, Solanaceae, *Pilea*, and other herbs. L.O.I. 550° C is a measure of organic matter content. L.O.I. 600-990° C represents the weight lost from carbonates (as CO_2) during combustion.

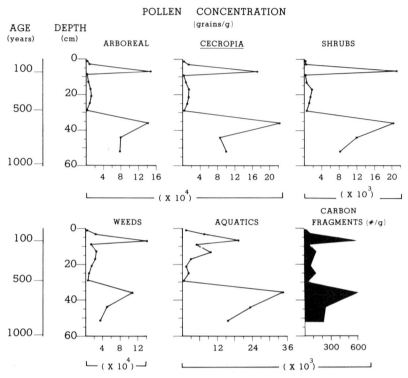

Figure 3. Concentrations of pollen and carbon fragments in Lake Miragoane core 17-VII-85-1. Concentrations are reported as number per gram dry sediment. The "arboreal" and "weed" categories include taxa as in Fig. 2, except that *Pinus* is included in the "arboreal" sum. *Cecropia* is plotted separately. "Shrubs" include pollen of Melastomataceae and Euphorbiaceae. "Aquatics" include *Cladium, Typha, Potamogeton, Nymphaea, Nelumbo*, and less common types.

QUATERNARY BIOGEOGRAPHICAL HISTORY OF LAND SNAILS IN JAMAICA

Glenn A. Goodfriend[1]

Abstract

A reconstruction of the history of regional endemism of land snails within the island of Jamaica is presented, based on a review of four Quaternary fossil deposits. The deposits represent three different faunal regions of the island: the central interior plateau, the western interior plateau, and the north coast. Three of the deposits contain material from the late Wisconsin and the Holocene, while the fourth is probably of Middle Pleistocene age. The predominant pattern at the interior sites is the similarity of the Pleistocene faunas to the modern ones at each site. In the great majority of cases, the endemics which are characteristic of each of the regions today also occur in the Pleistocene record at each site. However, some taxa (e.g., chondropomids) show greater stability of ranges through time than others (e.g., *Sagda*). Deposits from the north coast show a pattern different from the interior deposits, in that numerous local extinctions are recorded. These occurred around latest Wisconsinan to early Holocene time, apparently in relation to the warming trend, and again in latest Holocene time, apparently in relation to habitat disturbance by humans. The Wisconsinan faunas of the north coast deposits have no modern analogs, as is generally the case with Wisconsinan land snail faunas in the temperate zone. The fossil record does not support the Pleistocene refugium hypothesis as a cause of regional diversity of the Jamaican land snail faunas.

Introduction

Biogeography is essentially a historical science. Nevertheless, the data base most commonly used in biogeographical studies is the modern distributions of species. The temporal component is introduced through a hypothesis as to what influences were responsible for the development of the particular pattern of distribution. It is only through the fossil record that facts to replace hypotheses can be obtained. The shortcomings of the fossil record are well known. In general, speciation events are very poorly recorded and records of immigration events are nonexistent. However, the fossil record does tell us what taxa existed at a particular place at a particular time and this can be of considerable value in evaluating biogeographical scenarios.

One such biogeographical scenario amenable to evaluation through the fossil record is the Pleistocene refugium hypothesis (Haffer 1969). This holds that regional

---

[1] Dr. Goodfriend is in the Isotope Department, Weizmann Institute of Science, 76100 Rehovot, Israel.

endemism in the tropics results from differentiation of forms in isolated wetter patches (refugia), where forest persisted during dry glacial times. During wetter interglacial times (such as the present Holocene) these forests expanded to form a continuum, with the ranges of forest organisms expanding correspondingly and sometimes meeting to form the pattern presently seen. From this scenario, several predictions can be made which can be evaluated by the fossil record: 1) in most presently-moist areas, biotas of glacial periods should include or consist mostly of forms now inhabiting drier areas; 2) in a few presently-moist areas (especially the wettest ones), biotas of glacial periods should be similar to the present biotas; and 3) in drier areas, little difference should be seen between glacial and interglacial biotas.

These predictions are evaluated here based on four Quaternary fossil deposits of land snails from Jamaica. It is shown that none of the deposits is in agreement with the predictions of the Pleistocene refugium hypothesis. The fossil deposits come from three different faunal provinces in Jamaica, each representing a different climatic regime and having a distinct set of regionally endemic species. In this latter respect of showing high differentiation of faunas among regions, Jamaican land snails are classical tropical organisms (cf. various taxa in Prance 1982).

Central and western Jamaica consists of an interior plateau at c. 400-600 m elevation, with a narrow north coastal plain and a generally wider southern coastal plain (Fig. 1). The bedrock is limestone, except for a few areas where the limestone has been breached by erosion to expose non-calcareous strata. Rainfall is generally low on the coasts and higher in the interior, with the western interior plateau being especially wet (Fig. 1). Each of these regions possesses a land snail fauna containing species endemic to that region. The north coast fauna is represented by the deposits at Green Grotto Cave (Fig. 1), the central interior region by a cave deposit at Coco Ree, and the western interior region (Cockpit Country) by deposits at Sheep Pen Cave and Bonafide Cave.

Dating methods

The general approach to dating of the deposits involved the combined use of radiocarbon and amino acid epimer analysis (alloisoleucine/isoleucine ratio; hereafter A/I). The latter has the major advantage that it can be used to analyze individual shells, since only a small amount is required for analysis. This permits identification of mixed-age deposits and determination of ages of individual shells within such deposits (Goodfriend 1987b; Goodfriend and Mitterer 1988; Goodfriend and Mitterer ms). The ease of these analyses also permits large numbers of samples to be measured. However, the amino acid epimer ratios provide only a relative time scale. To obtain absolute age estimates, the amino acid time scale must be calibrated by another dating method (here, radiocarbon dating is used). The calibration differs among sites according to their temperatures, since the rate of epimerization is highly temperature dependent (it doubles with every c. 4°C increase; Mitterer 1975).

Radiocarbon dates on the carbonate of land snail shells from limestone areas require a correction for an age anomaly which results from incorporation of old carbonate carbon into the shell (Goodfriend and Stipp 1983). Limestone (lacking in ^{14}C) is ingested by the snails and digested to CO_2, which becomes dissolved in the bicarbonate pool of body fluids from which the shell carbonate is precipitated (Goodfriend and Hood 1983).

The age anomalies range up to a maximum of c. 3,000 yr (Goodfriend and Stipp 1983). The dates reported here have already been corrected for this age anomaly, as well as for isotopic fractionation (Goodfriend 1987a).

The fossil deposits
Summary of chronologies and land snail faunas

Coco Ree

The deposits at Coco Ree (Goodfriend and Mitterer ms) comprise a meter of mostly unconsolidated sediments within an unnamed cave in the interior plateau region (Fig. 1). Radiocarbon ages of bulk samples of land snails range from 13,400 yBP at 10-20 cm, to 30,100 yBP at 55-65 cm, to 36,300 yBP (but with a large uncertainty, c. ±4,000 yr) at 80-90 cm. Amino acid epimer analyses reveal that each layer contains a mixture of shells of quite variable ages, so the ^{14}C dates represent only average ages. However, below 55 cm depth, all measured A/I ratios indicate that shells are of Wisconsinan age, so glacial-interglacial faunal comparisons may be made between the faunas from these layers and the fauna presently inhabiting the area.

The Wisconsinan fauna in the deposits (31 species) is nearly identical with the modern fauna of the area. One species (*Helicina jamaicensis* Sow.) present from Wisconsinan to mid-Holocene times now occurs only to the west of the area; *Pleurodonte lucerna* (Müller) was replaced by its sister semispecies *P. sublucerna* (Pils.) in the late Wisconsin and now occurs 4 km to the northeast: *Pleurodonte aspera* (Fér.) was introduced into the area recently, perhaps on banana trees on which it lives. Apart from these, no faunal changes are seen between the Wisconsin and the present.

Bonafide

Deposits in Bonafide Cave, in the western interior plateau region (Fig. 1), were excavated by Mr. Fred Grady et al. of the U.S. National Museum to a depth of c. 40 cm (pit A). A description of the cave is given by Baker et al. (1986). Thus far, only preliminary analyses have been carried out. The deeper layers consist of shells of nearly uniform age (G. Goodfriend unpublished data), dated by radiocarbon at c. 13,000 yBP (RT-808). A few older shells also occur within these layers. The upper layer (0-15 cm) consists mostly of late Holocene shells, of variable ages, and contains also some shells of the same age as those in the deeper layers. The older shells are generally recognizable by their bleached appearance and the presence of patches of chalky texture (where the shell carbonate has partially broken down).

A tremendous difference in species diversity is seen between the upper and lower layers, with the upper 15 cm containing 69 species and the lower 15 cm containing 32 species. Most of this difference is accounted for by the complete absence of very small species (< 2 mm minimum dimension) in the lower layers (the upper layer contains 20 such species) and the rarity of shells in the 2-3 mm range in the lower layers. Their absence appears to be due to dissolution - the same process which produced partial breakdown of shell carbonate in the larger species. Considering only the larger shells (species > 3 mm), the lower 15 cm still has c. 20% fewer species than the upper 15 cm. However, a second factor must be taken into account in comparing these: the lower layers contain considerably fewer specimens than the upper. Consequently, the rarer species are more likely to be missing from the lower layer than from the upper, due to sampling error. Of the abundant species in the upper 15 cm (represented by > 10 specimens), most (13 out of 15) also occur in the lower 15 cm. Missing are *Stauroglypta anthoniana* (C.B. Ads.), a rather thin-shelled species which may have been lost below due to dissolution, and *Helicina jamaicensis*. This latter species interestingly shows quite the opposite of the pattern displayed at Coco Ree, where it is abundant in the Wisconsinan layers but absent from the area today. A recent shift in the range of this species may have occurred.

Thus the Wisconsinan fauna at Bonafide seems to represent a subset of the Holocene fauna, with missing taxa probably resulting from dissolution of shells and smaller sample number. No large differences between the faunas are indicated, although a statistical analysis is required to objectively and quantitatively assess the sample size effects.

Sheep Pen

The deposit at Sheep Pen Cave, located in the western interior region northeast of Bonafide Cave (Fig. 1), consists of a hard, fully indurated fissure fill breccia containing land snail shells, bones, limestone fragments, and soil peds (Goodfriend 1986a). The land snail shells are in most cases completely dissolved away and represented only by external and internal molds. The external molds retain in beautiful detail the sculptural features of the surfaces of the shells, which permitted even most small fragments of shells to be identified to species. Dating of the deposit is problematic. Two fragments of original shells yielded A/I ratios higher than the equilibrium value (R.M. Mitterer, pers. comm.), whereas bones and even tooth enamel were found to be totally lacking in amino acids (J.L. Bada, pers. comm.). This, combined with lithological evidence indicating a history of several dissolution-precipitation cycles (Goodfriend 1986a), suggests a very old age for the deposit, probably Middle Pleistocene or perhaps earlier.

The land snail fauna is quite rich (48 species; 47 reported in Goodfriend 1986a, plus the subsequent discovery of *Carychium jamaicense* Pils.). The assemblage is very similar to that of the same area today, but contains three species which appear to be extinct. Several species in the deposit are outside of their present ranges but no consistent geographic shift is seen - the species now occupy ranges not far away to the north (drier), to the south (wetter), and to the west (wetter). Given the considerable age of the deposit, the resemblance of the faunal assemblage to the modern fauna argues for considerable faunal stability.

Green Grotto

Land snails in six deposits from in and around Green Grotto Cave in the north coast region (Fig. 1) were studied (Goodfriend and Mitterer 1988). Five of the deposits come from fills of small solution holes, exposed by blasting away of the cliff next to the main cave entrance, while the sixth came from a thin layer of deposits in an entrance chamber of the cave. Two of the samples are breccias, with an indurated matrix, and date to 30,000 yBP (radiocarbon date) and c. 41,000 yBP (age estimate from A/I ratio, based on calibration from other breccia sample). Two deposits with mostly friable sediments have average ages of c. 16,000 yBP and 19,000 yBP (age estimate from A/I ratios, based on calibration from breccia sample) but the individual shells within them show considerable variation in age. The other two deposits consist of friable sediments. One is of late

Holocene age, whereas the other (from the cave chamber) is an age mixture containing mostly Holocene shells and a few late Wisconsinan shells.

In contrast to the other deposits, the Green Grotto deposits indicate considerable differences between the Pleistocene faunas and the present fauna. The differences are primarily the result of local extinctions of species present in the late Wisconsinan deposits. Of the 36 species present in the deposits, only 12 live at the site today. These extinctions appeared to have occurred mainly during two periods: latest Wisconsin to early Holocene and latest Holocene (last millenium). The former period of extinction seems to have been the result of the warming trend at the end of the Pleistocene. Several of the species that disappeared around that time are now restricted to or occur predominantly in the cooler interior areas. The latest Holocene extinctions may be attributed primarily to human disturbance of the habitat (e.g., selective cutting of trees).

History of patterns of regional endemism

One characteristic common to all the fossil sites studied is the presence during the Pleistocene of species presently endemic to the different regions represented by the fossil sites. In each case, the Pleistocene faunas suffice to unambiguously identify the region from which the deposit came because of their similarity to the modern faunas.

Endemic species presently characteristic of the faunal regions represented by the fossil sites, and appearing also in Pleistocene deposits at the sites, are shown in Fig. 2. The Wisconsinan deposits at Green Grotto on the north coast contain two species restricted to the north coast today and which do not appear in any of the other deposits (*Eutrochatella costata* (Sow.) and *Colobostylus albus* (Sow.); see Figs. 3b, 4 for *C. albus*). Two species presently characteristic of the north-central region appear in both the Coco Ree and Green Grotto Wisconsinan deposits but not in the western interior deposits. Both species live at both north-central sites today. The Wisconsinan layers at Coco Ree contain a number of species endemic to that area today and which do not occur in any other deposit (e.g., *Annularia pulchrum* (Wood), Figs. 3d, 4).

Similarly, the two deposits in the western interior contain a large number of species endemic to that region today, and which do not occur in either of the north-central region deposits. Interestingly, differences in the presence of local endemics between the two western interior deposits are also seen, despite their proximity. The modern land snail fauna of the area around Sheep Pen Cave has been extensively sampled by the author and is quite well known. Unfortunately, the modern fauna of the Bonafide Cave area is very poorly known at present. However, most of the species in the late Wisconsinan layers which are unique to this deposit have been recorded from adjoining areas to the west and south, where limited sampling has been done. Still, enough is known to reveal some interesting local patterns. For example, the local endemics *Colobostylus humphreyanus* (Pfr.) (Figs. 3c, 4) and *Urocoptis hollandi* (C.B. Ads.), abundant in the Wisconsinan layer at Bonafide, do not occur in the Sheep Pen deposit and do not live in the Sheep Pen area. Similarly, in the Pleistocene deposit of Sheep Pen, there occur species characteristic of that area today and which do not occur in the Bonafide deposits (Wisconsinan and Holocene) nor (apparently) live in the Bonafide area today (e.g., *Adamsiella miranda* (C.B. Ads.), Figs. 3a, 4).

Thus the pattern emerges clearly that today's pattern of regional endemism was already well established in the Wisconsin. The Sheep Pen deposit suggests a pre-Middle Pleistocene origin for these patterns.

Differences in geographic stability among taxa

Although the predominant historical biogeographic pattern is one of stability of geographic ranges through time, some differences are apparent among higher taxa (families, genera) with respect to the degree of stability of their ranges. Two examples - the family Chondropomidae and the genus *Sagda* (in the Sagdidae) will suffice to show the range of these patterns.

The Chondropomidae (=Annulariidae) is a very diverse family of Neotropical prosobranchs and is represented in Jamaica by four genera. Most of the species are quite localized in their distributions (e.g., Fig. 4; Jarvis 1903). In general, the Pleistocene and modern chondropomid faunas are very similar to each other at each of the sites (Table 1). One species, *Colobostylus banksianus* (Sow.), occurring in Pleistocene deposits at both of the Cockpit Country (western interior) sites, is not known to occur at either site today but occupies a range in the southern Cockpit Country, some kilometers south of Bonafide Cave. A *Parachondria* sp. at Sheep Pen is not known from modern collections and is apparently extinct. At Bonafide, two species occur as single specimens in the top (0-15 cm) layer, which is mostly of Holocene age but are not present in the late Wisconsinan layers. However, because of their rarity and the small sample size of these layers (see above), this cannot be taken to indicate their absence from the area during the Wisconsin. Apart from these few examples, no changes are seen in the chondropomid faunas at the sites from Pleistocene to the present.

A different situation is seen in the pulmonate genus *Sagda*. The genus is endemic to Jamaica and is represented by over 25 species. Most *Sagda* spp., like the chondropomids, have quite restricted distributions within the island (e.g., *S. bondi* Vanatta, Fig. 3e, and *S. montegoensis* Pils. & Brown; Goodfriend 1986b). Examination of the Pleistocene fossil record of *Sagda* indicates that considerable faunal changes have occurred at most sites since then (Table 2). In the Green Grotto area, only one *Sagda* species exists today (*S. centralis* Goodfriend) but three other species, not occurring in the area today, occur there in the late Wisconsinan deposits. *Sagda centralis* appeared at the site in the late Holocene but subsequently became extinct there, probably due to habitat disturbance (Goodfriend and Mitterer 1987). It occurs today elsewhere in the vicinity (Goodfriend 1986b). At Sheep Pen, only one of the three fossil species occurs in the area at present. However, the modern fauna includes another species, *S. grandis* Pils. & Brown (Fig. 3f), which does not appear in the fossil record. A similar pattern occurs at the other Cockpit Country site, Bonafide Cave, where *S. grandis* (as well as another unnamed species) occurs in the mostly Holocene upper layer (several specimens) but not in the Wisconsinan layers. In the Wisconsinan layers are two species not present in the upper layer and which are not known to inhabit the area today. Two other species occur throughout the deposits and the cave is within the known range of one of them (*S. foremaniana* (C.B. Ads.)); the other is known only from this cave deposit but will probably be found living in the vicinity of the cave when proper investigation is made. At Coco Ree, no faunal change is seen, with *S. spei* Pils. & Brown occurring throughout the deposits as well as living in the area at present.

Thus it is seen that, while some groups display a striking stability of their geographical ranges over time, others show considerable changes. While differences in their ecology must be involved, what these differences may be is not at all clear. Both *Sagda* spp. and the chondropomid spp. occupy similar microhabitats: most species occur (when inactive) under or among rocks, while some occur both among leaf litter as well as rocks (see Goodfriend 1986b for *Sagda* ecology).

The evolution of regional endemism in Jamaican land snails

The land snail faunas from the fossil sites permit evaluation of the applicability of the Pleistocene refugium hypothesis to Jamaican land snails, through testing of some of the predictions of the hypothesis (see Introduction). The major prediction of this hypothesis is that most sites in the moist interior should have experienced invasion of species from drier coastal areas during glacial periods; and that forms now characteristic of the interior should disappear from most sites. At Coco Ree, neither pattern is seen: all the present endemics characteristic of the area also appear in the Wisconsinan fossil record and no species characteristic today of drier areas appears in the Wisconsin. For Bonafide, a similar pattern is seen. No dry-area species entered the area during the late Wisconsin, and most of the present endemics are present also in the Wisconsinan layers. However, some of the present endemics are not present. As pointed out above, further study is needed to ascertain whether this is simply a problem of statistical sampling error or represents a real phenomenon. The ancient Sheep Pen fauna also argues for long-term stability of endemics and non-occurrence of dry-area species. However, it is not known whether this represents an interglacial fauna, a glacial fauna, or a mixture of the two. These faunal results are consistent with the Pleistocene refugium hypothesis only if each site happens to be in an area that was a refugium during the Pleistocene. This is unlikely simply for statistical reasons. But the sites also represent unlikely locations for refugia: none occurs in a particularly wet area. The Coco Ree area is bordered on the east by the wetter Mount Diablo area, while both Cockpit Country sites lie north of the wettest part of the Cockpit Country (Fig. 1).

No particular change in the composition of the dry-area fauna is predicted from the Pleistocene refugium hypothesis. Yet it is the opposite that actually occurred: it is the drier coastal fauna which was most affected by the glacial-interglacial transition, with species now limited to the interior disappearing from the north coast, apparently as a result of increased temperature. Contraction rather than expansion of ranges occurred. The occurrence of species endemic to the north coast is also not consistent with the Pleistocene refugium hypothesis, since this area has clearly been relatively dry throughout the Late Quaternary (Goodfriend and Mitterer 1988). The local endemics remained in place throughout the climatic changes which occurred during this time.

For the most part, species presently endemic to other parts of the north coast never occurred at Green Grotto in the Late Quaternary. To the east are endemics of the St. Mary coast (e.g., *Adamsiella pulchrius* (C.B. Ads.), *Parachondria aurora* (C.B. Ads.), and an undescribed *Poteria*) and an undescribed *Alcadia* from the Roaring River area of the St. Ann coast (24 km east of Green Grotto). To the west are coastal endemics of St. James (e.g., *Colobostylus chevalieri* (C.B. Ads.) and *Urocoptis gravesi* (C.B. Ads.)) and Trelawny (e.g., *Pleurodonte bronni* (Pfr.), known from 10 km W of Green Grotto). Two local north coast endemics, *Urocoptis brevis* (Pfr.) and *U. hendersoni* Pils., which now occur only outside the Green Grotto area, do occur in the Wisconsinan deposits at the cave. These appear to have occupied expanded ranges during glacial times, as did some of the interior species which then had ranges extending down to the coast.

Because only a few fossil sites have been studied, the details of the ranges of Jamaican land snail species during glacial times are not yet known. However, evidence from these sites is sufficient to show that the Pleistocene refugium hypothesis does not explain the patterns of regional endemism in Jamaican land snails. There is considerable evidence for dry conditions in the tropics during glacial times (although little for Neotropical lowland areas; Colinvaux 1987) and indeed there is some evidence for Jamaica itself (for the north coast; Goodfriend and Mitterer 1988). But conditions were probably not dry enough in Jamaica to cause replacement of forests by grasslands. Temperature changes seem to have played a more important role than moisture changes in the biogeographical histories of the snails, at least along the north coast.

No single scenario of regional differentiation appears to be applicable to the Jamaican land snail fauna as a whole. Different taxa show different patterns of changes or stability of ranges through time. The various regions of endemism tend to be characterized by different climates. This suggests that regional differentiation of faunas might have occurred in relation to climatic gradients, as was suggested by Endler (1982) for Amazonian biotas. The lack of barriers to dispersal in the mostly limestone western two-thirds of the island (rivers mostly limited to the coastal areas) would seem to role out allopatric models. However, it is questionable whether these climatic gradients have been stable enough in Jamaica during the Quaternary to produce such a geographical fine-tuning of the land snail fauna. But until the climatic changes are understood quantitatively and the geographical ranges of snail species through time are known in detail, this scenario cannot be ruled out.

Differentiation of faunas in relation to more stable aspects of the environment, such as substrate, offers a possible explanation for the tendency for geographical stability of the ranges through time. Geographically varying lithologies of the limestone, for example, provide different sizes and shapes of crevices and holes, which provide resting and activity sites for rock-dwelling species. *Annularia pulchrum* (Fig. 3d) appears to be an inhabitant of large crevices in limestone cliffs. Its distribution (Fig. 4) could be related to the distribution of appropriately-sized crevices. However, the same patterns of regional endemism are seen also in species that are leaf litter dwellers, e.g., *Zaphysema tenerrima* (C.B. Ads.) in the central interior region and *Z. tunicata* (C.B. Ads.) in the western interior region. And the rock-dwellers of the genus *Sagda* show among the least stable ranges of any taxa. The search for a single, generally applicable theory of regional endemism appears to be futile. Different taxa have their own intricate relations to the environment and respond differently to changes in various aspects of the environment.

Pleistocene - Holocene land snail faunal transitions --
A global overview

Is there anything distinctive about the Quaternary history of land snails in the tropics, as compared to the rest of the world, which might account for their high regional endemism? This question can be addressed for the best known and best dated period, which encompasses the late Wisconsin and the Holocene.

In northern areas of North America and Eurasia, land snails obviously became extinct in glacial times as glaciers advanced over the area. In regions south of the ice sheets, the Wisconsinan faunas were generally composed of a mixture of species - some which still live in the same areas and others which now occur only in other areas either to the north or at higher elevations. Examples include glacial faunas of southern Illinois (Frye et al. 1974), southwestern Kansas and northwestern Oklahoma (Miller 1975), Kentucky (Browne and McDonald 1960; Browne and Bruder 1968), Britain (Kerney 1977), and Central Europe (Lozek 1986). In nearly every case, the combinations of species in the glacial period faunas do not occur together anywhere today.

Well south of the ice sheets in the southwestern U.S., the Wisconsinan faunas were also composed of mixtures of species still occurring in the areas and species now living outside the areas, to the north or at higher elevations (Metcalf 1967, 1970). In the subtropical coastal plain of Israel, the Wisconsinan faunas were similar to the modern faunas, except for an additional species (*Theba pisana* (Müller)) in the modern fauna, which was introduced in historic times (Heller and Tchernov 1978) and the presence in some Wisconsinan faunas of one species (*Trochoidea langloisiana* (Bourg.)) which presently inhabits drier areas to the east (Brunnacker et al. 1981). In the Negev Desert of southern Israel, Pleistocene faunas contain no species that did not also occur in the area sometime during the Holocene (Goodfriend unpub. data). However, large faunal changes occurred during the Holocene, primarily in relation to changing moisture conditions (Goodfriend 1986c). On the tiny subtropical island of Bermuda, the only faunal changes recorded during the Late Quaternary were extinctions of certain species (Gould 1969).

Thus the predominant pattern of Wisconsinan faunas from the temperate zone to the tropical north coast of Jamaica (and appearing also in a preliminary analysis of a high plateau fauna from tropical Mexico; Taylor 1967) is the combination of species still living in the same areas with species now occurring only at higher latitudes or elevations. Exceptions occur on small, low islands (e.g., Bermuda) where immigration of species from nearby areas is not possible; in some semiarid areas (e.g., Israel) where the effects of changes in rainfall predominate over those of temperature; and in moister tropical areas (with Jamaica as the sole example) where climatic changes were perhaps not of sufficient magnitude to cause major faunal changes. Studies of glacial period faunas from other tropical areas are needed to assess the generality of this latter pattern.

Acknowledgements

Mr. Peter ("Tattered Dress") Clarke assisted in the field on numerous occasions. Most of the amino acid epimer data were provided by Dr. R.M. Mitterer. Access to material in the Florida State Museum was provided by Dr. F.G. Thompson. Mr. K. Auffenberg (FSM) patiently arranged numerous loans of specimens and assisted in the enumeration of material. The Bonafide Cave material was provided by Mr. F. Grady and Dr. R. Hershler of the U.S. National Museum.

Literature cited

Baker, L.L., E.A. Devine and M.A. DiTonto. 1986. Jamaica: the 1985 expedition of the NSS Jamaica Cockpits project. National Speleological Society News 44(1):4-15.

Browne, R.G. and D.E. McDonald. 1960. Wisconsin molluscan faunas from Jefferson County, Kentucky. Bulletins of American Paleontology 41(189):165-183.

_____, and P.M. Bruder. 1968. Wisconsin molluscan faunas from Henderson County, Kentucky. Bulletins of American Paleontology 54(241):191-275.

Brunnacker, K., H. Schütt and M. Brunnacker. 1981. ?ber das Hoch- und Spätglazial in der Küstenebene von Israel. Pp. 61-79 in W. Frey and H.-P. Uerpmann (eds.). Beiträge zur Umweltgeschichte des Vorderen Orients. Beihefte zum Tübinger Atlas des Vorderen Orients, reihe A (Naturwiss.), no. 8. Ludwig Reichert Ver lag, Wiesbaden.

Colinvaux, P. 1987. Amazon diversity in light of the paleoecological record. Quaternary Science Reviews 6:93-114.

Endler, J. 1982. Pleistocene forest refugia: fact or fancy? Pp. 641-657 in G.T. Prance (ed.). Biological Diversification in the Tropics. Columbia University Press, New York.

Frye, J.C., A.B. Leonard, H.B. Willman, H.D. Glass and L.R. Follmer. 1974. The late Woodfordian Jules Soil and associated molluscan faunas. Circular of the Illinois State Geological Survey (486):1-11.

Goodfriend, G.A. 1986a. Pleistocene land snails from Sheep Pen Cave in the Cockpit Country of Jamaica. Proceedings of the 8th International Malacological Congress. Hungarian Natural History Museum, Budapest:87-90.

_____. 1986b. Radiation of the land snail genus *Sagda* (Pulmonata: Sagdidae): comparative morphology, biogeography and ecology of the species of north-central Jamaica. Zoological Journal of the Linnean Society 87:367-398.

_____. 1986c. Holocene shifts in rainfall in the Negev Desert, as inferred from isotopic, morphological, and faunal analysis of fossil land snails. Pp. 122-124 in W.H. Berger and L.D. Labeyrie (eds.). Book of Abstracts and Reports, Conference on Abrupt Climatic Change. Scripps Institut of Oceanography, La Jolla, CA. SIO ref. ser., no. 86-88.

_____. 1987a. Radiocarbon age anomalies in shell carbonate of land snails from semi-arid areas. Radiocarbon 29(2):159-167.

_____. 1987b. Chronostratigraphic studies of sediments in the Negev Desert, using amino acid epimerization analysis of land snail shells. Quaternary Research 28:374-392.

_____, and D.G. Hood. 1983. Carbon isotope analysis of land snail shells: implications for carbon sources and radiocarbon dating. Radiocarbon 25(3):810-830.

_____, and R.M. Mitterer. 1988. Late Quaternary land snails from the north coast of Jamaica: local extinctions and climatic change. Palaeogeography, Palaeoclimatology, Palaeoecology 63:293-311.

_____, and R.M. Mitterer. ms. Late Quaternary land snail faunal history and chronostratigraphy of cave sediments at Coco Ree, Jamaica.

_____, and J.J. Stipp. 1983. Limestone and the problem of radiocarbon dating of land-snail shell carbonate. Geology 11:575-577.

Gould, S.J. 1969. Land snail communities and Pleistocene climates in Bermuda: a multivariate analysis of microgastropod diversity. Proceedings of the North American Paleontological Convention:486-521.

Haffer, J. 1969. Speciation in Amazonian forest birds. Science 165:131-137.

Heller, J. and E. Tchernov. 1978. Pleistocene landsnails from the coastal plain of Israel. Israel Journal of Zoology 27:1-10.

Jarvis, P.W. 1903. Distribution of Jamaican species of *Colobostylus*. Nautilus 17(6):62-65.

Kerney, M.P. 1977. British Quaternary non-marine Mollusca: a brief review. Pp. 31-42 in F.W. Shotton (ed.). British Quaternary Studies: Recent Advances. Clarendon Press, Oxford.

Lozek, V. 1986. Quaternary malacology and fauna genesis in Central Europe. Proceedings of the 8th International Malacological Congress. Hungarian Natural History Museum, Budapest:143-145.

Metcalf, A.L. 1967. Late Quaternary mollusks of the Rio Grande Valley, Caballo Dam, New Mexico to El Paso, Texas. Science Series, no. 1. Texas Western Press, University of Texas at El Paso:1-62.

_____. 1970. Late Pleistocene (Woodfordian) gastropods from Dry Cave, Eddy County, New Mexico. Texas Journal of Science 22(1):41-46.

Miller, B.B. 1975. A sequence of radiocarbon-dated Wisconsinan nonmarine molluscan faunas from southwestern Kansas-northwestern Oklahoma. Pp. 9-18 in G.R. Smith and N.E. Friedland (eds.). Studies on Cenozoic Paleontology and Stratigraphy. University of Michigan, Museum of Paleontology, Papers on Paleontology, no. 12.

Mitterer, R.M. 1975. Ages and diagenetic temperatures of Pleistocene deposits of Florida based on isoleucine epimerization in *Mercenaria*. Earth and Planetary Science Letters 28:275-282.

Prance, G.T. (ed.). 1982. Biological Diversification in the Tropics. Columbia University Press, New York.

Taylor, D.W. 1967. Late Pleistocene nonmarine mollusks from the state of Puebla, Mexico. Annual Report of the American Malacological Union (1967):76-78.

Table 1. Occurrence of Chondropomidae in the Pleistocene at fossil sites and modern occurrence in area.

Site	Pleistocene fossils only	Pleistocene fossils and living in area	No Pleistocene fossils, living in area
Coco Ree		*Adamsiella grayana* sp. *Annularia pulchrum* *Colobostylus thysanoraphe* *Parachondria columna*	
Bonafide	*Colobostylus banksianus*	*Adamsiella moribunda* *Colobostylus humphreyanus*	*Adamsiella ignilabre*[1] *variabilis*[1]
Sheep Pen	*Colobostylus banksianus* *Parachondria* sp [2]	*Adamsiella ignilabre miranda* *Annularia fimbriatulum* *Colobostylus thysanoraphe*	
Green Grotto		*Colobostylus albus* *Parachondria fecunda*	

[1] Represented by one specimen in mostly Holocene 0-15 cm layer.
[2] Species now apparently extinct.

Table 2. Occurrence of *Sagda* spp. in the Pleistocene at fossil sites and modern occurrence in area.

Site	Pleistocene fossils only	Pleistocene fossils and living in area	No Pleistocene fossils, living in area
Coco Ree		*S. spei*	
Bonafide	*S. ? alveare*[1] *S.* sp[1]	*S.* sp[2] *S. foremaniana*	*S. grandis*[3] *S.* sp[3]
Sheep Pen	*S. connectens* *S. montegoensis*	*S. pila*	*S. grandis*
Green Grotto	*S. bondi* *S. montegoensis* *S. spei*		*S. centralis*[4]

[1] Not present in mostly Holocene 0-15 cm layer and outside the poorly-known present range of the species.
[2] Present in all layers but species not known from modern collections.
[3] Present only in mostly Holocene 0-15 layer.
[4] Late Holocene records; now extinct at site but living nearby.

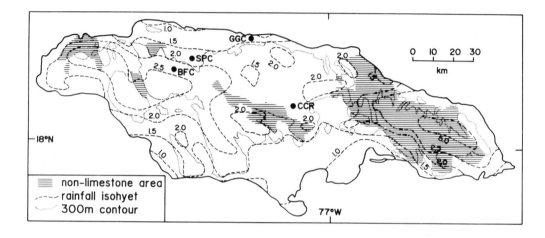

Figure 1. Map of Jamaica, showing mean annual rainfall isohyets (in m), 300 m elevational contours, non-limestone areas, and locations of fossil deposits. GGC = Green Grotto Cave, SPC = Sheep Pen Cave, BFC = Bonafide Cave, CCR = cave at Coco Ree.

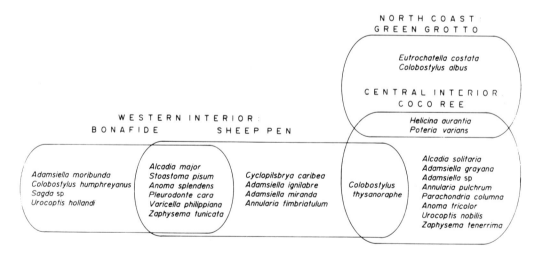

Figure 2. Pleistocene records of regionally endemic land snail species in the four fossil deposits.

Figure 3. Some Jamaican Chondropomidae and *Sagda*. A). *Adamsiella miranda*. B). *Colobostylus albus*. C). *Colobostylus humphreyanus*. D). *Annularia pulchrum*. E). *Sagda bondi*. F). *Sagda grandis*. Scale line in 5 mm.

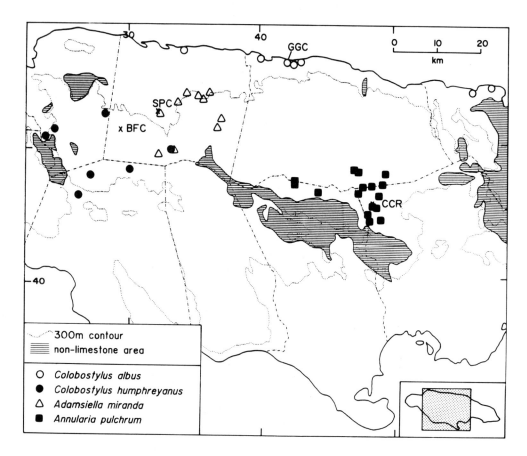

Figure 4. Modern distributions of some species of Chondropomidae, endemic to different regions of Jamaica.

THE POTENTIAL USE OF AMBER FOSSILS IN THE STUDY OF THE BIOGEOGRAPHY OF SPIDERS IN THE CARIBBEAN WITH THE DESCRIPTION OF A NEW SPECIES OF *LYSSOMANES* FROM DOMINICAN AMBER (ARANEAE: SALTICIDAE)

Jonathan Reiskind[1]

Abstract

Using the genus *Lyssomanes*, the value of Dominican amber to describe the biogeographic history of Caribbean spiders is considered. *Lyssomanes galianoae*, an amber fossil, is described for the first time. A probable history of the genus in the Greater Antilles is proposed utilizing cladistic analyses and geological evidence.

Introduction

The geographic distribution of spiders, as is the case in most organisms, is the consequence of a combination of both dispersal and vicariant events.

Sometimes it is possible to attribute the distribution entirely to dispersal--as when spiders are found on newly emerged land, either extensions of continental masses or new islands. This is indeed the origin of most of the spider fauna of peninsular Florida since vast expanses of it were submerged during the interglacial periods of the early Pleistocene. The source of the Florida fauna is predominantly North American, only about five percent attributable to Caribbean sources. The mode of spider dispersal in this case was a combination of both terrestrial and aerial locomotion--the latter being the ballooning unique to spiders. Ballooning along contiguous land was likely the most common form of dispersal during land recolonization. The ability of spiders to cross wide expanses of water in this way is limited to certain groups and should be considered rare in any case (Platnick 1976).

While only about 15 percent of the Florida fauna might be considered tropical in origin, only one-third of that component appears to have come via the West Indies, predominantly from Cuba and the Bahamas. The remainder came from Mexico probably via the widened coastal plain during glacial periods when the sea level was low.

But there is a vicariant element in the origin of the peninsular Florida spider fauna as well. This is reflected in its autochthonous element. Almost 20 percent of the spider fauna are species that probably arose on the islands of the Floridian archipelago of the early Pleistocene.

---

[1] Dr. Reiskind is an Associate Professor of Zoology, University of Florida, Gainesville, FL 32611.

The present distribution of spiders is more than the result of dispersal with or without subsequent vicariance. It is also the consequence of survival (that is, the avoidance of extinction).

This is especially true in insular situations where the chance of extinction may be significantly higher than on the mainland, as suggested in the theory of island biogeography (MacArthur and Wilson 1967).

Of course we might well ignore extinction if the only fauna for biogeographic consideration is the extant one. Who cares about what isn't there? You'll never know, so why worry about it? But there is an abundant number of past forms--both ancestral and collateral to our present organisms. And some of them are known--as fossils.

Although some consider it best to discount the importance of fossils in phylogenetic studies--that is, treat them merely as extant species for all intents and purposes--their fascination as potential ancestors, as well as their impact on other lineages, gives them a very special place in our imaginations and in our historical considerations.

When it comes to spiders there is, unfortunately, not a single known fossil of them in all of Florida. But in the West Indies such is not the case. There we have a narrow window into the obscurity of the past spider fauna, in the form of Dominican amber inclusions, and they aid us in the description of the historical basis for the distribution of spiders in the West Indies.

Discussion

Large and extensive deposits of amber are found in the Dominican Republic, on the island of Hispaniola, and they are conservatively estimated to be of Miocene age (about 20 million years before the present) (Wilson 1986).

There have been numerous publications on the spider and other arachnid inclusions of this amber by the German arachnologists Schawaller and Wunderlich and the latter is presently in the midst of publishing a major monograph on the fauna. Both have commented on the close similarity of the fossil fauna to the modern day fauna (at the generic level), yet the species are generally distinct.

The hope is that fossils will prove to be ancestral to modern forms. A cladistic analysis allows us to estimate the likelihood that a certain fossil is on the lineage, the end point of which we can examine in the present fauna.

This generates the question of whether the Miocene Dominican amber spider fossils are ancestral to the present day fauna of Hispaniola and the West Indies. One can never answer this question with absolute confidence or even with a good deal of confidence--for several reasons:

First, we have but a single window of perhaps a few million years, about 20 million years in the past, in which to observe the spider fauna--hardly the continuous fossil record we would like.

Second, the fossil record is a chancy one and there can be little confidence that the "collecting" is anywhere near comprehensive.

But we may be able to address the more reasonable question of whether it is probable that a certain fossil species is ancestral to one or more extant West Indian species. This is a far more restricted question and requires a case by case examination of both the fossils and their close living relatives.

Specifically I want to examine one case, answer that question, and propose a reasonable scenario for its present and past distribution. The case is that of the genus *Lyssomanes* (the only New World member of the salticid subfamily Lyssomaninae). This jumping spider genus is found throughout tropical America as well as in the southern United States (Galiano 1962, 1980).

Of the approximately 60 species in this genus, three are found in the West Indies. No island, large or small, has more than one extant species present, if one is present at all. When a fossil of a mature male *Lyssomanes* was found in the Dominican amber recently it offered a unique opportunity to examine the group's phylogeny and biogeography.

The fossil *Lyssomanes* belongs to the same species group to which all the West Indian species belong--the *antillanus* group. This is one of 12 species groups erected by Maria Elena Galiano (1980) and contains four species: *L. antillanus* Peckham and Wheeler, found on Cuba, Jamaica and Hispaniola, *L. portoricencis* Petrunkevitch (Puerto Rico and the Virgin Islands), *L. michae* Brignoli (Montserrat, in the northern Lesser Antilles), and the fossil species from the Miocene of Hispaniola, *L. galianoae* n. sp., which is described at the end of this paper.

This species group is surely monophyletic. The male genitalia is distinctive, with its pointed intromittent portion (the embolus) and a strong, angular median apophysis, making it an outstanding synapomorphy.

To ascertain the relationships among the members of the species group--especially that between the two Hispaniola species--a cladistic analysis was made establishing the likely sequence of speciation events.

A cladistic analysis of the group required the establishment of homologous traits and their polarity. The Central American *viridis* group was used as an outgroup for determination of the latter. This group has more than morphological resemblance to recommend it as the outgroup. The origin of the Greater Antilles was a result of movements of the Caribbean plate with respect to the North and South American plates in the early and mid-Tertiary. The proximity of the future Greater Antilles and Central America made the latter a likely source of the *antillanus* group.

The ancestral (plesiomorphic) and derived (apomorphic) character states are listed in Table 1. From them cladograms were generated.

The cladogram requiring the least homoplasy (the least convergences or character reversals), i.e. the most parsimonious one, shows a situation in which the fossil species, *L. galianoae* is not the ancestor to *L. antillanus*, the extant species on Hispaniola (Fig. 1). Rather *L. antillanus* is the sister-group to the other species of *Lyssomanes*, including the fossil.

If the fossil is not ancestral can it still give some insight into the evolution and biogeography of this group?

There are three main possibilities (of many) involving the two Hispaniola species:

1) Only one lineage of *Lyssomanes* has ever existed on Hispaniola and *L. galianoae* is ancestral to *L. antillanus*.
2) At least two lineages of *Lyssomanes* have coexisted on Hispaniola in the past but only one does today.
3) At least two lineages of *Lyssomanes* have existed, sequentially, on Hispaniola.

Let's look at each of these three scenarios:

The ancestor-descendent scenario

To assume that *L. galianoae* is ancestral to *L. antillanus* would require a modification of the cladogram, in which three additional character reversals would be required--in spination, cymbial width and the loss and gain again of the cheliceral promarginal tooth. This is too much to ask and, thus, *L. galianoae* being ancestral to *L. antillanus* is considered very unlikely.

The sympatric-synchronic scenario

That two lineages coexisted on Hispaniola in the past but only one is found there today would be consistent with the cladogram, not requiring any modifications. Here we return to the observation that, at present, only one species of *Lyssomanes* is found on any one island in the West Indies, even the largest, Cuba.

The reason for this is not clear, perhaps it is historical accident. But a more likely explanation is that the ecological requirements of members of this distinct genus are sufficiently restricted to result in the limitation of only one member of this homogeneous species group to any one island.

Little is known of the natural history of the genus in the West Indies, but the Florida species appears to have a preference for living under the leaves of magnolia (*Magnolia grandiflora* L.) (Richman & Whitcomb 1981.)

If we make this reasonable assumption of ecological exclusion then it is unlikely that the two lineages coexisted and we look to the third of our major scenarios.

The sympatric-allochronic scenario

That two lineages existed, sequentially, on Hispaniola is also consistent with the cladogram. This would require the initial existence of the ancestor to *L. antillanus* on, say, Cuba, and a member of the sister-group (that is, *L. galianoae*) on Hispaniola. And, indeed, by 20 million years ago Northern Hispaniola and Cuba had been separated for some 20 million years (Buskirk 1985) (Fig. 2). Then *L. galianoae* would be replaced by the *L. antillanus* lineage later on. But how? It could have been by dispersal from Cuba to Hispaniola.

But a more likely explanation is afforded by several geological models recently reviewed by Buskirk (1985). It appears likely that in the late Miocene, some 10 million years ago, Cuba and Northern Hispaniola formed a temporary super island (Wadge and Burke 1983). Applying our ecological exclusion principle only one species could and would remain. One species, presumably *L. galianoae*, went extinct resulting in a single and common species of *Lyssomanes* remaining on this island, the descendents of which exist as a single species, *L. antillanus*, today both in Cuba and Hispaniola.

Incidentally, *L. antillanus* is also found today on Jamaica, although it is morphologically distinct from the Cuban and Hispaniolan members of the species, having a somewhat reduced cheliceral tooth in the males. The origin of this geographic race surely is the consequence of dispersal since Jamaica was entirely submerged for over 20 million years from the upper Eocene until the early Miocene, and has remained a separate island ever since (Buskirk 1985).

This scenario best fits both the most likely phylogenetic and plate tectonic hypotheses as well as sits well with our ecological assumption.

The remaining members of the *antillanus* group, *L. portoricensis* and *L. michae*, apparently arose from a lineage of Puerto Rican origin having come from *L. galianoae* or its immediately ancestral lineage. This, too, agrees well with the recently hypothesized land-mass relationships in which a single land-mass splits into Cuba and Northern Hispaniola/Puerto Rico some 40 million years ago and then the latter island split into its two portions about 5 million years later (Buskirk 1985).

Conclusions

The case I have presented is not easily extended to some general statement of biogeographic application to all organisms, or even to all spiders. To say that each case is unique and must be explained with knowledge of the organism's cladistic relationships, ecology, and, if you are lucky, knowledge of its past (in the form of fossils), is not an admission of failure but rather an acceptance of the reality of a complex world, a world in which there is more than simple vicariance, a world in which dispersal and extinction also exist and may play important parts in reasonable explanations for the distribution of organisms on earth.

Lyssomanes galianoae n. sp.

MATERIAL EXAMINED. One male from Dominican amber (deposited in the Florida State Collection of Arthropods).

DESCRIPTION. Measurements (mm): Prosoma 2.10 long, 1.50 wide; chelicera 0.92 long; pedipalp cymbium 1.32 long, genital bulb 0.68. Indices listed in Table 2.

Male genitalia with a long, very thin cymbium and a strong, curved median apophysis typical of the *antillanus* species group and a simple embolus curved medially (Fig. 3).

The chelicerae somewhat divergent, moderately long and lacking any retromarginal apical tooth or boss. Legs with small, thin spines.

DIAGNOSIS. Table 2 shows a comparison of measurements and indices of several characters. Note the intermediate value of the Cheliceral Index. *L. galianoae* can be distinguished from all other members of the *antillanus* group of *Lyssomanes* by its long, thin cymbium, the absence of an apical retromarginal cheliceral tooth, and the reduced leg spination.

ETYMOLOGY. This species is named for Maria Elena Galiano, the Argentine arachnologist who has made important contributions to our knowledge of *Lyssomanes*.

Literature cited

Brignoli, P. M. 1984. On some West Indian *Mimetus* and *Lyssomanes* (Araneae: Mimetidae, Salticidae). Bulletin British Arachnological Society 6(5):200-204.

Buskirk, R. E. 1985. Zoogeographic patterns and tectonic history of Jamaica and the northern Caribbean. Journal of Biogeography 12:445-461.

Galiano, M. E. 1962. Redescripciones de especies del genero *Lyssomanes* Hentz, 1845, basadas en los ejemplares tipicos. Descripcion de una especie nueva (Araneae, Salticidae). Acta Zoologica Lilloana 18:45-97.

_____. 1980. Revision del genero *Lyssomanes* Hentz, 1845 (Araneae, Salticidae). Opera Lilloana 30:1-104.

MacArthur, R. H., and E. O. Wilson 1967. The Theory of Island Biogeography. Princeton University Press, Princeton, NJ. 203 pp.

Platnick, N. I. 1976. Drifting spiders or continents?: vicariance biogeography of the spider subfamily Laroniinae (Araneae: Gnaphosidae). Systematic Zoology 25:101-109.

Richman, D. B. and W. H. Whitcomb 1981. The ontogeny of *Lyssomanes viridis* (Walckenaer) (Araneae: Salticidae) on *Magnolia grandiflora* L. Psyche 88(1-2):127-133.

Wadge, G. & Burke, K. 1983. Neogene Caribbean plate rotation and associated Central American tectonic evolution. Tectonics 2:633-643.

Wilson, E. O. 1986. Ants of the Dominican Amber (Hymenoptera: Formicidae). 1. Two new myrmicine genera and an aberrant *Pheidole*. Psyche 92(1):1-9.

Table 1. Characters used in the cladistic analysis of the species in the *antillanus* group of *Lyssomanes*. The numbers correspond to the apomorphies noted on the cladogram in Fig.1.

ANCESTRAL (Plesiomorphic)	DERIVED (Apomorphic)
1. Median apophysis of male genitalia short and weak	Median apophysis of male genitalia long and strong
2. Chelicerae sexually dimorphic	No sexual dimorphism in chelicerae
3. Retromarginal apical tooth present on chelicerae	Retromarginal apical tooth absent
4. Leg spination robust	Leg spination weak
5. Cymbium thick	Cymbium thin
6. Male chelicerae long and divergent	Male chelicerae shorter and less divergent
7. Cymbium long	Cymbium short
8. Embolus curved medially	Embolus curved laterally

Table 2. Measurements and Indices of males of the *antillanus* group of *Lyssomanes*. (Measurements of *L. michae* from Brignoli, 1984.)

	antillanus	portoricensis	michae	galianoae
Prosoma length (mm)	1.90-2.40	1.75-2.10	2.62	2.10
Prosoma width (mm)	1.50-1.80	1.20-1.50	1.88	1.50
Prosomal Index (width/length x 100)	72-76	68-76	72	70
Chelicera length (mm)	1.25-1.81	0.52-0.81	--	0.92
Cheliceral Index (Chel. length/prosoma width x 100)	86-115	42-55	--	62
Cymbium length (mm)	0.90-1.23	0.71-0.86	1.60	1.32
Genital bulb length (mm)	0.49-0.58	0.46-0.54	0.69	0.68
Genital Index (bulb length/cymbium length x 100)	47-55	61-70	43	51

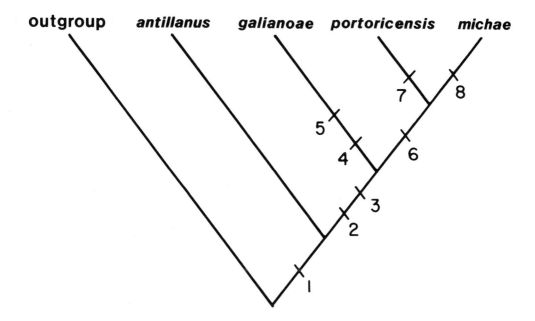

Figure 1. A cladogram of the *antillanus* species group of the genus *Lyssomanes*. The numbers refer to the apomorphic characters listed in Table 1.

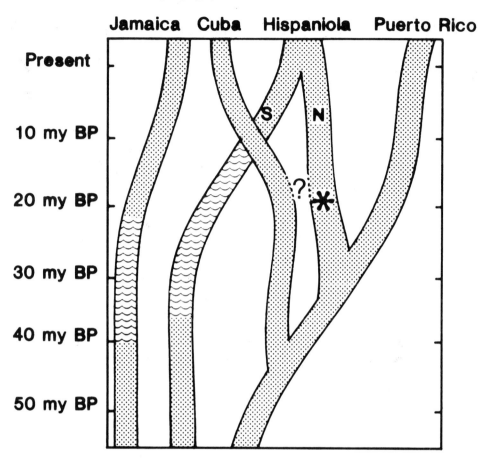

Figure 2. An area cladogram derived from Buskirk (1985) showing the physical relationships among the Greater Antilles over the past 50 million years. Abbreviations are: myBP = million years before the present; N and S are the northern and southern portions of Hispaniola; the asterisk indicates the location of *Lyssomanes galianoae*; the question mark indicates the close proximity of Northern Hispaniola to Cuba; the wavy pattern in Jamaica and Southern Hispaniola represents the inundation of both during the mid-Cenozoic.

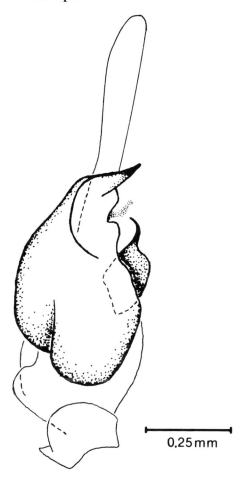

Figure 3. *Lyssomanes galianoae* n.sp. Ventral view of right male pedipalp.

Biogeography of the West Indies, 1989:229-262

THE BIOGEOGRAPHY OF WEST INDIAN BUTTERFLIES (LEPIDOPTERA:PAPILIONOIDEA, HESPERIOIDEA): A VICARIANCE MODEL

Lee D. Miller and Jacqueline Y. Miller[1]

Abstract

Previous biogeographical theories and models for West Indian butterfly biogeography are discussed. The age of butterflies, endemism of the West Indian species, and their propensities for or difficulties with dispersal are enumerated. A combination vicariance/dispersal model for the evolution of the Antillean butterfly fauna is proposed within the general constraints of current geological evidence. Jamaica is postulated to have occupied a more westerly position during the early Tertiary than is usually documented, and the southern Hispaniolan block is thought to have been closer to Yucatan during the Eocene than previously believed.

Introduction

Early biogeographic hypotheses were based on the belief that landmasses and sea floors have long been in their present positions. Wallace (1876), Matthew (1915), Simpson (1952), Darlington (1957; 1965) explain the arguments for this interpretation in detail. Organisms were distributed on these landmasses or oceans by the action of random dispersal through time. Still other biogeographers have felt it necessary to construct land bridges to distribute faunal and floral elements that could not be easily explained by other means (Schuchert 1935). Others were not so certain. The paleogeographical reconstructions of Wegener (1915) and du Toit (1927; 1937), which utilized moveable landmasses and sea floors, spurred the interest of a number of geologists (Carey 1958; Wilson 1963) and biologists (Cain 1944; Croizat 1958; Cracraft 1973; Raven and Axelrod 1974; Shields 1979).

Croizat's contribution during the 1950's and later, in particular, was the realization that entire biotas evolved and might be distributed in patterns that were not necessarily as random as had previously been assumed. Croizat postulated "tracks" along which entire biotas could be shown to have moved and evolved. By careful plotting of the ranges of diverse organisms, he determined that some of these tracks showed frequent congruent distribution patterns in such diverse groups as insects, trees, fresh-water fish and reptiles. From this began to emerge certain broad, repeated distributional patterns

---

[1] Drs. Lee and Jacqueline Miller are curators at the Allyn Museum of Entomology of the Florida Museum of Natural History, 3621 Bay Shore Road, Sarasota, FL 34234.

that he referred to as "generalized tracks". Croizat ct al. (1974) stressed that all species are components of biotas and that the generalized track estimates the composition and distribution of the ancestral biota before it subdivided into descendant biotas.

The arguments for vicariance biogeography in its various forms are to be found in Hennig (1966); Brundin (1972; 1981); Raven and Axelrod (1974); Rosen (1976); Savage (1973; 1974; 1983); Patterson 1981). The general vicariance rationale is well articulated by Savage (1983: 491-496).

It is generally agreed (Coney 1982; D. L. Smith 1985; and others) that the opening of the Gulf of Mexico and the Atlantic dates roughly from the Jurassic, and the floors of both were comprised of marine crustal oceanic basalt by the end of that period (Case et al. 1983; Perfit and Heezen 1978; Perfit et al. 1980). Further, North and South America were totally separated early in the Cretaceous, but a connection reformed later in that period from a series of volcanic islands connecting southern Mexico and northern South America. Guyer and Savage (1986), utilizing the current geological data summarized by Pindell and Dewey (1982) and a cladistic analysis of the animals, present a very plausible explanation for the distribution of anoles in the Americas. In this model the proto-Greater Antilles were part of the Caribbean plate that formed the early Central American connection between North and South America in the Paleocene, fragmenting away in the Eocene. These fragments drifted northeastward to become the Greater Antilles. These islands, however, may be more closely related to North than to South America by their original proximity to Yucatan and later by collision with the North American plate underlying what is now Florida and the Bahamas Rise. The Greater Antilles have undergone considerable modification during the Tertiary through the complex actions of many faults and by the independent movement of several smaller plates. This fragmentation of the proto-Greater Antilles and subsequent accretion of parts elsewhere results in at least three separate blocks forming Cuba and three or four additional separate blocks accreting to form Hispaniola, each contributing its own elements to the fauna. Parts of Hispaniola were attached either to Puerto Rico or to parts of Cuba during the Tertiary. The Lesser Antilles apparently either were offshoots of northern South America which rotated more or less clockwise around Trinidad, or they were an island arc between Central and South America, beginning in about the Eocene. The biota of these islands retains more similarity to South America than to any other landmass. Modern Middle America grew southward from southwestern Mexico as a complex volcanic arc-trench system, finally closing the gap between North and South America during the Pliocene, thus clearing the way for the "Great American faunal interchange" (Stehli and Webb 1985).

Previous biogeographical studies of butterflies

It is informative to examine previous studies on butterfly biogeography in light of the various geological and biogeographical models. The advent of Darlington's (1957) general treatise on zoogeography provided the impetus for several papers on the biogeography of the West Indian butterflies. Like the Darlington book, most of the butterfly studies to date have been based on the dispersalist model: lepidopterists have been slow to adopt the concepts of plate tectonics and vicariance.

One of these early studies was by Fox (1963) on the few species of West Indian Ithomiidae. He showed that these were most closely related to Central American species, but since they are not vagile, he resorted to Schuchert's (1935) land bridges to transport them onto the Greater Antilles. Fox, like Schuchert, assigned these land bridges to the Tertiary and gave very little significance to the Pleistocene.

Shortly thereafter, Clench (1964) published an analysis of the West Indian Lycaenidae. This was a purely Matthewian treatment ascribing virtually all of the West Indian fauna to dispersal events during Pleistocene interglacials with contraction and subsequent extinction of the fauna during glacial maxima and the concomitant lowering of temperature. The species themselves were largely ascribed to the Pleistocene, so practically all of the evolution of the family was of that age.

L. Miller (1965) examined the West Indian *Choranthus* (Hesperiidae) and its relatives. These he considered to have dispersed onto the Greater Antilles largely during the Tertiary, but he did not invoke land bridges, relying rather on rare, chance dispersals. The single Antillean species of *Paratrytone* was attributed to invasion from approximately present-day Honduras. He (L. Miller 1968) briefly mentioned the West Indies in his treatment of the zoogeography of the world Satyridae, again utilizing a dispersalist model. The only satyrid genus, *Calisto* Huebner, was thought to have been derived from a Middle American member of a wholly Neotropical tribe.

Brown and Heineman (1972) discussed Jamaica's butterflies on a classical dispersalist model, but they attributed some of the dispersal to events during the Tertiary. Scott (1972) listed the West Indian butterflies and made a few comments on their distribution. This, too, was a dispersalist discussion and strongly influenced by events of the Pleistocene. He (Scott 1986) again listed the West Indian butterflies and commented more widely on their postulated dispersal from the American continent.

Riley (1975) in his Field Guide offered some speculations on the origins of parts of the West Indian butterfly fauna, including a possible African link for one or two genera. His was not a vicariance study, however, and one is left with the impression that he was referring to long-distance dispersal, including Africa to the Antilles, to account for these anomalies.

Brown (1978) presented a paper on Antillean butterfly distributions during the same symposium in which Rosen (1976) presented his classic vicariance model for the Antillean fish fauna. Brown raised the possibility of vicariance, though he carried it no further, hinting at it only in connection with the danaid genus *Anetia*. He also mentioned *Calisto* as a possible link to Africa, basically repeating Riley's information.

Shields and Dvorak (1979) present an interesting, though controversial, vicariance model based on the late Jurassic to early Cretaceous separation of the Americas, Africa and the Caribbean. It ascribes most of the evolution of the butterflies to this time period. This model was totally at odds with the conventional dispersalist one, and since some of the taxonomic relationships that were postulated are questionable, it has not achieved much acceptance. It was, however, the first attempt at a vicariance study of butterflies.

Most of the emphasis has been placed on the Pleistocene by all of these authors with the exception of Fox (1963), L. Miller (1965), Shields and Dvorak (1979), and to a lesser degree, Brown (1978). This emphasis is inherent in adoption of the Matthew-Darlington model of biogeography. It requires the long distance transport of organisms across barriers to establish themselves in new territories. Perhaps nowhere are the difficulties of such transport more apparent than in colonization of islands, and why some islands are the recipients of various faunal elements while others are not.

Several factors need to be considered to determine the feasibility of the dispersalist model. First, one must examine the organisms themselves and their endemicity. It is also necessary to analyze the fossil record to determine the actual documented ages of butterflies. Since there are very few fossils available, what may we infer from those that are extant? Other questions that must be addressed involve dispersal itself. Are butterflies all "good" at dispersal, or not? In the case of disparate groups of butterflies, are they strong fliers? What are their lifestyles that would either facilitate dispersal or act against it? Is a sedentary butterfly much less likely to leave its home area? Some species are well known migrants, and their dispersal capabilities should be virtually unlimited. If one seeks to postulate dispersal by hurricanes, how do the organisms in question behave during storms (or even at the threat of them)? We will examine these aspects in some detail.

Endemism of the West Indian butterfly fauna

Endemism is a well-known phenomenom in insular butterflies (Holloway 1979), and those of the West Indies are no exception (Table I). Over half of the species are found exclusively in the West Indies. This statement is based on the assumption that the southern tip of Florida and the Keys are Antillean, rather than continental (Brown 1979; Scott 1972; 1986; Riley 1975; L. Miller and D. S. Smith in prep.), because some of the Cuban and Bahaman "endemics" occur there, too. Nearly one of every eight genera of West Indian butterflies is endemic.

It is necessary to examine the foodplant data for the butterflies that are in the West Indies (Brown and Heineman 1972; Scott 1972; Riley 1975; Shields and Dvorak 1979, etc). The foodplants either had to be on the islands when the butterflies arrived or they had to have arrived on the islands by vicariance along with their butterflies.

The endemic genera and the species are clearly derived from several different faunal sources (Table 2). Most of the butterflies of the Greater Antilles are most closely related to those of Central America and Mexico. This portion of the fauna may be explained either as dispersal from the mainland (most previous butterfly biogeographic studies) and/or by invoking a model similar to that proposed by Pindell and Dewey (1982) as a basis for vicariance. The clearly North American component of the Antillean butterfly fauna seems to be best explained by dispersal. The fauna of the southern Lesser Antilles is most closely related to South America via Trinidad, and this perhaps is best explained by dispersal. The Virgin Islands and northern Greater Antilles are faunistically related to Puerto Rico, again probably through dispersal. To determine which of the biogeographical models is most plausible, one must examine the present ranges of the butterflies that inhabit the islands (Table 2) and one must also determine the dispersal potential of the butterflies themselves.

Cuba and Hispaniola harbor the greatest percentage of endemics, but this might be a reflection of their status as the largest Antillean islands. It can equally well be a function of the islands as points of accretion of various small plates during the Tertiary, each harboring distinctive faunas, thus conforming to a vicariance model.

One would expect, on a purely dispersalist model, that Cuba would have a large component of species in common with the Yucatan, since the over water distance is only 220 km across the Yucatan Channel, but this is not so. Prevailing winds would tend to pass insects from Cuba to Yucatan, rather than vice versa. The only instance that has been reported in recent years (Shields 1985) of a Cuban species being taken on the Mexican (or Texas) mainland is of the libytheid *Libytheana motya* (Boisduval and LeConte). At the same time, there are some well documented invasions of Cuba by Mexican species, most notably *Hamadryas amphinome mexicana* (Lucas), which appeared in Pinar del Rio, Cuba in the 1860's and was reported by H. W. Bates. Numerous fresh specimens were taken western Cuba in 1930's (M. Bates 1936; de la Torre y Callejas 1954), indicating that the species was established there. Perhaps this was the result of several introductions. Although there are some old reports of Antillean species being taken in Honduras (Evans 1952), these are unverified and probably due to mislabelling.

There should have been a great influx of Cuban species on the Florida Keys, a distance of only 150 km, no matter which model is employed; and in fact, this has happened. There are a number of butterflies that can best be explained by waif dispersal (the lycaenid *Electrostrymon angelia angelia* (Hewitson) and the skipper *Asbolis capucinus* (Lucas) that have become well established in the Keys and South Florida in recent years (Klots 1951; Anderson 1974)). Sightings or captures have established that *Strymon limenia* (Hewitson), a lycaenid, *Anartia chrysopelea* (Huebner), a nymphalid (both Anderson 1974), the pierid *Aphrissa orbis* (Poey) (Scott 1986) and the Swallowtail *Eurytides celadon* (Lucas) (C. V. Covell, Jr. pers. comm.) have visited the Florida Keys recently. Many such reports involve strong-flying, often migratory, species.

Endemic species are usually not shared between islands, and those that are generally are shared only with nearby islands, for example, *Pyrgus crisia* (Herrich-Schaeffer) is found both in Cuba and Hispaniola (perhaps Puerto Rico), but not in Jamaica or Florida.

Most of the evidence points to these insular faunas having evolved in situ. The question must be answered as to whether they are the result of chance dispersals from the mainland or passive vicariance based on the mobile history of the islands, or both.

The age of butterflies and its biogeographical implications

The question of how old the butterflies are has vexed workers for many years. The problem becomes more complex when one realizes that there are fewer than three dozen unquestioned butterfly fossils known, and these span most of the Tertiary (Brown 1976; Whalley 1986). Because of their delicate wings and chitinous exoskeletons which require very special conditions for preservation, butterflies do not fossilize well. Hence, we have a few fragments with which to work. These are often too badly broken to even be recognizable as Lepidoptera. Even fewer more or less complete adult casts exist, and these are the basis of practically all of our knowledge about these insects. Since there are no fossil Lepidoptera known from the West Indies, it is necessary to extrapolate butterfly distributions with those of animal groups having congruent distributions and with better fossil records.

The majority of the lepidopterous fossils are of Oligocene age, chiefly from the Aix formation of France and the Florissant beds of Colorado. These fossils are mostly of Satyridae, Nymphalidae and Pieridae, the first two of which are considered among the most derived families in the Lepidoptera. Most of the fossils may be assigned with ease to extant genera, and in at least one instance, the fossil is almost indistinguishable from a present-day species (J. Miller and Brown 1989). One of the other fossils, *Doxocopa willmetae* Cockerell, from the Florissant, is closely allied to extant West Indian species. A fossil pierid, *Oligodonta florissantensis* Brown is very close to the present-day Andean genus *Leodonta* Butler, and the satyrid *Prodryas persephone* Scudder is a member of an extant, largely Old World tribe, the Pararginae (L. Miller, in prep.). These fossils are even older than the Miocene fossil fish that are referable to a present-day species reported by Lundberg, et al. (1986). Since fossils can be used only to give a *minimum* age for a taxon (Patterson 1981), any species could in fact be much older than indicated by its fossils.

Judging by the fact that there are very few genera of terrestrial animals that have persisted from at least Oligocene time to the present, the conclusion that at least some lineages of butterflies, even among the most apotypic families, are bradytelic is inescapable. This tends to confirm the postulated antiquity of Lepidoptera suggested by Carpenter (1930) and Forbes (1932), and presumably followed by Shields and Dvorak (1979), who assumed that the Lepidoptera arose or differentiated in the Jurassic. Whalley (1986) documents the lepidopteran fossils and describes one microlepidopteron from the Jurassic. He states that although Triassic and earlier fossils have been reported, these have proven to be cicadas or the like, but he does not discount Lepidoptera being extant in the Triassic. Based on geographic patterns, L. Miller (1968), in a dispersalist model, assigned the origin of the Satyridae to the Cretaceous. All of the above cited reports place the origin of the Lepidoptera farther in the past than that of various orders of mammals or birds: in fact, the situation much more closely parallels that of reptiles, amphibians and fish.

It is here postulated that the butterflies probably arose contemporaneously with the angiosperms (Smart and Hughes 1973; Raven and Axelrod 1974). The latter, long thought to be restricted to the early Cretaceous sediments of Laurasia, have recently been found in comparable deposits in southern South America (Romero and Archangelsky 1986), thereby establishing their presence in at least West Gondwanaland. Clearly the early angiosperms were much more widely distributed than previously believed.

From the above, it is evident that the Lepidoptera are much older than earlier authors thought. They are old enough that they were present on both parts of the divided Pangaea during the Mesozoic, and at least the Satyridae are beginning to show patterns of distribution and evolution that are consistent with vicariance models proposed for other heterothermic animals (L. Miller, in prep.). It must be noted here that the satyrids are considered to be "advanced" compared to some other butterfly groups; and if this interpretation is correct, then most butterfly groups should have been extant in the Cretaceous. It follows that all butterfly distribution in the West Indies is potentially explainable by a modification of the Pindell and Dewey (1982) model, as utilized by Guyer and Savage (1986) in their biogeographic explanation for anoles.

Are butterflies effective dispersalists?

The simple fact that butterflies have wings has led many to the conclusion that these insects must be accomplished flyers capable of long flights over water. Arguments for the classical dispersalist model for populating islands are eloquently discussed by Carlquist (1974). Over-water dispersal is surely true for a few species in selected genera, such as

Phoebis and some *Danaus*, and certainly this was how *some* Lepidoptera reached the Antilles.

This explanation for populating the Antilles long has been employed by zoogeographers (Clench 1964; L. Miller 1965; Scott 1972, for example), but it is necessary to examine individual butterfly species' "life styles" on a case-by-case basis to determine the accuracy of this argument.

Endler (1982:644, table 35.1) lists a number of animals, including a few butterflies, with their estimated dispersal potential gleaned from various published sources. The Lepidoptera that he mentioned have dispersal distances calculated at between 10 m and 5 km. Naturally, many butterflies engage in longer flights, and the possibility for some butterflies to disperse in this way cannot be discounted. Still, not all butterflies are capable of such flights, especially across water barriers.

The Satyridae, for example, are very sedentary (Endler 1982 lists the European species *Maniola jurtina* (Linnaeus) as having an effective dispersal distance of 10 m), and so are the Ithomiidae (Fox 1963). The latter author employed land bridges from the mainland to the Antilles to account for the spread of these sedentary ithomiids, but if we reject Schuchert's (1935) land bridges, we must explain how these non-vagile insects reached the islands.

Many butterflies take refuge at the slightest hint of inclement weather, and the probability of their dispersal by hurricanes is thus minimized (Fox 1963). While such species may disperse slowly over the relatively benign land given enough generations, it is extremely unlikely that they could have made long, over water flights. From personal observations, it is unlikely that most Lycaenidae could accomplish this feat, in contradiction to the dispersalist model proposed by Clench (1964).

The answer to the question "are butterflies effective dispersalists?", then is "it depends." Some butterflies have excellent dispersal potential and surely arrived in the Antilles by this means. In other instances, dispersal cannot have accounted for all of the Antillean distributions. For these species we must look to another model, perhaps a composite one incorporating both dispersal and vicariance, to explain lepidopteran distributions rationally.

Present study

The present study seeks to ascribe distribution of butterflies in the Caribbean to both vicariance and dispersalist events. Much of the distributional data are selected from museum collections and taken from Riley (1975), Clench (1977), Clench and Bjorndal (1980), J. Miller (in prep.), Scott (1986), Simon and L. Miller (1986), and L. Miller and D. S. Smith (in prep.) along with new data collected by the authors from field work and museum sources. The tectonic model employed is basically that of Pindell and Dewey (1982), as employed by Smith (1985) and Guyer and Savage (1986). Classical dispersalist explanations are roughly similar to those outlined by Darlington (1957) and modified by Clench (1964).

The dispersalists

There are many butterfly species that are adapted to long-range flight, migration, and these could be expected to have colonized the islands often. Many of these are well documented migrants, whose flights in the continental landmasses cover thousands of kilometers every year. If the dispersalist model is correct, there might have been a continuous interchange of migrants that would tend to facilitate genic exchange between island and mainland populations and, therefore, to make such species phenotypically similar, whether they came from the Antilles or from the mainland. Still other, not quite so vagile insects, may have dispersed to the islands and then subspeciated (perhaps speciated), but these should still be considered as dispersalists. Much of the dispersal should be of Pleistocene age (Clench 1964), and their affinities should be clear. Those species listed here are assigned to geographic areas from whence they came to the islands as follows: NA = those species that entered from North America; Mex = those which entered via Mexico; CA = those species which were derived from Central America; and SA = those species that entered from South America via Trinidad and/or the Lesser Antilles. The affinities of the West Indian genera are shown in Table 2. For further information on the Antillean distribution of the species listed below, see Riley (1975).

Many danaids are well documented dispersalists. The Antillean species of *Danaus* Kluk (mostly NA) quite possibly dispersed to the islands, usually subspeciated and often are involved in at least some inter-island movement (Simon and L. Miller 1986). *Lycorea* Doubleday (NA, but see below) has undergone a similar dispersal, and the Lesser Antilles are regularly visited by the South American *L. ceres atergatis* Doubleday (Riley 1975:39).

Several nymphalids best can be ascribed to dispersal from mainland populations. Examples include *Doxocopa laure* (Drury) (CA), *Marpesia petreus* (Cramer) and *chiron* (Fabricius) (CA), *Colobura dirce* (Linnaeus) (CA), *Historis odius* (Fabricius) and *acheronta* (Fabricius) (CA), at least *Hamadryas amphinome mexicana* (Lucas) (CA), *Dynamine* species (Mex or CA), *Eunica* species (CA, (except *heraclitus* (Poey)), *Adelpha iphicla* (Linnaeus) (Mex), *Hypolimnas misippus* (Linnaeus) (SA?), *Junonia* species (NA), some *Anartia* species (Mex, SA), *Biblis hyperia* (Cramer) (SA), *Siproeta stelenes* (Linnaeus) (Mex or CA), *Phyciodes phaon* W. H. Edwards (NA), *Eresia frisia* (Poey) (CA), *Vanessa* species (NA), *Euptoieta* species (NA, Mex), *Philaethria dido* (Clerck) (CA), *Agraulis vanillae* (Linnaeus) (NA, SA), *Dione juno* (Cramer) (Mex or CA), possibly *Dryas iulia* (Fabricius) (SA or CA), perhaps *Eueides melphis* (Godart) (CA), and perhaps *Heliconius charitonius* (Linnaeus) (all of the above regions).

Within the Lycaenidae, the following are almost certainly attributable to dispersal: *Pseudolycaena marsyas* (Linnaeus) (SA), *Chlorostrymon* species (Mex, CA), *Leptotes cassius* (Cramer) (probably SA), and *Hemiargus hanno* (Stoll) (CA).

Most of the Pieridae, except *Dismorphia* species, a few *Eurema* species, and perhaps *Melete salacia* (Godart) appear to be dispersalists from various sources.

Heraclides thoas (Linnaeus) (CA) and *cresphontes* (Cramer) (NA, Mex, or CA), *Papilio polyxenes* Fabricius (NA), *Pterourus palamedes* (Drury) and *troilus* (Linnaeus) (NA), and probably the *Eurytides* species (derived from the NA *E. marcellus* (Cramer)) are the West Indian Papilionidae whose distributions are probably referable to dispersal.

The distributions of several Hesperiidae best explained by a dispersalist model include *Phocides pigmalion* (Cramer) (CA), *Proteides* species (SA?), *Epargyreus* species (NA), *Polygonus* species (CA), *Aguna asander* (Hewitson) (CA), most *Urbanus* species (most areas), *Autochton* species (NA, Mex), *Cabares potrillo* (Lucas) (CA), *Achlyodes mithridates* (Fabricius) (CA), *Timochares* species (CA), *Grais* species (CA), *Gesta gesta* (Herrich-Schaeffer) (CA or SA), *Chiomara* species (SA), *Erynnis zarucco* (Lucas) (NA or Mex), *Pyrgus oileus* (Linnaeus) (Mex or CA), *Perichares philetes* (Gmelin) (Mex), *Synapte malitiosa* (Herrich-Schaeffer) (Mex), *Polites* species (NA or Mex), *Hylephila phyleus* (Drury) (any of the regions), *Atalopedes* species (NA or Mex), *Calpodes ethlius* (Stoll) (any of the regions), *Panoquina* species (NA, Mex, CA), *Nyctelius nyctelius* (Latreille) (CA or SA), *Lerodea eufala* (W. H. Edwards) (NA or Mex), and *Saliana esperi* Evans (CA).

Theoretically, the West Indian butterfly fauna in general has been well documented (Riley 1975; Scott 1972; 1976), but some of these records are based on older specimens. We require additional data for several species perhaps recorded in error from the Antilles, to fully assess the fauna of the islands and its possible origin. Additionally, several new insects are being described almost yearly. Regretfully, then, it must be admitted that lepidopterists often are still in an alpha taxonomic position, and biogeographic knowledge is equally fragmentary, so the full extent of the ranges of many butterflies has not yet been established, thereby making the task of delimiting either centers of origin or the assessment of potential vicariance difficult.

A vicariance/dispersal model for the biogeography of West Indian butterflies

There are numerous problems associated with the biogeographic analysis of butterflies. In the light of phylogenetics, there is uncertainty about the status of the classification (K. Johnson in litt.). Much of the taxonomy above the species level requires refinement (L. Miller and Brown 1983). To date, very few modern taxonomic and biogeographic revisions have been done, and even fewer have been published. The biogeographic analysis which follows is therefore presented in chronological sequence, and the examples employed are drawn from unrelated butterfly taxa whose geographical distributions show congruence with other animal groups that have usable fossil records.

We, therefore, accept the plate tectonic model of Pindell and Dewey (1982), with minor modifications, as explained below.

Late Mesozoic to Cretaceous

During this period Africa and South America were still in contact (Brundin 1981), but the continents were in the process of separating. The North and South American continents were connected during most of the Cretaceous (Pindell and Dewey 1982), and the Antilles did not exist in their present form.

To understand the biogeography of West Indian butterflies, one must consider the Mesozoic era even before the formation of the proto-Greater Antilles. During the Jurassic and the early Cretaceous, Pangaea split into a northern Laurasia and a southern continent, Gondwanaland, separated by the Tethys Sea. Later in the Cretaceous, Laurasia and Gondwanaland themselves began to fragment, and the parts to move toward their present positions. It is here that our story begins. Because of the probable age of butterflies (Whalley 1986), we cannot accept the Jurassic-early Cretaceous vicariance of butterfly groups postulated by Shields and Dvorak (1979), but their explanations might be adequate for some primitive moth groups.

Butterflies are believed to have been established during the Cretaceous (L. Miller 1968; in prep.; Whalley 1986); some groups appear to be basically Laurasian and others Gondwanian (L. Miller in prep.). During part of this time Africa and South America were in contact (Brundin 1981) and shared at least parts of the same fauna. Several butterfly groups fall into this category, including the satyrid sister groups *Manataria* (South America) + (the African *Aeropetes* + *Paralethe*) (L. Miller in prep.) These satyrids are too fragile to have been involved in significant over-water dispersal of the magnitude necessary to explain these distributions by more recent dispersal.

There are African affinities in a very small proportion of the West Indian butterfly fauna. Four butterfly genera, about 4.5% of the fauna, definitely have their sister groups in the Ethiopian region (see below). Such affinities, while rare, are by no means unique to butterflies. Flint (1977) commented on West Indian Odonata and Trichoptera and reported a small African influence on these, and recently Liebherr (1986) has described a genus of West Indian carabid beetles, *Barylaus*, from Hispaniola and Puerto Rico, whose nearest relatives are from Africa, Madagascar and Central America.

Although it has African ancestral relationships, the nymphalid genus *Eunica* is Neotropical with three species represented in the West Indies. Two of these, *monima* (Cramer) and *tatila* Herrich-Schaeffer, are more or less widely distributed in Central America and given to moderate mass movements on the continent (Howe, 1975). These species appear to be candidates for dispersal to the islands and are virtually indistinguishable on the islands from continental examples. The third species, however, *E. heraclitus* (Poey), known only from Cuba, is aligned with the south Brasilian species, *E. macris* (Godart). The two species superficially (and structurally) bear similarity to African *Sallya*, the sister genus of *Eunica* (Jenkins in prep.). All maintain the "primitive nymphaloid pattern" of Schwanwitsch (1924). Jenkins (pers. comm.) has derived a preliminary cladogram of *Eunica* and its relatives that places *Eunica* and the African *Sallya* as sister groups. This cladogram is more or less congruent with that of Liebherr (1986) for the Carabidae. This suggests a late Mesozoic vicariance separating the two genera on Africa and America followed by vicariance of *heraclitus* onto what is now Cuba rather early in the Tertiary.

An even more dramatic example of this biogeographic pattern involves the endemic *Archimestra teleboas* (Menetries) from Hispaniola. The sister group of this monobasic genus is *Neptidopsis* from Africa, and the closest outgroups for these two genera are the Neotropical genera *Mestra* and *Vila* (Jenkins in prep.). *Archimestra* and *Neptidopsis* are illustrated in Fig. 1. *Mestra* is represented on the islands by a pair of species, one in Jamaica and a second, an apparent dispersalist from South America on the Lesser Antilles. The implications are clear: the ancestral member of this group must have been rather similar to *teleboas*, which participated in the African-South American vicariance late in the Mesozoic, and then vicariated onto Hispaniola in the earliest Tertiary.

The nymphalid genera *Archaeoprepona* (Antilles and mainland tropical America) and *Prepona* (Central and South America), along with *Agrias* (American tropics) and *Charaxes* (Old World tropics), show a comparable distributional pattern. *Charaxes* and *Archaeoprepona* are sister genera and they in turn are the sister group of the more derived *Prepona* + *Agrias* (K. Johnson pers. comm.). It is interesting that the most plesiomorphic American genus in this cluster is the one that is represented in the West Indies, a situation that is comparable to the *Barylaus* model (Liebherr 1986). *Archaeoprepona demophoon* (Huebner), the West Indian species, has subspecies in Hispaniola, Cuba and Puerto Rico.

A similar pattern is evident in the lycaenid genus *Brephidium*, represented by three species: one from the southwestern U. S. to Venezuela and the Greater Antilles (Riley 1975), one from Florida and as a probable stray on the Bahamas (Riley 1975), and a third from South Africa (Dickson and Kroon 1978; Eliot 1973). The sister group of this genus, and the only other member of the *Brephidium* section of the Polyommatini of Eliot (1973), is the monobasic South African genus *Oraidium*. The male genitalia (Fig. 2) are illustrated for *Brephidium* and *Oraidium* to demonstrate their very close relationship: superficial characters also show their similarity. Here, too, is a very old, now relict African-Neotropical vicariance pattern that is congruent with Liebherr's pattern in beetles. In this genus, however, one must postulate extensive extirpation of these insects on intervening land masses between the Cretaceous and the present.

A possible relationship has been postulated between the American sister genera *Eretris* and *Calisto*, which is West Indian, and their putative African relatives, the *Dira* complex (Riley 1975; Brown 1978), but this relationship is much less close and dramatic as are the ones cited above. The possible African affinities of *Calisto* led Brown (1978:16) to state that it is "equally at home in the African tribe Dirini as in Pronophilini of the Andes." Reexamination of *Dira* and the Pronophilini in relation to *Calisto* does not support the close alignment of the African and West Indian insects shown in the examples given above: the latter are perfectly good members of the Pronophilini, a tribe that has not only Andean representatives but also members distributed in the lowlands from Arizona to Patagonia. If there is an African element here, it is not prepossessing. Members of *Calisto* are all endemic to the Greater Antilles (Riley 1975).

Late Cretaceous to Eocene

During this time, the first breakup of the old Central American connection began with great faults cutting across the southern boundary of Yucatan and the northwest corner of South America. The resulting block formed the proto-Greater Antilles, and by the end of this period the block was completely separate from South America, but still in approximation to the Yucatan peninsula. The Jamaican and southern Hispaniolan blocks were further to the west than the remainder of the proto-Greater Antilles and either still adherent to Central America, or at least approximate to it (Perfit and Heezen 1978; Case et al. 1984).

Dispersal between North and South America during the Cretaceous took place across a new connection that was to become in turn the proto-Greater Antilles (Savage 1983; Guyer and Savage 1986), resulting in an extensive feeder population of organisms for the original vicariance of the Antillean fauna. Pindell and Dewey (1982) and Smith (1985) propose that the proto-Greater Antilles began moving eastward relative to Mexico in the late Cretaceous, though significant movement of the Caribbean plate is questioned by Donnelly (1985). This proto-Antillean block was finally severed from the Yucatan during the Eocene. Several butterfly groups seem to have been part of this vicariance event, and these vicariants are usually species represented today on Cuba, Hispaniola, and Puerto Rico.

Guyer and Savage (1986) have correlated these vicariance events with the present-day distribution of anoles, and their patterns show congruence with some patterns of butterfly distributions. The primitive anole genera *Chamaeleolis* and *Chamaelinorops* are found exclusively on Cuba and Hispaniola, respectively, and are most closely allied to the sub-Andean genus *Phenacosaurus* (Guyer and Savage 1986:524-528). Those authors ascribed these distributions to these late Mesozoic and early Tertiary vicariance events.

A parallel vicariance pattern is evident in several butterfly genera. A few groups, however, such as *Atlantea* (Nymphalidae), *Calisto* (Satyridae), the *Lycorea* (Danaidae, cleobaea), *Heraclides* (Papilionidae), *Nesiostrymon* (Lycaenidae), and *Wallengrenia* (Hesperiiidae) are known from all four islands and must date from the early vicariance, even though Jamaica was more closely positioned against Central America (Perfit and Heezen 1978; Case et al. 1984) at about the time that the respective genera divided from their mainland sister groups. For example, the closest relatives of the strange and beautiful papilionid *Parides gundlachianus* (C. and R. Felder) are based in South America, though the genus is distributed now from Mexico to Argentina, but not on other Antillean islands. The vicariance of *Parides* (Papilionidae) is postulated as a Cretaceous or Paleocene event with further dispersal taking place in the Tertiary on continental landmasses. Dating the Papilionidae from this time is not unprecedented: Durden and Rose (1978) described a recognizable papilionid from the middle Eocene.

A similar scenario will explain the presence of the single member of the Riodinidae from the islands, though that species, *Dianesia carteri* (Holland), is also found on at least Andros and New Providence in the Bahamas (Harvey and Clench 1980). While we postulate short-range dispersal for this species across the Old Bahamas Channel to emergent land now submerged on the Great Bahamas Bank, perhaps during the Pleistocene, the butterfly originally vicariated to Cuba. Almost the exact same pattern (New Providence not represented, but present on Great Abaco) is shown within the genus *Eumaeus* Huebner (Lycaenidae) with some colonization, perhaps from Andros, of the southeasternmost Florida peninsula (Holland 1931 and others). We doubt that the situation is as complicated as indicated by Shields and Dvorak (1979), who postulated a Jurassic vicariance of this genus. It is more easily explained with reference to its Central American sister-species, *E. toxea* (Godart) and vicariance during the early Tertiary onto Cuba. The Cuba-Andros-south Florida connection will be discussed in L. Miller and D. S. Smith (in prep.)

Oligocene to Pliocene

During this period the proto-Greater Antillean block fragmented, and the resulting small blocks moved about and accreted onto the islands as we know them today. Most of this accretion was completed by the end of the Pliocene. Both Hispaniola and Cuba are products of accretion; in one case, both islands are the beneficiaries of the same early block (the eastern Cuba-northern Hispaniola block of Pindell and Dewey 1982), which fragmented, one part going to Hispaniola and one to Cuba, apparently in the Oligocene (for details see Guyer and Savage 1986). Most of Puerto Rico evidently became isolated early, but a small block that later split off that island accreted onto eastern Hispaniola.

Jamaica and southern Hispaniola separated from Nuclear Central America at a later date probably in the late Oligocene or early Miocene, (Pindell and Dewey 1982). Contrary to earlier speculations, the Blue Mountains of Jamaica and the La Selle massif of Haiti apparently were continuously emergent from the time of their separation from Central America (Case et al. 1984; Perfit pers. comm.) and were possible vehicles of vicariance.

Because Hispaniola and Cuba are far more complex geologically than other Greater Antillean islands, they have greater diversity of the fauna than expected. The complex satyrid genus *Calisto* is found on all of the Greater Antilles and on some of the Bahamian islands, with the greatest species diversity on Hispaniola followed by Cuba (Fig. 3). A single species of *Calisto* inhabits Jamaica, another is found on Puerto Rico, three or four in Cuba, two of which are reported from the northwestern Bahamas, and more than 20 species on Hispaniola. Each island has one (two in Hispaniola) large species with a more pronounced hindwing tornal lobe. These are most closely allied to the basically subandean *Eretris*, which has representatives in the mountains of Central America, one of which is today found in Guatemala and Chiapas. *Calisto* split into two lineages on Hispaniola and Cuba. The species most like *Eretris* have tornal lobes on the hindwings and genitalia that are considered plesiomorphous; they inhabit the lowlands and appear to have evolved there. Only Cuba and Hispaniola have smaller, more round-winged species, which appear to be a later development and do not so closely approximate *Eretris*. These more apomorphic *Calisto* have no tornal lobe, genitalia that are similar to each other, and are usually found in the mountains. Some of them have reinvaded the lowlands, but generally they are more forest insects than are the primitive species, and they are found nowhere else in the Antilles. This species diversity leads to the conclusion that most of the evolution in the genus took place on the several separated plates that later accreted to form Cuba and Hispaniola. Certainly the complex geological history of Cuba and Hispaniola accounts in part for the greater species diversity on those islands, but the presence of the small, apomorphic forms, evolving on separate blocks that accreted to form both islands, indicates additional speciation must have taken place on these formerly separated islands as well.

Evolution on the eastern Cuban-central Hispaniolan block could account for the present distribution of several organisms shared only by Cuba and Hispaniola. Examples in addition to the already mentioned group of *Calisto* include two species of *Anetia* (Danaidae), *Lucinia sida* (Huebner) (Nymphalidae), *Melete salacia* (Godart) and *Aphrissa orbis* ((Poey) (Pieridae), and *Astraptes xagua* (Lucas) and *habana* (Lucas), and *Polites baracoa* (Lucas) (Hesperiidae).

Many butterfly genera show a generalized distribution within the Greater Antilles that has been ascribed to past dispersal, but they are just as easily explained by late Cretaceous to Paleocene vicariance events (see the maps in Guyer and Savage 1986:527, fig. 10). Such an explanation accounts for much of the genus *Heraclides* (Papilionidae) in the islands. We have already suggested that *H. thoas* and *cresphontes* might be dispersalists, though the former could as easily be a vicariant, perhaps dating from the Eocene (Guyer and Savage 1986:527, fig. 10, top right map). Perhaps this model answers Riley's (1975: 143) comment that "It is curious that this widespread species should have reached none of the West Indies other than Cuba and Jamaica." Other *Heraclides* also fit a vicariance model. *H. aristodemus* (Esper) is a species restricted to Cuba, Hispaniola, Puerto Rico, Little Cayman, the Bahamas, and the Florida Keys. *H. andraemon* (Huebner) occurs naturally in Cuba, the Caymans, and was introduced into Jamaica. By contrast, *H. machaonides* (Esper) is restricted to Hispaniola and Puerto Rico, *thersites* is known from Jamaica, *aristor* (Godart) occurs in Hispaniola, *oxynius* (Huebner) is from Cuba, *pelaus* (Fabricius) is found on all of the Greater Antilles, and *caiguanabus* (Poey) is known from eastern Cuba. The distributions given for these species approximates ones already mentioned. All of these insects are rutaceous feeders and all were derived from the immigrant *thoas* (or *cresphontes*) stocks. A preliminary cladogram of these insects suggests that *machaonides*, *andraemon*, and *aristodemus* represent early branchings of the *Heraclides* stock in the Antilles, whereas *oxynius* and perhaps *pelaus* represent a somewhat later development. Finally, the more derived *oxynius* and *caiguanabus* evolved, the former in the bulk of Cuba and the latter probably on the eastern Cuba plate after it split off northern Hispaniola. The evolution and the complex biogeographic history of this genus in the West Indies is currently undergoing further study, but the data appear congruent with a vicariance model with some relatively modern inter-island dispersal having taken place. This is especially true in *H. aristodemus* in the Bahamas.

The papilionid species *Battus polydamas* (Linnaeus) basically is monomorphic on the mainland, but south Florida and the Antilles have 13 different subspecies (Fig. 4, and Riley 1975: 141-143). This butterfly is not an especially good flyer, as witness the lack of occasional records outside the normal range of even any of the Antillean subspecies (for example, St. Lucia and St. Vincent have different subspecies although they are in physical proximity). These subspecies are postulated to be the result of the initial breakup of the ancient Central America, though the situation may be complicated by a possibly dispersalist South American origin of most of the Lesser Antilles subspecies. The vicariant scenario would require complete separation of the subspecies other than the Lesser Antillean ones by about the Oligocene-Miocene boundary.

A complimentary pattern is observed in the papilionids *Battus devilliers* (Godart) from Cuba and the Bahamas which approaches the *Chamaeleolis* pattern, and *B. zetides* Munroe which approximates the *Chamaelinorops* pattern of Guyer and Savage (1986). Both of these species are probably most closely allied to the Central and North American *B. philenor* (Linnaeus). *Battus zetides* is now found only in portions of Hispaniola that were parts of the southern Hispaniolan block of Pindell and Dewey (1982), which block split away from Central America during the Eocene and accreted to Hispaniola in the Pliocene or Pleistocene (Case et al. 1984; Burke et al. 1984). A congruent pattern is shown in the castniid moth *Ircila hecate* (Herrich-Schaeffer), a species endemic to the southern Hispaniolan block and related to Mexican and Central American members of *Athis* (J. Miller 1986). The same pattern is roughly that shown in *Myscelia aracynthia* (Dalman) and its sister taxon *M. cyaniris* Doubleday (Jenkins 1984; pers. comm.). Strangely, there is not an endemic *Battus* on Jamaica, other than a subspecies of *polydamas* (see above), nor is any castniid or *Myscelia* recorded from there. This suggests that perhaps the relative position of Jamaica depicted in the Pindell and Dewey (1982) reconstruction might be erroneous: we suggest that Jamaica in the Eocene and the Oligocene may have been positioned against the area that later drifted southeastward as Central America (Fig. 6). This shift places Jamaica nearer its present position relative to the southern Hispaniolan block.

A pattern that approximates the situation with *Battus zetides* involves the only Antillean member of the genus *Paratrytone*. Insects of this genus are montane or sub-montane throughout most of their present range on the continent (L. Miller, 1964). Representative genitalia of *Paratrytone* are shown in Fig. 5. The West Indian species had been placed in the Antillean genus *Choranthus* until separated from that genus by L. Miller (1964). The distribution suggests that *Paratrytone* were in the contiguous parts of Nuclear Central America and what has drifted southward to become present-day Central America as shown in Fig. 6, along with the isolated southern Hispaniolan block. At present *P. batesi* (Bell) is found in the La Selle mountains of Haiti (Riley 1980:190) and in mountains in the southwestern Dominican Republic (Schwartz pers. comm.), both areas that are associated with the southern Hispaniolan block (Case et al. 1984). The derivation of this genus seems clear if one accepts closer proximity of the southern Hispaniolan block to Nuclear Central America during the Eocene-Oligocene, as suggested above.

The vicariance of the two ithomiids (perhaps they are merely subspecies) known from the Antilles, *Greta diaphana* (Drury) and *cubana* (Herrich-Schaeffer), as previously discussed, is somewhat different. These insects are represented on Jamaica, Hispaniola and Cuba, but not Puerto Rico. They are extremely sedentary and not subject to dispersal. The closest relatives are *Greta* from Mexico and northern Central America, and they appear to have been part of the early Tertiary vicariance of the entire protoAntillean block. Perhaps these insects were simply extirpated from Puerto Rico, but the evidence suggests that they were never there since Puerto Rico appears to have been the southern part of the proto-Greater Antillean block.

The genus *Euphyes* is represented by two species in the Greater Antilles, excluding Puerto Rico, and on a few Bahamian islands. They are most closely related to the continental *E. peneia* (Godman) and more distantly allied to several strictly South American congenors (Shuey 1986:104). That author ascribes the distribution of the group to "an old vicariant event between Cuba and Central America with subsequent speciation and dispersal..." (Shuey 1986:111).

Still other organisms are restricted to Jamaica and Hispaniola only, almost certainly derived from the late Eocene-Oligocene approximation of both the Jamaican and southern Hispaniolan blocks to the Mexican mainland; organisms fitting this pattern include *Danaus cleophile* (Danaidae) and *Aphrissa godartiana* (Pieridae).

Clench (1964) claims that Puerto Rico has a "predominantly Hispaniolan character", and goes on to postulate very cold Pleistocene temperatures that may have extirpated all of the Lycaenidae existing there before the Pleistocene. A few butterflies are apparently characteristic of the Puerto Rico-central Hispaniola block and are found nowhere else in the Antilles, including *Heraclides machaonides* (Papilionidae), *Dismorphia spio* (Pieridae), and *Pseudochrysops bornoi* (Lycaenidae) (Riley 1975). There may have been other species that today are found only on Hispaniola, and Clench's (1964) assumptions about Pleistocene extirpation in Puerto Rico might be correct for these organisms, but it probably is unnecessary to invoke Pleistocene mechanisms. As additional collecting is being done in Puerto Rico, more Hispaniolan species that might have been expected from there have been collected (D. Smith pers. comm.). The apparent lack of species in common between Hispaniola and Puerto Rico may be simply a lack of collecting on the latter.

The distribution of *Pyrgus crisia* (Herrich-Schaeffer), by contrast does suggest that this species was part of the Puerto Rico-Central Hispaniola plate fauna and probably dispersed over Hispaniola and to Cuba when those islands were in closer proximity than at present (Fig. 6). Few other species of butterflies show this distributional pattern, and the relationships are obscure. It is also possible that *P. crisia* was part of the Eastern Cuba-Northern Hispaniola plate and simply dispersed to Puerto Rico, perhaps *via* Mona Island (D. S. Smith pers. comm.).

The primitive danaid genus *Anetia* displays a distributional pattern that conforms to the pattern of *Battus* (Fig. 7). The only mainland *Anetia* has two subspecies, one in Mexico and northern Central America (a Nuclear Central American one) and a derived one from Costa Rica and Panama; all *Anetia* are associated with montane to sub-montane habitats. Cuba and Hispaniola have one endemic species each, and each shares two species; there are none known from Jamaica or Puerto Rico. This distribution suggests that the genus arose on the northern end of the proto-Greater Antilles, but excluding Jamaica and Puerto Rico. Perhaps most of them were isolated on the eastern Cuban-central Hispaniolan block during the Oligocene and evolved on the fragments when that block split.

Geological and further biogeographical evidence suggests that the evolution of the Jamaican fauna is the result of its relatively late connection with Nuclear Central America. Jamaica, the west and central Cuban plates, and the southern Hispaniolan plate were all approximate or adjacent to Nuclear Central America in the Eocene (Fig. 6). The evolution of the most spectacular of the Antillean butterflies, *Pterourus homerus* (Fabricius), should be discussed here. The sister taxon of *homerus* is *P. garamas* (Huebner), a butterfly found today chiefly in the mountains of western Mexico and Central America (Jordan 1908 [1907-1909]). *Pterourus homerus* is found in the submontane forests of Jamaica (Brown and Heineman 1972). The valvae of the male genitalia of *Pterourus homerus* and its near congeners are shown in Fig. 8. A plausible model demands that the progenitors of *homerus* and its sister group be in the west of Mexico-Central America as shown in Pindell and Dewey (1982) and that Jamaica be to the west of the position shown by those authors, perhaps accreted to this western spur in Eocene-Oligocene times. This is in agreement with the positions suggested by Burke et al. (1984), and it requires that at least part of Jamaica remain emergent, as suggested by Case et al. (1984) and Perfit (pers. comm.). There are no other *Pterourus* in the Antilles (Hancock 1983), other than *P. palamedes* and *troilus*, apparent vagrants on Cuba (Riley 1975). *Pterourus homerus* is presently found both east and west of the Blue Mountains and is generally associated with shales or calcareous rocks (Turner pers. comm.), but it is quite possible that its progenitors were part of a vicariance event centered on that part of the Blues that was emergent during Jamaica's early geological history. If this scenario were not true, one would have to require long-distance dispersal and evolution from the Miocene onwards to account for *homerus*, and the question of why the insect is present on Jamaica, rather than Cuba must be addressed.

The hesperiine genus, *Wallengrenia*, is widely distributed throughout North and South America and the Western Antilles and is an example of a number of the vicariant events previously discussed. Two species, *otho* (J. E. Smith) and *egeremet* (Scudder) (after Burns, 1985) are commonly found in the U. S. The neotropical species, including those represented in the Caribbean are currently under revision (J. Miller in prep.). Based on comparative morphological examination, the genus *Wallengrenia* includes two species in eastern South America, *premnas* (Wallengren) *otho*, which in turn is divided into two subspecies (*curassavica* (Snellen) and *sapuca* Evans). Three subspecies of *otho* are known from Central America. The greatest species diversity within the genus is in the West Indies, with four species and one new subspecies represented (Fig. 9). The darker species of these, *misera* (Lucas) is from Cuba and the northern Bahamian Islands. *W. druryi* (Latreille) is restricted to Hispaniola and Puerto Rico and the southern Bahamas. The bright fulvous species, *W. ophites* (Mabille), is found in the Lesser Antilles southward to Trinidad.

The present geographical range of *Wallengrenia* in Central America, and particularly in Mexico, is quite complex. *W. otho curassavica* (Snellen) inhabits on the western slopes of the Sierra Madre Occidental, with nominate *otho* in the east. The other Caribbean species, *W. vesuria* (Ploetz), is endemic to Jamaica and more closely resembles western Mexican *o. curassavica* than any other Antillean population, thus tending to support our modified Jamaican biogeographic model.

The distributional pattern of *Nesiostrymon* (Lycaenidae) is complex. This genus was initially thought to be monobasic, but it is now known to have at least one Mexican species (K. Johnson, pers. comm.). Its sister group, a new genus, contains a number of mainland species and a recently discovered, as yet unnamed Hispaniolan member. The sister group of these two members contains only mainland taxa. Johnson (pers. comm.) postulates a late Mesozoic vicariance of the *Nesiostrymon* and its sister genus from Central America and subsequent evolution on the islands during the Tertiary, not unlike the pattern shown for other genera.

Pliocene to Holocene

There seem no indication of recent vicariance events. The only possible ones involve the Bahamas, and these can as easily be ascribed to island-hopping dispersal.

The Lesser Antilles

These small islands apparently arose as a volcanic arc (the Aves Arc), a subduction zone (Pindell and Dewey 1982; Case et al. 1984; Burke et al. 1984).

Many of the islands are not yet well collected, so in many cases the real distribution of some of the endemic species is unknown. The most obvious members of the fauna are some dispersalists that doubtless came *via* Trinidad (Riley 1975; Scott 1986), and the few possible vicariants seem to have their roots in northern South America. We suspect dispersal in most cases for the Lesser Antilles.

Summary

Previous biogeographical theories are discussed, and previous studies on the biogeography of West Indian butterflies are enumerated. Most of these models have been of a dispersalist type and heavily influenced by Pleistocene events, with the exception of the model postulated by Shields and Dvorak (1979), which referred a number of butterfly distributions to the late Jurassic early Cretaceous opening of the Atlantic. We accept the findings of Whalley (1986) that the butterflies were not present much before the Cretaceous.

A synopsis of the evidence on the age of butterflies is provided. This demonstrates that some lineages are much older than previously believed; for example, most of the Florissant (Oligocene) fossils are congeneric with and/or very near existing species. These data suggest that butterflies probably were well differentiated by the late Cretaceous to earliest Tertiary, and therefore more easily explained by a vicariance model than previously thought.

Additional data suggesting that butterflies are not uniformly good dispersalists are presented. Data are given on the endemicity of West Indian butterflies, which suggest that the fauna is rather old.

A vicariance/dispersal model is proposed for the biogeography of West Indian butterflies, conforming in most respects to the Pindell and Dewey (1982) geologic reconstruction of the area. While there was probably some long-range dispersal, most of the butterfly stocks can be better ascribed to a vicariance model with shorter distance dispersal interspersed with the vicariance events. Most butterfly groups are considered unlikely candidates for long-distance dispersal because of their fragility and lack of vagility. Many of the original butterfly stocks were on the proto-Greater Antilles and evolved in situ within the islands. A chronology is given for the evolution and vicariance of several groups of Antillean butterfly groups.

Based on present taxonomic relationships in different butterfly families and the present biogeographical study, it is postulated that Jamaica may have occupied a more westerly position, against the extension of western Central America, and that the southern Hispaniolan block might have been in closer contact with the Yucatan peninsula during the Eocene than is shown in the Pindell and Dewey (1982) reconstruction. Areas of continuous emergence are postulated to be the Blue Mountains of Jamaica and the Massif de La Selle of Hispaniola, potential refugia for the butterflies of the parts of their respective islands. This reconstruction is in closer harmony with the positions of these island masses during the Oligocene and the later positions of the two landmasses. This better explains some of the more obvious similarities between the fauna of these insular areas and Central America than are shown on most other segments that formed the western Greater Antilles.

Acknowledgments

Several people have cheerfully provided data for this paper, often from unpublished studies. We are very grateful to Drs. Dale Jenkins of this institution, John Shuey of the Department of Entomology, Ohio State University, C. V. Covell, Jr., University of Louisville, and Kurt Johnson of the Department of Entomology, American Museum of Natural History. Dr. Thomas W. Turner, Clearwater, Florida, and lately of Jamaica, provided valuable data on the distribution and ecology of *Pterourus homerus*.

Special thanks are due Dr. Michael R. Perfit, Department of Geology, University of Florida, for discussing geological constraints with us, for providing literature, and generally for his enthusiastic help in the preparation of the manuscript.

Additional thanks are due Dr. Gerardo Lamas of the Museum Javier Prado in Lima, Peru, and Dr. Johnson for discussions on diverse biogeographical theories with us at some length, and Dr. Jenkins and Stephen Steinhauser of this institution, and Deborah L. Matthews, Department of Entomology, University of Florida, who critiqued this manuscript.

Finally, we thank Drs. Charles Woods and Hugh Genoways who convened the symposium in which these views were offered.

J. Miller's research on castniid taxonomy and biogeography was funded in part by National Science Foundation grant BSR-12437.

Literature cited

Anderson, R. A. 1974. Three new U. S. records and other unusual captures from the lower Florida Keys. Journal of the Lepidopterists' Society 28:354-358.

Bates, M. 1936. The butterflies of Cuba. Bulletin of the Museum of Comparative Zoology, Harvard, 78:63-258.

Brown, F. M. 1976. *Oligodonta florissantensis*, gen. n., sp. nov. (Lepidoptera: Pieridae). Bulletin of the Allyn Museum (37):4 pp.

_____. 1978. The origins of the West Indian butterfly fauna. Pp. 5-30 in F. B. Gill (ed.). Zoogeography in the Caribbean. Academy of Natural Sciences Philadelphia, Special Publication 13.

Brown, F. M. and B. Heineman. 1972. Jamaica and Its Butterflies. E. W. Classey, London, 478 pp.

Brown, F. M. and J. Y. Miller. in prep. *Vanessa amerindica*, a new Oligocene fossil butterfly from Florissant, Colorado (Lepidoptera: Nymphalidae) [tentative title].

Brundin, L. Z. 1972. Phylogenetics and systematics. Systematic Zoology 21:69-79.

_____. 1981. Croizat's panbiogeography versus phylogenetic biogeography. Pp. 94-138 in G. Nelson and D. E. Rosen (eds.). Vicariance Biogeography: A Critique. Columbia University Press, New York.

Burke, K., C. Cooper, J. F. Dewey, P. Mann, and J. L. Pindell. 1984. Caribbean tectonics and relative plate motions. Geological Society of America Memoir 162:31-63.

Burns, J. M. 1985. *Wallengrenia otho* and *W. egeremet* in Eastern North America (Lepidoptera: Hesperiidae: Hesperiinae). Smithsonian Contributions to Zoology 423:1-39.

Cain, S. A. 1944. Foundations of Plant Geography. Harper & Brothers, New York, 556 pp.

Carey, S. W. 1958. A tectonic approach to continental drift. Pp. 177-355 in S. W. Carey (convenor). Continental Drift, A Symposium. Hobart University, Tasmania.

Carpenter, F. M. 1930. A review of our present knowledge of the geological history of insects. Psyche 37:15-34.

Carlquist, S. J. 1974. Island Biology. Columbia University Press, New York and London, 660 pp.

Case, J. E., T. L. Holcombe, and R. G. Martin. 1984. Map of geological provinces in the Caribbean region. Geological Society of America Memoir 162:1-30.

Coney, P. J. 1983. Plate tectonic constraints on biogeographic connections between North and South America. Annals of the Missouri Botanical Garden 69:432-443.

Cracraft, J. 1973. Continental drift, paleoclimatology, and the evolution and biogeography of birds. Journal of Zoology (London) 169:455-545.

Clench, H. K. 1964. A synopsis of the West Indian Lycaenidac, with remarks on their zoogeography. Journal of Research on the Lepidoptera 2:247-270.

———. 1977. A list of the butterflies of Andros, Bahamas. Annals of the Carnegie Museum 46:173-194.

Clench, H. K. and K. A. Bjorndal 1980. Butterflies of Great and Little Inagua, Bahamas. Annals of the Carnegie Museum 49: 1-30.

Croizat, L. 1958. Panbiogeography. Private publication. Caracas, 2780 pp.

———, G. Nelson, and D. E. Rosen. 1974. Centers of origin and related concepts. Systematic Zoology 23:265-287.

Darlington, P. J. 1957. Zoogeography: The Geographical Distribution of Animals. John Wiley, New York, 675 pp.

———. 1965. Biogeography of the Southern End of the World: Distribution and History of the Far-Southern Life and Land, with an Assessment of Continental Drift. Harvard University Press, Cambridge, MA. 229 pp.

Dickson, C. G. C. and D. M. Kroon (eds.). 1978. Pennington's Butterflies of Southern Africa. A. Donker, Johannesburg and London, 669 pp.

Donnelly, T. W. 1985. Mesozoic and Cenozoic Plate Evolution of the Caribbean region. Pp. 89-121 in F. G. Stehli and S. D. Webb (eds.). The Great American Biotic Interchange. Plenum Press, New York.

Durden, C. J. and H. Rose. 1978. Butterflies from the middle Eocene. The earliest occurrence of fossil Papilionidae. Pearce-Sellard Series, Texas Memorial Museum 28:1-25.

du Toit, A. L. 1927. A geological comparison of South America with South Africa. Publication of the Carnegie Institute of Washington 381:1-158.

———. 1937. Our Wandering Continents: An Hypothesis of Continental Drift. Oliver & Boyd, Edinburgh, UK, 366 pp.

Eliot, J. N. 1973. The higher classification of the Lycaenidae (Lepidoptera): a tentative arrangement. Bulletin of the British Museum (Natural History), Entomology, 28:371-505.

Endler, J. A. 1982. Pleistocene forest refuges: fact or fancy? Pp. 641-657 in G. T. Prance (ed.). Biological Diversification in the Tropics. Columbia University Press, New York.

Evans, W. H. 1937. A Catalogue of the African Hesperiidae in the British Museum (Natural History). British Museum Trustees, London, 212 pp.

———. 1949. A Catalogue of the Hesperiidae Europe, Asia and Australia in the British Museum (Natural History). British Museum Trustees, London, 502 pp.

———. 1952. A Catalogue of the American Hesperiidae in the British Museum (Natural History). British Museum Trustees, London part 2, 178 pp.

———. 1955. A Catalogue of the American Hesperiidae in the British Museum (Natural History). British Museum Trustees, London part 4, 499 pp.

Flint, O. S., Jr. 1977. Probable origins of the West Indian Trichoptera and Odonata faunas. Pp. 215-223 in M. I. Crichton (ed.). Proceedings of the Second International Symposium on Trichoptera, Reading, UK.

Forbes, W. T. M. 1932. How old are the Lepidopera? American Naturalist 66:452-460.

Fox, R. M. 1963. Affinities and distribution of Antillean Ithomiidae. Journal of Research on the Lepidoptera 2:173-184.

Guyer, C. and J. M. Savage. 1986. Cladistic relationships among anoles (Sauria: Iguanidae). Systematic Zoology 35:509-531.
Hancock, D. L. 1983. Classification of the Papilionidae (Lepidoptera): a phylogenetic approach. Smithersia, (2):48 pp.
Harvey, D. J. and H. K. Clench. 1980. *Dianesia*, a new genus of Riodinidae from the West Indies. Journal of the Lepidopterists' Society 34:127-132.
Hennig, W. 1966. Phylogentic Systematics. University of Illinois Press, Urbana, IL, 263 pp.
Holloway, J. D. 1979. A survey of the Lepidoptera, Biogeography and Ecology of New Caledonia. W. Junk, The Hague, 588 pp.
Howe, W. H. 1975. Eunica. Pp.126-127 in W. H. Howe, (ed.). The Butterflies of North America. Doubleday and Co., New York.
Jenkins, D. W. 1984. Neotropical Nymphalidae. II. Revision of Myscelia. Bulletin of the Allyn Museum (87):64 pp.
_____. (in prep.). The phylogeny and biogeography of the Eurytelinae (Nymphalidae). [tentative title].
Jordan, K. 1908 (1907-1909). Papilionidae. Pp. 11-51 in A. Seitz (ed.). The Macrolepidoptera of the World, vol. 5. A. Kernan, Stuttgart.
Klots, A. B. 1951. A Field Guide to the Butterflies of North America, East of the Great Plains. Houghton Mifflin Co., Boston, MA, 349 pp.
Liebherr J. K. 1986. *Barylaus*, a new genus (Coleoptera: Carabidae) endemic to the West Indies with Old World affinities. Journal of the New York Entomological Society 94:83-97.
Lundberg, J. G., A. Machado-Allison, and R. F. Kay. 1986. Miocene characid fishes from Colombia: evolutionary stasis and extirpation. Science 234:208-209.
Matthew, G. A. 1915. Climate and evolution. Annals of the New York Academy of Science 24:171-318.
Miller, J. Y. 1986. The Taxonomy, Phylogeny and Zoogeography of the Neotropical Moth Subfamily Castniinae (Lepidoptera: Castnioidea: Castniidae). Unpublished Ph. D. dissertation, University of Florida, Gainesville, FL, 571 pp.
_____. (in prep.). Review of the neotropical species of Wallengrenia Berg (Lepidoptera: Hesperiidae) with description of two new taxa.
Miller, L. D. 1965. A review of the West Indian "Choranthus". Journal of Research on the Lepidoptera 4:259-274.
_____. 1968. The higher classification, phylogeny and zoogeography of the Satyridae (Lepidoptera). Memoirs of the American Entomological Society 24:1-174.
_____. in prep. The higher classification, phylogenetics, and biogeography of the Satyridae (Lepidoptera): a reappraisal. [tentative title].
_____, and F. M. Brown. 1983. Butterfly taxonomy: a reply. Journal of Research on the Lepidoptera 20:193-198.
_____, and D. Spencer-Smith, in prep. Why South Florida is best considered part of the West Indies. [tentative title].
Patterson, C. 1981. Methods of paleobiogeography. Pp. 446-489 in G. Nelson & D. E. Rosen (eds.). Vicariance Biogeography: A Critique. Columbia University Press, New York.

Perfit, M. R., and B. C. Heezen. 1978. The geology and evolution of the Cayman Trench. Geological Society of America Bulletin 89:1155-1174.

Pindell, J. and J. F. Dewey. 1982. Permo-Triassic reconstruction of western Pangea and the evolution of the Gulf of Mexico/Caribbean region. Tectonics, 1:179-211.

Raven, P. H. and D. I. Axelrod. 1974. Angiosperm biogeography and past continental movements. Annals of the Missouri Botanical Garden 61:539-673.

Riley, N. D. 1975. A Field Guide to the Butterflies of the West Indies. Collins Sons & Co., Ltd., London, 224 pp.

Romeo, E. J. and S. Archangelsky. 1986. Early Cretaceous Angiosperm leaves from southern South America. Science 234: 1580-1582.

Rosen, D. E. 1976. A vicariance model of Caribbean biogeography. Systematic Zoology 24:431-464.

_____. 1985. Geological hierarchies and biogeographic congruence in the Caribbean. Annals of the Missouri Botanical Garden 72:636-659.

Savage, J M. 1973. The geographic distribution of frogs: patterns and predictions. Pp. 349-445 in J. L. Vial (ed.). Evolutionary Biology of the Anura. University of Missouri Press, Columbia, MO.

_____. 1974. The isthmian link and the evolution of Neotropical mammals. Contributions in Science of the Natural History Museum of Los Angeles County 260:1-51.

_____. 1983. The enigma of the Central American herpetofauna: Dispersals or vicariance? Annals of the Missouri Botanical Garden 69:464-547.

Schuchert, G. 1935. Historical Geology of the Antillean-Caribbean Region or Lands Bordering the Gulf of Mexico and the Caribbean Sea. John Wiley & Sons, New York, 811 pp.

Schwanwitsch, B. N. 1924. On the ground-plan of wing-pattern in nymphalids and certain other families of rhopalocerous Lepidoptera. Proceedings of the Zoological Society London 1924:509-528.

Schwartz, A. 1983. Haitian Butterflies. Special Publication of the Museo Nacional de Historia Natural, Santo Domingo, Dominican Republic, 69 pp.

Scott, J. A. 1972. Biogeography of the Antillean Butterflies. Biotropica 4:32-45.

_____. 1986. Distribution of Caribbean Butterflies. Papilio (new series) 3:1-26.

Shields, O. 1979. Evidence for initial opening of the Pacific Ocean in the Jurassic. Palaeogeography, Palaeoclimatology, Palaeoecology 26:181-220.

_____. 1985. Zoogeography of the Libytheidae (Snouts or Breaks [sic!]) Tokurana 9:1-58.

Shields, O. and S. K. Dvorak. 1979. Butterfly distribution and continental drift between the Americas, the Caribbean and Africa. Journal of Natural History 13:221-250.

Shuey, J. A. 1986. The Ecology and Evolution of Wetland Butterflies, with Emphasis on the Genus *Euphyes* (Lepidoptera: Hesperiidae). Unpublished Ph. D. dissertation, Ohio State University, Columbus, OH, 145 pp.

Simon, M. J., and L. D. Miller. 1986. Observations on the butterflies of Great Inagua Island, Bahamas, with records of three species new to the island. Bulletin of the Allyn Museum (105):14 pp.

Simpson, G. G. 1952. Probabilities of dispersal in geologic time. Bulletin of the American Museum of Natural History 99:163-176.

_____. 1956. Zoogeography of West Indian mammals. American Museum Novitates 1759:1-28.

Smart, J. and N. F. Hughes. 1973. The insect and the plant: progressive palaeoecological integration. Pp. 143-155 in H. F. van Emden (ed.). Insect/Plant Relationships. Symposium of the Royal Entomological Society of London No. 6.
Smith, D. L. 1985. Caribbean plate relative motions. Pp. 17-48 in F. G. Stehli and S. D. Webb (eds.). The Great American Biotic Interchange. Plenum Press, New York.
Stehli, F. G. and S. D. Webb (eds.). 1985. The Great American Biotic Interchange. Plenum Press, New York, 532 pp.
de la Torre y Callejas, S. L. 1954. An annotated list of the butterflies and skippers of Cuba. Journal of the New York Entomological Society 62:1-25, 113-128, 189-192, 207-249.
Wallace, A. R. 1876. The Geographical Distribution of Animals. Harpers, New York. vol. 1, 503 pp.; vol 2, 607 pp.
Wegener, A. 1915. Die Enstehung der Kontinente und Ozeane. Sammlung Vieweg, 23: F. Vieweg und Sohn, Brauschweig, 144 pp.
Whalley, P. 1986. A review of the current fossil evidence of Lepidoptera in the Mesozoic. Biological Journal of the Linnean Society 28:253-271.
Wilson, J. T. 1963. Hypothesis of earth's behavior. Nature (London) 198:925-929.

Table 1. Endemicity of Antillean butterfly fauna. Note: southern Florida is included in the West Indies for purposes of establishing endemicity.

Family	Genera No.	Endemic No.	Pct. End.	Species No.	Endemic No.	Pct. End.
Danaidae	3	0	0.0	9	5	55.6
Ithomiidae	1	0	0.0	2	2	100.0
Satyridae	1	1	100.0	25+	25+	100.0
Nymphalidae	34	4	11.8	65	26	40.0
Libytheidae	1	0	0.0	3	3	100.0
Riodinidae	1	1	100.0	1	1	100.0
Lycaenidae	13	1	7.7	32	22	68.8
Pieridae	13	0	0.0	50	24	48.0
Papilionidae	5	0	0.0	22	15	68.2
Hesperiidae	49	7	14.3	92	47	51.1
Total	121	14	11.6%	301	170	56.5%

Table 2. Affinities of West Indian Butterfly Genera. Endemic genera are set off by an asterisk (*) and placed with nearest relative(s); number of Antillean speciers is in parentheses after generic name.

Widespread genera	Affinity with North America	Affinity with Mexico and/or Central America	Affinity with Central and South America
Danaus (4)	Asterocampa (1)	Anetia (4)	Calisto (25+)
Lycorea (1)	Basilarchia (1)	Greta (2)	Archaeprepona (1)
Doxocopa (2)	Phyciodes (1)	Anaea (1)	Myscelia (1)
Hypna (1)	Calephelis (1)	Siderone (1)	Eunica (3)
Memphis (4)	Eumaeus (1)	Hamadryas (3)	Siproeta (1)
Marpesia (3)	Atlides (1)	Dynamine (2)	Eresia (1)
Colobura (1)	Parrhasius (1)	Lucinia* (2)	Eueides (1)
Historis (2)	Ministrymon (1)	Adelpha (2)	Philaethria (1)
Archimestra* (1)	Pontia (1)	Junonia (3)	Pseudolycaena (1)
Anartia (4)	Pieris (1)	Atlantea* (4)	Cyanophrys (1)
Biblis (1)	Nathalis (1)	Antillea* (2)	Electrostrymon (4)
Vanessa (3)	Colias (1)	Hypanartia (1)	Pseudochrysops* (1)
Euptoieta (2)	Eurytides (4)	Dianesia* (1)	Melete (1)
Heliconius (1)	Pterourus (pt.) (2)	Allosmaitia (2)	Aphrissa (4)
Dryas (1)	Papilio (1)	Chlorostrymon (2)	Chiomara (2)
Agraulis (1)	Parides (1)	Nesiostrymon (1)	Pheraeus (1)
Libytheana (4)	Phocides (2)	Dismorphia (2)	Saliana (1)
Strymon (11)	Autochton (2)	Ganyra (2)	
Leptotes (2)	Erynnis (1)	Kricogonia (1)	
Brephidium (2)	Oarisma (2)	Zerene (1)	
Hemiargus (4)		Pterourus (pt.) (1)	
Ascia (1)		Aguna (1)	
Appias (2)		Polythrix (1)	
Eurema (23)		Astraptes (pt.) (3)	
Anteos (2)		Burca* (4)	
Phoebis (6)		Timochares (1)	
Heraclides (11)		Ephyriades (3)	
Battus (3)		Pyrrhocalles (2)	
Epargyreus (3)		Perichares (1)	
Polygonus (2)		Vettius (1)	
Chioides (4)		Synapte (1)	
Urbanus (6)		Rhinthon (2)	
Astraptes (pt.)(3)		Holguinia* (1)	
Cogia (1)		Polites (2)	
Nisoniades (1)		Atalopedes (4)	
Cabares (1)		Parachoranthus* (1)	

Table 2. Affinities of West Indian Butterfly Genera (Concluded).

Widespread genera	Affinity with North America	Affinity with Mexico and/or Central America	Affinity with Central and South America
Antigonus (1)		*Choranthus** (6)	
Achlyodes (1)		*Paratrytone* (1)	
Grais (1)		*Euphyes* (3)	
Gesta (1)		*Asbolis** (1)	
Ouleus (1)			
Heliopetes (1)			
Pyrgus (2)			
Cymaenes (1)			
Wallengrenia (4?)			
Hylephila (1)			
Panoquina (6)			
Nyctelius (1)			
Lerodea (1)			

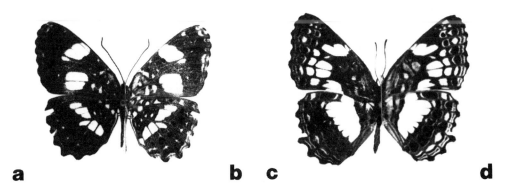

Figure 1. Facies of *Archimestra teleboas* (a = upper side; b = under side) from the Dominican Republic, and *Neptidopsis ophione* (c = upper side; d = under side) from Ghana. Structural features also support the resemblance as discussed in the text.

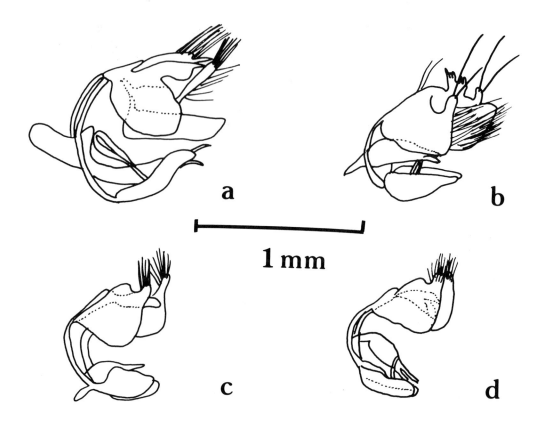

Figure 2. Structural examples of the African affinities in the West Indian butterflies are found in the left lateral views of male genitalia of *Oraidium* and *Brephidium* species. (a) *Oraidium barberae*, South Africa; (b) *Brephidium metophis*, Africa; (c) *B. isophthalma*, Florida; (d) *B. exilis*, Arizona.

Figure 3. Distribution of the genera *Calisto* (heavy stippling) and *Eretris* (vertical crosshatching) in the West Indies and continental America. *Eretris* is widely distributed and found southward in the Andes to Bolivia.

Figure 4. Distribution of *Battus polydamas* in the West Indies. The stippled area is inhabited only by *B. p. polydamas* (Linnaeus). Island subspecies include: (a) *lucayus* (Rothschild and Jordan); (b) *cubensis* (Du Frane); (c) *jamaicensis* (Rothschild and Jordan); (d) *polycrates* (Hopffer); (e) *thyamus* (Rothschild and Jordan); (f) *antiguus* (Rothschild and Jordan); (g) *christopheranus* (Hall); (h) *neodamas* (Lucas); (i) *dominicus* (Rothschild and Jordan); (j) *xenodamas* (Huebner); (k) *lucianus* (Rothschild and Jordan); (l) *vincentius* (Rothschild and Jordan); (m) *grenadensis* (Hall). See text for further discussion.

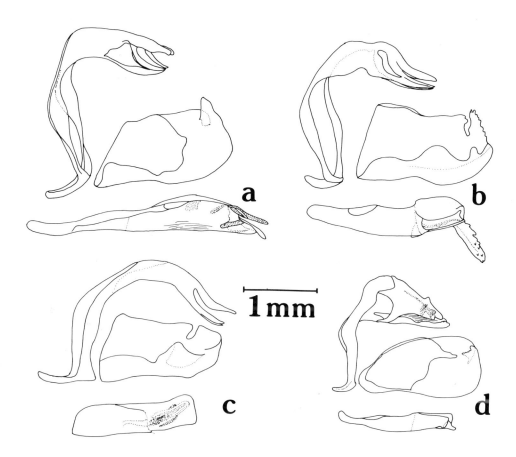

Figure 5. The potential evolutionary affinities of the hesperiid genera *Paratrytone* and *Choranthus* are apparent in the male genitalia of selected species: (a) *P. rhexenor*: Mexico; (b) *P. niveolimbus*: Guatemala; (c) *P. batesi*: Haiti; (d) *C. radians*: Cuba.

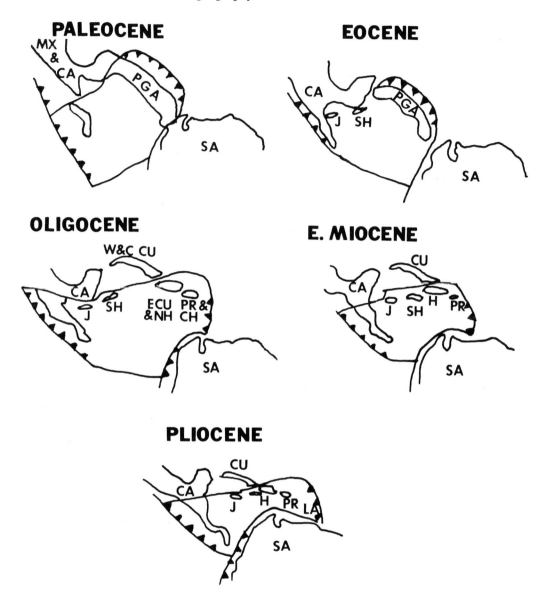

Figure 6. Modified Pindell and Dewey (1982) model of the evolution of the West Indies. Abbreviations: MX = Mexico; CA = Central America; PGA = preoto-Greater Antilles; SA = South America; J = Jamaica; SH = southern Hispaniola; W & C CU = western and central Cuba; ECU & NH = Eastern Cuba and northern Hispaniola; PR & CH = Puerto Rico and central Hispaniola; H = Hispaniola; PR = Puerto Rico; CU = Cuba; LA = Lesser Antilles. Exceptions to the Pindell and Dewey model involve particularly the relative Eocene placement of Jamaica and the southern Hispaniola blocks to conform with the biogeography of butterflies.

West Indian butterflies 259

Figure 7. The geographical distribution of the primitive danaid genus *Anetia* in Mexico, Central America, and the West Indies (stippled areas) reflects its past geological history. See text for further details.

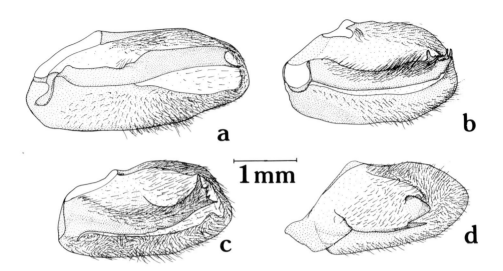

Figure 8. The possible evolutionary affinities of the Jamaican fauna are observed in the male valvae of selected *Pterourus* species; (a) *P. homerus*: Jamaica; (b) *P. cleotas*: Colombia; (c) *P. garamas*: Mexico; (d) *P. esperanza*: Mexico. For details of the relationships see text.

Figure 9. The distribution of the hesperine genus *Wallengrenia* (after J. Miller, in prep.) incorporates a number of vicariant events. *W. otho* (o) is widely distributed, as is *W. premnas* in South America. Antillean taxa include *W. misera* (vertical crosshatching and m), *druryi* (light stippling and d), *vesuria* (heavy stippling and v), and *ophites* (op). For additional discussion, see text.

Note added in proof

Since the account was written, two significant geological papers (Burke 1988; Donnelly 1988) have appeared which more accurately depict the geological history of the Caribbean than did previous accounts. The reader is referred to these and the Biogeography section of the forthcoming Smith, et al.

Bibliography

Burke, K. 1988. Tectonic evolution of the Caribbean. Ann. Rev. Earth Planet. Sci. 16: 201-230.
Donnelly, T.W. 1988. Geologic constraints on Caribbean biogeography. Pp. 15-37 *in* Liebherr, J.K. (ed.) Zoogeography of Caribbean Insects. Ithaca, Comstock: 285 pp.
Smith, D.S., L.D. Miller, and J.Y. Miller, in prep. Butterflies of the West Indies and South Florida. Oxford, Oxford Univ. Press.

ZOOGEOGRAPHY OF THE ANTILLEAN FRESHWATER FISH FAUNA

George H. Burgess and Richard Franz[1]

Abstract

The freshwater fish fauna of the Antilles (including the Lesser Antilles south to Grenada) consists of 71 mostly endemic species in the families Lepisosteidae, Bythitidae, Poeciliidae, Cyprinodontidae, Fundulidae, Rivulidae, Atherinidae, Synbranchidae, and Cichlidae. The family Poeciliidae with five genera and 46 species dominates the fauna. The Antillean ichthyofauna is derived from North American, South American, and marine protofaunas. The following patterns are suggested: (1) All island faunas have their closest relationships with Cuba; (2) Cuba's greatest affinity is with Hispaniola; and (3) after Cuba, the greatest affinities are Hispaniola and Jamaica with each other, the Bahamas with the Caymans, and the Caymans equally with the Bahamas, Hispaniola, and Jamaica. Cuba and Hispaniola, with their greater surface areas, host 89% of the fauna. Other island areas support from four to six species, except Puerto Rico, which totally lacks native freshwater fishes. Specific distributional patterns within Hispaniola and their relationships to past geological events and drainage histories are discussed. Historically proposed land bridge, dispersal, and continental drift (vicariance) models inadequately account for the presence of the freshwater fauna in the Antilles. Based on the most recent geological literature and modern fish distributions, we propose a zoogeographic model that merges elements of dispersal and vicariance to explain Antillean fish distributions.

Introduction

In 1938 Myers recognized that zoogeographers were divided into three schools of thought regarding insular zoogeography: bridge builders, dispersalists, and proponents of continental drift (vicariance). Following the consensus of the day, he did not pursue the latter option in detail (although he expressed the opinion that the Wegnerian school might eventually prevail) and proceeded to weigh bridges vs. dispersal in his treatment of West Indian freshwater fish zoogeography. Fifty years later, zoogeographers are still debating the subject, although vicariance has largely replaced bridge building in opposing dispersal as the prevailing zoogeographic philosophy. Students of West Indian freshwater fishes have played key roles in this continuing controversy (Table 1), and

---

[1] George H. Burgess is Senior Biologist in Ichthyology and Richard Franz is Associate in Ecology at the Florida Museum of Natural History, University of Florida, Gainesville, FL 32611 USA

philosophical proponents of all three schools continue to battle (Rosen 1976, 1985; Briggs 1984, 1987; Rivas 1958, 1986).

As we hope to show in subsequent discussions, it is our belief that attempts to explain Antillean zoogeography on the basis of the single philosophy are unsatisfactory. We suggest that Antillean freshwater fish zoogeography is best addressed by a merger of elements of dispersal and vicariance philosophies when the latest geological evidence is considered.

We dedicate this chapter to the memory of Luis Rene Rivas, friend and colleague. Luis dedicated most of his scientific life to studies of Antillean fishes and his publications form the basis of our understanding of the ichtyofaunal relationships and zoogeography of the region. Through interactions with Luis over many years we were privy to many of his unpublished ideas, but regrettably his death robbed us of much more.

Scope and derivation of the study

Our consideration of the freshwater fish fauna of the West Indies includes the geographic area encompassing the Bahamas, Cuba and Isla de Juventud (= Isle of Pines), the Cayman Islands, Hispaniola, Puerto Rico, and the Lesser Antilles southward to Grenada (Fig. 1). We exclude Trinidad, Tobago, Margarita, and the Netherland Antilles from our discussion since these islands are geologically and zoogeographically linked to South America. Their native freshwater fishes are typical of the Brazilian subregion of the Neotropics, but the fauna is depauperate. Only six species are present: two rivulid killifishes (*Rivulus harti*, *R. ocellatus*), two poeciliid killifishes (*Poecilia reticulata*, *P. vivipara*), and two synbranchid eels (*Synbranchus marmoratus*, *Ophisternon aenigmaticum*).

Distributional data were gathered from a variety of sources. In addition to the collections housed in the Florida Museum of Natural History, we utilized a series of faunistic surveys and regional synopses, including Alayo (1973), Breder (1932), Caldwell (1966), Eigenmann (1903), Evermann and Marsh (1900), Fowler (1915a, 1915b, 1928, 1938, 1952), Hildebrand (1935), Lee, Plantania and Burgess (1983), McIndoo (1906), Miller (1982), Nichols (1912, 1915, 1929), Ortiz Carrasquillo (1980), and Vergara (1980). Our taxonomic arrangements and understanding of phylogenetic relationships were derived from Cohen and Nielson (1978), Cohen and Robins (1970), Fink (1971a, 1971b), Fowler (1939), Franz and Burgess (1983), Greenfield (1983), Loiselle (1980), Miller (1962), Myers (1928), Parenti (1981), Relyea (1983), Rivas (1944, 1948, 1958, 1963, 1969, 1971, 1978, 1980, 1982), Rivas and Fink (1970), Rivas and Myers (1950), Regan (1905), Roloff (1938), Rosen and Bailey (1959, 1963), Rosen and Greenwood (1976), Rosen and Rumney (1972), Seegers and Huber (1981), Tee-Van (1935), Wiley (1976), and Vergara (1981). In addition, we have freely drawn upon Luis Rivas' unpublished thoughts regarding the relationships of the poeciliid genus *Limia*, a group critical to the zoogeography of the region. We follow Rivas' perceptions of *Limia* interrelationships throughout the text, but note that independent verification must await our completion of the revision of *Limia* initiated by Rivas. Our own unpublished data on Hispaniolan *Limia* species, some as yet undescribed, are also utilized in our synthetic zoogeographic discussions. We have chosen to follow Parenti's (1981) classification scheme for cyprinodontiform killifishes in the absence of published refutation, but are cognizant that this system has not been uniformly adopted.

Characterization of the Antillean freshwater fish fauna

The Antillean freshwater fish fauna includes 71 species in nine families (Table 2).

The presence of three species from scattered localities on Puerto Rico and the Lesser Antilles is problematic since introductions are likely in some, if not all, cases. Two live-bearing killifishes (Poeciliidae) native to northern South America are reported from the more northerly islands, but are thought to represent introductions (Myers 1938; Rosen and Bailey 1963). *Poecilia vivipara* is known from Puerto Rico, Martinique and some of the southern Lesser Antilles. It tolerates sea water well, commonly occurs around mangrove prop roots, and may have arrived fortuitously in the bilge water of ships or rafted with slow moving sailboats or canoes "island hopping" their way northward from the South American coast. It is known from Puerto Rico since at least the turn of the century (Evermann and Marsh 1900). The fancy guppy, *P. reticulata*, is discontinuously distributed on St. Thomas, Dominica, Barbados and Antigua (as well as on Cuba, Hispaniola, and Jamaica). This species is a popular aquarium fish and introduction seems highly likely. Species of the African cichlid *Tilapia* have often been taken syntopically, providing additional circumstantial evidence for introduction. The swamp eel *Synbranchus marmoratus* (Synbranchidae) has been take on St. Lucia, as well as natively on Trinidad, Tobago and in northern South America. While it is possible that *S. marmoratus* is native to St. Lucia, swamp eels are perhaps the most likely candidates for aboriginal introduction since they are capable of withstanding prolonged withdrawals from water if kept damp. Harris (1965) discussed numerous Carib and Arawak Indian introductions in the Lesser Antilles, and swamp eel introductions are documented or suspected in other parts of the world. We choose to omit these species from our zoogeographical considerations on the assumption that they are introductions, noting, however, that if native they would represent a significant northward extension of the Brazilian neotropical ichthyofauna.

Invading marine species constitute an important component of the ichthyofauna but are zoogeographically uninformative since high dispersal abilities have allowed them to become distributionally widespread. Many marine species spend at least portions of their lives, often as juveniles, in the brackish or freshwater lower reaches of insular rivers and streams. Seven species stand out from this group because they fully penetrate the most upstream areas and characteristically represent ecologically significant elements of the freshwater communities: *Anguilla rostrata* (Anguillidae), *Gobiesox nudus* (Gobiesocidae), *Strongylura notata* (Belonidae), *Agonostomus monticola* and *Joturus pichardi* (Mugilidae), *Gobiomorus dormitor* (Eleotridae), and *Sicydium plumieri* (Gobiidae). In addition, there are numerous established exotic species (Table 3) that have been intentionally or accidentally released into native waters from aquarium or culture stocks. Exotics and marine invaders will not be considered in our subsequent zoogeographic discussions.

Approximately 89% of the Antillean freshwater fish fauna occurs on Cuba and Hispaniola and their offshore islets. Cuba, with its 33% larger land surface, has fewer species (28) than Hispaniola (34), but has more families (nine vs. four) and genera (13 vs. seven) (Table 4). The high species diversity on Hispaniola is largely due to the explosive speciation of a single genus, *Limia* (Poeciliidae). The faunas of Jamaica, the Bahamas, and the Cayman Islands are small with five to six species each. Puerto Rico, despite having the fourth largest insular land mass, totally lacks a native freshwater ichthyofauna, but does support 24 established introductions as well as peripheral marine invaders. The Lesser Antilles collectively host only two species of freshwater fishes.

The Antillean fauna is dominated by cyprinodontiform fishes of the families Poeciliidae, Cyprinodontidae, Fundulidae, and Rivulidae (Fig. 2, 3). Eighty-five percent of the total fauna are cyprinodontoids with the poeciliids (65%) the largest family. The remainder of the fauna is confined to a gar (Lepisosteidae), four brotulas (Bythitidae), one silverside (Atherinidae), a swamp eel (Synbranchidae), and four cichlids (Cichlidae) (Fig. 2,3). All members of the ichthyofauna are secondary or peripheral freshwater species (Myers 1938, 1951). As a group, Antillean fishes are ecologically tolerant, with most species capable of living in marginal habitats including brackish water. They include both egg-laying species and live-bearers, many of which are reproductively prolific the year round.

The fauna is derived from North American, South American and marine protofaunas (Table 5). The North American (seven genera, 27 species) and South American (five genera, 39 species) components are essentially equidominant. The marine-derived component (two genera, five species) does not have New World affinities, but rather has ties to the Pacific or eastern Atlantic (Cohen and Nielson 1978, B. Chernoff, pers. comm.) suggesting an archaic Gondwanaland derivation.

Most of the fauna are Antillean endemics. Six of the 14 genera (*Lucifuga*, *Girardinus*, *Limia*, *Quintana*, *Cubanichthys*, and *Alepidomus*) are restricted to one or more islands. All but six taxa (*Gambusia p. puncticulata*, *Gambusia rhizophorae*, *Cyprinodon variegatus*, *Fundulus grandis saguanus*, *Rivulus marmoratus*, *Ophisternon aenigmaticum*) are endemic to a single island or island group; of these exceptions, at least two, *Cyprinodon variegatus* and *O. aenigmaticum*, probably will eventually warrant specific or subspecific recognition and *R. marmoratus* is certainly in need of critical examination from throughout its range. All of the widespread species except *O. aenigmaticum* frequent brackish water coastal habitats.

Seven genera are shared by two or more islands or island groups (Table 6). *Gambusia* and *Cyprinodon* are found on all five island groups, as well as in southern Florida. *G. puncticulata*, *C. variegatus*, and *R. marmoratus* (or their complexes) are widespread, occurring from the Bahamas to Yucatan. *G. puncticulata* and *R. marmoratus* have not yet been taken from Hispaniola, but are expected there.

Figure 4 presents tracks of shared presumed monophyletic lineages between islands or island groups and the closest continental coast, i.e. southern Florida. We emphasize that most of these relationships have not been proposed in phylogenies nor rigorously tested; thus this analysis is preliminary at best. Nevertheless, patterns do exist, including the following:

A. All island faunas have their closet relationships with Cuba, which has the largest land mass, largest familial and generic diversity, and is centrally located.

B. Cuba's greatest affinity is with Hispaniola, the second largest island.

C. After Cuba, the greatest affinities are Hispaniola and Jamaica with each other, the Bahamas with the Caymans, and the Caymans equally with the Bahamas, Hispaniola, and Jamaica.

Cuba

Cuba has the most distinctive freshwater fish fauna of the region with 34 species and subspecies in 13 genera and nine families (Table 2, 7). Among these are the Antilles' only gar (*Atractosteus tristoechus*); three cave-dwelling species of marine-derived brotulas (*Lucifuga*); an endemic genus of freshwater silverside (*Alepidomus evermanni*); a synbranchid eel (*Ophisternon aenigmaticum*) shared with Central and northeastern South America; two endemic genera of poeciliids, the monotypic *Quintana* and related *Girardinus*, which has radiated extensively (seven species); a representative of the Greater Antillean endemic genus *Cubanichthys* (*cubensis*); two species shared with southern Florida, *Fundulus grandis saguanus* and *Gambusia rhizophorae*; plus endemic species of *Cichlasoma* (2), *Rivulus* (3), *Limia* (1), and *Gambusia* (1). Distinctive subspecies or morphs of the widespread *Gambusia puncticulata* and *Cyprinodon variegatus* are also present.

One quarter of the fishes are widespread on the island (Table 7). The bulk of the fauna (59%) is confined to western Cuba, including Isla de la Juventud. Only one species (*Rivulus insulaepinorum*) is restricted to Isla de la Juventud; the taxonomic status of this species and *R. garciae*, endemic to western Cuba, are in need of review. Approximately 16% of the fishes are endemic to eastern Cuba (including three subspecies of *Gambusia puncticulata*). The low level of endemism in eastern Cuba is enigmatic since this area has the greatest topographic relief on the island and presumably is where one would expect the greatest range fragmentation and speciation. To our knowledge there are no freshwater fishes reported from the smaller islets associated with the southern shore of Cuba.

The majority of the freshwater fish fauna occurs in coastal regions in ponds, lakes and low gradient streams or rivers. At least four species (*Girardinus creolus*, *G. microdactylus*, *G. serripenis*, *G. uninotatus*) occur in upland streams in western Cuba and two species (*Girardinus denticulatus*, *Cichlasoma ramsdeni*) in eastern Cuba may represent a montane element similar to that reported for Hispaniola.

The marine-derived fauna is restricted to western Cuba. *Alepidomus* has been recorded from San Cristobal and Pinar del Rìo (Eigenmann 1903), while the three members of *Lucifuga* occur in subterranean freshwater of west central Cuba. The only other member of the highly adapted cavernicolous genus *Lucifuga* occurs in a brackish water cave system on New Providence Island in the Bahamas. All *Lucifuga* show reductions in pigmentation and eye structure, and modifications in reproductive behavior and physiology typical of highly adapted cave dwellers.

Hispaniola

The freshwater fish fauna of Hispaniola consists of 35 species belonging to seven genera in four families (Table 2). One family, Poeciliidae, with 28 species (four species of *Gambusia*, 21 species of *Limia*, and three species of *Poecilia*) dominates the fauna. Étang de Miragôane on the Tiburon Peninsula of Haiti hosts an extensive series of endemics, including eight species of *Limia* and one *Gambusia*. Other distinctive elements in the Hispaniolan ichthyofauna include an undescribed species of *Cubanichthys*; three species of *Poecilia*, a genus otherwise absent from the Antilles (see earlier remarks concerning *Poecilia vivipara* and *P. reticulata*); the lacustrine endemic *Cyprinodon bondi*; plus endemic species of *Rivulus* (2), *Cichlasoma* (2), and *Gambusia* (2-3). The Hispaniolan fish fauna has exploited most of the available aquatic habitats, from hypersaline lakes and coastal mangrove sites to montane streams.

Most Hispaniolan species have very restricted ranges and specific habitat requirements, e.g. *Limia (Limia) sulphurophilia* is confined to one sulphur spring which joins Lago Enriquillo in Republica Dominicana (RD); seven *Limia (Odontolimia)*, one *Limia (Limia)*, and one *Gambusia* from Étang Miragôane in Haiti; *Gambusia pseudopunctata*, *Limia (Limia) dominicensis*, *Limia* species D from isolated streams on the extreme western end of the Tiburon Peninsula; *Limia (Limia) rivasi* from a small mangrove area at Anse-a-Galet on Ile de la Gonâve. Only *Cichlasoma haitiensis* is found nearly island-wide, although this fish has been used by humans for food and may have been transported to many sites.

The Hispaniolan fauna has been divided into three distinct groups (Franz and Thompson 1978). Their distributions correspond with (1) northern Hispaniola, (2) the Tiburon and Barahona peninsulas in southern Hispaniola, and (3) the Cul-de-Sac/Valle-de-Neiba Plain (Fig. 5). Areas (1) and (2) have been referred to in the literature as the "North and South islands." Area (3) is a large tectonic valley that separates the north and south areas. This valley was periodically flooded to form a seaway during periods of higher sea levels.

The North and South islands are composed of a series of NW-SE oriented mountain blocks separated by extensive valleys, many of which also have tectonic origins. Maurrasse (1981a) considered these mountain blocks to have remained above water as separate islands during late Tertiary and Pleistocene times. Rivas (1982) used this model to explain the distribution of the genus *Poecilia* in northern Hispaniola.

The fishes associated with the North and South islands and the Cul-de-Sac/Valle-de-Neiba Plain are listed in Table 8. Examination of this table shows the presence of high levels of endemism within each region. The endemic fauna of the South Island includes two *Gambusia*, seven *Limia (Odontolimia)*, and three *Limia (Limia)* (the latter in two species groups), while that of the North Island includes three *Poecilia*, five *Limia (Limia)* (three species groups), one *Cubanichthys*, two *Rivulus* (including *R. heyei* from Isla de la Saona), and *Cichlasoma vombergi* (from the Rìo Yaque del Sur drainage). Only the widespread *Cichlsoma haitiensis* and the *dominicensis* species group of *Limia (Limia)* (Fig. 6) are shared by both islands, as well as the Cul-de-Sac/Valle-de-Neiba region. We define the Cul-de-Sac/Valle-de-Neiba faunal region as including the spring at Puerto Alejandro at the base of the Sierra Martin Garcia, lower portions of the Rìo Yaque del Sur, Lago Enriquillo, Laguna del Rincon, Laguna Oviedo, and their springs in the RD and Étang Saumâtre, Trou Caiman and their springs in Haiti (Fig. 5). Its fauna includes an endemic *Gambusia*, *Limia (Limia) tridens* (which occurs widely on the South Island, as well as in coastal rivers in western Haiti), four endemic *Limia (Limia)* (two species groups), and the unique *Cyprinodon bondi*. Portions of this fauna have also invaded the lower Rivière l'Artibonite in northern Haiti and eastern portions of the Barahona Peninsula in southwestern RD. The Cul-de-Sac/Valle-de-Neiba Plain seems to have its greatest faunal affinities with the South Island. *Limia (Limia) rivasi*, an Ile de la Gonâve endemic, probably best fits with this fauna.

Two distinct faunal elements occur in the North Island. A lowland element (*Limia*, *Cubanichthys*, and two species of *Cichlasoma*) inhabits warm streams and larger rivers at elevations less than 200 m while a montane element (*Poecilia dominicensis*, *P. elegans*, *P. hispaniolana*, and *Rivulus roloffi*) occurs in the mid-to high elevation streams in the Cordillera Central. The main centers for montane fishes are the headwaters of the Rìo

Yaque del Norte and Rìo Yuna in LaVega and Peravia provinces in the eastern part of the Cordillera Central (RD); all species inhabit riffles and pools in clear, cool streams (Franz and Rivas 1983a; Franz and Rivas 1983b; Rivas and Franz 1983). The distributions of *Poecilia elegans* and *Rivulus roloffi* are restricted to this region while the ecologically flexible *P. dominicensis* and *P. hispaniolana* are more widespread (Fig. 7). There are no records of the montane ichythyofauna from the Cordillera Septentrionale, north of the Valle de Cibao. The relictual nature of the North Island montane element suggests an archaic assemblage that retreated upstream with changing environments and/or competition with a more ecologically tolerant lowland element.

The South Island fauna includes only a lowland element which consists of *Gambusia pseudopunctata*, *G. beebei*, *Limia dominicensis*, *L. tridens*, and *Limia* (*Odontolimia*) species. There is no evidence of a montane element, similar to that on the North Island, despite extensive sampling in the Massif de la Selle and Massif de la Hotte.

Table 9 provides a list of fishes by drainage (see also drainage map, Fig. 8). Presumably, the presence of naturally-occurring species or groups of species in more than one drainage indicates some level of faunal exchange. These movements probably resulted from captures in headwater streams (for montane species), lowland streams, and tributaries in lower floodplains; escapes during catastrophic floods; and movements along coastal swamp corridors, or across narrow salt water channels (for lowland species). The following avenue for dispersal are postulated (see Fig. 9):

A. Cul-de-Sac Plain to lower l'Artibonite. *Gambusia hispaniolae*, *Limia melanonotata* to the l'Artibonite via low elevation stream captures in mountain areas separating these two areas, rather than dispersal through precipitous coastal corridors around the western edge of the Chaines dux Matheux.

B. Lower Rivière l'Artibonite to northwest Haiti. *Limia tridens* to Les Trois Rivière through low elevation captures between the lower l'Artibonite, Rivière la Quinta and Les Trois Rivière (at Gros Morne).

C. Rìo San Juan to upper l'Artibonite and Les Trois Rivières. *Poecilia hispaniolana* to Les Trois Rivières through headwater captures between the upper San Juan and eastern tributaries of the upper l'Artibonite, then between Rivière Canot, possibly via Riviere d'Ennery (upper tributary of Rivière la Quinta) to Les Trois Rivières.

D. Upper and lower Rivière l'Artibonite. The faunas between the upper and lower parts of this river are distinct which suggests that these two areas were only recently connected to one another probably via the water gap at Mirebalais.

E. Upper l'Artibonite to Rivière du Massacre. *Poecilia hispaniolana* to the Massacre via headwater captures.

F. Upland streams of the eastern Cordillera Central. Montane fauna invading headwater streams of south slope (Rìo Ocoa, Haina, and Ozama).

G. Rìo Yuna to streams on the Samana Peninsula. *Poecilia dominicensis* to streams near Sanchez by stream captures through lowland areas during a glacial period. Presumably streams were cooler than present which would allow for the movements of montane fishes across areas no longer suitable for habitation.

H. Rìo Yuna to northeast RD. *Limia zonata* to the Samana Peninsula and to streams in the vicinity of Nigua and Cabrara across coastal corridors during periods of low sea levels.

I. Rìo Yaque del Sur to Lago Enriquillo spring. *Poecilia hispaniolana* to spring-fed stream near La Descubierta via headwater capture with Rìo Cano (tributary of Rìo San Juan).

J. Valle-de-Neiba Plain to western slope of Sierra Martin Garcia and eastern Barahona Peninsula. *Limia* species A to the spring at Puerto Alejandro and *Gambusia hispaniolae, Limia tridens, Limia* species B to Lago Oviedo via coastal corridors.

K. Western Tiburon Peninsula. Dispersal of *Gambusia pseudopunctata, Limia dominicensis, L. tridens* between the northern drainages (vicinity of Jeremie) and the Les Cayes basin through coastal corridors.

Jamaica

Jamaica hosts six native species of freshwater fishes (Table 2, 4). Four of these are endemic: *Cubanichthys pengelleyi, Limia melanogaster, Gambusia melapleura,* and *Gambusia wrayi*. The genus *Cubanichthys* is shared with Cuba and Hispaniola. *L. melanogaster* belongs to the same subgenus (*Limia*) represented on Cuba, Hispaniola and the Cayman Islands, but is placed in its own species group. *G. melapleura* and *G. wrayi* are members of the *nicaraguensis* species group, and have affinities with Central America. Two widespread forms, *G. puncticulata* and *Cyprinodon variegatus*, complete the fauna.

Bahamas

The Bahamas, despite their relatively large collective land mass (Table 3,4), almost totally lack bodies of pure fresh water. Hence, the five "freshwater" species (Table 1,3) are all brackish water forms capable of existing in fluctuating salinities. *Lucifuga (Stygicola) spelaeotus*, endemic to New Providence Island, is the most distinctive form. The subgenus *Stygicola* also includes the western Cuban species *dentatus* and *simile*. Also endemic is *Cyprinodon laciniatus*; non-endemics include the widespread *Cyprinodon variegatus, Gambusia p. puncticulata*, and *Rivulus marmoratus*.

Cayman Islands

The most distinctive elements of the Cayman ichthyofauna are two endemics, *Limia caymanensis* and *Gambusia xanthosoma* (Table 2,4) which are apparently restricted to Grand Cayman. Populations of *G. p. puncticulata* occur on all three islands and there are subtle pigmentation differences suggesting at least some level of differentiation, particularly on Cayman Brac. Single members of the *Cyprinodon variegatus* and *Rivulus marmoratus* complexes are also present. As in the Bahamas, true bodies of fresh water are virtually non-existent in the Caymans and the "freshwater" fishes are actually euryhaline species.

Lesser Antilles

Aside from *Poecilia vivipara, P. reticulata,* and *Synbranchus marmoratus*, all thought to be introductions, only two species of secondary freshwater fishes are known from the Lesser Antilles: *Rivulus cryptocallus* and *R. ocellatus*. *R. cryptocallus*, endemic to Martinique, and *R. marmoratus*, widespread throughout the Caribbean, are highly tolerant of salt water.

An overview of Antillean biogeography

As noted in our introductory remarks, there are three schools of thought regarding Caribbean insular biogeography. The land bridge hypothesis, first proposed by Spencer (1895) and promoted by Barbour (1916) and his followers during the first half of this century, has fallen into disfavor in recent years. Rivas (1958, 1986), discussing Cuban poeciliid fish zoogeography, steadfastly embraced this hypothesis despite the absence of confirmatory geologic evidence. As Briggs (1984, 1987) has pointed out, there is also "the problem of having to destroy [a land bridge] before too many other species get across" and the question of why primary freshwater fishes are totally absent from the Antilles. For these reasons we reject the land bridge hypothesis as a viable alternative.

Briggs (1984, 1987) advocated a pure dispersal hypothesis based on fortuitous overseas transport of organisms over a long period of time, probably beginning in the mid-Tertiary. His explanation was predicated on a lack of confidence in the prevailing geologic (plate tectonic) concensus summarized below, specifically the placement and timing of land emergences. We do not totally share his appraisal of this consensus and believe that there is adequate evidence of the uninterrupted aerial existence of some islands to invoke some early vicariant events between Central America and the emerging Antilles.

The vicariant approach espoused by Rosen (1976, 1985) proposed migrating islands with nearly intact biotas moving in a northeastward path from present-day Central America to their current positions. Rosen's hypothesis, however, inadequately considered the problems associated with uplift and subsistence of insular land masses, and eustatic sea level changes. We join Briggs (1987) in objecting to the concept of *entire* biotas being transported on intact islands because the geologic record clearly indicates that relative sea-levels have played important roles in Antillean biogeography. However, we differ from Briggs in our belief that certain large islands (Cuba, northern Hispaniola) were sufficiently well emerged during the Tertiary to have accepted biotic elements; that these islands remained adequately emergent since then, supporting a continuous, albeit fluctuating, biota; and that subsequent dispersal events have occurred that allowed colonization and recolonization of other smaller islands that emerged at later times. Thus, we propose a biogeographic model that blends portions of the vicariance and dispersal hypotheses.

Geological interpretation

We accept the continental drift interpretation of tectonic evolution of the Caribbean as the geological basis for our zoogeographical analysis. Plate tectonics was initially applied to the Caribbean region by Malfait and Dinkelman (1972) and Tedford (1974). Numerous subsequent studies (cited and summarized by Hedges 1982) expanded upon these, and most recently Pindel and Dewey (1982), Sykes et al. (1982), Coney (1982), Wadge and Burke (1983), Burke et al. (1984), Mann and Burke (1984), and Smith (1985), among others, have furthered our understanding of the orogeny of the region. The Willis (permanent pre-Jurassic deep sea basin) and Suess (subsided continental terrain) schools of thought (Burke et al. 1984) on Caribbean tectonic evolution are rejected by the overwhelming weight of the aforementioned Wegnerian (continental separation) evidence. Exact timing of activities, derivations and temporal placements of the land masses are still in a state of flux, but the consensus is that:

A. The Greater Antilles arose in the area currently occupied by lower Central America, either attached to the continental land mass or (more likely) as a series of subduction-derived islands. Cuba, Puerto Rico and northern Hispaniola are the oldest permanently exposed insular land masses; Jamaica and southern Hispaniola are younger in age (relative to continuous land emergence).

B. Subsequent eastward and northeastward movements brought these islands to their current positions. Southern Hispaniola, of independent origin from northern Hispaniola, has collided with northern Hispaniola (which may itself represent the merger of two islands of similar origin and geologic history).

C. Eustatic sea level changes and localized uplift and subsidence have greatly influenced the relative vertical positions of the islands. Jamaica may only have been continuously emergent since the early Miocene, and with southern Hispaniola has exhibited the greatest recent uplift.

D. The Lesser Antilles are more recently derived as a volcanic island arc with no past continental connections.

At odds with this consensus are two alternative tectonic reconstructions. One, proposed by Salvador and Green (1980) and Anderson and Schmidt (1983), advocates a pre-Cenozoic north-south expansion of the Caribbean-Gulf of Mexico region as North and South America underwent latitudinal displacements. The Antilles are viewed as having had volcanic origins along the north coast of South America. The second (Donnelly 1985) has the Greater Antilles originating not afar from their present location and emphasizes vertical rather than horizontal insular movements. Maurrasse (1982a) presents a somewhat similar scenario of essentially in-place origins of Jamaica and southern Hispaniola via island arc development and asymmetric back arc spreading. In the absence of additional support for these alternatives, we choose to refer to the above consensus in our subsequent zoogeographic discussions.

The exact early geologic histories of Jamaica and southern Hispaniola are still not clear and are critical to the origin of the faunas of these islands. The geological consensus has them as part of a primitive island arc with Cuba, northern Hispaniola and Puerto Rico that moved northeasterly during the Cretaceous. One school of thought (as espoused by Pindel and Dewey 1982 and others), hereafter referred to as "Geology A", has Jamaica colliding with or passing closely to the Guatemala-Yucatan platform in the late Cretaceous; southern Hispaniola was concurrently located farther to the northeast. An alternative hypothesis ("Geology B") has Jamaica and southern Hispaniola arising close to the southern shore of Cuba in the late Cretaceous or Paleocene (Stephan 1982; Thierry 1983). Both Jamaica and southern Hispaniola are thought to have inititally emerged in the Paleocene as small islands, grown larger in the Eocene and then diverged drastically in their geologic histories. Maurrasse (1982a, b) inferred that after initial emergence at least parts of southern Hispaniola remained continuously above sea level. Jamaica, on the other hand, is thought to have undergone total subsistence from the late to mid Eocene until it re-emerged in the early Miocene (Buskirk 1985). This view, however, is tempered by later statements by Buskirk "that there was *little*, if any, emergent land" (emphasis ours) during the Oligocene, and by Arden (1975), who hypothesized almost complete submergence except for a few small islands, suggesting the possibility that elements of an archaic freshwater fish fauna could have survived after innoculation in the Paleo-Eocene.

Proposed zoogeographical model

This discussion addresses only freshwater fishes, although the story appears consistent with other biotic groups. Differences in dispersal abilities and evolutionary rates between groups are immense, and should be carefully considered when comparing proposed zoogeographies involving evolutionarily divergent groups.

Ancient South American and North American elements dispersed into Central America during the late Cretaceous or Paleocene (Bussing 1976, 1985). Among these were the ancestors of poeciliids (*Gambusia, Limia, Poecilia, Girardinus, Quintana*), *Cichlasoma, Rivulus, Ophisternon*, cyprinodontoids (*Cubanichthys, Cyprinodon, Fundulus*), all South American in origin; and *Atractosteus* of North American origin.

Central and North America were isolated from South America by a seaway during much of the Tertiary. During the early Tertiary, Cretaceous island arc features (proto-Cuba, proto-northern Hispaniola, and possibly proto-Puerto Rico, proto-Jamaica and proto-southern Hispaniola) moving through the seaway or uplifting from the seafloor were inoculated with the ancient South American and North American elements and with ancient marine elements. These included the prototypic South American elements *Cichlasoma, Cubanichythys, Limia*, and *Rivulus* (Cuba and/or northern Hispaniola); *Ophisternon, Quintana, Girardinus*, and the *Gambusia punctatus* group (Cuba); and *Poecilia* (northern Hispaniola). *Atractosteus*, of North American origin, and *Lucifuga* and *Alepidomus*, of marine origin, also invaded Cuba. Some of these colonizing forms undoubtedly were also present initially on other islands but eventually disappeared as a result of extinction; regrettably our nearly total lack of freshwater fish fossils forces a less than desirable zoogeographical reconstruction based totally on present-day distributional evidence. The only known freshwater fish fossil from the Antilles, *Cichlasoma woodringi*, came from Tertiary beds in Haiti (Cockerell 1924). Thought to be Miocene in age, *C. woodringi* is apparently indistinguishable from extant *C. haitiensis* (L.R. Rivas, pers. comm.).

Inoculation was most likely through dispersal among nearby islands via contiguous brackish to marine shallows. The most compelling evidence for this is the domination of secondary freshwater fishes and the total lack of primary freshwater fishes in the Antilles, despite their acknowledged inclusion (Bussing 1976, 1985) as members of the Old South America colonizing element in Central America. Geologic evidence, as noted in Hedges (1982), is also available that argues against these islands having been contiguously attached to the Central American continental land mass. Similar distributional patterns involving marine-derived archaic fishes (Bussing 1982) and aquatic invertebrates (Holsinger 1986) presently established in the Yucatan Peninsula, thought once to be an island itself (Pindel and Dewey 1982), and the Antilles is additional evidence of early inoculation.

Subsequent eustatic sea level changes (Vail and Hardenbol 1979; Vail et al. 1977) altered the topography of Cuba and northern Hispaniola, alternately fragmenting each of these land masses into smaller islands and re-uniting them, thereby influencing the zoogeography and evolution of their archaic faunas. In the late Miocene to early Pliocene the Bolivar Seaway was closed by the isthmus, connecting North and South America; New South America faunistic elements invaded Central America from the south, and North and Old South America elements moved southward (Bussing 1985). Southern

Hispaniola connected with northern Hispaniola at about 10 myBP (Pindel and Dewey 1982), allowing *Limia* (proto-*Odontolimia*) to invade the southern peninsula from the north. Late Pliocene sea level fluctuations greatly influenced speciation events on Hispaniola, especially within *Limia* (*Odontolimia*).

The uninterrupted post-Eocene aerial existence of proto-Jamaica and proto-southern Hispaniola is important because it would allow an early vicariant origin for three groups of fishes on these islands (assuming Geology A) rather than subsequent dispersal events. Acceptance of Geology A allows *Cubanichthys*, and *Limia* (*Limia*) predecessors to colonize Jamaica and a *Gambusia nicaraguensis* antecedent to invade both Jamaica and southern Hispaniola in the Eocene, providing a simpler explanation than subsequent longer lengthened dispersals to these islands during the Miocene or Pleistocene from Cuba or northern Hispaniola (*Cubanichthys* and *Limia*) or from Central America via the Nicaragua Rise (*Gambusia nicaraguensis*).

According to Maurrasse (1982a), the earliest portion of southern Hispaniola to arise was the eastern half of the present-day Tiburon Peninsula. Presumably this is where progenitor *G. nicaraguensis* would have settled. Ancestral *G. nicaraguensis* is thought to be closest to the extant *G. nicaraguensis*, commonly found in lower Central American marine and brackish water environments; it is also found on San Andres Island, located well offshore of the mainland Nicaraguan coast (Fink 1971a). It seems reasonable to assume that ancestral *G. nicaraguensis* could have survived in early marine to brackish water environments on the eastern island of southern Hispaniola and on Jamaica. Later, probably in the late Miocene, as the western island(s) of southern Hispaniola emerged and during a low stand of sea level, members of the *G. punctata* group arrived via dispersal onto the westernmost island of southern Hispaniola from Cuba. High stands of sea level eventually isolated this group from the *G. nicaraguensis* stock already established on the eastern island of southern Hispaniola and eventually split the western island into two islands, facilitating *G. punctata* differentiation into *G. pseudopunctata* (western island) and *G. beebei* (Étang de Miragôane on the central island). In the Plio-Pleistocene, *G. nicaraguensis* stock (perhaps by now differentiated into *G. hispaniolae*) spread off the easternmost island of southern Hispaniola onto northern Hispaniola, utilizing brackish water (Mann et al. 1984) routes to colonize the Cul-de-Sac/Valle-de-Neiba Plain depressions where there were no *Gambusia* competitors. Dispersal to the west by the *nicaraguensis* stock was curtailed by the already established *G. punctata* morphs.

This scenario also explains the presence of *Cubanichthys* and *Limia* on Jamaica. Otherwise dispersal from Cuba or Hispaniola to Jamaica must be invoked during a low stand of sea level during the late Miocene or Pleistocene. This would require a long swim over deep waters by a small fish, an even longer voyage than at present if late Miocene Jamaica was located 200 km farther to the southwest as suggested in Geology A. *Cubanichthys* presently occupies streams, but often is found in waters up to full strength sea water in Cuba and has presumably dispersed onto Isla de la Juventud. At least one *Limia* (*Limia*) species (*rivasi*) is known to prefer brackish waters and others (e.g. *caymanensis* and *vittata*) are commonly found in brackish waters. Thus, prototypic *Cubanichthys* and *Limia* should have been able to survive on a small Jamaican island in brackish to marine situations.

Alternatively, if Jamaica became completely submerged, *Cubanichthys*, *Limia* (*Limia*), and ancestral *Gambusia nicaraguensis* probably had to arrive via dispersal during a low stand of sea level at a later time (early Miocene onwards). *Cubanichthys* and *Limia* (*Limia*) probably immigrated from Cuba; dispersal would have been greatly facilitated by the emergence of Jamaica in the position advocated by Geology B. *G. nicaraguensis* stock arrived via the exposed Nicaragua Rise (Figure 1). This explanation would predict that Jamaican members of the *nicaraguensis* complex (*wrayi*, *melapleura*) would be among the most derived members of the species group. If southern Hispaniola was not continuously above sea level or never was initially inoculated with ancestral *G. nicaraguensis* in the Eocene, then its form (*hispaniolae*) would also be expected to be derived. Our understanding of the phylogenetic relationships of this group is poor, with Fink's (1971a) single sentence, "*G. nicaraguensis* seems to be direct descendent of the progenitor stock of the species group", our only clue to the relationship.

During Pleistocene glacial periods, sea levels dropped markedly, exposing many modern banks (Fig. 1) and facilitating dispersal by reducing distances and providing "stepping stones" between land masses. We hypothesize that several dispersal events could have occurred during this epoch:

A. *Limia* (*Limia*) from Cuba to the Cayman Islands. *L. caymanensis*, like its presumed sister species, *L. vittata*, frequents brackish water coastal mangrove lagoons, as well as freshwater situations and should have been able to withstand sea dispersal. Numerous banks off southern Cuba undoubtedly were exposed, including Pickle Bank, which offered a convenient mid-way refugium between Cuba and the Caymans.

B. *Gambusia p. puncticulata* from Yucatan to Cuba, thence to the Cayman Islands, Jamaica and the Bahamas. Fink (1971b) considered *G. p. puncticulata* the direct descendent of the ancestral stock of the species complex. This subspecies prefers brackish water, but is also found in full strength fresh and salt water. Other insular subspecies (*monticola*, *bucheri*, *baracoana* in Cuba and *manni* in the Bahamas) apparently evolved after subsequent isolation events.

C. *Gambusia rhizophorae* from Cuba to southern Florida. This dispersion could be a relatively recent event since *G. rhizophorae* is unlikely to have survived lowered south Florida water temperatures associated with the last glacial period. *G. rhizophorae* is a member of the *G. punctata* species group which is otherwise confined to Cuba and the Tiburon Peninsula of Hispaniola (Rivas 1969). In Florida it is restricted to red mangrove shorelines (Getter 1980). Its evolution in warm Antillean waters plus its minimal northward range extension in peninsular Florida, where red mangroves abound (Odum et al. 1982), suggest that temperature has played a key role in affecting distribution.

D. *Fundulus grandis* (and possibly *Cyprinodon variegatus* complex) to Cuba and *C. variegatus* complex to Hispaniola from southern Florida by way of the exposed Cay Sal Bank (Fig. 1) of the Bahamas. This event occurred during a glacial period utilizing a pathway similar to the one that brought *Opsanus* to the Bahamas (Walters and Robins 1961). *F. grandis*, confined to the north-central Cuban coast, is considered a glacial relict; its extinction from Hispaniola and the Bahamas is likely the result of thermal intolerance and the submergence of the Bank.

C. variegatus was able to survive in Hispaniola and the Bahamas because of higher thermal tolerance and greater habitat flexibility; subsequent speciation events resulted in the Bahaman *C. laciniatus*, the Hispaniolan *C. bondi*, and the nominal *variegatus* morphs in the Bahamas and Cuba (and an unnamed form from Ile de la Gonave, Hispaniola). Jamaican *C. variegatus* possibly arrived during this period from Central America via the exposed Nicaragua Rise. Cayman and Cuban *C. variegatus* could also have dispersed from Yucatan using the route proposed above for *Gambusia puncticulata*.

E. *Rivulus marmoratus* complex from Yucatan or from the Lesser Antilles towards the central Antilles, Bahamas and Florida. As noted earlier, this species warrants a detailed systematic study, and direction of dispersal is purely conjectural at this time.

Puerto Rico is ichthyologically depauperate, possibly because it 'missed out' during initial inoculation and subsequently was the farthest removed zoogeographically of all the Antilles from Central and South American dispersal centers of freshwater fishes. Puerto Rico's ichthyofaunal fate was probably sealed early if the island was located too far offshore or was submerged during the critical time period when ancient South and North American elements dispersed into Central America.

The geologic history (in-place origin and relatively young age) and limited availability of freshwater habitats explain the virtual absence of freshwater fishes in the Lesser Antilles. Dispersal of the salt tolerant genus *Rivulus* from the northern South American coast probably occurred during the Pleistocene.

Summary

The native Antillean freshwater fish fauna is comprised of 71 species in nine families. In addition, invading marine fishes form an important component of the ichthyofauna and at least 27 exotic species have become established, including 24 in Puerto Rico. Cuba and Hispaniola have the largest land masses, most complex topographic relief and habitat diversity, and host the greatest number of families, genera, and species of freshwater fishes. Other Antillean islands have fish diversities that are approximately proportional to their land masses, except Puerto Rico, which enigmatically lacks freshwater fishes despite ranking fourth in land area. The Antillean ichthyofauna is dominated by secondary freshwater species, with 85% of the fauna cyprinodontoid fishes of the families Poeciliidae, Cyprinodontidae, Rivulidae, and Fundulidae. The fauna is derived from North American, South American, and marine protofaunas. Six genera and all but six species are Antillean endemics; there are no endemic families.

Although the freshwater fishes are salt tolerant, we do not believe that these species or their predecessors were able to cross long reaches of salt water (as they exist today) to colonize the Antilles. We propose a zoogeographic model that blends elements of dispersal and vicariance with recent geological evidence to explain their presence on the islands. Following initial dispersal into Central America during the late Cretaceous or Paleocene, ancient South and North American elements invaded emerging Cretaceous island arc features moving through a seaway or uplifting from the ocean floor. Inoculation was most likely through dispersal among nearby islands via contiguous brackish to marine shallows. Subsequent eastward and northeastward movements of the islands, plus eustatic sea level changes and localized uplift and subsidence, have greatly influenced the evolution and zoogeography of the founding ichthyofauna. Distinctive ichthyofaunas developed on each island or island group as a result of isolation and periodic fragmentation of ranges caused by high and low stands of sea levels. During Pleistocene glacial periods many banks were exposed as sea levels dropped, allowing small-scale dispersal events. Puerto Rico is thought to be ichthyologically depauperate because it missed early inoculation and was subsequently far removed from dispersal centers. The ichthyofauna of Hispaniola is composed of distinctive North Island, South Island, and Cul-de-Sac/ Valle-de-Neiba Plain assemblages. Recent dispersals of species between drainages have occurred via headwater and lowland stream captures, flooding, coastal swamps, and short-distance marine routes.

Our zoogeographic model is highly conjectural because of the nearly total absence of fossil fishes, an imprecise knowledge of the origins and timing of orogenic events, and the paucity of detailed phylogenies for critical fish groups. Confirmation, refutation or refinement of our model awaits advances in our understanding of the paleontology, geology, and evolutionary relationships of the fishes of the region.

Acknowledgments

We thank J. E. Davis for his support of our collecting efforts in the Caymans; Francisco Gerard, Fred G. Thompson and Charles A. Woods for support of field work in the Republica Dominicana and Haiti; Jackie Belwood, Ronald Crombie, Dan Cordier, Francisco Gerard, Gary Morgan, Jose Ottenwalder and Fred G. Thompson for field assistance in Republica Dominicana, Haiti, Jamaica, and the Cayman Islands; Barry Chernoff, Walter R. Courtenay, Gary Morgan, Michael L. Smith and Ricardo Vergera R. for shared information and discussions; David S. Lee and Renaldo Kuhler for allowing use of illustrations; and Mandy Garcia for preparation of Figures 1 and 2. We especially appreciate Barbara Stanton's patience during repeated typings and revisions.

Literature cited

Alayo D., P. 1973. Lista de peces fluviatiles de Cuba. Torreia 29, 55 pp.

Anderson, T.H., and V.A. Schmidt. 1983. The evolution of Middle America and the Gulf of Mexico-Caribbean Sea region during Mesozoic time. Geological Society of America Bulletin 94:941-966.

Arden, D.D. 1975. Geology of Jamaica and the Nicaraguan Rise. Pp. 617-661 *in* Navin, A.E. and F.G. Stehli (eds.). The Ocean Basins and Margins, Plenum Press, New York.

Barbour, T. 1916. Some remarks upon Matthew's "Climate and Evolution." Annals of the New York Academy of Sciences 27:1- 15.

Breder, C.M. 1932. An annotated list of fishes from Lake Forsyth, Andros Island, Bahamas with the description of three new forms. American Museum of Natural History Novitates 551, 7 pp.

Briggs, J.C. 1984. Freshwater fishes and biogeography of Central America and the Antilles. Systematic Zoology 33(4):428-435.

_____. 1987. Biogeography and Plate Tectonics. Elsevier Science Publishing Company, Inc., New York, 204 pp.

Burke, K., C. Cooper, J.F. Dewey, P. Mann, and J.L. Pindel. 1984. Caribbean tectonics and relative plate motions. Geological Society of America Memoir 162:31-63.

Bussing, W.A. 1976. Geographic distribution of the San Juan ichthyofauna of Central America with remarks on its origin and ecology. Pp. 157-175 *in* T.B. Thorson (ed.). Investigations of the Ichthyofauna of Nicaraguan Lakes, University of Nebraska Press, Lincoln.

Bussing, W.A. 1985. Patterns of distribution of the Central American ichthyofauna. Pp. 453-473 *in* Stehli, F.G. and S.D. Webb (eds.). The Great American Biotic Interchange, Plenum Press, New York.

Caldwell, D.K. 1966. Marine and freshwater fishes of Jamaica. Bull. Inst. Jamaica, Sci. Ser. 17, 120 pp.

Cockerell, T.B.A. 1924. A fossil cichlid fish from the Republic of Haiti. Proceedings of the United States National Museum 63(7):1-2.

Cohen, D.M., and J.G. Nielson. 1978. Guide to the identification of genera of the fish order Ophidiiformes with a tentative classification of the order. United States National Marine Fisheries Service, NOAA Technical Report NMFS Circular 417, 72 pp.

Cohen, D.M., and C.R. Robins. 1970. A new ophidioid fish (genus *Lucifuga*) from a limestone sink, New Providence Island, Bahamas. Proceedings of the Biological Society of Washington 83(11):133-144.

Coney, P.J. 1982. Plate tectonic constraints of the biogeography of Middle America and the Caribbean region. Annals of the Missouri Botanical Garden 69:432-443.

Darlington, P.J. 1938. The origin of the fauna of the Greater Antilles, with discussion of dispersal of animals over water and through the air. Quarterly Review of Biology 13:274-300.

_____. 1957. Zoogeography: The Geographical Distribution of Animals. John Wiley and Sons, New York, 675 pp.

de Beaufort, L.F. 1951. Zoogeography of the Land and Inland Waters. Sidgwick and Jackson, London.

Donnelly, T.W. 1985. Mesozoic and Cenozoic plate evolution of the Caribbean region. Pp. 89-121 *in* Stehli, F.G. and S.D. Webb (eds.). The Great American Biotic Interchange, Plenum Press, New York.

Eigenmann, C.H. 1903. The fresh-water fishes of western Cuba. Bulletin of the United States Fishery Commission 22:213-236 + 2 pl.

Evermann, B.W. and M.C. Marsh. 1900. The fishes of Porto Rico. Pp. 49-350 *in* Fishes and Fishery of Porto Rico, United States Fishery Commission, Washington, D.C.

Fink, W.L. 1971a. A revision of the *Gambusia nicaraguensis* species group (Pisces: Poeciliidae). Publications of the Gulf Coast Research Laboratory Museum 2:47-77.

_____. 1971b. A revision of the *Gambusia puncticulata* complex (Pisces: Poeciliidae). Publications of the Gulf Coast Research Laboratory Museum 2:11-46.

Fowler, H.W. 1915a. Cold-blooded vertebrates from Florida, the West Indies, Costa Rica, and eastern Brazil. Proceedings of the Academy of Natural Sciences of Philadelphia 67:244-269.

_____. 1915b. The fishes of Trinidad, Grenada, and St. Lucia, British West Indies. Proceedings of the Academy of Natural Sciences of Philadelphia 67:520-546.

_____. 1928. Fishes from Florida and the West Indies. Proceedings of the Academy of Natural Sciences of Philadelphia 80:451-473.

_____. 1938. A small collection of fresh-water fishes from eastern Cuba. Proceedings of the Academy of Natural Sciences of Philadelphia 90:143-147.

_____. 1939. Notes on the fishes from Jamaica with descriptions of three new species. Academy of Natural Sciences of Philadelphia, Notulae Naturae 35, 16 pp.

Fowler, H.W. 1952. The fishes of Hispaniola. Memorias de la Sociedad Cubana de Historia Natural 21(1):83-115 + 7 pl.

Franz, R., and G.H. Burgess. 1983. A new poeciliid killifish, *Limia rivasi*, from Haiti. Northeast Gulf Science 6(1):51- 54.

_____, and L.R. Rivas. 1983a. *Poecilia dominicensis*. P. 48. in Lee, D.S., S.P. Platania, and G.H. Burgess (eds.). Atlas of North American Freshwater Fishes, 1983 Supplement. Freshwater Fishes of the Greater Antilles. North Carolina State Museum of Natural History, Raleigh.

_____, and L.R. Rivas. 1983b. *Poecilia elegans*. P. 49. in: Lee, D.S., S.P. Platania, and G.H. Burgess (eds.). Atlas of North American Freshwater Fishes, 1983 Supplement. Freshwater Fishes of the Greater Antilles. North Carolina State Museum of Natural History, Raleigh.

Franz, R. and F.G. Thompson. 1978. Distributional patterns of Hispaniolan poeciliid fishes. Abstracts of 58th Annual Meeting, American Society of Ichthyologists and Herpetologists, Tempe, Arizona.

Getter, C.D. 1980. *Gambusia rhizophorae*. P. 545. in Lee, D.S., C.R. Gilbert, C.H. Hocutt, R.E. Jenkins, D.E. McAllister, and J.R. Stauffer (eds.). Atlas of North American Freshwater Fishes, North Carolina State Museum of Natural History, Raleigh.

Greenfield, D. 1983. *Gambusia xanthosoma*, a new species of poeciliid fish from Grand Cayman Island, BWI. Copeia 1983(2):457-464.

Harris, D.R. 1965. Plants, animals, and man in the outer Leeward Islands, West Indies. University of California Publication in Geography 18, 164 pp. + 18 pl.

Hedges, S.B. 1982. Caribbean biogeography: implications of recent plate tectonic studies. Systematic Zoology 31(4):518-522.

Hildebrand, S.F. 1935. An annotated list of fishes of the fresh waters of Puerto Rico. Copeia 1935(2):49-56.

Holsinger, J.R. 1982. Zoogeographic patterns of North American subterranean amphipod crustaceans. Pp. 85-106 in Gore, R.H. and K.L. Heck (eds.). Crustacean Biogeography, A.A. Balkema, Boston.

Lee, D.S., S.P. Plantania, and G.H. Burgess (eds.). 1983. Atlas of North American Freshwater Fishes, 1983 Supplement. Freshwater Fishes of the Greater Antilles. North Carolina State Museum of Natural History, Raleigh, 67 pp.

Loiselle, P.V. 1980. Giant predatory cichlids: the true guapotes. Freshwater and Marine Aquarium 3(8):39-47, 71-74.

Malfait, B.T., and M.G. Dinkelman. 1972. Circum-Caribbean tectonic and igneous activity and the evolution of the Caribbean plate. Geological Society of America Bulletin 83:251-272.

Mann, P. and K. Burke. 1984. Neotectonics of the Caribbean. Review of Geophysics and Space Physics 22:309-362.

_____, F.W. Taylor, K. Burke, and R. Kulstad. 1984. Subaerially exposed coral reef, Enriquillo Valley, Dominican Republic. Geological Society of America Bulletin 95:1084-1092.

Maurrasse, F.J.-M.R. 1982a. Relations between the geologic settings of Hispaniola and the origin and evolution of the Caribbean. Pp. 246-264 in Maurasse, F.J.-M.R. (ed.). Presentations, Transactions du ler Colloque sur la Geologie d'Haiti, Port-au-Prince.

_____. 1982b. Survey of the geology of Haiti. Guide to the Field Excursions in Haiti, March 3-8, 1982. Miami Geological Society, 103 pp.

McIndoo, N.E. 1906. On some fishes of western Cuba. Proceedings of the Academy of Natural Sciences of Philadelphia 58:484- 488.

Miller, R.R. 1962. Taxonomic status of *Cyprinodon baconi*, a killifish from Andros Island, Bahamas. Copeia 1962(4):836- 837.

_____. 1982. Pisces. Pp. 486-501. *in* Hurlbert, S.H. and A. Villalobos (eds.). Aquatic Biota of Mexico, Central America and the West Indies, San Diego State University, San Diego, California.

Myers, G.S. 1928. The existence of cichlid fishes in Santo Domingo. Copeia 1928(167):33-36.

_____. 1938. Fresh-water fishes and West Indian zoogeography. Smithsonian Report for 1937:339-364 + 3 pl.

_____. 1951. Fresh-water fishes and East Indian zoogeography. Stanford Ichthyological Bulletin 4:11-21.

Nichols, J.T. 1912. Notes on Cuban fishes. Bulletin of the American Museum of Natural History 31(18):179-194.

_____. 1915. Fishes new to Porto Rico. Bulletin of the American Museum of Natural History 34(7):141-146.

_____. 1929. Scientific survey of Porto Rico and the Virgin Islands. The fishes of Porto Rico and the Virgin Islands. Branchiostomidae to Sciaenidae. New York Academy of Science 10(2):1-295.

Odum, W.E., C.C. McIvor, and T.J. Smith. 1982. The ecology of the mangroves of south Florida: A community profile. United States Fish and Wildlife Service, Biological Services Program FWS/OBS-81/24, 144 pp.

Ortiz Carrasquillo, W. 1980. Resumen historico de la introduccion de los peces de agua dulce en los lagos artifiales de Puerto Rico desde 1915 hasta 1975. Science-Ciencia 7(3):95-107.

Parenti, L.R. 1981. A phylogenetic and biogeographic analysis of cyprinodontiform fishes (Teleostei, Atherinomorpha). Bulletin of the American Museum of Natural History 168(4):335-557.

Pindel, J. and J.F. Dewey. 1982. Permo-Triassic reconstruction of Western Pangea and the evolution of the Gulf of Mexico/Caribbean region. Tectonics 1(2):179-211.

Regan, C.T. 1905. A revision of the fishes of the American cichlid genus *Cichlasoma* and of the allied genera. Annals Magazine of Natural History, Series 7, 16:60-77, 225-243, 316-340, 433-445.

Relyea, K. 1983. A systematic study of two species complexes of the genus *Fundulus* (Pisces: Cyprinodontidae). Bulletin of the Florida State Museum, Biological Sciences 29(1):1-64.

Rivas, L.R. 1944. Contribuciones al estudio de los peces Cubanos de la familia Poeciliidae. II. *Glaridichthys atherinoides*, nueva especie de la Provincia de Camaguey. Contribuciones Ocasionales del Museo Historia Natural del Colegio De La Salle 2:1-7.

_____. 1948. Cyprinodont fishes of the genus *Fundulus* in the West Indies, with description of a new subspecies from Cuba. Proceedings of the United States National Museum 98(3299):215-221.

Rivas, L.R. 1958. The origin, evolution, dispersal, and geographical distribution of the Cuban poeciliid fishes of the tribe Girardinini. Proceedings of the American Philosophical Society 102(3):281-320.

_____. 1963. Subgenera and species groups in the poeciliid fish genus *Gambusia* Poey. Copeia 1963(2):331-347.

_____. 1969. A revision of the poeciliid fishes of the *Gambusia punctata* species group, with descriptions of two new species. Copeia 1969(4):778-795.

_____. 1971. A new subspecies of poeciliid fishes of the genus *Gambusia* from eastern Cuba. Publications of the Gulf Coast Research Laboratory Museum 2:5-9.

_____. 1978. A new species of poeciliid fish of the genus *Poecilia* from Hispaniola, with reinstatement and redescription of *P. dominicensis* (Evermann and Clark). Northeast Gulf Science 2(2):98-112.

_____. 1980. Eights new species of poeciliid fishes of the genus *Limia* from Hispaniola. Northeast Gulf Sciences 4(1):28-38.

_____. 1982. Character displacement and coexistence in two poeciliid fishes of the genus *Poecilia* (*Mollienesia*) from Hispaniola. Northeast Gulf Science 5(2):1-24.

_____. 1986. Comments on Briggs (1984): Freshwater fishes and biogeography of Central America and the Antilles. Systematic Zoology 35(4):633-639.

Rivas, L.R. and W.L. Fink. 1970. A new species of poeciliid fish of the genus *Limia* from the Island of Grand Cayman, B.W.I. Copeia 1970(2):270-274.

_____, and R. Franz. 1983. *Poecilia hispaniolana*. p. 50. in Lee, D.S., S.P. Platania, and G.H. Burgess (eds.). Atlas of North American Freshwater Fishes, 1983 Supplement. Freshwater Fishes of the Greater Antilles. North Carolina State Museum of Natural History, Raleigh.

_____, and G.S. Myers. 1950. A new genus of poeciliid fishes from Hispaniola, with notes on genera allied to *Poecilia* and *Mollienesia*. Copeia 1950(4):288-294.

Roloff, E. 1938. *Rivulus roloffi* Trewavas 1938. Wochenschrift fur Aquarien und Terrarienkunde 35:597-598.

Rosen, D.E. 1976. A vicariance model of Caribbean biogeography. Systematic Zoology 24(4):431-464.

_____. 1985. Geological hierarchies and biogeographical congruence in the Caribbean. Annals of the Missouri Botanical Garden 72:636-659.

_____, and R.M. Bailey. 1959. Middle-American poeciliid fishes and the genera *Carlhubbsia* and *Phallichthys*, with descriptions of two new species. Zoologica, New York 44(1):1-44 + 6 pl.

_____, and R. M. Bailey. 1963. The poeciliid fishes (Cyprinodontiofrmes), their structure, zoogeography, and systematics. Bulletin of the American Museum of Natural History 126:1-176.

Rosen, D.E. and P.H. Greenwood. 1976. A fourth neotropical species of synbranchid eel and the phylogeny and systematics of synbranchiform fishes. Bulletin of the American Museum of Natural History 157(1):1-70.

_____, and A. Rumney. 1972. Evidence of a second species of *Synbranchus* (Pisces, Teleostei) in South America. American Museum of Natural History Novitates 2497, 45 pp.

Salvador, A., and A.R. Green. 1980. Opening of the Caribbean Tethys (origin and development of the Caribbean and the Gulf of Mexico). Memoire Bureau de Recherche Geologie et Minieres 115:224-229.

Schuchert, C. 1935. Historical Geology of the Antillean- Caribbean Region. John Wiley and Sons, New York, 811 pp.

Seegers, L., and J.H. Huber. 1981. *Rivulus cryptocallus* n.sp. von der Insel Martinique (Pisces: Atheriniformes: Cyprinodontidae). Senckenbergiana Biologie 61(3/4):169-177.

Smith, D.L. 1985. Caribbean Plate relative motions. Pp. 17-48 *in* Stehli, F.G. and S.D. Webb (eds.). The Great American Biotic Interchange, Plenum Press, New York.

Spencer, J.W. 1985. Reconstruction of the Antillean continent. Geological Society of America Bulletin 6:103-140.

Stephan, J.F. 1982. Evolution geodynamique du domaine caraibe Andes el chaine sur la transversale de Barquisimeto (Venezuela). These d'Etat, Paris.

Sykes, L.R., W.R. McCann and A.L. Kafka. 1982. Motion of Caribbean Plate during last 7 million years and implications for earlier Cenozoic movements. Journal of Geophysical Reserach 87(B13):10656-10676.

Tedford, R.H. 1974. Marsupials and the new paleogeography. Pp. 109-126 *in* Ross, C.A. (ed.). Paleogeographic provinces and provinciality, Society of Economic Paleontologists Special Publication 21.

Tee-Van, J. 1935. Cichlid fishes in the West Indies with especial reference to Haiti, including the description of a new species of *Cichlasoma*. Zoologica, New York 10(2):281- 300.

Thierry, C. 1983. Contribution a l'Etude Geologique du Massif de Macaya (Sud-Ouest d'Haiti, Grandes Antilles). Sa Place dans l'Evolution de l'Orogene Nord-Caraibe. Universite de Pierre et Marie Curie, Paris, These de Doctorat, 163pp. + 4 Appendices.

Vail, P.R. and J. Hardenbol. 1979. Sea-level changes during the Tertiary. Oceanus 22(3):71-79.

____, R.M. Mitchum and S. Thompson. 1977. Seismic stratigraphy and global changes of sea level, Part 4. Global cycles of relative changes of sea level. American Association of Petroleum Geologists Memoir 26:83-97.

Vergara R., R. 1976. Factors of distribution of Cuban fresh- water fishes. XXIII International Geographical Congress, Moscow, Add. Vol.:I54-I57.

____. 1980. Principales caracteristicas de la ictiofauna dulceacuicola cubana. Ciencas Biologicas 5:95- 106.

____. 1981. Estudio filogenetico de los peces ciegos del genero *Lucifuga* (Pisces: Ophidiidae). II. Biogeografia filogenetica. Revision Ciencias Biologicas, Havana 12(1):99-107.

Wadge, G. and K. Burke. 1983. Neogene Caribbean plate rotation and associated Central American tectonic evolution. Tectonics 2:633-643.

Walters, V., and C.R. Robins. 1961. A new toadfish (Batrachoididae) considered to be a glacial relict in the West Indies. American Museum of Natural History Novitates 2047, 24 pp.

Wiley, E.O. 1976. The phylogeny and biogeography of fossil and recent gars (Actinopterygii: Lepisosteidae). University of Kansas Museum of Natural History Miscellaneous Publication 64, 111 pp.

TABLE 1. Schools of thought on Antillean ichthyofaunal zoogeography.

LAND BRIDGE HYPOTHESIS

    Schuchert (1935)
    de Beaufort (1951)
    Rivas (1958)
    Bussing (1976)[a]
    Rivas (1986)

PURE DISPERSAL HYPOTHESIS

    Darlington (1938)
    Myers (1938)
    Darlington (1957)
    Rosen and Bailey (1963)
    Vergara (1976)
    Vergara (1980)
    Briggs (1984)
    Briggs (1987)

PURE VICARIANCE HYPOTHESIS

    Rosen (1976)
    Rosen (1985)

---

[a] Bussing's "land connection between the Antilles and Central America" may also be interpreted as an affirmation of Rosen's (1976) proto-Antilles vicariant school of thought.

TABLE 2. List of freshwater fishes of the Bahamas and the Antilles.

Order LEPISOSTEIFORMES
    Family LEPISOSTEIDAE
        Genus *Atractosteus* (western Cuba and Isla de la Juventud)
            1. *A. tristoechus* (Bloch & Schneider)

Order OPHIDIIFORMES
    Family BYTHITIDAE
        Genus *Lucifuga*
            Subgenus *Stygicola*
                2. *L. dentatus* (Poey) (western Cuba)
                3. *L. spelaeoteus* Cohen & Robins (New Providence Island, Bahamas)
                4. *L. simile* Nalbant (western Cuba)

            Subgenus *Lucifuga*
                5. *L. subterraneus* Poey (western Cuba)

Order CYPRINODONTIFORMES
    Family POECILIIDAE
        Genus *Gambusia*
            Nicaraguensis Species Group
                6. *G. hispaniolae* Fink (Cul-de-Sac/ Valle-de-Neiba Plain of Hispaniola, peripheral on North and South Islands)
                7. *G. melapleura* (Gosse) (Bluefields and Shrewsberry rivers, Jamaica)
                8. *G. wrayi* Regan (Jamaica)

            Nobilis Species Group
                9. *G. dominicensis* Regan (listed as "Haiti")
            Punctatus Species Group
                10. *G. beebei* Myers (Étang de Miragôane, South Island of Hispaniola)
                11. *G. pseudopunctata* Rivas (South Island of Hispaniola)
                12. *G. punctata* Poey (Cuba and Isla de la Juventud)
                13. *G. rhizophorae* Rivas (southern Florida, northern Cuba)

            Puncticulata Species Group
                14. *G. puncticulata* Poey (includes the following subspecies: *bucheri* Rivas from Rio Moa system in eastern Cuba, *baracoana* Rivas from Rio Baracoa system in eastern Cuba, *manni* Hubbs from New Providence Island in the Bahamas, *monticola* Rivas from Rio Cauto system in eastern Cuba, *puncticulata* Poey from Cuba, Isla de la Juventud, Jamaica, Cayman Islands, Bahamas)

TABLE 2. List of freshwater fishes (continued 2).

Order CYPRINODONTIFORMES
   Family POECILIIDAE
     Genus *Gambusia*
        Unassigned Species Group
           15. *G. xanthosoma* Greenfield (Grand Cayman)

     Genus *Girardinus* (confined to Cuba and Isla de la Juventud)
           16. *G. creolus* Garman (extreme western Cuba)
           17. *G. cubensis* (Eigenmann) (extreme western Cuba)
           18. *G. denticulatus* Garman (eastern 2/3 of Cuba) (includes *G. d. denticulatus* from central and eastern Cuba; *G. d. ramsdeni* Rivas from Rio Guaso and Rio Yateras of southeastern Sierra Maestra in Oriente Province)
           19. *G. falcatus* (Eigenmann) (western 3/4 of Cuba and Isla de la Juventud)
           20. *G. metallicus* Poey (island-wide but not on Isla de la Juventud)
           21. *G. microdactylus* Rivas (extreme western Cuba and Isla de la Juventud)
           22. *G. serripenis* Rivas (Rio Taco Taco drainage, western Cuba)
           23. *G. uninotatus* Poey (extreme western Cuba) (includes *G. u. uninotatus* from east of Rio Guama; *G. u. torralbasi* Rivas from west of this river)

     Genus *Limia*
       Subgenus *Odontolimia* (confined to Étang de Miragôane, South Island of Hispaniola)
         Grossidens Species Group
           24. *L. grossidens* Rivas

         Ornata Species Group
           25. *L. ornata* Regan
           26. *L. fuscomaculata* Rivas
           27. *L. garnieri* Rivas
           28. *L. immaculata* Rivas
           29. *L. miragoanensis* Rivas

       Subgenus *Limia* (Hispaniola, Ile de la Gonave, Cuba, Isla de la Juventude, Jamaica, Cayman Islands)
         Nigrofasciata Species Group
           30. *L. nigrofasciata* Regan (Étang de Miragôane drainage, South Island of Hispaniola)

TABLE 2. List of freshwater fishes (continued 3).

Order CYPRINODONTIFORMES
    Family POECILIIDAE
        Genus *Limia*
            Subgenus *Limia*
                Vittata Species Group
                      31. *L. vittata* (Guichenot) (Cuba, Isla de la Juventud)
                      32. *L. caymanensis* Rivas and Fink (Cayman Islands)
                      33. *L. melanonotata* Nichols and Myers (Cul-de-Sac/ Valle-de-Neiba Plain, peripheral on North Island of Hispaniola)
                      34. *L. perugiae* (Evermann and Clark) (North Island of Hispaniola)

                Versicolor Species Group
                      35. *L. versicolor* (Gunther) (North Island of Hispaniola)
                      36. *L. zonata* (Nichols) (North Island of Hispaniola)

                Melanogaster Species Group
                      37. *L. melanogaster* (Gunther) (Jamaica)

                Dominicensis Species Group
                      38. *L. dominicensis* (Valenciennes) (South Island of Hispaniola)
                      39. *L. pauciradiata* Rivas (Grande Riviere du Nord, North Island of Hispaniola)
                      40. *L. rivasi* Franz and Burgess (Ile de la Gonave)
                      41. *L. sulphurophila* Rivas (sulfur spring in Valle-de-Neiba Plain, Hispaniola)
                      42. *L. tridens* (Hilgendorf) (South Island, Cul-de-Sac/ Valle-de-Neiba Plain, peripheral on North Island of Hispaniola)
                      43. *L.* yaguajali Rivas (Rio Yaguajali drainage, North Island of Hispaniola)

                Unassigned Species Groups
                      44. *L.* new species A (spring at Puerto Alejandro, North Island of Hispaniola)
                      45. *L.* new species B (Lago Oviedo, South Island of Hispaniola)
                      46. *L.* new species C (Étang de Miragôane, South Island of Hispaniola)
                      47. *L.* new species D (Riviere Cavaillon drainage, South Island of Hispaniola)

TABLE 2. List of freshwater fishes (continued 4).

Order CYPRINODONTIFORMES
    Family POECILIIDAE
        Genus *Poecilia* (North Island of Hispaniola)
            Subgenus *Poecilia*
                48. *P. dominicensis* (Evermann and Clark)
                49. *P. elegans* (Trewavas)
                50. *P. hispaniolana* Rivas

        Genus *Quintana* (extreme western Cuba and Isla de la Juventud)
            51. *Q. atrizona* Hubbs

    Family CYPRINODONTIDAE
        Genus *Cubanichthys*
            52. *C. cubensis* (Eigenmann) (western Cuba and Isla de la Juventud)
            53. *C. pengelleyi* (Fowler) (Jamaica)
            54. *C.* new species (Rio Yaque del Sur drainage, North Island of Hispaniola)

        Genus *Cyprinodon*
            55. *C. bondi* Myers (Cul-de-Sac/Valle-de-Neiba Plain in Hispaniola)
            56. *C. laciniatus* Hubbs & Miller (New Providence Island, Bahamas)
            57. *C. variegatus* complex (Antillean forms includes *baconi* from the Bahamas, *jamaicensis* from Jamaica, *riverendi* and *felicianus* from Cuba, unnamed populations in Cayman Islands and Hispaniola).

    Family FUNDULIDAE
        Genus *Fundulus*
            58. *F. grandis saguanus* Rivas (southern Florida, northern Cuba)

    Family RIVULIDAE
        Genus *Rivulus*
            59. *R. cylindraceus* Poey (Cuba and Isla de la Juventud)
            60. *R. garciae* de la Cruz & Dubitsky (western Cuba)
            61. *R. heyei* Nichols (Saona Island, Hispaniola)
            62. *R. insulaepinorum* de la Cruz & Dubitsky (Isla de la Juventud)
            63. *R. roloffi* Trewavas (North Island of Hispaniola)
            64. *R. cryptocallus* Seagers and Huber (Martinique)
            65. *R. ocellatus* Hensel (= *R. marmoratus* Poey (Cuba, Bahamas, Cayman Islands, and many islands in Lesser Antilles).

TABLE 2. List of freshwater fishes (concluded)

Order ATHERINIFORMES
   Family ATHERINIDAE
      Genus *Alepidomus*
            66. *A. evermanni* (Eigenmann) (western Cuba)

Order SYNBRANCHIFORMES
   Family SYNBRANCHIDAE
      Genus *Ophisternon*
            67. *O. aenigmaticum* Rosen & Greenwood (Cuba, Middle and South America)

Order PERCIFORMES
   Family CICHLIDAE
      Genus *Cichlasoma*
            68. *C. haitiensis* Tee-Van (Hispaniola)
            69. *C. ramsdeni* Fowler (extreme eastern Cuba)
            70. *C. tetracanthus* (Valenciennes) (Cuba)
            71. *C. vombergi* Ladiges (North Island of Hispaniola)

TABLE 3. Established exotic freshwater fishes of the Antilles.*

SPECIES	WHERE ESTABLISHED
Dorosoma cepedianum	Puerto Rico
Barbus conchonius	Puerto Rico
Carassius auratus	Puerto Rico
Ctenopharyngodon idella	Cuba
Cyprinus carpio	Cuba
Pimephales promelas	Puerto Rico
Ictalurus catus	Puerto Rico
Ictalurus nebulosus	Puerto Rico
Ictalurus punctatus	Puerto Rico
Gambusia affinis	Hispaniola, Puerto Rico
Poecilia helleri	Hispaniola, Jamaica, Puerto Rico
Poecilia maculata	Jamaica, Puerto Rico
Poecilia reticulata	Cuba, Hispaniola, Puerto Rico, some Lesser Antilles
Poecilia vivipara	Puerto Rico, some Lesser Antilles
Ophisternon aenigmaticum	St. Lucia
Lepomis auritus	Puerto Rico
Lepomis gulosus	Puerto Rico
Lepomis macrochirus	Cuba, Puerto Rico
Micropterus coosae	Puerto Rico
Micropterus salmoides	Cuba, Puerto Rico
Astronotus ocellatus	Puerto Rico
Cichla ocellaris	Puerto Rico
Tilapia aurea	Puerto Rico
Tilapia honororum	Puerto Rico
Tilapia melanopleura	Cuba
Tilapia mossambica	Cuba, Hispaniola, Jamaica, Puerto Rico
Tilapia nilotica	Cuba, Puerto Rico
Tilapia rendalli	Puerto Rico

* Status and specific identity of some introduced *Tilapia* in Lesser Antilles uncertain at this time.

TABLE 4. Numbers of freshwater fish species with representatives in freshwater habitats in the Antilles.

	Area (km^2)	Families	Genera	Species
Cuba	114,524	9	13	28
Hispaniola	76,190	4	7	32
Jamaica	11,425	2	4	6
Bahamas	13,935	4	3	5
Puerto Rico	8,697	0	0	0
Martinique	1,101	1	1	1
Cayman Islands	216	2	4	4

TABLE 5. Derivation of the Antillean freshwater fish fauna. Numbers of species in parentheses after each genus.

NORTH AMERICA	SOUTH AMERICA	MARINE
Atractosteus (1)	*Limia* (24)	*Lucifuga* (4)
Gambusia (10)	*Poecilia* (3)	*Alepidomus* (1)
Girardinus (8)	*Rivulus* (7)	
Quintana (1)	*Ophisternon* (1)	
Cyprinodon (3)	*Cichlasoma* (4)	
Cubanichthys (3)		
Fundulus (1)		

TABLE 6. Number of freshwater fish species within genera shared by islands of the Antilles.

Genus	Cuba	Hispaniola	Jamaica	Cayman Is.	Bahamas
Lucifuga	3	--	--	--	1
Limia	1	21	1	1	--
Gambusia	3^a	4	3^a	2^a	1^a
Cubanichthys	1	1	1	--	--
Cyprinodon	1^b	2^b	1^b	1^b	1^b
Rivulus	4^c	2	--	1^c	1^c
Cichlasoma	2	2	--	--	--

(a-c) = includes the widespread, highly euryhaline species *Gambusia puncticulata* (a), *Cyprinodon variegatus* (b), and *Rivulus marmoratus* (c). All three species may represent species complexes.

TABLE 7. Distributional patterns of Cuban freahwater fishes.

RESTRICTED DISTRIBUTION	WIDESPREAD DISTRIBUTION
W CUBA AND ISLA DE LA JUVENTUD	**E, W CUBA AND ISLA DE LA JUVENTUD**
Atractosteus tristoechus	*Gambusia puncticulata puncticulata*
Gambusia punctatus	*Limia vittata*
Girardinus falcatus	*Rivulus cylindraceus*
Girardinus microdactylus	*Rivulus marmoratus*
Quintana atrizona	*Cyprinodon variegatus*
Cubanichthys cubensis	*Cichlasoma tetracanthus*
ISLA DE LA JUVENTUD	**E AND W CUBA**
Rivulus insulaepinorum	*Girardinus metallicus*
	Ophisternon aenigmaticum
W CUBA	
Lucifuga dentatus	
Lucifuga simile	
Lucifuga subterraneus	
Gambusia rhizophorae	
Girardinus creolus	
Girardinus cubensis	
Girardinus serripenes	
Girardinus uninotatus uninotatus	
Girardinus uninotatus torralbosi	
Fundulus grandis.saguanus	
Rivulus garciae	
Alepidomus evermanni	
E CUBA	
Gambusia puncticulata baracoana	
Gambusia puncticulata bucheri	
Gambusia puncticulata monticola	
Girardinus denticulatus denticulatus	
Girardinus denticulatus ramsdeni	
Cichlasoma ramsdeni	

TABLE 8. Hispaniola fish assemblages.

I. NORTH ISLAND ASSEMBLAGE

CHARACTERISTIC SPECIES

Limia pauciradiata
Limia perugiae
Limia versicolor
Limia yaguajali
Limia zonata
Limia new species A
Poecilia dominicensis
Poecilia elegans
Poecilia hispaniolana
Cubanichthys new species
Rivulus roloffi
Cichlasoma vombergi

WIDESPREAD SPECIES

Limia tridens
Cichlasoma haitiensis

PERIPHERAL SPECIES (from Cul-de-Sac/ Valle-de-Neiba Assemblage)

Gambusia hispaniolae
Limia melanonotata

II. CUL-DE-SAC/ VALLE-DE-NEIBA ASSEMBLAGE

CHARACTERISTIC SPECIES

Gambusia hispaniolae
Limia melanonotata
Limia sulphurophila
Cyprinodon bondi

WIDESPREAD SPECIES

Limia tridens
Cichlasoma haitiensis

PERIPHERAL SPECIES

Poecilia hispaniolana (from North Island Assemblage)

TABLE 8. Hispaniola fish assemblages (concluded).

III. SOUTH ISLAND ASSEMBLAGE

CHARACTERISTIC SPECIES

*Gambusia beebei**
Gambusia pseudopunctata
Limia dominicensis
*Limia fuscomaculata**
*Limia garnieri**
*Limia grossidens**
*Limia immaculata**
Limia miragoanensis
*Limia nigrofasciata**
*Limia ornata**
Limia new species B
Limia new species C*
Limia new species D

WIDESPREAD SPECIES

Limia tridens
Cichlasoma haitiensis

PERIPHERAL SPECIES

Gambusia hispaniolae (from Cul-de-Sac/ Valle-de-Neiba Plain Assemblage)

IV. COASTAL ISLAND ASSEMBLAGES

CHARACTERISTIC SPECIES

Limia rivasi (Ile de la Gonâve)
Cyprinodon cf. *variegatus* (Ile de la Gonâve)
Rivulus heyei (Isla Saona)

---

* Endemic to Étang de Miragôane.

TABLE 9. Hispaniola freshwater fishes listed by drainage basin.

I. NORTH ISLAND DRAINAGES

HAITI, Northwestern Drainages (includes Les Trois Rivières): *Limia tridens, Poecilia hispaniolana.*

HAITI, Northeastern Drainages (includes Rivière du Limbe, Grande Rivière du Nord): *Limia pauciradiata.*

HAITI, Lower l'Artibonite Basin: *Gambusia hispaniolae, Limia melanonotata, Limia tridens.*

HAITI and REPUBLICA DOMINICANA, Upper l'Artibonite Basin: *Poecilia hispaniolana.*

HAITI and REPUBLICA DOMINICANA, Rivière du Massacre Drainage: *Poecilia dominicensis, Poecilia hispaniolana.*

REPUBLICA DOMINICANA, Puerto Plata Drainages (includes Rìo Bajabonico, Rìo Yasica): No fish reported.

REPUBLICA DOMINICANA, Rìo Yaque del Norte Drainage: *Limia yaguajali, Limia zonata, Poecilia dominicensis, Poecilia elegans, Poecilia hispaniolana, Rivulus roloffi, Cichlasoma haitiensis.*

REPUBLICA DOMINICANA, Rìo Yuna Drainage: *Limia zonata, Poecilia dominicensis, Poecilia elegans, Poecilia hispaniolana, Rivulus roloffi, Cichlasoma haitiensis.*

REPUBLICA DOMINICANA, Nagua-Cabrera Drainages (includes Rìo Bacu, Rìo Boba, Rìo Nagua): *Limia zonata.*

REPUBLICA DOMINICANA, Samana Peninsula Drainages (includes north and south drainages): *Limia zonata.*

REPUBLICA DOMINICANA, Seibo Peninsula Drainages: No fishes are reported for this large area.

REPUBLICA DOMINICANA, Rìo Ozama Drainage: *Poecilia dominicensis.*

REPUBLICA DOMINICANA, Bani Drainages (includes Rìo Haina, Rìo Nizao, Rìo Ocoa, Rìo Jura, Rìo Tabara): *Limia perugiae, Limia versicolor, Poecilia dominicensis, Poecilia elegans, Poecilia hispaniolana, Rivulus roloffi, Cichlasoma haitiensis.*

REPUBLICA DOMINICANA, Rìo Yaque del Sur Drainage: *Limia melanonotata, Limia perugiae, Limia tridens, Poecilia hispaniolana, Cubanichthys* new species, *Cichlasoma vombergi.*

REPUBLICA DOMINICANA, Spring at Puerto Alejandro: *Limia* new species A.

TABLE 9. Hispaniola freshwater fishes listed by drainage basin (concluded).

II. CUL-DE-SAC/VALLE-DE-NEIBA DRAINAGES

HAITI, Cul-de-Sac Lake Drainages (includes Étang Saumâtre, Trou Caiman): *Gambusia hispaniolae, Limia melanonotata, Limia tridens, Cyprinodon bondi, Cichlasoma haitiensis.*
REPUBLICA DOMINCANA, Lago Enriquillo Drainages: *Gambusia hispaniolae, Limia melanonotata, Limia sulphurophila, Limia tridens, Poecilia hispaniolana, Cyprinodon bondi, Cichlasoma haitiensis.*

III. SOUTH ISLAND DRAINAGES

REPUBLICA DOMINICANA, Barahona Peninsula (includes Lago Oviedo drainage): *Gambusia hispaniolae, Limia* new species B.
HAITI, Eastern Tiburon Peninsula Drainage (includes Rivière de Jacmel): *Limia tridens.*
HAITI, Étang de Miragôane Basin: *Gambusia beebei, Limia fuscomaculata, Limia garnieri, Limia grossidens, Limia immaculata, Limia miragoanensis, Limia nigrofasciata, Limia ornata, Limia* new species C, *Cichlasoma haitiensis.*
HAITI, Rivière Cavaillon: *Limia* new species D.
HAITI, Les Cayes Basin (includes Rivière du Sud, Rivière Torbeck, Rivière Duclerc): *Gambusia pseudopunctata, Limia dominicensis, Limia tridens.*
HAITI, Jérémie Drainages (includes Rivière de la Grande Anse, Rivière des Roseaux): *Gambusia pseudopunctata, Limia tridens.*

IV. COASTAL ISLANDS

HAITI, Ile de la Gonâve: *Limia rivasi, Cyprinodon* cf. *variegatus.*
REPUBLICA DOMINICANA, Isla Saona: *Rivulus heyei.*

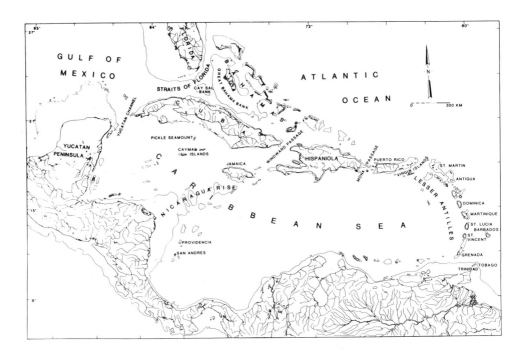

Figure 1. The Caribbean region showing position of islands and important shallow banks discussed in text. Two hundred meter isobath defines banks and shelf zones.

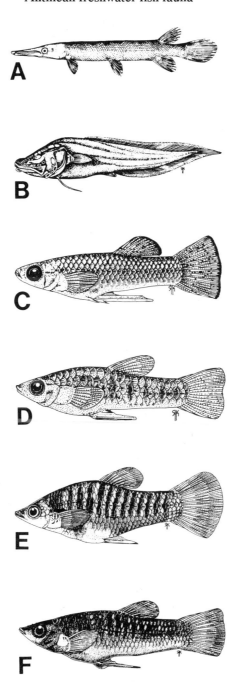

Figure 2. Representative native Antillean freshwater fishes. (A) Lepisosteidae: *Atractosteus tristoechus*. (B) Bythitidae: *Lucifuga (Stygicola) spelaeotes*. (C) Poeciliidae: *Gambusia puncticulata bucheri*. (D) Poeciliidae: *Girardinus creolus*. (E) Poeciliidae: *Limia (Odontolimia) grossidens*. (F) *Limia (Limia) pauciradiata*.

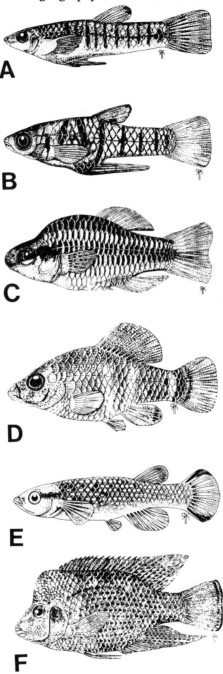

Figure 3. Representative native Antillean freshwater fishes. (A) Poeciliidae: *Poecilia (Poecilia) hispaniolana*. (B) Poeciliidae: *Quintana atrizona*. (C) Cyprinodontidae: *Cubanichthys pengelleyi*. (D) Cyprinodontidae: *Cyprinodon variegatus* "baconi". (E) Rivulidae: *Rivulus cylindraceus*. (F) Cichlidae: *Cichlasoma ramsdeni*.

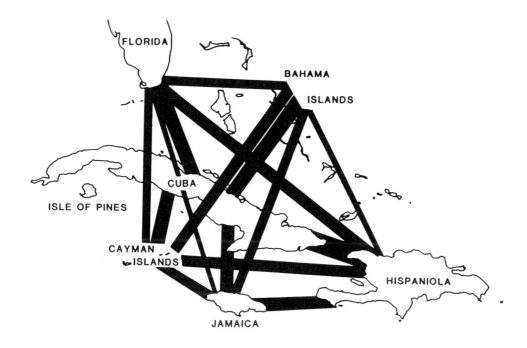

Figure 4. Tracks of shared presumed monophyletic lineages between islands or island groups and the closest continental coast. Widths of tracks are proportional to numbers of shared lineages.

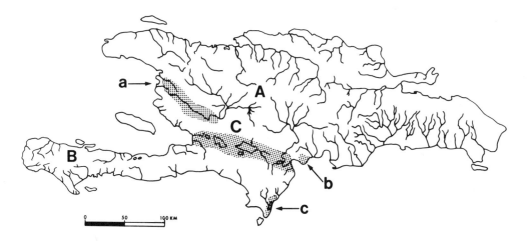

Figure 5. North Island (A), South Island (B), and Cul-de-Sac/Valle-de-Neiba Plain (C). The faunal area of C also extends to the lower Rivière l'Artibonite (a), a spring at the base of the Sierra Martin Garcia (b), and Laguna Oviedo (c).

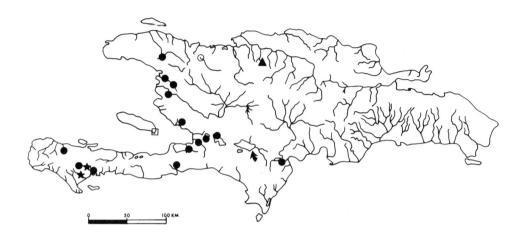

Figure 6. Distribution of species in the *Limia dominicensis* group. *L. tridens* (solid circle), *L. pauciradiata* (hollow circle), *L. yaguajali* (solid triangle), *L. rivasi* (hollow square), *L. sulphurophila* (arrow), *L. dominicensis* (star).

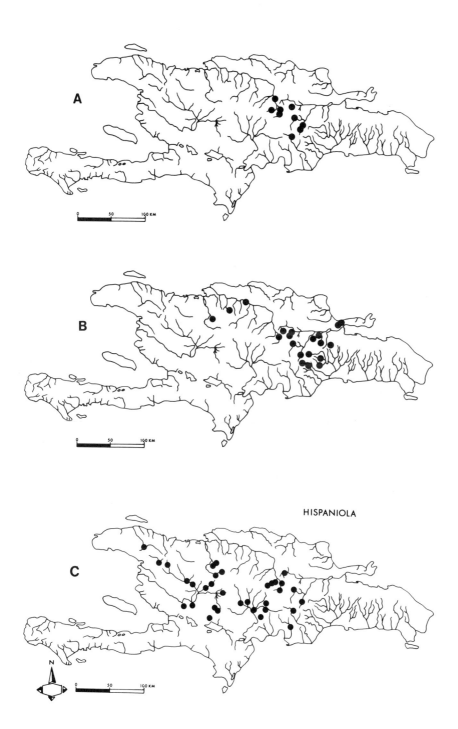

Figure 7. Distributions of *Poecilia dominicensis* (a), *P. elegans* (b) and *P. hispaniolana* (c). Maps adapted from Lee et al. (1983).

304 Biogeography of the West Indies

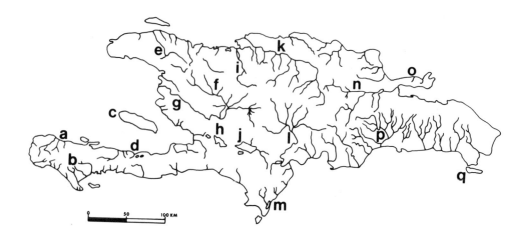

Figure 8. Drainage map of Hispaniola. Jeremie drainages (a), Les Cayes basin (b), Ile de la Gonâve (c), Étang de Miragane (d), Northwestern drainages (e), upper Rivière l'Artibonite (f), lower Rivière l'Artibonite (g), Cul-de-Sac drainages (h), Rìo Massacre (i), Lago Enriquillo drainages (j), Rìo Yaque del Norte (k), Rìo Yaque del Sur (1), Laguna Oviedo (m), Rìo Yuna (n), Samana Peninsula drainage (o), Rìo Ozama (p), Isla Saona (q).

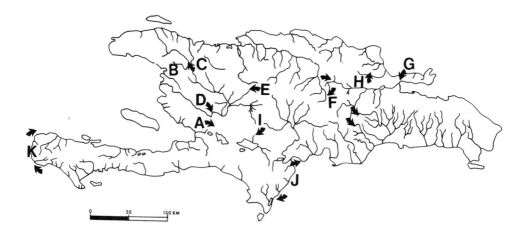

Figure 9. Postulated dispersal routes of fishes in Hispaniola (refer to text, p. 269).

EVOLUTION AND BIOGEOGRAPHY OF WEST INDIAN FROGS OF THE GENUS *ELEUTHERODACTYLUS*: SLOW-EVOLVING LOCI AND THE MAJOR GROUPS

S. Blair Hedges[1]

Abstract

A new approach to of electrophoretic analysis is presented in which only slow-evolving protein loci are examined, permitting large numbers of species to be compared in a single study. Using this method, 82 West Indian species of the leptodactylid frog genus *Eleutherodactylus* were compared with sequential electrophoresis at six loci. The initial number of alleles detected after one condition (113) was nearly doubled (223) after varied conditions were applied. Genetic distance analyses defined four major groups: 1) all native Jamaican species; 2) the majority of Hispaniolan South Island species; 3) most species previously placed in the *auriculatus* group from the Hispaniolan North Island, Puerto Rico, and the Lesser Antilles; and 4) species previously placed in the *inoptatus* group (Hispaniola). Variation in two morphological characters, liver shape and vocal sac condition, showed congruence with the allozyme groupings. Character analyses (using parsimony) of those four major groups defined subgroups, many of which were also supported by morphology and geography.

A revised classification is presented whereby four genera (*Ladailadne*, *Sminthillus*, *Syrrhophus*, and *Tomodactylus*) are placed in the synonymy of the genus *Eleutherodactylus* (ca. 450 sp.) and five subgenera are recognized: *Craugastor* (a Middle American clade of 68 sp.), *Eleutherodactylus* (presumably a paraphyletic taxon containing ca. 275 sp. distributed mostly in South America but with an eastern Caribbean clade), *Euhyas* (a western Caribbean clade of 78 sp.), *Pelorius* new subgenus (a Hispaniolan clade of six sp.), and *Syrrhophus* (a southern North American clade of 24 sp. previously placed in the genera *Syrrhophus* and *Tomodactylus*). Within the tribe Eleutherodactylini, which contains genera with direct development, the genus *Eleutherodactylus* is defined by a single synapomorphy: T-shaped terminal phalanges. Three subgenera, one section, five series, and 15 species groups are recognized for the West Indian species.

An hypothesized biogeographic history of West Indian *Eleutherodactylus* begins with dispersal from South America into a proto-Antillean land mass in the late Cretaceous or early Tertiary. The subsequent break-up of the proto-Antilles probably isolated the subgenus *Craugastor* on the Chortis Block and southern North America, the subgenus *Euhyas* on Cuba, and the subgenus *Eleutherodactylus* on the Hispaniolan North

---

[1] Blair Hedges completed his Ph.D. at the Department of Zoology, University of Maryland, College Park, MD 20742

Island and/or Puerto Rico, although the latter group may have dispersed to the Antilles in the Paleocene or Eocene. Late Eocene or early Oligocene dispersal of Cuban *Euhyas* to southern North America apparently led to the subgenus *Syrrhophus*. Jamaica and the Hispaniolan South Island (separate tectonic blocks) were submerged until the Miocene, at which time they probably were colonized by *Euhyas* from Cuba. Collision of the South Island and the North Island in the late Miocene resulted in limited overland dispersal of the subgenus *Euhyas* to the North Island and the subgenus *Eleutherodactylus* to the South Island. The striking morphological similarity of many West Indian species is considered to be convergence as a result of independent island radiations, as in the anoline lizards. Ecomorphs (*sensu* Williams) of West Indian *Eleutherodactylus* are described for concordant variation in morphology and ecology of species on different islands.

Introduction

Over the past decade, attention has focused on two competing theories of Caribbean biogeography: vicariance and dispersal (Rosen 1976, 1978, 1985; MacFadden 1980, 1981; Pregill 1981; Armas 1982; Hedges 1982; Coney 1982; Savage 1982; Briggs 1984; Buskirk 1985). The vicariance theory suggests that the Greater Antilles once were part of a late Mesozoic/early Cenozoic land mass, and that the relationships of the present day biota reflect those ancient land connections (Rosen 1976, 1985). Dispersalists contend that the West Indian biota is the result of overwater transport from the mainland during the Cenozoic (Matthew 1918; Simpson 1956; Pregill 1981; Briggs 1984). Three critical elements are needed to test these theories: 1) a geological history; 2) the distribution of the groups; and 3) the phylogenetic history of the groups.

The geological history of the Caribbean region is becoming well known, and most recent syntheses are in agreement with a proto-Antillean land mass similar to that proposed by Rosen (Pindell and Dewey 1982; Sykes et al. 1982; Duncan and Hargraves 1984; Mann and Burke 1984; Buskirk 1985). Also, the geographic distributions of most major extant groups in the Caribbean are well documented, although paleo-distributions are poorly known due to the scant Tertiary fossil record (Williams this volume). A major obstacle now remaining to test these competing theories of Caribbean biogeography is the lack of accurate phylogenetic reconstructions for the groups. This study represents an effort at obtaining these phylogenetic data for a major vertebrate group in the Caribbean region, leptodactylid frogs of the genus *Eleutherodactylus*.

The genus *Eleutherodactylus*

With over 400 described species, *Eleutherodactylus* is the largest vertebrate genus. It is a Neotropical group with two major centers of species diversity: northwestern South America and the West Indies. Nearly all species share two characteristics: "T-shaped" terminal phalanges and direct development (Lynch 1971). The T-shaped terminal phalanx probably is an adaptation for climbing, since it is best developed in arboreal species with expanded digital pads. It is also present in terrestrial species (although reduced) indicating that the ancestral *Eleutherodactylus* likely was a species adapted for climbing. However, direct development, allowing for reproduction away from water and the subsequent exploitation of new and diverse habitats, probably is largely responsible for the enormous success of this genus (Lynch 1971; Bogart 1981b).

About one third (ca. 130) of *Eleutherodactylus* species occur in the West Indies, where they are the dominant amphibian group. No single species is naturally found on more than one of the four Greater Antillean Islands, and most are restricted to small areas within an island. Also, no species are endemic to the Bahamas, and only five are known from the Lesser Antilles (Schwartz 1978; Schwartz and Henderson 1985). The recent discovery of an Eocene *Eleutherodactylus* on Hispaniola (Poinar and Cannatella 1987) confirms that this group has been evolving for at least 40 million years in the West Indies. Its wide distribution, high endemism, and long period of residence in the West Indies makes *Eleutherodactylus* an ideal group for the study of Caribbean biogeography.

No comprehensive phylogenetic study has been done on the entire genus *Eleutherodactylus*, and only one unambiguous synapomorphy is known that defines a major division: the "E" condition of the mandibular ramus of the trigeminal nerve defining the Middle American Clade (Lynch 1986). Most other divisions of the genus have been assigned the rank species group, which are largely phenetic assemblages defined primarily by skin texture, digital pad size, finger length, and vomerine odontophore length (Lynch 1976; Schwartz 1978). Some species groups, such as the *unistrigatus* group (>155 species), have more taxa than most anuran families and contain species with considerable morphological diversity.

In the West Indies, about 130 species are placed in seven species groups (Schwartz 1978, 1985), mostly in the *auriculatus* and *ricordii* groups. This study examines the West Indian species using six slow-evolving allozyme loci in an attempt to identify the major lineages and their relationships. In turn, the phylogenetic data are used to develop a new hypothesis for the evolution and biogeography of West Indian *Eleutherodactylus*.

Slow-evolving loci

Soon after natural populations of organisms were examined at multiple electrophoretic loci (Hubby and Lewontin 1966; Lewontin and Hubby 1966), interlocus variation was found in the number of allelomorphs (hereafter referred to as alleles). Avise (1975) noted the usefulness of this variation for systematics, and Sarich (1977) introduced the terms "slow-" and "fast-evolving" loci. Sarich considered plasma proteins (albumin, esterases, hemoglobin, transferrin) to be fast-evolving and intracellular enzymes to be slow-evolving, with a tenfold difference in substitution rate. More recent studies have not supported a bimodality in allelic variation (Avise and Aquadro 1982; Hedges 1986:fig. 5). Instead, it appears that variability in protein loci spans a continuum, from slow-, through intermediate- to fast-evolving loci. Such a finding is not unexpected since variability in the amino acid substitution rate of proteins also spans a continuum (Dayoff 1978; Nei 1987). However, these terms have proven useful in describing loci that have relatively few (=slow-evolving) or many (=fast-evolving) alleles, a usage that is followed here.

The importance of electrophoresis in systematics and its widespread use is largely a result of this continuum of variability in protein loci. If all loci had the same electrophoretically detectable substitution rate, then the resolution of a phylogeny would be very limited. For example, if the average amino acid substitution rate for every locus was one in 20 million years, it would be difficult to resolve the relationships of a group of species resulting from a radiation in the late Pliocene, two million years before present (myBP), since most species would appear identical in a set of 30 loci. Likewise, if the average substitution rate was one in two million years, then species that diverged in the Miocene (20 myBP) would not be expected to have any alleles in common. The systematic value of electrophoresis derives from variability in substitution rate: different loci often resolve different portions of a phylogeny. A suite of loci, with a continuum of variability, has the potential for resolving the relationships of a group spanning a considerable amount of evolutionary time (ca. 1-30 my).

Although the differential resolving power of electrophoresis has been known for some time (Avise 1975; Sarich 1977; Berlocher 1984), few systematists have taken advantage of it (e.g. Lanyon 1985). An exception is contact zone studies where fast-evolving loci are routinely examined (Barton and Hewitt 1983). One reason for this has been the strong reliance on the electrophoretic "clock" (Thorpe 1982) and the necessity to randomly sample as many loci as possible to obtain genetic distances that are comparable from study to study. Although many valuable studies have resulted, the full potential of electrophoretic loci as characters in a character analysis has not been fully realized. In addition, large amounts of time and effort have been expended comparing numerous alleles at fast-evolving loci only to find in the end that all of the species in the group have a different allele! The technical aspect of comparing large numbers of alleles at a particular locus also has limited the number of species that can be compared in any one study. Additional species require a disproportionately larger amount of laboratory work and thus electrophoretic studies rarely involve more than 25 species (Avise and Aquadro 1982). By using only slow-evolving loci, however, large numbers of species can be examined in one study because there usually are fewer alleles to compare.

Another advantage of using slow-evolving loci is that heterozygosity generally is much lower. Intraspecific polymorphism, a major problem in standard character analysis of electrophoretic data, is only rarely encountered at most slow-evolving loci. Differences between loci usually are "fixed" (with no heterozygotes) and thus the coding of alleles as character states is identical to the coding of morphological character states. Trees can then be generated by standard parsimony programs. Once the major groups have been identified by slow-evolving loci, individual studies can be conducted on each of those groups using faster-evolving loci.

In this study, I selected six slow-evolving loci to examine the major groups of West Indian *Eleutherodactylus*. These loci were selected by first running all species (84) at a large number of loci (>40) and choosing those loci that had the fewest number of alleles.

Materials and methods

Frogs were collected over a period of five years during 16 trips to the West Indies. A total of 84 species of *Eleutherodactylus* was obtained: all 17 native Jamaican species, 47 of 54 from Hispaniola, 12 of 15 from the Puerto Rican Bank, and four of five from the Lesser Antilles (Appendix 1). Access to Cuban species was available only through Guantanamo Bay Naval Station (*atkinsi*) and introduced populations in Jamaica (*planirostris*). Thus only two of the 33 Cuban *Eleutherodactylus* were obtained, a major limitation of the study. Two non-West Indian species were also included: *bransfordii* from Costa Rica, and *fenestratus* from Peru. If samples were available from multiple localities within a species, the locality nearest to the type locality was used (for taxonomic purposes).

Specimens were processed in the field and tissue samples were transported to College Park in a liquid nitrogen tank, or live frogs were brought to the laboratory for processing. Processing included weighing and photographing each species, obtaining blood for microcomplement fixation studies, removing viscera (primarily heart, liver, and kidney) and leg muscle (or an entire leg) for electrophoresis, preparing intestines and testes for chromosome analysis (in specimens injected with colchicine), removing one finger and fixing in 2.5% glutaraldehyde for scanning electron microscopy of the digital pad, and preserving the carcass in 10% formalin (transferred later to 70% ethanol) as a voucher for deposition in the United States National Museum of Natural History (USNM). Tissue samples were stored in ultracold freezers (-75°C) in the laboratory until needed. Samples were homogenized in distilled water at a ratio of 5:1 (distilled water:tissue). Homogenates were then refrozen, thawed, and centrifuged at 2°C for 20 min and 10,000 rpm. The aqueous protein supernatants were stored again at -75°C until use.

Electrophoretic differences between species and species groups usually are "fixed" with no heterozygotes (Avise 1975; Gorman and Renzi 1979). Thus, for the purpose of this study, very little would be gained by having more than one individual per species and therefore the sample size used was one.

Sequential electrophoresis

For systematic purposes, it is important to know whether two or more alleles with the same mobility are homologous or convergent. Apparently, convergence of electromorphs is a technique problem, related to the efficiency of electrophoresis in detecting protein variants, and not a result of adaptation to similar environments. This efficiency has been examined in two proteins: hemoglobin (Ramshaw et al. 1979) and myoglobin (McLellan 1984). In both cases, standard electrophoresis (one condition) detected about 40% of the protein variants of known amino acid sequence. By using different electrophoretic conditions successively, a procedure termed "sequential electrophoresis" (Coyne 1982), 85-93% of the protein variants could be detected. Additional evidence from sequencing studies of natural populations suggests that sequential electrophoresis can detect all or nearly all amino acid substitutions (Lewontin 1985).

Despite the obvious implications for systematics in terms of increasing resolution and reducing homoplasy (Coyne et al. 1979), very few researchers (e.g. Aquadro and Avise 1982; Lanyon 1985) have used sequential electrophoresis in their systematic studies. Some use multiple conditions simultaneously but this is inefficient since an electrophoretic difference defined at one condition needs no additional confirmation. Also, unless alleles are characterized by the conditions under which they were detected, later comparisons (running samples side-by-side with known standards) are difficult or impossible since allelic differences detected under one condition may not be detected under another. Therefore, the use of multiple conditions simultaneously is usually not equivalent to sequential electrophoresis.

An even greater error in the estimate of allelic variation results if comparisons are not performed. Ratios of the distance travelled by an electromorph relative to a standard sometimes are used in place of side-by-side comparisons. However, this likely will result in errors, such as missing small differences or in scoring differences when none exist, since no two electrophoretic runs are completely identical.

In order to reduce the amount of homoplasy (e.g. allelic convergence) in my data, I used sequential electrophoresis. The primary variable chosen was buffer type, since it has been shown to have substantial effects on mobility (Coyne 1982). After experimentation, I found that many loci were not resolvable on all buffer systems or in all taxa. Typically, only two or three buffer systems resulted in gels that were fully scorable at a particular locus. Thus, I used either one, two, or three conditions with the following six protein loci: *Acp, Ck, Icd-1, Lgl, Pgm*, and *Pt-3* (Table 1).

Horizontal starch gel electrophoresis was employed using Connaught starch at a concentration of 12.5%. Buffers were prepared following the methods of Selander et al. (1971). Assays and references for the proteins are given in Hedges (1986).

Differences and similarities in electrophoretic mobility were confirmed in comparison runs. To ensure detection of very small differences, samples representing the same presumed allele were alternated on the same gel (e.g. Coyne 1982:fig. 1). This procedure was repeated for all pairs of samples representing the same presumed allele. Initial experimenting confirmed that more differences could be detected using this "alternating" method of comparison over one involving samples run side-by-side.

Alleles and multiple loci are ordered from cathode to anode. Alleles detected during the first electrophoretic run are assigned lower-case letters. If additional alleles were detected during the second and third runs, they are assigned numbers and upper-case letters, respectively. This is done in a "nested" fashion so that subdivided allelomorphs retain their initial designation, but are uniquely defined by their second and/or third additional designations (Appendix 2). In the case of multilocus systems, protein homology was assessed by the methods described in Hedges (1986).

The electrophoretic data were analyzed by three different methods. Two involve the use of genetic distances and the third is a character analysis.

Genetic distance analyses

A UPGMA phenogram was produced using a modified Cavalli-Sforza distance (D_C; Nei et al. 1983), and a distance Wagner tree was generated with the Cavalli-Sforza and Edward's (1967) chord distance (D_A). A fuller discussion of the use of these distances and methods is presented elsewhere (Hedges 1986). The distance Wagner tree was rooted with *E. fenestratus* as the outgroup since that species is not believed to be phylogenetically close to any of the West Indian species (Lynch 1976). All trees using genetic distance data were produced with BIOSYS-I (Swofford and Selander 1981), modified to incorporate the Cavalli-Sforza distance used by Nei et al. (1983).

Character analyses

Character analyses were performed on the allelic data using PAUP (Phylogenetic Analysis Using Parsimony) computer software (Swofford 1985). Each locus was treated as a character and alleles as unordered character states. In cases where heterozygotes were encountered that had one allele shared with other species and another that was unique (autapomorphic), the unique allele was not considered since it was of no value in tree construction. In three cases, both alleles of the heterozygote were shared with two or more species. Since frequency coding was not possible, it was determined by outgroup analysis, in two of those cases, which allele was derived and the species were coded as possessing only that allele, thus minimizing the loss of cladistic information. In the third case, (*Lgl*), the locus had to be omitted from the analysis of that group (IV).

Global branch-swapping was used to find the most-parsimonious tree (MPT) and in all cases more than one tree was found (in many cases, this was due to forced dichotomous resolution of polychotomies by PAUP). If a large number of MPT's exist, then PAUP only stores the first 100, which will be a biased sample dependent on the initial ordering of species in the data file. In order to eliminate or reduce this bias in those cases, I obtained 11 consensus trees (Adams 1972), each constructed from 100 MPT's generated by a random reordering of the species. A majority-rule consensus tree (Margush and McMorris 1981) was then constructed from those 11 initial consensus trees, thus representing 1100 MPT's.

Cladograms showing allelic changes were constructed using the topology of the final consensus tree except in some cases where aspects of one MPT were favored by data external to the study (e.g. morphology). In all cases, the character-state cladogram was slightly longer than any single MPT due to conflicting character state distributions generated in the consensus process.

Morphology

In an effort to find some morphological traits of systematic value in West Indian *Eleutherodactylus*, I examined preserved specimens in the United States National Museum of Natural History and in my own collection. Additional data were extracted from literature accounts. Emphasis was placed on traits that were not obviously correlated with environmental variables to reduce the likelihood of convergence. Three useful characters were identified that showed variation concordant with other sources of data and were thus surveyed in a majority of West Indian species: liver shape (small and/or with rounded lobes vs large and with a long pointed left lobe), testis color (black or pigmented vs white or unpigmented), and vocal sac condition (single, paired, or absent; internal or external).

In the case of liver shape, the lateral incision normally used to sex anurans was sufficient to allow scoring of that character. In some individuals, liver shape could be seen (without dissection) by holding the specimen in front of a strong light source. No vocal sac was ever found, upon dissection, in an adult male frog lacking vocal slits. Therefore, the absence of vocal slits in species not dissected was taken as evidence for the absence of a vocal sac. In the case of testis color, species polymorphic (usually >80% pigmented) for this character were treated as having pigmented testes since this method of coding resulted in a higher degree of concordance with other types of data.

Results

The number of alleles per locus ranged from 15 to 54 with an average of 37 and a total of 223 alleles detected (Appendix 2). Before sequential electrophoresis, only 113 alleles were detected and thus the use of additional buffer systems nearly doubled the total number of alleles. Heterozygosity averaged 5% (SE= ± 1%) among all loci (6) and species, which is about average for anurans and other amphibians and reptiles (Nevo 1984). However, since sequential electrophoresis significantly increases heterozygosity estimates to the more correct value (Coyne 1982; Lewontin 1985), and since slow-evolving loci are expected to have lower heterozygosity, comparison with other studies may be misleading.

Genetic distance analyses

A UPGMA phenogram of 84 *Eleutherodactylus* species (Fig. 1) has a cophenetic correlation coefficient of 0.81, and a Prager and Wilson (1976) F-value of 4.73. Several large groups are defined. One contains all native Jamaican species (Group I), except for *nubicola*. Two others (Groups II and III) correspond, in general, to the *ricordii* and *auriculatus* groups of Dunn (1926a) and Schwartz (1958, 1969, 1978). Of two smaller groups, one (Group IV) is equivalent to the *inoptatus* group (Schwartz 1965, 1976; Hedges and Thomas 1987). The other is a cluster of six species morphologically allied to the *ricordii* group (Schwartz 1976), and which share an allele ($Pt\text{-}3^{i2}$) with Group II. The remaining six species (*counouspeus*, sp. nov. N, *glanduliferoides*, *richmondi*, *bransfordii*, and *fenestratus*) have either one or none of their alleles in common with other species at any of the loci and are therefore the most divergent.

Geographically, the species in Groups I and II are western Caribbean, occurring primarily on Jamaica, Cuba, and the South Island of Hispaniola. Species in Group III are eastern Caribbean, occurring primarily on the North Island of Hispaniola, Puerto Rico, and the Lesser Antilles. Group IV is restricted to Hispaniola.

A distance Wagner tree (Fig. 2) has a cophenetic correlation coefficient of 0.57 and a Prager and Wilson's (1976) F-value of 44.1 (0.80 and 5.77, respectively, after branch-length optimization). In general, the groupings are similar to those in the phenogram. The one Jamaican species omitted from Group I in the phenogram (*nubicola*) now associates with that group. Also, the two separate units of Group II in the phenogram form one cluster in the distance Wagner tree. However, Group III is broken into several units, Groups I and II do not associate, and Group IV clusters with Group I.

A character analysis of the entire group of 84 *Eleutherodactylus* species was not performed because of the absence of shared alleles between the ingroup and outgroup species (*bransfordii* and *fenestratus*), and the low levels of allelic similarity among the major subdivisions (Groups I-IV) of the ingroup. However, examination of the allelic data (Appendix II) reveals that each of the four groups is defined by one or more unique alleles (not necessarily possessed by all of the species in a group): Icd^{q2} (Group I), Pgm^{jB} (Group II), Icd^{f1} (Group III), and Icd^{p5}, Lgl^{a1}, and Pgm^{o} (Group IV). In a cladistic (character) analysis, only shared derived character states (synapomorphies) are used to cluster species whereas phenetic analyses use both synapomorphies and symplesiomorphies (shared primitive traits). Although some or all of the alleles defining Groups I-IV could be symplesiomorphic (i.e undesirable), the congruence with morphology (see below), immunology (Hass and Hedges, unpubl. MS), and geography suggests that they are synapomorphies defining monophyletic groups. For that reason, the four allozyme groups were treated separately in the following character analyses.

Character analyses

Individual character analyses (using PAUP) were performed on the four groups defined in the genetic distance analyses. Each tree was rooted by using a composite outgroup of all West Indian species not present in the ingroup under consideration. This composite outgroup was the same for all four ingroups, and consisted of alleles shared between those groups: Acp^g, Lgl^{g2B}, and $Pt-3^{k1}$. Primitive alleles were therefore those alleles present in both the ingroup and the outgroup. There were no shared alleles at the other three loci. In two cases (Acp^o and Lgl^{13B}), ingroup/outgroup convergences were detected by examining outgroup topology, and those alleles were treated as apomorphies in their respective ingroups.

Group I.-- Numerous MPT's were generated, all with a length of 29 and a consistency index (CI) of 1.00 (i.e. no homoplasy). A majority rule consensus tree representing 1100 MPT's (Fig. 3) shows strong support for several clusters (*cavernicola* and *cundalli*; *fuscus* and *pentasyringos*; *luteolus*, *grabhami*, and *sisyphodemus*) and weaker support for others (*nubicola* and *andrewsi*; *cavernicola*, *cundalli* and *glaucoreius*; *junori*, *gossei*, *pantoni*, *fuscus*, and *pentasyringos*). These groupings are similar to those found in a more complete study of 29 loci in the Jamaican species (Hedges, in press). A cladogram of allelic changes (Fig. 4) has a length of 31 and a CI of 0.94. Allele $Icd-1^{q2}$ is a probable synapomorphy for the group, although the primitive state cannot be identified. It is present in 15 of the 17 species and is not found in any other West Indian species examined (Fig. 5). No ancestral allele could be determined for *Ck* and *Pgm*.

Several groups are indicated on the tree of allelic changes. A group of three species in western Jamaica (*grabhami*, *luteolus*, and *sisyphodemus*) is defined by alleles Pgm^{1B} (absent in *sisyphodemus*) and $Pt-3^g$. Another group, consisting of five morphologically similar terrestrial species (*fuscus*, *gossei*, *junori*, *pantoni*, and *pentasyringos*), has the allele $Pt-3^{k2}$. Three arboreal and rock-dwelling species with long limbs and large eyes (*cavernicola*, *cundalli*, and *glaucoreius*) form a group based on alleles Ck^{s2D} and Pgm^{fc}, and another trio of species (*alticola*, *andrewsi*, and *nubicola*) restricted to the Blue Mountains share alleles Ck^{s2B} (absent in *andrewsi*) and Pgm^{1a} (absent in *alticola*). The remaining three species (*griphus*, *orcutti*, and *jamaicensis*) do not show affinities with the other groups.

Group II.-- A majority-rule consensus tree representing 1100 MPT's of this group (Fig. 6) shows considerable resolution. All MPT's were of length 64 and had a CI of 0.95. A cladogram of allelic changes (Fig. 7) has a length of 65 and a CI of 0.94. Allele Pgm^{jB} (Fig. 8) is a probable synapomorphy for the group, although the primitive state cannot be identified. No ancestral allele could be determined for *Ck* and *Icd*.

A group of seven largely arboreal or rock-dwelling species (*amadeus*, *bakeri*, *eunaster*, *glanduliferoides*, *glaphycompus*, *heminota*, and sp. nov. P) is defined by $Pt-3^{l3}$. Eight species (*furcyensis*, *grahami*, *pictissimus*, *probolaeus*, *rhodesi*, *rufifemoralis*, *schmidti*, and *weinlandi*) form another group based on Ck^{m2A} (absent in *rufifemoralis* and *schmidti*). Within that group, two species with red flash marks and a dorsal pattern of reverse parentheses (*furcyensis* and *rufifemoralis*) also share allele Icd^i. Another subgroup is formed of five allopatric species (*grahami*, *pictissimus*, *probolaeus*, *rhodesi*, and *weinlandi*) that share allele Pgm^{s2A}.

Two species with a pattern of diagonal leg barring and restricted to the Massif de La Hotte (*glandulifer* and *sciagraphus*) also share allele Ck^r. Another pair of species (*apostates* and *oxyrhyncus*) have a similar robust appearance (stout-bodied with long thick legs), extreme sexual dimorphism in body size, and share allele Ck^{kC}. The remaining species in this assemblage do not associate into groups.

Group III.-- A majority-rule consensus tree representing 1100 MPT's (Fig. 9) resolves only one large group of Puerto Rican and Lesser Antillean species, and three small groups of Hispaniolan species. All MPT's were of length 71 and had a CI of 0.97. A cladogram of allelic changes (Fig. 10) has a length of 74 and a CI of 0.95. Allele Icd^{f1} (Fig. 5) is a probable synapomorphy for Group III (present in 25 species; absent only in *minutus, montanus, poolei, richmondi,* and *unicolor*) and not found in any other West Indian species, although the primitive state cannot be identified. No ancestral allele could be determined for *Ck* and *Pgm*.

Allele Acp^d defines the large group of Puerto Rican and Lesser Antillean species: *antillensis, barlagnei, cochranae, cooki, coqui, eneidae, johnstonei, martinicensis, pinchoni, portoricensis, richmondi, unicolor,* and *wightmanae* (the primitive allele is Acp^g). It is absent only in *richmondi*, but that species shares allele Ck^{g2} with three of those species (*antillensis, cochranae,* and *unicolor*). Within that group, three of the Lesser Antillean species (*johnstonei, martinicensis,* and *pinchoni*) form a subgroup defined by Ck^{g3A} and Pgm^{n2B} (absent in *johnstonei*). A fourth species, *barlagnei*, is also included in that subgroup based on the presence of dorsal chevrons and its morphological resemblance to *pinchoni*. Although data are included only for the introduced Jamaican *johnstonei*, specimens from Guadeloupe were also compared and found to be similar to the Jamaican sample at most loci examined.

All other species occur in Hispaniola and form a weakly defined group based on allele Ck^{f2B} (present in seven of 14 species). Within that group are three well-defined subgroups. One contains a trio of largely bromeliad-dwelling species (*fowleri, lamprotes,* and *wetmorei*) defined by alleles Acp^o, Lgl^{l5A} (absent in *wetmorei*) and Pgm^{u3B} (absent in *lamprotes*). Two species with notched digital pads, *poolei* and *flavescens*, share allele Lgl^{l2}, and two nearly identical montane species, *montanus* and *patricae*, share allele Pt-3^{l2}. A group of three, small high elevation species (*audanti, haitianus,* and *minutus*) share allele Ck^{f3B}.

Group IV.-- In this group, one locus (*Lgl*) had to be omitted from the parsimony analysis because it was polymorphic, with heterozygotes and homozygotes of both alleles present, and thus could not be coded for entry in PAUP. The parsimony analysis of five loci yielded three MPT's of length 10 and with a CI of 1.00. The cladogram of allelic changes (Fig. 11) also represents a strict consensus tree of the three MPT's (length=11, CI=0.91). Although not used in the parsimony analysis, *Lgl* is included on the tree.

Two groups of three species each are defined by unique alleles. The burrowing species (*hypostenor, parapelates,* and *ruthae*) form a group defined by alleles Acp^b and Ck^{j1A} (absent in *parapelates*). The three large terrestrial and arboreal species (*chlorophenax, inoptatus,* and *nortoni*) share alleles Ck^{j2} and Pt-3^{l1}.

Morphology

Variation in liver shape, testis color, and vocal sac condition was scored for 113 species of West Indian *Eleutherodactylus* (Appendix 3), although in many cases, data were unavailable for one or two of the characters. No information was available for the following 12 species: *albipes, cubanus, delacruzi, emiliae, guanahacabibes, gundlachi, intermedius, zeus, darlingtoni, lucioi, neodreptus,* and *warreni*.

Liver shape (Fig. 12) showed a strong correlation with the allozyme groupings. With few exceptions, species having livers with long and pointed left lobes belong to allozyme Groups I and II whereas species with short and rounded left lobes belong to Groups III and IV. Exceptions are four Hispaniolan species (*abbotti*, *audanti*, *minutus*, and *parabates*) and one Puerto Rican species (*locustus*) belonging to Group III but which have long and pointed left lobes. The livers of five species (*poolei*, *cooki*, *karlschmidti*, *pinchoni*, and *wightmanae*) were intermediate in shape. In the case of two species, *johnstonei* (N=95, small liver with short and rounded left lobe) and *gossei* (N=31, large liver with long and pointed left lobe), large series were examined and no significant intraspecific variation in liver shape was found. Poorly preserved specimens or those with empty and contracted digestive tracts (indicating that the specimen was starved prior to preservation) were not scored for liver shape.

Of particular interest is the apparent lack of association between liver shape and ecology. Although Group III species are primarily arboreal, the arboreal species of Groups I and II (e.g. *jamaicensis*, *armstrongi*, *bakeri*, and *heminota*) have livers with long and pointed left lobes like other members of those allozyme groups. Also, the terrestrial Puerto Rican Bank species, *richmondi* and *lentus*, always associated with the *ricordii* group (Schwartz 1976), have a liver shape characteristic of the *auriculatus* group, thus in agreement with the allozyme data (no allozyme data were available for *lentus*). Also, there was no apparent correlation between liver shape and body size or altitude. Using liver shape, species for which allozyme data are unavailable can now be associated with allozyme groupings. Thus the Cuban species *Sminthillus limbatus*, associated with the *ricordii* group by Bogart (1981a) based on chromosomes, has a liver shape characteristic of that group (Groups I and II).

Testis color did not show the major geographic patterns that liver shape exhibited, except that all 17 species in Group I (Jamaica) and all six in Group IV (Hispaniola) had unpigmented testes whereas pigmented testes were found in all four species from Guadeloupe. Several of the smaller clades defined by allozyme data were supported by testis color (Fig. 7). There was no apparent association between testis color and either body size, altitude, or habitat.

Vocal sac condition showed a pattern similar to liver shape in exhibiting a dichotomy between Groups I & II and Group III. Nearly all Group III species have single external submandibular vocal sacs, as noted by Schwartz (1969) for the *auriculatus* group. Most species in Groups I and II either lack a vocal sac, or have an internal one. A clade of Hispaniolan (South Island) species have paired vocal sacs (Fig. 6). The condition of the vocal sac showed support for the allozyme data in several cases where the latter disagreed with previous species group allocations (e.g. *parabates* in Group III, and *armstrongi*, *bakeri*, and *heminota* in Group II). In two species, *amadeus* and *bakeri*, the vocal sac was absent in some adult males and present in others.

In an effort to establish polarity for the variation found in these three characters, preliminary information was obtained for several "outgroup" species from the mainland and from other leptodactylid genera. The same variation present in the ingroup was also found in the outgroup indicating that a more extensive survey of the genus *Eleutherodactylus* (and related genera) will be necessary to determine polarity for these characters.

Discussion

In comparison with other vertebrate groups, anurans are morphologically conservative (Wilson et al. 1977; Cherry et al. 1978) thus limiting the number of useful morphological characters for phylogenetic analysis. In the speciose genus *Eleutherodactylus*, the problem is compounded in that similar morphologies have appeared in many unrelated lineages.

For example, both terrestrial and arboreal habits probably have evolved multiple times in the genus *Eleutherodactylus*. Since most terrestrial species have small digital pads and nearly all arboreal species have enlarged digital pads (presumably to aid in climbing), the use of such a highly adaptive trait as digital pad size may result in homoplasy in a phylogenetic analysis. Other characteristics that are correlated with the environment or ecology of the animal include ventral skin texture (rough or areolate in arboreal species) and interdigital webbing (present in aquatic species). All three of these characters have been used in previous systematic studies of *Eleutherodactylus* (Schwartz 1958; Shreve and Williams 1963; Lynch 1976).

In contrast, electrophoretic alleles (and most other types of molecular data) essentially are free from the problem of adaptive convergence, and therefore should be better indicators of phylogeny than morphology. Even if very strong directional selection for a specific enzyme function in two unrelated lineages existed, two identical convergent amino acid sequences would not likely result. Differences would nearly always occur at "functionally neutral" sites. Since electrophoresis, if properly applied using the sequential method (Coyne 1982), can detect most single amino acid substitutions, then in theory, alleles shared among taxa should be identical (or nearly so) in sequence. In practice, however, a small percentage of allelic differences likely will go undetected even using sequential electrophoresis (due to chance convergence), but this should not pose serious problems in a phylogenetic analysis.

In this study, slow-evolving electrophoretic loci provide useful phylogenetic information in West Indian *Eleutherodactylus*. Four major groups and many smaller clades are defined by allelic data. At present, the alleles defining the major groups can only be treated as "presumed synapomorphies" until mainland species (and other related genera) are surveyed at these loci and polarities are established. However, the general concordance of these allelic groups with morphology (liver shape and vocal sac condition, and to a lesser degree, external morphology) and geography suggests that they are monophyletic.

Systematics of the genus *Eleutherodactylus*

The basic systematic category within the genus *Eleutherodactylus* has been the species group, of which more than 30 are currently recognized (Lynch 1985; Schwartz 1985), if "assemblies" (Lynch 1980, 1981) are included. The first attempt at a higher-level organization of the genus was by Lynch (1971), who proposed an Alpha/Beta dichotomy on osteological grounds. The Alpha division included most of the West Indian species and "parts of the Andean system," *Syrrhophus*, and *Tomodactylus*. The Beta division included all other *Eleutherodactylus* examined. Later, Lynch (1976) proposed four "infrageneric units" for the South American species groups based on variation in the relative lengths of the two inner fingers, and ventral skin texture.

Recently, Lynch (1986) defined a Middle American clade of *Eleutherodactylus* based on a shared derived trait, the "E" condition of the mandibular ramus of the trigeminal nerve. This clade, which also extends into southern North America and northern South

America, comprises 68 species of *Eleutherodactylus* (including three species previously placed in the genus *Hylactophryne*). The defining character is the position of the mandibular ramus of the trigeminal nerve relative to the *M. adductor mandibulae externus superficialis* (jaw muscle). The nerve passes either lateral ("S") or medial ("E") to the muscle, with the former condition being primitive based on outgroup comparisons (Lynch 1986:fig. 1).

Lynch (1986) suggested that if the Middle American Clade were to be recognized taxonomically, it would take the generic or subgeneric name *Craugastor* Cope 1862. I propose that it be recognized as a subgenus (type species =*Hylodes fitzingeri* by subsequent designation [Dunn and Dunn 1940]).

Savage (1987) recently presented a phylogenetic scheme for *Eleutherodactylus* and closely related genera using variation in jaw musculature and chromosomes. His analysis, based primarily on the work of Starrett (1968) for jaw musculature, and Bogart (1970, 1981a) and Deweese (1976) for chromosomes, largely conflicts with the results of this study and previous morphological studies (Lynch, 1970, 1971, 1976; Heyer, 1975).

Savage (1987) recognized Lynch's Middle American Clade but also discussed data on variation in the depressor mandibulae muscle, which appears to define three lineages within that clade. However, only one condition of the depressor muscle (DFSQdAT) is present in species possessing the "S" condition of the trigeminal nerve. For those species and closely related eleutherodactyline genera outside of the Middle American Clade (ca. 80% of the genus *Eleutherodactylus*), Savage uses only chromosome number and chromosome arm number to define relationships, assuming 2n=26 is primitive and all other numbers (lower and higher) are derived. This approach resulted in the following three groups (Savage 1987:fig. 28): (I) one lineage containing *Tomodactylus* and *Ischnocnema* (sister genera), and a second lineage containing *Euparkerella* and *Holoaden* (sister genera), Cuban *E. auriculatus* group, and the *E. diastema* group (the most distant taxon of the second lineage); (II) *Syrrhophus* as a sister group to a lineage containing *Sminthillus* and the combined *E. ricordii* and *E. unistrigatus* groups); and (III) *E. altae* and Puerto Rican *E. auriculatus* group.

The genera *Tomodactylus* and *Syrrhophus* are very similar morphologically (osteology, external morphology) and have largely parapatric distributions in Mexico suggesting a close relationship (Lynch 1970). Previous workers have considered them to be sister genera (Dixon 1957; Lynch 1970, 1971; Heyer 1975) and related to West Indian *Eleutherodactylus* (Lynch 1971; Bogart 1981a). Specifically, they share with the West Indian species several osteological traits: fusion of the frontoparietals, degree of overlap of the parasphenoid alae and median rami of the pterygoids, and median separation of the prevomers (Lynch 1971). Therefore the placement of these two genera in different lineages by Savage (1987), and the grouping of *Tomodactylus* with *Ischnocnema* seems unlikely. The latter genus occurs in the upper Amazon basin and in southeastern Brazil, distant from the range of *Tomodactylus*. The placement of Cuban *auriculatus* group species with *Euparkerella* and *Holoaden*, two genera found in southeastern Brazil, seems even more improbable based on geography and morphology (Lynch 1971). Finally, the splitting of the Cuban and Puerto Rican *auriculatus* group species, which are similar in external morphology (Dunn 1926; Schwartz 1969) and osteology (Lynch 1971; Joglar 1986), and the clustering of the *ricordii* and *unistrigatus* groups, which differ considerably in external morphology and osteology (Lynch 1976; Joglar 1986), are also unlikely arrangements. I suggest that these anomalous results can be explained by the use of too few data (karyotypes of 65 out of 400+ *Eleutherodactylus* species were used), and the finding that chromosome evolution in *Eleutherodactylus* apparently is too rapid (Bogart 1981a; Bogart and Hedges, in press) to be very useful in defining major groups due to the high probability of convergence in chromosome number.

Since only one condition of the depressor jaw musculature (DFSQdAT) appears to be present in 80% of *Eleutherodactylus* species (i.e. those outside of the Middle American Clade), that character is not very useful in defining major groups within the genus (although many species have yet to be examined). On the other hand, chromosome data appear to be too variable to be used for this purpose, except in conjunction with other types of data where they can be placed in a phylogenetic framework. Instead, the rapid rate of chromosome evolution in *Eleutherodactylus* is ideally suited for resolving lower-level relationships (species groups and smaller clades). Although slow-evolving electrophoretic loci have proven useful in defining major groups of West Indian *Eleutherodactylus*, they also may be too variable to establish relationships of major divisions within the genus. The technique of micro-complement fixation of albumin, which can be used to examine relationships extending into the Cretaceous (Maxson and Maxson 1986), shows promise for unravelling the phylogenetic history of the major groups of *Eleutherodactylus* (Hass and Hedges, unpubl. MS).

The genera *Sminthillus*, *Syrrhophus* and *Tomodactylus* have been recognized as offshoots of the *Eleutherodactylus* radiation for some time, based on osteology, but have been retained as genera in anticipation of further splitting of the genus *Eleutherodactylus* (Lynch 1971). If the major monophyletic groups within the genus *Eleutherodactylus* are treated as subgenera, as advocated here, then it becomes undesirable to retain those three as genera. To eliminate this paraphyletic situation, I place *Sminthillus*, *Syrrhophus*, and *Tomodactylus* in the synonymy of the genus *Eleutherodactylus*. A welcome result of this change is that it aids in defining the genus *Eleutherodactylus*, now comprising about 450 described species. Although other genera in the tribe Eleutherodactylini have direct development, only species of the genus *Eleutherodactylus* (including species previously placed in the genera *Sminthillus*, *Syrrhophus*, and *Tomodactylus*) possess T-shaped terminal phalanges (Lynch 1971). Nonetheless, several other eleutherodactyline genera lacking T-shaped terminal phalanges (*Adelophryne*, *Euparkerella*, *Ischnocnema*, *Phyllonastes*, *Phyzelaphryne*, and *Phrynopus*) are believed to be part of the *Eleutherodactylus* radiation (Lynch 1971, 1986; Savage 1987) and thus eventually may be synonymized in the genus *Eleutherodactylus* when additional information becomes available. I concur with Lynch's (1986) placement of *Hylactophryne* in the synonymy of the genus *Eleutherodactylus*.

Since the species previously placed in the genera *Syrrhophus* and *Tomodactylus* appear to form a monophyletic group united by several osteological characters (Lynch 1971; Joglar 1986), I propose that those species be placed in the subgenus *Syrrhophus* Cope 1878 (the older name), with the type species *Syrrhophus marnockii* Cope 1878 by monotypy. In so doing, it becomes necessary to erect two new categories to include the species previously placed in those two genera. For the 15 species (5 species groups) previously placed in the genus *Syrrhophus* (Lynch 1970), I propose the *longipes* series. This series includes the following species groups: *longipes*, *leprus*, *marnockii*, *modestus*, and *pipilans*. For the nine species (no recognized species groups) previously placed in the genus *Tomodactylus* (Lynch 1971), I propose the *nitidus* series. However, one species in the *nitidus* series, *fuscus* Davis and Dixon 1955, becomes a junior secondary homonym of *Eleutherodactylus fuscus* Lynn and Dent 1943 and therefore a replacement name must be found. I suggest the name *Eleutherodactylus (Syrrhophus) maurus* (nomen novum) as a replacement, continuing allusion to the dark coloration of this species. Diagnoses and definitions for these two series are given in Lynch (1968, 1971) under their previous generic names. Albumin immunological distances (Hass and Hedges, unpubl. MS) suggest that the subgenus *Syrrhophus* is closer to West Indian Groups I and II (a separate subgenus defined below) than to III and IV (two additional subgenera), a finding which is concordant with their terrestrial habits, presence of glandular areas, and distribution (western Caribbean).

A revised classification of West Indian *Eleutherodactylus*

The taxonomic framework for West Indian *Eleutherodactylus* initiated by Dunn (1926a) and developed extensively by Schwartz (1985) is largely supported by the electrophoretic data presented herein. In particular, the morphological division between two major assemblages, the *auriculatus* and *ricordii* groups, also is reflected in major allelic differences. However, some significant changes in taxonomy are suggested, and a more refined estimate of the relationships within the major groups is presented. Therefore, it is desirable to have a classification which more accurately reflects our current knowledge of the relationships of West Indian *Eleutherodactylus*.

A considerable body of molecular and chromosomal data now exist for the West Indian species, but comparable data are lacking for most of the mainland taxa. Therefore, the term "presumed synapomorphy" will be used in the systematic account below to refer to those shared derived characters (including those of morphology) in which polarity is suggested by only a limited sampling of mainland representatives and outgroups, or to those unique alleles which show concordance with other data sets. A summary of the following classification is presented in Table 2. Distributions do not include introduced populations.

Genus *Eleutherodactylus* Dumeril and Bibron 1841
Subgenus *Euhyas* Fitzinger 1843

TYPE SPECIES.-- *Hylodes ricordii* Dumeril and Bibron 1841

DEFINITION.-- Long vomerine odontophores, a large liver with a long and pointed left lobe, inguinal glands, and the absence of vocal slits (and vocal sac) are presumed synapomorphies, although not all of these traits are possessed by all species. Primarily terrestrial or rock-dwelling frogs with smooth or weakly rugose (or weakly areolate) ventral skin.

CONTENT.-- 78 species, 11 species groups, and 3 series.

DISTRIBUTION.-- Cuba and Bahamas (27 sp.), Jamaica (17 sp.), Hispaniola (33 sp.), and Mona Island (1 sp.).

REMARKS.-- This subgenus corresponds to allozyme Groups I and II defined in this study, and largely contains species previously placed in the *ricordii* group (Dunn 1926a; Schwartz 1958, 1965b, 1973, 1976, 1985). Although long vomerine odontophores are believed to be a synapomorphy (Joglar 1986), species with short vomerine odontophores can be placed confidently in this subgenus based on a combination of other characteristics (morphological and molecular; see below) indicating that there have been multiple reversals in this character. Those species are *amadeus*, *bakeri*, *brevirostris*, *cubanus*, *delacruzi*, *eunaster*, *glanduliferoides*, *heminota*, *orcutti*, sp. nov. P (Hedges, unpubl. MS), *rufifemoralis*, *sciagraphus*, *semipalmatus*, *symingtoni*, *thorectes* (Hedges 1988), *turquinensis*, *varleyi*, *ventrilineatus*, and *zeus*. It is possible that the size of the vomerine odontophores (and hence number of vomerine teeth) is correlated with feeding habits: short for soft-bodied prey such as Diptera and Lepidoptera, long for hard-bodied prey such as Orthoptera and Coleoptera. This would explain why most arboreal species (those that would encounter prey such as Diptera and Lepidoptera more frequently) have short odontophores and most terrestrial species (which would encounter hard-bodied prey more often) have long odontophores. Preliminary data on stomach contents lends initial support to that hypothesis but a much more extensive survey is warranted.

Of 46 species in this subgenus that have been karyotyped, 35 are either 2N=30 or 32 (Bogart 1981a; Bogart and Hedges, in press; and J. Bogart, pers. comm.). Diploid chromosome numbers in the other major West Indian assemblage, the *auriculatus* section of the subgenus *Eleutherodactylus*, are 2N=28 or fewer except for three Puerto Rican species which are 2N=30 (*karlschmidti*, *richmondi*, and *unicolor*). A few mainland species are 2N=32 and *marnockii* is 2N=30 (Bogart 1970, 1981a), although, except for the latter species, none appear to be close to the subgenus *Euhyas*. Bogart's (1981a) inclusion of *Sminthillus limbatus* in the *ricordii* group (=subgenus *Euhyas*) of *Eleutherodactylus* based on chromosome data is supported by its liver shape (long and pointed left lobe).

luteolus series

DEFINITION.-- Allele *Icd-1*^{q2} (absent in *fuscus* and *nubicola*) is a presumed synapomorphy (there are no known morphological synapomorphies). These are primarily terrestrial species but are morphologically and ecologically diverse. This series includes all native Jamaican *Eleutherodactylus*.

CONTENT.-- 17 species; 5 species groups.

DISTRIBUTION.-- Jamaica

REMARKS.-- Additional synapomorphic alleles are presented elsewhere (Hedges, in press). All species possess relatively large, white testes and lack vocal slits (and vocal sacs), but these probably are plesiomorphic traits within the subgenus *Euhyas*. Diploid chromosome numbers in this series are 24, 26, 28, 30, and 32 (Bogart and Hedges, unpubl. MS). Albumin immunological distance data support the monophyly of this series (Hass and Hedges, in press), although an analysis of morphological variation in the Jamaican species by Flores (1984) does not.

luteolus group
Figure 13

DEFINITION.-- Alleles *Pgm*^{1B} (absent in *sisyphodemus*) and *Pt-3*^{g} are synapomorphies.

CONTENT.-- 3 species: *grabhami*, *luteolus*, and *sisyphodemus*.
DISTRIBUTION.-- Western Jamaica.
REMARKS.-- Dunn (1926b) used the "*luteolus* group" in a broader sense, referring to most Jamaican species known at the time. The group defined here is a restricted version primarily based on allozyme characters but with a geographic cohesiveness: all three species occur in karst areas of Western Jamaica. They appear to be morphologically dissimilar, although *luteolus* and *sisyphodemus* are both very small (13-16 mm SVL, males). There is chromosomal evidence for a relationship between *grabhami* and *sisyphodemus* (Bogart and Hedges, in press).

gossei group
Figure 14

DEFINITION.-- Allele $Pt\text{-}3^{k2}$ is a synapomorphy for the group. Stout-bodied and short-legged terrestrial frogs with a smooth dorsum and small digital pads.
CONTENT.-- 5 species: *fuscus*, *gossei*, *junori*, *pantoni*, and *pentasyringos*.
DISTRIBUTION.-- Jamaica.
REMARKS.-- Goin (1954) constructed the *gossei* group to accommodate most Jamaican species previously placed in Dunn's *luteolus* group (excluding *luteolus*), and it has been used in a similar manner since that time (Goin 1960; Schwartz and Fowler 1973; Crombie 1977, 1986). It is used here in a restricted sense to define a monophyletic subset of those species. Schwartz and Fowler (1973) described *pentasyringos* as a subspecies of *pantoni*, but I regard it as a distinct species based on call differences (Crombie 1986), a different chromosome number (Bogart and Hedges, in press), allozyme differences (Hedges, in press), and an apparent lack of intergradation (R. Crombie, pers. comm.).

This is a morphologically well-defined group (Flores 1984). Chromosome, immunological, and additional allozyme data supporting this group will be presented elsewhere (Bogart and Hedges, in press; Hass and Hedges, unpubl. MS; Hedges, in press).

cundalli group
Figure 15

DEFINITION.-- Alleles Ck^{s2D} and Pgm^{fC} are synapomorphies. Long-limbed species with large eyes, a rugose or tuberculate dorsum, and large digital pads.
CONTENT.-- 3 species: *cavernicola*, *cundalli*, and *glaucoreius*.
DISTRIBUTION.-- Jamaica
REMARKS.-- Synapomorphic alleles at six additional loci will be presented elsewhere (Hedges, in press). The species *glaucoreius* was described as an eastern subspecies of *cundalli* by Schwartz and Fowler (1973) based on smaller body size and shorter vomerine odontophores. It can be distinguished from both *cavernicola* and *cundalli* by electrophoretic (Hedges in press) and chromosomal (Bogart and Hedges, in press) differences. All three taxa are allopatric with non-adjoining ranges. Since *glaucoreius* has differentiated to at least the same degree as *cavernicola* (from *cundalli*), I regard it as a distinct species. All previous workers on Jamaican *Eleutherodactylus* have considered *cundalli* (and by association, *cavernicola* and *glaucoreius*) and *grabhami* to be closer to Cuban and Hispaniolan species of the *ricordii* group (=subgenus *Euhyas*) than to other Jamaican

species based on external morphology and osteology (Dunn 1926b; Lynn 1940; Goin 1954; Schwartz and Fowler 1973; Crombie 1977; Flores 1984). However, one morphological trait is in agreement with the molecular data showing a relationship between the *cundalli* group and other Jamaican *Eleutherodactylus*. The "picket" dorsal pattern apparently is a rare variant only known in seven Jamaican species (*cundalli, glaucoreius, fuscus, gossei, pantoni, pentasyringos,* and *sisyphodemus*) and one Middle American species of *Eleutherodactylus* (Goin 1960; Schwartz and Fowler 1973; Crombie 1977).

jamaicensis group
Figure 16

DEFINITION.-- Allele Ck^{s2f}, and an areolate venter are autapomorphies for the single bromeliad-dwelling species in this group.
CONTENT.-- 1 species: *jamaicensis*.
DISTRIBUTION.-- Jamaica.
REMARKS.-- Dunn (1926b) and Schwartz (1969) associated *jamaicensis* with the *auriculatus* group (=*auriculatus* section). Schwartz and Fowler (1973) considered it an aberrant *gossei* group (=*luteolus* series) species, and Crombie (1977) placed it in its own group (as a separate invasion to Jamaica). Flores (1984) considered it distant from all other Jamaican species based on osteological characters. Although considered here to be part of the Jamaican radiation (*luteolus* series), it is retained in a separate group.

nubicola group
Figure 17

DEFINITION.-- A diploid chromosome number of 32 (Bogart and Hedges, in press) and absence of inguinal glands are synapomorphies. Stout-bodied terrestrial species with small digital pads (except *orcutti*) and a smooth dorsum (except *alticola*).
CONTENT.-- 5 species: *alticola, andrewsi, griphus, nubicola,* and *orcutti*.
DISTRIBUTION.-- Jamaica.
REMARKS.-- Except for *griphus* (Crombie 1986), all are restricted to the Blue Mountains of eastern Jamaica. Flores (1984) considered the absence of inguinal glands in this group to be a synapomorphy. Although no allelic synapomorphies were detected in this study, chromosome and allozyme data to be presented elsewhere support the monophyly of this group (Bogart and Hedges, in press; Hedges, in press).

bakeri series
Figure 18

DEFINITION.-- Allele Pt-3^{13} (absent in *glaphycompus*; unknown in *semipalmatus* and *thorectes*), a paired vocal sac (vocal sac absent in *glanduliferoides* and *semipalmatus*), and enlarged digital pads (reduced in *glanduliferoides*) are synapomorphies. Primarily arboreal and rock-dwelling species with rugose or areolate venters.
CONTENT.-- 9 species: *amadeus* (Hedges et al. 1987), *bakeri, eunaster, glandulife-*

roides, glaphycompus, heminota (Fig. 18), sp. nov. P (Hedges, unpubl. MS), *semipalmatus* and *thorectes* (Hedges 1988).

DISTRIBUTION.-- Hispaniola (South Island).

REMARKS.-- This series has invaded the arboreal adaptive zone normally occupied by species of the *auriculatus* section of the subgenus *Eleutherodactylus* (see below). The paired vocal sac is internal in *heminota*, external in *eunaster*, *glaphycompus*, and sp. nov. P, absent in *glanduliferoides*, *semipalmatus*, and *thorectes*, and either internal or absent (i.e. polymorphic) in *amadeus* and *bakeri*. Most are climbing species (*amadeus*, *bakeri*, *eunaster*, and *heminota* are arboreal; *glaphycompus* and sp. nov. P are rock-dwelling) although two species (*glanduliferoides* and *thorectes*) are found on or near the ground. The poorly known *semipalmatus* is believed to inhabit streams based on its digital fringe and webbing, although some examples of the bromeliad-dwelling *heminota* also possess these traits (to a lesser extent). Several additional undescribed species belonging to this series are present in collections from the Massif de la Hotte (Hedges, unpubl. MS).

pictissimus series

DEFINITION.-- Allele Ck^{m2A} (absent in *rufifemoralis* and *schmidti*) is a synapomorphy. Terrestrial species with reduced digital pads and a common dorsal pattern of mottling and/or dorsolateral stripes (or reverse parentheses).

CONTENT.-- 12 species; 3 species groups.

DISTRIBUTION.-- Hispaniola and Mona Island.

REMARKS.-- Aside from the co-occurrence of *paulsoni* and *pictissimus*, and possibly *schmidti* and *weinlandi*, all of the species in this series are allopatric.

rufifemoralis group
Figure 19

DEFINITION.-- Allele $Icd\text{-}1^i$, red "flash" marks on the concealed portions of the thigh and groin, and a "reverse parentheses" dorsal pattern are synapomorphies.

CONTENT.-- 2 species: *furcyensis* and *rufifemoralis* (Fig. 19).

DISTRIBUTION.-- Hispaniola (South Island), in the Sierra de Baoruco and Massif de la Selle.

REMARKS.-- These two allopatric species differ greatly in body size: *furcyensis* = 20 mm (male), 37 mm (female) SVL; *rufifemoralis* = 15 mm (male), 19 mm (female) SVL.

schmidti group
Figure 20

DEFINITION.-- Allele Ck^{kA}, webbing (slight) on hind feet, and large body size (to 58 mm SVL) are autapomorphies.

CONTENT.-- 1 species: *schmidti*.

DISTRIBUTION.-- Hispaniola (North Island).

REMARKS.-- Geographic variation in this stream-associated species was detailed by Schwartz (1971).

pictissimus group
Figure 21

DEFINITION.-- Allele Pgm^{s2A} (unknown in *lucioi*, *monensis*, and *warreni*) is a synapomorphy. Most species possess vocal slits and an internal submandibular vocal sac (absent in *lucioi*, *monensis*, and *probolaeus*). Dorso-ventrally flattened terrestrial species with a dorsal pattern of mottling or dorsolateral stripes.

CONTENT.-- 8 species: *grahami*, *lucioi*, *monensis*, *pictissimus*, *probolaeus*, *rhodesi*, *warreni*, and *weinlandi* (Fig. 21).

DISTRIBUTION.-- Hispaniola and Mona Island.

REMARKS.-- Morphological variation in this complex of allopatric species was discussed by Schwartz (1976), who proposed a phylogeny based on the presence or absence of glandular areas and digital pad size. Two species considered by Schwartz to be closely related to this assemblage (*richmondi* and *alcoae*) were not found to be close based on allozyme data.

Species unassigned to species group:

A single species, *paulsoni*, is placed in the *pictissimus* series based on its morphological resemblance to those species (Schwartz 1964), but is unassignable to species group (no allozyme data are available). It is restricted to the Tiburon peninsula of Hispaniola (South Island) and adjacent areas.

Species groups unassigned to series:
emiliae group

DEFINITION.-- A smooth dorsum, vocal slits (unknown in *emiliae*), and an absence of glandular areas are presumed synapomorphies.

CONTENT.-- 3 species: *albipes*, *emiliae*, and *intermedius*.

DISTRIBUTION.-- Cuba.

REMARKS.-- This group originally was designated as the *dimidiatus* group by Dunn (1926a), but Schwartz and Fowler (1973) removed *dimidiatus* (which they considered close to Jamaican species) and erected the *emiliae* group for the remainder of the species. Of the three Hispaniolan species previously placed in this group, I consider one (*parabates*) to belong to the *auriculatus* section (subgenus *Eleutherodactylus* [discussed below]) and the other two (*jugans* and *ventrilineatus*) to be unassigned to a series or group. Both *jugans* and *ventrilineatus* have a rugose and tuberculate (not smooth) dorsum, and at least *jugans* has glandular areas (supraxillary and inguinal). Also, both species lack vocal slits whereas at least two of the three Cuban members of the *emiliae* group possess vocal slits (not examined in *emiliae*). Thus, the two Hispaniolan species, although morphologically similar to one another, do not agree with the *emiliae* group characteristics as defined by Shreve and Williams (1963) or here, and are therefore removed from that group. No allozyme data are available for this group of Cuban species.

symingtoni group

DEFINITION.-- Short vomerine odontophores (a reversal), large body size, and a very rugose and tuberculate dorsum are synapomorphies.

CONTENT.-- 3 species: *delacruzi, symingtoni,* and *zeus*.

DISTRIBUTION.-- Cuba; the western province of Pinar del Rio (and possibly Matanzas).

REMARKS.-- This group recently was reviewed by Estrada et al. (1986). Although no allozyme data are available, I place the *symingtoni* group in the subgenus *Euhyas* based on the absence of vocal slits (in at least *symingtoni*), and the presence of a rugose dorsum and smooth venter (weakly rugose in *delacruzi* and *zeus*). Also, the species appear to have terrestrial or rock-dwelling habits, characteristic of species in this subgenus. Joglar (1986) places *symingtoni* and *zeus* in the *unistrigatus* group (=*auriculatus* section) based on short vomerine odontophores. I interpret the short vomerine odontophores in this group as a reversal to the plesiomorphic state.

varleyi group

DEFINITION.-- Short vomerine odontophores (a reversal), and small body size are presumed synapomorphies.

CONTENT.-- 2 species: *cubanus* and *varleyi*.

DISTRIBUTION.-- Cuba.

REMARKS.-- I place the Hispaniolan species *eunaster* and *glanduliferoides* in the *bakeri* series, rather than in this group (Schwartz 1985), based on morphology. The two Cuban species lack glandular areas (Shreve and Williams 1963) whereas both Hispaniolan species possess supraxillary, inguinal, and postfemoral glands. Also, *eunaster* has a paired vocal sac like most other species in the *bakeri* series whereas *varleyi* possesses a single pectoral vocal sac (vocal sac is absent in *glanduliferoides* and no information is available for *cubanus*). It is possible that the *varleyi* group belongs in the *bakeri* series, but there is little morphological justification, aside from short vomerine odontophores, and no allozyme data are available for the Cuban species.

Species unassigned to series:

The following species are placed in the subgenus *Euhyas* but are unassignable to series: Cuba (19) - *acmonis, atkinsi, bresslerae, cuneatus, dimidiatus, etheridgei, greyi, guanahacabibes, gundlachi, klinikowskii, limbatus, pezopetrus, pinarensis, planirostris, ricordii, sierramaestrae, thomasi, turquinensis,* and *zugi*; Hispaniola (13) - *alcoae, apostates, armstrongi, brevirostris* (Fig. 22), sp. nov. C (Hedges unpubl. MS), *darlingtoni, glandulifer, jugans, leoncei, neodreptus, oxyrhyncus* (Fig. 23), *sciagraphus,* and *ventrilineatus*).

Subgenus *Eleutherodactylus* Dumeril and Bibron 1841

TYPE SPECIES.-- *Hylodes martinicensis* Tschudi 1838
CONTENT.-- ca. 275 species.

DISTRIBUTION.-- South America, Middle America, and the West Indies (excluding Jamaica).

REMARKS.-- Presumably a paraphyletic assemblage which includes more than half of the named species in the genus, mostly distributed in northwestern South America. Only one division of this subgenus will be considered here: the *auriculatus* section. The South American species groups are reviewed by Lynch (1976), and the current composition of these groups can be found in Lynch (1985), with some recent changes (Lynch 1986).

auriculatus section

DEFINITION.-- Allele $Icd\text{-}1^{f1}$ (Fig. 5) is a presumed synapomorphy, present in 25 of 32 species examined (absent in *counouspeus*, *minutus*, *montanus*, sp. nov. N, *poolei*, *richmondi*, and *unicolor*; 12 species not examined electrophoretically). Primarily an arboreal group distinguished from other West Indian species of the genus *Eleutherodactylus* by a combination of short vomerine odontophores, areolate venter, external submandibular vocal sac, and enlarged digital pads.

CONTENT.-- 44 species, 2 series.

DISTRIBUTION.-- Cuba (6 sp.), Hispaniola (16 sp.), Puerto Rican Bank (17 sp.), and the Lesser Antilles, Trinidad, and northeastern South America (5 sp.).

REMARKS.-- Previously referred to as the *auriculatus* group (Dunn 1926a; Schwartz 1969), this section is morphologically similar to the *unistrigatus* group (Lynch 1976; Joglar 1986). However, in the *auriculatus* section, the frontoparietal and otoccipital bones normally are fused (not fused in most *unistrigatus* group species) and the median ramus of the pterygoid does not overlap the parasphenoid alae (usually overlaps in *unistrigatus* group species) (Lynch 1976). Also, some differences exist in the tympanum, hyoid processes, and presence or absence of nuptial pads, but none are diagnostic (Joglar 1986). Until a critical examination of the mainland *Eleutherodactylus* (especially the *unistrigatus* group) reveals that the *auriculatus* section is not monophyletic, I prefer to continue regarding it as such given the limited data available. Considering the close association between geography and phylogeny in other *Eleutherodactylus* groups, and the relatively high degree of morphological similarity among most *auriculatus* section species, it is likely that this geographic unit is monophyletic. Since *counouspeus* has a small liver with rounded lobes and an external submandibular vocal sac, it tentatively is placed in this section, but its chromosome number (2N=32; J. Bogart, pers. comm.), rock-dwelling habits, and distribution (South Island) would otherwise associate it with the subgenus *Euhyas*.

martinicensis series
Figures 24-25

DEFINITION.-- Allele Acp^d (absent in *brittoni*, *gryllus*, *locustus*, and *richmondi*) is a synapomorphy. There are no known morphological synapomorphies.

CONTENT.-- 22 species; 1 species group.

DISTRIBUTION.-- The Puerto Rican Bank, Lesser Antilles, and northeastern South America.

REMARKS.-- Although allele Acp^d is absent in *brittoni, gryllus, locustus,* and *richmondi*, other data suggest that these species are associated with the *martinicensis* series. Allele Pgm^{u4B} is unique to *locustus* and a species in the *martinicensis* series, *antillensis*. Also, *richmondi* has an allele, Ck^{g2}, found only in species of this series. In an electrophoretic study of eight proteins in seven Puerto Rican *Eleutherodactylus*, Smith et al. (1981) found *brittoni* and *locustus* to be close to other species on the island (*antillensis* and *eneidae*, respectively). Since the synapomorphic allele Acp^d is polymorphic in at least two species (*coqui* and *portoricensis*), then it is possible that it was not found in those four species due to the small sample size (one).

No allozyme data are available for *hedricki* and *karlschmidti*, but chromosomal data suggest an association with other Puerto Rican species (Bogart 1981a). The inclusion of *karlschmidti* and *unicolor* in the *ricordii* group (=subgenus *Euhyas*) by Savage (1987:table 3) based on chromosome number (2N=30) apparently was an error.

The ovoviviparous species *jasperi* recently was placed in a separate genus, *Ladailadne* (Dubois 1986) primarily based on its unique mode of reproduction. However, ovoviviparity has not been found in any other *Eleutherodactylus* species and is a derived trait. Since it is therefore an autapomorphy, it conveys no information concerning the relationship of *jasperi* to other *Eleutherodactylus* species. No electrophoretic data are available for *jasperi*, but it has a small liver with a rounded left lobe and an external submandibular vocal sac, two traits which associate it with the *auriculatus* section of the subgenus *Eleutherodactylus*. Additionally, it has a diploid chromosome number of 26 (Drewry and Jones 1976) like most other Puerto Rican species (Bogart 1981a), although a few Hispaniolan species in the *auriculatus* section also have that number (Bogart and Hedges, unpubl. data). There is no indication that *jasperi* is other than a member of the *auriculatus* section (based on morphology, geography, and chromosome number) and thus I place the genus *Ladailadne* in the synonymy of the genus *Eleutherodactylus*.

martinicensis group

DEFINITION.-- Alleles Ck^{g3A} (absent in *barlagnei*), Pgm^{n2B} (absent in *barlagnei* and *johnstonei*), and a common pattern of dorsal chevrons are synapomorphies.

CONTENT.-- 4 species: *barlagnei, johnstonei, martinicensis,* and *pinchoni*.

DISTRIBUTION.-- Lesser Antilles.

REMARKS.-- Schwartz (1967:58) suggested that these four Lesser Antillean species might share a common ancestor based on a pattern variant of one or two dorsal chevrons present in all. Two of the four, *barlagnei* and *pinchoni*, are endemic to Guadeloupe and all four occur on that island (*johnstonei* recently was introduced). Based on geography, *urichi*, which occurs in the southern Lesser Antilles and northern South America, may also be a member of this group, although no allozyme data are available for that species.

Species unassigned to species group:

The following species in the *martinicensis* series are unassignable to species group: *antillensis, brittoni, cochranae, cooki* (Fig. 24), *coqui* (Fig. 25), *eneidae, gryllus, hedricki, jasperi, karlschmidti, lentus, locustus, portoricensis, richmondi, schwartzi, unicolor, urichi,* and *wightmanae*.

montanus series
Figures 26-27

DEFINITION.-- Alleles Icd^{f2B}, Lgl^{l3A}, and Pgm^{p3B} are presumed synapomorphies, but see remarks below.
CONTENT.-- 15 species; 1 species group.
DISTRIBUTION.-- Hispaniola.
REMARKS.-- This series is weakly defined since out of 14 species examined, only seven have allele Icd^{f2B}, only five have Lgl^{l3A}, and only two have allele Pgm^{p3B} (9 species have at least one of the three). However, all occur on Hispaniola and most have chromosome numbers that are lower than those in the *martinicensis* series (Bogart 1981a; Bogart and Hedges, unpubl. data). It is possible that the six Cuban *auriculatus* section species are associated with this series since the two species that have been karyotyped (*auriculatus* and *varians*; Bogart 1981a) also have a low number, $2N = 18$, which is the modal number in the *montanus* series.

Among the unassigned species, several groups are suggested by the allelic data. Two species with "notched" digital pads, *flavescens* and *poolei*, share the unique allele Lgl^{i2} and are allopatric in distribution. Also, two virtually identical species, *montanus* (Fig. 26) and *patricae* (Schwartz 1965c), share the unique allele $Pt\text{-}3^{l2}$. Three small high elevation species, *audanti*, *haitianus*, and *minutus*, share the unique allele Ck^{f3B}.

wetmorei group
Figure 27

DEFINITION.-- Alleles Acp^o, Lgl^{l5A} (absent in *wetmorei*), and Pgm^{u3B} (absent in *lamprotes*) are synapomorphies. Moderate to large-sized bromeliad-dwelling species with large circular digital pads.
CONTENT.-- *fowleri*, *lamprotes*, and *wetmorei* (Fig. 27).
DISTRIBUTION.-- Hispaniola.
REMARKS.-- Allele Acp^o also occurs in *glanduliferoides* (subgenus *Euhyas*, *bakeri* series), but is interpreted as a convergence based on other allelic data, chromosomes, and morphology (see discussion of *bakeri* series above). The two allopatric South Island species, *fowleri* and *lamprotes*, are relatively large, have greatly expanded digital pads, and live almost exclusively in bromeliads. The third species, *wetmorei*, occurs on both the North and South Islands and is ecologically more variable, but is also commonly found in bromeliads.

Species unassigned to species group:

The following species in the *montanus* series are unassignable to species group: *abbotti*, *audanti*, *auriculatoides*, *flavescens*, *haitianus*, *minutus*, *montanus*, *parabates*, *patricae*, *pituinus*, *poolei*, and sp. nov. N (Sierra de Neiba).

Species unassigned to series:

The following species of the *auriculatus* section are unassignable to series: *auriculatus*, *bartonsmithi*, *counouspeus* (Fig. 28), *eileenae*, *leberi*, *ronaldi*, and *varians*.

Pelorius new subgenus

Cornufer Tschudi 1838 [Type species by monotypy, *Cornufer unicolor* Tschudi 1838 (=*Eleutherodactylus inoptatus* Barbour 1914; see Zweifel 1967). The specific name *Cornufer unicolor* Tschudi 1838 was placed on the Official Index of Rejected and Invalid Specific Names in Zoology (Bulletin of Zoological Nomenclature 34:267).]

TYPE SPECIES.-- *Leptodactylus inoptatus* Barbour 1914

DEFINITION.-- Alleles Icd^{p5} (absent in *inoptatus*), Lgl^{a1} (absent in *parapelates*), and Pgm^o (absent in *chlorophenax* and *nortoni*), a relatively long first finger (longer than second finger in *inoptatus*, *parapelates*, and *ruthae*), and large body size (50-90 mm SVL) are presumed synapomorphies. These are robust species with a relatively wide head and a smooth to weakly areolate venter.

CONTENT.-- 6 species; 2 species groups.

DISTRIBUTION.-- Hispaniola.

ETYMOLOGY.-- From the Greek, *Pelorios*, meaning huge, prodigious, awe-inspiring; referring to the large size of the species in this group, and the striking appearance of some (e.g. *nortoni*).

REMARKS.-- Schwartz (1965a) separated the *inoptatus* group (=subgenus *Pelorius*) from the *auriculatus* group (=*auriculatus* section) citing differences in size (larger in the *inoptatus* group), vomerine odontophore length (short, but not "patch-like"), vocal sac (not external), vocalization, and calling site (low to the ground or underground). Hedges and Thomas (1987) found that all six species have a single, internal, submandibular vocal sac and discussed species differences in vocalization and calling sites.

inoptatus group
Figure 29

DEFINITION.-- Alleles Ck^{j2} and $Pt\text{-}3^{11}$, and very large body size (66-90 mm SVL) are synapomorphies.

CONTENT.-- 3 species - *chlorophenax* (Fig. 29), *inoptatus*, and *nortoni*.

DISTRIBUTION.-- Hispaniola.

REMARKS.-- Two of the three species, *chlorophenax* and *nortoni*, form a subgroup based on synapomorphic allele Pgm^q. Both species also have greatly enlarged digital pads and a similar rising call (Hedges and Thomas 1987).

ruthae group
Figure 30

DEFINITION.-- Alleles Acp^b and Ck^{j1A} (absent in *parapelates*), a protruding snout with cornified skin at tip, chevron-shaped shank bars, and a laterally extended vocal sac are synapomorphies. Moderate-sized (ca. 50 mm SVL) burrowing species.

CONTENT.-- 3 species - *hypostenor*, *parapelates*, and *ruthae* (Fig. 30).

DISTRIBUTION.-- Hispaniola.

REMARKS.-- In addition to a cornified snout, these three allopatric species have unusually large subarticular tubercles, both probably adaptations for burrowing (I have observed captive *ruthae* use all four limbs and snout while burrowing). Considerable geographic call variation in *ruthae* (Schwartz 1965a; Hedges and Thomas 1987) may indicate that there are additional undescribed species.

BIOGEOGRAPHY

The results of this study indicate that overwater dispersal has not been a major factor in the recent evolutionary history of the genus *Eleutherodactylus* in the West Indies. This is suggested by the high degree of intra-island similarity, and the nearly complete absence of shared alleles between species on different islands. Others have assigned a more important role to dispersal, based on the morphological resemblance of species on different islands (Shreve and Williams 1963; Schwartz 1978). However, the allelic data indicate that intra-island radiations accompanied by morphological convergence, as in the anoline lizards (Williams 1969, 1983), has been the major theme in West Indian *Eleutherodactylus* evolution and biogeography.

The relevance of vicariance to *Eleutherodactylus* evolution in the Caribbean region depends strongly on the time frame involved. If the times of divergence of the major groups in the West Indies post-date the early Tertiary breakup of the proto-Antilles, then the vicariance theory (Rosen 1976, 1985) is not supported for this genus, and dispersal must have occurred. Information on times of divergence can come from fossil data, or from molecular data such as albumin immunological distances calibrated with geological time.

FOSSIL RECORD.-- An Upper Eocene *Eleutherodactylus* recently was discovered in Dominican Republic amber (Poinar and Cannatella 1987). It is the only unquestionable pre-Quaternary fossil *Eleutherodactylus* (Lynch 1971). This important find establishes the presence of the genus (in addition to *Anolis* and *Sphaerodactylus*; see Williams, this volume) in the Antilles 35-40 myBP. However, the description of the frog is insufficient to associate it with one of the five subgenera of *Eleutherodactylus*, making any other biogeographic interpretations difficult.

Notwithstanding, Poinar and Cannatella (1987) asserted that the presence of the Eocene amber fossils supports the vicariance model (Rosen 1976, 1985) in that they occur at a much earlier time than predicted by the dispersal model (Oligocene or Miocene at earliest). However, the mid-Tertiary arrival of Antillean groups in the dispersal model was based on the earliest known fossils of those groups, all from the mainland (Pregill 1981). That some of the earliest fossils are now from Hispaniola does not counter the dispersal model but simply extends the minimum age for these groups. Dispersal may have occurred during the Eocene or before.

Also, Poinar and Cannatella suggested that the relationships of the amber *Anolis* (Rieppel 1980), also from the North Island, supports the vicariance model, since it belongs to a subgroup of the genus which is distributed primarily on Cuba and Hispaniola (Williams 1976). However, this in fact supports dispersal, since most of the twenty Hispaniolan species of the subgroup (*carolinensis* subsection) occur on the South Island of Hispaniola (Schwartz 1980, Henderson and Schwartz 1984), which was separated from the North Island by over 1000 km of ocean in the Upper Eocene (Buskirk 1985).

Finally, a test was proposed by Poinar and Cannatella whereby vicariance is disproven if the fossil *Eleutherodactylus* (North Island) eventually is found to be more closely related to present-day South Island *Eleutherodactylus* than to North Island *Eleutherodactylus*. One difficulty with this test is the extremely low probability of rejecting the vicariance hypothesis. Only a complicated and unlikely biogeographic scenario would result in the Eocene North Island species being more closely related to present-day South Island species than to North Island species. Secondly, geologic data (Buskirk 1985) suggest that Jamaica and the South Island of Hispaniola were submerged during the mid-Tertiary. If true, then the present biota of those two islands must have arrived by dispersal. Although the Eocene amber fossils from Hispaniola establish the antiquity of those lineages in the West Indies, they do not presently favor one model over the other.

DISTRIBUTION.-- The five subgenera of *Eleutherodactylus* (*Craugastor*, *Eleutherodactylus*, *Euhyas*, *Pelorius*, and *Syrrhophus*) all occur in the Caribbean region (including adjacent mainland areas) and have largely allopatric distributions (Figs. 31-32). Although both subgenera *Eleutherodactylus* and *Euhyas* occur on Cuba and Hispaniola, the number of species per island of each subgenus indicates that the former taxon primarily is an eastern Caribbean group whereas the latter taxon is a western Caribbean group (Fig. 33). All Jamaican *Eleutherodactylus* belong to the subgenus *Euhyas* whereas all Puerto Rican Bank and Lesser Antillean species are in the subgenus *Eleutherodactylus*. The subgenus *Pelorius* is restricted to Hispaniola where it occurs with both subgenera (*Eleutherodactylus* and *Euhyas*).

In Middle America, the subgenus *Syrrhophus* has a northern distribution, occurring in Mexico, Belize, and northern Guatemala. Although the subgenus *Craugastor* is sympatric with *Syrrhophus* in northern Middle America, it is more widely distributed and has its highest diversity on the Chortis Block (southern Guatemala, Honduras, Nicaragua, and northern Costa Rica). The subgenus *Eleutherodactylus* (excluding the *auriculatus* section) extends into southern Middle America but has its highest diversity in South America.

BIOGEOGRAPHIC HISTORY.-- The distribution and relationships of the major groups of the genus *Eleutherodactylus* in the Caribbean region suggest a biogeographic history that includes both vicariance and dispersal. The timing of the events is inferred from albumin immunological data (Hass and Hedges, unpubl. MS, unpubl. data) using the albumin clock (Wilson et al. 1977). Since molecular data are lacking for most mainland species, the following biogeographic scenario focuses primarily on Antillean events.

The genus *Eleutherodactylus* likely arose in South America (Lynch 1971) and dispersed across the proto-Antilles to reach southern North America and the Chortis Block in the late Cretaceous. Perfit and Williams (this volume) argue that the proto-Antilles was a discontinuous chain of islands and not a continuous land mass. Geologic data are insufficient to distinguish clearly between these two alternatives, although in contrast to Williams, I believe that the latter is more likely. First, the proto-Antillean volcanic arc was already about 50-60 million years old by the late Cretaceous (Donnelly 1985). It is possible that this long period of activity resulted in a larger accretion of land than would a newly formed island arc system, although this has not been reflected in the sedimentary strata (T.W. Donnelly pers. comm.). Also, the poor overwater dispersal ability of *Eleutherodactylus* indicated by this study suggests that the connection between North and South

America at the end of the Cretaceous was relatively continuous for the South American *Eleutherodactylus* to have reached the Chortis block and southern North America.

The breakup of the proto-Antilles probably was the vicariant event that isolated the subgenus *Craugastor* on the Chortis block (and possibly southern North America) in the early Tertiary (60 myBP). Likewise, the subgenus *Euhyas* became isolated on Cuba. Jamaica and the South Island of Hispaniola may have carried an *Eleutherodactylus* fauna at this time but these islands later were completely submerged in the Oligocene (Buskirk 1985). It is unclear when the subgenera *Eleutherodactylus* and *Pelorius* entered the West Indies, or what vicariant event (if any) led to their divergence. If they diverged in the Oligocene as suggested by immunological data (Hass and Hedges, unpubl. data), then the Eocene amber frog (Poinar and Cannatella 1987) likely was a member of the ancestral stock of those two groups, based on their current distributions (Figs. 32-33). This ancestral stock may have been present on the North Island of Hispaniola and/or Puerto Rico during the initial breakup of the proto-Antilles, or dispersed there from South America sometime before the late Eocene.

Following the breakup of the proto-Antilles, Cuba, the North Island of Hispaniola, and Puerto Rico moved northeastward relative to North and South America, eventually colliding with the Bahamas platform in the Eocene (Pindell and Dewey 1982; Sykes et al. 1982; Duncan and Hargraves 1984). In the late Eocene or early Oligocene (35-40 myBP), dispersal from Cuba (subgenus *Euhyas*) to nearby southern North America may have led to the establishment of the subgenus *Syrrhophus* in what is now southern Mexico, Guatemala, and Belize. Since all species of that subgenus lack vomerine teeth (Joglar 1986), the dispersal event probably was from Cuba to North America and not in the other direction, as most Cuban *Eleutherodactylus* have vomerine teeth and it is less likely that this trait would reappear after being lost. A resident North American *Eleutherodactylus* fauna (the subgenus *Craugastor*) may have posed an ecological barrier (e.g. Williams 1969) to dispersal. However, species of the subgenera *Euhyas* and *Syrrhophus* are more terrestrial than those of the subgenus *Craugastor* and thus less likely to compete. Also, the subgenus *Craugastor* may have been restricted to the Chortis Block at that time, in which case the proto-*Syrrhophus* colonist would not have faced a potential competitor.

At about this time (late Eocene), a fault zone developed in the northern Caribbean, extending from southern Mexico and northern Guatemala and through the present day Cayman Trough (Burke et al. 1978; Pindell and Dewey 1982; Sykes et al. 1982). Jamaica and the South Island of Hispaniola since have moved eastward along this fault zone.

Thick limestone sequences lacking terrestrial sediments were deposited throughout Jamaica during the Oligocene indicating that it was completely submerged (Robinson et al. 1970; Horsfield 1973; Comer 1974; Horsfield and Roobol 1974; Arden 1975; Kashfi 1983; Wadge and Dixon 1984). Since limestone does not presently cover the Blue Mountains, presumably having been eroded away, it cannot be proven that all of Jamaica was submerged. However, the purity of the limestone immediately adjacent to the Blue Mountains argues against emergence even in that region (Horsfield and Roobol 1974). Also, there was no orogenic activity (uplift) occurring in Jamaica during most of the Oligocene. Instead, up to 3650 m of subsidence (Kashfi 1983) further suggests that there was no emergent land other than coral atolls, unlikely to support a continuous lineage of *Eleutherodactylus*. The South Island of Hispaniola also apparently was submerged at that time (Bowin 1975; Maurrasse 1982), although the geological evidence is not as strong due to extensive late Cenozoic uplift and erosion of mid-Tertiary limestones.

During the Oligocene, the subgenus *Euhyas* probably was restricted to Cuba where it was evolving in isolation. This is suggested by its present distribution and the fact that Jamaica and the South Island of Hispaniola probably were submerged at the time.

By the early Miocene (20 myBP), Jamaica and the South Island had moved further eastward relative to Cuba, and were now emergent (Fig. 34:top). Dispersal of the subgenus *Euhyas* from Cuba to Jamaica and the South Island of Hispaniola probably occurred at this time, based on albumin immunological distances (Hass and Hedges, unpubl. data). The latter two islands initially were low and flat with a blanket of highly dissected limestone (Comer 1974: fig. 5). Major uplift since the late Miocene (10 myBP) resulted in the Blue mountains in Jamaica and the three South Island ranges: the Massif de la Hotte, Massif de la Selle, and the Sierra de Baoruco (Horsfield 1973; Burke et al. 1980). It was during this geologically active time that most of the speciation on Jamaica and the South Island probably occurred. This can be inferred by the fact that many species from these two islands presently are restricted to upland areas, where they presumably evolved.

Pregill and Olson (1981) suggested that the large radiation of South Island *Eleutherodactylus* mainly was the result of Pleistocene sea level and climatic changes. However, the levels of allozyme divergence among South Island species examined in this study (some with different alleles at most of the six slow-evolving loci) indicate a longer period of evolution. This is also supported by the high degree of morphological and ecological differentiation of the species, with at least 20 sympatric at one site in the Massif de la Hotte (Schwartz 1973; Hedges and Thomas 1987).

After the South Island collided with the North Island in the late Miocene (10 myBP; Sykes et al. 1982), one lineage of the subgenus *Euhyas* (*pictissimus* series) dispersed northward and one lineage of the *auriculatus* section of the subgenus *Eleutherodactylus* (*wetmorei* group) dispersed southward (Fig. 34:bottom). These overland dispersal events probably occurred relatively recently (late Pliocene) since each led to a small radiation of closely related species, mostly allopatric and morphologically similar. The origin of the Cuban *auriculatus* section species is less clear, although their morphological and chromosomal similarity to North Island species suggests a relatively recent (late Miocene or Pliocene) overwater dispersal from the North Island.

Two *auriculatus* section species which are islandwide on Hispaniola, *abbotti* and *audanti*, probably dispersed from the North Island to the South Island in the Quaternary. The spotty distribution of *audanti* on the South Island (absent from many undisturbed areas in the Massif de la Hotte) further suggests it is a recent arrival.

The present distribution of the subgenus *Pelorius* throughout Hispaniola implies that it dispersed from the North to the South Island after the collision. Another possibility is that it dispersed to the South Island soon after the emergence of that island (late Oligocene?), evolved along with the subgenus *Euhyas* during the Miocene, and dispersed (*inoptatus* and *ruthae*) back to the North Island after collision. That would provide a vicariance explanation for the origin of *Pelorius*. There are other possibilities, but whichever is correct, the large body size characteristic of the subgenus *Pelorius* probably has facilitated coexistence with the other two subgenera (*Eleutherodactylus* and *Euhyas*) on Hispaniola.

The preceeding scenario differs substantially from previous explanations for North vs

South Island faunal distribution patterns (Mertens 1939; Williams 1961, 1965; Schwartz 1978, 1980; Pregill and Olson 1981). Those authors suggested that sea level and climatic changes, especially those that occurred during the Pleistocene, were responsible for the isolation of populations on either side of the Cul de Sac/Valle de Neiba trough. While such a mechanism may explain North vs South Island species pairs (e.g. the *Anolis chlorocyanus* species group; Williams 1965), the more trenchant differences between the North and South Island *Eleutherodactylus* faunas revealed by this study are better explained by the tectonic history of the two islands outlined above.

Island radiations and convergence

Broadly defined, almost any monophyletic group of organisms can be referred to as an evolutionary radiation. When most or all of the species of a taxon inhabiting an island form a monophyletic group, it can be referred to as an island radiation (Williams 1983). An adaptive radiation (Osborn 1902) is an evolutionary radiation believed to be the result of the filling of newly available ecological niches through adaptation (Romer 1966).

Raup (1984) considers the term adaptive radiation to be a tautology because "a group that suddenly increases in diversity does so for reasons of adaptive success," and therefore he prefers the term evolutionary radiation. I maintain the distinction here since I believe that cladogenesis and diversity can increase without "adaptive success". This could occur when a widespread species is fragmented into geographical isolates, each resulting in a new species (vicariance) but with similar ecologies.

An example of such a nonadaptive radiation is the *pictissimus* group of *Eleutherodactylus* (subgenus *Euhyas*). It includes eight allopatric Hispaniolan species similar in habitus and ecology. In this case, cladogenesis probably was caused by Pleistocene climatic changes (Pregill and Olson 1981) resulting in range fragmentation of a widespread ancestral species. Over time, a nonadaptive radiation may lead to an adaptive radiation if species become sympatric and develop morphological and ecological differences. Thus, evolutionary radiations (or island radiations) may include both adaptive and nonadaptive radiations.

One of the best known examples of island radiations involves the West Indian anoline lizards of the family Iguanidae. Although the systematics of the Cuban and Hispaniolan species are not well known, island radiations probably occurred on each of the four Greater Antilles (Williams 1976, 1983). This has resulted in numerous cases of morphological and ecological convergence, and formed the basis for the concept of ecomorph (Williams 1972, 1983).

In contrast, the morphological and ecological diversity of West Indian *Eleutherodactylus* are not equally represented among the major lineages (subgenera). The frogs of the subgenus *Euhyas* generally are terrestrial in habits whereas the subgenus *Eleutherodactylus* (*auriculatus* section) mainly is composed of arboreal species. The six species in the subgenus *Pelorius* span a variety of ecological types, although burrowing species are unique to this group. Thus it appears that the ecological habits of the ancestors of these groups have been largely maintained in the descendants, which may explain why the correlated morphological traits (small digital pads and smooth venter in terrestrial species; large digital pads and areolate or rugose venter in arboreal species) generally have proven to be useful diagnostic characters.

Nonetheless, island radiations have occurred in West Indian *Eleutherodactylus*, but within the phylogenetic and ecological context of the two major subgenera. Thus, the radiations on Cuba, Jamaica, and the South Island of Hispaniola largely involve terrestrial species (subgenus *Euhyas*), whereas arboreal species dominate the radiations on the North Island of Hispaniola, Puerto Rico, and the Lesser Antilles (subgenus *Eleutherodactylus*).

One example of convergence that has resulted from these independent island radiations involves the aquatic or semiaquatic species found on Cuba (*cuneatus, sierramaestrae,* and *turquinensis*), Jamaica (*orcutti*), Hispaniola (*semipalmatus*), and Puerto Rico (*karlschmidti*). Shreve and Williams (1963) erected the *orcutti* group for this assemblage based on the presence of toe webbing, but subsequent authors (Schwartz and Fowler 1973; Schwartz 1967; Crombie 1977) have considered those species (and *barlagnei* of Guadeloupe) to be convergent. Although only *barlagnei* and *orcutti* were examined in this study, the results also indicate convergence in morphology, with toe webbing being an adaptation to an aquatic lifestyle.

In addition to the aquatic ecomorph, at least two other widespread ecomorphs of West Indian *Eleutherodactylus* can be recognized: rock/cave and bromeliad. The rock/cave ecomorph includes species with long limbs, large eyes, and large, truncated (or notched) digital pads. Those species are: *greyi, guanahacabibes, thomasi,* and *zeus* (Cuba); *cavernicola* and *cundalli* (Fig. 15; Jamaica); *counouspeus* (Fig. 28), *glaphycompus,* and sp. nov. P (Hispaniola-South Island); *pituinus* (Hispaniola-North Island), and *cooki* (Fig. 24; Puerto Rico). The bromeliad ecomorph includes dorsoventrally flattened species with an areolate venter (smooth in *heminota*) and with large, rounded or circular digital pads. They are: *jamaicensis* (Fig. 16; Jamaica); *fowleri, heminota* (Fig. 18), and *lamprotes* (Hispaniola-South Island); *auriculatoides* (Hispaniola-North Island); and *gryllus* and *jasperi* (Puerto Rico).

In Hispaniola, the largely independent evolution of the North Island (subgenus *Eleutherodactylus*) and South Island (subgenus *Euhyas*) *Eleutherodactylus* faunas has led to some striking examples of convergence (Fig. 35; Table 3). At least seven ecomorphs can be recognized, six of which involve species from both subgenera and presumably separated for most of the Cenozoic (60 my) based on albumin immunological distance data (Hass and Hedges unpubl. MS).

One example of the remarkable similarity in some of these convergent species pairs involves *jugans, ventrilineatus,* and *parabates*. Schwartz (1964) gave the North Island species its name, *parabates* (meaning "transgressor"), in allusion to its resemblance to South Island *jugans* and *ventrilineatus,* presuming that *parabates* had transgressed the Valle de Neiba from the South. However, allozyme data (Fig. 1) and ecological observations suggest that *parabates* is a North Island *auriculatus* section member, convergent with *jugans* and *ventrilineatus*. The latter two species are terrestrial in habits as are most species in the subgenus *Euhyas*. Despite its very stocky and short-legged appearance (Fig. 35), *parabates* males call from arboreal sites, like nearly all other species in the *auriculatus* section (Hedges and Thomas, unpubl. data).

Very little is known about the ecology and behavior of most West Indian *Eleutherodactylus* and thus the ecomorph categories described here are likely to change in definition and composition. Also, the underlying assumption with the concept of ecomorph, that island radiations are also adaptive radiations, is a difficult hypothesis to test but one which agrees with the nonrandom associations between morphology and ecology.

Acknowledgments

This study would not have been possible without the large number of Hispaniolan *Eleutherodactylus* species (47) collected in six months of field work, largely a result of the assistance, encouragement, and companionship of Richard Thomas. Others who generously provided assistance in the field were: M. Coggiano, D. Hardy, C. A. Hass, R. Highton, M. Londner, C. Mayer, J. Piñero, K. and S. Schindler, and M. and W. Stephenson. Collecting permits and general support were received from P. Fairbairn and A. Haynes (Jamaica), G. Hermatin, E. Magny, P. Paryski, R. Pierre-Louis, and F. Sergile (Haiti), S. and Y. Incháustegui (Dominican Republic), E. Cardona (Puerto Rico) and J. Fifi (Guadeloupe). The *bransfordii* sample was kindly provided by S. Werman, *atkinsi* by Brian Crother, and *fenestratus* was donated by R. Heyer and R. Cocroft. I am particularly grateful to H. Dowling for encouragement, and the opportunity to conduct field work from the Dowling House, Jamaica. BIOSYS-I and PAUP computer programs generously were made available by D. Swofford; G. Pereira assisted with PAUP. The examination of preserved specimens was made possible by R. Crombie, R. Heyer, and G. Zug (Smithsonian Institution); W. Auffenberg, D. Auth, and R. Franz (Florida State Museum); and Albert Schwartz (private collection). Discussions with J. Bogart, R. Crombie, C. A. Hass, R. Highton, J. Lynch, D. Swofford, R. Thomas, D. Townsend, and J. Wright were particularly beneficial. J. Coyne, C. A. Hass, R. Heyer, R. Highton, G. Mayer, C. Mitter, R. McDiarmid, M. Reaka, G. Vermeij, E. Williams, and A. Wilson offered helpful suggestions on the manuscript (or portions thereof). Financial support was provided by C. J. Hass, The University of Maryland Computer Science Center and Graduate School, and the National Science Foundation (grants BSR 83-07115 to R. Highton and BSR 89-0635 to SBH). This paper was submitted to the University of Maryland in partial fulfillment of the requirements for the degree of Doctor of Philosophy.

Literature cited

Adams, E.N. 1972. Consensus techniques and the comparison of taxonomic trees. Systematic Zoology 21:390-397.

Aquadro, C.F., and J.C. Avise. 1982. Evolutionary genetics of birds. VI. A reexamination of protein divergence using varied electrophoretic conditions. Evolution 36:1003-1019.

Arden, D.D. 1975. The geology of Jamaica and the Nicaraguan Rise. Pp. 617-661 in Nairn, A.E.M. and F.G. Stehli (eds.). Ocean Basins and Margins, Vol. 3, Gulf Coast, Mexico, and the Caribbean. Plenum Press, New York.

Armas, L.F. de. 1982. Algunos aspectos zoogeographicos de la escorpionfauna antillana. Poeyana 238:1-17.

Avise, J.C. 1975. Systematic value of electrophoretic data. Systematic Zoology 23:465-481.

_____, and C.F. Aquadro. 1982. A comparative summary of genetic distances in the vertebrates. Evolutionary Biology 15:151-185.

Barbour, T. 1914. A contribution to the zoogeography of the West Indies, with special reference to amphibians and reptiles. Memoirs of the Museum of Comparative Zoology 44:209-359.

Barton, N.H., and G.M. Hewitt. 1983. Hybrid zones as barriers to gene flow. Pp. 341-359 in Oxford, G.S., and D. Rollinson (eds.). Protein Polymorphism: Adaptive and Taxonomic significance. Academic Press, London.

Berlocher, S.H. 1984. Insect molecular systematics. Annual Review of Entomology 29:403-433.

Bogart, J.P. 1970. Los cromosomas de anfibios anuros del genero *Eleutherodactylus*. Actas IV Congresso Latinoamericano de Zoologia 1:65-78.

_____. 1981a. Chromosome studies in *Sminthillus* from Cuba and *Eleutherodactylus* from Cuba and Puerto Rico (Anura: Leptodactylidae). Royal Ontario Museum Life Sciences Contributions (129):1-22.

_____. 1981b. How many times has terrestrial breeding evolved in anuran amphibians? Monitore Zoologico Italiano, N.S. Supplemento 15:29-40.

Bowin, C. 1975. The geology of Hispaniola. Pp. 501-522 in Nairn, A.E.M. and F.G. Stehli (eds.). Ocean Basins and Margins, Vol. 3, Gulf Coast, Mexico, and the Caribbean. Plenum Press, New York.

Briggs, J.C. 1984. Freshwater fishes and biogeography of Central America and the Antilles. Systematic Zoology 33:428-434.

Burke, K., P. J. Fox, and A. M. C. Sengor. 1978. Bouyant ocean floor and the evolution of the Caribbean. Journal of Geophysical Research 83:3949-3954.

Burke, K., J. Grippi, and A.M.C. Sengor. 1980. Neogene structures and the tectonic style of the northern Caribbean plate boundary zone. Journal of Geology 88:375-386.

Buskirk, R. 1985. Zoogeographic patterns and tectonic history of Jamaica and the northern Caribbean. Journal of Biogeography 12:445-461.

Cavalli-Sforza, L.L., and A.W.F. Edwards. 1967. Phylogenetic analysis: Models and estimation procedures. Evolution 21:550-570.

Cherry, L.M., S.M. Case, and A.C. Wilson. 1978. A frog perspective on the morphological difference between humans and chimpanzees. Science 200:209-211.

Comer, J.B. 1974. Genesis of Jamaican bauxite. Economic Geology 69:1251-1264.

Coney, P.J. 1982. Plate tectonic constraints on the biogeography of Middle America and the Caribbean region. Annals of the Missouri Botanical Garden 69:432-443.

Cope, E.D. 1862. Catalogues of the reptiles obtained during the explorations of the Parana, Paraguay, Vermejo and Uruguay rivers, by Capt. Thos. J. Page, U.S.N.; and those procured by Lieut. N. Michler, U.S. Top. Eng., Commander of the expedition conducting the survey of the Atrato River. Proceedings of the Academy of Natural Sciences, Philadelphia 14:346-359.

_____. 1878. New genus of Cystignathidae from Texas. American Naturalist 12:252-253.

Coyne, J.A. 1982. Gel electrophoresis and cryptic protein variation. Isozymes: Current Topics in Biological and Medical Research 6:1-32.

_____, W.F. Eanes, J.A.M. Ramshaw, and R.K. Koehn. 1979. Electrophoretic heterogeneity of α-glycerophosphate dehydrogenase among many species of *Drosophila*. Systematic Zoology 28:164-175.

Crombie, R.I. 1977. A new species of frog of the genus *Eleutherodactylus* from the Cockpit Country of Jamaica. Proceedings of the Biological Society of Washington 90(2):194-204.

———. 1986. Another new forest-dwelling frog (Leptodactylidae: *Eleutherodactylus*) from the Cockpit Country of Jamaica. Transactions of the San Diego Society of Natural History 21:145-153.

Davis, W.B., and J.R. Dixon. 1955. Notes on Mexican toads of the genus *Tomodactylus* with descriptions of two new species. Herpetologica 11:154-160.

Dayoff, M.O. 1978. Survey of new data and computer methods of analysis. Pp. 7-16 in Dayoff, M.O. (ed.). Atlas of Protein Sequence and Structure, vol. 4. National Biomedical Research Foundation, Silver Spring, MD.

DeWeese, J.E. 1976. The Karyotypes of Middle American Frogs of the Genus *Eleutherodactylus* (Anura: Leptodactylidae): Case Study of the Significance of the Karyologic Method. Unpublished Ph.D. dissertation, University of Southern California, Los Angeles, CA. 210p.

Dixon, J.R. 1957. Geographic variation and distribution of the genus *Tomodactylus* in Mexico. Texas Journal of Science 9:379-409.

Donnelly, T.W. 1985. Mesozoic and Cenozoic plate evolution of the Caribbean region. Pp. 89-121 in Stehli, F.G. and S.D. Webb (eds.). The Great American Biotic Interchange. Plenum Press, New York.

Drewry, G.E., and K.L. Jones. 1976. A new ovoviviparous frog, *Eleutherodactylus jasperi* (Anura, Leptodactylidae), from Puerto Rico. Journal of Herpetology 10:161-165.

Dubois, A. 1986. Miscellanea taxinomica batrachologica (I). Alytes 5:7-95.

Duméril, A.M.C., and G. Bibron. 1841. Erpétologie Général ou Histoire naturelle complete des reptiles. Roret, Paris, 792 pp.

Duncan, R.A., and R.B. Hargraves. 1984. Plate tectonic evolution of the Caribbean region in the mantle reference frame. Geological Society of America, Memoirs 162:81-93.

Dunn, E.R. 1926a. Additional frogs from Cuba. Occasional Papers of the Boston Society of Natural History 5:209-215.

———. 1926b. The frogs of Jamaica. Proceedings of the Boston Society of Natural History 38:111-130.

———, and M.T. Dunn. 1940. Generic names proposed in herpetology by E.D. Cope. Copeia 1940:69-76.

Estrada, A.R., J.N. Rodriguez, and L.V. Moreno. 1986. Las ranas del grupo *symingtoni*, genero *Eleutherodactylus* (Anura: Leptodactylidae) de Cuba. Poeyana 329:1-14.

Fitzinger, L.J.F.J. 1843. Systema Reptilium. Vindobonae, Braumuller und Seidel, Vienna, 106 pp.

Flores, G. 1984. Comparative Osteology, Relationships, and Evolution in Jamaican Frogs of the Genus *Eleutherodactylus*. Unpublished B.A. thesis, Harvard University, Cambridge, MA. 122 pp.

Goin, C.J. 1954. Remarks on the evolution of color pattern in the *gossei* group of the frog genus *Eleutherodactylus*. Annals of the Carnegie Museum 33(10):185-195.

———. 1960. Pattern variation in the frog *Eleutherodactylus nubicola* Dunn. Bulletin of the Florida State Museum 5(5):243-258.

Gorman, G.C., and J. Renzi, Jr. 1979. Genetic distance and heterozygosity estimates in electrophoretic studies: Effects of sample size. Copeia 1979:242-249.

Hedges, S.B. 1982. Caribbean biogeography: implications of recent plate tectonic studies. Systematic Zoology 31:518-522.

_____. 1986. An electrophoretic analysis of Holarctic hylid frog evolution. Systematic Zoology 35:1-21.

_____. 1988. A new diminutive frog from Hispaniola (Leptodactylidae, *Eleutherodactylus*). Copeia 1988:636-641.

_____, and R. Thomas. 1987. A new burrowing frog from Hispaniola with comments on the *inoptatus* group of the genus *Eleutherodactylus* (Anura, Leptodactylidae). Herpetologica 43:269-279.

_____, R. Thomas, and R. Franz. 1987. A new species of *Eleutherodactylus* (Anura, Leptodactylidae) from the Massif de La Hotte, Haiti. Copeia 1987:943-949.

Henderson, R.W., and A. Schwartz. 1984. A guide to the identification of amphibians and reptiles of Hispaniola. Milwaukee Public Museum Special Publications, Biology and Geology 4:1-70.

Heyer, W.R. 1975. A preliminary analysis of the intergeneric relationships of the frog family Leptodactylidae. Smithsonian Contributions to Zoology 199:1-55.

Horsfield, W.T. 1973. Late Tertiary and Quaternary crustal movements in Jamaica. Journal of the Geological Society of Jamaica 13:6-13.

_____, and M.J. Roobol. 1974. A tectonic model for the evolution of Jamaica. Journal of the Geological Society of Jamaica 14:31-38.

Hubby, J.L., and R.C. Lewontin. 1966. A molecular approach to the study of genic heterozygosity in natural populations. I. The number of alleles at different loci in *Drosophila pseudoobscura*. Genetics 54:577-594.

Joglar, R.L. 1986. Phylogenetic Relationships of the West Indian Frogs of the Genus *Eleutherodactylus*. Unpublished Ph.D dissertation. University of Kansas, Lawrence, KA, 142 pp.

Kashfi, M.S. 1983. Geology and hydrocarbon prospects of Jamaica. Bulletin of the American Association of Petroleum Geologists 67:2117-2124.

Lanyon, S.M. 1985. Molecular perspective on the higher-level relationships in the Tyrannoidea (Aves). Systematic Zoology 34:404-418.

Lewontin, R.C. 1985. Population genetics. Annual Review of Genetics 19:81-102.

_____, and J.L. Hubby. 1966. A molecular approach to the study of genic heterozygosity in natural populations. II. Amount of variation and degree of heterozygosity in natural populations of *Drosophila pseudoobscura*. Genetics 54:595-609.

Lynch, J.D. 1968. Genera of leptodactylid frogs in Mexico. University of Kansas Publications, Museum of Natural History 17:503-515.

_____. 1970. A taxonomic revision of the leptodactylid frog genus *Syrrhophus* Cope. University of Kansas Publications, Museum of Natural History 20:1-45.

_____. 1971. Evolutionary relationships, osteology, and zoogeography of leptodactyloid frogs. Miscellaneous Publications of the University of Kansas Museum of Natural History 53:1-238.

_____. 1976. The species groups of South American frogs of the genus *Eleutherodactylus* (Leptodactylidae). Occasional Papers of the Museum of Natural History, University of Kansas 61:1-24.

———. 1980. The *Eleutherodactylus* of the Amazonian slopes of the Ecuadorian Andes (Anura: Leptodactylidae). Miscellaneous Publications of the University of Kansas Museum of Natural History 69:1-86.

———. 1981. Leptodactylid frogs of the genus *Eleutherodactylus* in the andes of northern Ecuador and adjacent Colombia. Miscellaneous Publications of the University of Kansas Museum of Natural History 72:1-46.

———. 1985. *Eleutherodactylus* (part). Pp. 265-331 in Frost, D.R. (ed.). Amphibian Species of the World. Allen Press and the Association of Systematics Collections, Lawrence, KS.

———. 1986. The definition of the Middle American clade of *Eleutherodactylus* based on jaw musculature (Amphibia: Leptodactylidae). Herpetologica 42:248-258.

Lynn, W.G. 1940. I. Amphibians. Pp. 2-60 in Lynn, W.G., and C. Grant. The Herpetology of Jamaica. Bulletin of the Institute of Jamaica No. 1.

———, and J.N. Dent. 1943. Notes on Jamaican amphibians. Copeia 1943:234-242.

MacFadden, B. 1980. Rafting mammals or drifting islands?: biogeography of the Greater Antillean insectivores *Nesophontes* and *Solenodon*. Journal of Biogeography 7:11-22.

———. 1981. Comments on Pregill's appraisal of historical biogeography of Caribbean vertebrates: vicariance, dispersal, or both? Systematic Zoology 30:370-372.

Mann, P., and K. Burke. 1984. Neotectonics of the Caribbean. Reviews of Geophysics and Space Physics 22:309-362.

Margush, T., and F.R. McMorris. 1981. Consensus n-trees. Bulletin of Mathematical Biology 43:239-244.

Matthew, N.D. 1918. Affinities and origin of the Antillean mammals. Bulletin of the Geological Society of America 29:657-666.

Maurrassee, F.J. 1982. Survey of the Geology of Haiti. Miami Geological Society, Miami, FL, 103 pp.

Maxson, R.D., and L.R. Maxson. 1986. Micro-complement fixation: A quantitative estimator of protein evolution. Molecular Biology and Evolution 3:375-388.

McLellan, T. 1984. Molecular charge and electrophoretic mobility in Cetacean myoglobins of known sequence. Biochemical Genetics 22:181-200.

Mertens, R. 1939. Herpetologische Ergebnisse einer Reise nach der Insel Hispaniola, Westindien. Abh. Senckenberg Naturf. Ges. 449:1-84.

Nei, M. 1987. Molecular Evolutionary Genetics. Columbia University Press, New York, 512 pp.

———, F. Tajima, and Y. Tateno. 1983. Accuracy of estimated phylogenetic trees from molecular data. Journal of Molecular Evolution 19:153-170.

Nevo, E. 1984. The evolutionary significance of genetic diversity: Ecological, demographic and life history correlates. Pp. 13-213 in Mani, G.S. (ed.). Evolutionary Dynamics of Genetic Diversity. Lecture Notes in Biomathematics, vol. 53.

Nomenclature Committee of the International Union of Biochemistry. 1984. Enzyme Nomenclature 1984. Academic Press, New York. 646 pp.

Osborn, H.F. 1902. The law of adaptive radiation. American Naturalist 36:353-363.

Pindell, J., and J.F. Dewey. 1982. Permo-triassic reconstruction of western Pangaea and the evolution of the Gulf of Mexico/Caribbean region. Tectonics 1:179-211.

Poinar, G.O., Jr., and D.C. Cannatella. 1987. An Upper Eocene frog from the Dominican Republic and its implication for Caribbean biogeography. Science 237:1215-1216.

Prager, E.M., and A.C. Wilson. 1976. Congruency of phylogenies derived from different proteins. A molecular analysis of the phylogenetic position of cracid birds. Journal of Molecular Evolution 9:45-57.

Pregill, G.K. 1981. An appraisal of the vicariance hypothesis of Caribbean biogeography and its application to West Indian terrestrial vertebrates. Systematic Zoology 30:147-155.

_____, and S.L. Olson. 1981. Zoogeography of West Indian vertebrates in relation to Pleistocene climatic cycles. Annual Review of Ecology and Systematics 12:75-98.

Ramshaw, J.A.M., J.A. Coyne, and R.C. Lewontin. 1979. The sensitivity of gel electrophoresis as a detector of genetic variation. Genetics 93:1019-1037.

Raup, D.M. 1984. Evolutionary radiations and extinctions. Pp. 5-14 in Holland, H.D. and A.F. Trendall (eds.). Patterns of Change in Earth Evolution. Springer-Verlag, Berlin.

Rieppel, O. 1980. Green anole in Dominican amber. Nature 286:486-487.

Robinson, E., J.F. Lewis, and R.V. Cant. 1970. Field guide to aspects of the geology of Jamaica. Pp. 3-9 in Donnelly, T.W. (ed.). International Field Institute Guidebook to the Caribbean Island-Arc System. American Geological Institute/National Science Foundation.

Romer, A.S. 1966. Vertebrate Paleontology. University of Chicago Press, Chicago, IL. 468 pp.

Rosen, D.E. 1976. A vicariance model of Caribbean biogeography. Systematic Zoology 24:431-464.

_____. 1978. Vicariant patterns and historical explanation in biogeography. Systematic Zoology 27:159-188.

_____. 1985. Geological hierarchies and biogeographic congruence in the Caribbean. Annals of the Missouri Botanical Garden 72:636-659.

Sarich, V.M. 1977. Rates, sample sizes, and the neutrality hypothesis for electrophoresis in evolutionary studies. Nature 265:24-28.

Savage, J.M. 1982. The enigma of the Central American herpetofauna: Dispersals or vicariance? Annals of the Missouri Botanical Garden 69:464-547.

_____. 1987. Systematics and distribution of the Mexican and Central American rainfrogs of the *Eleutherodactylus gollmeri* group (Amphibia: Leptodactylidae). Fieldiana, Zoology (new series) 33:1-57.

Schwartz, A. 1958. Four new frogs of the genus *Eleutherodactylus* (Leptodactylidae) from Cuba. American Museum of Natural History Novitates 1873:1-20.

_____. 1964. Three new species of frogs (Leptodactylidae, *Eleutherodactylus*) from Hispaniola. Museum of Comparative Zoology, Breviora 208:1-15.

_____. 1965a. Variation and natural history *Eleutherodactylus ruthae* on Hispaniola. Bulletin of the Museum of Comparative Zoology 132:481-508.

_____. 1965b. Geographic variation in two species of Hispaniolan *Eleutherodactylus*, with notes on Cuban members of the *ricordi* group. Studies of the Fauna of Curaçao and Other Caribbean Islands 22(86):98-123.

_____. 1965c. Two new species of *Eleutherodactylus* from the eastern Cordillera Central of the Republica Dominicana. Caribbean Journal of Science 4:473-484.

_____. 1967. Frogs of the genus *Eleutherodactylus* in the Lesser Antilles. Studies of the Fauna of Curaçao and Other Caribbean Islands 91:1-62.

_____. 1969. The Antillean *Eleutherodactylus* of the *auriculatus* group. Studies of the Fauna of Curaçao and Other Caribbean Islands 30:99-115.

_____. 1971. The subspecies of *Eleutherodactylus schmidti* Noble (Anura: Leptodactylidae). Caribbean Journal of Science 10(3-4):109-118.

_____. 1973. Six new species of *Eleutherodactylus* (Anura: Leptodactylidae) from Hispaniola. Journal of Herpetology 7:249-273.

_____. 1976. Variation and relationships of some Hispaniolan frogs (Leptodactylidae: *Eleutherodactylus*) of the *ricordi* group. Bulletin of the Florida State Museum 21:1-46.

_____. 1978. Some aspects of the herpetogeography of the West Indies. Pp. 31-51 in Gill, F.B. (ed.). Zoogeography in the Caribbean. Special Publications of the Academy of Natural Sciences of Philadelphia No. 13.

_____. 1980. The herpetogeography of Hispaniola, West Indies. Studies on the Fauna of Curaçao and Other Caribbean Islands 189:86-127.

_____. 1985. *Eleutherodactylus* (part). Pp. 265-331 in Frost, D.R. (ed.). Amphibian Species of the World. Allen Press and the Association of Systematics Collections, Lawrence, KS.

_____, and D. Fowler. 1973. The anura of Jamaica: a status report. Studies of the Fauna of Curaçao and Other Caribbean Islands 43(142):50-142.

_____, and R.W. Henderson. 1985. A Guide to the Identification of the Amphibians and Reptiles of the West Indies Exclusive of Hispaniola. Milwaukee Public Museum, Milwaukee, WI. 165 pp.

_____, and R. Thomas. 1975. A checklist of West Indian amphibians and reptiles. Special Publication of the Carnegie Museum of Natural History 1:1-216.

_____, R. Thomas, and L.D. Ober. 1978. First supplement to a checklist of West Indian amphibians and reptiles. Special Publication of the Carnegie Museum of Natural History 5:1-35.

Selander, R.K., M.H. Smith. S.Y. Yang, W.E. Johnson, and J.B. Gentry. 1971. Biochemical polymorphism and systematics in the genus *Peromyscus*. I. Variation in the old-field mouse (*Peromyscus polionotus*). Studies in Genetics VI. University of Texas Publications 7103:49-90.

Shreve, B., and E.E. Williams. 1963. The herpetology of the Port-au-Prince region and Gonave Island, Haiti. Part II. The frogs. Pp. 302-342 in Bulletin of the Museum of Comparative Zoology (129):293-342.

Simpson, G.G. 1956. Zoogeography of West Indian land mammals. American Museum Novitates 1759:1-28.

Smith, M.H., D. Straney, and G. Drewry. 1981. Biochemical similarities among Puerto Rican *Eleutherodactylus*. Copeia 1981(2):463-466.

Starrett, P.H. 1968. The phylogenetic significance of jaw musculature in anuran amphibians. Unpublished Ph.D dissertation, University of Michigan, Ann Arbor, MI. 179 pp.

Swofford, D.L. 1985. Phylogenetic Analysis Using Parsimony (PAUP), version 2.4 (computer software). Illinois Natural History Survey, Champaign, IL.

_____, and R.B. Selander. 1981. BIOSYS-1: A FORTRAN program for the comprehensive analysis of electrophoretic data in population genetics and systematics. Journal of Heredity 72:281-283.

Sykes, L.R., W.R. McCann, and A.L. Kafka. 1982. Motion of Caribbean plate during last 7 million years and implications for earlier Cenozoic movements. Journal of Geophysics Research 87:10656-10676.

Thorpe, J.P. 1982. The molecular clock hypothesis: Biochemical evolution, genetic differentiation and systematics. Annual Review of Ecology and Systematics 13:139-168.

Tschudi, J.J. 1838. Classification der Batrachier, mit Berucksichtigung der Fossilien Thiere dieser Abtheilung der Reptilien. Neuchatel, 100 pp.

Wadge, G., and T.H. Dixon. 1984. A geological interpretation of SEASAT-SAR imagery of Jamaica. Journal of Geology 92:561-581.

Williams, E.E. 1961. Notes on Hispaniolan herpetology. 3. The evolution and relationships of the *Anolis semilineatus* group. Museum of Comparative Zoology, Breviora 136:1-8.

_____. 1965. The species of Hispaniolan green anoles (Sauria, Iguanidae). Museum of Comparative Zoology, Breviora 227:1-16.

_____. 1969. The ecology of colonization as seen in the zoogeography of anoline lizards on small islands. Quarterly Review of Biology 44:345-389.

_____. 1972. Origin of faunas: evolution of lizard congeners in a complex island fauna - a trial analysis. Evolutionary Biology 6:47-89.

_____. 1976. West Indian anoles: a taxonomic and evolutionary summary. I. Introduction and a species list. Museum of Comparative Zoology, Breviora 440:1-21.

_____. 1983. Ecomorphs, faunas, island size, and diverse end points in islands radiations of *Anolis*. Pp. 326-370 in Huey, R.B., E. R. Pianka, and T.W. Schoener, (eds.). Lizard Ecology. Harvard University Press, Cambridge, MA.

Wilson, A.C., S.S. Carlson, and T.J. White. 1977. Biochemical evolution. Annual Review of Biochemistry 46:573-639.

Zweifel, R.G. 1967. Identity of the frog *Cornufer unicolor* and application of the generic name *Cornufer*. Copeia 1967:117-121.

Table 1. Protein loci and electrophoretic conditions.

Protein[1]	Locus	Enzyme Commission[1] Number	Electrophoretic Conditions[2]		
			First	Second	Third
1. Acid phosphatase	*Acp*	3.1.3.2	E	-	-
2. Creatine kinase	*Ck*	2.7.3.2	C	D	E
3. Isocitrate dehydrogenase (NADP$^+$)	*Icd-1*	1.1.1.42	A	B	-
4. Lactoylglutathione lyase	*Lgl*	4.4.1.5	F	D	C
5. Phosphoglucomutase	*Pgm*	5.4.2.2	B	C	A
6. Protein 3	*Pt-3*	-	C	D	-

[1] Nomenclature Committee of the International Union of Biochemistry (1984)

[2] (A)Tris-citrate pH 8.0, 130v, 6h; (B)Tris-citrate pH 6.7, 150v, 6h; (C)Poulik, 300v, ca. 7h; (D)Lithium hydroxide, 400v, ca. 8h; (E)Tris-versene-borate, 250v, 6h; (F)Tris-HCl, 250v, 4h.

Table 2. A revised classification of West Indian *Eleutherodactylus* (summary).

Genus *Eleutherodactylus* (ca. 450 sp.; 128 sp. in West Indies) - South America, Central America, North America, West Indies

    Subgenus *Euhyas* (78 sp.) - Bahamas, Cuba, Jamaica, Hispaniola, Mona Island

        luteolus series (17 sp.) - Jamaica
            luteolus group (3 sp.) - Jamaica
            gossei group (5 sp.) - Jamaica
            cundalli group (3 sp.) - Jamaica
            jamaicensis group (1 sp.) - Jamaica
            nubicola group (5 sp.) - Jamaica
        bakeri series (9 sp.) - Hispaniola (South Island)
        pictissimus series (12 sp.) - Hispaniola
            rufifemoralis group (2 sp.) - Hispaniola (South Island)
            schmidti group (1 sp.) - Hispaniola (North Island)
            pictissimus group (8 sp.) - Hispaniola
            unassigned species (1) - Hispaniola (South Island)
        unassigned groups:
            emiliae group (3 sp.) - Cuba
            symingtoni group (3 sp.) - Cuba
            varleyi group (2 sp.) - Cuba
        unassigned species (32) - Bahamas, Cuba, Hispaniola

    Subgenus *Eleutherodactylus* (ca. 275 sp.) South America, Central America, West Indies

        auriculatus section (44 sp.) - Cuba, Hispaniola, Puerto Rican Bank, Lesser Antilles, NE South America
        martinicensis series (22 sp.) - Puerto Rican Bank, Lesser Antilles, NE South America
            martinicensis group (4 sp.) - Lesser Antilles
            unassigned species (18) - Puerto Rican Bank, Lesser Antilles, NE South America
        montanus series (15 sp.) - Hispaniola
            wetmorei group (3 sp.) - Hispaniola
            unassigned species (12) - Hispaniola
        unassigned species (7) - Cuba, Hispaniola (South Island)

    Subgenus *Pelorius* (6 sp.) - Hispaniola
        inoptatus group (3 sp.) - Hispaniola
        ruthae group (3 sp.) - Hispaniola

Table 3. Convergence in Hispaniolan *Eleutherodactylus*. North and South Island refer to paleoislands presently separated by the arid Cul de Sac (Haiti) and Valle de Neiba (Dominican Republic). Species are listed in order of increasing body size (maximum female snout-vent length in mm), indicated in parentheses.

ECOMORPH	SOUTH ISLAND[1]	NORTH ISLAND[2]	MORPHOLOGY	ECOLOGY
Small Terrestrial Montane	*thorectes* (17)	sp. nov. N (15) *haitianus* (17)	short limbs; small digital pads; rugose dorsum	high elevation; ground and leaf litter
Small Arboreal Montane	*amadeus* (25)	*minutus* (19) *audanti*[3] (25)	moderately developed digital pads	high elevation; low vegetation
Intermediate Terrestrial Montane	*ventrilineatus* (25) *jugans* (33)	*parabates* (24)	short snout; stocky habitus; small digital pads	moderate to high elevation; ground and low vegetation
Rock/cave	*glaphycompus* (29) sp. nov. P (33)	*pituinus* (29)	long limbs; large, notched digital pads; large eyes	rock- and cave-dwelling
Bromeliad	*lamprotes* (28) *heminota* (30) *fowleri* (33)	*auriculatoides* (33)	dorsoventrally flattened; large, rounded digital pads	bromeliad-dwelling

Table 3. Convergence in Hispaniolan *Eleutherodactylus* (concluded).

ECOMORPH	SOUTH ISLAND[1]	NORTH ISLAND[2]	MORPHOLOGY	ECOLOGY
Large Arboreal Montane	*bakeri* (37) *armstrongi* (43)	*patricae* (35) *montanus* (45)	large digital pads; little or no sexual dimorphism in size[4]	arboreal generalists; whistle call
Large Terrestrial Montane	*apostates* (44) *oxyrhyncus* (55)	*schmidti* (58)	robust habitus; large limbs; small digital pads	terrestrial; ravine and stream associated

[1] subgenus *Euhyas*, except *fowleri* and *lamprotes* (subgenus *Eleutherodactylus*).
[2] subgenus *Eleutherodactylus*, except *schmidti* (subgenus *Euhyas*).
[3] also occurs on the South Island.
[4] except *armstrongi*.

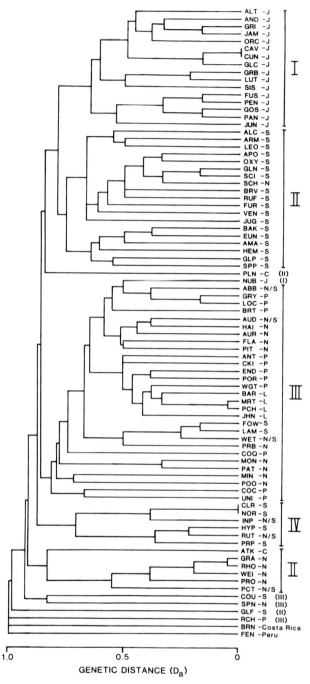

Figure 1.-- Phylogenetic tree of 84 *Eleutherodactylus* species constructed by UPGMA clustering of modified Cavalli-Sforza distances. Prager and Wilson's F value = 4.73. Species abbreviations defined in Appendix 1, island abbreviations are: C, Cuba; N, Hispaniola - North Island; S, Hispaniola - South Island; J, Jamaica; L, Lesser Antilles; P, Puerto Rico.

Evolution and biogeography of *Eleutherodactylus*

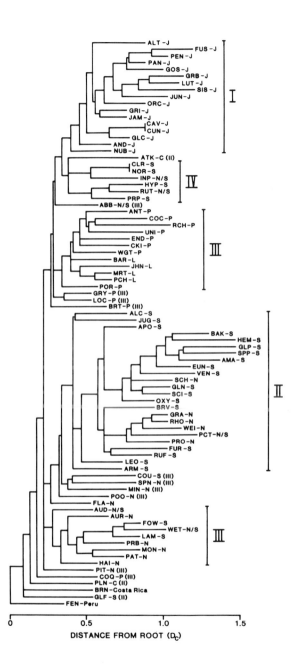

Figure 2.-- Phylogenetic tree of 84 *Eleutherodactylus* species constructed by the distance Wagner method using Cavalli-Sforza and Edwards (1967) chord distance and rooted with *fenestratus* (FEN). Prager and Wilson's F value = 44.1 (5.77 with branch-length optimization).

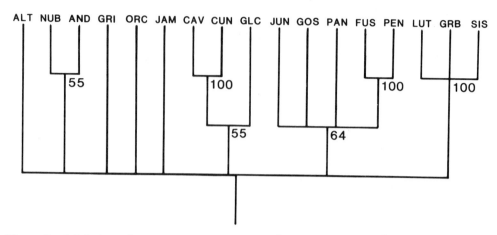

Figure 3.-- Majority-rule consensus tree representing 1100 most-parsimonious trees (each of length = 29, CI = 1.00) of Group I (subgenus *Euhyas*, *luteolus* series; = all 17 native Jamaican species). Numbers refer to percent of trees defining a particular group.

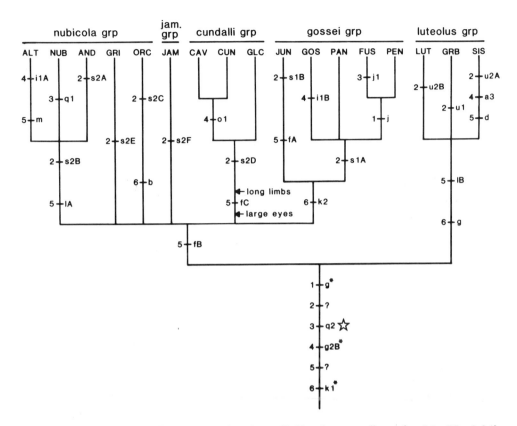

Figure 4.-- Cladogram of Group I showing allelic changes (length=31, CI=0.94). Number at left of tick mark is locus: 1) *Acp*, 2) *Ck*, 3) *Icd-1*, 4) *Lgl*, 5) *Pgm*, and 6) *Pt-3*. The combination of number and letters at right identifies the allele (Appendix 2) possessed by all taxa above the tick mark (unless additional changes at that locus are indicated). The tree is rooted by a composite outgroup (see text). A star indicates a presumed synapomorphy for the group, asterisks are plesiomorphic alleles (shared with outgroup), and loci where the primitive allele cannot be determined are indicated by question marks. Morphological changes are also indicated but were not used in the analysis.

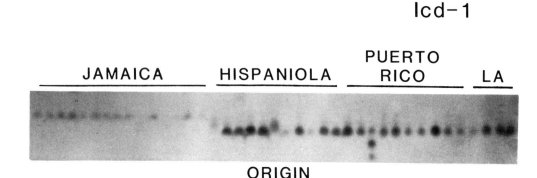

Figure 5.-- Allelic differences at a slow-evolving locus, isocitrate dehydrogenase (*Icd*) illustrating the separation of Jamaican species of the genus *Eleutherodactylus* (allele q2) from species on other islands (allele f1). The species (one per slot) are arranged in the following order (see Appendix 1 for abbreviations): Jamaica - ALT, AND, CAV, GLC, CUN, GOS, GRB, GRI, JAM, JUN (h/q2 heterozygote), LUT, ORC, PAN, PEN, SIS; Hispaniola - ABB (f1/m2 heterozygote), AUD, AUR, FLA, FOW, HAI (f1/m1 heterozygote), LAM, PRB, PAT, PIT, WET; Puerto Rico - ANT, BRT, COC (a/f1 heterozygote), CKI, COQ, END, GRY, LOC, POR, WGT; Lesser Antilles - BAR, JHN, MRT, and PCH. Note that no alleles are shared between the Jamaican species and species from other islands (species possessing other alleles are not shown). In addition to the mobility difference, note the difference in intensity: the Jamaican species are more weakly staining. (Gel BH-1165; Tris-citrate pH 8.0.)

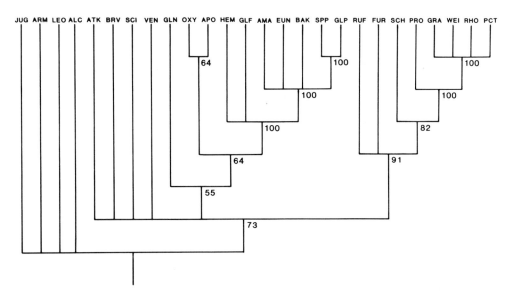

Figure 6.-- Majority-rule consensus tree representing 1100 most-parsimonious trees (each of length = 64, CI = 0.95) of Group II (subgenus *Euhyas*, part; = 25 Hispaniolan and one Cuban species, ATK).

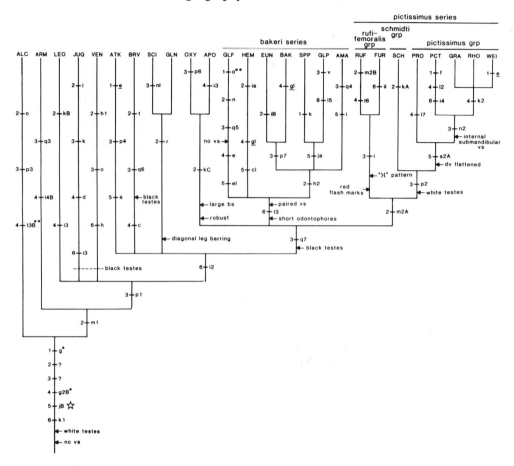

Figure 7.-- Cladogram of Group II showing allelic changes (length = 65, CI = 0.94). Morphological changes are also indicated but were not used in the analysis (BS = body size, DV = dorsoventrally, VS = vocal sac). Underlined alleles are those convergent within the group; two asterisks indicate allelic convergence with species outside of the group. Other symbols as in Fig. 4.

Evolution and biogeography of *Eleutherodactylus* 355

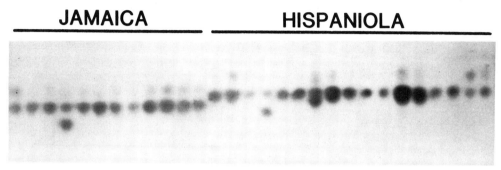

Figure 8.-- Allelic differences at a slow-evolving locus, phosphoglucomutase (*Pgm*), illustrating the separation of Jamaican and Hispaniolan (South Island, except SCH) species of the genus *Eleutherodactylus*. Although only two major allelic classes (f and j) are resolved on this gel, tris-citrate pH 6.7, additional hidden allelic variation within these classes (fA, fB, fC; jA, jB) was detected on tris-citrate pH 8.0. The species are arranged in the following order: Jamaica - ALT (fB/m heterozygote), CAV (fC), CUN (fC), FUS (a1/fB heterozygote), GLC (fC), GOS (fB), GRI (fB), JAM (fB), JUN (fA), ORC (fB), PAN (fB), PEN (fB); Hispaniola - ALC (jB), APO (jB), ARM (jB), BAK (c2/jB heterozygote), BRV (jB), EUN (jB), FUR (g1/jB heterozygote), GLN (jB), GLP (jA), JUG (jB), LEO (jB), OXY (jB), SPP (h/jA heterozygote), RUF (jB), SCI (jB), VEN (jB/s2B heterozygote), and SCH (jB). Note that no alleles are shared between Jamaican and Hispaniolan species (species possessing other alleles are not shown). (Gel BH-1164.)

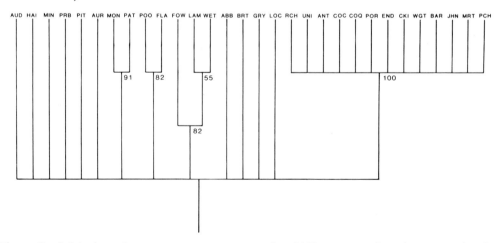

Figure 9.-- Majority-rule consensus tree representing 1100 most-parsimonious trees (each of length=71, CI=0.97) of Group III (subgenus *Eleutherodactylus*, *auriculatus* section).

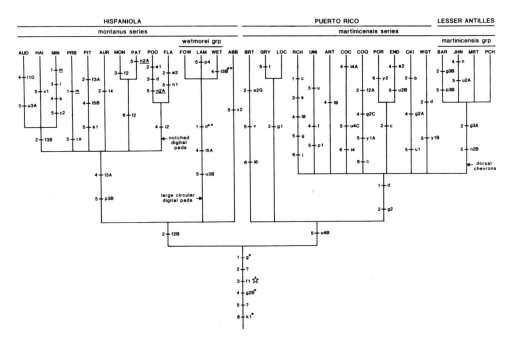

Figure 10.-- Cladogram of Group III showing allelic changes (length=74, CI=0.95). Morphological changes and geographic associations are also indicated but were not used in the analysis. Symbols as in Figures 4 and 6.

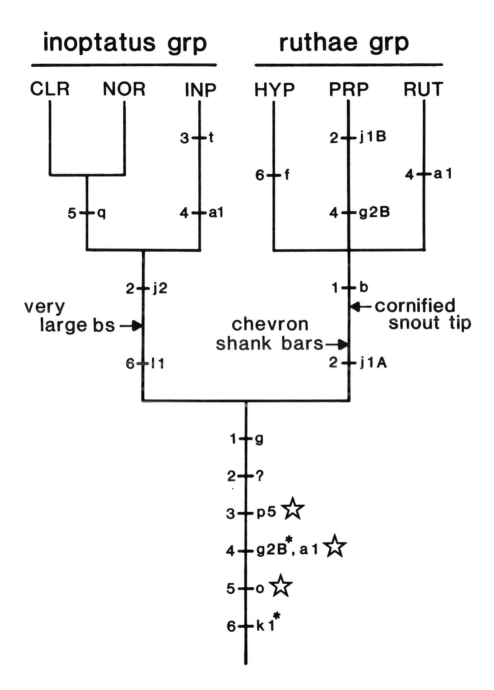

Figure 11.-- A strict consensus tree (length=11, CI=0.91) of the three most-parsimonious trees (length=10, CI=1.00) of Group IV (subgenus *Pelorius*; = six Hispaniolan species) showing allelic changes. Morphological changes are also indicated but were not used in the analysis. Symbols as in Figures 4 and 6.

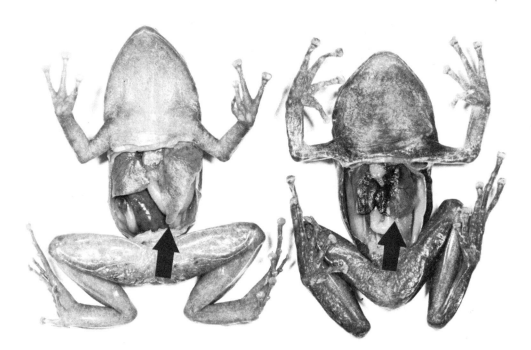

Figure 12.-- *Eleutherodactylus armstrongi* (left) and *Eleutherodactylus montanus* (right) with venters dissected away to show two major differences in liver shape among West Indian *Eleutherodactylus*: long and pointed left lobe (*armstrongi*) vs short and rounded left lobe (*montanus*). Arrows point to left lobe of liver. Although both species have been placed in the same species group based on similarities in large digital pad size, arboreality, and call, allozyme data and liver shape show that each belongs to one of two major West Indian groups and that the external morphological similarities are due to convergence.

Figures 13-18: West Indian frogs of the genus *Eleutherodactylus*. Figure 13 (upper left).-- *Eleutherodactylus* (*Euhyas*) *luteolus* of the *luteolus* group and *luteolus* series (Jamaica: St. James; 2.4 km W Mocho). Figure 14 (upper right).-- *Eleutherodactylus* (*Euhyas*) *gossei* of the *gossei* group and *luteolus* series (Jamaica: St. James; 3.2 km W Mocho). Figure 15 (middle left).-- *Eleutherodactylus* (*Euhyas*) *cundalli* of the *cundalli* group and *luteolus* series (Jamaica: Trelawny; 0.8 km N Burnt Hill). Figure 16 (middle right).-- *Eleutherodactylus* (*Euhyas*) *jamaicensis* (dark phase) of the *jamaicensis* group and *luteolus* series (Jamaica: Trelawny; 0-11 km NNW Quick Step). Figure 17 (lower left).-- *Eleutherodactylus* (*Euhyas*) *nubicola* of the *nubicola* group and *luteolus* series (Jamaica: St. Andrew; 1.3 km W Hardwar Gap). Figure 18 (lower right).-- *Eleutherodactylus* (*Euhyas*) *heminota* of the *bakeri* series (Haiti: Grande Anse; 17.6 km N Camp Perrin).

Figures 19-24: West Indian frogs of the genus *Eleutherodactylus*. Figure 19 (upper left).-- *Eleutherodactylus (Euhyas) rufifemoralis* of the *rufifemoralis* group and *pictissimus* series (Dominican Republic: Barahona; 15 km SSW La Guazara). Figure 20 (upper right).-- *Eleutherodactylus (Euhyas) schmidti* of the *schmidti* group and *pictissimus* series (Dominican Republic: Elias Piña; Loma Nalga de Maco). Figure 21 (middle left).-- *Eleutherodactylus (Euhyas) weinlandi* of the *pictissimus* group and *pictissimus* series (Dominican Republic: El Seibo; 22 km WNW El Valle). Figure 22 (middle right).-- *Eleutherodactylus (Euhyas) brevirostris* (Haiti: Grande Anse; 11.2 km S, 1.9 km E Marché Léon). Figure 23 (lower left).-- *Eleutherodactylus (Euhyas) oxyrhyncus* (Haiti: Grande Anse; 9.0-9.7 km S Marché Léon). Figure 24 (lower right).-- *Eleutherodactylus (Eleutherodactylus) cooki* of the *martinicensis* series and *auriculatus* section (Puerto Rico: 2.3 km SW Yabucoa).

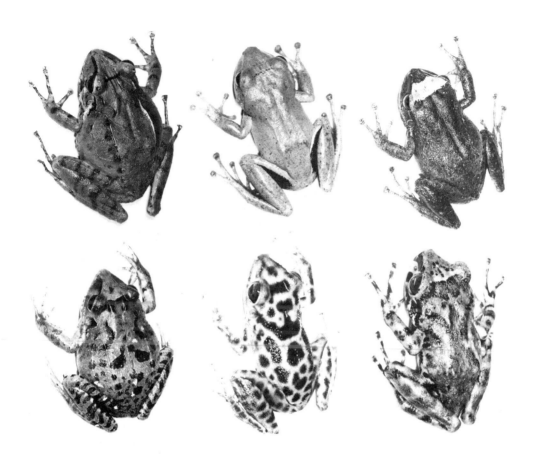

Figures 25-30: West Indian frogs of the genus *Eleutherodactylus*. Figure 25 (upper left).-- *Eleutherodactylus (Eleutherodactylus) coqui* of the *martinicensis* series and *auriculatus* section (Puerto Rico: El Yunque peak). Figure 26 (upper right).-- *Eleutherodactylus (Eleutherodactylus) montanus* of the *montanus* series and *auriculatus* section (Dominican Republic: La Vega; 13 km NW La Horma). Figure 27 (middle left).-- *Eleutherodactylus (Eleutherodactylus) wetmorei* of the *wetmorei* group and *auriculatus* section (Haiti: Grande Anse; 9.0-9.7 km S Marché Léon). Figure 28 (middle right).-- *Eleutherodactylus (Eleutherodactylus) counouspeus* (Haiti: Sud; 13.5 km N Camp Perrin). Figure 29 (lower left).-- *Eleutherodactylus (Pelorius) chlorophenax* of the *inoptatus* group (Haiti: Sud; Plain Formon). Figure 30 (lower right).-- *Eleutherodactylus (Pelorius) ruthae* of the *ruthae* group (Haiti: Sud; ca. 5-6 km NW Les Platons).

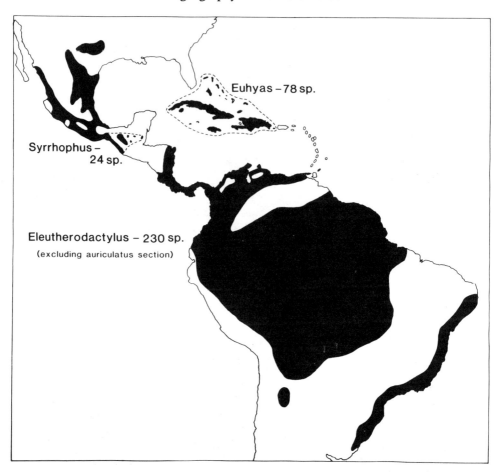

Figure 31.-- Map showing the distribution of the subgenera *Syrrhophus*, *Euhyas*, and *Eleutherodactylus* (excluding *auriculatus* section) based on Lynch (1970, 1976), Schwartz and Thomas (1975), Henderson and Schwartz (1984), and Schwartz and Henderson (1985).

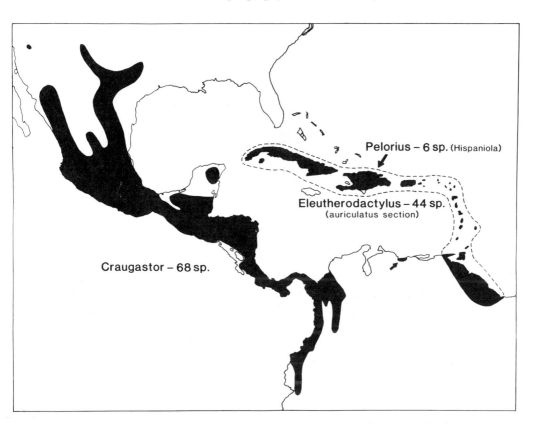

Figure 32.-- Map showing the distribution of the subgenera *Craugastor*, *Eleutherodactylus* (*auriculatus* section), and *Pelorius* based on Lynch (1986), Schwartz and Thomas (1975), Schwartz et al. (1978), Henderson and Schwartz (1984), and Schwartz and Henderson (1985).

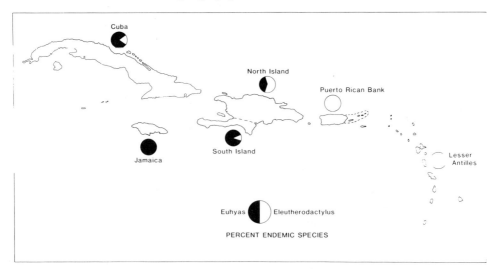

Figure 33.-- Map showing the percentage of species of the subgenera *Eleutherodactylus* and *Euhyas* endemic to each island or island group in the West Indies. The number of species (*Euhyas/Eleutherodactylus*) are: 27/6 (Cuba); 17/0 (Jamaica); 25/3 (South Island); 7/10 (North Island); 0/17 (Puerto Rican Bank); and 0/5 (Lesser Antilles). Four islandwide Hispaniolan species (*abbotti, audanti, pictissimus,* and *wetmorei*) and one species from Mona Island (*monensis*) are not included.

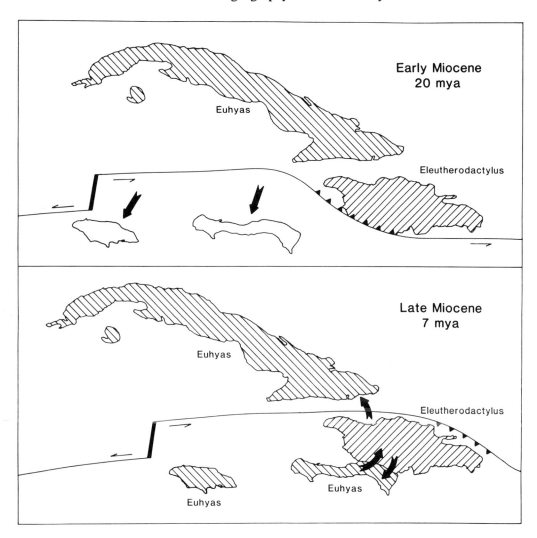

Figure 34.-- Hypothesized biogeographic history of *Eleutherodactylus* (subgenera *Euhyas* and *Eleutherodactylus*) in the Western Caribbean during the Miocene. Paleoreconstructions are based on Sykes et al. (1982). Spreading zone is indicated by wide line, narrow lines are plate boundaries, small arrows show relative direction of plate movement, large arrows indicate dispersal, and subduction is indicated by tooth marks on overriding plate. Early Miocene - Jamaica and the South Island arise above water and are colonized by Cuban frogs of the subgenus *Euhyas* while the North Island is occupied by the subgenus *Eleutherodactylus*. Late Miocene - The South Island has collided with the North Island, halting subduction, and a new fault zone forms to the North of Hispaniola. Overland dispersal of one South Island lineage (subgenus *Euhyas*, *pictissimus* series) to the North Island and one North Island lineage (subgenus *Eleutherodactylus*, *wetmorei* group) to the South Island occurs. Overwater dispersal of North Island *Eleutherodactylus* to Cuba may have occurred at this time.

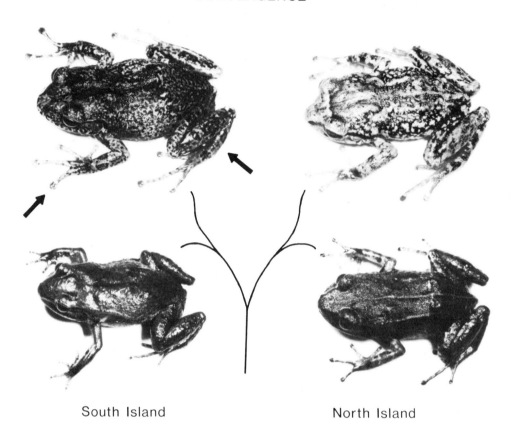

Figure 35.-- Morphological convergence in Hispaniolan *Eleutherodactylus*. Shown are two examples of apparent convergence between North and South Island species. TOP: species of the <u>large</u> <u>arboreal</u> <u>montane</u> ecomorph (*bakeri*, South Island; *patricae*, North Island) show convergence in digital pad size (large) and relative leg length (long). BOTTOM: species of the <u>intermediate</u> <u>terrestrial</u> <u>montane</u> ecomorph (*jugans*, South Island; *parabates*, North Island) have small digital pads and relatively short legs.

APPENDIX 1
Localities and Voucher Specimens

CUBA.-- *atkinsi* (ATK) - Guantanamo Bay Naval Station (University of Southern California 7516); *planirostris* (PLN) - Jamaica (introduced), St. Mary, 2.9 km NW Port Maria (tissue voucher only, but from same population as USNM 266461-465).

JAMAICA.-- *alticola* (ALT) - St. Thomas, Blue Mountain Peak, 1980-2256 m (USNM 269234); *andrewsi* (AND) - St. Andrew, 1.3 km W Hardwar Gap, ca. 1200 m (USNM 269235); *cavernicola* (CAV) - Clarendon, ca. 1.6 km ESE Jacksons Bay, 15 m (USNM 266359); *cundalli* (CUN) - Trelawny, ca. 11 km NNW Quick Step, ca. 450 m (USNM 266362); *fuscus* (FUS) - St. James, 3.2 km W Mocho (USNM 266381); *glaucoreius* (GLC) - St. Andrew, 1.3 km W Hardwar Gap, ca. 1200 m (tissue voucher only, but from same population as USNM 266374-375); *gossei* (GOS) - St. James, 3.2 km W Mocho (USNM 269236); *grabhami* (GRB) - Trelawny, ca. 11 km NNW Quick Step, ca. 450 m (USNM 269237); *griphus* (GRI) - Trelawny, ca. 11 km NNW Quick Step, ca. 450 m (USNM 269238); *jamaicensis* (JAM) - St. Andrew, 2.4 km NW Hardwar Gap, ca. 1200 m (tissue voucher only); *junori* (JUN) - Trelawny, 7.7 km WNW Troy, 625 m (USNM 269239); *luteolus* (LUT) - St. James, 2.4 km W Mocho, 640 m (USNM 269246); *nubicola* (NUB) - St. Andrew, vicinity of Hardwar Gap, 1220 m (USNM 269247); *orcutti* (ORC) - Portland, 0.8 km S Section (USNM 269248); *pantoni* (PAN) - Trelawny, 10.1 km NW Troy (USNM 269254); *pentasyringos* (PEN) - Portland, 2.3 km S Fellowship (USNM 266460); *sisyphodemus* (SIS) - Trelawny, ca. 11 km NNW Quick Step, ca. 450 m (USNM 266468).

HISPANIOLA.-- *abbotti* (ABB) - Dom. Rep., El Seibo, ca. 10 km W Sabana de la Mar (airline), 5 m (USNM 269255); *alcoae* (ALC) - Dom. Rep., Barahona, Los Patos, 0 m (USNM 269256); *amadeus* (AMA) - Haiti, Sud, N slope of Morne Formon, 1700 m (USNM 258691); *apostates* (APO) - Haiti, Grande Anse, 11.7 km S, 1.7 km E Marché Léon (airline), 1480 m (USNM 269257); *armstrongi* (ARM) - Dom. Rep., Barahona, 15.3 km S, 3.2 km E Cabral (by road), 1220 m (USNM 269258); *audanti* (AUD) - Haiti, Ouest, 5 km S Furcy, 1520 m (USNM 269259); *auriculatoides* (AUR) - Dom. Rep., La Vega, 19 km E El Río, 1140 m (USNM 269260); *bakeri* (BAK) - Haiti, Sud, crest of Formon Ridge, 1840-1880 m (USNM 269261); *brevirostris* (BRV) - Haiti, Sud, 2.6 km N, 15.1 km W Camp Perrin (airline), 1650 m (USNM 269262); *chlorophenax* (CLR) - Haiti, Sud, vic. of Plain Formon, 1000 m (USNM 257729); *counouspeus* (COU) - Haiti, Sud, 13.5 km N Camp Perrin, 750 m (USNM 269263); *eunaster* (EUN) - Haiti, Grande Anse, 9.0-9.7 km due S of Marché Léon (airline), 1030-1090 m (USNM 269264); *flavescens* (FLA) - Dom. Rep., Altagracia, ca. 2 km N Boca de Yuma (on new road), (USNM 269265); *fowleri* (FOW) - Dom. Rep., Pedernales, Los Arroyos, 1180 m (USNM 269266); *furcyensis* (FUR) - Haiti, Sud Est, 8.4 km SW Seguin, 1040 m, (USNM 269267); *glandulifer* (GLN) - Haiti, Sud, 2.6 km N, 15.1 km W Camp Perrin (airline), 1650 m (USNM 269268); *glanduliferoides* (GLF) - Haiti, Ouest, 5 km S Furcy, 1520 m (USNM 269269); *glaphycompus* (GLP) - Haiti, Grand Anse, 9.0-9.7 km due S of Marché Léon (airline), 1030-1090 m (USNM 269270); *grahami* (GRA) - Haiti, Artibonite, 10.4 km NW Ça Soleil, 130 m (USNM 269272); *haitianus* (HAI) - Dom. Rep., La Vega, ca. 37 km SE Constanza (via new road), 2300 m (USNM 269273); *heminota* (HEM) - Haiti, Ouest, Furcy (USNM 269274); *hypostenor* (HYP) - Dom. Rep., Barahona, 19.5 km SW Barahona, 880 m (USNM 257734); *inoptatus* (INP) - Dom. Rep., Pedernales, Los Arroyos, 1180 m (USNM 269275); *jugans* (JUG) - Haiti, Sud'Est, 8 km NW Seguin, 1850 m (USNM 269276, 269279); *lamprotes* (LAM) - Haiti, Grande Anse, 9.0-9.7 km due S of Marché Léon (airline), 1030-1090 m (USNM 269280); *leoncei* (LEO) - Haiti, Sud'Est, 8 km NW Seguin, 1850 m (USNM 269281); *minutus* (MIN) - Dom. Rep., La Vega, 14.2 km SE Constanza (via new road), 2000 meters (USNM 269282); *montanus* (MON) - Dom. Rep., La Vega, 18 km SE Constanza, 1770 m (USNM 269283); sp. nov. N (SPN) - Dom. Rep., Independencia, 7 km N Cacique Enriquillo, 1640 m (USNM 269316); *nortoni* (NOR) - Haiti, Sud, vic. of Plain Formon, 1000 m (USNM 257744); *oxyrhyncus* (OXY) - Haiti, Grande Anse, 9.5 km S, 0.6 km W Marché Léon (airline), 1030 m (USNM 269284); sp. nov. P (SPP) - Haiti, Sud, vicinity of Plain Formon, 1000 m (USNM 269271); *parabates* (PRB) - Dom. Rep., Elias Piña, 13 km N Cacique Enriquillo, 1870 m (USNM 269285); *parapelates* (PRP) - Haiti, Grande Anse, 7.8 km S, 0.3 km E Marché Léon (airline), 960 m (USNM 257726); *patricae* (PAT) - Dom. Rep., La Vega, ca. 37 km SE Constanza, 2300 m (USNM 269286); *pictissimus* (PCT) - Dom. Rep., Barahona, Los Patos, 0 m (tissue voucher only, but from same population as USNM 266310-314); *pituinus* (PIT) - Dom. Rep., Peravia, 10.5 km NW La Horma, 1645 m (USNM 269287); *poolei* (POO) - Haiti, Nord, Citadel of King Christophe, 600 m (USNM 269288); *probolaeus* (PRO) - Dom. Rep., Altagracia, 2 km N Boca de Yuma (on old road) (USNM 269290); *rhodesi* (RHO) - Haiti, Nord Ouest, 1.0 km N Balladé, (USNM 269291); *rufifemoralis* (RUF) - Dom. Rep., Barahona, ca. 15 km SSW La Guazara, 1036-1219 m, (USNM 269292); *ruthae* (RUT) - Haiti, Sud, ca. 5-6 km NW Les Platons, ca. 900 m (USNM 257751); *schmidti* (SCH) - Dom. Rep., Elias Piña, 3.2 km S, 4.0 km E Rio Limpio (CREAR), ca. 1270 m (USNM 269293); *sciagraphus* (SCI) - Haiti, Grande Anse, 10.7 km S, 1.6 km E Marché Léon (airline), 1270 m (USNM 269294); *ventrilineatus* (VEN) - Haiti, Sud, crest of Formon Ridge, 1840-1880 m (USNM 269295); *weinlandi* (WEI) - Dom. Rep., El Seibo, 22 km WNW El Valle, 76 m (USNM 269296); *wetmorei* (WET) - Haiti, Grande Anse, 9.0-9.7 km due S of Marché Léon (airline), 1030-1090 m (USNM 269297).

PUERTO RICO.-- *antillensis* (ANT) - 2.2 km S Palmer on route 191, (USNM 269298); *brittoni* (BRT) - 4.2 km E Catalina in Luquillo National Forest, (USNM 269299); *cochranae* (COC) - 2.2 km S Palmer on route 191, (USNM 269300); *cooki* (CKI) - 2.3 km SW Yabucoa (USNM 269301); *coqui* (COQ) - El Yunque (near peak), 1000 m (USNM 269302); *eneidae* (END) - El Yunque (near peak), 1000 m (USNM 269303); *gryllus* (GRY) - El Yunque (near peak), 1000 m (USNM 269304); *locustus* (LOC) - El Yunque (near peak), 1000 m (USNM 269305); *portoricensis* (POR) - El Yunque (near peak), 1000 m (USNM 269306); *richmondi* (RCH) - 3.2 km S Campamento Guavate, (USNM 269307); *unicolor* (UNI) - El Yunque (near peak), 1040 m (USNM 269308); *wightmanae* (WGT) - 1.3 km S, 1.1 km E of El Yunque peak, (USNM 269309).

LESSER ANTILLES.-- *barlagnei* (BAR) - Guadeloupe, Basse Terre, 4 km E Marigot, 120 m (USNM 269310); *martinicensis* (MRT) - Guadeloupe, Basse Terre, 5 km W St. Sauveur, 390 m (USNM 269313-314); *pinchoni* (PCH) - Guadeloupe, Basse Terre, 5 km W St. Sauveur, 390 m (USNM 269318); *johnstonei* (JHN) - Jamaica (introduced), Trelawny, 8.0-8.9 km NW Troy, 610-640 m (USNM 269312), and Guadeloupe, Basse Terre, Pointe de la Grande Anse (near Trois Rivieres), 5 m (USNM 269311).

COSTA RICA.-- *bransfordii* (BRN) - Heredia, Finca La Selva, 60 m (USNM 266332).

PERU.-- *fenestratus* (FEN) - Madre de Dios, Tambopata, near Puerto Maldonado (USNM 268941).

APPENDIX 2
Allelic Variation

Allelic variation is presented for 84 *Eleutherodactylus* species at six slow-evolving protein loci: *Acp*, *Ck*, *Icd-1*, *Lgl*, *Pgm*, and *Pt-3* (alleles listed in that order). In all cases, sample size is one and heterozygotes are indicated by parentheses. Alleles are designated by the conditions under which they were detected (Table 1): lower-case letters, numbers, and upper-case letters refer to first, second, and third conditions, respectively. The total number of alleles at each locus and each condition (cumulative) are: *Acp* (15), *Ck* (21,33,49), *Icd-1* (24,41), *Lgl* (15,33,41), *Pgm* (26,41,54) and *Pt-3* (12,23); total=223 alleles.

CUBA.-- *atkinsi* - e,m1,p4,g2B,a2,i2; *planirostris* - g,p,j2,o2,e2,i5.

JAMAICA.-- *alticola* - g,s2B,q2,i1A,(fB,m),k1; *andrewsi* - g,s2A,q2,g2B,1A,k1; *cavernicola* - g,s2D,q2,o1,fC,k1; *cundalli* - g,s2D,q2,o1,fC,k1; *fuscus* - j,s1A,j1,g2B,(a1,fB),k2; *glaucoreius* - g,s2D,q2,g2B,fC,k1; *gossei* - g,s1A,q2,i1B,fB,k2; *grabhami* - g,u1,q2,(g2B,j),lB,g; *griphus* - g,s2E,q2,g2B,fB,k1; *jamaicensis* - g,s2F,q2,g2B,fB,k1; *junori* - g,s1B,(h,q2),g2B,fA,k2; *luteolus* - g,u2B,q2,g2B,lB,g; *nubicola* - g,s2B,q1,g2B,1A,k1; *orcutti* - g,s2C,q2,g2B,fB,b; *pantoni* - g,s1A,q2,g2B,fB,k2; *pentasyringos* - j,s1A,q2,g2B,fB,k2; *sisyphodemus* - g,u2A,q2,(a3,g3),d,g.

HISPANIOLA.-- *abbotti* - g,f2B,(f1,m2),g2B,(x2,z1),k1; *alcoae* - g,o,p3,l3B,jB,(k1,m); *amadeus* - g,h2,q4,g2B,i,l3; *apostates* - g,kC,q7,i3,jB,i2; *armstrongi* - g,m1,q3,l4B,jB,k1; *audanti* - g,f3B,f1,l10,u3A,k1; *auriculatoides* - g,f4,f1,l3A,p3B,k1; *bakeri* - g,h2,p7,g1,(c2,jB),l3; *brevirostris* - g,t,q6,c,jB,i2; *chlorophenax* - g,j2,p5,(a1,g2B),q,l1; *counouspeus* - a,f5,j3,l5C,p2,k1; *eunaster* - g,iB,p7,g2B,jB,l3; *flavescens* - g,e2,f1,(b,i2),n1,k1; *fowleri* - o,f2B,f1,l5A,u3B,k1; *furcyensis* - g,m2A,i,(g2B,k1),(g1,jB),i1; *glandulifer* - g,r,p1,g2B,jB,i2; *glanduliferoides* - o,n,q5,e,e1,l3; *glaphycompus* - g,h2,v,g2B,jA,l5; *grahami* - g,m2A,(n2,w),k2,s2A,i2; *haitianus* - g,f3B,(f1,m1),l3A,x1,k1; *heminota* - g,iA,q7,g1,c1,l3; *hypostenor* - b,j1A,p5,(a1,g2B),o,f; *inoptatus* - g,j2,t,a1,o,l1; *jugans* - g,l,k,d,jB,i3; *lamprotes* - o,f2B,f1,l5A,p4,k1; *leoncei* - g,kB,p1,i3,jB,k1; *minutus* - m,f3B,l,b,z2,k1; *montanus* - g,f2B,f2,l3A,p3B,l2; sp. nov. N - i,f1,(b,e),l1,rB,k1; *nortoni* - g,j2,p5,(a1,g2B),q,l1; *oxyrhyncus* - g,kC,p6,g2B,jB,i2; sp. nov. P - k,h2,q7,g2B,(h,jA),l3; *parabates* - m,f2B,f1,l3A,rA,k1; *parapelates* - b,j1B,p5,g2B,o,k1; *patricae* - g,f2B,f1,l3A,n2A,l2; *pictissimus* - f,m2A,n2,l2,s2A,i4; *pituinus* - g,f3A,f1,l5B,s1,k1; *poolei* - g,e1,(d,r),(b,i2),n2A,k1; *probolaeus* - g,m2A,p2,l7,s2A,i2; *rhodesi* - g,m2A,n2,k2,s2A,i2; *rufifemoralis* - g,m2B,i,l6,jB,i2; *ruthae* - b,j1A,p5,a1,o,k1; *schmidti* - g,kA,p2,g2B,jB,i2; *sciagraphus* - g,r,n1,g2B,jB,i2; *ventrilineatus* - g,h1,o,g2B,(jB,s2B),h; *weinlandi* - e,m2A,n2,k2,s2A,i2; *wetmorei* - o,f2B,f1,l3B,u3B,k1.

PUERTO RICO.-- *antillensis* - d,g2,f1,l9,u4B,k1; *brittoni* - g,s2G,f1,g2B,v,l6; *cochranae* - d,g2,(a,f1),l4A,u4C,l4; *cooki* - d,b,f1,g2A,u1,k1; *coqui* - (d,g),f2A,f1,g2C,y1A,c; *eneidae* - d,c,f1,a2,u2B,k1; *gryllus* - g,g1,f1,g2B,t,k1; *locustus* - g,g1,f1,g2B,u4B,k1; *portoricensis* - (d,g),c,f1,g2B,y2,k1; *richmondi* - c,g2,s,l8,g2,j; *unicolor* - d,g2,u,f,p1,k1; *wightmanae* - d,d,f1,(g2B,i4),y1B,k1.

LESSER ANTILLES.-- *barlagnei* - d,g3B,f1,g2B,p3A,k1; *martinicensis* - d,g3A,f1,g2B,(n2B,u4A),k1; *pinchoni* - d,g3A,f1,g2B,n2B,k1; *johnstonei* - d,g3A,f1,n,u2A,k1.
COSTA RICA.-- *bransfordii* - l,q,j4,h,b,a.
PERU.-- *fenestratus* - (h,n),a,g,m,k,(d,e).

APPENDIX 3
Morphological Variation

Variation in liver shape, testis color, and vocal sac condition (in that order) is presented for 113 species of West Indian frogs of the genus *Eleutherodactylus*. Abbreviations are as follows: liver shape (L=long and pointed left lobe, S=short and rounded left lobe, I=intermediate condition), testis color (U=unpigmented, P=pigmented or polymorphic for pigmented and unpigmented), and vocal sac (I=internal, X=external, A=absent; P=paired, S=single). A question mark ("?") indicates that no data were available.

Cuba.-- *acmonis* - L,?,A; *atkinsi* - L,U,IS; *auriculatus* - S,P,XS; *bartonsmithi* - ?,?,XS; *bresslerae* - ?,?,A; *cuneatus* - L,U,IS; *dimidiatus* - L,U,A; *eileenae* - ?,?,XS; *etheridgei* - ?,U,IS; *greyi* - L,?,A; *klinikowskii* - ?,P,?; *leberi* - ?,?,XS; *limbatus* - L,?,?; *pezopetrus*-?,?,A; *pinarensis* - L,P,A; *planirostris* - L,U,A; *ricordii* - ?,?,A; *ronaldi* - ?,?,XS; *sierramaestrae* - L,P,A; *symingtoni* - ?,?,A; *thomasi* - L,U,IS; *turquinensis* - L,P,A; *varians* - S,?,XS; *varleyi* - L,U,XS; *zugi* - ?,P,A.

Jamaica.-- *alticola* - L,U,A; *andrewsi* - L,U,A; *cavernicola* - L,U,A; *cundalli* - L,U,A; *fuscus* - L,U,A; *glaucoreius* - L,U,A; *gossei* - L,U,A; *grabhami* - L,U,A; *griphus* - L,U,A; *jamaicensis* - L,U,A; *junori* - L,U,A; *luteolus* - L,U,A; *nubicola* - L,U,A; *orcutti* - L,U,A; *pantoni* - L,U,A; *pentasyringos* - L,U,A; *sisyphodemus* - L,U,A.

Hispaniola.-- *abbotti* - L,P,XS; *alcoae* - ?,U,A; *amadeus* - L,P,IP; *apostates* - L,P,A; *armstrongi* - L,U,A; *audanti* - L,P,XS; *auriculatoides* - S,U,XS; *bakeri* - L,P,IP; *brevirostris* - L,P,A; *chlorophenax* - ?,U,IS; *counouspeus* - S,U,XS?; *eunaster* - L,P,XP; *flavescens* - S,U,XS; *fowleri* - ?,U,XS; *furcyensis* - L,P,A; *glandulifer* - L,U,?; *glanduliferoides* - L,P,A; *glaphycompus* - L,P,XP; *grahami* - L,U,IS; *haitianus* - S,P,XS; *heminota* - L,P,IP; *hypostenor* - ?,U,IS; *inoptatus* - S,U,IS; *jugans* - L,P,A; *lamprotes* - ?,U,XS; *leoncei* - L,U,A; *minutus* - L,P,XS; *montanus* - S,U,XS; sp. nov. "N" - ?,P,XS; *nortoni* - S,U,IS; *parabates* - L,P,A; *parapelates* - ?,U,IS; *patricae* - S,U,XS; *paulsoni* - L,?,A; *pictissimus* - L,U,IS; *pituinus* - ?,U,XS; *poolei* - I,P,XS; *probolaeus* - L,U,A; *rhodesi* L,U,IS; *rufifemoralis* - L,P,A; *ruthae* - S,U,IS; *schmidti* - L,U,A; *sciagraphus* - ?,U,A; *semipalmatus* - L,P,A; *thorectes* - L,P,A; *ventrilineatus* - L,P,A; *weinlandi* - L,U,IS; *wetmorei* - S,P,XS.

Mona Island.-- *monensis* - L,U,A.

Puerto Rican Bank.-- *antillensis* - S,P,XS; *brittoni* - S,P,XS; *cochranae* - S,P,XS; *cooki* - I,U,XS; *coqui* - S,U,XS; *eneidae* - S,P,XS; *gryllus* - ?,P,?; *hedricki* - ?,?,XS; *jasperi* - S,P,XS; *karlschmidti* - I,U,XP?; *lentus* - S,U,IS; *locustus* - L,P,XS; *portoricensis* - S,P,XS; *richmondi* - S,U,XS; *schwartzi* - S,U,XS; *unicolor* - ?,U,?; *wightmanae* - I,U,XS.

Lesser Antilles.-- *barlagnei* - S,P,XS; *johnstonei* - S,P,XS; *martinicensis* - S,P,XS; *pinchoni* - I,P,XS; *urichi* - S,U,XS.

PHYLOGENETIC RELATIONSHIPS OF THE WEST INDIAN FROGS OF THE GENUS *ELEUTHERODACTYLUS*: A MORPHOLOGICAL ANALYSIS

Rafael L. Joglar[1]

Introduction

The herpetofauna is one of the most prominent elements of the West Indian biota. In order to interpret the origin and evolution of the Antillian biota it is necessary to understand evolutionary relationships of component species of its herpetofauna. Species of the leptodactylid frog genus *Eleutherodactylus* constitute 25 percent of the Antillean herpetofauna (87% of the West Indian amphibians). An understanding of the phylogenetic relationships of this large group of frogs will contribute to a better understanding of the entire West Indian fauna.

The members of the genus *Eleutherodactylus* are a highly diverse group of leptodactylid frogs characterized by direct development of terrestrial eggs rather than a free-living larval stage. Frost (1985) listed 404 described species of *Eleutherodactylus*. Since then, the total has increased to 432 species with the descriptions of 28 new species (W.E. Duellman pers. comm.). More than one fourth of the total number of species (119) occur in the West Indies. Unlike mainland taxa of *Eleutherodactylus* that have been placed in a reletively few species groups (Lynch 1976), at least 16 species groups of West Indian *Eleutherodactylus* have been recognized by different authors (Dunn 1925, 1926; Goin 1954; Schwartz 1958, 1969, 1976, 1978; Shreve and Williams 1963; Crombie 1977; Joglar 1983). Recognition of these phenetic groups has not contributed to an understanding the phylogenetic relationships of the West Indian *Eleutherodactylus* .

The principal characters used to define species groups in the West Indian *Eleutherodactylus* are those proposed by Dunn (1926): 1) texture of the skin on the venter; 2) length of the vomerine series; 3) presence or absence of vocal sac; and 4) size of the digital discs. The texture (smooth versus granular) of the venter shows enormous intraspecific variation. The length of the vomerine series is a problematic character because the terms "long" and "short" have never been defined. The vocal sac is present or absent in various groups. The size of the digital disc is correlated with habitat, and expanded digits occur in members of different groups.

Schwartz (1978) recognized six "natural" phenetic groups of *Eleutherodactylus* in the West Indies: 1) *E. auriculatus* group (42 species); 2) *E. gossei* group (12 species); 3) *E.*

---

[1] Dr. Joglar is a professor at the University of Puerto Rico, Rio Piedras, PR. 00931

emiliae group (6 species); 4) *E. inoptatus* group (5 species); 5) *E. symingtoni* group (2 species); and 6) *E. ricordii* group (43 species). Furthermore, no evidence has been presented to determine that any of these groups is monophyletic. Also, several authors (e.g. Lynch 1976; Rivero and Solano 1977) have suggested that the *E. auriculatus* group in the West Indies may not be separable from the South American *E. unistrigatus* group.

Some of the taxonomic work related to the West Indian *Eleutherodactylus* has dealt only with the species on a single island (Dunn 1926; Goin 1954; Shreve and Williams 1963; Crombie 1977; Joglar 1983; Flores 1985). By comparing only the species on one island, these authors failed to recognize that the sister taxa may occur on other islands or on the mainland. Moreover, different suits of characters have been used to define species groups of West Indian *Eleutherodactylus* from those in Central and South America. Consequently, meaningful comparisons of West Indian groups with Central and South American groups has not been possible.

If the large assemblage of species of *Eleutherodactylus* in the West Indies is going to contribute to an understanding of the West Indian biota, it is desirable to present testable hyphotheses of relationships of the West Indian and mainland species. Such hypotheses can be based on different suites of characters. I have undertaken an analysis of morphological characters. A comparable analysis of allozymes is being carried out by S. Blair Hedges (University of Maryland).

The purposes of this study are to: 1) present an analysis of morphological characters of the West Indian species of *Eleutherodactylus*; 2) hypothesize the phylogenetic relationships of the species of *Eleutherodactylus* in the West Indies; 3) define natural (monophyletic) groups and identify phenetic clusters of species; 4) examine the relationships of the West Indian species with the mainland species of *Eleutherodactylus*; 5) explore the relationships of the West Indian *Eleutherodactylus* with the genera *Sminthillus*, *Syrrhophus*, and *Tomodactylus*; 6) present a biogeographic scenario based on the hypothesized relationships of West Indian *Eleutherodactylus*.

Materials and methods

In order to investigate the phylogenetic relationships among West Indian *Eleutherodactylus*, I gathered data on 52 characters (21 external, 28 osteological, 1 myological, 2 reproductive) for 179 taxa. The complete data matrix of character-states for each taxon is given in Appendix I. The ingroup includes 118 species of West Indian *Eleutherodactylus*. For outgroup comparison, I examined representatives of the subgenus *Craugastor* (Lynch 1986; 19 species), the mainland *E. unistrigatus* group (13 species), *Leptodactylus* (10 species), *Adenomera* (3 species), *Lithodytes* (1 species), *Cycloramphus* (6 species), *Ischnocnema* (2 species), *Syrrhophus* (5 species), *Tomodactylus* (3 species), and the monotypic genus *Sminthillus*. Osteological data were obtained from 229 dried skeletons and alizarin-or alizarin-alcian--stained skeletons (Dingerkus and Uhler 1977; Cannatella pers. comm.). A total of 1,007 specimens was examined. For a complete list of specimens examined see Joglar 1986. The only two species of West Indian *Eleutherodactylus* that were not examined were *Eleutherodactylus guanahacabibes* Estrada and Novo Rodriguez (1986) from Cuba and *E. parapelates* Hedges and Thomas (1987) from Hispaniola.

Polarity of the transformation series of each of the characters was determined by outgroup comparison. Phylogenetic analysis of the taxa was accomplished by the PAUP

(Phylogenetic Analysis Using Parsimony) software (Swofford 1984). Missing data and character-states that represented logical impossibilities, were coded as "9" in the input data matrix. An example of a character that is a logical impossibility is the length of the odontophores in taxa that have no odontophores (i.e. *E. jasperi*, *Syrrhophus*, and *Tomodactylus*).

Characters

In the following list of characters, state 0 is the primitive state, and sequential integers are used for consecutively derived states. For a complete character description see Joglar 1986.

1. Tympanic annulus- 0 = distinct; 1 = indistinct.
2. Supraocular tubercles- 0 = present; 1 = absent.
3. Head width- 0 = narrow; 1 = wide.
4. Cranial Crests- 0 = absent; 1 = present.
5. Vocal slits- 0 = present; 1 = absent.
6. Fingers I and II- 0 = I > II; 1 = II > I.
7. Finger discs- 0 = absent; 1 = present.
8. Shape of subarticular finger tubercles- 0 = oval; 1 = round.
9. Height of subarticular tubercles of the finger- 0 = raised; 1 = low.
10. Nuptial pads- 0 = absent; 1 = present.
11. Accessory palmar tubercles- 0 = present; 1 = absence.
12. Inner tarsal fold- 0 = present; 1 = absent.
13. Toe discs.; 0 = absent; 1 = present.
14. Shape of subarticular tubercles of the toe- 0 = oval; 1 = round.
15. Height of subarticular tubercles of the toe- 0 = raised; 1 = low.
16. Plantar tubercles- 0 = present; 1 = absent.
17. Toe webbing; 0 = absent; 1 = present.
18. Dorsolateral glandular ridges- 0 = present; 1 = absent.
19. Ventral skin- 0 = smooth to weakly granular; 1 = coarsely granular.
20. Dorsal skin- 0 = smooth to granular; 1 = rugose.
21. *Ricordii* dorsal pattern- 0 = absent; 1 = present.
22. Nasals- 0 = average size; 1 = small; 2 = large.
23. Median separation of nasals- 0 = narrow; 1 = wide; 2 = almost in contact; 3 = in contact.
24. Nasals-frontoparietals- 0 = widely separated; 1 = narrowly separated; 2 = almost in contact.
25. Frontoparietal-otoccipital bones- 0 = not fused; 1 = fused.
26. Parasphenoid-pterygoid- 0 = no overlap; 1 = overlap.
27. Pterygoid-maxilla articulation- 0 = broad; 1 = narrow
28. Length of zygomatic ramus of squamosal- 0 = not greatly elongated; 1 = greatly elongated
29. Robustness of zygomatic ramus of squamosal- 0 = robust; 1 = slender
30. Otic ramus of squamosal- 0 = not reduced; 1 = reduced

31. Length of lateral palatine process of the pars palatina of the premaxilla- 0 = moderately developed and distinct from palatal shelf; 1 = greatly developed and long
32. Orientation of the lateral palatine process of the pars palatina of the premaxilla- 0 = process oriented posterolaterally; 1 = process oriented posteromedially.
33. Premaxillary-maxillary separation- 0 = narrow; 1 = wide.
34. Anterior expansion of palatal shelf of maxilla- 0 = absent; 1 = present.
35. Posterior expansion of palatal shelf of maxilla- 0 = absent; 1 = present.
36. Palatine-maxilla- 0 = in contact or narrowly separated; 1 = widely separated.
37. Vomers- 0 = not reduced in size; 1 = reduced in size.
38. Presence of odontophores- 0 = present; 1 = absent.
39. Width of odontophores (= length of vomerine series, Dunn 1926)- 0 = short; 1 = long.
40. Width of odontophores- (= length of vomerine series, Dunn 1926)- 0 = very short; 1 = short; 2 = long; 3 = very long (This character was used to construct a cladogram for the *E. ricordii* group, but it was never used together with character 39).
41. Vomerine teeth- 0 = present; 1 = absent.
42. Ornamentation of skull- 0 = absent; 1 = present.
43. Anterior process of hyale- 0 = absent; 1 = present.
44. Anterolateral process of hyoid plate- 0 = present; 1 = absent.
45. Sternum- 0 = biforcated posteriorly; 1 = elongated rectangular; 2 = pendulum shaped; 3 = anchor shaped.
46. Terminal phalange of third finger- 0 = knobbed; 1 = T-shaped.
47. Terminal phalange of third toe- 0 = knobbed; 1 = T-shaped.
48. Relative lengths of Fingers I and II (osteological data)- 0 = I > II; 1 = I = II; 2 = I < II.
49. Fusion of presacral vertebrae- 0 = absent; 1 = present.
50. Mandibular ramus of the trigeminal nerve- 0 = "s" condition; 1 = "e" condition.
51. Egg deposition- 0 = eggs laid directly on water; 1 = eggs deposited in foam nest; 2 = eggs laid on moist, terrestrial situation; 3 = ovoviviparity.
52. Larvae- 0 = aquatic larvae; 1 = non-aquatic larvae; 2 = direct development, no larvae.

Phylogenetic analysis

The initial data base consisted of 179 taxa and 52 characters (28 osteological, 21 external, 1 myological, and 2 reproductive). Fifty-three taxa are outgroups from five genera and two species groups of *Eleutherodactylus*. External data were obtained from all taxa. Osteological data were available for only 141 taxa.

A number of trees were generated to determine the capacity of the initial data base to resolve the phylogeny of the West Indian *Eleutherodactylus*. Two major problems were obvious in these preliminary analyses. 1) The number of taxa exceeded the efficiency of the PAUP program, and 2) there were numerous homoplasious characters. Consistency indices (CI) range from 0 to 1, with a value of 1 indicating no homoplasy (convergences, parallelism or reversal). As the amount of homoplasy increases, the consistency index decreases and approaches 0. In order to correct for these two problems the initial data base was modified by deleting homoplasious characters and deleting certain taxa.

Taxa were deleted at the outgroup and ingroup levels. The outgroup taxa were reduced by choosing a plesiomorphic species to be representative of each outgroup. At the ingroup level taxa were deleted in the following two ways: 1) all taxa without osteological data were deleted; and 2) species were collapsed into monophyletic groups or higher taxonomic levels. Following David Swofford's suggestions (pers. comm.), species were collapsed into species groups and/or genera by coding all characters they share in common and coding as missing values (= 9) those characters that show variation within each taxonomic level. This procedure was followed to obtain one taxon from each of the following: *E. inoptatus* group, *E. ricordii* group, *E. auriculatus* group, *Syrrhophus* and *Tomodactylus*. The trees that were generated in the preliminary analysis, showed two monophyletic groups within the West Indies: the *E. ricordii* group and the *E. inoptatus* group. The *E. ricordii* group is characterized by long odontophores, a derived state not found in any other *Eleutherodactylus*. Osteological data show that the members of the *E. inoptatus* have the first finger longer than the second, a condition not shared by any other West Indian *Eleutherodactylus*. The remaining of the West Indian species were not separable from the species of the *E. unistrigatus* and for that reason were collapsed into the *E. auriculatus* group.

The modified data base (taxa = 12, number of chararters = 50) was used to generate a cladogram. The resulting cladogram (68 steps; CI = 0.691) is reprented in Fig. 1.

Following is a discussion of the phylogeny presented in the cladogram in Fig. 1. In Appendix II characters are listed numerically with their consistency ratio, direction of change, and branches along which shifts in character state occur. A summary of character-state changes at each branch (internode) is given in Appendix III.

The phylogeny represented in Fig. 1 is moderately well supported; the consistency index is 0.69. Nodes A-F, and the terminal nodes (= outgroup taxa) Leptodactylinae, *Cycloramphus*, *Hylactophryne*, *Ischnocnema*, and *Craugastor* are well supported by synapomorphies and a few reversals. Two of these outgroups, *Craugastor* and *Hylactophryne* require some discussion. The subgenus *Craugastor* includes 56 species of Central American *Eleutherodactylus* (Lynch 1986). These species share a synapomorphy, the "e" condition of the mandibular ramus of the trigeminal nerve (50). This condition also is present in the three species of *Hylactophryne*. According to Lynch (1986) the genus *Hylactophryne* is a member of a subset of *Eleutherodactylus* and cannot be recognized as a genus if classifications are to reflect logic. His argument for this hypothesis is that the synapomorphy of the "e" condition evolved only once and not repeatedly. I examined 17 species of the subgenus *Craugastor* of *Eleutherodactylus* and two species of *Hylactophryne*. These 19 taxa share the "e" condition but the 17 species of the *Eleutherodactylus* share another synapomorphy not present in *Hylactophryne*. In *Hylactophryne*, the medial separation of nasals is wide, which is the plesiomorphic condition. In the 17 species of *Eleutherodactylus*, the medial separation is narrow in few species, and in most species the nasals are in contact or nearly so. If the phylogeny in Fig. 1 is correct, the "e" condition of the mandibular ramus of the trigeminal nerve evolved twice, and not once, as stated by Lynch (1986).

Node F bears three synapomorphies: large nasals (22), narrow pterygoid-maxilla articulation (27), and presence of anterior expansion of palatal shelf of maxilla (34), and two reversals: rugose dorsal skin (20) and no overlap of parasphenoid-pterygoid (26). The monophyly of the *E. inoptatus* group is supported by two characters: 1) Finger I longer than II; 2) presence of a broad or wide head (more than 40% or 45% of SVL, respectively).

There are two synapomorphies at Node G: 1) fusion of the frontoparietal-otoccipital bones (25); and 2) Finger II is longer than I (48). The *E. auriculatus* group, the *E. unistrigatus* group, the *E. ricordii* group, and the genera *Syrrhophus* and *Tomodactylus* form an unresolved polytomy. The *E. unistrigatus* group is defined in the cladogram by two synapomorphies and one reversal. These three characters are present in the species I used as representative of the group, but they are not present in all of the species in the group. In this cladogram there are no characters supporting the *E. auriculatus* group, but the group could be defined with the same characters used to define the *E. unistrigatus* group. I consider the *E. auriculatus* group to be the West Indian representatives of the mainland *E. unistrigatus* group. I used the *E. unistrigatus* group as one of my close outgroups, and for this purpose, I assumed that the group is monophyletic. This assumption might be incorrect. Lynch (1976) did not provide evidence to support the monophyly of this group.

Despite the fact that *Sminthillus limbatus* is one of the smallest of all frogs and has four apomorphies, it does not deserve generic recognition. *Sminthillus* was named by Barbour and Noble (1920) on the basis of partial fusion of the epicoracoids. Noble (1922, 1931) claimed that the cartilages were united for more than half their length. Griffiths (1959) stated that the epicoracoidal condition of *S. limbatus* does not differ substantially from that of a generalized leptodactyloid. The epicoracoidal condition of *S. limbatus* is not different from that of other West Indian *Eleutherodactylus*. My data, which support previous morphological comparisons (Griffiths 1959; Lynch 1971), in addition to chromosomal information (Bogart 1981) justifies inclusion of *S. limbatus* in the genus *Eleutherodactylus*. I consider *E. limbatus* to be a member of the West Indian subset of the *E. unistrigatus* group.

It has been suggested that the genera *Syrrhophus* and *Tomodactylus* are most closely related to the West Indian *Eleutherodactylus* (Lynch 1971). The dorsal pattern of *Syrrhophus* and *Tomodactylus* is shared by the *E. ricordii* group. If these two genera had long odontophores, I would not hesitate to assign them to the *E. ricordii* group. The possibility of using the length of the odontophores to ally *Syrrhophus* and *Tomodactylus* to the *E. unistrigatus* or the *E. ricordii* group is unfeasible because these two genera lack odontophores (38). According to Lynch (1971), within the West Indies the *E. auriculatus* group (= *E. unistrigatus* group) most closely approaches *Syrrhophus* and *Tomodactylus* in morphology. In the West Indies, *E. jasperi* is the only species that is like *Syrrhophus* and *Tomodactylus* in lacking odontophores. It is feasible that *Syrrhophus* and *Tomodactylus* are closely related to the West Indian subset of the *E. unistrigatus* group or the *E. ricordii* group, but other possibilities should be explored. It is also possible that these two genera are most closely related to the mainland subset of the *E. unistrigatus* group, as hypothesized by Rivero and Solano (1977). The northernmost distribution of the *E. unistrigatus* group approaches the southernmost distribution of *Syrrhophus* and *Tomodactylus* (Guatemala). Actually, *E. ridens*, a member of the *E. unistrigatus* group, occurs as far north as Honduras; this species lacks odontophores.

My analysis provides no characters to separate the genera *Syrrhophus* and *Tomodactylus* from one another. *Syrrhophus* has been distinguished from *Tomodactylus* by the presence of lumbar glands in *Tomodactylus* and by the arrangement of the supernumerary plantar tubercles (Lynch 1968, 1970, 1971). The latter character is homoplasious. Lumbar glands occur in *E. darlingtoni* and other West Indian *Eleutherodactylus*. In a cladogram of *Syrrhophus* and *Tomodactylus* (Fig. 2), *T. albolabris* is the sister taxon of *S. leprus*. The separation of *Syrrhophus* and *Tomodactylus* therefore is highly questionable, but this taxonomic issue is beyond the scope of this paper. For a complete description of the characters and their consistancy ratios as well as the character state changes at each branch of Fig. 2 (see Joglar 1986).

The *E. ricordii* group is a natural group characterized by the unique condition of long odontophores (39). The absence of this condition in *Eleutherodactylus* from the mainland indicates that the condition evolved in the West Indies. I consider the West Indian subset of the *E. unistrigatus* group to be the sister group of the *E. ricordii* group. Species such as *E. klinikoskii*, *E. orcutti* and/or *E. zugi* could be considered intermediate between the West Indian subset of the *E. unistrigatus* group and the *E. ricordii* group. Also it is possible that these three species, together with some other species with short odontophores, are members of the *E. ricordii* group, in which the condition of long odontophores reversed to the short condition. Flores (1984) argued that heterochrony could be the mechanism responsible for the evolution of one species with short odontophores (*E. orcutti*) from a species with the long condition (*E. nubicola*).

I do not recognize the *E. gossei* and *E. emiliae* groups of Schwartz (1978). These are phenetic clusters of species based on a few external, homoplasious characters. The species with long odontophores that were previously included in the *E. gossei* and *E. emiliae* groups are now assigned to the *E. ricordii* group.

Taxonomy

On basis of a cladistic analysis of morphological characters, I recognize three major lineages of *Eleutherodactylus* in the West Indies. In the following definition of these lineages, characters are followed by abbreviations of species groups, if any, that share the same character: u = *E. unistrigatus* group, i = *E. inoptatus* group, r = *E. ricordii* group, and S/T = *Syrrhophus* and *Tomodactylus*. A list of species within each lineage and distribution of species per island are included under Content and Distribution, respectively.

All the the groups share the following: 51) eggs laid in moist, terrestrial situations; and 52) direct development, no larvae.

Eleutherodactylus inoptatus group

DEFINITION.-- Apomorphic characters include: 3) head width "relatively" wide (>40%) or wide; 48) Fingers I longer than II; and large body size.

This species group shares with other West Indian species groups the following characteristics: 1) tympanum distinct (r, u, S/T); 2) supraocular tubercles present (r, S/T); 4) cranial crest absent (r, u, S/T); 5) vocal slits present; 7) finger discs present (r, u); 8) shape of subarticular tubercles of fingers oval (S/T); 9) height of subarticular tubercles of fingers raised (S/T); 10) nuptial pads absent (r, u, S/T); 11) accessory palmar tubercles present (r, S/T); 12) inner tarsal fold absent or present (u); 13) toe discs present (r); 14) shape of subarticular toe tubercles oval (r, S/T); 15) height of subarticular toe tubercles raised (S/T); 16) plantar tubercles present; 17) toe webbing absent (r, S/T); 18) dorsolateral glandular ridges present; 19) ventral skin smooth to weakly granular or coarsely granular (u); 20) dorsal skin smooth to granular or rugose (r, u); 21) *ricordii* dorsal pattern absent; 22) nasals of average size; 23) medial separation of nasals narrow, almost in contact or in contact (r); 24) nasals and frontoparietals widely separated, narrowly separated, or almost in contact (S/T); 25) frontoparietal-otoccipital bones unfused; 26) parasphenoid and pterygoid with or without overlap; 27) pterygoid-maxilla articulation narrow or broad (r, u); 28) zygomatic ramus of squamosal variable in length (r, u); 29) breadth of zygomatic ramus of squamosal narrow or broad (r, u); 30) otic ramus of squamosal not reduced (r, u, S/T); 31) lateral process of premaxilla well developed and long (r, u, S/T); 32) lateral process of premaxilla with posteromedial orientation (S/T); 33) premaxilla-maxilla separation wide (S/T); 34) anterior expansion of palatal shelf of maxilla present; 35) posterior expansion of palatal shelf of maxilla present or absent (u, r); 36) palatine-maxilla in contact or narrowly separated (r); 37) vomers not reduced in size (r); 38) odontophores present (r); 39) odontophores short (u); 41) vomerine teeth present (r); 42) ornamentation of skull present or absent (u); 43) anterior process of hyale present (r, u, S/T); 44) anterolateral process of hyoid plate present (r, S/T); 45) sternum bifurcate posteriorly; 46) terminal phalange of third finger T-shaped (S/T); 49) fusion of presacral vertebrae absent (S/T); 50) "S" condition of the mandibular ramus of the trigeminal nerve (r, u, S/T).

CONTENT.-- 5 species: *E. chlorophenax*, *E. hypostenor*, *E. inoptatus*, *E. nortoni* and *E. ruthae*.

DISTRIBUTION.-- The *E. inoptatus* group is confined to Hispaniola.

REMARKS.-- In *E. hypostenor*, the snout shape (dorsal view) is more narrow and pointed in males than in females. A similar situation occurs in *E. unicolor* (Rivero 1978; Joglar 1983), a species with habits similar to those of *E. hypostenor*.

Eleutherodactylus ricordii group

DEFINITION.-- Apomorphic characters include: 21) *ricordii* dorsal pattern present in most species; 40) odontophores long.

This group also shares with other West Indian species groups the following characters: 1) tympanum distinct (u, i, S/T); 2) supraocular tubercles present (i, S/T); 3) head width narrow (u, S/T); 4) cranial crest absent (u, i, S/T); 5) vocal slits absent or present (u, S/T); 7) finger discs present (u, i); 8) shape of subarticular tubercles of fingers oval or round (u, i); 9) height of subarticular tubercles of fingers raised or low (u); 10) nuptial pads absent (u, i, S/T); 11) accessory palmar tubercles present (i, S/T); 12) inner tarsal fold absent (i, S/T); 13) toe discs present (i); 14) subarticular toe tubercles oval (i, S/T); 15) height of subarticular toe tubercles raised or low (u); 16) plantar tubercles present or absent (u); 17) toe webbing absent (i, S/T); 18) dorsolateral glandular ridges absent or present (u); 19) ventral skin smooth to weakly granular (S/T); 20) dorsal skin smooth to granular or rugose (u, i); 22) nasals large or average size (S/T); 23) medial separation of nasals absent, almost in contact or narrow (S/T); 24) nasals and frontoparietals in contact, almost in contact, narrowly separated, or widely separated; 25) frontoparietal-otoccipital bones fused or not (u, S/T); 26) parasphenoid and pterygoid lacking overlap (u, S/T); 27) pterygoid-maxilla articulation narrow or broad (u, i); 28) zygomatic ramus of squamosal variable in length (u, i); 29) breadth of zygomatic ramus of squamosal variable (u, i); 30) otic ramus of squamosal not reduced (u, i, S/T); 31) lateral process of premaxilla well developed and long (u, i, S/T); 32) lateral palatine process of the pars palatina of the premaxilla with posterolateral orientation (u); 33) premaxillary-maxillary separation wide or narrow (u); 34) anterior expansion of palatal shelf of maxilla present or absent (u, S/T); 35) posterior expansion of palatal shelf of maxilla absent or present (u); 36) palatine-maxilla in contact or narrowly separated (i); 37) vomers not reduced in size (i); 38) odontophores present (i); 41) vomerine teeth present (i); 42) ornamentation of the skull absent (S/T); 43) anterior process of hyale present (u, i, S/T); 44) anterolateral process of hyoid plate present (i, S/T); 45) sternum bifurcate posteriorly, pendulum-shaped or elongated rectangular (u, S/T); 46) terminal phalange of third finger T-shaped or knobbed (u); 47) terminal phalange of third toe T-shaped or knobbed (u); 48) Fingers II longer than I or II equal to I; 49) fusion of presacral vertebrae absent or present (u); 50) "s" condition of the mandibular ramus of the trigeminal nerve (u, i, S/T).

CONTENT.-- 52 species: *E. acmonis, E. albipes, E. alcoae, E. alticola, E. andrewsi, E. apostates, E. armstrongi, E. atkinsi, E. bresslerae, E. cavernicola, E. cundalli, E. cuneatus, E. darlingtoni, E. dimidiatus, E. emiliae, E. furcyensis, E. fuscus, E. glandulifer, E. glaphycompus, E. gossei, E. grabhami, E. grahami, E. greyi, E. guanahacabibes, E. gundlachi, E. intermedius, E. jamaicensis, E. jugans, E. junori, E. lentus, E. leoncei, E. lucioi, E. luteolus, E. monensis, E. nubicola, E. oxyrhyuchus, E. pantoni, E. paulsoni, E. pezopetrus, E. pictissimus, E. pinarensis, E. planirostris, E. probolaeus, E. rhodesi, E. richmondi, E. ricordii, E. schmidti, E. sierramaestrae, E. sisyphodemus, E. thomasi, E. warreni* and *E. weinlandi*.

DISTRIBUTION.-- Cuba--17 species; Jamaica--18 species; Hispaniola--19 species; Puerto Rico--3 species.

Eleutherodactylus unistrigatus group

DEFINITION.-- No apomorphic characters. This group shares with other West Indian species groups the following characteristics: 1) tympanum distinct (r, i); 2) supraocular tubercles present or absent; 3) head width narrow (r, S/T); 4) cranial crest absent (r, i, S/T); 5) vocal slits present or absent (r, S/T); 6) finger discs present (r, i); 8) shape of subarticular tubercles of fingers variable (r, i); 9) height of subarticular tubercles of fingers variable (r); 10) nuptial pads absent (r, i, S/T); 11) accessory palmer tubercles present or absent; 12) inner tarsal fold absent or present (i); 13) toe discs present or absent (S/T); 14) shape of subarticular toe tubercles variable 15) height of subarticular toe tubercles variable (r); 16) plantar tubercles present or absent (r); 17) toe webbing absent or present; 18) dorsolateral glandular ridges absent or present (r); 19) ventral skin smooth to weakly granular or coarsely granular (i); 20) dorsal skin smooth to granular or rugose (r, i); 21) *ricordii* dorsal pattern absent or present (r, S/T); 22) size of nasals variable; 23) medial separation of nasals variable (S/T); 24) nasals and frontoparietals widely separated, narrowly separated, almost in contact, or in contact (r); 25) frontoparietal-otoccipital bones fused or not (r, S/T); 26) parasphenoid and pterygoid without overlap (r, S/T); 27) pterygoid-maxilla articulation narrow or broad (r, i); 28) zygomatic ramus of squamosal variable in length (r, i); 29) breadth of zygomatic ramus of squamosal narrow or broad (r, i); 30) otic ramus of squamosal not reduced (r, i, S/T); 31) lateral process of premaxilla well developed and long (r, i, S/T); 32) lateral palatine process of the pars palatina of the premaxilla with posterolateral or posteromedial orientation (r); 33) premaxilla-maxilla separation variable (r); 34) anterior expansion of palatal shelf of maxilla present or absent (r, S/T); 35) posterior expansion of palatal shelf of maxilla absent or present (r, i); 36) palatine-maxilla in contact or narrowly separated, or widely separated; 37) vomers not reduced in size or reduced in size; 38) odontophores present or absent; 39) odontophores short (i); 41) vomerine teeth present or absent; 42) ornamentation of skull absent or present (i); 43) anterior process of hyale present (r, i, S/T); 44) anterolateral process of hyoid plate present or absent; 45) sternum bifurcate posteriorly, elongated rectangular or pendulum shaped (r, S/T); 46) terminal phalange of third finger T-shaped or knobbed (r); 47) terminal phalange of third toe T-shaped or knobbed (r); 48) Finger II longer than I (S/T); 49) fusion of presacral vertebrae absent or present (r); 50) "S" condition of the mandibular ramus of the trigeminal nerve (r, i, S/T).

CONTENT.-- 61 species in the West Indies: *E. abbotti*, *E. antillensis*, *E. audanti*, *E. auriculatoides*, *E. auriculatus*, *E. bakeri*, *E. barlagnei*, *E. bartonsmithi*, *E. brevirostris*, *E. brittoni*, *E. cochranae*, *E. cooki*, *E. coqui*, *E. counouspeus*, *E. cubanus*, *E. eileenae*, *E. eneidae*, *E. etheridgei*, *E. eunaster*, *E. flavescens*, *E. fowleri*, *E. glanduliferoides*, *E. gryllus*, *E. haitianus*, *E. hedricki*, *E. heminota*, *E. jasperi*, *E. johnstonei*, *E. karlschmidti*, *E. klinikowskii*, *E. lamprotes*, *E. leberi*, *E. locustus*, *E. martinicensis*, *E. minutus*, *E. montanus*, *E. neodreptus*, *E. orcutti*, *E. parabates*, *E. patriciae*, *E. pinchoni*, *E. pituinus*, *E. poolei*, *E. portoricensis*, *E. ronaldi*, *E. rufifemoralis*, *E. schwartzi*, *E. sciagraphus*, *E. semipalmatus*, *E. symingtoni*, *E. turguinensis*, *E. unicolor*, *E. urichi*, *E. varians*, *E. varleyi*, *E. ventrilineatus*, *E. wetmorei*, *E. wightmanae*, *E. zeus*, *E. zugi*, *Sminthillus limbatus*.

DISTRIBUTION.-- Throughout forested Colombia and Ecuador as well as into the high altitude grasslands; along the western edge of the Amazon Basin in Peru and adjacent Bolivia ; the northern Amazon Basin in Brazil and Venezuela; Guyanas; Merida Andes and the costal Range of Venezuela; In Central America as far north as Honduras. In the West Indies, members of this group are distributed as follows: Cuba--15 species; Hispaniola--25 species; Jamaica--1 species; Lesser Antilles--5 species; Puerto Rico--15 species.

REMARKS.-- There are no apomorphic characters to define the *E. auriculatus* group as recognized formerly. The tympanum is indistinct or distinct in the *E. unistrigatus* group but always distinct in the *E. auriculatus* group. Nuptial pads are present or absent in the *E. unistrigatus* group but always absent in the *E. auriculatus* group. The absence of the anterolateral process of the hyoid plate is more widespread in the *E. unistrigatus* group than in the *E. auriculatus* group. Except for these three minor differences, the *E. auriculatus* group is indistinguishable from the *E. unistrigatus* group of South America.

Therefore, I consider the species listed above to be the West Indian representatives of the *E. unistrigatus* group.

Biogeography

The historical geology of the Caribbean region has been the subject of considerable controversy. The model of plate tectonics proposed by Malfait and Dinkelman (1972) and refined by Perfit and Heezen (1978) is generally accepted, but there are many gaps in our knowledge of the region (see Coney 1982 for review). According to those authors, a volcanic arc formed a land extension eastward from nuclear Central America in the late Cretaceous. Toward the eastern part of this arc were areas that subsequently became three islands: 1) Jamaica; 2) the eastern part of Cuba; and 3) the western part of Hispaniola. The rest of Cuba already existed as a separate land mass. In the early Tertiary (pre-Oligocene), the land connection with Central America subsided. In the late Tertiary (post-Oligocene), the Greater Antilles moved eastward on the Caribbean plate, eastern Cuba united with the rest of the island, and the Lesser Antilles were formed as volcanos along the eastern edge of the Caribbean plate.

Prior to the evidence in support of the theory of plate tectonics, explanations of the distribution of animals in the West Indies were based on over-water dispersal (e.g. Simpson 1956). Rosen (1975, 1985), proposed a model of vicariance biogeography that involved the eastward movement of a "proto-Antillean archipelago" from a position in the region of present lower Central America to the present position of the Greater Antilles. MacFadden's (1980) interpretation of the biogeography of insectivores in the Greater Antilles was based on this model. Pregill (1981b) argued that Rosen's vicariance model departed critically from the geological evidence, and he concluded that neither the patterns of distribution in the West Indies nor evidence on the ages of most of the taxa support Rosen's model. On the other hand, the vicariance model of Crother et al. (1986) for xantusiid lizards on Cuba and in Central America conforms to the geological evidence. According to Guyer and Savage (1986) the presence of the anoline lizards in the West Indies resulted from vicariant events by means of a "continuous" land connection between Central America and the proto-Greater Antillean block. According to these authors, only three anoline groups could have resulted from overwater dispersal (two groups in Lesser Antilles and one group in Cuba).

Other workers have interpreted the biogeography of groups of amphibians and reptiles in the West Indies as the result of dispersal. Trueb and Tyler (1974) proposed that the hylid frogs resulted from multiple invasions from the mainland. Pregill (1981a) considered that the bufonid genus *Peltophryne* was the result of a single invasion, and Cadle (1985) suggested two invasions for xenodontine snakes. Savage (1982) favored the distribution of some groups by vicariance and others by dispersal.

In his summary of the herpetofauna of Middle America, Savage (1982) emphasized the importance of the land connection between Central America and South America in the late Cretaceous and early Paleocene. During the time of this land connection early eleutherodactyline frogs could have dispersed from South America into Central America. Subsequent separation of the land masses during most of the Tertiary would have resulted in separate lineages of *Eleutherodactylus* in Central America and South America represented today by the *Craugastor* and *E. unistrigatus* clades, respectively.

Based on this assumption and the results of my phylogenetic analysis, I offer the following biogeographic scenario for the West Indian *Eleutherodactylus*:

1. During the late Cretaceous and very early Tertiary, *Eleutherodactylus* in Middle America and its West Indian extension differentiated into three lineages: 1) *Craugastor* in Middle America; 2) undifferentiated *E. unistrigatus* clade in Middle America and West Indian extension; and 3) the *E. ricordii-Syrrhophus-Tomodactylus* lineage in Middle America and West Indian extension.

2. During the early Tertiary subsequent to the subsidence of the land connection between Middle America and the West Indies and the fragmentation of the Greater Antilles, differentiation occurred that resulted in: 1) isolation of *Craugastor* in Middle America; 2) differentiation of the West Indian *E. unistrigatus* clade into *E. inoptatus* group in Hispaniola and *E. ricordii* group in Cuba and Hispaniola; and 3) differentiation of the *E. unistrigatus* clade into *Syrrhophus-Tomodactylus* in Middle America.

3. During the late Tertiary, the *E. ricordii* and *E. unistrigatus* lineages dispersed and differentiated throughout the Greater Antilles, and members of the South American clade of the *E. unistrigatus* group dispersed from the mainland into the Lesser Antilles.

Thus, if eleutherodactyline frogs existed in the late Cretaceous, the differentiation and basic distribution of clades could have resulted from vicariance associated with the breakup of the West Indian land extension from nuclear Central America. However, the distribution and differentiation of some of the species in the Greater Antilles seems to have resulted from dispersal within the Greater Antilles and dispersal from South America into the Lesser Antilles. Any further biogeographic refinement would exceed the limits of available data.

Summary

Using cladistic methods on morphological characters of frogs of the genus *Eleutherodactylus*, three clades (groups) are recognized in the West Indies---*E. inoptatus* group, *E. ricordii* group, and *E. unistrigatus* group. The latter may not be monophyletic. The genus *Sminthillus* is considered to be a synonym of *Eleutherodactylus*, and *E. limbatus* is included in the West Indian subset of the *E. unistrigatus* group.

The separation of the genera *Syrrhophus* and *Tomodactylus* is questioned. These two genera may be related to the mainland subset of the *E. unistrigatus* group and therefore not closely related to the West Indian taxa. The inclusion of *Hylactophryne* in the subgenus *Craugastor* (Lynch 1986) is questioned.

Acknowledgements

Numerous people and institutions assisted generously in various aspects of this research. For the loan of specimens, permission to prepare skeletons, and patience in answering my numerous letters and loan requests, I am grateful to the following persons and their institutions: Charles W. Myers, American Museum of Natural History (AMNH); C. J. McCoy, Carnegie Museum of Natural History (CM); William E. Duellman, Museum of Natural History, University of Kansas (KU); Pere Alberch and Jose Rosado, Museum of Comparative Zoology, Harvard University (MCZ); Arnold G. Kluge, University of Michigan Museum of Zoology (UMMZ); Juan A. Rivero, Universidad de Puerto Rico, Recinto de Mayaguez (UPR-M); George R. Zug, National Museum of Natural History (USNM); and Yale University Peabody Museum of Natural History (YPM).

I thank the members of my graduate committee, Drs. William E. Duellman, Edward O. Wiley, Linda Trueb, and John D. Lynch for their advice and critical reviews of my dissertation, on which this paper is based. I thank the Division of Herpetology, Museum of Natural History for assisting in various aspects of this study. Special thanks go to Darrel Frost, David Cannatella, and Rebecca Pyles for numerous stimulating discussions on amphibian systematics, cladistics, and computer science. I thank Dr. Charles A. Woods and the Florida State Museum for organizing this wonderful symposium. This paper is dedicated to my parents, Rafael Joglar and Olga Jusino de Joglar, and my two sisters, Marichel Joglar de Muñoz and Teresita Joglar.

Literature cited

Barbour, T., and G. K. Noble. 1920. Some amphibians from northwestern Peru, with a revision of the genera *Phyllobates* and *Telmatobius*. Bulletin of the Museum of Comparative Zoology 63:393-427.

Bogart, J.P. 1981. Chromosome studies in *Sminthillus* from Cuba and *Eleutherodactylus* from Cuba and Puerto Rico (Anura: Leptodactylidae). Life Sciences Contributions Royal Ontario Museum (129):1-22.

Cadle, J.E. 1985. The Neotropical colubrid snake fauna (Serpentes: Colubridae): lineage components and biogeography. Systematic Zoology 34(1):1-20.

Coney, P.J. 1982. Plate tectonics constraints on biogeographic connections between North and South America. Annals of the Missouri Botanical Garden 69:432-443

Crombie, R.I. 1977. A new species of frog of the genus *Eleutherodactylus* (Amphibia: Leptodactylidae) from the cockpit country of Jamaica. Proceedings of the Biological Society of Washington 90(2):194-204.

Crother, B.I., M.M. Miyamoto, and W.F. Presh. 1986. Phylogeny and biogeography of the lixard family Xantusiidae. Systematic Zoology 35(1):37-45.

Dingerkus, G., and L.D. Uhler. 1977. Enzyme clearing of alcian blue stained whole small vertebrates for demonstration of cartilage. Stain Techology 52:229-232.

Dunn, E.R. 1925. New frogs from Cuba. Occassional Papers of the Boston Society of Natural History 5:163-166.

_____. 1926. Additional frogs from Cuba. Occassional Papers of the Boston Society of Natural History 5:209-215.

Estrada, A., and N. Rodriguez. 1985. Nueva especie de *Eleutherodactylus* del grupo *ricordii* (Anura: Leptodactylidae) del occidente de Cuba. Poeyana (303):1-10.

Flores, G. 1985. Comparative osteology, relationships, and evolution in Jamaican frogs of the genus *Eleutherodactylus*. Unpublished B.S. thesis. Harvard University, Cambridge, MA.

Frost, D. (ed.). 1985. Amphibians Species of the World. Association of Systematic Col lections, Lawrence, KS. 732pp.

Goin, C.J. 1954. Remarks on evolution of color pattern in the *gossei* group of the frog genus *Eleutherodactylus*. Annals of the Carnegie Museum 33(10):185-195.

Griffiths, I. 1959. The phylogeny of *Sminthillus limbatus* and the status of the Brachycephalidae (Amphibia, Salienta). Proceedings of the Zoolological Society of London 132(3):457-487.

Guyer, C., and J.M. Savage. 1986. Cladistic relationships among anoles (Sauria: Iguanidae). Systematic Zoology 35(4):509-531.

Hedges, S. B. 1982. Caribbean biogeography: implications of recent plate tectonic studies. Systematic Zoology 31(4):518-522.

____, and R. Thomas. 1987. A new burrowing frog from Hispaniola with comments on the *inoptatus* group of the genus *Eleutherodactylus* (Anura: Leptodactylidae). Herpetologica 43(3):269-279.

Joglar, R. 1983. Estudio fenetico del genero *Eleutherodactylus* en Puerto Rico. Caribbean Journal of Science 19(3-4):33-40.

____. 1986. Phylogenetic relationships of the West Indian frogs of the genus *Eleutherodactylus*. Unpublished Ph.D. dissertation, University of Kansas, Lawrence, KS. 142pp.

Lynch, J.D. 1968. Genera of leptodactylid frogs in Mexico. University of Kansas Publications of the Museum of Natural History 17:503-515.

____. 1970. A taxonomic revision of the leptodactylid frog genus *Syrrhophus* Cope. University of Kansas Publications of the Museum of Natural History 20:1-45.

____. 1971. Evolutionary Relationships, Osteology, and Zoogeography of Leptodactyloid Frogs. Miscelaneous Publications of the University of Kansas Museum of Natural History 53:1-238.

____. 1976. The species groups of the South American frogs of the genus *Eleutherodactylus* (Leptodactylidae). Occassional Papers of the Museum of Natural History of the University of Kansas (61):1-24.

____. 1986. The definition of the Middle American clade of *Eleutherodactylus* based on jaw musculature (Amphibia: Leptodactylidae). Herpetologica 42(2):248-258.

MacFadden, B.J. 1980. Rafting mammals or drifting islands?: biogeography of the Greater Antillean insectivores *Nesophonthes* and *Solenodon*. Journal of Biogeography 7:11-22.

Malfait, B.T., and M.G. Dinkelman. 1972. Circum-Caribbean tectonic and igneous activity and the evolution of the Caribbean plate. Bulletin of the Geological Society of America 83:251-272.

Noble, G.K. 1922. The phylogeny of the Salienta I. - The osteology and the thigh musculature; their bearing on classification and phylogeny. Bulletin of the American Museum of Natural History 46:1-87.

____. 1931. The Biology of the Amphibia. McGraw-Hill Inc., New York. 577 pp.

Perfit, M.R., and B.C. Heezen. 1978. The geology and the evolution of the Cayman Trench. Bulletin of the Geological Society of America 89:1155-1174.

Pregill, G.K. 1981a. Cranial morphology and the evolution of the West Indian toads (Salienta: Bufonidae): resurrection of the genus *Peltophryne* Fitzinger. Copeia (2):273-285.

____. 1981b. An appraisal of the vicariance hypothesis of Caribbean biogeography and its application to West Indian terrestrial vertebrates. Systematic Zoology 30(2):147-155.

Rivero, J.A. 1978. Los anfibios y reptiles de Puerto Rico. Universidad de Puerto Rico Editorial Universitaria 1-152 + lams.

____, and H. Solano. 1977. Origen y evolucion de los *Eleutherodactylus* (Amphibia: Leptodactylidae) de los Andes Venezolanos. Separata Memoria Sociedad de Cienias Naturales de La Salle (108):249-263.

Rosen, D.E. 1975. A vicariance model of Caribbean biogeography. Systematic Zoology 24(4):431-464.

____. 1985. Geological hierarchies and biogeographic congruence in the Carribbean. Annals of the Missouri Botanical Garden 72:636-659.

Savage, J.M. 1982. The enigma of the Central American herpetofauna: dispersals or vicariance? Annals of the Missouri Botanical Garden 69:464-547.

Schwartz, A. 1958. Four new frogs of the genus *Eleutherodactylus* (Leptodactylidae) from Cuba. Amererican Museum of Natural History Novitates, (1873):1-20.

____. 1969. The Antillean *Eleutherodactylus* of the *auriculatus* group. Studies on the fauna of Curacao and other Caribbean Islands 30:100-115.

____. 1976. Variation and relationships of some Hispaniolan frogs (Leptodactylidae, *Eleutherodactylus*) of the *ricordi* group. Bulletin of the Florida State Museum 21(1):1-46.

____. 1978. Some aspects of the herpetogeography of the West Indies. Pp. 31-51 in F.B. Gill (ed.). Zoogeography in the Caribbean, The 1975 Leidy Medal Symposium. Special Publication (No. 13) of the Academy of Natural Sciences, Philadelphia,

Shreve, B., and E.E. Williams. 1963. The frogs. Pp. 302-342 in E.E. Williams, B. Shreve, and P.S. Humphrey (eds.). The herpetology of the Port-au-Prince region and Gonove Island, Haiti. Parts. I-II. Bulletin of the Museum of Comparative Zoology 129:291-342.

Simpson, G.G. 1956. Zoogeography of the West Indian land mammals. American Museum of Natural History Novitates 1759:1-28.

Swofford, D.L. 1984. PAUP--Phylogenetic Analysis Using Parsimony, Version 2.2, User's Manual, Privately Published.

Trueb, L., and M.J. Tyler. 1974. Systematics and evolution of the Greater Antillean hylid frogs. Occassional Papers of the Museum of Natural History, University of Kansas (24):1-60.

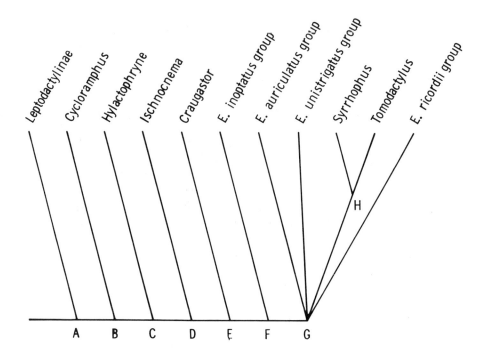

Figure 1. Hypothesized phylogenetic arrangement of the West Indian *Eleutherodactylus*. Letters at nodes refer to characters and their changes as given in Appendices II and III.

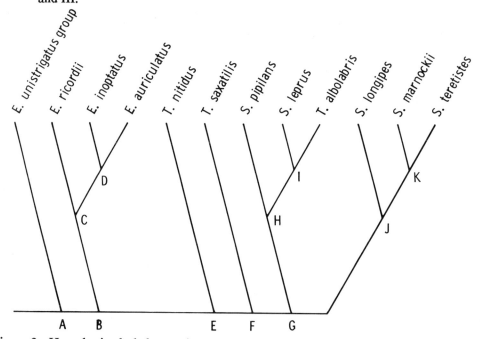

Figure 2. Hypothesized phylogenetic relations of *Syrrhophus* and *Tomodactylus*. Letters at node refer to characters and their changes as given in Joglar, 1986.

APPENDIX 1. Data matrix for phylogenetic analysis with 134 taxa and 52 characters. Within the matrix a "9" represents missing data or that the datum is logically impossible.

	1 2
	1 2 3 4 5 6 7 8 9 0 1 2 3 4 5 6 7 8 9 0

Ancestor	0 0 0 0 0 0 0 0 0 0 0 0 0 0 0 0 0 0 0 0 0 0 0 0 0 0 0 0 0 0 0 0 0 0 0 0 0 0 0 0 0 0 0 0 0 0 0 0 0 0 0 0
Craugastor	0 0 1 0 0 0 1 0 0 0 0 1 0 0 1 1 1 0 1 0 2 3 2 0 1 0 0 0 0 1 0 1 0 1 0 0 0 0 1 0 0 1 0 0 1 1 0 0 1 2 2
Cycloramphus	1 0 1 0 0 1 0 0 0 0 0 0 0 0 0 1 1 0 1 0 1 0 0 0 1 0 1 0 1 1 0 9 0 1 0 0 0 0 0 0 0 0 0 0 0 0 2 0 0 2 1
E. unistrigatus group	1 0 0 0 0 1 1 0 0 0 0 0 1 0 0 0 0 1 0 0 0 0 0 0 9 0 1 0 0 0 1 1 1 1 9 0 0 0 0 0 0 0 1 1 0 1 1 2 0 0 2 2
E. abbotti	0 0 0 0 0 1 1 0 0 0 0 1 1 0 0 0 0 1 1 0 0 0 3 3 1 0 1 0 1 0 1 1 1 0 0 0 0 0 0 1 0 0 9 9 9 1 1 2 0 0 2 2
E. acmonis	0 0 0 0 1 1 1 0 0 0 0 1 1 0 0 0 0 1 0 0 1 9 9 9 9 9 9 9 9 9 9 9 9 9 9 9 1 0 9 2 0 9 9 9 9 9 9 9 9 0 9 9
E. albipes	0 0 0 0 0 1 1 1 1 0 0 0 1 0 9 0 0 0 0 0 0 9 9 9 9 9 9 9 9 9 9 9 9 9 9 9 1 0 9 3 0 9 9 9 9 9 9 9 9 0 9 9
E. alcoae	0 0 0 0 1 1 1 0 0 0 0 1 1 0 0 0 0 1 0 0 1 9 9 9 9 9 9 9 9 9 9 9 9 9 9 9 1 0 9 2 0 9 9 9 9 9 9 9 9 0 9 9
E. alticola	0 0 0 0 1 1 1 0 1 0 0 1 1 0 1 0 0 1 0 0 0 0 0 0 1 0 1 1 1 0 1 1 1 1 0 0 0 0 0 3 0 0 1 0 1 1 1 2 0 0 2 2
E. andrewsi	0 0 0 0 1 1 1 0 0 0 0 1 1 0 0 0 0 9 0 0 1 0 0 1 1 9 1 0 1 0 1 1 1 1 0 0 0 0 0 3 0 0 1 0 1 1 1 2 0 0 2 2

APPENDIX 1. Data matrix for phylogenetic analysis with 134 taxa and 52 characters. Within the matrix a "9" represents missing data or that the datum is logically impossible.

	1 2
	1 2 3 4 5 6 7 8 9 0 1 2 3 4 5 6 7 8 9 0

E. antillensis	0 1 0 0 0 1 1 1 1 0 0 1 1 1 1 9 0 1 1 0 0 2 3 0 1 0 1 0 1 0 1 1 1 1 0 0 0 0 0 1 0 0 1 0 0 1 1 2 0 0 2 2
E. apostates	0 0 0 0 1 1 1 0 0 0 0 1 1 0 0 0 0 0 0 0 1 9 9 9 9 9 9 9 9 9 9 9 9 9 9 9 1 0 9 3 0 9 9 9 9 9 9 9 9 0 9 9
E. armstrongi	0 0 0 0 1 1 1 0 0 0 0 1 1 0 0 0 0 1 0 0 1 2 3 3 1 0 1 0 1 0 1 1 1 1 0 0 0 0 0 3 0 0 9 9 9 1 1 2 0 0 2 2
E. atkinsi	0 0 0 0 0 1 1 0 0 0 0 1 1 0 0 0 0 1 0 0 1 2 3 0 0 0 1 0 1 0 1 1 1 1 0 0 0 0 0 3 0 0 1 0 0 1 1 2 0 0 2 2
E. audanti	0 0 0 0 0 1 1 0 0 0 0 1 1 1 1 0 0 1 0 0 0 0 0 0 1 0 1 0 1 0 1 1 1 1 0 0 0 0 0 0 0 0 1 0 0 1 1 2 0 0 2 2
E. auriculatoides	0 0 0 0 0 1 1 0 0 0 0 1 1 0 0 0 0 1 1 0 0 0 1 0 0 0 1 0 1 0 1 1 1 1 0 0 0 0 0 1 0 0 1 0 2 1 1 2 0 0 2 2
E. auriculatus	0 0 0 0 0 1 1 0 0 0 0 1 1 0 0 0 0 1 0 0 0 0 0 0 9 0 1 0 1 0 1 1 1 1 0 0 1 0 0 0 0 0 1 0 2 1 0 2 0 0 2 2
E. bakeri	0 0 0 0 9 1 1 0 0 0 0 1 1 0 0 0 0 1 0 0 0 2 3 0 1 0 1 0 1 0 1 1 0 1 0 0 0 0 0 1 0 0 1 0 1 1 1 2 0 0 2 2
E. barlagnei	0 0 0 0 9 1 1 1 1 0 0 0 1 1 1 1 1 1 0 9 0 0 0 0 1 0 1 0 1 0 1 1 0 1 0 0 0 0 0 0 0 0 1 0 1 1 1 2 0 0 2 2
E. bartonsmithi	0 0 0 0 0 1 1 0 0 0 0 1 1 0 0 0 0 1 0 0 1 9 9 9 9 9 9 9 9 9 9 9 9 9 9 9 0 0 9 0 0 9 9 9 9 9 9 9 9 0 9 9

APPENDIX 1. Data matrix for phylogenetic analysis with 134 taxa and 52 characters. Within the matrix a "9" represents missing data or that the datum is logically impossible.

	1 2 1 2 3 4 5 6 7 8 9 0 1 2 3 4 5 6 7 8 9 0
E. bresslerae	0 0 0 0 1 1 1 0 0 0 0 1 1 0 0 0 0 9 0 0 1 9 9 9 9 9 9 9 9 9 9 9 9 9 9 9 1 0 9 3 0 9 9 9 9 9 9 9 9 0 9 9
E. brevirostris	0 0 0 0 9 1 1 1 1 0 0 1 1 1 1 0 0 1 0 0 0 0 0 1 0 0 1 0 0 0 1 1 0 1 0 0 0 0 0 1 0 0 9 9 9 1 1 2 0 0 2 2
E. brittoni	0 0 0 0 0 1 1 1 1 0 0 1 1 1 1 0 0 1 0 0 0 1 1 0 1 0 1 0 1 0 1 1 1 1 0 1 1 0 0 0 0 0 9 9 1 0 0 2 0 0 2 2
E. cavernicola	0 0 0 0 1 1 1 0 0 0 0 1 1 0 0 0 0 1 0 0 1 9 9 9 9 9 9 9 9 9 9 9 9 9 9 9 1 0 9 2 0 9 9 9 9 9 9 9 9 0 9 9
E. chlorophenex	0 0 0 0 0 1 1 0 0 0 0 0 1 0 0 0 0 0 1 0 0 9 9 9 9 9 9 9 9 9 9 9 9 9 9 9 0 0 9 1 0 9 9 9 9 9 9 9 9 0 9 9
E. cochranae	0 0 0 0 0 1 1 1 1 0 0 1 1 1 1 9 0 1 0 0 0 0 0 0 1 0 1 0 1 0 1 1 1 1 0 0 0 0 0 0 0 0 1 0 0 1 1 2 0 0 2 2
E. cooki	0 1 0 0 0 1 1 1 1 0 0 1 1 1 1 1 0 1 0 0 0 0 3 0 1 0 1 0 1 0 1 1 1 1 0 0 0 0 0 1 0 0 1 0 0 1 1 2 0 0 2 2
E. coqui	0 0 0 0 0 1 1 0 0 0 0 9 1 0 0 0 0 1 0 0 0 2 2 0 0 0 0 1 0 0 1 1 1 1 1 0 0 0 0 1 0 0 1 0 0 1 1 2 0 0 2 2
E. counouspeus	0 0 0 0 0 0 1 0 0 0 0 1 1 0 0 0 0 1 0 0 1 9 9 9 9 9 9 9 9 9 9 9 9 9 9 9 0 0 9 1 0 9 9 9 9 9 9 9 9 0 9 9
E. cubanus	0 1 0 0 0 1 1 0 1 0 0 1 1 9 9 0 0 1 0 0 0 1 1 0 1 0 1 0 1 0 1 1 1 0 0 0 0 0 0 0 0 0 9 9 9 1 1 2 0 0 2 2

APPENDIX 1. Data matrix for phylogenetic analysis with 134 taxa and 52 characters. Within the matrix a "9" represents missing data or that the datum is logically impossible.

	1　　　　　　　　2　　　　 1 2 3 4 5 6 7 8 9 0 1 2 3 4 5 6 7 8 9 0
E. cundalli	0 0 0 0 1 1 1 0 0 0 0 1 1 0 0 0 0 1 0 0 1 2 3 2 9 0 1 0 1 0 1 1 1 1 0 0 0 0 0 3 0 0 1 0 2 1 1 2 0 0 2 2
E. cuneatus	0 0 0 0 0 1 1 0 0 0 0 1 1 0 0 0 0 9 0 0 1 2 3 3 1 0 1 0 1 0 1 1 1 1 0 0 0 0 0 3 0 0 1 0 0 1 1 2 0 0 2 2
E. darlintoni	0 0 0 0 9 1 1 0 0 9 0 1 1 0 1 0 0 1 0 0 0 0 0 1 1 0 1 0 1 0 1 1 1 1 0 0 0 0 0 1 0 0 9 9 9 0 0 2 0 0 2 2
E. dimidiatus	0 0 0 0 0 1 1 0 0 0 0 1 1 0 0 0 0 0 0 0 1 2 2 2 0 0 0 0 0 0 1 0 0 1 9 0 0 0 0 3 0 0 1 0 0 1 1 2 0 0 2 2
E. eileenae	0 0 0 0 0 1 1 0 0 0 0 1 1 0 0 0 0 9 0 0 0 0 0 0 1 0 1 0 1 0 1 1 1 1 0 0 0 0 0 0 0 0 1 0 0 1 1 2 0 0 2 2
E. emiliae	0 0 0 0 9 1 1 0 1 0 0 1 1 0 0 0 0 0 0 0 1 2 0 1 1 0 1 0 1 0 1 0 0 1 0 0 0 0 0 3 0 0 1 0 1 0 0 1 0 0 2 2
E. eneidae	0 0 0 0 0 1 1 0 0 0 0 1 1 0 0 0 0 1 0 0 0 0 2 0 1 0 1 0 1 0 1 1 1 1 0 0 0 0 0 1 0 0 9 9 9 1 1 2 0 0 2 2
E. etheridgei	0 0 0 0 0 1 1 0 0 0 0 1 1 0 0 0 0 0 0 0 0 2 2 0 1 0 1 0 1 0 1 1 1 0 0 1 0 0 0 1 0 0 1 0 0 1 1 2 0 0 2 2
E. euraster	0 0 0 0 0 1 1 0 1 0 0 1 1 1 1 9 0 1 0 0 0 9 9 9 9 9 9 9 9 9 9 9 9 9 9 9 0 0 9 1 0 9 9 9 9 9 9 9 9 0 9 9
E. flavecens	0 0 0 0 0 1 1 1 0 0 0 1 1 0 0 0 0 1 0 0 0 0 0 0 1 0 1 0 0 0 1 1 1 1 0 0 0 0 0 1 0 0 9 9 9 1 1 2 9 0 2 2

APPENDIX 1. Data matrix for phylogenetic analysis with 134 taxa and 52 characters. Within the matrix a "9" represents missing data or that the datum is logically impossible.

	1 2
	1 2 3 4 5 6 7 8 9 0 1 2 3 4 5 6 7 8 9 0
E. fowleri	0 0 0 0 0 1 1 0 0 0 0 1 1 0 0 0 0 1 0 0
	0 9 9 9 9 9 9 9 9 9 9 9 9 9 9 9 0 0 9 1
	0 9 9 9 9 9 9 9 9 0 9 9
E. furcyensis	0 0 0 0 9 1 1 0 0 0 0 1 1 0 0 0 0 0 0 0
	1 2 0 3 1 0 1 0 1 0 1 1 1 1 0 0 0 0 0 3
	0 0 9 9 1 1 1 2 0 0 2 2
E. fuscus	0 0 0 0 1 0 1 0 0 0 0 1 1 0 0 0 0 9 0 0
	0 9 9 9 9 9 9 9 9 9 9 9 9 9 9 9 1 0 9 3
	0 9 9 9 9 9 9 9 9 0 9 9
E. glandulifer	0 0 0 0 0 1 1 0 0 0 0 1 1 0 0 0 0 1 0 0
	9 0 0 1 0 0 1 0 1 0 1 1 1 1 0 0 0 0 1 2
	0 0 9 9 9 1 1 2 0 0 2 2
E. glanduliferoides	0 1 0 0 9 1 1 1 1 0 0 1 1 1 1 0 0 1 0 0
	1 9 9 9 9 9 9 9 9 9 9 9 9 9 9 9 0 0 9 0
	0 9 9 9 9 9 9 9 9 0 9 9
E. glaphycampus	0 0 0 0 0 1 1 0 0 0 0 1 1 0 0 0 0 1 0 0
	1 2 2 3 1 0 1 0 1 0 1 1 1 1 1 0 0 0 0 3
	0 0 1 0 2 1 1 2 0 0 2 2
E. gossei	0 0 0 0 1 1 1 0 0 0 0 1 1 0 0 0 0 0 0 0
	0 2 3 2 1 0 1 0 0 0 1 1 0 1 1 0 0 0 0 3
	0 0 1 0 2 1 1 2 0 0 2 2
E. grabhami	0 0 0 0 1 1 1 0 0 0 0 1 1 0 0 0 0 1 0 0
	1 2 3 3 1 0 1 0 1 0 1 1 1 1 0 0 0 0 0 2
	0 0 1 0 1 1 1 2 0 0 2 2
E. grahami	0 1 0 0 0 1 1 1 0 0 0 1 1 0 0 0 0 1 0 0
	1 9 9 9 9 9 9 9 9 9 9 9 9 9 9 9 1 0 9 2
	0 9 9 9 9 9 9 9 9 0 9 9
E. greyi	0 0 0 0 9 1 1 0 0 0 0 1 1 0 0 0 0 1 0 0
	1 2 0 2 0 0 1 0 1 0 1 1 1 1 0 0 0 0 0 3
	0 0 1 0 0 1 1 2 0 0 2 2

APPENDIX 1. Data matrix for phylogenetic analysis with 134 taxa and 52 characters. Within the matrix a "9" represents missing data or that the datum is logically impossible.

	1 2 1 2 3 4 5 6 7 8 9 0 1 2 3 4 5 6 7 8 9 0
E. gryllus	0 0 0 0 0 1 1 0 1 0 0 1 1 1 0 1 0 1 0 0 0 9 9 9 9 9 9 9 9 9 9 9 9 9 9 9 0 0 9 0 0 9 9 9 9 9 9 9 9 0 9 9
E. guanahacabibes	0 9 0 0 9 1 1 9 9 0 9 9 1 0 0 9 0 9 0 0 1 9 9 9 9 9 9 9 9 9 9 9 9 9 9 9 1 0 9 2 0 9 9 9 9 9 9 9 9 0 9 9
E. gundlachi	0 0 0 0 0 1 1 0 0 0 0 1 1 0 0 0 0 1 0 0 0 9 9 9 9 9 9 9 9 9 9 9 9 9 9 9 1 0 9 3 0 9 9 9 9 9 9 9 9 0 9 9
E. haitianus	0 0 0 0 9 1 1 1 1 0 0 1 1 1 1 0 0 1 0 0 0 9 9 9 9 9 9 9 9 9 9 9 9 9 9 9 0 0 9 0 0 9 9 9 9 9 9 9 9 0 9 9
E. hedricki	0 0 0 0 0 1 1 1 1 0 0 1 1 1 1 0 0 1 1 0 0 2 3 1 1 0 1 1 0 0 1 1 1 1 1 0 0 0 0 0 0 0 9 9 0 1 1 2 9 0 2 2
E. heminota	0 0 0 0 9 1 1 0 0 0 0 1 1 0 0 0 0 1 0 0 0 2 0 0 1 0 1 0 1 0 1 1 0 1 0 0 0 0 0 1 0 0 1 0 1 1 1 2 0 0 2 2
E. hypostenor	0 0 1 0 0 1 1 0 0 0 0 1 1 0 0 0 0 9 0 0 0 0 0 0 0 0 1 0 1 0 1 1 1 1 0 0 0 0 0 1 0 0 1 0 0 1 1 0 0 0 2 2
E. inoptatus	0 0 1 0 0 0 1 0 0 0 0 1 1 0 0 0 0 0 0 0 0 0 3 2 0 1 0 0 0 0 1 1 1 1 0 0 0 0 0 1 0 1 1 0 0 1 1 0 0 0 2 2
E. intermedious	0 1 0 0 0 1 1 1 1 0 0 1 1 0 0 0 0 9 1 0 0 9 9 9 9 9 9 9 9 9 9 9 9 9 9 9 1 0 9 2 0 9 9 9 9 9 9 9 9 0 9 9
E. jamaicensis	0 0 0 0 1 1 1 0 1 0 0 1 1 0 1 9 0 1 0 0 0 0 0 0 1 0 1 0 1 0 1 1 1 1 0 0 0 0 0 2 0 0 1 0 1 1 1 2 0 0 2 2

APPENDIX 1. Data matrix for phylogenetic analysis with 134 taxa and 52 characters. Within the matrix a "9" represents missing data or that the datum is logically impossible.

	1 2
	1 2 3 4 5 6 7 8 9 0 1 2 3 4 5 6 7 8 9 0
E. jasperi	0 0 0 0 0 1 1 1 1 0 1 1 1 1 1 1 0 1 1 0
	0 2 0 0 1 0 1 0 1 0 1 1 1 1 0 0 0 1 0 9
	1 0 1 0 0 0 0 2 0 0 3 2
E. johnstonei	0 0 0 0 0 1 1 0 0 0 0 1 1 0 0 0 0 0 1 0
	0 0 3 0 1 0 1 0 1 0 1 1 1 1 0 0 0 0 0 1
	0 0 1 0 2 0 0 2 0 0 2 2
E. jugans	0 0 0 0 1 1 1 0 0 0 0 1 1 0 0 0 0 0 0 0
	0 0 3 2 0 0 0 0 1 0 1 1 1 1 1 0 0 0 0 3
	0 0 9 9 9 1 1 2 0 0 2 2
E. junori	0 0 0 0 9 1 1 1 1 0 0 1 1 0 0 0 0 1 0 0
	0 9 9 9 9 9 9 9 9 9 9 9 9 9 9 9 1 0 9 2
	0 9 9 9 9 9 9 9 9 0 9 9
E. karlschmidti	0 0 0 0 0 1 1 0 1 0 0 0 1 0 1 1 1 1 0 0
	0 2 2 0 0 0 1 0 0 0 1 0 1 1 0 0 0 0 0 1
	0 0 1 0 0 1 1 2 0 0 2 2
E. klinikoskii	0 0 0 0 9 1 1 0 0 0 0 1 1 0 0 0 0 1 0 0
	1 2 3 1 1 0 1 0 1 0 1 1 1 1 0 0 0 0 0 1
	0 0 1 0 2 1 1 2 0 0 2 2
E. lamprotes	0 0 0 0 0 1 1 0 1 0 0 1 1 1 0 0 0 1 1 0
	0 9 9 9 9 9 9 9 9 9 9 9 9 9 9 9 0 0 9 0
	0 9 9 9 9 9 9 9 9 0 9 9
E. leberi	0 0 0 0 0 1 1 0 0 0 0 1 1 0 0 0 0 1 0 0
	1 9 9 9 9 9 9 9 9 9 9 9 9 9 9 9 0 0 9 0
	0 9 9 9 9 9 9 9 9 0 9 9
E. lentus	0 0 0 0 0 1 1 0 0 0 0 1 1 0 0 0 0 0 0 0
	1 2 2 1 1 0 1 0 1 0 1 1 1 1 1 0 0 0 0 3
	0 0 1 0 0 1 1 2 0 0 2 2
E. leonci	0 0 0 0 1 1 1 0 0 0 0 1 1 0 0 0 0 1 0 0
	1 0 0 1 1 0 1 0 1 0 1 1 1 1 0 0 0 0 0 2
	0 0 9 9 9 1 1 2 0 0 2 2

APPENDIX 1. Data matrix for phylogenetic analysis with 134 taxa and 52 characters. Within the matrix a "9" represents missing data or that the datum is logically impossible.

	1 2 1 2 3 4 5 6 7 8 9 0 1 2 3 4 5 6 7 8 9 0
E. locustus	0 0 0 0 0 1 1 0 0 0 0 1 1 0 0 1 0 1 1 0 0 0 1 0 1 0 1 0 0 0 1 1 1 1 0 0 0 0 0 1 0 0 9 9 9 1 1 2 0 0 2 2
E. lucioi	0 1 0 0 1 1 1 0 0 0 0 1 1 0 0 0 0 1 0 0 0 9 9 9 9 9 9 9 9 9 9 9 9 9 9 9 1 0 9 3 0 9 9 9 9 9 9 9 9 0 9 9
E. luteolus	0 0 0 0 9 1 1 0 0 0 0 1 1 0 0 0 0 0 0 0 0 2 3 3 1 0 1 0 1 0 1 1 1 1 0 0 0 0 0 3 0 0 1 0 2 1 1 2 0 0 2 2
E. martinicensis	0 0 0 0 0 1 1 0 0 0 0 1 0 0 0 0 0 1 0 0 0 0 3 0 1 0 1 0 1 0 1 1 1 1 0 0 0 0 0 1 0 0 1 0 0 1 1 2 0 0 2 2
E. minutus	0 0 0 0 0 1 1 1 0 0 0 1 1 0 0 0 0 1 0 0 0 1 1 3 1 0 1 0 1 0 1 1 0 1 1 0 0 0 0 0 0 0 9 9 9 9 9 2 0 0 2 2
E. monensis	0 0 0 0 1 1 1 0 0 0 0 1 1 0 0 0 1 1 0 0 1 2 3 3 1 0 0 0 1 0 1 1 1 1 1 0 0 0 0 3 0 0 1 0 2 1 1 2 0 0 2 2
E. montanus	0 0 0 0 9 1 1 1 1 0 0 1 1 1 1 0 0 1 0 0 0 0 2 2 1 0 1 0 0 0 1 1 0 1 0 0 0 0 0 1 0 0 9 9 1 1 1 2 0 0 2 2
E. neodreptus	0 1 0 0 9 1 1 1 1 0 0 1 1 1 1 0 0 1 0 0 0 9 9 9 9 9 9 9 9 9 9 9 9 9 9 9 0 0 9 0 0 9 9 9 9 9 9 9 9 0 9 9
E. nortoni	0 0 0 0 0 1 1 0 0 0 0 0 1 0 0 0 0 0 0 0 0 9 9 9 9 9 9 9 9 9 9 9 9 9 9 9 0 0 9 1 0 9 9 9 9 9 9 9 9 0 9 9
E. nubicola	0 0 0 0 1 1 1 0 0 0 0 1 1 0 0 0 0 0 0 0 0 2 3 3 1 0 1 1 0 0 1 1 0 1 0 0 0 0 0 3 0 0 1 0 2 1 1 2 0 0 2 2

APPENDIX 1. Data matrix for phylogenetic analysis with 134 taxa and 52 characters. Within the matrix a "9" represents missing data or that the datum is logically impossible.

	1 2
	1 2 3 4 5 6 7 8 9 0 1 2 3 4 5 6 7 8 9 0
E. orcutti	0 0 0 0 1 1 1 0 1 0 0 0 1 1 1 0 1 1 0 9
	0 2 0 0 1 0 1 0 1 0 1 1 1 1 0 0 1 0 0 1
	0 0 1 0 1 1 1 2 0 0 2 2
E. oxyrhyncus	0 1 0 0 1 1 1 0 1 0 0 1 1 0 0 0 0 0 0 0
	0 9 9 9 9 9 9 9 9 9 9 9 9 9 9 9 1 0 9 3
	0 9 9 9 9 9 9 9 9 0 9 9
E. pantoni	0 0 0 0 1 1 1 0 0 0 0 1 1 0 0 0 0 0 0 0
	0 2 3 3 1 0 1 0 1 0 1 1 1 1 0 0 0 0 0 3
	0 0 1 0 2 1 1 2 0 0 2 2
E. parabates	0 0 0 0 9 1 1 0 0 0 0 1 1 0 0 0 0 1 0 0
	0 9 9 9 9 9 9 9 9 9 9 9 9 9 9 9 0 0 9 1
	0 9 9 9 9 9 9 9 9 0 9 9
E. patriciee	0 0 0 0 0 1 1 0 0 0 0 1 1 0 0 0 0 1 1 9
	0 0 1 0 1 0 1 0 1 0 1 1 1 1 0 0 0 0 0 0
	0 0 1 0 0 1 1 2 0 0 2 2
E. paulsoni	0 0 0 0 1 1 1 0 0 0 0 1 1 0 0 0 0 1 0 0
	1 9 9 9 9 9 9 9 9 9 9 9 9 9 9 9 1 0 9 2
	0 9 9 9 9 9 9 9 9 0 9 9
E. pezopetrus	0 0 0 0 1 1 1 0 0 0 0 1 1 0 0 0 0 1 0 0
	1 2 3 1 1 0 1 0 1 0 1 1 1 1 0 0 0 0 0 2
	0 0 1 0 9 1 1 2 0 0 2 2
E. pictissimus	0 0 0 0 0 1 1 0 0 0 0 1 1 0 0 0 0 1 0 0
	1 2 0 0 1 0 1 0 1 0 1 1 1 1 0 0 0 0 0 2
	0 0 1 0 0 1 1 2 0 0 2 2
E. pinarensis	0 0 0 0 1 1 1 1 1 0 0 1 1 0 0 0 0 1 0 0
	1 2 3 3 0 0 1 0 1 0 1 1 1 1 9 0 0 0 1 3
	0 0 1 0 0 1 1 2 0 0 2 2
E. pinchoni	0 0 0 0 9 1 1 0 1 0 0 1 1 0 1 1 0 1 0 0
	0 0 3 1 1 0 1 0 1 0 1 1 1 1 0 0 0 0 0 1
	0 0 1 0 2 0 0 2 0 0 2 2

APPENDIX 1. Data matrix for phylogenetic analysis with 134 taxa and 52 characters. Within the matrix a "9" represents missing data or that the datum is logically impossible.

	1 2 1 2 3 4 5 6 7 8 9 0 1 2 3 4 5 6 7 8 9 0
E. pituinus	0 0 0 0 0 1 1 1 1 0 0 9 1 0 0 0 0 1 0 0 0 9 9 9 9 9 9 9 9 9 9 9 9 9 9 9 0 0 9 1 0 9 9 9 9 9 9 9 9 0 9 9
E. planirostris	0 0 0 0 1 1 1 0 0 0 0 1 1 0 0 0 0 1 0 0 1 2 2 3 1 0 1 0 1 0 1 1 1 1 0 0 0 0 0 2 0 0 1 0 0 1 1 2 0 0 2 2
E. poolei	0 0 0 0 0 1 1 0 0 0 0 1 1 0 0 0 0 1 0 0 1 0 0 9 1 0 1 0 1 0 1 1 1 0 0 0 0 0 0 0 0 0 9 9 1 1 1 2 0 0 2 2
E. portoricensis	0 0 0 0 0 1 1 1 1 0 0 1 1 1 1 0 0 1 1 0 0 0 3 0 1 0 1 0 1 0 1 1 1 1 0 0 0 0 0 1 0 0 1 0 0 1 1 2 0 0 2 2
E. probolaeus	0 0 0 0 0 1 1 0 0 0 0 1 1 0 0 0 0 1 0 0 1 9 9 9 9 9 9 9 9 9 9 9 9 9 9 9 1 0 9 2 0 9 9 9 9 9 9 9 9 0 9 9
E. rhodesi	0 1 0 0 9 1 1 0 0 9 0 1 1 0 0 0 0 1 0 0 0 9 9 9 9 9 9 9 9 9 9 9 9 9 9 9 1 0 9 3 0 9 9 9 9 9 9 9 9 0 9 9
E. richmondi	0 0 0 0 0 1 1 0 0 0 0 1 1 0 0 1 0 1 0 0 1 0 2 0 1 0 1 0 1 0 1 1 1 0 0 0 0 0 0 3 0 0 1 0 0 1 1 2 1 0 2 2
E. ricordii	0 0 0 0 1 1 1 0 0 0 0 1 1 0 0 0 0 9 0 0 1 2 2 3 0 0 1 0 1 0 1 1 1 1 0 0 0 0 1 3 0 0 1 0 0 1 1 2 0 0 2 2
E. ronaldi	0 0 0 0 0 1 1 0 0 0 0 1 1 0 0 0 0 1 0 0 0 0 0 0 1 0 1 0 1 0 1 1 1 1 0 0 0 0 0 1 0 0 1 0 2 1 1 2 1 0 2 2
E. rufifemoralis	0 0 0 0 9 1 1 0 0 0 0 1 1 0 0 0 0 9 0 0 0 2 2 1 1 0 1 0 1 0 1 1 1 0 0 0 0 0 0 1 0 0 9 9 9 9 9 2 0 0 2 2

APPENDIX 1. Data matrix for phylogenetic analysis with 134 taxa and 52 characters. Within the matrix a "9" represents missing data or that the datum is logically impossible.

	1 2 1 2 3 4 5 6 7 8 9 0 1 2 3 4 5 6 7 8 9 0
E. ruthae	0 0 1 0 0 0 1 1 0 0 0 1 1 0 0 0 0 0 0 0 0 0 2 1 0 0 1 1 1 0 1 1 1 1 1 0 0 0 0 1 0 1 1 0 0 1 1 0 0 0 2 2
E. schmidti	0 0 0 0 1 1 1 0 0 0 0 9 1 0 0 0 0 1 0 0 1 2 3 1 0 0 1 0 1 0 1 1 1 1 0 0 0 0 0 2 0 0 1 0 0 1 1 0 0 0 2 2
E. schwartzi	0 0 0 0 0 1 1 0 0 0 0 1 1 0 0 0 0 1 0 0 0 9 9 9 9 9 9 9 9 9 9 9 9 9 9 9 0 0 9 1 0 9 9 9 9 9 9 9 9 0 9 9
E. sciagraphus	0 0 0 0 1 1 1 0 1 0 0 1 1 1 1 0 0 1 0 0 0 9 9 9 9 9 9 9 9 9 9 9 9 9 9 9 0 0 9 1 0 9 9 9 9 9 9 9 9 0 9 9
E. semipalmatus	0 0 0 0 0 1 1 1 1 0 0 0 1 1 1 9 1 0 0 0 0 0 2 1 1 0 1 0 1 0 1 1 1 1 1 0 0 0 0 1 0 0 1 0 1 1 1 2 0 0 2 2
E. sierramaetrae	0 0 0 0 1 1 1 1 1 0 0 1 1 0 0 0 0 0 0 0 1 0 2 0 0 0 1 0 1 0 1 1 1 1 0 0 0 0 0 2 0 0 1 0 0 1 1 2 0 0 2 2
E. sisyphodemus	0 0 0 0 1 1 1 1 1 0 0 1 1 0 1 0 0 0 0 0 0 9 9 9 9 9 9 9 9 9 9 9 9 9 9 9 1 0 9 2 0 9 9 9 9 9 9 9 9 0 9 9
E. symingtoni	0 0 1 0 1 9 1 1 0 0 0 0 1 0 0 0 0 1 0 1 0 2 3 2 0 0 0 0 0 0 1 1 1 1 1 0 0 0 0 1 0 1 1 0 1 1 1 2 1 0 2 2
E. thomasi	0 0 0 0 0 1 1 0 0 0 0 1 1 0 0 0 0 9 0 0 1 2 3 1 1 0 1 0 1 0 1 1 1 1 1 0 0 0 0 3 0 0 1 0 0 1 1 2 0 0 2 2
E. turquinensis	0 0 0 0 1 1 1 0 1 0 0 0 1 1 1 0 1 1 0 0 0 9 9 9 9 9 9 9 9 9 9 9 9 9 9 9 0 0 9 1 0 9 9 9 9 9 9 9 9 0 9 9

APPENDIX 1. Data matrix for phylogenetic analysis with 134 taxa and 52 characters. Within the matrix a "9" represents missing data or that the datum is logically impossible.

	1 2
	1 2 3 4 5 6 7 8 9 0 1 2 3 4 5 6 7 8 9 0
E. unicolor	0 0 0 0 0 1 1 1 1 0 1 1 0 1 1 1 0 1 0 0
	0 1 1 2 0 0 1 0 1 0 1 1 0 1 0 0 0 0 0 1
	0 0 1 0 1 0 0 2 1 0 2 2
E. urichi	0 0 0 0 0 1 1 0 0 0 0 1 1 0 0 0 0 1 0 0
	0 0 0 0 0 0 1 0 1 0 1 1 1 1 0 0 0 0 0 0
	0 0 1 1 0 1 1 2 0 0 2 2
E. varians	0 0 0 0 0 1 1 0 0 0 0 1 1 0 0 0 0 1 0 0
	0 9 9 9 9 9 9 9 9 9 9 9 9 9 9 9 0 0 9 0
	0 9 9 9 9 9 9 9 9 0 9 9 0 0
E. varleyi	0 0 0 0 0 1 1 0 0 0 0 1 1 0 0 0 0 9 0 0
	0 2 3 0 1 0 1 0 1 0 1 1 1 1 0 0 0 0 0 1
	0 0 1 0 2 1 1 2 0 0 2 2
E. ventrilineatus	0 1 0 0 1 0 1 1 1 0 0 1 0 1 1 0 0 1 0 0
	0 1 0 1 0 0 1 0 1 0 1 1 1 1 0 0 0 0 0 1
	0 0 9 9 9 1 1 2 0 0 2 2
E. warreni	0 0 0 0 0 1 1 0 0 0 0 1 1 0 0 0 0 1 0 0
	1 9 9 9 9 9 9 9 9 9 9 9 9 9 9 9 1 0 9 3
	0 9 9 9 9 9 9 9 9 0 9 9
E. weinlandi	0 0 0 0 0 1 1 0 0 0 0 1 1 0 0 0 0 1 0 0
	1 2 3 1 1 0 1 0 1 0 1 1 1 1 0 0 0 0 0 3
	0 0 1 0 0 1 1 2 0 0 2 2
E. wetmori	0 1 0 0 0 1 1 0 0 0 0 1 1 0 0 0 0 1 0 0
	0 0 0 0 1 0 1 0 1 0 1 1 1 1 0 0 0 0 0 0
	0 0 1 0 2 1 1 2 0 0 2 2
E. wightmanae	0 0 0 0 0 1 1 1 1 0 0 1 1 1 1 9 0 1 0 0
	0 0 0 1 1 0 1 0 1 0 1 1 1 1 0 0 0 0 0 1
	0 0 1 0 2 1 1 2 0 0 2 2
E. zeus	0 0 1 0 9 1 1 1 0 0 0 1 1 1 1 9 0 9 0 0
	0 2 3 1 0 0 0 0 1 0 1 1 1 1 1 0 0 0 0 1
	0 1 1 0 1 1 1 2 1 0 2 2

APPENDIX 1. Data matrix for phylogenetic analysis with 134 taxa and 52 characters. Within the matrix a "9" represents missing data or that the datum is logically impossible.

	1 2
	1 2 3 4 5 6 7 8 9 0 1 2 3 4 5 6 7 8 9 0

E. zugi	0 0 0 0 1 1 1 0 0 0 0 1 1 0 0 0 0 1 0 0 1 0 3 3 9 0 1 0 1 0 1 1 1 1 0 0 0 0 0 1 0 0 9 9 9 1 1 2 0 0 2 2
Hylactophryne	0 0 9 0 0 0 0 0 0 0 1 0 0 0 1 0 9 0 9 0 9 9 0 0 9 0 0 0 0 1 1 1 1 9 0 0 0 0 0 0 0 0 1 0 0 0 0 0 0 1 2 2
Ischnocnema	0 0 0 0 1 0 0 0 0 0 0 1 0 0 0 0 0 1 0 1 0 2 2 0 0 1 0 0 1 0 0 1 1 0 0 0 0 0 0 1 0 0 1 0 0 1 1 0 0 0 2 2
Leptodactylinae	0 1 0 0 0 0 0 0 0 0 0 1 0 1 0 0 0 0 0 0 0 0 1 0 0 0 0 0 0 1 0 0 0 0 0 0 0 0 0 1 1 0 0 0 3 0 0 0 0 0 1 0
S. leprus	0 0 0 0 1 1 1 0 0 0 0 1 1 0 0 0 0 1 0 0 1 2 3 2 1 0 1 0 1 0 1 1 1 1 0 1 1 1 9 9 1 0 1 0 2 1 1 2 0 0 2 2
S. lorgipes	0 0 0 0 1 1 1 0 0 0 0 1 1 0 0 0 0 1 0 0 1 2 0 0 1 0 1 0 1 0 1 1 1 1 0 1 1 1 9 9 1 0 1 0 0 1 1 2 9 0 2 2
S. marnokii	0 0 0 0 1 1 1 0 0 0 0 1 1 0 0 0 0 1 0 0 1 2 2 0 1 0 1 0 1 0 1 1 1 1 0 1 1 1 9 9 1 0 1 0 0 1 1 2 0 0 2 2
S. pipilans	0 0 0 0 0 1 1 0 0 0 0 1 1 0 0 0 0 1 0 0 1 2 3 0 1 0 1 0 1 0 1 1 1 1 0 1 1 1 9 9 1 0 1 0 0 1 1 2 0 0 2 2
S. teretistas	1 0 0 0 0 1 1 0 0 0 0 1 0 0 0 0 0 1 0 0 1 2 2 0 0 0 1 0 1 0 1 1 1 0 0 1 1 1 9 9 1 0 1 0 1 1 1 2 0 0 2 2
T. albolabris	0 0 0 0 0 1 1 0 0 0 0 1 0 0 0 0 0 1 0 0 1 2 3 1 1 0 1 0 1 0 1 1 1 1 0 1 1 1 9 9 1 0 1 0 0 1 1 2 0 0 2 2

APPENDIX 1. Data matrix for phylogenetic analysis with 134 taxa and 52 characters. Within the matrix a "9" represents missing data or that the datum is logically impossible.

	1 2
	1 2 3 4 5 6 7 8 9 0 1 2 3 4 5 6 7 8 9 0
T. nitidus	0 0 0 0 0 1 0 0 0 0 0 1 0 0 0 0 0 1 0 0 0 0 1 0 1 0 1 0 1 0 1 1 1 1 0 1 1 1 9 9 1 0 1 0 0 1 1 2 0 0 2 2
T. saxatilis	0 0 0 0 0 1 1 0 0 0 0 1 1 0 0 0 0 1 0 0 1 0 0 0 1 0 1 0 1 0 1 1 1 1 0 1 1 1 9 9 1 0 1 0 1 1 1 2 0 0 2 2
S. limbatus	0 0 0 0 0 1 1 0 0 0 0 1 1 0 0 0 0 1 0 0 0 0 1 0 0 0 1 0 1 0 1 1 1 1 0 1 1 0 0 0 1 0 1 1 9 1 1 2 0 0 2 2

APPENDIX II. List of characters, direction of change of states, branches along which the changes occur, and consistency ratios on the cladogram of relationships in Fig.1. Deleted characters (*) not included in length calculations.

CHARACTER	CHANGED FROM	TO	ALONG BRANCH	CONSISTENCY
1	0	1	G ---> *E. unistrigatus* group	0.500
	0	1	B ---> *Cycloramphus*	
2	0	1	A ---> Leptodactylinae	1.000
3	0	1	B ---> *Cycloramphus*	1.000
4	CONSTANT CHARACTER			0
5	0	1	D ---> *Ischnocnema*	1.000
6 *	0	1	E ---> F	0.500
	0	1	B ---> *Cycloramphus*	
7	0	1	D ---> E	1.000
8	CONSTANT CHARACTER			0
9	CONSTANT CHARACTER			0
10	CONSTANT CHARACTER			0
11	CONSTANT CHARACTER			0
12	0	1	B ---> C	0.333
	1	0	G ---> *E. unistrigatus* group	
	1	0	E ---> *Craugastor*	
13	0	1	D ---> E	1.000
14	CONSTANT CHARACTER			0
15	CONSTANT CHARACTER			0

APPENDIX II. (continued).

CHARACTER	CHANGED FROM	TO	ALONG BRANCH	CONSISTENCY
16	0	1	ANCESTOR ---> A	0.33
	1	0	E ---> *Craugastor*	
	1	0	B ---> *Cycloramphus*	
17	0	1	E ---> *Craugastor*	0.500
	0	1	B ---> *Cycloramphus*	
18	0	1	B ---> C	0.500
	1	0	F ---> *E. inoptatus* group	
19	CONSTANT CHARACTER			0
20	0	1	A ---> B	0.500
	1	0	E ---> F	
21	0	1	G ---> H	1.000
22	0	2	A ---> B	0.500
	2	0	E ---> F	
	0	2	G ---> H	
	2	1	B ---> *Cycloramphus*	
23	0	3	E ---> *Craugastor*	1.000
	0	2	D ---> *Ischnocnema*	
	0	1	A ---> Leptodactylinae	
24	0	2	E ---> *Craugastor*	1.000
25	0	1	G ---> H	1.000

APPENDIX II. (continued).

CHARACTER	CHANGED FROM	CHANGED TO	ALONG BRANCH	CONSISTENCY
26	0	1	A ---> B	0.500
	1	0	E ---> F	
27	0	1	E ---> F	1.000
28	0	1	B ---> *Cycloramphus*	1.000
29	0	1	G ---> H	0.500
	1	0	D ---> *Ischnocnema*	
30	0	1	ANCESTOR ---> A	0.500
	1	0	B ---> C	
31	0	1	A ---> B	0.500
	1	0	D ---> *Ischnocnema*	
32	0	1	B ---> C	0.500
	1	0	E ---> *Craugastor*	
33	0	1	A ---> B	1.000
34	0	1	E ---> F	0.500
	0	1	C ---> *Hylactophryne*	
35	0	1	E ---> *Craugastor*	0.500
	0	1	B ---> *Cycloramphus*	
36	0	1	G ---> H	1.000
37	0	1	G ---> H	1.000
38	0	1	G ---> H	1.000
39	0	1	G ---> *E. ricordii* group	1.000

APPENDIX II. (concluded).

CHARACTER	CHANGED FROM	TO	ALONG BRANCH	CONSISTENCY
40 *	0	1	D ---> E	0.500
	1	2	G ---> *E. ricordii* group	
	1	0	G ---> *E. unistrigatus* group	
	0	1	A ---> Leptodactylinae	
41	0	1	G ---> H	0.500
	0	1	A ---> Leptodactylinae	
42	CONSTANT CHARACTER			0
43	0	1	B ---> C	1.000
44	0	1	G ---> *E. unistrigatus* group	1.000
45	0	3	A ---> Leptodactylinae	.000
46	0	1	C ---> D	1.000
47	0	1	C ---> D	1.000
48	0	2	F ---> G	0.500
	0	2	B ---> *Cycloramphus*	
49	CONSTANT CHARACTER			0
50	0	1	C ---> *Hylactophryne*	0.500
	0	1	E ---> *Craugastor*	
51	0	2	ANCESTOR ---> A	1.000
	2	1	A ---> Leptodactylinae	
52	0	1	A ---> B	1.000
	1	2	B ---> C	

Appendix III---List of occurence of preferred character-state transitions on the cladogram of relationships in Fig. 1. Deleted characters (*) not included in length calculations.

Node	Ancestor	Character	Ancestral State	Derived State
Leptodactylinae	A	2	0	1
		23	0	1
		40 *	0	1
		41	0	1
		45	0	3
		51	2	1
Cycloramphum	B	1	0	1
		3	0	1
		6 *	0	1
		16	1	0
		17	0	1
		22	2	1
		28	0	1
		35	0	1
		48	0	2
Ischnocnema	D	5	0	1
		23	0	2
		29	0	1
		31	1	0

Appendix III---(continued).

Node	Ancestor	Character	Ancestral State	Derived State
Craugastor	E	12	1	0
		16	1	0
		17	0	1
		23	0	3
		24	0	2
		32	1	0
		35	0	1
		50	0	1
E. unistrigatus group	G	1	0	1
		12	1	0
		40 *	1	0
		42	0	1
E. auriculatus group	G	[None]		
E. inoptatus group	F	9	1	0
E. ricordii group	G	39	0	1
		40 *	1	0
Syrrhophus	H	[None]		
Tomodactylus	H	[None]		
Hylactophryne	C	34	0	1
		50	0	1

Appendix III---(continued).

Node	Ancestor	Character	Ancestral State	Derived State
H	G	21	0	1
		22	0	2
		29	0	1
		36	0	1
		37	0	1
		38	0	1
		41	0	1
G	F	25	0	1
		48	0	2
F	E	6 *	0	1
		20	1	0
		22	2	0
		26	1	0
		27	0	1
		34	0	1
E	D	7	0	1
		13	0	1
		40 *	0	1
D	C	46	0	1
		47	0	1

Appendix III---(concluded).

Node	Ancestor	Character	Ancestral State	Derived State
C	B	12	0	1
		18	0	1
		30	1	0
		32	0	1
		43	0	1
		52	1	2
B	A	20	0	1
		22	0	2
		26	0	1
		31	0	1
		33	0	1
		52	0	1
A	ANCESTOR	16	0	1
		30	0	1
		51	0	2

THE RELATIONSHIPS OF ANTILLEAN *TYPHLOPS* (SERPENTES: TYPHLOPIDAE) AND THE DESCRIPTION OF THREE NEW HISPANIOLAN SPECIES

Richard Thomas[1]

Abstract

Three new species of *Typhlops* are described from Hispaniola (*T. schwartzi*, *T. tetrathyreus*, and *T. titanops*). A phylogeny is proposed in which two groups occupy the West Indies, one with three members, which Dixon and Hendricks (1978) included in their Caribbean Arc Groups with representatives in Central and South America, and another, the Major Antillean Radiation (MAR), with 16 species. The MAR species show synapomorphic connectivity between various Antillean islands: The Caymans-Cuba-Bahamas, Cuba-Hispaniola, Puerto Rico-Bahamas and, less securely: Jamaica-Northern Lesser Antilles. Certain groupings within the MAR are unresolved multifurcations. The MAR appears to share its defining synapomorphies with the West African *Typhlops caecatus* rather than with any New World species.

Introduction

In the West Indies, species of *Typhlops* are found on all major and most smaller islands of the Greater Antilles and irregularly in the Lesser Antilles (south of the Anegada Passage). They appear to be absent from the northernmost of the Lesser Antilles, the Anguilla Bank, and Saba, but are present, almost continuously, from the Antigua-Barbuda bank and the St. Kitts-Nevis Bank south to Dominica. Their apparent absence on La Desirade, Marie-Galante, and Les Saintes may only reflect insufficient field effort. Between Dominica and Grenada, including Barbados, the genus is unknown.

In the following list of West Indian *Typhlops* I have elevated *T. biminiensis epactius* and *T. capitulatus gonavensis* to specific rank. Both are insular forms that are relatively well differentiated from their closest relatives. The islands are listed west to east and them north to south with allowances for contiguity of faunas.

Cuba

    lumbricalis (Linnaeus) (also some peripheral islands of Cuba)
    biminiensis Richmond

---

[1] Richard Thomas is Associate Professor of Biology, University of Puerto Rico, Río Piedras, PR 00931.

Bahamas
    lumbricalis (Linnaeus) (Great and Little Bahama Banks)
    biminiensis Richmond (Great Bahama Bank and Great Inagua)
    richardi Duméril and Bibron (Turks and Caicos)

Cayman Islands
    caymanensis Sackett (Grand Cayman)
    epactius Thomas, new combination (Cayman Brac)

Jamaica
    jamaicensis (Shaw)

Navassa
    sulcatus Cope

Hispaniola
    capitulatus Richmond
    gonavensis Richmond
    hectus Thomas
    pusillus Barbour
    sulcatus Cope
    syntherus Thomas
    schwartzi, new species
    tetrathyreus, new species
    titanops, new species

Mona
    monensis (Meerwarth)

Puerto Rico Bank
    granti Ruthven and Gaige (Puerto Rico and Caja de Muertos only)
    platcephalus Duméril and Bibron (Puerto Rico only)
    richardi Duméril and Bibron
    rostellatus Stejneger (Puerto Rico only)
    sp. (Puerto Rico only)

St. Croix
    richardi Duméril and Bibron

Antigua-Barbuda Bank, St. Eustatius, St. Kitts, Nevis, and Montserrat
    monastus Thomas

Guadeloupe and Dominica
    dominicanus Stejneger

Grenada
    tasymicris Thomas

Some groupings of Antillean *Typhlops* have been recognized. Ruthven and Gaige (1935) designated a "*jamaicensis* group" that included *Typhlops jamaicensis*, *richardi*, *granti*, and *monastus* (not then names). Richmond (1955) noted that in *T. biminiensis*, the preocular contacts supralabials 2 and 3 and does not conform to the condition in the majority of Antillean species in which the preocular contacts supralabial 3 only. I recognized (Thomas 1968) a "*biminiensis* group" that included *caymanensis*, *biminiensis*, and *epactius* (then described as a subspecies of *biminiensis*). I later (Thomas 1976) included all Antillean species except those of the *biminiensis* group in a Major Antillean Radiation and elaborated on the features characterizing it. Dixon and Hendricks (1979) recognized a Caribbean Arc group that includes the *biminiensis* group plus the Central American species, *costaricensis*, *tenuis*, and *microstomus*, and from northwestern South America and Grenada, *lehneri*, *trinitatus*, and *tasymicris*. The four remaining South American species (*brongersmianus* Vanzolini, *minisquamus* Dixon and Hendricks, *paucisquamis* Dixon and Hendricks, and *reticulatus* Linnaeus) they relegated to a third group having preocular contact with supralabials 2 and 3 and a semidivided nasal scale.

Before proceeding with a discussion of the relationships of the Antillean species of *Typhlops* it is necessary to deal with some unresolved problems: the taxonomy and nomenclature of the populations now included under the name "*lumbricalis*." The type of Linnaeus's *Anguis lumbricalis* was in the collection of Gronovius and is apparently no longer extant (A. C. Wheeler in litt.). Cochran (1924) concluded that Linnaeus's brief description (1758) was sufficient to restrict the name *Typhlops lumbricalis* to those West Indian populations having low longitudinal scale counts (around 300 or less) and 20 scale rows on the anterior part of the body. She considered the name *lumbricalis* to apply to populations inhabiting Cuba, the Bahamas, and Hispaniola. I concluded (Thomas 1974) that some of these *Typhlops* of southwestern Hispaniola having 20 scale rows and moderate longitudinal counts pertained to a distinct species and proposed the name *Typhlops hectus* for them but continued to apply the name *lumbricalis* to other Hispaniolan populations of 20-18 row snakes with low middorsal-counts, although noting that the various populations differ among themselves and from those of Cuba and the Bahamas. With the acquisition of considerably more material of the Hispaniolan "*lumbricalis*," it has become possible to better discriminate the taxa. To avoid formal attribution of names in a dissertation (Thomas 1976), I designated two unnamed Hispaniolan species *Typhlops* I and *Typhlops* II. The former I considered to be represented by two subspecies, one in eastern and one in western Hispaniola. More material and further assessment of their characteristics lead me to believe that these eastern and western populations are best treated as separate species.

I have argued (Thomas 1976) that because of the low longitudinal counts given in Linnaeus's description (1758), the most likely provenance for the type of *Typhlops lumbricalis* is the Bahamas. Since New Providence has long been the commercial and political center of the Bahamas and since *Typhlops lumbricalis* occurs there, it is a reasonable source for Linnaeus's type, and it is appropriate to restrict the type-locality of *Typhlops lumbricalis* to the island of New Providence, Bahama Islands. Since my dissertation (Thomas 1976) was not published, the present statement may be considered a formal restriction of the type-locality. On this premise the name *lumbricalis* is attributable to populations of 20-row *Typhlops* (reducing to 18 rows posteriorly) occurring on many islands of the Little and Great Bahama Banks, Cuba, some of its offshore islands, and the Isla de la Juventud (Isle of Pines). Within *lumbricalis*, as I envision it, there is some, largely clinal, geographic variation, but the species is not well-represented in collections from some areas of Cuba and the variation remains to be assessed. (See Schwartz and Thomas 1975, for a complete listing of islands of occurrence.) Remaining on Hispaniola are the three allopatric populations of 20-18 row *Typhlops* with low middorsal counts and relatively unspecialized anterior head scales but differing from one another in a number of characters. It is these populations that I describe in this paper.

Materials and methods

The descriptions of the species follow a format used by Thomas (1976) with some changes. Asterisks in the diagnoses indicate autapomorphies. Linear measurements of head scales were taken with an ocular micrometer in a dissecting microscope. Skeletal data were taken from cleared and stained specimens, supplemented by X-rays to ascertain variation in some features.

The following conventions and abbreviations are used in the descriptions: RL, rostral length (from internasal suture to tip); RW, rostral width (widest part in dorsal aspect); TL, total length; TA, tail length; MBD, midbody diameter. Sinuosity of the anterior edge of the ocular (= the ocular-preocular suture) is the ratio of the smallest length of the ocular to the greatest length of the ocular. The posterior nasals are described as being divergent, if their edges diverge anterior to posterior, parallel, or calyculate if they are bowed (parenthesis-like); APNW is the anterior postnasal width, the distance between the apices of the preoculars; PPNW is the posterior postnasal width, the distance between the edges of the postnasals near their ends. A comparison of these two measurements (as in Fig. 6) indicates amount of divergence, although it will not separate parallel from calyculate conditions. For the most part the osteological descriptions are self-explanatory. For the foramina of cranial nerve V, I have used the term "lappet" for the ventrad projection of the prootic and the term "tongue" for the anteriad projection of the prootic (Fig. 4). Figures 3 and 4 show the range of conditions in order to make the textual descriptions meaningful and do not necessarily illustrate specimens or species mentioned in the text.

The cladogram (Fig. 7) was constructed by searching a character matrix for less and less inclusive synapomorphies and for apparent character transformation series. A number of characters used in the descriptions were eliminated from consideration in forming the cladogram, because their pattern of occurrence was unorderly and because their polarity could not be assessed with even slight confidence. Collections cited: KU, University of Kansas, Museum of Zoology; MCZ, Museum of Comparative Zoology, Harvard; MNHNSD, Museo Nacional de Historia Natural, Santo Domingo, República Dominicana; UF, Florida State Museum, University of Florida, Gainesville; USNM, United States National Museum; RT, Richard Thomas personal collection.

Taxonomic descriptions

The first of these three species I take great pleasure in naming for Dr. Albert Schwartz both for his enormous contributions to our knowledge of the Antillean fauna and, personally, for his stimulus and encouragement to me throughout my biological career. Since this species is one of the largest and most spectacular of the Hispaniolan *Typhlops* it is particularly appropriate that it be named after him; furthermore, he was personally responsible for obtaining a majority of the specimens.

Typhlops schwartzi, new species

HOLOTYPE.--KU 208752, one of a series collected 1.5 km W Jayaco, La Vega Province, República Dominicana, 244 m, on 11 August 1973 by native collectors.

PARATYPES. (all from República Dominicana).--Azua Prov.: KU 208753, 2.9 km W, thence 16.4 km N, Azua. Puerto Plata Prov.: USNM 10276, Puerto Plata. Duarte Prov.: MNHNSD 165, San Pedro de Macorís, Ingenio Colén, August 1973, Zacarias Ceara Aybar. La Vega Prov.: KU 208754, 4 km S La Vega, 20 July 1968, Albert Schwartz; RT 3633, 2 km W Jayaco, 20 August 1975, native collector; KU 208755-58, 208759-70, 208771-74, 208775, 208776, same data as holotype; KU 208777-80, 208781-83, 208784-86, same locality as holotype, 1-2 January 1975; KU 208787-90, 208791-92, 1 km W Jayaco, 900 feet (274 m), 2 January 1973, Mark D. Lavrich, natives. Samaná Prov.: USNM 55298, Sánchez, October, 1916, W. L. Abbott. San Pedro de Macorís Prov.: AMNH 13630, San Pedro de Macorís, G. K. Noble. La Altagracia Prov.: RT 3601, 3 km SW Higüey, 17 August 1975, R. Thomas.

DIAGNOSIS.--A very large, stout *Typhlops* (to 326 mm TL; Fig. 2) with extensive pigmentation on the facial region and venter; no pigment collar; low to moderate number of middorsal scales (237-282); scale rows 20 reducing to 18 at 46-67% TL; rostral a narrow oval to parallel sided (W/L = .47-.56), extending posteriorly to level of eyes; postnasal pattern divergent (Fig.1); preocular contacting only 3rd of supralabials, broadly angled, rounded apically; upper and lower sutures with postnasal not strongly curved. Parietals single, greatly extended laterally and short (major axis transverse), blade-like, extending ventrad along posterior edge of ocular to the level of the lower edge of the eye; postoculars single, high and short; prominent blade-like atlantal hypapophyses; pelvic moieties with ilium ischium and pubis fused (hatchet-shaped).

DESCRIPTION.--1) Snout rounded. 2) Rostral in dorsal aspect a narrow oval to parallel sided (RW/RL 0.43-0.60), oval (mode); slightly flared on apex; labial margin slightly flared. 3) Preocular subtriangular, broadly angled (50-80 degrees), apex rounded; lower portion contacting only 3rd of upper labials. 4) Ocular length about 2/3 height, sinuosity 0.16-0.07. 5) Postnasal pattern divergent. 6) Postocular single (strong mode), higher than long. 7) First parietal greatly extended laterally and blade-like (major axis transverse), extending along ocular to below level of eye. 8) Second parietal spanning 2 scale rows or absent. 9) TL to 326 mm (Fig. 2). 10) TL/TA: males 22-37, females 26-41. 11) TL/MBD: 23-38. 12) Middorsal scales 237-282 ($\bar{x}$ = 265.1, SE = 1.38, N = 52). 13) Scale rows 20-18 with reduction occurring at about midbody (46-67 percent TL). 14) Coloration: dark brown dorsal pigmentation extending onto ventrolateral surface and fading onto venter; facial pigmentation dark and extensive, extending over sides of head onto ventrolateral surface; rostral pigmented over nearly all of its length. 15) Rectal Caecum present. 16) Hemipenes expanded, apical region oblique in completely everted organs. 17) Cranium broad, sides of parietals tapering very slightly. 18) Premaxilla broad, about 40 percent of width across prefrontals, slightly convex, not protuberant; posteroventral edged transverse, making a right-angle juncture with narrow blade. 19) Nasals without lateral angle. 20) Septomaxilla with silver of bone extending along lateral margin of naris, anterior portion tapered. 21) Frontal-parietal suture transverse, slightly sinuous. 22) Frontal with mostly unfused anterior ventral blade-like process (Fig. 3A). 23) Optic foramen canalicular (Fig. 3B-D). 24) Postorbital process of parietal prominent. 25) Foramen of cranial nerve V: lappet of prootic very prominent, in broad contact with sphenoid and parietal (apparently fused with tongue; tongue not visible); prominent secondary foramen isolated (similar to Fig. 4A). 27) Supraoccipitals unfused, in broad median contact (ASFS V26874 with left supraoccipital very reduced). 28) Exoccipitals not fused with prootics. 29) Angular not sliver-like. 30) Dorsal process of quadrate hooked. 31) Atlantal hypapophysis prominent, bladelike; 5-6 total hypapophyses. 32) Hyoid U-shaped, composed of 2 fused ceratobranchials; basihyal absent. 33) Pelvic moieties: in females, absent or composed of 2 small rodlike ischia; in males, prominent and hatchet-shaped with broad pubic process and more slender ilium and ischium. 34) Eye moderate with narrow orbital space.

DATA ON HOLOTYPE.--An adult female, TL 265 mm, tail 10.9 mm; MBD 8.6; middorsal scales 256; ventrals 244; scale rows 20, reducing to 19 at midventral scale 139 and to 18 at midventral 149 (135 and 146 mm posterior to snout, respectively); final reduction step (19 to 18 rows) occurring at 55 % TL; parietals 1/1, both extended; postoculars 1/1; pigmented scale rows 15.

DISTRIBUTION.--*Typhlops schwartzi* occurs in eastern Hispaniola, where it is known from the eastern two-thirds of the Dominican Republic, from northwest of Azua east to Higüey and north to Puerto Plata and Sánchez at the base of the Samaná Peninsula (Fig. 5).

Typhlops tetrathyreus, new species

HOLOTYPE.--KU 208793, an adult female taken 3 mi. (4.8 km) N Pétionville, Dépt. de l'Ouest, Haiti, by "Marcellus" on 28 November 1970.

PARATYPES.--Haiti: Dépt. du Centre: KU 208794-95, Fond Michelle, 545 m. KU 208796, 0.5 km S Terre Rouge, 545 m, 26 May 1974, native collector. Dépt. de l'Ouest: USNM 117270-71, Trou Caïman, 1945-46, Anthony Curtiss; KU 208797-98, 2.7 km E Trou Caïman (village), 18 May 1974, native collectors; MCZ 62637, CM 38886, Manneville, April 1963, George Whiteman; MCZ 81150, KU 208799-298800, 208801, Manneville, 20 February 1966, native collectors; BMNH 1948.1.6.63-34 (2 specimens), Pont Beudet, 12 July 1937, Ivan Sanderson; KU 208802, Tabarre, 5.1 km SE François Duvalier Airport, 28 October 1973, native collectors; USNM 75893, Port-au-Prince, 1928, J. S. C. Boswell; USNM 123792, Port-au-Prince, 1945-46, Anthony Curtiss; MCZ 51426, Port-au-Prince, September 1950, Anthony Curtiss; MCZ 62631-33, Port-au-Prince, June 1960, Luc Whiteman; UF 32179, Port-au-Prince, 29 August 1962, Peter C. Drummond; MCZ 65812, near Port-au-Prince, June 1960, Luc Whiteman; KU 208803-06, LDO 7-6464-69 (6 specimens), same data as holotype; RT 7628-30, ca. 3 km (airline) NE Pétionville, vicinity of Château Blond, 30 July 1979, Luís Rivera, R. Thomas; KU 208807-10, Château Blond, 6.4 km NE Pétionville, 160 m, 2 March 1966, native collectors, R. Thomas; RT 4916, Dumay, east side Rivière Grise (9 km airline E Pétionville), 100 m, 10 June 1978, R. Thomas; RT 5043-44, Dumay, east side Rivière Grise (9 km airline E Pétionville, 100 m, 1 July 1978, R. Thomas; KU 208811, 13.1 km E Croix des Bouquets, 27 June 1976, A. Schwartz.

DIAGNOSIS.--A large, stout *Typhlops* (to 273 mm total length). Tan to medium brown dorsal pigmentation ending midlaterally. Middorsal scale counts moderate (246-285; X = 273.67); scale rows 20, reducing to 18 at 46 to 66 percent of total length. Rostral broadly oval (Fig. 1), width 52 to 66 percent of length from internasal suture, extending posteriorly to level of ocular-preocular suture; rostronasal pattern calyculate (Fig. 1); posterior nasal angle [angle formed by edges of postnasals] 0 (parallel)-20 degrees; preocular contacting only 3rd of supralabials, moderately angled anteriorly, angle 48-78 degrees. Parietals double on each side*, not greatly expanded transversely. Postoculars 2, cycloid.

DESCRIPTION.--1) Snout rounded. 2) Rostral oval in dorsal aspect (RW/RL .52-.66), extending posteriad to a level with the preocular-ocular suture; not flared on apex; slightly flared at labial end. 3) Preoculars angulate (48-78 degrees), apex rounded, lower end contacting only 3rd of upper labials. 4) Ocular length about 2/3 height. 5) Postnasal pattern calyculate (sides of upper limb of postnasal parenthesis-like, often with a slight break rather than smoothly curved). 6) Postoculars 2, cycloid, normal size of body scales. 7) First parietal standard, spanning 2 scale rows. 8) Second parietal standard, spanning 2 scale rows; first and second parietals co-occurring bilaterally in 84% of the specimens, unilaterally in 8%, single in 8%. 9) TL to 273 mm (Fig. 2). 10) T</TA: males: 23-37 ($\bar{x}$ = 27.42, SD = 4.06, N = 19); females: 28-41 ($\bar{x}$ = 34.69, SD = 3.67, N = 16). 11) TL/MBD = 25-35. 12) Middorsal scales between rostral and caudal occurring at about midbody (29-66%TL, $\bar{x}$ = 52%). 14) Coloration bicolor, dorsum tan to medium brown, fading over a narrow midlateral zone to unpigmented venter; pigmentation fading out on snout, dorsal part of rostral pigmented; no collar. 15) Rectal caecum present. 16) Hemipenes expanded, apical region oblique. 17) Cranium broad, sides of parietals tapering only slightly. 18) Premaxilla broad, about 40% of width across prefrontals, slightly convex, not protuberant; posteroventral edges transverse, making a right-angle juncture with blade; blade narrow. 19) Nasals without lateral angles. 20) Septomaxilla with silver, anterior portion tapered. 21) Frontal-parietal suture transverse, slightly sinuous. 22) Frontal with unfused anterior ventral blade-like process (Fig. 3A). 23) Optic foramen canalicular (Fig. 3B-D). 24) Postorbital process of parietal prominent. 25) Temporal ridge of parietal present. 26) Lappet of prootic very prominent, in broad contact with sphenoid and parietal, apparently fused with tongue; tongue not visible; prominent secondary foramen isolated (similar to Fig. 4A). 27) Supraoccipitals unfused, in broad median contact. 28) Exoccipitals not fused with prootics. 29) Angular not sliver-like. 30) Dorsal process of quadrate hooked. 31) Atlantal hypapophysis prominent, blade-like; 5-6 total hypapophyses. 32) Hyoid U-shaped, composed of 2 fused ceratobranchials; basihyal absent. 33) Pelvic moieties L-shaped. 34) Eye moderate with narrow orbital space.

DATA ON HOLOTYPE.--An adult male, TL 240 mm; tail length 7.8 mm; MBD 8.0 mm; middorsal scales 267; midventral scales 260; scale rows 20, reducing to 19 at midventral scale 146 and to 18 at scale 154 (120 and 128 mm posterior to snout, respectively)_; final reduction step (19 to 18) occurring at 53% TL; parietals 2/2; postoculars 2/2. Coloration bicolored with 11 pigmented scale rows at midbody.

DISTRIBUTION.--Principally the Cul-de-Sac Plain of Haiti, with some records from the north slope of the La Selle immediately to the south and the south slopes of the Montagnes de Trou d'Eau to the north (Fig. 5).

ETYMOLOGY.--*Tetrathyreus* is from the Greek, *tetra*, four, and *thyreos*, shield, in reference to the characteristic condition of four parietals.

Typhlops titanops, new species

HOLOTYPE.--KU 208812, an adult female from El Mulito on Río El Mulito, 18 km N Pedernales, Pedernales Province, República Dominicana, 500 feet (152 m), one of a series taken 5 August 1975, by Richard Thomas.

PARATYPES.--Haiti: Dépt. de l'Ouest (the localities reckoned from Découzé are near the boundary between Dépts. de l'Ouest and Sud Est but appear to be in the former.): KU 208813, 2.5 kn N Découzé, 727 m, 6 July 1979, natives; KU 208814, 2.3 kn N Découzé, 727 m; RT 5548, 1.2 kn N Découzé, 591 m, 10 July 1978, native, RT 7477, Vendal, 1.5 km N Découzé, 636 m, 16 July 1979, native. Dépt. du Sud Est: MCZ 68571, Colombier near Saltrou; RT 7310, 5.3 k by road W Thiote (= Colombier) 485 m, 11 July 1979, native. República Dominicana, Pedernales Province: KU 208815, 11 k SW Los Arroyos, 394 m, 27 June 1964, R. Thomas; USNM 266301, 20 km N Pedernales, 274 m; KU 208816, 21 kn N Pedernales, 242 m, 29 June 1964, R. Thomas; RT 3440, 3442-44, 3569, same data as holotype; USNM 266302, 22 km N Pedernales, Río Mulito, 274 m.

DIAGNOSIS.--A moderate-sized species of *Typhlops* (to 216 mm total length) having large eyes (orbits)*, truncate preoculars*, a narrow constricted (waisted) rostral*; parallel to weakly divergent postnasals; low middorsal scale counts (231-264); reduction from 20-18 scale rows occurring relatively far posteriorly (61-70% total length); pigmentation bicolor, snout unpigmented, no pigment collar; dorsal process of quadrate hooked; anteroventral process of frontal broad, wedge-like and fused with overlying part of frontal; frontal-parietal suture V-shaped, optic foramen canalicular; atlantal hypapophysis rounded, tab-like; pelvic moieties L-shaped.

DESCRIPTION.--1) Head rounded. 2) Rostral narrow in dorsal aspect (RW/RL .42-.55), parallel, waisted to oval, not flared on apex and with minimal labial flaring. 3) Preocular angle 59-83 degrees, apex truncate modally or rounded; lower portion contacting only 3rd of upper labials. 4) Ocular length approximately 2/3 height, sinuosity .17-.05. 5) Postnasal pattern parallel to divergent. 6) Postoculars 2 (cycloid) with 1 or 3 as variants. 7) First parietal standard, width spanning 2 scale rows (mode) or as a variant, blade-like and extending laterally along posterior edge of ocular to below eye. 8) Second parietal present and equal in size to first (but never extended) or absent. 9) TL to 216 mm (Fig. 2). 10) TL/TA (tail length): males 21-23, females 26-34. 11) TL/MBD 25-30. 12) Middorsal scales 231-264; $\bar{x}$ = 243.4; SE = 1.97, N = 14. 13) Scale rows 20-18 with reduction posterior to midbody (61-68% TL). 14) Coloration bicolor with dorsal pigmentation (gray-brown to tan) fading over a midlateral zone to an unpigmented venter; pigment not extending onto snout region (rostral, postnasals, anterior parts of preoculars); no collar of pigment across throat. 15) Rectal caecum present. 16) Hemipenes expanded, no apparent strongly differentiated apical region; fully everted organ probably with oblique apical region. 17) Cranium broad, width across prefrontals 88 percent of width across prootics. 18) Premaxilla moderate in width (33 percent of width across prefrontals), slightly concave anteriorly and slightly protuberant; posteroventral edges transverse, forming slightly greater than right-angle juncture with blade; blade broad, subtriangular. 19) Nasals without lateral angle, narrowly bordering narial opening. 20) Septomaxilla with sliver, anterior portion tapered. 21) Frontal-parietal suture forming shallow V. 22) Frontal with broad, wedge-shaped, anteroventral blade-like process partly fused with overlying portion of frontal (Fig. 3B). 23) Optic foramen canalicular (Fig. 3B). 24) Postorbital process of parietal moderate. 25) Parietal without temporal ridges. 26) Foramen of cranial nerve V: lappet of prootic small, slightly overlapping moderately long tongue; tongue not reaching parietal (similar to Fig. 3C). 27) Supraoccipitals unfused, in broad median contact. 28) Exoccipitals not fused with prootic. 29) cipitals unfused, in broad median contact. 28) Exoccipitals not fused with prootic. 29) Angular not sliver-like. 30) Dorsal process of quadrate hooked. 31) Atlantal hypapophysis tab-like; total hypapophyses 5. 32) Hyoid U-shaped with joined ceratobranchials, no basihyal. 33) Pelvic moieties L-shaped with prominent ischia and ilia, pubic processes reduced; ilia with cartilaginous process (or tendon?) oriented dorsoposteriorly. 34) Orbit large with a large pale conjunctival space.

DATA ON HOLOTYPE.--An adult female, TL 166 mm, tail length 8.0; MBD 5.5; middorsal scales 236; ventral scales 232; scale rows 20, reducing to 19 at midventral scale 144 and to 18 at midventral scale 150 (95 and 100 mm posterior to snout respectively), final reduction occurring at 60% SVL. Parietals 2/2, postoculars 2/2. Pigmentation bicolored with 13 pigmented rows at midbody.

DISTRIBUTION AND COMMENTS.--*Typhlops titanops* is known from a limited extent of the southern slopes of the Massif de La Selle-Sierra de Baoruco montane chain between the region about Découzé (north of Jacmel, Haiti) in the west and the area below Los Arroyos (Dominican Republic) in the east (Fig. 5). It appears to be restricted to an elevational zone between about 240 and 727 meters, the latter being near the upper limit for *Typhlops* in Hispaniola. At both extremes of its known range it is replaced at lower elevations by different sets of species: In the west it is replaced on the northern slopes of the Tiburon Peninsula by *T. hectus* and *T. capitulatus* and on the southern slopes around Jacmel by *T. pusillus*, *T. hectus*, *T. capitulatus*, and in some areas, *T. sulcatus*. In the east, it is replaced in the xeric lowlands of the Barahona Peninsula by *T. sulcatus* and *T. syntherus*.

ETYMOLOGY.--From the Greek, *titanos*, pale, *Titan*, a reference to large size, and *ops*, eye, refers to the eyes with large, pale conjunctival spaces.

COMPARISON.--*Typhlops schwartzi* and *titanops* need comparison principally with the 20-row species having low middorsal scale counts: *T. lumbricalis* of Cuba and the Great and Little Bahama banks and *T. hectus* of southern Hispaniola. *T. lumbricalis* itself shows a considerable amount of geographical variation in external features. Of the four, *lumbricalis* has more strongly divergent postnasals and greater preocular angle, and the rostral is broadly oval, even more than in *T. tetrathyreus* (Fig. 6). *T. lumbricalis* has but one parietal, the four-parietal condition of *tetrathyreus* appearing as a variant in only 3% of the specimens; the parietal is never extended laterally as in *schwartzi*. Postoculars are two with only rare fusions. Most populations of *lumbricalis* are bicolor with the pigment fading over a narrow midlateral zone; at least one population has extensive ventral pigmentation. A collar of pigment across the throat is a unique feature of *lumbricalis* that would probably approach 100% incidence in fresh (unfaded) material. *T. lumbricalis* is distinctly smaller than *schwartzi*, and in most of the populations individuals average smaller than *titanops* (Fig. 2).

Populations of *Typhlops lumbricalis* from the Bahamas have narrower rostrals, less divergent postnasals, and smaller preocular angles; the size is even smaller than the Cuban and Isla de la Juventud snakes (Fig. 2); middorsal counts are also low.

Typhlops titanops is extreme among these four species in its narrow rostrals, large eyes, reduced head pigmentation, low middorsal scale counts, and far posterior scale-row reduction. Parietals are usually single but may be double bilaterally as in *tetrathyreus* (2 specimens), and 7 specimens have elongated parietals similar to *schwartzi* (only 2 bilaterally). Postoculars are usually two but the single condition occurs with expanded parietals.

The postnasals vary from nearly parallel to divergent but never assume the extreme parenthesis-like condition of *tetrathyreus*. Thus, aside from its unique or extreme features, *T. titanops* shares features with *schwartzi*, *tetrathyreus*, and *hectus*. The expanded first parietal of *T. schwartzi* is probably the result of fusion of the parietal with the uppermost postocular. The single postocular is higher and shorter than is typical of the other species.

T. titanops agrees with *schwartzi* and *lumbricalis* in having a broad, non-tapered cranium, but in the one skeleton examined, the structures of the prootic that form the trigeminal foramina are closest to the condition in *T. hectus*: the lappet is small and slightly overlaps the long tongue, which does not reach the parietal (similar to Fig. 4D). Both *titanops* and *schwartzi* have the dorsal process of the quadrate hooked, whereas in *lumbricalis* it is low and triangular; the optic foramen is slitlike in *lumbricalis*, canalicular in *titanops* and *schwartzi*. In *schwartzi* the atlantal hypapophysis is prominent and blade-like; in *lumbricalis* and *titanops* it is small and rounded (tablike). Similarly, the pelvic moieties are heavy and hatchet-shaped in *schwartzi*, (ilium and pubic wide and strongly fused to ischium), L-shaped (lacking the prominent ilial extensions) in *lumbricalis* and *titanops*. In *lumbricalis* the moieties are only weakly L-shaped.

Typhlops hectus is a 20-row species that occurs at least macrosympatrically with *tetrathyreus* and *titanops*, and along with these two was once considered to be conspecific with *lumbricalis*. (It appears to be syntopic with *titanops*, at least at the localities near Decouze in Haiti). It has the calyculate postnasal configuration (parenthesis-like postnasals, clavate rostral, and small preocular angle), higher middorsal scale counts than *schwartzi* or *tetrathyreus* (284-328), and farther posterior reduction (possibly not true for all populations) or no reduction at all (see Thomas, 1976 for reduction characteristics of the different populations). It is the least highly modified member of a narrow-skulled group that I postulate to have evolved from the group to which *lumbricalis*, *schwartzi*, and *titanops* belong (Fig. 7). The postnasal configuration of *tetrathyreus* appears to be an early stage in the evolution of the more extreme condition of *hectus* and other relatives (*syntherus*, *capitulatus*, *pusillus*, and *rostellatus*).

KEY TO WEST INDIAN *TYPHLOPS*

A.	Preocular contacting supralabials 2 and 3	B
A'.	Preocular contacting supralabial 3 only (of suparlabial series) .... E	
B.	Rostral scale relatively broad (RW/RL .85-.95)	C
B'.	Rostral scale narrower (RW/RL .64-.80)	D
C.	Scale rows 22 (mode) or 24 anteriorly, 22 posteriorly, dorsum uniform brown or tan	*Typhlops bimimiensis*
C'.	Scale rows 20, no posterior recution, dorsum lineate .... *Typhlops tasymicris*	
D.	Middorsal scales 473-505	*Typhlops epactius*
D'.	Middorsal scales 351-408	*Typhlops caymanensis*

E.	Two preoculars	*Typhlops pusillus*
E'.	One preocular	F

F.	Scale rows 24 anteriorly, 22 posteriorly	*Typhlops dominicana*
F'.	Scale rows 22 with no posterior reduction	G

G.	Middorsal scales 299-353, rostral broad (RW/RL .39-.47)	*Typhlops syntherus*
G'.	Middorsal scales 379-448, rostral broad (RW/RL .58-.81)	*Typhlops jamaicensis*

H.	Scale rows 22 anteriorly with posterior reduction	I
H'.	Scale rows less than 22 anteriorly	J

I.	Hemipenes attenuate	*Typhlops richardi*
I'.	Hemipenes expanded apically	*Typhlops monastus*

J.	Scale rows 20 anteriorly	K
J'.	Scale rows 18 anteriorly	*Typhlops granti*

K.	Scale rows 20-20 (no posterior reduction)	L
K'.	Scale rows 20-18	M

L.	Middorsal scales over 350	N
L'.	Middorsal scales 284-350	*Typhlops hectus* (part)

M.	Pigmentation uniformly dark brown with white spots at mouth and vent, postnasal pattern calyculate	*Typhlops rostellatus*
M'.	Pigmentation bicolor, postnasal pattern divergent	P

N.	Head relatively small, body slender (TL/MBD 46-57) body pigmented dorsally and ventrally (reddish brown)	M
N'.	Head about same width as body, body not so slender (TL/MBD 37-44), pigmentation sharply bicolored (gray to brown above)	*Typhlops suclcatus*

O.	Middorsal scales 358-406, preocular acuminate, rostronasal pattern parallel	*Typhlops gonavensis*
O'.	Middorsal scales 398-456, preocular well extended but not acuminate, postnasal pattern calyculate	*Typhlops capitulatus*

P.	Hemipenes attenuate, snout shape in dorsal aspect ogival	*Typhlops monensis*
P'.	Hemipenes expanded, snout shape rounded in dorsal aspect	Q

Q.	Reduction from 20 to 18 scale rows occurring far posteriorly on body (69-93% TL), middorsal scales 284-324	*Typhlops hectus* (part)
Q'.	Scale rows reducing from 20 to 18 farther anteriorly (29-68% TL)	R

R.	A collar of pigment across the throat	*Typhlops lumbricalis*
R'.	No collar of pigment across the throat	S
S.	Postnasal pattern divergent	T
S'.	Postnasal pattern calyculate	*Typhlops tetrathyreus*
T.	Orbit large, pale; rostral a narrow oval; preocular smoothly angled; size large (to 326 mm TL)	*Typhlops schwartzi*
T'.	Orbit smaller, pigmented; rostral narrow, waisted; preocular truncate on apex; size small (to 216 mm TL)	*Typhlops titanops*

Relationships of the Antillean species

Figure 7 presents my conclusions about the phylogenetic relationships of the West Indian *Typhlops*. In order to determine the character states in putative out-groups I have relied on the literature for head scale, skeletal, and hemipenial features in non-West Indian typhlopoids (Hemipenes: Branch 1986. Head scales: Dixon and Hendricks 1979 for 8 non-Antillean, New World *Typhlops*; Roux-Esteve, 1974 and Dunn 1944 for *Typhlopis, Helminthophis, Liotyphlops*, and *Anomalepis*; Kinghorn 1961 and McDowell 1974 for 35 Australasian *Ramphotyphlops*; Roux-Esteve 1974 for 21 African *Typhlops* and 25 *Rhinotyphlops*. Skeleton: Hass, 1964, 1968; List 1966; McDowell and Bogert 1954 for 11 typhlopids and 1 anomalepid). In addition I have examined external morphology of a variety of non-West Indian typhlopoids in various collections. I have used *T. biminiensis* as an outgroup representative for the Major Antillean Radiation (MAR), especially for characters not described adequately in the literature.

The following characters have been used to define synapomorphic groupings of Antillean *Typhlops*:

Preocular scale.--The synapomorphies that define the Major Antillean Radiation of *Typhlops* (MAR) are (1) that of the preocular contacting supralabial 3 only rather than 2 and 3 and (2) the preocular with a pronounced anteriad extension, either sharply angular or rounded. Richmond (1955) stated that two mainland species of *Typhlops* have preocular contact with labials 2 and 3, but in Dixon's and Hendricks' review of mainland species (1979) they found none with this feature. (They could not verify as being New World six species that have long been regarded as South or Central American). Among most species of typhlopids throughout the world the preocular-nasal contact differs significantly from that of the Antillean species. Roux-Estève (1974) noted preocular contact with labial 3 alone in three African species; one, *caecatus* of the Ivory Coast and Ghana, is like the Antillean (MAR) species in also having an angular preocular. Thus one African species shares both defining synapomorphies of the MAR. The extended preocular, although uncommon, occurs in typhlopids in other parts of the world, including *Ramphotyphlops*; the difference in labial contact is a matter of a small shift in position of the postnasal scale. Before being certain of a synapomorphic connection with African species, I would want to find more synapomorphies. The African *T. leucostictus* and *T. socotranus* have extended preoculars but not the MAR-style labial contact (Roux-Estève 1974). In *T. pusillus* the preocular is double and in *T. syntherus* it is greatly extended.

Both *T. syntherus* and *T. hectus* have or often have a step (inflexion) in the lower margin of the preocular. I interpret this as transitional to the split preocular (*T. hectus*, in fact, occasionally has either a marked eye-level notch or an incompletely or completely divided preocular.

Postnasal configuration.--I identify two states: divergent, which is plesiomorphous, and calyculate, which is apomorphous. The calyculate postnasal configuration is associated with a narrowed rostral (dorsal portion), in association with an extended preocular (and consequently a small preocular angle) and bowed but not diverging lateral edges to the dorsal rami of the postnasals. The most extreme development of this character complex is found in *pusillus, syntherus, hectus, capitulatus, gonavensis,* and *rostellatus*. It is well-developed in *Typhlops tetrathyreus*, whereas *schwartzi* and *titanops* show tendencies toward this condition. *T. hectus* seems "annectant" between these three and the others; because of this character progression I unite the entire group as depicted in Figure 7.

Snout shape.--Among the species with the divergent postnasals, three have rostrals that are wide and somewhat protuberant, a condition that I call ogival, since the outline is like an ogive arch. It is apparently a synapomorphy for the Puerto Rican populations of *T. richardi* and *T. monesis*. *T. jamaicensis* also has this snout shape along with a wider rostral; because the hemipenes shape differs from that of *richardi* and *monensis*, I presume that this snout shape is convergent in *jamaicensis*.

Cranial shape.--Broad (plesiomorphic), in which the sides of the parietals are virtually parallel, and tapered (apomorphic) in which the sides of the parietals taper noticeably from posterior to anterior.

Premaxilla.--Broad (plesiomorphic), in which the width is 30-40% the maximum width of the rostrum and narrow (apomorphic), in which the width is 40-60% of the maximum width.

Septomaxilla.--Hooked and tapered (plesiomorphic), with a hook-like lateral process around the edge of the naris and with a distinct narrowing just posterior to the hook versus the apomorphic non-hooked, broad, blunt triangular process forming the posterolateral edge of the naris and not narrowing just posterior to this process.

Nasal.--With the lateral sides smoothly curved (plesiomorphic) or with a distinct angle (apomorphic).

Blade-like process of frontal.--On the ventral aspect of the frontal a blade-shaped process forms the floor of the optic foramen. When the blade is not fused with the overlying part of the frontal, the optic foramen is a slit. In most osteological descriptions of typhlopids, this region has not been shown or described in detail; or, when shown (e.g., McDowell and Bogert 1954), none of the complexity seen in the Antillean species is evident. Most cleared and stained material does not permit a clear view, especially if the specimen is not skinned (removal of the mandible may be necessary). I regard the unfused condition as plesiomorphic because this is the condition in the outgroup member (*T. biminiensis*). I have seen no evidence of an ontogenetic progression from unfused (slit-like) to fused (canalicular) or the reverse. The blade shows various degrees of anterior fusion with the overlying part of the frontal (in which case the optic foramen is a closed tubular canal); with fusion (apomorphic) the remnants of the blade may be visible in their entirety, as wedge-like process on the ventromedial edge of the frontal, or completely absent. *Typhlops monastus* and *jamaicensis* appear to share a synapomorphy --

complete fusion, with disappearance of the blade-like process of the frontal. I regard this synapomorphy uniting the two species as very provisional. *T. capitulatus* also has this condition, but because of strong synapomorphies with other species I interpret it as an independent acquisition. The degree of fusion does not seem to be related to overall ossification (slender vs. heavy crania).

Frontal-parietal suture.--In dorsal aspect, this may be transverse (pleisomorphic) or broadly V-shaped (apomorphic).

Quadrate.--The quadrate has a dorsal process that may be low and wedge-like (triangular) or narrow and hooked (apomorphic).

Basihyal.--The plesiomorphic condition is a Y-shaped, cartilaginous basihyal with bony rodlike ceratohyals attached to each leg of the Y. Various stages of reduction in the basihyal occur to complete absence in which the ceratohyals are joined to form a U-shaped unit. In some species the basihyal is absent, and the two bony ceratohyals are not untied.

Hemepenes.--Within the MAR, attenuate hemipenes (versus apically expanded organs) provide an apparent synapomorphy uniting several species. Although *T. biminiensis*, not a MAR species and therefore a putative outgroup species, appears to have attenuate hemipenes, the conditions reported for non-Antillean species are all expanded in some fashion, or at least not attenuate (Branch 1986; Dixon and Hendricks 1979; McDowell 1974). At any rate, within the MAR, attenuate versus expanded hemipenes serves to define two groups. Expanded hemipenes appear to be of at least three types: Simply expanded and sac-like, trumpet-shaped having a flattened apex with a circumferential sulcus, and diagonal in which the expanded apex is at an angle to the axis of the organ. For some species I have not seen everted organs and have assessed the condition by dissection.

Scutellation.--I have not used scale-row data or middorsal scale counts to define synapomorphies when they conflicted with other apparent synapomorphies. High middorsal counts (above 350) and primary (=before reduction) scale rows of 22 or over I regard as plesiomorphic. These meristic characters are very variable intraspecifically in many species; however, some conditions seem to characterize major groups of species. Although many species are constant in scale row characteristics, in some species the zones of scale-row fusion can vary markedly in position. The phylogeny that I present necessitates three reversals in middorsal counts (Fig. 7, arrows) and two in scalerows.

Biogeography

Although the Caribbean Arc Group is not unequivocally based on synapomorphies (Dixon and Hendricks 1979), I have associated the four members of th Caribbean Arc Group as shown in Figure 7, since *tasymicris* occurs at the extreme southeastern part of the Caribbean and shares a number of features with *trinitatus* and *lehneri* of northeastern South America. It is obviously an insular derivative of these species. I found no obvious synapomorphies that resolve the trichotomy of *caymanensis-epactius-biminiensis*. *T. biminiensis* and *epactius* are similar in middorsal scale counts and scale rows, hence I show them to be the more closely related pair (*T. epactius* was originally described as a subspecies of *biminiensis*).

The *biminiensis* group occupies the geologically associated area of the Caymans,

Cuba, and the Bahamas. The MAR, on the other hand, is diverse in Hispaniola and Puerto Rico with representatives on Cuba (1 species), Jamaica (1 species), and in the Lesser Antilles (2 species). Synapomorphic connectivity (tracks) exists between: Cuba and the Bahamas (both *biminiensis* and MAR), Cuba and the Caymans (*biminiensis* group), Hispaniola and Cuba (MAR), and Cuba and Puerto Rico (MAR). Less certain is the apparent connectivity between Jamaica and the northern Lesser Antilles (Antigua-Barbuda, St. Kitts-Nevis, and Montserrat Banks). Connectivity of such remote islands is not unknown in Antillean biogeography, however; it is paralleled in the *bilineatus* group of leptotyphlopid snakes, which has members only on Hispaniola and the southern Lesser Antilles (Thomas et. al. 1985). I cannot resolve the polychotomy involving the southernmost Lesser Antillean MAR species, *T. dominicanus*; the connectivity could be with Puerto Rico or Hispaniola.

Typhlops tracks in which I have the most confidence:

1. Puerto Rico<--->Turks and Caicos (*richardi*)
2. Puerto Rico<--->Mona (*platycephalus, richardi, monensis*)
3. Hispaniola<--->Navassa (*sulcatus*)
4. Hispaniola<--->Cuba (*lumbricalis, schwartzi, tetathyreus, titanopis*)
5. Cuba<--->Little and Great Bahama Banks (*lumbricalis*)
6. Cuba<--->Bahamas (Cay Sal, Great Bank, Inagua)<--->Caymans (*biminiensis, epactius, caymanensis*)

The following two tracks implied by the synapomorphies, Jamaica<--->Northern Leeward Islands (*jamaicensis, monastus*) and Puerto Rico<--->Southern Windward Islands (*richardi, dominicanus*) are tentative and require corroboration. Likewise, the apparent synapomorphy linking *Typhlops caecatus* of Africa with the Major Antillean Radiation requires further investigation using other character suites.

For the evolution of *Typhlops* in the West Indies, I suggest the following events: 1) The evolution of the MAR species group in the Cretaceous prior to the separation of Africa and South America and the formation of the Caribbean Basin. 2) Continued diversification in the Central Greater Antilles -- Hispaniola and Puerto Rico -- with derivatives Jamaica, Greater Cuba, and some islands of the Lesser Antilles (1 species only to island banks inhabited). The Bahamas are populated from two sources: via Cuba to the Great and Little Bahama Banks, and from Puerto Rico to the Turks and Caicos banks. 3) Peripheral invasion by mainland taxa: Greater Cuba (Caymans, Cuba, the Bahamas) by *biminiensis* and *caymanensis* from Central America and the southernmost Lesser Antilles (Grenada, at least) by *tasymicris* from norther South America. The connectivities presented above imply the possibility of more complex dispersal (or vicariant histories) relative to the central Antilles (Hispaniola and Puerto Rico) than a simple central-to-peripheral one. However, the details of position and emergence of the islands during the evolution of the present Antillean physiography are not sufficiently well-known to test cladistic pattern against history.

With little doubt, overwater dispersal has been important in establishing parts of the present pattern. The Hispaniola-Cuba-Bahamas pattern strongly bespeaks this, as does the Puerto Rico-Turks and Caicos connection. The Lesser Antilles likewise have been accessible only via waif dispersal.

If the MAR *Typhlops* are indeed as old as the above scenario states and have one or more African representatives, then the group is not a strictly Antillean radiation. It could be a relict assemblage, whose New World members are now found only in the West Indies (this could be true even without an African representative). For example, the MAR *Typhlops* might be only a remnant of a once larger clade. It would then be erroneous to assume that a correct phylogeny would necessarily correspond in an interpretable way to Antillean geological history.

Acknowledgments

My major debt of gratitude is to Albert Schwartz for reasons already mentioned.At an early stage in this study Douglas A. Rossman contributed greatly to the development of my ideas about West Indian *Typhlops*. Luis Rivera and Blair Hedges were outstanding rock-turners at one stage or another of the field work and have my great appreciation. Blair Hedges further was instrumental in initiating some of the field work, some of which was done under the aegis of National Science Foundation Grant BSR-83-7115 to Richard Highton. Some of my field work critical to this study was supported by NSF Grant SER 77-04629. Thomas G. Gush read a draft of the manuscript, and David Auth helpfully reviewed the manuscript.

Literature cited

Branch, W. R. 1986. Hemipenial morphology of African snakes: A taxonomic review. Part. 1. Scolecophidia and Biodae. Journal of Herpetology 20(3):285-299.

Cochran, D. M. 1924. *Typhlops lumbricalis* and related forms. J. Washington Academy of Sciences 14(8):174-177.

Dixon, J. R., and F. S. Hendricks. 1979. The wormsnakes (family Typhlopidae) of the Neotropics, exclusive of the Antilles. Zooligsche Verhandlingen 173:3-39

Dunn, E. R. 1944. A review of the colombian snakes of the families Typhlopidae and Leptotyphlopidae. Cladasia 3:47-55.

Haas, G. 1964. Anatomical observations on the head of *Liotyphlops albirostris* (Typhlopidae, Ophidia). Acta Zoologica 45:1-62.

———. 1968. Anatomical observations on the head of *Anomalepis aspinosus* (Typhlopidae, Ophidia). Acta Zoologica 49:1-77.

Kinghorn, J. R. 1929. The snakes of Australia. Angus and Robertson. 197 pp.

Linnaeus, C. 1758. Systema naturae per Regnum Tria Naturae. . . Ed. 10. Jarrold & Sons, Norwich and London. 824 pp.

List, J. C. 1966. Comparative osteology of the snake families Typhlopidae and Leptotyphlopidae. University of Illinois Biological Monographs 38:1-112.

McDowell, S. B., and C. M. Bogert. 1954. The systematic position of *Lanthanotus* and the affinities of the anguinomorphan lizards. Bulletin American Museum of Natural History 105(1):1-142.

———. 1974. A catalogue of the snakes of New Guinea and the Solomons, with special reference to those in the Bernice P. Bishop Museum. Part. I. Scholecophidia. Journal of Herpetology 8(1):1-57.

Richmond, N. D. 1955. The blind snakes (*Typhlops*) of Bimini, Bahama Islands, British West Indies, with description of a new species. American Museum Novitates (1734): 1-7.

Rosen, D. 1985. Geological hierarchies and biogeographic congruence in the Caribbean. Annals of the Missouri Botanical Garden 72:636-659.

Roux-Estève, R. 1974. Revision systematique des Typhlopidae d'Afrique Reptilia-Serpentes. Memoires Museum National d'Histoire Naturelle, Nouvelle Serie, Serie A, Zoologie 87:1-313.

Ruthven, A. G., and H. T. Gaige. 1935. Observations of *Typhlops* from Puerto Rico and some of the adjacent islands. Occasional Papers Museum of Zoology, University of Michigan 307:1-12.

Schwartz, A., and R. Thomas. 1975. A check-list of West Indian amphibians and reptiles. Carnegie Museum Special Publications 1:1-216.

Thomas, R. 1968. The *Typhlops biminiensis* group of Antillean blind snakes. Copeia 1968(4):713-722.

———. 1974. A new species of *Typhlops* (Serpentes: Typhlopidae) from Hispaniola. Proceedings of the Biological Society of Washington 87:11-18.

———. 1976. Systematics of Antillean blind snakes of the genus *Typhlops* (Serpentes: Typhlopidae). Unpublished dissertation, Louisiana State University, Baton Rouge, LA.

———. R. W. McDiarmid, and F. G. Thompson. 1985. Three new species of thread snakes (Serpentes: Leptotyphlopidae) from Hispaniola. Proceedings of the Biological Society of Washington 98(1):204-220.

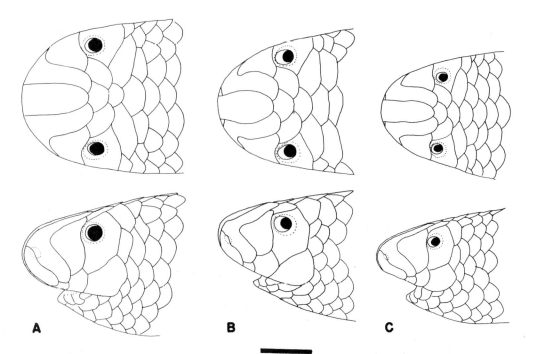

Figure 1. Dorsal and lateral views of the heads of A, *Typhlops schwartzi* (ASFS V27866); B, *Typhlops tetrathyreus* (ASFS V22440, holotype); and C, *Typhlops titanops* (ASFS V2604).

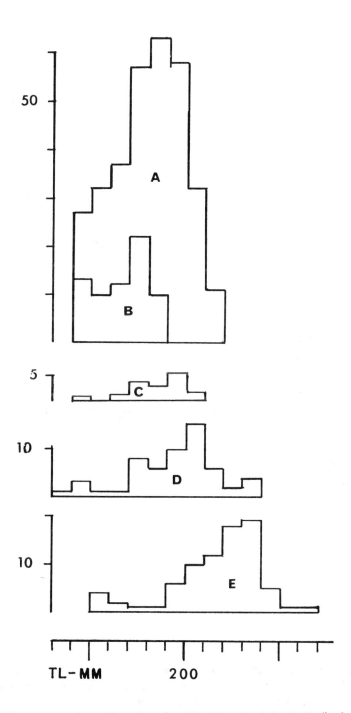

Figure 2. Histograms of total lengths of A, *Typhlops lumbricalis*; B (included histogram), Bahamian *T. lumbricalis*; C, *T. schwartzi*; D *T. tetrathyreus*; and E *T. titanops*. Vertical scales indicate numbers of individuals.

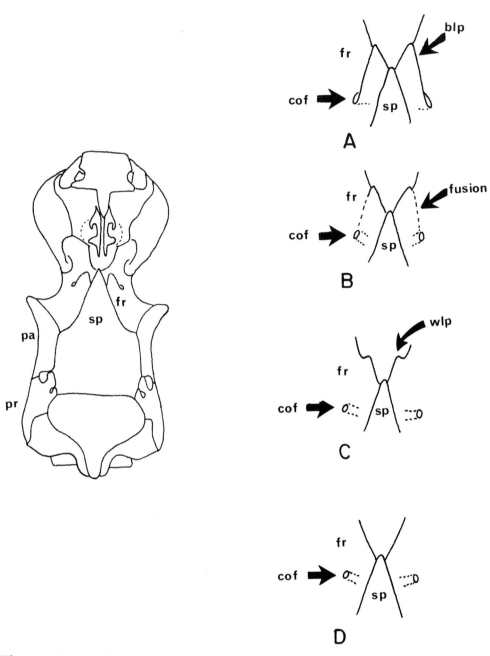

Figure 3. Range of conditions seen in ventral aspect of the frontal (fr) of *Typhlops* crania (A-D). Ventral view of entire cranium (*T. schwartzi*) is shown for localization of structures in A-D and those in Fig.4. BP = blade-like process; FR = frontal; PA = parietal; PR = prootic; SOF = slit-like optic foramen; COF = canalicualar optic foramen; SP = sphenoid. Dotted lines indicate fusion or features beneath the surface.

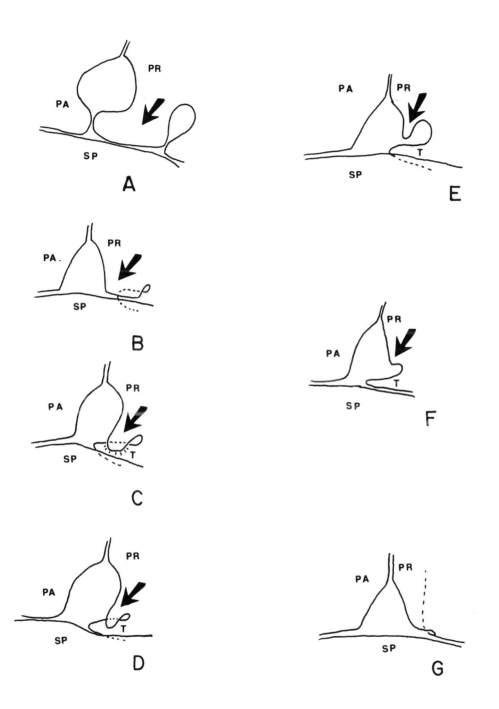

Figure 4. Range of conditions seen in the foramina of cranial nerve V in *Typhlops* crania. PA = parietal bone; PR = prootic bone; SP = sphenoid bone; Arrow indicates "lappet"; T indicates "tongue." See Fig. 3 for localization of the structures.

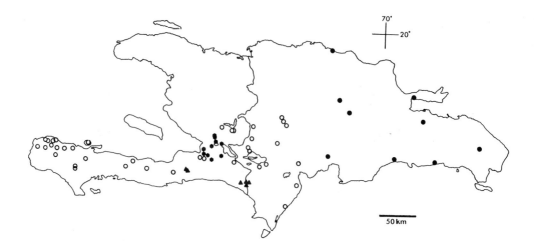

Figure 5. Map of Hispaniola showing localities for *Typhlops hectus* (open circles), *Typhlops schwartzi* (solid circles), *Typhlops tetrathyreus* (solid rhombs), and *Typhlops titanops* (solid triangles).

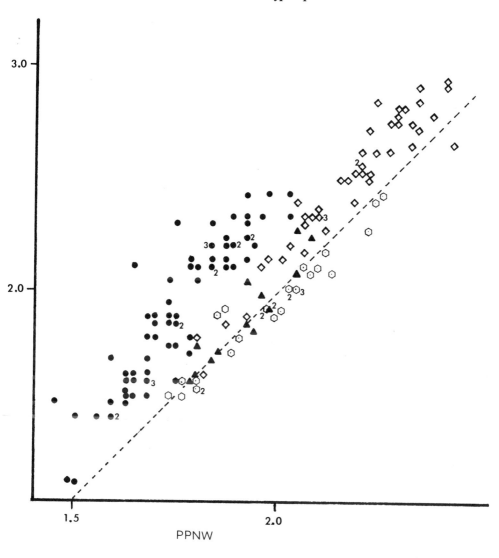

Figure 6. Scatter diagram of anterior postnasal width (APNW) vs. posterior postnasal width (PPNW) for *T. lumbricalis* (open circles), *T. schwartzi* (solid triangles), *T. tetrathyreus* (open rhombs), and *T. titanops* (hexagons).

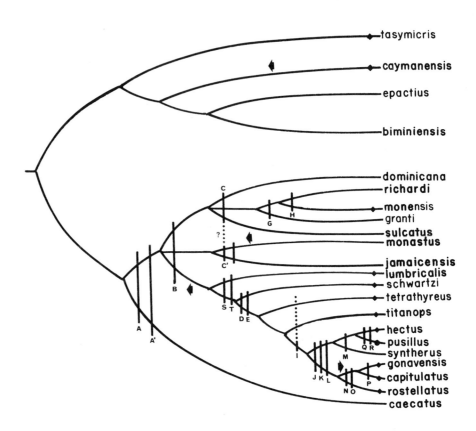

Figure 7. Cladogram of relationships of Antillean *Typhlops*. A = extended (angular) preocular; A' = Preocular contact with supralabial 2 only; B = sinuous Preocular-ocular suture; C = attenuate hemipenes; C' = expanded, non-oblique hemipenes; D = hooked quadrate; E = narrow premaxilla; F = completely fused blade-like process of frontal; G = completely fused blade-like process of frontal (putative second occurrence); H = ogival head shape; I = calyculate rostronasal complex, progression series; J = tapered crania; K = broad, non-hooked septomaxilla; L = narrow premaxilla; M = stepped preocular; N = reduced caudal spine; O = complete pigmentation; P = shape similarities of head scales; Q = scale rows (some populations only of *pusillus*;) R = divided preocular (a variant in *hectus*); S = reduced basihyal; T = decrease in middorsal scales. Arrows indicate major increase (right) or decrease (left) in middorsal scale counts. Rombs mark species with 20 scale rows anteriorly (includes 20-20 and 20-18 conditions); *pusillus* (circle) has both conditions; all others have 22 or more scale rows initially.

A CRITIQUE OF GUYER AND SAVAGE (1986): CLADISTIC RELATIONSHIPS AMONG ANOLES (SAURIA: IGUANIDAE): ARE THE DATA AVAILABLE TO RECLASSIFY THE ANOLES?

Ernest E. Williams[1]

Abstract

The attempted cladistic analysis of anoline lizards by Guyer and Savage (1986) is here criticized on the basis of serious errors and confusions. It is recommended that neither the data bases nor the taxonomic conclusions be accepted without re-examination of primary evidence. The claim is made that even corrected data bases that are now available are not adequate to reclassify the anoles.

Introduction

I concur with Guyer and Savage's first statement in the paper here reviewed: "The iguanid lizard genus *Anolis* (*sensu lato*), among the most diverse (approximately 250 recognized species) vertebrate genera, continues to challenge attempts by systematic herpetologists to analyze its phylogeny." I can also applaud their intent: "Our purpose is to review previously proposed phylogenies and classification against the data of osteology, karyology and albumin immunology in a revaluation of relationships based upon cladistic analysis." I am persuaded, however, that they have taken upon themselves a very large task and done so too cavalierly. They have tried to work with old data, incomplete data, misunderstood data, bad data, and have compounded already great difficulties with errors, large and small, of their own. In the end, except for resurrected generic names, there is very little novelty in their results, and, in fact, more agreement with my own former opinions than may accord with reality. There is more ground for uncertainty and controversy than they admit. In my opinion to alter formal classification at this time is indiscreet and unwarranted.

Of the three data sets that Guyer and Savage analyze, only the osteological data set has any pretension to completeness. The taxonomic samples for both karyology and immunology are inadequate, as Guyer and Savage are aware. They are aware also of the problems in analysis that make these non-morphological sets of limited value. In fact, after description, discussion, and analysis of each of the three data sets, they come to an

---

[1] Dr. Williams is Curator Emeritus of Herpetology at the Museum of Comparative Zoology, Harvard University, Cambridge, MA. 02138

admission that, in their taxonomic decisions, they have abandoned two of them (p. 522-523): "Since the evolution of karyotypes cannot be inferred independently and since the immunological data have serious methodological problems, we have based our reclassification primarily on what osteology tells us about the evolution of anoles."

Guyer and Savage's paper is peculiarly unfortunate because it has the manner though not the substance of a modern systematic review. The superficial reader will assume it to be not only the latest but the best word on its topic. Its taxonomic decisions have been adopted in Schwartz and Henderson's 1988 "West Indian Amphibians and Reptiles: A Check-List" and are, I am told, being utilized in a number of collections. I want here to demonstrate the unwisdom of such action. I cannot recommend that any biogeographer reorder his thoughts on the basis of this paper nor that any curator rearrange his collection on the basis of the new classification. The publication of this paper in its present form is an egregious failure of the review process.

I will not discuss here cladistic method or analytical theory. Cannatella and de Queiroz (1988) will elsewhere criticize the analytical methods employed by Guyer and Savage. I will myself discuss the value and validity of the underlying data. There will be first some necessary comment on the taxonomic units Guyer and Savage use in their analysis. I will then discuss in some detail the three data bases that Guyer and Savage reexamined, elaborating on my comment above that none of them is really adequate to the revisionary task, and pointing out the deficiencies in the evidence that must be remedied before that task can be successfully accomplished. I next discuss briefly their biogeography and its geological base. I review finally their definitions of genera. Under every one of these topics I will show that Guyer and Savage have committed confusions and errors that, quite apart from any problems of method, vitiate their analysis. In quotations from Guyer and Savage I will underline passages that I regard as requiring special comment. To avoid the issue presented by the controversial genera recognized by Guyer and Savage, I shall in the relevant cases use only specific names without a generic prefix.

The units of analysis

The units of analysis of Guyer and Savage - the terminal taxa of their cladograms - are "series"; they nowhere descend to subseries or species groups, superspecies, nor to discussion of variation between or within species. Unfortunately, they introduce the terminal taxa that they use throughout the paper almost covertly in table 1, which is said to report "the widely accepted classification of *Anolis* lizards based on sections, subsections and series recognized by Etheridge (1960) and Williams (1976a, b)." The result is serious taxonomic confusion at the very beginning of their paper. Their table 1 is a strange hybrid list of taxa new and old that needs careful examination and explanation.

The "series" category was introduced by Etheridge (1960). Etheridge did not go below this category, except to list the species he had himself examined, and even so left a number of species that he had examined unplaced in any series. In 1976 I employed the series category in my West Indian checklist (Williams 1976a) but also for that region employed the categories subseries, species groups and superspecies. In a second checklist for South America (Williams 1976b) I used only the species group category. (In South America I was very unsure how to combine species groups into any higher units - even more so for the beta than the alpha anoles.) In both checklists I spelled out the content of each of my groups. (A printer's error in table 2 of 1976b resulted in *sulcifrons* and *ibague* being inadvertently referred to the *petersi* species group (=*biporcatus* species group of my present usage). I have never provided a checklist or a series or even species group level classification for species restricted to Central America or Mexico. I have had almost no first hand knowledge of any of these animals in life; I have not even reviewed them more than casually as museum specimens.

In purporting to record "the widely accepted classification of *Anolis*," Guyer and Savage's table 1 attempts an impossible task, since there are, in fact, two classifications. That of Williams is a modification of that proposed by Etheridge - but with substantive differences. The two sections, first proposed by Etheridge - alpha and beta - are the same in both systems. The two subsections of the alpha section - *punctatus* and *carolinensis* - represent a formalization by Williams of groupings recognized by Etheridge. The problems enter with the category "series" - the category that Guyer and Savage thereafter use almost exclusively. Unfortunately Etheridge and Williams had quite different ideas of the contents of their series even when they used the same series names. There has, in other words, never been a complete classification of the anoles at the series level, and, thus, never a single classification that could be "widely accepted."

That the actual tabulation of series provided by Guyer and Savage is a failed hybrid of two partial classifications might be a trivial error or at least an unimportant one. Their table 1, however, is not intended as just an historical record. Its real significance is as the list of the operational units which Guyer and Savage analyze in the whole remainder of the paper. Because of this, the confusions in table 1 taint the remainder of the paper.

Twenty one series are listed. Only two series - *grahami* and *sagrei* series - can from the text be confidently identified as used in Etheridge's sense. Five series - *alutaceus, cuvieri, lucius, monticola,* and *occultus* - are clearly those first named by Williams. Of these the *alutaceus, lucius,* and *monticola* series were included in Etheridge's *carolinensis* series. The giant Hispaniolan and Puerto Rican species placed by Williams in the *cuvieri* series were regarded by Etheridge as problematic and not placed in any series. Neither *occultus* (nor *sheplani,* included by Williams in the *occultus* series) had been described at the time of Etheridge's study.

Five series cited by Guyer and Savage - the *chlorocyanus, cybotes, equestris, meridionalis,* and *onca* series - were never recognized at the series level by either Etheridge or Williams. All were listed at the species group level by Williams, *cybotes* also as a subseries of the *cristatellus* series. Etheridge included *chlorocyanus* in his *carolinensis* series, *cybotes* in his *cristatellus* series, left *equestris* and *meridionalis* unplaced, and retained the genus *Tropidodactylus* for *onca*.

The erratic choice of series from two incompatible classifications and the raising of other taxa to series level with neither justification or comment is unfortunate, but not necessarily disastrous. My own evaluation is that it was both correct and shrewd of Guyer and Savage to treat groups of questionable affinities such as the *chlorocyanus, cybotes, equestris, meridionalis,* and *onca* assemblages as independent units during their re-analysis. Some of the most dubious and controversial decisions of Etheridge and of Williams certainly involved these taxa. How closely, for example, the *cybotes* complex is related to the *cristatellus* radiation may long remain a much disputed question (see Gorman and others, several papers, for views very different from the decision finally made by Guyer and Savage). The action of Guyer and Savage in treating these two units as separate series and therefore operational units in their analysis is unquestionably useful in making re-evaluation of that supposed relationship a genuinely open question. It is their introduction of new series without prelude or discussion that is confusing and dismaying. Guyer and Savage conceal their own most perceptive insights under an actually fictitious "widely accepted classification."

Some of their inventiveness in regard to classification, however, is not nearly so fortunate as their erection of a *cybotes* series. In three cases it is possible to discern from clues in the text what the content of certain series is. In the case of the *latifrons* series in most of the text this series is understood in an Etheridgean sense. It is, in fact, on p. 522 directly equated with *Dactyloa, sensu* Guyer and Savage. Unhappily, almost immediately opposite on p. 523 *Dactyloa* is divided into six series, only one of which is a *latifrons* series. In this instance species groups of Williams 1976b have been raised to series rank.

In the case of the *cristatellus* and *bimaculatus* series it is possible to determine the content intended by Guyer and Savage not by any explicit statement but from the chromosome formulae given in their table 4, the citation in fig. 7 of *evermanni* as a member of the *cristatellus* series, and the inclusion of Hispaniola and the northern Bahamas in the distribution of the *cristatellus* series in fig. 9 (implying inclusion of the *distichus* complex in the *cristatellus* series). Their usage is clearly not that of Etheridge 1960 or Williams 1976a. Etheridge's division, followed faithfully by Williams, was based on parasternal (=inscriptional) formulae (see below) and separated a group with two attached and two free inscriptional rib pairs - his *cristatellus* series (= the *cybotes*+*pulchellus*+*cristatellus* species groups of Williams 1976a) from a group with three attached and one free pair - his *bimaculatus* series (= the *stratulus* subseries + *bimaculatus* subseries of Williams 1976a). Guyer and Savage's *cristatellus* series, according to the karyotypes listed for it in Table 4, clearly contains the *stratulus* subseries of Williams 1976a as well as the *cristatellus* series, *sensu* Etheridge or Williams. Just as clearly the *bimaculatus* series of Guyer and Savage, again inferred from listed karyotypes, contains only the *bimaculatus* species group and the *wattsi* species group *sensu* Williams 1976a.

It is in six cases impossible to infer from the text what the content of their "series" are:

1) An *angusticeps* series was recognized by Etheridge; however, it contained in addition to *angusticeps* the species *argillaceus* and *loysiana*. Williams did not accept Etheridge's *angusticeps* series, merging all three species in his *carolinensis* series, placing *angusticeps* as a subgroup of the *carolinensis* species group, while separating *argillaceus* and *loysiana* as a species group of their own. It is not possible to tell whether Guyer and Savage's *angusticeps* series is strictly equivalent to Etheridge's or not.

2) Since the *chlorocyanus* series is regarded by them as distinct, Guyer and Savage's *carolinensis* series clearly does not include *chlorocyanus* itself, as both Etheridge and Williams suggested. But *coelestinus* is nowhere mentioned in Guyer and Savage's text. Etheridge regarded it as belonging to a distinct series. At an another extreme, Williams tentatively placed *coelestinus* and *chlorocyanus* in the same superspecies.

3) It is, on its face, an irresolvable question whether Guyer and Savage's *carolinensis* series is strictly synonymous with Williams' *carolinensis* species group or is more extensive. It is, for example, uncertain whether the *hendersoni* superspecies is a member of the *carolinensis* or the *chlorocyanus* series - or merely forgotten.

4) The *auratus* series of Guyer and Savage is undefined, No *auratus* series was

recognized by Etheridge. The *auratus* species group of Williams (1976b) did not include any Central American species. Perhaps Guyer and Savage's *auratus* series is Etheridge's *chrysolepis* series renamed. If so, the karyotypes attributed to the *auratus* series in table 4 are not those published by Gorman (1973) for species belonging to Etheridge's *chrysolepis* series. No clues are given in Guyer and Savage's text.

5) The *fuscoauratus* series of Guyer and Savage is again undefined. In table 4 the karyotypes attributed to the *fuscoauratus* series include counts and xy sex chromosome data that in the karyotype tree (fig. 6) on the next page are attributed to the *auratus* series. Again I find no clues in Guyer and Savage's text.

The *darlingtoni* series of Guyer and Savage is a different case; it is not a simple failure to expressly identify included species. It is an even more elementary taxonomic error - a confusion of names. There is a nomenclatorial tangle that it is necessary to unravel: Cochran endeavored to honor P.J. Darlington by describing two anoles with the trivial name *darlingtoni*. One, described in 1935, she placed in the genus *Xiphocercus*, the other, not described until 1939, in *Anolis*. Etheridge (1960) synonymized *Xiphocercus* with *Anolis*, thus creating homonymy between Cochran's two names. He disavowed proposing a substitute name for the later described species, and, instead, employed consistently in both the text and appendix of his thesis the usage "*Anolis [Xiphocercus] darlingtoni*" for the 1935 taxon and "*A. darlingtoni*" for that of 1939. In 1962 Williams erected the name *etheridgei* for *darlingtoni* Cochran, 1939 preoccupied.

Etheridge assigned "[*Anolis*] *darlingtoni*" (by implication the 1939 species) to his *carolinensis* series. Williams (1976a) referred the same species, as *etheridgei*, to his *monticola* series. Etheridge did not designate any series for "*Anolis [Xiphocercus] darlingtoni*," since he considered it as one of a number of forms of uncertain position. I quote his complete remarks regarding that species:

> "With the exception of the absence of functional autotomic caudal vertebrae, there is no feature of the skeletal morphology of *Anolis [Xiphocercus] darlingtoni* to distinguish it from the *coelestinus* series. The species bears some superficial resemblance to certain South American members of the *latifrons* series, and to *Phenacosaurus*, but differs from them in having a Y-shaped parietal, a flaring, half-funnel shaped occipital portion of the parietal, and the lateral processes of the interclavicle in contact with the expanded proximal portions of the clavicles."

The key points here are 1) that Etheridge placed *Anolis darlingtoni* Cochran 1939 (= *Anolis etheridgei* Williams 1962) in his *carolinensis* series, while affirming that only one character known to him kept him from referring *Anolis [Xiphocercus] darlingtoni* (Cochran, 1935) to his *coelestinus* series; 2) that Williams (1976), in erecting a *darlingtoni* series had in mind *Xiphocercus darlingtoni* Cochran 1935 and not *Anolis darlingtoni* Cochran 1939 - which he had renamed in 1962 precisely in order to avoid the sort of confusion into which Guyer and Savage have fallen.

Guyer and Savage's references to the characters of the *darlingtoni* series are only two: 1) p. 516, fig. 5 where in a "most parsimonious tree of relationships based on osteological data" the *darlingtoni* series is given a position that implies that it is extremely derived, adjacent to the *monticola* series; 2) p. 521 where they discuss their restricted concept of the genus *Anolis*:

> "The relationships of this genus to other anoles as determined in our reanalysis of the osteological data differ from those suggested by Etheridge (1960) and Williams (1969a). In our tree the number of lumbar vertebrae is used as the most stable character for determining relationships, whereas Etheridge and Williams relied on parasternal rib formulae, a character for which numerous homoplasies must be invoked. The principal difference in our tree is the placement of the *darlingtoni* series as highly derived based upon the presence of six lumbar vertebrae and the *carolinensis* series as more primitive based upon the presence of four lumbar vertebrae. Etheridge and Williams considered the former series to be more primitive and the latter to be more derived."

Both figure and text are badly confused. Even if Guyer and Savage are right about the value of the lumbar character, they have their facts awry. On page 147 of his thesis Etheridge does, indeed, give six as the "most usual number" of lumbar vertebrae in *darlingtoni*, but he is here referring to *Anolis darlingtoni* Cochran 1939. On page 155 he finds "no feature of the skeletal morphology" of *Anolis* [*Xiphocercus*] *darlingtoni* to distinguish it from his *coelestinus* series (*A. coelestinus* and its synonym *A. latirostris*). Reference to p. 148 discovers that the *coelestinus* series is characterized by four lumbar vertebrae. Thus *darlingtoni* Cochran 1939 = *etheridgei* Williams 1962 has six lumbar vertebrae like *monticola*, with which Williams will associate it in the *monticola* series in 1976, while *darlingtoni* Cochran 1935 has four lumbar vertebrae like *coelestinus*, *chlorocyanus*, and *carolinensis*, which Williams (1976), unlike Guyer and Savage, associated in a single group. Carelessness in compilation has resulted in an unfortunate error.

The data bases
1. The osteological data set

Guyer and Savage present no new data of their own. Most of their osteological data comes from Etheridge's 1960 thesis. Since that time a great number of new species have been described: some of these are clearly of phylogenetic and biogeographic importance: *occultus, insolitus, sheplani, fowleri, eugenegrahami*. None of these have been described in the osteological detail that they deserve. Guyer and Savage have had to omit them from their study. Indeed, of these more recently described species they mention only *occultus* - but only as a series in table 1 and again in their statement on p. 523 of the content of the residual genus *Anolis* as recognized by them. They present no data for *occultus*, and it does not appear in their tables or cladograms. Apart from these omissions, there is a serious bias in the skeletal sampling on which they rely. The relatively complete survey of anole species - 175 species of 5 genera - that was fundamental to Etheridge's groundbreaking study was permitted by a technique - the use of soft X-rays. He had fewer dry skeletons - 67 species of 4 genera - and nothing cleared and stained. X-ray images necessarily are most useful for superficial or uncomplicated structures. Etheridge expressly states (1960:120) that characters were chosen because "they are easy to study, and because they showed promise of being important taxonomic characters." It is, therefore, not surprising that many of the 15 characters that Guyer and Savage extract from Etheridge's thesis for their reanalysis deal with vertebrae or superficial portions of the skull. A thorough comparative analysis of anole osteology was not achieved by Etheridge (1960). It has not been attempted by anyone since.

Fifteen characters of the skeleton are listed by Guyer and Savage in their table 2. The characters are those used by Etheridge in his description of the series of *Anolis* (*sensu lato*), that he recognized in his 1960 thesis. He was very consistent in utilizing these characters comparatively in the description of series. It is important, however, that he did not use these characters routinely or comparatively in the description of the genera that he retained as separate from *Anolis*, nor, as I will have multiple occasion to repeat below, in the description of species that he felt unable to place with confidence in his series.

I have explained above the factors of easy visibility in X-rays or in dry skeletons that led to Etheridge's choice of characters. It will be useful to review these selected characters and Guyer and Savage's coding of them seriatim. I cite **in bold face** my preferred coding when it differs from that of Guyer and Savage, and I have not hesitated to add new information when it is relevant. Skeletons of the species mentioned in the case of new observations are in the collections of the Museum of Comparative Zoology (MCZ), except for the species *darlingtoni*, a skeleton belonging to the National Museum of Natural History (USNM), and *solitarius*, a skeleton belonging to the Instituto de Ciencias Naturales, Bogota (ICN), and the skeleton of the paratype of a new species of *Phenacosaurus* belonging to the Museo Ecuatoriano de Ciencias Naturales (MECN). (I am not, however, as this time, attempting a comprehensive new survey.)

1. Palatine Teeth: Absent (0), Present (1).

Teeth on the palatine bones are unusual in lizards, including iguanids, but primitive if *Sphenodon* or fossil lepidosauromorphs are used as outgroups. The possibility has been raised (Estes et al. 1988) that absence is primitive for squamates, with the known occurrences within squamate lineages (most conspicuously in snakes) reversals.

In anoles palatine teeth are known only in *Chamaeleolis*. They are absent in *Polychrus* which Guyer and Savage have chosen as their outgroup - hence the coding in table 2. The text (p. 519), however, describes *Chamaeleolis* as the most primitive of the anoles and says: "This placement of the genus is based primarily on the presence of angular bone and the presence of palatine teeth."

Guyer and Savage are in their text following the preferred judgment of Etheridge, who regards *Chamaeleolis* as the sister group of the remaining anoles. The choice of coding in the table 2, with the condition in *Polychrus* as the chosen outgroup defining the polarity of the character, has the unfortunate result that the palatine teeth of *Chamaeleolis* are regarded as an autapomorphic reversal in that genus and phylogenetically meaningless.

I would myself provisionally code the presence of palatine teeth as primitive. However, I recognize, in agreement with Estes et al. (1988), that reversal is this case quite possible; the only requirement is the extension forward of an ontogenetic field. (The palatine teeth in *Chamaeleolis* are at the rear of the bone, near and in line, or almost in line, with the pterygoid teeth.) I have not, in this case, ventured to call Guyer and Savage wrong.

2. <u>Angular</u>: Present (0), Absent (1).

The presence of an angular is certainly a primitive feature whatever outgroup is chosen. In anoles the bone or bones in the splenial-angular area are vestigial. Only in *Chamaeleolis* are there two. The more posterior element in *Chamaeleolis* is especially small but constantly present and distinct and has been identified as the angular. As the angular it is, as the text of Guyer and Savage states, one of the two arguments for regarding *Chamaeleolis* as especially primitive, and for regarding this genus as the sister group of all other anoles. There are, however, data in regard to the element ordinarily identified as the splenial in other anoles (see below) that may complicate this story.

3. <u>Splenial</u>: Present (0), Absent (1).

Again presence of a splenial is primitive by outgroup comparison. Distribution within the anoles is in the *latifrons* series *sensu* Etheridge (=*Dactyloa* of Guyer and Savage) and in the *cuvieri, cybotes, cristatellus*, and *bimaculatus* series *sensu* Guyer and Savage - it is not constant in all of these - but also in the *equestris* complex and *hendersoni* superspecies, and in *darlingtoni*, but not *insolitus*, of the *darlingtoni* series. All beta anoles, except a few of the most primitive (*biporcatus, parvauritus, loveridgei*) and the other series in Guyer and Savage's residual *Anolis* lack the bone (except that Etheridge (1960:144) mentions its occurrence in *Anolis longiceps* of Navassa, a member of the *carolinensis* series). It is difficult to make anything of this distribution except the expected statement that presence is primitive, and that the bone is absent in dwarf species belonging to primitive lineages. The ontogeny of this bone or bones has not been reported.

What is consistently called the splenial in anoles is, when best developed, a long and narrow sliver extending far posteriorly to extend onto the inferior surface of the mandible (Fig. 1b). In its posterior portion its position more nearly corresponds to the position of the angular than to any part of the splenial of other iguanids. Recognition of the very small posterior bone in *Chamaeleolis* as an angular is based, as Etheridge has reminded me, on the presence of the posterior mylohyoid foramen within it. I quote his statement (pp. 9-10 of his 1966 thesis): "The presence of a tiny element penetrated by the posterior mylohyoid foramen in *Chamaeolis* and the location of this foramen in other anoles between the dentary and the remaining element [= splenial fide Etheridge], rather than through this element, is interpreted as evidence that the angular is missing." I formerly accepted this view, but examination of the variation of the "splenial" in more anole taxa has raised questions in my mind. Most impressively for me, in the *cuvieri* series (*cuvieri, ricordii, baleatus, barahonae*) only an ovoid or trapezoidal anterior portion of the "splenial" (centered underneath the coronoid process) is present (Fig. 1a). In the *bimaculatus* series it has long been known that one to two or even three slivers of bone may be found in the position of the "splenial." In an undescribed giant species of *Phenacosaurus* I find well within the posterior "splenial" a foramen that appears to be the posterior mylohyoid foramen. This, however, occurs on one side only. (The posterior mylohyoid foramen is not always present in adult *Chamaeleolis*, although the two bones always are; Fig 1C shows the two elents and the foramen in the "angular" in a juvenile *Chamaeleolis chamaeleonides*.) The interpretation of vestigial bones of this sort is clearly not easy.

A correction of the printed record is required: when he was about to describe *eugenegrahami*, Albert Schwartz urgently asked me to examine its osteology. I did so too hastily, reporting a large splenial, mistaking a fragment of tissue for a suture, and an arrow-shaped interclavicle, misinterpreting, as I now believe, a variant of T-shape of the interclavicle. Both statements are wrong; *eugenegrahami* is not a member of the *bimaculatus* series as I stated in Schwartz (1978), but instead a member of the *carolinensis* subsection (*sensu* Williams 1976a). There is no splenial and a T-shaped interclavicle. Guyer and Savage would refer the species to their residual genus *Anolis*.

4. Jaw Sculpturing: Absent (0), Present (1).

This character - called by Etheridge (1960) "excavations in the lower jaw" is described only in males of some West Indian species and occurs in three types - called by Etheridge the *cristatellus* type, the *krugi* type and the *cybotes* type - which are quite different from one another and very characteristic. Etheridge (1960) regarded the three types in spite of differences as homologous and partly on the basis of this character, placed the species showing these in a single series, the *cristatellus* series (= *cybotes* + *cristatellus* series of Guyer and Savage). Etheridge (1964) noted sculpturing in *leachii*, quite different in configuration, increasing with size and becoming "grotesque, almost pathological" in very large fossil specimens. He also mentioned at this time sculpturing in *evermanni*. This feature has not been sufficiently examined in other anoles, and other non-homologous states may exist. It is known that rugosities, greater or lesser in extent, may occur on the lower jaw in *Chamaeleolis* and the *equestris* series. That these rugosities are sexually dimorphic is not yet ascertained.

5. Pineal Foramen: **U-shaped notch at the fronto-parietal suture (0), Within the parietal (1).**

Etheridge (1960:121) discusses this character at length both in regard to ontogeny and phylogeny. Ontogenetically the foramen begins as a large mediodorsal exposure of the brain that is then reduced by growth inward of the parietal from the two sides. Primitively and in most anoles a median u-shaped notch is retained in the parietal at the frontoparietal margin. In some forms - always in the *grahami* series and almost always in the *cristatellus* series - inward growth of the anterior margins of the parietal toward each other may result in a foramen well within the parietal. Sometimes a narrow open groove may connect a foramen lying within the parietal with the anterior frontoparietal suture. There is individual variation - specifically mentioned by Etheridge (p. 151) for the species *cristatellus*.

Guyer and Savage distinguish three states of this character: "fronto-parietal suture (0), the anterior edge of the parietal (1), parietal (2)." I find the distinction between the first two to be oversubtle. The anterior edge of the parietal cannot help but be at the fronto-parietal suture. Some justification for the Guyer and Savage coding can be found in the slightly different phraseology used by Etheridge in his descriptions under the several series. My own reaction, however, is that there has been overinterpretation here. I have coded only the two states above.

6. Parietal shape: Trapezoidal (0), Triangular (1).

This should be called "parietal roof shape," not "parietal shape"; the parietal is a moderately complex bone, and its ventrolateral components enter in no way into the character here described. On outgroup comparison there is no doubt about the polarity in this case.

7. Parietal crest: None or U-shaped (0), **V-shaped (1), Y-shaped (2).**

The term "crest" was used by Etheridge (1960) for the low and simple keels that define parietal roof shape in anoles. It is an unfortunate term - all too reminiscent of the high vertical crests of basiliscines and unfortunate in still another way; it prevents recognition of the difference between "casquing" - the impressive phenomenon seen in *Chamaeleolis* and a few other primitive anoles - and the mere ridges or keels present only in the advanced anoles. I shall discuss casquing below, after dealing with "crests" as understood by Guyer and Savage.

There is, not surprisingly, a close relationship between the shape of the parietal roof and the "cresting" that bounds it. Etheridge (1960) in his description of series did not treat parietal shape as an independent character. Cannatella and de Queiroz (1988) treat characters 6, 7, and also 8 of Guyer and Savage as synonymous. I have retained the supposed distinctions in order to be closer to Guyer and Savage's presentation. There is, in fact, a distinction to be made that is, to some degree, useful. U-shape or triangular shape for "parietal shape," understood as parietal roof shape, does describe a reality different from that seen in the pattern of ridging or keels found in the Y-shape of the parietal. In particular the part of the crest that is the tail of the Y is not part of a horizontal parietal roof, which is only the triangular area contained within the anterior V of the Y.

The coding for "parietal crest" given in table 2 of Guyer and Savage is different from that provided above and is erroneous. Etheridge (1960) was able, with the aid of a large series of *Anolis carolinensis* showing growth stages, to provide an ontogenetic criterion for the polarity of this character, well shown in his fig. 9. The condition with no or U-shaped parietal cresting (Fig. 2a) is primitive on both outgroup and ontogenetic criteria. V-shaped is less derived than Y-shaped (Fig. 2b) on ontogenetic criteria.

The effect of Guyer and Savage's coding in tables 2 and 3 is to make the two island beta series (*grahami* and *sagrei*) appear more primitive than they, in fact, are, and the mainland betas less primitive than they are. This is true also of the statement in the text, which, however, ascribes a different parietal cresting for the two series. Thus Guyer and Savage assert (p. 521) that "the two island series of *Norops* (*grahami* and *sagrai*[sic] series) appear at the base of a single *Norops* lineage due to their primitive V-shaped parietal cresting." In fact the two island beta series have in common the most derived condition a Y-shaped parietal crest - as Etheridge stated. The condition of this character in all the mainland betas I have checked is more primitive than that found in the island series - a V sometimes tending toward a U.

Ignored in Guyer and Savage's discussion is the phenomenon of casquing. Casquing is found in its clearest and most extreme form in *Chamaeleolis* but is present in a much less impressive state in *Phenacosaurus* and in the *equestris* complex. Ignoring casquing results in the coding of the extraordinary parietal roof in *Chamaeolis* by Guyer and Savage as (1) under character 6 (=triangular) and (1) under character 7 (=Y), as though *Chamaeolis* were not different from *grahami* or *sagrei*. *Phenacosaurus* is coded for the two characters (0),(0), and the *equestris* series as (1),(1). None of these codings would aid anyone to visualize the conditions actually found.

The parietal casque in *Chamaeleolis* is U-shaped and expanded laterally as well as very extensively posteriorly. In profile the casque totally transforms the head, making it markedly oblique dorsally - high posteriorly, much lower anteriorly. Similar but less striking conditions occur in the *equestris* series (Fig. 2c) and in *Phenacosaurus* (Fig. 2d).

It is possible to imagine the development of the casque from outward - lateral and posterior - growth of the primitive trapezoidal state of the parietal roof. However, the U to V to Y evolutionary trend of the parietal ridges of the advanced anoles is in a very different direction from that manifested by the casque margins of *Chamaeleolis*, *Phena-*

cosaurus, and *equestris*. Parietal casquing and parietal "cresting" are two divergent evolutionary tracks. Etheridge's description of "crests" in his series is accurate and pertinent; he did not, however, - and very rightly - endeavor to compare what he described as "a large shield" in *Chamaeleolis* and "more or less of a casque" in *Phenacosaurus* with the "crests" of *Anolis* (*sensu lato*).

8. "Occipital": "Exposed" (0), "Half-funnel" (1).

Character 8 is strongly correlated with characters 6 and 7. Cannatella and De Queiroz (1988) therefore treat it as synonymous. Morphologically there is, in fact, some difference in the degree of development of the "half funnel" that is realized in the different taxa. I therefore discuss it as an independent character.

The term "occipital" is in error as to the element involved. It is the parietal, not any part of the occipital complex that can be described as "half-funnel." The character cited - the expanded and flared posterior margin of the parietal, turned upward and extended backward to partly overlap the occipital region - is another of the characters that I regard as a synapomorphy of all the advanced anoles. It is not a feature that Etheridge (1960) used in his formal definitions of series. He, however, faithfully mentioned it in every series description in which it was relevant. He did not explicitly mention it for *cuvieri*, *ricordii*, or *roosevelti* (the only species of the *cuvieri* group that he knew) because he did not allocate them to any series. (See below under the definitions of genera for the confusions in regard to this character that may have resulted from Etheridge's omission.)

While the half-funnel condition of the posterior parietal margin is very distinctive (Fig. 2b) and makes the skull of any anole having it immediately recognizable, the fact that it is a terminal state in an ontogenetic transformation series makes its phylogenetic interpretation at times difficult. It may not be clear when the absence of the half-funnel state is primitive and when it is paedomorphic. In, for example, the *darlingtoni* series (Williams 1976a) two species - *darlingtoni* and *insolitus* - have been associated on the basis of distinctive phenetically very similar squamation. One - *insolitus* - is a dwarf; the other - *darlingtoni* - is of moderate size. A skull of *darlingtoni* is now available; it can be confirmed - as Etheridge had already determined from X-rays - that a parietal half-funnel is well-developed. The half-funnel is quite absent in *insolitus*. In the original description (Williams and Rand 1969) *insolitus* was regarded as a primitive anole, partly on characters now thought to be paedomorphic or ecomorphic. It now seems probable that both the dwarfism and absence of the parietal half-funnel are paedomorphic in *insolitus*, and that *darlingtoni* and *insolitus* are, indeed, related, but are both relatively advanced anoles.

In table 3 Guyer and Savage correctly describe the condition in the *cuvieri* series (Fig. 2b) as (1) = "half funnel". This is in keeping with the ontogenetic and size related changes in the dorsal aspect of the parietal. In the advanced anoles the larger species show the derived features of the parietal best. While the *cuvieri* series is in several characters relatively primitive among advanced anoles, it is the derived character of giant size that is the most conspicuous feature of the series. Since both size and phylogenetic position are relevant for the parietal half-funnel, this character state will need careful checking in all anoles; there are ambiguous situations both structurally and phylogenetically.

9. Interclavicle: **T-shaped (0), Arrow-shaped (1).**

The descriptive epithets here are a bit too simplistic, as Etheridge long ago emphasized to me. There is variation in several features of the interclavicle and variation in the clavicle also that partly parallels that of the interclavicular arms. At issue also is polarity.

The polarity favored by Etheridge and by Guyer and Savage depends upon two observations: 1) that the interclavicle in *Chamaeleolis* is arrow-shaped, i.e., "its lateral processes only partly in contact with the slender proximal parts of the clavicles" (Etheridge 1960:82) and 2) that the interclavicle in the South American alpha anoles (*Dactyloa* of Guyer and Savage) is the same. The first of these observations I can confirm; the second I find to be true only for the *punctatus, tigrinus*, and *roquet* species groups, not true for the *latifrons* and *aequatorialis* species groups.

The observation that the South American alpha anoles are not a uniform group in terms of interclavicular structure is the most startling new information that I will present in this paper. It is based on dry skeletons of critical species in the Museum of Comparative Zoology. Dry skeletons can involve some distortion due to warping, but this should be least important in large species. *Latifrons, frenatus* and *aequatorialis* are in anole terms "giant" species, and each has an interclavicle conforming fully to Etheridge's definition of the T-condition - "lateral processes closely overlapped by the flattened proximal parts of the clavicles" (again Etheridge 1960:82). The interclavicles of the species *punctatus, solitarius*, or *roquet* are very different, although not literally arrow-shaped; their lateral processes, at about one-third of their length, sharply and abruptly diverge from the adjacent clavicles. The interclavicles of South American alpha anoles, in fact, segregate more sharply into two types than do some of the T-interclavicles of the *carolinensis* subsection from some of the arrow interclavicles of the *cuvieri* or *bimaculatus* series. The question is not one of the existence of two types within the South American alphas but rather whether similarly clear and homologous states can be recognized in the West Indian series - or in *Phenacosaurus* (Fig. 3e) and *Chamaelinorops* (fig. 4 in Case and Williams 1987). I provisionally classify the interclavicles of the two latter genera as T-shaped but I regard neither of them as so clearly T-shaped as the elements in the *latifrons* and *aequatorialis* series.

In view of the observations and the hesitations expressed earlier regarding the characters of palatine teeth and angular bone as evidence for the position of *Chamaeleolis* as the sister group of all other anoles, the case for the primitiveness of either of the two conditions of interclavicle is weak. Despite variation in each type, and some approach, particularly in the West Indian taxa, I do not see in the South American alphas any overlap. In the absence of unambiguous outgroup comparison (morunasaurines and oplurines with "arrow" interclavicles, *Polychrus* and leiosaurs with "T" interclavicles) I have ventured here to code the T-shaped interclavicle as primitive, because, as I interpret it, it is so very widespread in anole lineages that I believe to be independent - three of the genera - *Phenacosaurus, Chamaelinorops*, and the new genus that I will be proposing for the species *occultus*, the *latifrons* and *aequatorialis* series of mainland alpha anoles plus all beta anoles and all species of the *carolinensis* subsection. It seems most parsimonious to assume that a condition common to all of such a diverse assemblage is primitive than to assume that it has arisen independently in each genus or group independently. There is further citation of this character in the section on the definition of genera.

10. "Parasternal ribs"

There is here a perhaps minor point in morphological nomenclature. Guyer and Savage consistently employ the term "parasternal ribs." "Parasternum" was, indeed, used by Etheridge in 1960. However in 1965, in a paper not cited by Guyer and Savage, Etheridge corrected his nomenclature, noting that "parasternum" or "parasternalia" was a term coined for dermal abdominal structures such as those in crocodiles and *Sphenodon* and should not be used for the superficially similar serial ventral cartilaginous structures in lizards. He utilized the term "inscriptional ribs," specifically "postxiphisternal inscriptional ribs," and this has been the conventional language for the latter structures in anole literature ever since. Perhaps, however, Guyer and Savage had in mind Camp's 1923 use of "parasternum" and preferred this usage as classical for lizards.

a. Total Number of "Parasternal" (=Inscriptional) Ribs: >7 (0), 7 (1), 6 (2), 5 (3), 4 (4).

There is some question as to the polarities within this sequence. Guyer and Savage have selected *Polychrus* as an outgroup - hence their decision to call >7 the most primitive condition. They have also chosen to regard *Chamaeleolis* as the most primitive anole. For them this involves treating all of the characters of *Chamaeleolis* that are not obviously autapomorphic (e.g. the huge parietal crest/casque) as also the most primitive condition for anoles. *Chamaeleolis* has 6 total inscriptional ribs; hence Guyer and Savage postulate in their text that the condition of *Phenacosaurus*, which has 7, is a reversal. There is no question in any case that the general trend in anoles is to reduction in total number of inscriptional ribs, but the reduction has occurred in different lineages independently in both alpha and beta anoles.

Etheridge (1965) further reported some data that Guyer and Savage have not commented on, including totals of 6 and 7 inscriptional ribs in *Anolis* (*sensu lato*). He did not cite the specific species, but presumably these were variant conditions in South American alphas. His count for *Chamaelinorops* was a total of 6 as in *Chamaeleolis*. Forsgaard (1983) counted a total of only 5 in *Chamaelinorops*.

b. Number of Attached Ribs: >6 (0), 5 (1), 4 (2), 3 (3), 2 (4).

Etheridge (1960 and 1965), and I have followed him, has used a formula for inscriptional rib states that gives as the first number the number of attached inscriptional ribs, and as the second the number of free chevrons, thus 4:2, 3:1, 2:2. The total number of ribs and of free and attached elements is thus given simultaneously. Thus Etheridge's (1965) high count formulae for *Anolis* (*sensu lato*) were 5:2, 5:1, and 4:2, and for *Chamaeleolis* 4:2 and *Chamaelinorops* 4:2, 3:3. Forsgaard's (1983) formula for *Chamaelinorops* was 4:1.

Again the general trend is toward reduction of attached ribs, again in different lineages independently. The extreme reported by Etheridge (1960, 1965) was one - formula 1:3 - in the *nebulosus* complex of mainland betas. There is some tendency for related forms to have the same inscriptional rib formula, but this is not sufficiently consistent to be a useful clue to phylogeny. Etheridge (1960) made routine use of inscriptional formula in series definitions, but, at that time, overvalued this character. There is more individual variation, at least in taxa with higher counts, than Etheridge was aware of in 1960.

11. Number of Presacral Vertebrae: 25 (0), 24 (1), 23 (2), 22 (3).

This character does not appear to be of major taxonomic value within anoles. It seems most useful at the species group and species levels. Guyer and Savage list variation from the primitive count of 24 presacrals only for *Phenacosaurus* (content: three recognized species) 22 presacrals, for the *cristatellus* series (content: a species group) 23 presacrals, the *equestris* series (content: a superspecies) 23 presacrals, and for the *auratus* series (content: unknown) 23 presacrals. I do not know the source of the count for the *equestris* series, and Etheridge's count for the species *auratus* and most of the other members of his *chrysolepis* series was 24. Those species of Etheridge's *chrysolepis* series for which Etheridge (1960:120) did give a presacral count of 23 seem a rather miscellaneous set that I am unable to evaluate, except that most are Mexican. Several beta species show derived counts. *Crassulus* was reported by Etheridge as having a count of 22. In *humilis* Etheridge found a range of 22 to 24 in a large sample, with the mode at 23. In related *uniformis*, also in a large sample, the range was 23 to 24, with a very strong mode at 24. In the *nebulosus* complex counts ranged from 21 to 25. There is also variability within the *latifrons* series, *sensu* Etheridge, that Guyer and Savage do not mention. Etheridge cited counts of 23 for *jacare* and *transversalis* (and the synonym of the latter *buckleyi*).

12. Number of Lumbar Vertebrae: <2 (0), 3 (1), 3-4 (2), 4 (3), 5 (4).

The number of lumbar vertebrae undoubtedly increases within the anoles. On the basis of outgroup comparison, the primitive condition is none. The most primitive condition in anoles, 2 or less, and the most derived, 5 or more, are distinct and infrequent. This, however, does not suffice to separate out groups cleanly. Guyer and Savage's coding of 3 lumbar vertebrae, 3 to 4 lumbar vertebrae, and 4 lumbar vertebrae as three discrete character states is plain evidence of the difficulties attending use of the character. Etheridge's cautionary remarks (1960:127) are pertinent: "Three or four lumbar vertebrae are usually present; there may be as many as seven. There is some variation in all species, but the range of variation in any one is seldom more than two or three. Because of the small total range of variation in the entire group, the number of lumbar vertebrae is not a particularly useful taxonomic character."

It should be mentioned that that it is not always easy to determine the number of lumbar vertebrae on dry skeletons, particularly of small species. The articular facets for the last dorsal ribs may not be at all obvious. Presumably Etheridge (1960), dealing primarily with X-ray photographs of intact specimens, had better data.

13. Number of Aseptate Caudal vertebrae: >8 (0), 8 (1).

This is again a character routinely reported by Etheridge (1960) but not used by him in the definition of series because of individual variation and overlap between series. As he had with lumbar vertebrae, Etheridge records "the most usual number" - a category that is very unsatisfactory when sample size is small. Etheridge cites for *equestris* 10 anterior aseptate caudals and lists the range in his *latifrons* series as 7-11. The species names in the *latifrons* series utilized by Etheridge in his thesis have not been verified, except for those represented by specimens at the Museum of Comparative Zoology, but the following breakdown by species and series *sensu* Guyer and Savage seem likely to be correct: *laevis* series: *proboscis* 9; *latifrons* series: *fraseri* 8, *latifrons* 9, *frenatus* 10, *insignis* 10, *microtus* 10, *squamulatus* 11; *punctatus* series: *chloris* 7, *peraccae* 7, *punctatus* 9, *transversalis* 9; *roquet* series: *bonairensis* 7, *richardi*, *roquet*, and *trinitatis* 8; *tigrinus* series: *solitarius* 8. The West Indian series, other than the *roquet* and *equestris* series, are all reported by Etheridge as having 6 or 7 aseptate caudals. The West Indian betas fall in nearly the same range the *grahami* series with 5-7 aseptate caudals, The "most usual number" 7, the *sagrei* series 5 or 6, the most usual number 5. The *petersi* series of the mainland betas according to Etheridge show a greater range and higher, presumably more primitive, numbers: 7-11, loveridgei 7, *pentaprion* 8, *biporcatus* 8 or 9, *capito* 9, *petersi* 11. The more derived series - *fuscoauratus* and *chrysolepis* - are said to show a range of 6-8.

Etheridge's data suggest that this character may be interesting when compared between closely related series. It is also a character that can be assessed readily on dry skeletons. A proper survey has, however, not been done. It is unclear why Guyer and Savage chose to break the continuum at >8 contrasted with 8; the character might better have been discarded. It seems of no taxonomic utility at the level of analysis that they chose to make.

14. Caudal Autotomy Septa: Absent (0), Present (1).

The problem here is polarity. Having accepted *Polychrus* as their outgroup, Guyer and Savage are committed to accepting lack of autotomy as primitive in anoles. This is one of the two cases in which, solely on the basis of *Polychrus* as their outgroup, they reject polarity decisions by Etheridge. Etheridge's own arguments for his preferred polarity are two: 1) that on wider outgroup comparison caudal autotomy appears to be primitive for lizards; 2) reversal in an embryonic sense means re-invention of breakage planes and the considerable structural complexity that breakage planes entail. I will make an alternative suggestion of my own: *Chamaeleolis, Phenacosaurus, Chamaelinorops*, and a new genus that in my opinion (see further remarks below) should be erected for *occultus* have non-autotomic vertebrae. These might be relics of an ancient radiation for which non-autotomic vertebrae was primitive, while *Anolis* (*sensu lato*) could be another radiation in which tail autotomy was primitive. I suspect this problem to be unresolvable without further evidence.

In some anoles the absence of caudal autotomy is a derived state. I have discovered the highly derived beta anole *bombifrons*, which is only with some difficulty distinguished from its near relative *chrysolepis*, to have a non-autotomic tail. Another beta anole - *meridionalis* - has been long known to have a non-autotomic tail, but in this instance Etheridge (1960:175) apparently had evidence from X-rays that "the septa of the caudal vertebrae become fused and non-functional." The ontogeny of this character in *bombifrons* is not known.

Etheridge mentions several instances in beta anoles of partial fusion of the caudal septa. On p. 163 he specifies this for the species *petersi*; while on p. 164 he mentions that in several unspecified forms of his *fuscoauratus* series from Peru and Ecuador "fusion of the septa leaves only a few of the more anterior autotomic vertebrae functional." He was not aware of any such cases in his *chrysolepis* series, which he believed to be more derived.

15. Caudal Vertebrae:(for coding see below).

Of the two kinds of autotomic vertebrae that occur in anoles, only the alpha type - transverse processes absent - is paralleled in other lizards, but, according to Etheridge and de Queiroz 1988, even this type is non-homologous with the caudals without transverse processes of other lizards. The beta type differs from the autotomic vertebrae of other lizards with transverse processes in the orientation - anterolateral - and position - posterior to the breakage plane - of the transverse processes. The caudal vertebrae of *Chamaelinorops* are distinctive - non-autotomic and with transverse processes that in adults are oriented directly laterally and widened distally. Etheridge believes the transverse processes of *Chamaelinorops* to be derived from the beta condition.

The transformation series of the anole caudal sequences is controversial. Etheridge (1960) suggested the beta gave rise to alpha. An obvious alternative is the reverse. Williams (1977, and Williams in Peterson 1983) preferred the sequence: *Chamaelinorops* to alpha to beta. Cannatella and de Queiroz (1988) mention that there are other possibilities and cite independent origin of alphas and betas.

Guyer and Savage code three possibilities, taking *Polychrus* as the plesiomorphic state for two of them. Since *Polychrus* is already a non-autotomic alpha, "system 1" (=15a in table 2) might be interpreted to imply the transformation series non-autotomic alpha to autotomic beta to autotomic alpha, and "system 2" (=15b in table 2) non-autotomic alpha to autotomic alpha to autotomic beta, but the coding in table 3 appears not to agree with this: In both 15a and 15b *Polychrus* is indeed coded (0), but in 15a *Chamaeleolis* and *Phenacosaurus*, both non-autotomic alphas are coded (2) like the autotomic alpha series, but *Chamaelinorops* also non-autotomic but certainly <u>not</u> an alpha is also coded (2). In 15a of table 3 betas are consistently coded 1, including *meridionalis*, which under character 13 is coded as non-autotomic. In 15b, what seem to me to be similar errors are made, including again the coding of *Chamaelinorops* as an alpha!

"System 3" (=15c of table 2) purports to be the Williams hypothesis, but the sequence given is somewhat peculiar: "Alpha (1), *Chamaelinorops* (0), Beta (2)," perhaps intended to imply the independent origin of alpha and beta caudals from a *Chamaelinorops*-like type. However, the text below table 2 (p. 513) plainly is congruent only with a sequence for this system in which beta is derived from alpha. In table 3 the coding in the column for 15c is mostly irrelevant to System 3 of table 2: Both *Polychrus* and *Chamaelinorops* are coded (1), all alpha caudals are coded (0), and only beta caudal are coded correctly - (2). For character 15 it is doubtful that even table 2 has been fully worked out, but the errors in table 3 are blatant.

Evaluation of the osteological data set.--Setting aside the errors and confusions Guyer and Savage have introduced, there are evident intrinsic ambiguities and contradictions in even this minimal set of 15 characters. Several characters appear to be of little phylogenetic value. There are questions of polarity that are unresolved as well as observational gaps and problems of interpretation. In such a situation, there is clearly an urgent need for additional characters. Etheridge's 1960 data was an immensely useful first approximation. It needs to be superseded by a new still more comprehensive study. It is important that much of the material for such a study is now available.

2. The karyological data set

Guyer and Savage's count of "at least 96 species" of anole karyotypes includes species studied by Carl S. Lieb in his unpublished thesis (University of California, Los Angeles). Their table 4 intends to be a summary of the available data, listed by diploid number and presence or absence of sex chromosomes. Macro- and microchromosomes are recorded as separate numbers, but acrocentrics and metacentrics are not distinguished. Sex chromosomes, if present, are recorded as xy or xxy and counted with the macrochromosomes.

The evolution of anole karyotypes is inferred directly by Guyer and Savage from diploid chromosome <u>numbers</u>. This has to be deplored as a very unsafe procedure. Banding is almost unknown for any anole (see, however, the pioneering study of C-banding in *grahami* by Blake 1983b). Banding of any sort is itself is poorly understood and very subject to interpretation; chromosome homology without banding is a matter of guesswork.

I would emphasize that, in the absence of banding, I cannot regard it as confirmed that all 12+24 karyotypes are in fact the same. Even in this chromosome pattern, which does look to be relatively stereotyped, there may be significant differences in arm length between supposedly homologous macrochromosomes (Paull et al. 1976). I do, in discussions of phylogeny, assume that 12+24 karyotypes are both primitive and identical, but I do so hesitantly, as a heuristic first approach. Correspondingly, when other karyotypes with different numbers and differing arm lengths of macrochromosomes come under consideration, I regard the possibility of complex and non-homologous changes as dangerously great. Karyology is an evolutionary area in which change is characteristically saltatory and not easily traced without good markers. In anole species groups that on other grounds are known to be closely related, it may be possible to trace sequences of change with modest hope of approximating reality. In the case, however, of species distantly related or of uncertain affinities, I quite lack the confidence which Guyer and Savage display. I reiterate that until anole chromosome segments can be homologized with reasonable accuracy, anole chromosome evolution will be poorly understood.

On page 516 Guyer and Savage describe the mechanisms in terms of which they analyze the karyotypic evolution of anoles: "In determining evolutionary sequences from chromosome number, we assumed that macrochromosomes evolve from low to high numbers via fission (Webster et al. 1972) and microchromosomes evolve from high to low numbers via disappearance or fusion (Gorman 1973)." These mechanisms are certainly real for specific cases (e.g., fission of macrochromosomes in the *monticola* series and in *insolitus*, the taxa for which Webster et al. 1972 invoked it; disappearance or fusion of microchromosomes in the *roquet* superspecies of the *roquet* series (Gorman and Atkins 1967)). However, fission of macrochromosomes does not seem to be the most frequent mode of origin of macrochromosomes in anoles; instead, fusions of microchromosomes with each other are the most likely explanation of increase in macrochromosome number, as well as, in parallel, of decrease in microchromosome number (Webster et al.1972, Gorman 1973, Paull et al. 1976).

There are certainly also a plethora of chromosomal changes additional to and very different from the two - and only two - mechanisms on which Guyer and Savage attempt to build the whole evolutionary story in anoles.

But, apart from the serious problems always present in working out chromosome homology and chromosome evolution in any group, it is a conspicuous and grievous fact that in anoles the present karyological sample is woefully biased and inadequate. Ninety six taxa karyotyped may seem impressive, but most of these are West Indian, almost all the rest are Mexican or Central American. South America is very nearly a terra incognita: karyotypes for only nine species have thus far been published - *jacare*, *squamulatus*, *auratus*, *biporcatus*, *chrysolepis*, *fuscoauratus*, *gracilipes*, *lineatus*, and *tropidogaster*. Of Guyer and Savage's *Dactyloa* only *jacare* (Williams et al. 1970) and *squamulatus* of Venezuela, *frenatus* of Panama and *agassizi* of Malpelo Island (Stamm and Gorman 1975), and the *roquet* series of the Southern Lesser Antilles have been studied. *Agassizi*, *squamulatus*, and the *roquet* series have the primitive 12+24 karyotype. *Jacare*, despite minor differences, appears to differ significantly from the primitive condition only - as does the *roquet* species group - by loss of a pair of microchromosomes. *Frenatus*, however, has a karyotype that is distinctly derived: there is a size break between pairs 4 and 5, the macro- and microchromosomes are not sharply distinct; pair 7 is intermediate in size. There is undoubtedly some tendency to retain the primitive condition, but there are at least 45 species of South American alphas for which no karyotypes are known.

Because of the confusion as to which species are contained in which series, it is impossible to check the accuracy of most of the chromosome numbers assigned to series in table 4. The *carolinensis* subsection, the *cybotes* series, and *cuvieri* series have no karyotype numbers listed opposite them; I assume that the reader is supposed to infer - most often correctly - that their karyotypes are primitive. There is no mention at all of the *latifrons* series *sensu* Etheridge nor of the included *roquet* series. The chromosome counts given for *Chamaeleolis* and *Chamaelinorops* do not accord with the information available to me. (See further under the discussion of the definitions of these genera below.)

Their fig. 6 is said to present the most parsimonious karyotype tree. Taxa with the primitive 12+24 karyotype are not listed at all - on the excuse that their tree is rooted with this karyotype. However, the *monticola* series, which has several derived karotypes, is omitted also. The tree, in consequence, except for two minimally derived *latifrons* karyotypes at the base of the tree, portrays only karyotype evolution in the beta anoles and the *bimaculatus-cristatellus* series. It uses for this purpose chromosome number - macro- and micro- and sex chromosome data (none, xy or xxy) - in a very literal and simple numerical sequence. Several beta karyotypes - West Indian and mainland - are placed near the base of the tree prior to the node which divides beta from *bimaculatus-cristatellus* patterns. There is a conspicuous anomaly: on the mainly *bimaculatus-cristatellus* branch there is one node - 17,14 xxy - at which a *cristatellus* series (Puerto Rican alpha) karyotype is treated as equivalent to a *petersi* (mainland beta) karyotype with the same numbers and sex chromosome pattern.

The anomaly becomes especially awkward when the tree (fig. 6) is compared with table 4 which purports to be a "List of karyotypes known from *Anolis*." The *petersi* series is listed three times - with different karyotype numbers - in the tree, but only once - and for a single species - in table 4, where the sex chromosomes are, indeed, said to be xxy, but the macrochromosome-microchromosome count is given as 15,14. The entry in table 4 for the *petersi* series does correspond to Gorman's 1973 description of *biporcatus*, but is not reconcilable with the node on the tree in fig. 6.

On the basis of the tree of fig. 6 modal types are defined (p. 517). Type I is the primitive karyotype. Type II is the *cristatellus* series and Type III the *bimaculatus* series. Within the betas Type IV (modal 2N=30) and type V (modal 2N=40) are recognized. The conclusion is asserted that: "Thus the karyotype data support a single lineage of derived anoles that has undergone karyotypic evolution from Type I chromosomes to Type IV chromosomes - through a reduction of the microchromosomes to 16 and an increase in macrochromosomes to 14, followed by derivation of Type V chromosomes through a dramatic, possibly saltatory increase in macrochromosomes to 24. Types II and III (*cristatellus* and *bimaculatus* series respectively) differ from all the others by sharing derived xxy heterochromatism but are related to Type V by sharing the condition of 18 macrochromosomes."

It is very difficult to know what to make of either the tree or of the discussion of it. Both are simplistic to the point of absurdity. To reiterate just one critical point, chromosome number is no guarantee at all of homology and serves only to conceal both convergences that are only numerical and the real homologies of chromosome segments that might be revealed by banding. The discussion glosses over these and other difficulties with no indication of awareness.

The scheme of p. 517 is disavowed in the discussion of karyotypic data in the section on "Consensus among data sets" on p. 519. Guyer and Savage rid themselves of the most serious anomaly in the tree - the xxy pattern in the *petersi* series - by revealing that "Gorman (1973) noted that the condition in *A. biporcatus* is not homologous with that in the *bimaculatus* or *cristatellus* series and, therefore, it appears to be of independent origin." (Gorman disavows this in a recent letter to me, commenting that he "was then biased in favor of the alpha and beta dichotomy and hence dismissed all similarities between beta and *cristatellus* karotypes as convergent.") A second attempt of Guyer and Savage (p. 519) to rescue themselves from the "single lineage of chromosome evolution" is a singular mixture of naivete and error (my emphases):

> "The reductions of microchromosomes and additions of macrochromosomes needed to derive Types IV and V chromosomes are also found in some *carolinensis* subsection anoles (*monticola* species group; Williams and Webster, 1974). Therefore, we think it most likely that advanced karyotypes were derived independently in anoles of the beta section and those of the *bimaculatus-cristatellus* series."

There are here both factual and logical errors. It is, first, an error to assert that Williams and Webster (1974) found reductions in microchromosome number in the *monticola* species group. On the contrary, Webster in that paper recorded for *rupinae*, as a phenomenon not known in any other anole, an increase in microchromosome number from 12 to 13 pairs! Secondly, while it is certainly possible or even probable that the beta karyotypes and those of the *bimaculatus-cristatellus* series are independently evolved, the "therefore" in Guyer and Savage's remarks is, a total non sequitur. The increase in macrochromosome number in the *monticola* series is strictly by fission. In contrast, except in the most highly derived taxon in the *bimaculatus* series - the species complex *occulatus* - increase in macrochromosome number in the *bimaculatus-cristatellus* series, and much of this increase and all of the other changes parallel to the *bimaculatus-cristatellus* patterns in betas, i.e., the disappearance of a sharp distinction between macro- and microchromosomes, have nothing to do with fission. Instead, they involve, most probably, translocations between macrochromosomes and fusions of microchromosomes. The chromosomal changes in the *monticola* series are quite irrelevant to any evolutionary comparison of beta and *bimaculatus-cristatellus* karyotypes. All of this increase in total chromosome number by fissioning has without reasonable doubt arisen independently in highly derived very separate lineages.

A curious detail in Guyer and Savage's karyology is the use of the term "heterochromatism" for heteromorphic sex chromosomes. It is a usage unknown to me from a fairly extensive reading of the cytological literature. Perhaps Guyer and Savage's usage is only a lapsus; if so, it is a confusing and much repeated lapsus.

Guyer and Savage's ultimate conclusion (p. 522) is that "the evolution of karyotypes [in anoles] cannot be inferred independently." This is surely true - certainly for the anoles as a whole - but, given this harsh fact, it is difficult to defend their rather elaborate analytical exercise on earlier pages, which proceeded as if such independent inference were possible.

Evaluation of the karyological data set.--In the absence of banding techniques for anoles that are adequate to ascertain the homology of whole chromosomes, chromosome arms or segments, karyology in anoles must remain at a primitive level. Hypotheses may be suggested regarding possible modes of chromosome evolution, but they cannot at present be fully confirmed. Within species, i.e., for such variation as described by Blake 1983a, or between very closely related species, evolutionary hypotheses may have reasonable probability. However, when such hypotheses are extended beyond the level of the species group or series, and especially when phylogenetic relationships are founded on assumptions as unrealistic as those of Guyer and Savage, they must be regarded as very unsafe - if not outright erroneous.

Chromosome banding is possible in lizards (e.g. in geckos Moritz 1984), and Kasahara et al. (1987a and b) have achieved very satisfactory G-, C-, and R-banding, as well as NOR staining in an iguanid genus *Tropidurus*. Blake (1983b) has aready demonstrated C-banding in one anole species, *grahami*, but her pioneer effort has not been followed up in any other species.

Two things clearly are needed if there is to be such across the board phylogenetic interpretation of anole chromosomes as Guyer and Savage have here first attempted and then abandoned: 1) There must be consistent use of banding techniques to ascertain chromosome homologies. In effect, this means that all karyotypes now reported for anoles must be redone. 2) There must additionally be more complete sampling of anoles, i.e., a serious effort to fill all taxonomic gaps.

3. The immunological data set

In their section analyzing the immunological data base Guyer and Savage utilize only the data of Shochat and Dessauer (1981), dismissing almost without comment the substantial contribution by Gorman and collaborators and not even citing the paper by Gorman, Lieb, and Harwood 1984. (They do later discuss the taxonomic implications of Gorman's work, but only in relation to their genus *Ctenonotus*; see further below.)

The pioneering work of Shochat and Dessauer (1974, 1977, 1980) was based on 40 species (not six as one would infer from Guyer and Savage p. 517) compared against reciprocal antisera from seven taxa that represented all major anole groups (betas, *punctatus* subsection and *carolinensis* subsection). The six series phenogram of Shochat and Dessauer closely associated the *bimaculatus* series, the *cristatellus* plus *acutus* series, and the Jamaican betas and found the *carolinensis* series (represented by *carolinensis* itself) and the *latifrons* series *sensu* Etheridge (represented by the *roquet* series) to be distinctly more distant. One species - *cybotes* - associated by both Etheridge and Williams with the *cristatellus* series Shochat and Dessauer expressly excluded from that series on the basis of their immunological data. Two mainland betas (*chrysolepis* and *lineatus*) were grouped with the Jamaican betas. Shochat and Dessauer grouped the *cristatellus* (including *acutus*), *bimaculatus*, and *grahami* series as the "central-Caribbean-series-complex," referring to them by the acronym "CCSC."

Gorman et al. (1980) confirmed the crucial immunological similarity of the *cristatellus* (excluding *cybotes*) and *bimaculatus* series with the Jamaican betas and attempted use of immunological distance in cladistic analysis, adding other characters to demonstrate the plausibility of new definitions of series and subseries in the anoles of the eastern Caribbean. Wyles and Gorman (1980) extended the immunological sample (using *cybotes* antiserum) to include seven previously unstudied Hispaniolan anoles including the endemic genus *Chamaelinorops* and one Mexican species *gadovii*. They found greater immunological similarity among the Hispaniolan taxa (including *Chamaelinorops*) than between these and any other species compared. Gorman, Lieb, and Harwood (1984), using *gadovii* antisera, measured the immunological distances between *gadovii* and 20 other anole species, 14 of them Mexican or Central American betas and one - *frenatus* - a Panamanian alpha of the *latifrons* series (*sensu stricto*) one - *luteogularis* - a Cuban alpha of the *equestris* series, one - *garmani* - a Jamaican beta, one - *sagrei* - a Cuban beta, one - *evermanni* - a Puerto Rican alpha of the *bimaculatus* series, one - *extremus* - a *roquet* series alpha. Confirmed again was the relative similarity (smaller immunological distance) of some betas (*gadovii* and its close relatives and Jamaican *garmani*) to the alphas of the *bimaculatus* and *cristatellus* series and the greater difference (larger distance) between all these and other alphas. They were impressed that as great differences exist between *gadovii* and relatives and several of the other mainland betas as exist within the alphas. They interpret this finding as implying that there may be several major Central American beta lineages, and, stressing the similarity of Jamaican betas and the *cristatellus-bimaculatus* series alphas, suggest that at least one lineage of the mainland betas may have arrived from the West Indies. Gorman et al. (1984) conclude that "1. There is no apparent Alpha-Beta dichotomy; 2. closest relatives appear to be geographic near neighbors; 3. the genus contains a number of distantly related lineages of uncertain relationships. Moreover the data suggest the possibility of a West Indian origin for the large endemic radiation of Mexican anoles."

Despite the absence of sufficient reciprocal sera - Guyer and Savage's reason for disregarding Gorman's data - the consistent results of his independent trials and their strong concurrence with the results of Shochat and Dessauer are very suggestive of the validity at some level of all of the immunological information. Guyer and Savage, however, do not adequately discuss any of it. The end result of a rather elaborate analytical procedure (p. 517) is a tree (p. 518) said to be identical to Shochat and Dessauer's (1980) phenogram. The description of the tree is brief but correct: "In it, anoles of the beta section and *cristatellus* series cluster as sister groups; a more distant relationship is indicated between this pair and the *bimaculatus* series. The *carolinensis* subsection, as the sister group of these three groups, and the *latifrons* series, as the sister group of the four-group cluster lie at the base of the tree (fig. 7)." Comment on the tree is postponed to the consensus section (p. 519), where the remarks have an odd emphasis: "The best-fit albumin tree supports the karyotype tree by suggesting that the *carolinensis* subsection is more primitive than the *bimaculatus-cristatellus* series and the beta section. However, it conflicts with the karyotypic data by failing to confirm the sister-group relationship of the anoles in the *bimaculatus-cristatellus* series."

This comment fails to make a point that would certainly be seen as important by Shochat and Dessauer and by Gorman - that the reason that the immunological tree does not confirm the karyological tree is that the beta anoles find their place on the immunological tree between the *cristatellus* and *bimaculatus* series and therefore appear closer to these series than to the *carolinensis* subsection. Gorman certainly considers that the karyotypes, like the immunological data, imply relationship of the beta anoles to the *cristatellus* and *bimaculatus* series.

The dispute as to the relationships of the betas is certainly not an unimportant one. Guyer and Savage ultimately (p. 529) resolve the issue, rather tentatively, in favor of a sister group relationship between the *carolinensis* subsection (=*Anolis* in their restricted sense) and the betas (their *Norops*). They mention, only to discard, an early separation from the *latifrons* series (= their *Dactyloa*). They do not discuss, almost do not mention, the Schochat and Dessauer-Gorman hypothesis of relationship between Jamaican betas and the *bimaculatus-cristatellus* sister-series (= in part their *Ctenonotus*).

Guyer and Savage's basis for associating beta anoles and their restricted genus *Anolis* is morphological - the presence in both of T-shaped interclavicles and some skull features found in "advanced" anoles, that otherwise must be regarded as homoplasies. They are compelled by this decision to reject the immunological data as of lesser importance - as Gorman and others tend to denigrate morphological details that I and others have regarded as important.

There are, indeed, grounds for thinking that close affinity between Jamaican betas and the *bimaculatus-cristatellus* series - and especially Gorman's suggestion of invasion of the mainland from the West Indies - is implausible (see below under the definition of *Norops*). But again Guyer and Savage have neither confronted nor even mentioned the hypothesis of a origin of the betas - or of some of them - within the West Indies. They have apparently rejected the hypothesis out of hand. The misfortune in this case as in their preference of osteological over immunological data is not the decision itself, but that it is made with no discussion.

Evaluation of the immunological data set.--For my own part I am very reluctant to throw real data away. Immunological distance has for me definite flaws - difficulties in interpretation that go the heart of its use. As a systematist using conventional morphological data I have difficulty with immunological distance also because it has, at times, been accepted prima facie as automatically and necessarily better than any morphological data. In this regard Gorman et al. (1980) are to be complimented on attempting to reconcile morphological and immunological data and making a good case for their reinterpretation.

But there are cases for which such reconciliation will not be at all easy. As I would now see it, the flaw in Wyles and Gorman (1980) is not at all their suggestion - repeated with additional evidence in Gorman et al. (1984) - that taxa geographically close are closest relatives, but rather their failure to realize how extraordinarily strong a statement they were making. In associating *Chamaelinorops* with Hispaniolan alpha *Anolis*, Wyles and Gorman were disavowing the relevance of extreme multi-system differences in morphology. (See the documentation in Case and Williams 1988.). There is now a parallel case to make the same apparent point. Burnell and Hedges (unpublished MS), studying 49 West Indian anoles (including *Chamaelinorops barbouri*) electrophoretically, find discrepancies nearly or quite as great between the relationships inferred from morphology and from shared slow-evolving electromorphs, with the biochemical data again associating nearest neighbors in spite of very strong morphological diversity.

I have no instant solution for such contradictions, often blatant, between and within data sets. The biochemical and morphological contradictions are merely those most advertised at the moment. Similar contradictions occur between electrophoretic and immunological data (Case and Williams 1988) and within the morphological set. Despite my own disputes with Gorman and collaborators I am impressed by their data and by Shochat and Dessauer's. I would like to see an immunological data set for all anoles - certainly for as many as possible. I am dismayed only by the tendency of the immunologically adept to discard the morphological data. I want to keep both - and, at this moment, to do so means - very obviously.-.keeping some problems unsolved.

The vicariance biogeography of anoles

Guyer and Savage in addition to their cladistic analysis add a section on biogeography, in which their paleogeography is plate tectonics and their method vicariance biogeography.

It is clear from the accumulated data of geology (Perfit and Williams this volume) that there is no escape from plate tectonics. It is now the base from which we start - the given that constrains our speculation. But it is now equally clear that plate tectonics provides no simple biogeographic answers.

I have elsewhere (Williams in the introductory chapter this volume) commented that Guyer and Savage somewhat misrepresent the geological evidence on the Caribbean now available (see Perfit and Williams this volume for a summary), and in particular misconstrue the geological authorities (Pindell and Dewey 1982) on whom they purport to rely. The first two panels, allegedly "Paleocene" and "Eocene" of their fig. 10, which intends to summarize Caribbean geology, are clearly mislabelled when compared directly with the figures in Pindell and Dewey from which they are said to be adapted. However, there are sufficient uncertainties in all reconstructions of the early Caribbean that theirs can neither be confidently rejected nor confidently accepted. I shall here, for the purposes of discussion, assume the correctness of their geology. I shall argue that some novel insights but very little clarity is added to anole biogeography by their version of a plate tectonic model for the Caribbean.

It is a theoretically imperative initial assumption of vicariance biogeography that the taxon that is to undergo vicariance is initially widespread over the whole land mass that is later to be broken up by barriers. Guyer and Savage believe that they have geological and biological evidence for an isthmus - "more or less continuous land mass" - a "proto-Antillean block" - spanning the gap between North and South America during the late Cretaceous and early Tertiary. They assume that the ancestral stock of anolines was distributed on and at both ends of this isthmus. The breaking away of this isthmus at both ends and its movement eastward was the initial vicariant event in anole history. Guyer and Savage describe the further events: "The subsequent fragmentation of this block into the precursors of the Greater Antilles and their displacement northeastward by the movement of the Caribbean Plate appears to have provided the basis for cladogenesis in Antillean anoles." As panel 2 in fig. 10 ("Eocene") shows the situation, Jamaica and the future southwest peninsula of Hispaniola remain well to the west, Jamaica in contact with Central America, while the remainder of the proto-Greater Antilles (presumably Cuba+Hispaniola+Puerto Rico) are a single eastern mass.

Thus far the vicariant events described by Guyer and Savage might account for

Dactyloa in South America, *Norops* in Central America, and something, perhaps, in what will become the southwest peninsula of Hispaniola. Jamaica, according to Guyer and Savage, would have received *Norops* from Central America at this time. If anoles were differentiating at this time on proto-Cuba-Hispaniola-Puerto Rico, it was not, according to Guyer and Savage's scenario, by any macro-vicariant plate tectonic event.

The next panel is said to portray the Oligocene. Jamaica is shown as still in contact with Central America. The southwest peninsula of Hispaniola is shown as moving eastward along a fault line that splits proto-western and central Cuba from the remainder of the Antilles. The western tip of Cuba is shown as close to Yucatan. What will be eastern Cuba below the fault line is identified as a fragment of the old proto-Antilles coalesced (or still joined) to future northern Hispaniola. Somewhat to the southeast another fragment is labelled to represent Puerto Rico+Central Hispaniola. The vicariance events here identified (p. 528) are those that permitted the evolution of *Ctenonotus* and *Semiurus* (on the Central Hispaniola-Puerto Rican unit) and of *Anolis* (on the eastern Cuban-Northwest Hispaniolan unit).

In the succeeding (post-Oligocene) panels and in the relevant explanatory text Guyer and Savage report fragmentation of the eastern Cuban-Northwest Hispaniolan unit and of the Central Hispaniola-Puerto Rican unit, with eastern Cuba joining the rest of that modern island and Puerto Rico separating off to the east. Northwest Hispaniola and Central Hispaniola, for their part, joined "an insular core that had existed as a unit through the Cenozoic" to form the bulk of that present island and were later joined in their turn by the present southwest peninsula after a long eastward voyage.

In the latest stage of Antillean history Guyer and Savage admit to some dispersal - of *Dactyloa* from South America to the southern Lesser Antilles and of the *sagrei* series from Jamaica to Cuba.

There are some loose ends to the story. No vicariance events are invoked to explain the separation of *Phenacosaurus* and *Dactyloa* or of *Semiurus* and *Ctenonotus*. There is no discussion of these omissions at all.

There is, indeed, discussion of *Chamaeleolis* and *Chamaelinorops*. I find it unspecific and inadequate. Guyer and Savage (p. 528) tell us that their area cladogram

> "confirms the Williams (1969b, 1976a) hypothesis that the primitive anoles *Chamaeleolis* and *Chamaelinorops* are ancient inhabitants of the Antilles. Unlike Williams, we believe that their association with Cuba and Hispaniola, respectively, predates the formation of these islands; and the two genera represent derived relicts of the ancestral anoline stock that ranged throughout tropical America at the end of the Mesozoic. Unlike Williams, we do not believe that these stocks are invaders of the Greater Antilles by overwater dispersal. Instead, we propose that the two genera have evolved in situ on portions of the fragmented proto-Greater Antilles as these geographic units were brought to their present distributions through movement of the Caribbean Plate."

These remarks, although appropriately calling attention to plate tectonic theory and the movement eastward of the Caribbean Plate, do not address the reasons that Etheridge (1960) and Williams had for regarding *Chamaeleolis* and *Chamaelinorops* as possibly "ancient inhabitants of the Antilles." Etheridge (p. 186) spelled out his suggestion for *Chamaeleolis* very specifically: "there was either a very early split in the islands, isolating *Chamaeleolis* from the remaining island stock, or there were two arrivals, the first of which is represented by the monotypic *Chamaeleolis* and the second by the West Indian species of the Alpha Section. *Chamaeleolis*, after long separation, became specialized in the condition of the parietal and the form of the teeth, and in the loss of functional autotomy, while retaining the other primitive features of its mainland ancestors." His comments (pp. 190-191) on *Chamaelinorops* were conspicuously similar: "the position of *Chamaelinorops* in the division of anoles with primitive caudal vertebrae is comparable to the position of *Chamaeleolis* in the other. Both genera show primitive as well as highly specialized skeletal characteristics. Thus, *Chamaelinorops* may represent either an invasion separate from, and prior to, that which gave rise to the island series of the section, or an early division of a single island invasion."

My own contribution (Williams 1969) was to adopt the suggestion of Etheridge of the earlier temporal origin of *Chamaeleolis* and *Chamaelinorops*. I described the two genera as relicts of an early period of dispersal to the islands - a Stage I in anole history - and found in a mid-Tertiary restriction of area or subsidence of the islands, a geological phenomenon that terminated that first stage and left only two (or as I would now say - see above - three) genera as relicts. (This mid-Tertiary inundation, although I in 1969 derived my information from Schuchert (1935), is amply documented in more recent literature: Mattson 1984; Perfit and Williams this volume.) I inferred that it was during a second period of dispersal - a Stage II - after the Antillean islands regained their former or a greater area that the other island anoles, collectively called *Anolis* by Etheridge and myself, arrived.

For both Etheridge and myself the extraordinary specializations of *Chamaeleolis* and *Chamaelinorops* combined with retained primitive features had the probable implications of 1) a long period in which to evolve the extreme specializations, i.e., divergence at some relatively early time; 2) separation from other anole stocks at a period before the specializations of these other stocks were acquired, i.e., again early divergence. Both of us thus considered time a factor in the differentiation of these highly distinctive genera; both of us also were very aware of another possible factor - isolation - which might have accelerated that differentiation. This was Etheridge's alternative explanation for the evolution of both genera. My own explanation involved, in addition, two points not part of Etheridge's argument - the special degree of isolation enforced by the mid-Tertiary inundation of the Antilles plus another consideration derived from my earlier familiarity with vertebrate paleontology - a phenomenon often seen in the early phases of the evolution of a novel group - an explosive experimental phase (cf. Rensch 1960) - of which *Chamaeleolis* and *Chamaelinorops* were <u>relicts</u>.

Guyer and Savage deal with none of issues raised by Etheridge and myself. The questions regarding the evolution of these two genera that require answers are: when, where and how? Etheridge and I attempted answers to the question of "how?" To the question of "when?" we answered in vague terms - early. To "where?" we did not answer directly, but, by implication, and in ignorance of plate tectonics, our answer was the naive one - Cuba and Hispaniola.

Guyer and Savage specifically reject my 1969 hypothesis, but they themselves provide no specific temporal nor spatial explanation of the evolution of either genus nor any explanation of the mode of evolution of such specialized genera. The earliest date permissible under their scenario is given by the first fragmentation of their late Cretaceous-Paleocene isthmus. Only if the very first fragments of the isthmus to break off carried *Chamaeleolis* and *Chamaelinorops* are these genera genuinely "ancient inhabitants of the Antilles" - older. Which were these fragments - and where? Guyer and Savage say only that the "association [of *Chamaeleolis* and *Chamaelinorops*] with Cuban and Hispaniola, respectively, predates the formation of these islands." This answer lacks precision. Perhaps the southwest peninsula of Hispaniola was available for *Chamaelinorops*. This and Jamaica are, according to Guyer and Savage's fig. 10, the first islands to separate from the proto-Antillean block. But where was *Chamaeleolis*? If the relevant part of Cuba was any part of the massive island mass that they figure as the residue of the proto-Antillean block, in what sense - in particular what sense meaningful in vicariance biogeography - can *Chamaeleolis* be the most primitive of all anoles, the sister group to all the others? A plate tectonic solution as to this question seems at the moment obscure. In any case, Guyer and Savage do not confront the problem that impelled Etheridge and I to propose two stages in anole history. They attempt no explanation of the most conspicuous aspect of the two "ancient inhabitants of the Antilles" - their extreme and bizarre specialization and offer no alternatives to the Etheridge-Williams explanations based on greater age or greater isolation.

The most confused of Guyer and Savage's vicariance scenarios involves the origin of *Norops* and of *Anolis*. Guyer and Savage present two very distinct hypotheses of the origin of the beta anoles. One is given on pages 525-526. A second hypothesis is introduced on p. 529. The first would separate beta anoles (*Norops*) from South American alphas (*Dactyloa*) by the Panamanian sea portal at about the late Cretaceous to early Paleocene. This very early origin of *Norops* they describe as the most generally accepted view. They immediately discard it in favor of their plate tectonic hypothesis of an isthmian connection during the same late Cretaceous to early Tertiary time, that later fragments, with that fragmentation the "basis for cladogenesis" of all the West Indian genera and ultimately of *Norops*. "Ultimately" is here the appropriate word. According to the preferred hypothesis of Guyer and Savage the beta anoles would not be among the earliest differentiated of the anoline stocks, but among the latest. They suggest in fact, not vicariant separation from *Dactyloa* but from *Anolis*.

They arrive at this hypothesis by using their osteological cladogram (fig. 5) to construct an area cladogram (fig. 11) in which the latest and most derived branches are *Anolis* (with a primarily Cuban distribution) and *Norops* (with its distribution primarily in Central America). As they tell it (p. 529) this "branching pattern implies a common geographic antecedent area for Cuba and Central America persisting from a time after the separation of Hispaniola-Puerto Rico from Cuba, and suggests an Oligocene divergence time between *Anolis* and *Norops* (fig. 11). Possibly a portion of western Cuba had a close relationship to the Yucatan area (similar to the Jamaica-Central American one) during Eocene-Oligocene, if so, separation of this segment from Central America that led to the origin of *Anolis* (Cuba) and *Norops* (Central America)."

Given the "more or less continuous land mass" connecting North and South America that Guyer and Savage postulate, anoles of some sort were surely present ab initio in

Central America. A portion of western Cuba in "close relationship to Yucatan" might easily receive anoles from Central America. But this association of western Cuba with Yucatan would not be part of a proto-Antillean block (see Perfit and Williams this volume) and might be quite early. Separation and divergence might be much earlier than Oligocene. Whether this scenario can be squared with the presence in Hispaniola of relatively primitive series that Guyer and Savage would allocate to *Anolis* (e.g. *chlorocyanus*) cannot at present be determined. Guyer and Savage are hesitant in presenting it, using phrases such as "possibly" and "if so." The details needed to flesh out the hypothesis are missing.

There is a further difficulty. There are not merely two hypotheses about the origin of *Norops*, there are two stories for *Anolis*. On p. 528 it is said (my emphasis) that it is "a Cuban-northwest Hispaniolan island upon which *Anolis* evolved." On p. 529 - as the obverse of the origin of *Norops* - it was the separation of that portion of western Cuba that may have had a close relationship to the Yucatan area that might have provided the vicariant event that differentiated *Anolis* from *Norops*. A Hispaniolan involvement is not mentioned. The two hypotheses cannot be reconciled. Cuba, like Hispaniola, is a composite island. Its eastern and western portions are not geologically similar. Hispaniolan participation in the origin of *Anolis* either did or did not occur.

Again plate tectonics has not made the puzzle simpler. But this is not always true. The panel for the Oligocene in fig. 10 of Guyer and Savage suggests that one ancient island pattern that may make one specific distributional problem more readily explicable - and in a rather elegant way. The case in question is that of the modern island of Hispaniola interpreted in terms of its plate tectonic precursors.

The distributional puzzle is that, in terms of relationships, the anole fauna of the modern island of Hispaniola faces in two directions. Three of the anole series - those belonging to the *punctatus* subsection of Williams (1976a) - the *cuvieri*, *cybotes*, and *bimaculatus* series - have relatives in Puerto Rico, and none in Cuba. All the remaining anole series of Hispaniola are members of the *carolinensis* subsection and have relatives at the subsection level in Cuba - and, if, as I have suggested above, *occultus* is a distinct genus and not a member of the *carolinensis* subsection, none at all in Puerto Rico. There are thus, in effect, two anole faunas in Hispaniola, one with affinities with Puerto Rico, one with affinities with Cuba - two faunas not very readily derived one from the other, in whichever direction evolution might be assumed to have taken place.

This might be a matter of chance, and chance - after origin of the *carolinensis* subsection from the *punctatus* subsection within Hispaniola - was the rather troubling explanation that I, for a long period, accepted for this pattern. The discovery of the supposedly defining character of the *carolinensis* subsection - the T-shaped interclavicle - in some of the mainland alphas further shook my confidence in this too facile explanation. Now, however, with the revelation that, on geological evidence, Hispaniola is composite, that the southwest peninsula was a very late addition to the modern island, and that central fraction of the island was once united with Puerto Rico, and a northwest fraction with Cuba, I find another explanation more satisfying. The two faunas evolved on two different islands and came together only only when the parent islands fragmented, and two portions accreted to form the major portion of the island of Hispaniola, while the appropriate other fragments moved westward to join Cuba or stayed in place as modern Puerto

Rico. The present distributional patterns then match with a sequence of paleogeographic patterns that explain them.

The Guyer and Savage "genera"

I cannot subscribe to the generic definitions provided by Guyer and Savage, not even to those for genera that have long been recognized as valid. Again there are outright errors as well as poorly supported new conclusions. I deal below with the Guyer and Savage genera in the order of and initially in terms of the phyletic sequencing they provide on p. 523. I comment, however, on other statements about the genera wherever found, especially on the more detailed discussions and informal definitions found on pp.519-522. I underline for each genus the defining autapomorphy. I add information on the shoulder girdle where pertinent and discuss karyotypes under each genus.

1. *Chamaeleolis*.

"Six parasternal ribs, iliac process acute and elongate, peg-like teeth, unique parietal cresting, 2 spp."

Two complaints seem warranted. 1) The intention is said to be to employ phyletic sequencing. This should imply listing of two sorts of characters - 1) characters synapomorphic at this node of the cladogram and plesiomorphic at subsequent nodes; 2) characters autapomorphic for the genus branching off at this point. The iliac process seems inappropriate in either category. The only discussion of the iliac process (p. 520) is that it is blunt and rectangular in *Phenacosaurus* but is "acute and elongate in all other anoles surveyed by Lazell (1969)," a statement immediately challenged by Etheridge's quoted personal comment that the character is much more variable than Lazell implied. 2) The statement that there are two species in the genus conflicts with the four listed in the species column in table 4. The correct number of currently recognized species is three (Garrido 1982).

The content and validity of this genus is doubted by no one. It is the only anoline genus for which this is true. Guyer and Savage introduce some confusion even here.

Already mentioned above is Guyer and Savage's placement of *Chamaeleolis* (p. 519) as "the most primitive of the anoles" on the basis of two characters, one of which - the presence of palatine teeth - they have coded as derived in tables 2 and 3.

The unfortunate coding of the parietal crest/casque in table 3 that would appear to imply that the same state is found in *grahami* and *sagrei* has also been mentioned above. In striking contradiction to that implication, Guyer and Savage elsewhere (p. 520) make the extraordinary statement that in *Chamaeleolis* "the parietal crest extends far back on body" (my emphasis), contrasting this with the condition in other anoles, in which they report the "parietal crest, if present, restricted to head." As in many other examples the bizarre phrasing here corresponds to a correct statement on another page (p. 519): "a unique pattern of parietal cresting that extends far back over the neck."

The karyotype of *Chamaeleolis* is probably primitive - 12 macrochromosomes + 24 microchromosomes. It is so reported by Paull et al. (1976), but the only detailed description and figure is that of Gorman et al. (1969) (see also Gorman (1973), which reported 12 macrochromosomes plus "more than 20 microchromosomes." Guyer and Savage (Table 4) report the diploid count as "12, 22"; I do not know the source of their information.

2. *Phenacosaurus*.

"Seven parasternal ribs, 3 lumbar vertebrae, 4 attached ribs, 24 presacral vertebrae, swollen head scales, no palatine teeth, no angular, 2 sternal ribs, 7 species."

There are errors here. The number of presacral vertebrae is erroneously given as 24. It is correctly stated elsewhere in the text (p. 520) as 22 and are so coded in table 3. The character of "swollen head scales" must reflect a misunderstanding of the casque present in *Phenacosaurus*. (See further below.) The number of species reported in the definition - seven - is also wrong. Three is the present number of recognized species; this may be a simple lapsus: the number of species in *Polychrus* is seven. The species number of 24 given in table 4 may be a typographical error.

Only one of the three named species of *Phenacosaurus* is described osteologically. (I figure the interclavicle of a still unnamed species in Fig. 3e) The characters of two sternal ribs, 7 postxiphisternal inscriptional ribs and 22 (not 24) presacral vertebrae are therefore thus far verified only for *P. heterodermus*. One of the three named species, *P. orcesi*, described in Lazell's 1969 paper cited by Guyer and Savage, lacks the heterogeneous dorsal scutellation that two species (Guyer and Savage p. 520 say all), share with *Chamaeleolis*. The heterogeneous dorsal squamation of *Chamaeleolis* is, indeed, quite similar to that of the two *Phenacosaurus*, but a number of other anoles have more or less similar squamation - the *equestris* complex, Central American *sminthus* (as its synonym *heteropholis* implies), and *Chamaelinorops*. It can be confidently stated that not all of these conditions are homologous; it is not certain that any of them are.

Guyer and Savage (p. 520) describe "enlarged swollen circumorbital scales" as diagnostic of *Phenacosaurus*. All the head scales in *Phenacosaurus* tend to be enlarged, least so in *P. nicefori*, but the circumorbitals are not necessarily swollen. Contrary to Guyer and Savage's statement one other anole - *latifrons* - does have swollen circumorbitals as a species character. As I suggest above, Guyer and Savage must be referring to the swellings that mark the casque of *Phenacosaurus*, but the swollen area is parietal, not circumorbital.

Etheridge has believed that *Phenacosaurus* was derived within *Anolis* (*sensu lato*). If this were true and *Phenacosaurus* recognized, *Anolis* would be paraphyletic. Additional information on scales and skull in the genus will be put on record by Flores and Williams (in prep.).

As mentioned the sternal ribs in *Phenacosaurus* are two and the interclavicle T-shaped.

The karyotype of *P. heterodermus* is 12+24, the primitive condition. A figure was published in Gorman et al. (1969).

3. *Chamaelinorops*:

"Four parasternal ribs, no palatine teeth, no angular, thickened and fused pre- and postzygapophyses. 1 sp."

The character "four parasternal ribs" presumably means total number. If so, it is in error. Etheridge (1965) reported this number as 6 (rib formulae 4:2; 3;3). Forsgaard (1983) listed 5 (rib formula 4:1). The description of the diagnostic autapomorphy of *Chamaelinorops* - "thickened and fused pre- and postzygapophyses" is not very readily visualizable and erroneous if it applies - as it would appear it does - to all vertebrae. The two more elaborate descriptions of this character on p. 520 are serious misstatements twice repeated. The most detailed description - in the second column of p. 520 - is, in fact, doubly wrong, (my emphasis): "Vertebrae, including caudal vertebrae, with thick transverse processes formed from fused pre- and postzygapophyses that extend at right angles to the vertebral axis."

The vertebral column of *Chamaelinorops* has been well described by Forsgaard (1983) in a paper that Guyer and Savage list in their bibliography. Examination of Forsgaard's figures as well as of cleared and stained specimens and dry skeletons at the Museum of Comparative Zoology reveal a structure quite different from that described by Guyer and Savage: on the dorsal vertebrae of *Chamaelinorops* wide plates do extend along each side of the neural arches between the pre- and post- zygapophyses. However, well underneath these dorsal interzygapophysial plates are small cylindrical peg-like transverse processes on the centra that bear the dorsal ribs. In the dorsal region the there is no association between the interzygapophysial plates and the transverse processes that lie beneath them. The caudal vertebrae have no trace of interzygapophysial plates. The caudal pre- and postzygapophyses are in no way special; they are like the zygapophysial processes of any other lizard. The caudal transverse processes again are on the centra, and in adults they are, indeed, large stout tranverse structures wider distally. There is, again, no association between transverse processes and zygapophyses except presence on the same vertebrae.

Etheridge inclines to the opinion that *Chamaelinorops* is derivable from or closely related to the beta anoles - *Norops*, *sensu* Guyer and Savage. He (pers. comm.) defends this position on the basis of his observation of slight forward inclination of the caudal transverse processes in cleared and stained hatchling *Chamaelinorops*.

The interclavicle of *Chamaelinorops* has been figured in Case and Williams (1988). It has been called a T-interclavicle because the margins of the clavicle and interclavicle are parallel, but the interclavicular arms are not at right angles with the interclavicular shaft; it is not so literally T-shaped as those of *carolinensis* or *latifrons*. As in *Phenacosaurus*, there are two instead of the primitive three sternal ribs (Forsgaard 1983).

The diploid count for *Chamaelinorops* was cited as 36 by W.P. Hall in Paull et al. (1976) - a bare mention that so far as I know is the only published statement on the karyotype. This count and the primitive nature of the karyotype has been confirmed by Judith Blake (pers. comm.). A primitive karyotype is poorly congruent with the derivation from within the beta anole at one time suggested by Etheridge. The karyotypes of all unquestioned betas are highly derived (see below) and derived in ways that make reversal extraordinarily unlikely. The diploid count of "12,20" given by Guyer and Savage in table 4 appears to be mistaken.

4. *Dactyloa*.

"Caudal autotomy present, < 8 aseptate caudal vertebrae, 47 species."

Usage of "*latifrons* series" in Guyer and Savage has to my knowledge twice dismayed careful readers. From table 1 to and including page 522, where it is explicitly equated with *Dactyloa*, the *latifrons* series is employed consistently in the sense of Etheridge (1960). Almost immediately opposite on p. 523 the genus *Dactyloa* is said to be composed of six series, one of which is the *latifrons* series - clearly not *sensu* Etheridge. All the "series" in this list were "species groups" in Williams in 1976b.

The presence of caudal autotomy here is intended to distinguish *Dactyloa* from the preceding genera all of which have non-autotomic caudals. The character of 8 or fewer aseptate caudals likewise is supposed distinguish it from these genera since all of their caudals are aseptate. (As reported above, the actual range of aseptate caudals in *Dactyloa* is 7-11, with the type of *Dactyloa - punctatus* - having 9.) Neither character distinguishes *Dactyloa* against the four subsequently defined genera. In the absence of any synapomorphy mentioned for *Dactyloa* the implication would be that the genus is defined solely on plesiomorphic characters. However, a possible synapomorphy for the mainland species of *Dactyloa* is absence of the caudal breakage planes in adults of all species. Guyer and Savage in discussing *Dactyloa* comment that "caudal autotomy septa are present in all juvenile *Dactyloa*, although these septa may become fused in adults of some species (Etheridge, 1960)." This somewhat misstates the evidence. My own observation is that all mainland South American alphas (and their lower Central American extensions or derivatives) have non-autotomic caudals as adults. That juveniles have open septa is the statement that has to be made faith of Etheridge. His statement (in litt.), when asked by me, was that he no longer has any notes on the topic, but that, he now assumes, he saw evidence of breakage planes in X-rays of some species. I know of no other evidence or comment on the point. On the other hand, the *roquet* series in the southern Lesser Antilles and *agassizi* on Malpelo Island in the Pacific (both part of *Dactyloa* as understood by Guyer and Savage) do consistently retain septa throughout life, as specimens readily demonstrate. Etheridge's own view - and mine - of the evolutionary history of caudal autotomy in the alpha anoles has been that caudal autotomy is primitive, and that the *roquet* series have retained the primitive condition while all the mainland alphas of South and southern Central America have secondarily lost or fused the caudal breakage planes.

The new information on shoulder girdle structure (clavicular-interclavicular relationship) is a complication for the hypothesis the monophyly of the mainland alphas. Interclavicular shape and relationships appears to cleanly separate two groups within the South American alphas, one of the two also containing the *roquet* series of the Southern Lesser Antilles. The absence of caudal breakage planes in adults can clearly be interpreted as a synapomorphy uniting the mainland South American alphas as a single group. This would, however, exclude the *roquet* series. Guyer and Savage's suggestion that absence of autotomy for the anoline lineage is primitive does not simplify the situation.

It is tempting to associate as phyletic units the mainland and West Indian alphas with T-shaped interclavicles and, similarly, the West Indian and mainland alphas that have arrow-shaped interclavicles. There is even an apparent geographic correlation: the alpha series that have T-shaped interclavicles (*aequatorialis* and *latifrons*) are western in distribution - Colombia and Ecuador and adjacent southern Central America with one outlier: *squamulatus* in Venezuela. West Indian alpha series with T-shaped interclavicles are Cuban and Hispaniolan with one supposed outlier - *occultus* - in Puerto Rico. I have already stated above that I consider *occultus* as properly referred to an undescribed genus as distinct as *Chamaeleolis*, *Chamaelinorops*, and *Phenacosaurus*. In contrast, series with arrow-shaped interclavicles in the West Indies - *cuvieri*, *cybotes*, *cristatellus*, and *bimaculatus* - have an exclusively eastern distribution, none getting further west than Hispaniola. The mainland alpha series with arrow interclavicles - *punctatus* and *solitarius* - while occurring also in the west extend further east than the *latifrons* or *aequatorialis* series and have an extreme eastern relative as an outlier - the *roquet* series of the southern Lesser Antilles. It does not diminish the apparent geographic correlation that the beta section - *Norops* fide Guyer and Savage - which has T-shaped interclavicles, has its center of distribution to the west in Central America, and, while as widespread in South America as any alphas, is confined in the West Indies to Cuba and Jamaica.

While it is attractive to postulate that the cross resemblances between interclavicular types in South American and West Indian series are phylogenetically meaningful, these cross resemblances do not, at this time, appear to be supported by any other characters. The West Indian series, with the except of *roquet*, are strongly differentiated when compared with the mainland species. The inferred phylogenetic relationships may be real, but it is reasonable to suspect that the separation of the Greater Antillean and mainland stocks is old, certainly significantly older than that of the *roquet* series.

A distributional error here repeated may be a source of confusion. The genus *Dactyloa* is described (p. 523) as having a range in South America of Colombia, northwestern Venezuela and Ecuador, southward in upper Amazonia to eastern Peru, and southeastern Brazil. The type of *Dactyloa* is *Anolis punctatus*. The distribution of this species includes, in addition to the Guianas (= ? "upper Amazonia" of Guyer and Savage's statement of the range of *Dactyloa*), the whole of forested Brazil north of Santa Catarina, including the Atlantic Forest. The map given for *Dactyloa* in fig. 8 of Guyer and Savage includes the Guianas, but clearly excludes most of Amazonia; it is clearly a copy of map 2 in Etheridge (1960), which had previously been copied in Savage (1966, fig. 12). The initial error is Etheridge's, presumably because of his very small sample of *punctatus* (four specimens). (*Punctatus* is the only South American alpha anole that has so wide an eastern extension - all of Amazonia and the Guianas plus the Atlantic Forest of eastern Brazil.)

The karyotype of all species of the *roquet* series - the *Dactyloa* of the southern Lesser Antilles is known. There are two sublineages, one with a 12+24 karyotype, one with 12+22, i.e. the primitive karyotype or a simple derivative. The karyotype is known for only three mainland "*Dactyloa*" - *jacare*, *squamulatus*, and *frenatus*. Of these *frenatus* shows a derived condition that Gorman (in litt.) compares with that of the *cristatellus* series (although lacking sex chromosomes). *Agassizi* of Malpelo Island in the Pacific, like *squamulatus* retains the primitive karyotype. *Jacare* differs by the absence of two microchromosomes.

5. *Semiurus*.

"Triangular-shaped parietal, Y-shaped parietal cresting, 5 total parasternal ribs, <3 attached parasternal ribs, 5 spp."

Three diagnoses of *Semiurus* are given that appear verbally distinct, one given in the abstract at the beginning of the paper, another on p. 522, still another on p. 523. No two are in agreement. That in the abstract contains two errors. I underline the portion of the diagnosis in the abstract to which I wish to call attention):

"The *cuvieri* series (placed in *Semiurus*) shares with other advanced anoles the derived conditions for the shape of the parietal, parietal cresting, total number of parasternal ribs, and number of attached parasternal ribs, but lacks the derived condition of the occipital region of more advanced anoles."

Etheridge's own description of the three species of the *cuvieri* series (*sensu* Williams 1976a) that had been recognized in 1960 was the following: "*Anolis ricordi* and *roosevelti* are the only species of the Alpha Section with a parasternal formula of 3:2. Were it not for their parasternal formula these two species might be placed in the *bimaculatus* series on the basis of the presence of a splenial and the condition of the girdle. *Anolis cuvieri*, with its parasternal formula of 3:1 and condition of the pectoral girdle, might be placed in the *bimaculatus* series but for the absence of a splenial."

Comparison of Etheridge's statements with the definition of *Semiurus* in Guyer and Savage's abstract reveals the two discrepancies: 1) The total number of "parasternal" ribs in at least two species of the *cuvieri* series is 5, not 4 as in other advanced anoles. (In fact, Guyer and Savage elsewhere reveal awareness of this, giving in table 3 the coding of 3 to character 8a in the *cuvieri* series, implying that the whole series, or at least its primitive members had a total number of 5 "parasternal" ribs.) 2) Etheridge's *bimaculatus* series is included by Guyer and Savage in their *Ctenonotus*, for which the first phrase of the definition on p. 523 is "Funnel shaped occipital". Examination of skulls of *cuvieri* and *ricordi* shows that the "funnel" is especially well-developed in these species.

The apparent intention in the diagnosis in the text on p. 523 is to list the characters at a particular node in the implied cladogram. More derived characters first appearing at the next node are listed only there. If this interpretation is correct, the definition in mentioning "five parasternal ribs" has corrected one of the misstatements in the abstract, but in failing to mention "funnel-shaped occipital" is impliedly repeating the other.

There is still the third discussion of the genus on p. 522 (again I underline the pertinent part of the text):

> "This leaves the *cuvieri* series, a group that <u>shares</u> with <u>*Anolis, Ctenonotus* and *Norops* the: derived characteristics of shape of the occipital region</u>; shape of the parietal and parietal cresting; and <u>number of attached parasternal ribs</u>. However, this series shares no derived features that might link it with any of the other advanced anole genera. The stock has a more primitive total number of parasternal ribs, a primitive karyotype, and albumin proteins that are equally dissimilar with respect all other anoles. Unfortunately, the lineage possesses no autapomorphic osteological characters by which it might be diagnosed; however, these giant anoles appear to represent an independent lineage that should be recognized as a 'separate genus.'"

This is a correct, although certainly weak diagnosis of the genus *Semiurus*. Unfortunately it is contradicted by the two other definitions in the same paper.

I have previously mentioned that "occipital" is a misnomer for the structure referred to. Etheridge's 1960 description (e.g., p. 143) is both lucid and correct: "The occipital part of the parietal is half-funnel shaped, its posterior border extending backward over the supraoccipital."

The sternal ribs are three in the *cuvieri* series and the interclavicle arrow-shaped, but not always as distinctly so as in *punctatus, solitarius,* or *roquet*.

As mentioned, the known karotypes of the *cuvieri* series have the primitive formula 12+24.

6. *Ctenonotus*.

"Funnel-shaped occipital, 4 total parasternal ribs, jaw-sculpturing, small processes on caudal vertebrae, ca. 30 spp."

There are again questions to be raised about this definition, even as an expression of characters at a node in a cladogram. The character "funnel-shaped occipital" - corrected as to anatomical nomenclature - should have been cited at the previous node. "Four total parasternal ribs" is correct for this and subsequent genera, but there is homoplasy for this character in *cuvieri*, again at the previous node.

"Small processes on the caudal vertebrae" only occur in two of the series - *cybotes* and *cristatellus* - referred to *Ctenonotus*, never in the third series - *bimaculatus*. Unless this character is assumed to be a plesiomorphic feature secondarily lost in the *bimaculatus* series the monophyly of *Ctenonotus* would be in question. Fortunately Guyer and Savage (p. 522, see below) have made this assumption explicit.

In associating the three series they unite in *Ctenonotus* Guyer and Savage are in basic agreement with Etheridge (1960, see the dendrogram on p. 185) and Williams (1976a), with, however, the difference they separate a *cybotes* and a *cristatellus* series, whereas Etheridge, and, following him Williams, united these in a single *cristatellus* series. The real divergence of opinion, is negligible. Guyer and Savage (p. 522) state the arguments for and against this phylogeny well; it is worth quoting them in extenso:

> "The *bimaculatus* and *cristatellus* series are clearly sister taxa, since they share derived karyotypes, especially xxy heterochromatism. The *cybotes* series appears to be more closely related to the *bimaculatus-cristatellus* series than to any other group since they share with some *cristatellus*-series anoles the derived features of jaw sculpturing and small lateral projections on caudal vertebrae (reminiscent of transverse processes). Gorman et al. (1980) argued that these vertebral projections may represent an intermediate stage in the evolution of caudal vertebrae from the alpha to the beta condition. However, the shape and extent of these 'processes' do not support this hypothesis (Etheridge, pers. comm.), and their presence appears to be a distinctive condition of the caudal vertebrae that is synapomorphic for these two series....Since *bimaculatus* series anoles lack both jaw sculpturing and the small projections on the caudal vertebrae, characters that are primitive for this lineage, they must have been lost in that series. Gorman et al. (1980) argued for independent evolution of these two characters in anoles of the *cristatellus* and *cybotes* series due to differences in karyotype (advanced karyotypes in *cristatellus* and *bimaculatus* series; primitive in *cybotes* series) and albumin immunology (similar albumins in *cristatellus* and *bimaculatus* series; dissimilar albumins between these two series and *cybotes* series). While they argued correctly that the number of steps needed to evolve the two characters independently in the two series equals the number of steps needed to evolve the two characters in a *bimaculatus-cristatellus-cybotes* ancestor and then lose them in the *bimaculatus* stock, we maintain that it is easier to lose these characters than to gain them independently. Our suggestion that anoles of the *bimaculatus*, *cristatellus* and *cybotes* series belong to a single lineage is not incongruent with the tree published by Gorman et al. (1980:fig. 6). We merely suggest a resolution to their trichotomy."

It is notable that after this extensive discussion, more elaborate than for any other case, Guyer and Savage still entertain some apparent doubts. They admit parenthetically (p. 523) to "...some question about the appropriateness of including the *cybotes* series with *bimaculatus* and *cristatellus* series in the genus *Ctenonotus*."

This is a case in which karyotypes are, on the one hand a source of confidence, and on the other an occasion for doubt. A derived karyotype with differentiated sex chromosomes securely unites the *bimaculatus* and *cristatellus* series. All, except one species - *evermanni* - with xy sex chromosomes, have the xxy sex chromosome pattern. In contrast, the primitive 12+24 karotype of the *cybotes* series without differentiated sex chromosomes, along with immunological data, is a major reason for doubting the plausibility of its inclusion in *Ctenonotus*.

7. *Anolis*.
"T-shaped interclavicle, 4-6 lumbar vertebrae, ca. 55 spp."

On p. 523 the genus *Anolis* in the restricted sense is said to contain nine series - four are restricted to Cuba, three restricted to Hispaniola, one occurs on both Cuba and Hispaniola, and one on Puerto Rico. On p. 525 the genus *Anolis* is described as a Cuban radiation and "is represented in Hispaniola by three closely related species in the *alutaceus* series." The first statement is one of the possible interpretations of the genus as restricted; the second statement is literally true but ignores the *chlorocyanus*, *darlingtoni*, *monticola*, and *hendersoni* series.

Anolis in the restricted sense of Guyer and Savage is coterminous with the *carolinensis* subsection of Williams 1976a, which was at that time defined as *Anolis* (*sensu lato*) without caudal transverse processes, but with T-shaped interclavicles. I now think that this is a too inclusive category - a miscellany, not at all monophyletic. It is certainly no longer definable just on the T-shaped interclavicle now that we know that some *Dactyloa*, *sensu* Guyer and Savage, have T-shaped interclavicles. Nor is it possible to define it on a high number of lumbar vertebrae as Guyer and Savage insist. Four is a count of lumbar vertebrae that Etheridge (1960) reports for every one of his alpha series, including *latifrons* (=*Dactyloa*). There is, therefore, first, extreme overlap and second incomplete information. Etheridge did, indeed, say that his *carolinensis* series - which as he then defined it included the *alutaceus*, *carolinensis*, *chlorocyanus*, *hendersoni*, *lucius*, and *monticola* series of Guyer and Savage - had a range of 4 to 6 lumbar vertebrae. He also said that his *coelestinus* series had 4 and his *angusticeps* series had 4 to 5 lumbars. His phrase "most usual number" for the species in these series for which he reported four lumbars does clearly imply that any variation was in the direction of higher numbers. However, he did not report the lumbar count for *equestris* - then considered a single species, nor on "*Anolis* (*Xiphocercus*) *darlingtoni*," which he did not assign to a series. Nor, of course, was he able to report on still to be described *occultus*, *insolitus*, *sheplani*, *fowleri*, and *eugenegrahami*. I do not have counts in which I have confidence for any of the species that are missing from Etheridge's account. However, these are just the species whose assignment to a restricted genus *Anolis* I find most doubtful.

I would subtract, first of all, and with no hesitation, the only Puerto Rican species that remains in Guyer and Savage's new conception of *Anolis*. As I have already mentioned above, I consider *occultus* not a series within *Anolis*, whether understood in the restricted or the broad sense, but an undescribed genus.

Not as clear a case, but one in which I regard allocation to *Anolis* in any sense as suspect, is the *equestris* complex. The casquing of the parietal, the relatively large splenial, some curious resemblances in the squamation to *Chamaeleolis* lead me to question its affinities. I have always regarded the *equestris* complex as either the most primitive member of the *carolinensis* subsection (this is the conclusion implied by Etheridge 1960 fig. 10) or possibly not correctly assigned to *Anolis* at all.

As for the other more recently described species for which the lumbar character is not known, all are highly distinctive and insufficiently understood. A case for including them in a restricted genus *Anolis* is still to be made.

Even among those species for which Etheridge provided data there is striking diversity that ought to imply an old radiation. There is an assemblage that is close to *carolinensis* itself and another that is *chlorocyanus* and its sister species, large and small. Then there is the *lucius* series, remarkably divergent within itself and not obviously close to other series, the chromosomally and in other ways very singular *monticola* series, the aberrant *hendersoni* superspecies, which, unexpectedly, has retained the splenial, *argillaceus* and *loysiana*, curiously parallel ecomorphically to the distichoid radiation in Hispaniola, *darlingtoni* and *insolitus*, the relationships of which are doubtful, *sheplani*, once thought to be close to *occultus*, but the few resemblances are ecomorphic, *fowleri* and *eugenegrahami*, both chromosomally very unprimitive (J. Blake pers. comm.) and very different from all other species and from each other. The genus *Anolis* as conceived by Guyer and Savage is residual - those species left over after other better understood groupings have been separated out. If the other groupings are to be regarded as genera, then this residue needs to be reexamined - carefully.

As equivalence with the *carolinensis* subsection implies, all *Anolis* in the sense of Guyer and Savage have T-shaped interclavicles. All *Anolis*, sensu Guyer and Savage, have three sternal ribs, with the exception of *occultus*, which has only two sternal ribs like *Phenacosaurus* and *Chamaelinorops*. (This is one of the several reasons for my suggested separation of *occultus* at the generic level. Among the others are loss of a coronoid labial blade and of the angular process of the mandible (Fig. 4) and the presence of a high number of post-xiphisternal inscriptional ribs - 5 or 6, all attached - the latter count fide Webster et al. 1972.)

Karyotypes in *Anolis* as restricted are often 12+24, but many species have not been studied. The *monticola* series contains both species like *christophei, rimarum*, and *etheridgei* with 12+24 and species like *monticola, koopmani,* and *rupinae* in which some or all of the macrochromosomes have undergone fission, with the extreme resulting karyotype being 24+24. *Insolitus*, which has not been regarded as closely related, is also known to show macrochromosomal fission - 2n=44 (Webster et al. 1972). *Fowleri* and *eugenegrahami* have highly derived karyotypes that are incompletely analyzed. Webster (1974) reported a diploid count of 44 in *fowleri*. Macrochromosal fission is demonstrated, but the single karyotype known shows other still not understood peculiarities. Blake (pers. comm.) reports a beta/*cristatellus*-like karyotype in *eugenegrahami*.

8. *Norops*.

"T-shaped interclavicle, transverse processes on autotomic caudal vertebrae, ca. 120 spp."

The anoles with beta type caudal tranverse processes are, as compared with the alpha anoles, limited in diversity. Guyer and Savage have shown their recognition of this phenomenon by dividing the alpha anoles into four genera while retaining only one for the beta anoles.

The argument for the monophyly of the betas is morphological: the uniqueness of the character that defines them: caudal transverse processes that are directed obliquely anteriorly and arise posterior to the breakage plane dividing each vertebra. This is a very distinctive condition, unique within lizards. The immunological and karyotypic evidence would imply involvement of paraphyly or polyphyly.

Gorman and collaborators, following the leads provided by Shochat and Dessauer (1981), assume a West Indian origin for the betas, relying on the immunological and the highly derived karyotypic similarities between Jamaican and Puerto Rican anoles. Gorman et al. (1984) suggest successive dispersals by beta anoles from the West Indies into Central America. Again details are lacking: when and where in the West Indies did the betas evolve. What is now present on Jamaica is immunologically a single radiation; there has not been retained in the Antilles the diversity that the 1984 paper requires. This was the consideration that restrained Etheridge in 1960 (p. 194) from making a similar suggestion:

> "With the appearance of this series [the *carolinensis* series] the splenial has been lost and the pectoral girdle become specialized. The *grahami* and *sagrei* Series might then be derived from the *carolinensis* Series by the appearance of pseudodiapophyses on the autotomic caudal vertebrae. The structure of the diapophyses in members of the Beta Section is very unusual in comparison with those of all other lizards; and at one time I gave this interpretation serious consideration. The main objection to this idea is not the unlikelihood of the evolution of pseudodiapophyses, but rather the necessity for deriving all mainland species of the Beta Section from some Antillean ancestor."

Guyer and Savage's apparently preferred hypothesis of the origin of *Norops* resembles Etheridge's rejected hypothesis in associating *Norops* (the beta anoles and *Anolis* (the *carolinensis* subsection) but does so in a vicariance scenario not involving any dispersal between the mainland and the islands. Postulated is an early contact between western Cuba (then not yet joined to the rest of the modern island) and Yucatan, then an Oligocene separation as the vicariant event separating the populations that will become *Anolis* (in Guyer and Savage's restricted sense) on Cuba from the populations that will become *Norops* in Central America. Presumably the undivided population had already acquired the T-shaped interclavicle that *Anolis* and *Norops* share, but beta-type caudal transverse processes were only evolved after vicariance.

This scenario allows the major radiation of the betas to occur on the Central American mainland, spreading from there into Jamaica (and later Cuba) and into South America. There is no clear tectonic evidence for or against the contact between western Cuba and Central America - although a dry land connection might be unlikely. However, post-Oligocene spread of *Norops* to Jamaica or to South America does not seem possible without cross-water dispersal.

In the case of Jamaica the evidence is for submergence or extreme restriction of area from the Oligocene into the early Miocene. Even assuming prior dry land contact with the Yucatan area and restriction of area rather than total submergence, an Oligocene date of divergence poses difficulty for the evolution of the *sagrei* and *grahami* series except by relatively late invasion across a water gap.

A similar problem exists for dispersal into South America. The range of beta anoles in South America is essentially coterminous with that of the sympatric alpha anoles, which are clearly older in South America. It seems improbable that beta anoles have achieved so extensive a distribution and a considerable radiation in South America since the Pliocene closure of the Panamanian gap, as Guyer and Savage (in the abstract) assume. If the invasion was earlier, it had to be cross-water, probably by several invasions across a narrowing gap.

More difficult for the Guyer and Savage hypothesis of *Norops* and residual *Anolis* as the most derived anoles and therefore related is the accumulating evidence of primitive features in both groups and the further evidence that other features have to be considered widely homoplasious. The splenial has now been found in primitive mainland betas and in the *hendersoni* superspecies as well as the *equestris* series among the alphas. The T-interclavicle may be primitive, and, without question, the derived conditions of the parietal have evolved independently in Guyer and Savage's *Ctenonotus* (the West Indian component of my *punctatus* subsection).

The karotypes known in all betas thus far examined are highly derived and very different from the primitive $12+24$ pattern found in many of Guyer and Savage's residual *Anolis*. This not per se an objection to close relationship, but it is also not a resemblance. Some known beta karyotypes do show chromosomal fission, somewhat as in the *monticola* series, but the species in which this occurs are highly derived. The case for close phyletic affinity of these two assemblages has not been made.

Final comment

My name appears in the first sentence of the acknowledgments of the paper that I have spent so long commenting on. I am described as enhancing the authors' understanding of anoline lizards by discussions with them. After my description of so many errors and confusions, it will not be a surprise that I enter a disclaimer 1) denying that there have ever been any substantive discussions with either of the authors orally or in writing; 2) asserting that, if, in the view of the authors, significant discussions did occur, my contribution to the authors' understanding of the anoles was unsuccessful.

It has never been my opinion that the formal "widely accepted" classification of the anoline lizards, which Guyer and Savage would discard, is final and correct. The classical classification has always been for me a classification to be retained in the absence of anything better. I am not at this time very far from joining the 1980 opinion of Gorman et al. that all classification of *Anolis* (*sensu lato*) above the species group level needs serious re-examination. This opinion is already joined by Burnell and Hedges in the unpublished MS mentioned above.

Clearly that serious re-examination and better classification has not been provided by Guyer and Savage (1986). In the interim before something better appears, I do not at all object to tinkering with sections, subsections, series, or species groups, although I reserve my right to quarrel with such informal categories when I think them wrong, and I reserve also the right to change my own mind. At the informal level, experimental taxonomic changes based on new data and new analyses appear to me desirable; such suggestions tested by time and by rigorous, even aggressive, criticism will be the base for the improved classification that is so much to be desired. Gorman (in litt.) and Cannatella and de Queiroz (1988) express not dissimilar opinions.

Acknowledgments

I thank Charles Woods, Richard Etheridge, and George Gorman for encouragement in a task that I thought very necessary. They, Gregory Mayer, and Blair Hedges have made many useful comments. Others who have seen early versions of this critique have been as much impressed by my rhetoric as by my substantive comment. To these I say that I have modified the rhetoric but not the severity of my judgment.

There have been many who have enhanced my own knowledge of the taxonomy, morphology, karyology and immunology of the anoles. Richard Etheridge stands in first place; George Gorman is not far behind. There, has been a great deal of discussion and correspondence between us. We have at times collaborated, and we have often disagreed. My debt, however, is great. I do not want to disparage the many others with whom contact has not been as long or has been less continuous. I do not list them for fear that I would omit some worthy name. All have my heartfelt thanks.

Literature cited

Blake, J.A. 1983a. Complex chromosomal variation in natural populations of the Jamaican lizard *Anolis grahami*. Genetica 69:3-17.

_____.1983b. Chromosomal C-banding in *Anolis grahami*. Pp. 621-625 in A.G.C. Rhodin and K.Miyata (eds.). Advances in Herpetology and Evolutionary Biology. Museum of Comparative Zoology, Cambridge, MA.

Cannatella, D.C., and K. de Queiroz. 1988. Phylogenetic systematics of the anoles: Is a new taxonomy warranted? Systematic Zoology, in press.

Case, S.M., and E.E. Williams. 1988. The cybotoid anoles and *Chamaelinorops* lizards (Reptilia: Iguanidae): evidence of mosaic evolution. Zoological Journal of the Linnaean Society 91:325-341.

Cochran, D. 1935. New reptiles and amphibians collected in Haiti by P.J. Darlington. Proceedings of the Boston Society of Natural History 40:367-376.

_____. 1939. Diagnoses of three new lizards and a frog from the Dominican Republic. Proceedings of the New England Zoological Club 18:1-3.

Estes, R., K. de Queiroz, and J. Gauthier. 1988. Phylogenetic relationships within squamate reptiles. In R. Estes and G.K. Pregill (eds.). Phylogenetic Relationships of the Lizard Families. Stanford University Press, Palo Alto, CA.

Etheridge, R. 1960. The relationships of the anoles (Reptilia: Sauria: Iguanidae): An interpretation based on skeletal morphology. Unpublished Ph.D. dissertation, University of Michigan, Ann Arbor, MI.

_____. 1964. Late Pleistocene lizards from Barbuda, British West Indies. Bulletin of the Florida State Museum 9:43-75

_____. 1965. The abdominal skeleton of lizards in the family Iguanidae. Herpetologica 21:161-168.

Etheridge, R., and K. de Queiroz. 1988. A phylogeny of the Iguanidae. In R. Estes and G. K. Pregill (eds.) Phylogenetic Relationships of the Lizard Families. Stanford University Press, Palo Alto, CA.

Forsgaard, K. 1983. The axial skeleton of *Chamaelinorops*. Pp. 284-295 in A.G.J. Rhodin and K. Miyata (eds.). Advances in Herpetology and Evolutionary Biology. Essays in Honor of Ernest E. Williams. Museum of Comparative Zoology, Cambridge, MA.

Garrido, O. 1982. Descripcion de una nueva especie de *Chamaeleolis* (Lacertilia: Iguanidae) con notas sobre su comportamiento. Poeyana 236:1-25.

Gorman, G.C. 1973. Chromosomes of the Reptilia, a cytotaxonomic interpretation. Pp. 349-424 in A.B. Chiarelli and E. Capanna (eds.). Cytotaxonomy and Vertebrate Evolution. Academic Press, New York.

_____ and L. Atkins. 1967. The relationships of the *Anolis* of the *roquet* species group (Sauria: Iguanidae). II. Comparative chromosome cytology. Systematic Zoology 16:137-143.

_____, R.B. Huey, and E.E. Williams. 1969. Cytotaxonomic studies on some unusual iguanid lizards assigned to the genera *Chamaeleolis*, *Polychrus*, *Polychroides*, and *Phenacosaurus*, with behavioral notes. Museum of Comparative Zoology, Breviora 316:1-17.

_____, D.G. Buth, and J.S. Wyles. 1980. Anole lizards of the eastern Caribbean: A case study in evolution. III. A cladistic analysis of albumin immunological data and the definition of species groups. Systematic Zoology 29:143-158.

_____, C.S. Lieb, and Robert H. Harwood. 1984. The relationships of *Anolis gadovi*: Albumin immunological evidence. Caribbean Journal of Science 20:145-152.

Guyer, C. and J. M. Savage. 1986. Cladistic relationships among anoles (Sauria: Iguanidae). Systematic Zoology 35:509-531.

Kasahara, S., Y. Yonenaga-Yassuda, and M.T. Rodrigues. 1987a.. Geographical karyotypic variations and chromosome banding in *Tropidurus hispidus* (Sauria, Iguanidae) from Brazil. Caryologia 40:43-57.

_____. 1987b. Karyotype and evolution of the *Tropidurus nanuzae* species group (Sauria, Iguanidae). Revista Brasileira de Genetica 10:185-197.

Lazell, J. 1969. The genus *Phenacosaurus* (Sauria: Iguanidae). Breviora, Museum of Comparative Zoology 325:1-24.

Mattson, P.H. 1984. Caribbean structural breaks and plate movements. In W.E. Bonini, R.B. Hargraves, and R. Shagam (eds.). The Caribbean-South American Plate Boundary and Regional Tectonics. Memoir of the Geological Society of America 162:131-152.

Moritz, C. 1984. The evolution of a highly variable sex chromosome in *Gehyra purpurascens* (Gekkonidae). Chromosoma 90:111-119.

Paull, D., E.E. Williams, and W.P. Hall. 1976. Lizard karyotypes from the Galápagos Islands: Chromosomes in phylogeny and evolution. Museum of Comparative Zoology, Breviora 441:1-31.

Peterson, J. 1983. The evolution of the subdigital pad in *Anolis*. I. Comparisons among the anoline genera. Pp. 245-283 in A.G.J. Rhodin and K. Miyata (eds.). Advances in Herpetology and Evolutionary Biology. Essays in Honor of Ernest E. Williams. Museum of Comparative Zoology, Cambridge, MA.

Pindell, J. and J.F. Dewey. 1982. Permo-Triassic reconstruction of western Pangea and the evolution of the Gulf of Mexico/Caribbean region. Tectonics 1:179-211.

Rensch, B. 1960. Evolution Above the Species Level. Columbia University Press, New York. 419pp.

Savage, J.M. 1982. The enigma of the Central American herpetofauna: Dispersals or vicariance? Annals of the Missouri Botanical Garden 69:464-547.

Schuchert, C. 1935. Historical Geology of the Antillean-Caribbean region. John Wiley and Sons, New York. 811pp.

Schwartz, A. 1978. A new species of aquatic *Anolis* (Sauria, Iguanidae) from Hispaniola. Annals of Carnegie Museum 47:261-279.

_____ and Robert W. Henderson. 1988. West Indian Amphibians and Reptiles: A Checklist. Milwaukee Public Museum Contributions in Biology and Geology 74:1-264.

Shochat, D., and H.C. Dessauer. 1977. Report. Pp. 184-191 in E.E. Williams (ed.). The Third *Anolis* Newsletter. Museum of Comparative Zoology. Cambridge, MA.

_____. 1981. Comparative immunological study of albumins of *Anolis* lizards of the Caribbean islands. Comparative Biochemistry and Physiology 68A:67-73.

Stamm, B., and G.C. Gorman. 1975. Notes on the chromosomes of *Anolis agassizi* (Sauria: Iguanidae) and *Diploglossus millepunctatus* (Sauria: Anguidae) Smithsonian Contributions to Zoology 176:52-54.

Webster, T.P. 1974. Report Pp.1-4. E. E. Williams (ed.). The Second Anolis Newsletter. Museum of Comparative Zoology, Cambridge, Massachusetts. (Reports, alphabetically arranged by author, are separately paginated and without title.)

_____, W.P. Hall, and E.E. Williams. 1972. Fission in the evolution of a lizard karyotype. Science 177:611-613.

Williams, E.E. 1962. Notes on the herpetology of Hispaniola. 7. New material of two poorly known anoles, *Anolis monticola* Shreve and *Anolis christophei* Williams. Museum of Comparative Zoology, Breviora 164:1-11.

_____. 1969. The ecology of colonization as seen in the zoogeography of anoline lizards on small islands. Quarterly Review of Biology 44:345-389.

_____. 1976a. West Indian anoles: A taxonomic and evolutionary summary. 1. Introduction and a species list. Museum of Comparative Zoology, Breviora 440:1-21.

_____. 1976b. South American anoles: the species groups. Papeis Avulsos de Zoologia Sao Paulo 29:259-268.

_____. 1983. Ecomorphs, faunas, island size and diverse end points in island radiations of *Anolis*. Pp. 326-370 in R.B. Huey, E.R. Pianka, and T.W. Schoener (eds.). Lizard Ecology: Studies of a Model Organism. Harvard University Press, Cambridge, MA.

_____ and A.S. Rand. 1969. *Anolis insolitus*, a new dwarf anole of zoogeographic importance from the mountains of the Dominican Republic. Museum of Comparative Zoology, Breviora 326:1-21.

Wyles, J.S., and G.C. Gorman. 1980. The classification of *Anolis*: Conflict between genetic and osteological interpretation as exemplified by *Anolis cybotes*. Journal of Herpetology 14:149-153.

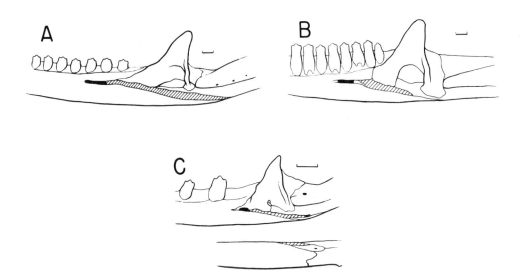

Figure 1. The "splenial" in some anoles. **A.** *"Anolis" noblei galeifer*, MCZ 127832 (*equestris* series) showing the elongate single bone (cross-hatched) of most anoles that have a "splenial." **B.** *"Semiurus" barahonae barahonae*, MCZ 173169 (*cuvieri* series) - the short single bone characteristic of this series. **C.** *Chamaeleolis chamaeleonides*, juvenile, MCZ 13328. **Above**: the two bones characteristic of this genus in lingual view, the "splenial" anterior, overlapping the more posteriorly and ventrally placed "angular." Below: the same two bones in ventral view, the posteriormost portion of the "splenial" seen at the upper central margin, the "angular" below it showing the posterior mylohyoid foramen.

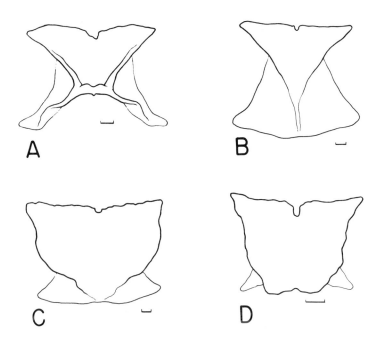

Figure 2. Anole parietals from above. **A.** *"Dactyloa" princeps*, MCZ 147444 (*latifrons* series) showing "U-shaped" parietal crests and no posterior parietal "half-funnel." **B.** *"Semiurus" ricordii ricordii*, MCZ 119715 (*cuvieri* series) - "Y-shaped" parietal crests and a strongly developed posterior parietal "half-funnel". **C.** *"Anolis" noblei galeifer*, MCZ 127832 (*equestris* series), showing "casquing." **D.** *Phenacosaurus heterodermus*, MCZ 17111 - "casquing."

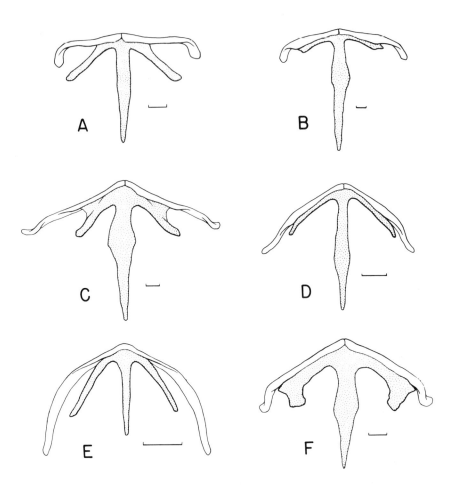

Figure 3. Anole interclavicles (stippled). **A.** *"Dactyloa" punctatus* (*punctatus* series) - an "arrow" interclavicle. **B.** *"Dactyloa" latifrons* (*latifrons* series) - a "T" interclavicle. **C.** *"Ctenonotus" leachii* MCZ 82110 (*bimaculatus* series) - an "arrow" interclavicle. **D.** *"Norops" onca*, MCZ 140263 (*onca* series) a "T" interclavicle. **E.** *Chamaeleolis chamaeleonides*, juvenile, MCZ 13328 - an "arrow" interclavicle. **F.** *Phenacosaurus* n.sp. MECN 0327 - a "T" interclavicle.

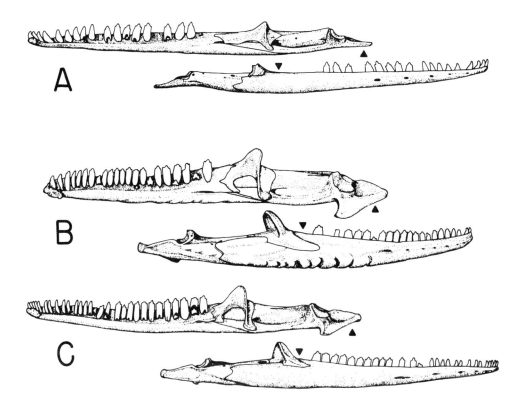

Figure 4. Anole lower jaws. lingual aspect above, labial aspect below. **A.** *"Anolis" occultus* Paratype MCZ 83660. **B.** *"Ctenonotus" gundlachi* MCZ 61848. **C.** *"Ctenonotus" evermanni* MCZ 61813. The black triangles point to the areas of the angular process and of the labial blade of the coronoid, both present in *gundlachi* and *evermanni*, both absent in *occultus*.

BIOGEOGRAPHIC PATTERNS OF PREDATION IN WEST INDIAN COLUBRID SNAKES

Robert W. Henderson and Brian I. Crother[1]

Abstract

Most West Indian colubrid snakes (41 species) share a number of morphological, ecological and behavioral characteristics, and dissections of about 2,700 snakes have indicated that a general dietary pattern is also present. Lizards accounted for 74.9 percent of all prey items (n = 707) found in the digestive tracts of West Indian colubrid snakes, followed by frogs (20.8%), snakes (1.7%), and mammals (1.6%). Combined, frogs and lizards accounted for 95.7 percent of all prey items. Lizards of the iguanid genus *Anolis* were most frequently exploited (56.8%), followed by leptodactylid frogs of the genus *Eleutherodactylus* (13.5%). Among lizard genera, 75.8 percent were *Anolis*, and among frog genera 64.6 percent were *Eleutherodactylus*. Birds and mammals together comprised 1.9 percent of the total prey items consumed by colubrids, compared to 70.9 percent for the boid genus *Epicrates*, and 100 percent for adults of West Indian representatives of the viperid genus *Bothrops*.

Within each of four geographic units of the West Indies, anoles accounted for 50-60 percent of the prey items taken by colubrids. Similarly, *Eleutherodactylus* and *Ameiva* also contributed to the diets of colubrids in each of the four geographic samples. Invertebrates appeared only in the diets of colubrid snakes (*Liophis*) in the Lesser Antilles.

Biogeographic patterns of predation exist among West Indian colubrids within the Antilles and also in relation to neighboring continents. Diets in general vary little among West Indian snakes according to distribution, habitat, and habitus, but on larger, physiographically complex islands, trophic specialization occurs. The snake-anole relationship in the West Indies is probably a unique predator-prey system.

Introduction

The palm-fringed islands of the West Indies have long been famous for their abundance of golden sunshine, smooth rum, beautiful beaches, and *Anolis* lizards, and scientists long ago recognized that because of their number, variation in area, and physiographic complexity, the Antilles would provide an excellent arena in which to study various aspects of ecology, behavior, evolution and biogeography. The latter has received much attention and probably has become the most controversial aspect to scientists

---

[1] R.W. Henderson is Curator of Vertebrate Zoology, Milwaukee Public Museum, Milwaukee, WI 53233; B.I. Crother, Department of Zoology, University of Texas, Austin, TX 78712.

working with the West Indian fauna (e.g., Matthews 1915; Barbour 1915; Rosen 1975, 1978; Pregill 1981; to cite only a few debated views).

Here, in a largely qualitative analysis, we present an overview of biogeographic patterns of predation in the West Indian snake fauna. Field work, coupled with the examination of about 2,700 museum specimens, has taught us a great deal about an ecologically intriguing fauna. Our emphasis is on the colubrid fauna (which is the richest with 41 species), and on the island of Hispaniola (which has the richest snake and colubrid faunas). The boid, tropidophiid, and viperid faunas are also discussed, but the leptotyphlopid (thread snakes) and typhlopid (blind snakes) faunas, because of their radically different modes of living (e.g., vestigial eyes; fossorial; diets consist entirely of invertebrates, primarily ants and termites), are not (but see Appendix 1).

The West Indies harbor a large snake fauna, consisting of 6 families, 21 genera and 88 described species (Table 1; Schwartz and Henderson 1985). This fauna is interesting in terms of predation patterns for several reasons: 1) the snakes' distributions vary from island-specific endemics with geographical ranges among the smallest known for vertebrates, to multiple-island distributions; 2) the snake fauna displays a large variation in morphology both inter- and intra-taxonomically; and 3) potential prey items are unusually ubiquitous and are found in sympatry and syntopy with the snakes throughout the islands in nearly all habitats. In addition, many West Indian snakes have incredibly small ranges, and coupled with habitat destruction and the introduction of exotic wildlife (i.e., the mongoose *Herpestes auropunctatus*), many species are at genuinely tenuous stages in their history; several species may already be extinct.

The obvious questions to ask then are, do diets vary according to distribution, habitat, and habitus of the snakes? Or do snakes prey on the ubiquitous prey items regardless of the latter three parameters? To understand this, more specific questions are addressed: 1) What kinds of prey do West Indian snakes eat in general? 2) Are certain prey types and taxa exploited more or less uniformly throughout the islands? 3) Why are some prey groups exploited heavily and others not? 4) What generalizations can be made regarding trophic niche breadth and foraging mode in regard to island size and complexity? 5) Do West Indian snakes share morphological, ecological and behavioral characteristics?

Materials and methods

The stomachs of about 2,700 West Indian snakes representing four families (Tropidophiidae, Boidae, Colubridae, Viperidae) were examined for prey remains. Collections from which specimens were necropsied were the American Museum of Natural History (AMNH), the Albert Schwartz Field Series in Miami (ASFS), the Museum of Comparative Zoology at Harvard University (MCZ), the Milwaukee Public Museum (MPM), the private collection of Richard Thomas (RT), the Florida State Museum at the University of Florida (UF), and the National Museum of Natural History (USNM). Methods of dissection, measurement and determination of individual prey volumes have been described in detail elsewhere (Henderson 1984a). Snake and prey snout-vent length is abbreviated SVL.

Some information on snake diets has been gleaned from the literature. Sources used are Arendt and Anthony (1986) Carey (1972), Crother (1986), Dixon and Soini (1986),

Duellman (1978), Grant (1940), Greene (1979, 1983a, 1983b), Hardy (1957a, 1957b), Iverson (1986), Lancini (1979), Lando and Williams (1969), Lazell (1964), Lynn and Grant (1940), Pendlebury (1974), Reagan (1984), Rodriguez and Reagan (1984), Schmidt (1928), Schwartz and Rossman (1976), Sheplan and Schwartz (1974), Stewart (1977), Stewart and Martin (1980) and Thomas (1966). In addition to the literature, food data for certain species were provided by Richard Franz (*Tretanorhinus variabilis*), James C. Gillingham (*Epicrates inornatus, Alsophis portoricencis, Arrhyton exiguum*), and Peter J. Tolson (*Epicrates monensis*). From these data general patterns of snake diets were determined and a qualitative island or island group description of snake diets was compiled.

Data for several islands were combined to increase sample size for our analysis (i.e., all of the Lesser Antilles; Cuba, Puerto Rico, Cayman Islands, Jamaica, and the Virgin Islands). Although the latter group is, admittedly, an unnatural geographic assemblage, it does provide another sample for geographic comparison.

The colubrid snake fauna has received most of our attention; it is the most diverse and the most intriguing. Colubrid snakes prey predominantly on small, ectothermic prey species. Most species of dwarf boas (*Tropidohis*) do likewise, and large members of the boid genus *Epicrates* and the viperid genus *Bothrops* prey predominantly on endothermic prey, but we did not examine large numbers of these snakes. Hispaniola, with its rich colubrid fauna, has also received a disproportionate amount of our focus. These biases are reflected in our analysis, but we feel that, overall, an accurate view of predation trends in West Indian snakes is available from our data.

Table 1 presents a species list and a summary of aspects of snake ecology and behavior, including prey genera exploited by each species. Appendices 1-3 offer brief synopses of 1) each genus of snakes that occurs in the West Indies; 2) each island's snake fauna; 3) the prey groups/genera most frequently exploited by West Indian snakes. Photographs of many of the species discussed here appear in Henderson and Schwartz (1984a) and Schwartz and Henderson (1985).

Results

Most West Indian colubrid snakes share a number of morphological, ecological, and behavioral characteristics in common: 1) small to moderate size (SVL of 1.0 m or less) (75.6% of 41 species), with a relatively slender habitus; 2) diurnal (95.1%); 3) prey is subdued by a method other than constriction (100%); 4) employ an active foraging mode (90.5%); and 5) are ground-dwelling (85.4%). All of the major islands have one or more ground-dwelling colubrids of small to moderate size that employ an active foraging mode. Only Hispaniola has colubrids that fit this pattern plus colubrids that diverge from it (e.g., *Hypsirhynchus, Uromacer*).

The dissections yielded 778 individual prey items for all snake taxa combined, with 707 of the prey items from the colubrid taxa. The following percentages are of total prey items yielded by colubrid taxa. In general, lizards (74.9%) were taken most frequently by colubrids, with frogs (20.8%) a distant second. Snakes (1.7%), mammals (1.6%), fishes (0.4%), birds (0.3%), invertebrates (0.3%), and turtles (0.1%) were taken much less frequently.

Prey frequency at the genus/group level for colubrids is a little more revealing.

Eleutherodactylus represents 13.5 percent and *Osteopilus* represents 7.2 percent of the prey consumed. *Anolis* lizards made up 56.8 percent of the prey consumed with *Ameiva* contributing 9.5 percent and *Leiocephalus* 4.4 percent. The other prey categories were too small to divide meaningfully. (Guyer and Savage [1986] have recently presented strong evidence in favor of partitioning *Anolis* into five genera.)

Clearly, the most frequently consumed prey items are lizards and, among the lizards, *Anolis* is the most heavily exploited at a 75.8 percent frequency, followed by *Ameiva* at 12.7 percent, *Leiocephalus* at 5.9 percent and *Sphaerodactylus* at 3.2 percent. Within frogs, *Eleutherodactylus* was the most commonly exploited at a 64.6 percent frequency, followed by *Osteopilus* at 34.7 percent. Frogs and lizards combined comprise 95.7 percent of the prey items in our sample, while birds and mammals account for only 1.9 percent of the prey items. By contrast, birds and/or mammals comprise 70.9 percent of the prey items (n = 72) found in the digestive tracts of members of the boid genus *Epicrates*, and frogs and lizards accounted for only 29.2 percent of the total sample (Fig. 1). For Hispaniolan material, the mean size (= volume) of 260 colubrid prey items was 3.6 ± 0.3 cm^3 and 53.3 ± 10.5 cm^3 for 33 boid prey items. Similarly, West Indian representatives of the viperid genus *Bothrops* exploited only rodents in our sample (n = 8), with a mean volume of 37.5 ± 9.7 cm^3.

The colubrids for which we have dietary data, or for which we are reliably certain what they do eat (38 species), were divided into six broad taxonomic assemblages based on diet. Results of this division are: 1) predominantly or exclusively fishes - 5.3 percent of the colubrids fed predominantly or exclusively on fishes (*Nerodia* and *Tretanorhinus*); 2) predominantly or exclusively frogs - 2.6 percent (*Darlingtonia haetiana*); 3) frogs and lizards - 50.0 percent; 4) predominantly or exclusively lizards - 10.5 percent; 5) snakes - 5.3 percent (*Arrhyton taeniatum* and *Ialtris agyrtes*); 6) all or most terrestrial vertebrates - 26.3 percent (all *Alsophis*).

The frequency with which prey groups/genera are exploited in comparing four areas of the West Indies: 1) Bahama Islands; 2) Cuba + Cayman Islands + Jamaica + Puerto Rico + Virgin Islands; 3) Hispaniola; and 4) Lesser Antilles) yielded interesting trends. In each area anoles accounted for 50-60 percent of the prey sample (Fig. 2). *Eleutherodactylus* and *Ameiva* contributed to snake diets in each area. Some prey genera did not occur in every area (e.g., *Osteopilus*, *Leiocephalus*), and made varying contributions to snake diets. Invertebrates were eaten only by snakes in the Lesser Antilles (*Liophis*), and fishes are consumed only by the Cuban and Cayman Island *Tretanorhinus variabilis*; they are undoubtedly eaten by the Cuban *Nerodia fasciata* as well.

Larger, more physiographically complex islands (e.g., Cuba, Jamaica, and Hispaniola) have larger frog and lizard faunas and, concomitantly, larger snake faunas (Fig. 3). They have a more diverse range of prey resources than do small islands. Exceptions to this are satellite islands associated with islands in the Greater Antilles. For example, Hispaniolan satellite islands of Isla Saona and Ile de la Gonave, despite having smaller areas and prey faunas, support richer colubrid snake faunas than either Jamaica or Puerto Rico. Large islands support the same "types" of snakes (e.g., *Alsophis*) as do small islands but, in addition, have more specialized species (*Arrhyton taeniatum*, *Ialtris agyrtes*, *Uromacer oxyrhynchus*). Cuba has a rich *Tropidophis* fauna (nine species) and six species of *Arrhyton*. Hispaniola has the richest colubrid fauna (11 species, including three genera with two or more representatives), only one species of *Tropidophis* (that also occurs on Cuba and Jamaica) and no *Arrhyton*. Only Hispaniola has colubrids that: 1) utilize a sit-and-wait foraging mode; 2) exploit the arboreal adaptive zone; and 3) belong to endemic genera (*Darlingtonia*, *Hypsirhynchus*, *Ialtris*, *Uromacer*). Jamaica has three species of *Arrhyton*. No other islands have co-occurring congenerics.

Discussion

A general dietary pattern for West Indian snakes is clearly present; most or all exploit *Anolis* lizards. This uniformity in diet may be a consequence of morphological and behavioral uniformity throughout the West Indian snake assemblage.

With few exceptions (e.g., *Uromacer*), West Indian colubrids tend to be generalized, exhibiting little in the way of morphological specialization. There is, likewise, a high degree of homogeneity in time of activity, foraging mode, prey capture and immobilization, prey size, and adaptive zone. Even those colubrids that have largely neotropical mainland distributions (e.g., *Clelia*, *Liophis*, *Mastigodryas*, *Pseudoboa*) but that have successfully colonized the Lesser Antilles, fit this homogeneous pattern (not unlike the characteristics shared by *Anolis* that have been successful colonizers, [Williams 1969]). Perhaps predictably, with knowledge of the lack of diversity within this suite of characteristics, West Indian colubrids also exhibit trophic uniformity.

Antillean snakes prey predominantly on frogs and lizards (95.7% by frequency), with lizards the most frequently consumed type of prey (74.9%). *Anolis* lizards constitute the single most frequently exploited prey genus (56.8%), followed distantly by the frog genus *Eleutherodactylus* (13.5%). These are the two most speciose vertebrate genera in the West Indies, with representatives on nearly every island and islet, and populations of some species at very high densities (Rubial and Philibosian 1974; Schoener and Schoener 1980; Stewart and Pough 1982; Turner and Gist 1970). Other lizard genera (e.g., *Sphaerodactylus*) may be important in the diets of newborn snakes (P.J. Tolson, pers. comm.). Perhaps they constitute a less "convenient" (i.e., inappropriate size; less ubiquitous) food source than *Anolis*, despite being found at high population densities (Cheng 1983; Schoener et al. 1982).

Mammals, especially rodents, are eaten by a wide variety of snakes, but in the West Indies only boids and viperids exploit them routinely; they cannot, on the basis of our data, be considered an important prey group for any species of colubrid snake that is endemic to the Antilles. Historically, species of rodents and insectivores that are now extinct were probably exploited by snakes in the West Indies before the arrival of Europeans and the introduction of *Mus* and *Rattus*. An hypothetical scenario has previously been presented for the boid *Epicrates striatus* on Hispaniola (Henderson et al. 1987b); it suggests a dietary transition from native insectivores (*Nesophontes*) and rodents (*Brotomys* and *Isolobodon*) to introduced rodents. Evidence indicates that representatives of these genera survived up until the early decades of the 20th Century, and simultaneously with *Mus* and *Rattus* (Allen 1942; Miller 1929, 1930). The dietary shift from endemic mammals to introduced rodents by *Epicrates striatus* would have increased gradually until the extirpation of the endemic species was complete, or at least until it became energetically disadvantageous to forage for them. Similar scenarios could likewise be projected for snakes (most boids, some colubrids, viperids) on other West Indian islands. See Morgan and Woods (1986) for a discussion of extinctions in West Indian land mammals.

Invertebrates, fishes, snakes, turtles, and birds, although eaten, are like, mammals, infrequently preyed upon by West Indian colubrids. The reason for the absence of these prey groups in the diets of Antillean colubrids may be many (e.g., the prevalence, density, and ubiquity of *Anolis*), but at least one may be particularly relevant: 1) Many Central and South American colubrids belong to the subfamily Xenodontinae as do most West Indian colubrids. According to Dowling and Duellman (1978), the Xenodontinae "appears to be a true phylogenetic entity," and "most feed on animals other than mammals and birds." 2) Most West Indian colubrids are small, energy-intensive foragers that prey on small frogs and lizards (generally < 3.0 cm^3). None preys heavily on birds or mammals. 3) Xenodontines may be poorly adapted to prey on birds and mammals, because they lack the means of efficiently subduing (by constriction or efficient envenomation) larger prey items. 4) Because the West Indies have a generally depauperate, non-volant, small mammal fauna, but do have rich frog and lizard faunas, xenodontine colubrids may have been the optimum "type" of snake for radiating in the West Indies: adapted morphologically, physiologically, ecologically and behaviorally to feed primarily on the kinds of vertebrate prey that occurred with the greatest diversity and at the highest population densities in the West Indies. Similarly, Nussbaum (1984) suggested that snakes that eat lizards and birds are more common on oceanic islands because lizards and birds are more common than other vertebrates on oceanic islands.

It might be expected that, in an historical context, predators would thoroughly exploit anoles as food and then undergo an adaptive radiation to utilize other available food sources, also. But because of their diversity, ecological ubiquity, and frequently high population densities, the availability of *Anolis* as a food source in the West Indies may preclude the necessity of intense exploitation of many other types of prey (e.g., invertebrates and *Eleutherodactylus* frogs). Antillean boids (Henderson et al. 1987b; Pendlebury 1974) and *Tropidophis* (Greene 1979, 1983a; Iverson 1986) prey frequently (*Epicrates striatus*, *Tropidophis greenwayi*) or almost exclusively (*Epicrates gracilis*) on anoles, even though they do not do so elsewhere in their range (i.e. *Corallus enydris* [Pendlebury 1974]). It is likely that about 94 percent of all West Indian boid, tropidophiid, and colubrid snakes (62 of 66 species) prey on anoles at some time during their life history. Possible exceptions are *Arrhyton taeniatum*, *Ialtris agyrtes*, *Nerodia fasciata* and *Tretanorhinus variabilis*; the latter two species occur on Cuba, are members of genera that are not West Indian endemics, and are probably fairly recent arrivals. It is possible that all West Indian snakes (exclusive of *Typhlops* and *Leptotyphlops*) prey on anoles on occasion.

On the neotropical mainland, anoles are not nearly as conspicuous, do not occur at such high densities, do not comprise such a major portion of the lizard fauna (Williams 1983), and do not contribute as much to the diets of most snake species as they do in the Antilles. Food data for colubrid snakes from the Upper Amazon Basin of Ecuador and Peru were gleaned from Duellman (1978) and Dixon and Soini (1986). Of 158 prey items, 48.1 percent were frogs, 21.5 percent were lizards and 13.3 percent were birds and mammals combined. Only 3.2 percent of the prey items were anoles, and they comprised 14.7 percent of the lizard prey. This is a significant reversal from the situation in the West Indies. At some localities on the neotropical mainland, however, anoles may be a major food source in the diets of some very specialized colubrids (e.g., *Oxybelis aeneus*; Henderson 1982). Although the number of species of *Eleutherodactylus* is greater than the number of species of *Anolis* in the Antilles, frogs generally reach their highest levels of species diversity at higher elevations than do most lizard and snake species. On Hispaniola, frog species diversity peaks at an elevation at which snake species diversity has already declined (Fig. 4).

According to Cadle (1985), Greater Antillean xenodontines (with the exception of *Tretanorhinus*) are phylogenetically related to members of the South American xenodontine clade rather than to Central American taxa. In the Lesser Antilles, the xenodontine fauna contains minor elements (*Clelia, Liophis, Pseudoboa*) distinct from Greater Antillean taxa and related to South American congeners. A Tertiary dispersal from South America and subsequent radiation within the Greater and Lesser Antilles has been hypothesized to explain the origin of the West Indian xenodontine fauna (Cadle 1985). Pregill (1981) has suggested that arrival of colubrids to the West Indies began in the late Miocene, and a fossil *Anolis* is known from the early Miocene on Hispaniola (Rieppel 1980).

Overwater dispersal is generally regarded as the most plausible explanation for the distribution of West Indian snakes (Pregill 1981; Tolson 1982; Cadle 1985), and Maglio has suggested that the present xenodontine fauna was derived through at least four oversea colonizations. He further suggested that there have been numerous combinations of inter-island migrations, especially to centrally located Hispaniola, and they seem to have been much more frequent than mainland-island migrations.

Hispaniola, with its geographically central position in the West Indies, led Schwartz (1978) to conclude that "without question, Hispaniola is the center of colubrid radiation at the generic level in the Antilles." According to Maglio (1970), "Its large size, varied habitats, complex physiography and history have provided an excellent opportunity for immigrants to differentiate into noncompetitive forms." An hypothetical historical scenario of a segment of the *cantherigerus* species assemblage (sensu Maglio 1970) on Hispaniola, based on morphological (Henderson et al. 1988), osteological (Maglio 1970), biochemical (Cadle 1984a, 1985), and ecological (Henderson et al. 1988) data may be constructive. According to Maglio (1970), "The assemblage appears to have been derived from an ancestral species not unlike *Alsophis cantherigerus* in its osteological, hemipenial, and external morphology." *Alsophis* is the most widespread colubrid genus in the West Indies, and representatives also occur in western South America and on the Galapagos Islands. *Alsophis cantherigerus* is a morphologically generalized colubrid, an active forager, and trophically euryphagous. Proto-*cantherigerus* gave rise to proto-*Alsophis anomalus*, and subsequent specialization resulted in *Hypsirhynchus* and *Uromacer*; *Alsophis* is considered the sister taxon to the *Hypsirhynchus-Uromacer* clade. It is possible, as has been suggested by Maglio (1970), that *Alsophis, Hypsirhynchus*, and the tree snake genus *Uromacer* "may have differentiated as a means of dividing up the habitat more efficiently" and thereby avoiding competition. Species of *Alsophis* are ground-dwelling, active foraging, trophic generalists; *Hypsirhynchus ferox* is a ground-dwelling, sit-and-wait forager that is a lizard specialist; *Uromacer catesbyi* is an arboreal, active/ambush forager, and trophic generalist that is osteologically close to *H. ferox*; the other two species of *Uromacer* (*frenatus* and *oxyrhynchus*) are highly arboreal, morphologically specialized, sit-and-wait foragers that are trophic specialists. The available evidence suggests that a generalized diet and an active foraging mode are primitive, and a specialized diet and sit-and-wait foraging mode are derived (Henderson et al. 1988).

Many factors may be relevant in determining species diversity or richness in a particular geographic area, but "The underlying cause of the apparent latitudinal, elevational, and/or habitat patterns in snake species density appears best explained on the basis of the number and abundance of prey types available" (Vitt 1987). The West Indian islands with the greatest area and most complex physiography (Cuba, Hispaniola, Jamaica) possess the richest frog, lizard, and snake faunas. We ran Spearman Rank Correlation Procedures for the number of colubrid snake species vs. island size, and number of colubrid snake species vs. number of species of *Anolis* occurring on the same island. Both correlation coefficients were highly significant ($P < .01$), but with snake diversity-island size ($r^s = .96$) somewhat higher than snake diversity-anole diversity ($r^s = .82$).

The relationship between snake prey diversity and the number of snake species has been noted in temperate (e.g., Arnold 1972) and tropical (e.g., Greene in press; Miyata 1980) snake faunas. Although not enough data for each snake species have accumulated, available evidence suggests that the narrowest (i.e., most specialized) trophic niche breadths occur on those islands with the richest endemic faunas (Hispaniola, Cuba, possibly Jamaica). Only Cuba has snakes that feed primarily (*Nerodia clarki*) or exclusively (*Tretanorhinus variabilis*) on fish. *Darlingtonia haetiana*, an *Eleutherodactylus* frog specialist, occurs on Hispaniola. Four species of strictly saurophagous snakes (that includes three colubrids (*Hypsirhynchus ferox, Uromacer frenatus, U. oxyrhynchus*) and a boid (*Epicrates gracilis*)) occur on Hispaniola. The only snakes that appear to specialize in eating other snakes occur on Cuba (*Arrhyton taeniatum*) and Hispaniola (*Ialtris agyrtes*); evidence indicates that both are stenophagous for *Typhlops*. Conversely, the most widespread colubrid genus in the West Indies is *Alsophis*, and its members are euryphagous vertebrate predators (Henderson 1984a; Henderson et al. 1988). Similarly, *Liophis* occurs on many islands in the Lesser Antilles, and available evidence suggests that it, too, is a trophic generalist (invertebrates, frogs and lizards).

Cuba, although larger than Hispaniola does not have so diverse a colubrid fauna. It does, however, have a rich *Tropidophis* fauna (nine species, seven of which are endemic); Hispaniola has but a single species (*T. haetianus*) which it shares with Cuba and Jamaica. The Cuban *Tropidophis* fauna includes species (*T. feicki, T. semicinctus*) that are morphological extremes compared to most dwarf boas. It has been suggested that, on Cuba, *Tropidophis* occupies niches filled by colubrids on Hispaniola (Greene 1979 and 1983a).

West Indian colubrids and boids have diets that are largely exclusive of one another (Fig. 1). Although members of both families eat anoles, evidence from the Hispaniolan snake fauna indicates that boas eat larger individuals of the same species exploited by the colubrids (Henderson et al. 1987b). The diets of West Indian *Tropidophis* and colubrids are much more similar (primarily small frogs and lizards; Fig. 1).

Various aspects of West Indian snake diets have been discussed and lead us to the point of the title. Biogeographic patterns of predation do exist for West Indian snakes within the Antilles and, also, relative to neighboring continents. In general, diets vary little among most West Indian colubrid snakes according to habitus, distribution, and habitat. This is especially true on small islands that harbor only one or two colubrids. Most species exploit the most ubiquitous prey, regardless of differences in the above three parameters, throughout the Antilles; that is, they eat anoles. If not anoles, then some other lizard species or leptodactylid frogs. But on the largest islands that exhibit the greatest physiographic complexity (i.e., Cuba, Hispaniola, Jamaica), morphological, behavioral, and trophic specializations are apparent within the boid, tropidophiid, and, especially, colubrid faunas.

The *Anolis* factor is of singular importance in the biogeography of predation in West Indian snakes. With over 90 percent of all West Indian boid, tropidophiid, and colubrid snakes exploiting anoles as food at some time in their life history, it is likely that no other vertebrate-eating snake fauna in the world preys to such a large extent on a single genus (*sensu lato*; see Guyer and Savage 1986) of prey organisms. For some snakes, anoles may be just one of many prey taxa exploited (e.g., Greater Antillean *Alsophis*). Others, especially boids (*Epicrates*), but some colubrids also (e.g., *Hypsirhynchus ferox*), consume anoles primarily when young (= small) and gradually shift their diets to other prey taxa (e.g., larger lizards or mammals and birds). Yet others exploit anoles as their primary food source throughout their lives (e.g., *Epicrates gracilis*, *Uromacer oxyrhynchus*).

Like xenodontine colubrids on the neotropical mainland, West Indian xenodontines prey predominantly on lizards and frogs, but even though the neotropical mainland harbors more species of anoles (albeit in a larger geographical area), the intensity of predation by the snake fauna does not begin to compare with that of West Indian snakes. Frogs comprise the most important prey group for the majority of neotropical mainland snakes (Duellman 1978; Dixon and Soini 1986; Greene 1988). Indeed, it is the snake-anole/predator-prey relationship throughout the Antilles that is so spectacular and that will generate provacative questions and stimulate future research on the ecology of West Indian snakes.

Acknowledgments

We are grateful to the personnel responsible for loaning the many specimens that were borrowed during the course of this ongoing project: Richard G. Zweifel (AMNH), Albert Schwartz (ASFS), José Rosado and Pere Alberch (MCZ), Richard Thomas (RT), Walter Auffenberg, David Auth, and Peter Meylan (UF), and George Zug (USNM). For critical comments on the manuscript we are indebted to G.S. Casper, R. de Sa, M.A. Donnelly, H.S. Fitch, P.N. Lahanas, J.C. Lee, M.A. Nickerson, T.A. Noeske -Hallin, R.A. Sajdak, A. Schwartz, R. Shine, J.B. Slowinski and C.A. Woods. Discussions of West Indian snake ecology with R. Sajdak, A. Schwartz and P. Tolson were especially helpful. Figure 5 was rendered by Rose M. Henderson, and we appreciate her efforts. Portions of the manuscript were typed by Patrice Lastufka and Karen LaVetter. Recent field work by Henderson on Hispaniola and in the Lesser Antilles has been generously funded by Friends of the Milwaukee Public Museum and the Institute of Museum Services. Craig Dethloff has been an enthusiastic assistant in dissecting snake stomachs; his help has been appreciated.

Literature cited

Allen, G.M. 1942. Extinct and vanishing mammals of the Western Hemisphere. Committee on International Wildlife Protection Special Publication 11:620.

Arendt, W.J. and D. Anthony. 1986. Bat predation by the St. Lucia boa (*Boa constrictor orophias*). Caribbean Journal of Science 22(3-4):219-220.

Arnold, S.J. 1972. Species densities of predators and their prey. American Naturalist 106:220-236.
Barbour, T. 1916. Some remarks upon Matthew's "Climate and Evolution." Annals of the New York Academy of Science 27:1-15.
____, and B. Shreve. 1935. Concerning some Bahamian reptiles, with notes on the fauna. Proceedings of the Boston Society of Natural History 40(5): 347-365.
Bogert, C.M. 1968. The variations and affinities of the dwarf boas of the genus *Ungaliophis*. American Museum of Natural History Novitates (2340):1-26.
____. 1969. Boas - a paradoxical family. Animal Kingdom 1969:18-25.
Brongersma, L.D. 1959. Some snakes from the Lesser Antilles. Studies on the Fauna of Curaçao and other Caribbean Islands 9:50-60.
Buden, D.W. 1966. An evaluation of Jamaican *Dromicus* (Serpentes, Colubridae) with the description of a new species. Breviora (238):1-10.
Cadle, J.E 1984a. Molecular systematics of neotropical xenodontine snakes: I. South American xenodontines. Herpetologica 40(1):8-20.
____. 1984b. Molecular systematics of neotropical xenodontine snakes: III. Overview of xenodontine phylogeny and the history of New World snakes. Copeia 1984:641-652.
____. 1985. The neotropical colubrid snake fauna (Serpentes: Colubridae): lineage components and biogeography. Systematic Zoology 34(1):1-20.
Carey, W.M. 1972. The herpetology of Anegada, British Virgin Islands. Caribbean Journal of Science 12(1-2):79-89.
Cheng, H.Y 1983. Predation, tail regeneration and reproduction in two sympatric geckos (*Sphaerodactylus*) on Hispaniola. Abstracts, Society for the Study of Amphibians and Reptiles/Herpetologists' League, p.52.
Crother, B.I. 1986. Diet record of *Alsophis cantherigerus*. Herpetological Review 17(2):47.
Dixon, J.R. 1980. The neotropical colubrid snake genus *Liophis*. The generic concept. Contributions in Biology and Geology, Milwaukee Public Museum (31):1-40
____. 1981. The neotropical colubrid snake genus *Liophis*: the eastern Caribbean complex. Copeia 1981(2):296-304.
____, and P. Soini. 1986. The reptiles of the Upper Amazon Basin, Iquitos Region, Peru. Milwaukee Public Museum, 154 pp.
Dowling, H.G. and W.E. Duellman. 1978. Systematic herpetology: a synopsis of families and higher categories. HISS Publication, New York.
Duellman, W.E. 1978. The biology of an equatorial herpetofauna in Amazonian Ecuador. University of Kansas Museum of Natural History Miscellaneous Publication (65):1-352.
Dunn, E.R. 1932. The colubrid snakes of the Greater Antilles. Copeia 1932 (2):89-92.
Garrido, O.H. and M.L. Jaume. 1984. Catalogo descriptivo de los anfibios y reptiles de Cuba. Donana, Acta Vertebrata 11(2): 5-128.
Grant, C. 1940. The herpetology of the Cayman Islands. Bulletin of the Institute of Jamaica, Science Series (2):1-65.
Greene, H.W. 1979. Evolutionary biology of the dwarf boas (Serpentes: Tropidophiidae). Yearbook of the American Philosophical Society 1979:206-207.

_____. 1983a. Dietary correlates of the origin and radiation of snakes. American Zoologist 23:431-441.

_____. 1983b. *Boa constrictor* (Boa, Bequer, Boa Constrictor). Pp. 380-382 in D.H. Janzen (ed.). Costa Rican natural history. University of Chicago Press, Chicago, IL.

_____. 1988. Species richness in tropical predators. In F. Almeda, and C.M. Pringle (eds.). Diversity and conservation of tropical rainforests. Memoirs of the California Academy of Science.

Guyer, C. and J.M. Savage. 1986. Cladistic relationships among anoles (Sauria: Iguanidae). Systematic Zoology 35(4):509-531.

Hardy, J.D 1957a. A note on the feeding habits of the Cuban racer, *Alsophis angulifer* (Bibron). Copeia 1957:49-50.

_____. 1957b. Bat predation by the Cuban boa, *Epicrates angulifer* Bibron. Copeia 1957(2):151-152.

Henderson, R.W 1982. Trophic relationships and foraging strategies of some New World tree snakes (*Leptophis*, *Oxybelis*, *Uromacer*). Amphibia-Reptilia 3:71-80.

_____. 1984a. The diets of Hispaniolan colubrid snakes I. Introduction and prey genera. Oecologia 62:234-239.

_____. 1984b. The diet of the Hispaniolan snake *Hypsirhynchus ferox* (Colubridae). Amphibia-Reptilia 5:367-371.

_____, B.I. Crother, T.A. Noeske-Hallin, A. Schwartz, and C.R. Dethloff. 1987a. The diet of the Hispaniolan colubrid snake *Antillophis parvifrons*. Journal of Herpetology 21(4):328-332.

_____, T.A. Noeske-Hallin, B.I. Crother, A. Schwartz. 1988. The diets of Hispaniolan colubrid snakes II. Prey species, prey size, and phylogeny. Herpetologica 44(1):55-70.

_____, T.A. Noeske-Hallin, J.A. Ottenwalder, and A. Schwartz. 1987b. On the diet of the boa *Epicrates striatus* on Hispaniola, with notes on *E. fordi* and *E. gracilis*. Amphibia-Reptilia 8:251-258.

_____, and R.A. Sajdak. 1986. West Indian racers: a disappearing act or a second chance? Lore 36(3):13-18.

_____, and A. Schwartz. 1984a. A guide to the identification of the amphibians and reptiles of Hispaniola. Milwaukee Public Museum Special Publication in Biology and Geology (4):1-70.

_____, and A. Schwartz. 1984b. *Uromacer frenatus*. Catalogue of American Amphibians and Reptiles (357):1-2.

_____, and A. Schwartz. 1986. The diet of the Hispaniolan colubrid snake, *Darlingtonia haetiana*. Copeia 1986(2): 529-531.

_____, A. Schwartz, and T.A. Noeske-Hallin. 1987c. Food habits of three colubrid tree snakes (genus *Uromacer*) on Hispaniola. Herpetologica. 43(2):241-248.

Iverson, J.B. 1986. Notes on the natural history of the Caicos Islands dwarf boa, *Trophidophis greenwayi*. Caribbean Journal of Science 22(3-4):191-198.

Jaume, M.L. 1965. La mas rara culebra de Cuba. Torreia, Nueva Serie 35: 3-8.

Jenner, J.V. 1981. A Zoogeographic Study and Taxonomy of the Xenodontine Colubrid Snakes. Unpublished Ph.D. dissertation. New York University.

_____, and H.G. Dowling. 1985. Taxonomy of American xenodontine snakes: the tribe Pseudoboini. Herpetologica 41(2):161-172.

Lancini V., A.R. 1979. Serpientes de Venezuela. Graficas Armitano, C.A., Caracas, Venezuela. 262 pp.

Lando, R.V. and E.E. Williams. 1969. Notes on the herpetology of the U.S. Naval Base at Guantanamo Bay, Cuba. Studies on the Fauna of Curaçao and Caribbean Islands 34:46-72.

Lazell, J.D., Jr. 1964. The Lesser Antillean representatives of *Bothrops* and *Constrictor*. Bulletin of the Museum of Comparative Zoology 132(3): 245-273.

Lynn, W.G. and C. Grant. 1940. The herpetology of Jamaica. Bulletin of the Institute of Jamaica 1:1-148.

Maglio, V.J. 1970. West Indian xenodontine colubrid snakes: their probable origin, phylogeny, and zoogeography. Bulletin of the Museum of Comparative Zoology 141(1):1-53.

Matthew, W.D. 1915. Climate and evolution. Annals of the New York Academy of Science 24:171-318.

Miller, G.S., Jr. 1929. A second collection of mammals from caves near St. Michel, Haiti. Smithsonian Miscellaneous Collection 81(9):1-30.

_____. 1930. Three small collections of mammals from Hispaniola Smithsonian Miscellaneous Collection 82(15):1-10.

Miyata, K.I. 1980. Patterns of Diversity in Tropical Herpetofaunas. Unpublished Ph.D dissertation, Harvard University, Cambridge, MA. 805 pp.

Morgan, G.S., and C.A. Woods. 1986. Extinction and zoogeography of West Indian land mammals. Biological Journal Linnaean Society 28:167-203.

Myers, C.W. 1974. The systematics of Rhadinaea (Colubridae), a genus of New World snakes. Bulletin of the American Museum of Natural History 153(1): 1-262.

Nussbaum, R.A. 1984. Snakes of the Seychelles. Pp. 361-377 In Stoddart, D.R. (ed.). Biogeography and ecology of the Seychelles Islands. W. Junk, Publisher, The Hague.

Parker, H.W. 1933. Some amphibians and reptiles from the Lesser Antilles. Annals of the Magazine of Natural History, series 10(11):155-158.

Pendlebury, G.B. 1974. Stomach and intestine contents of *Corallus enydris*; a comparison of island and mainland specimens. Journal of Herpetology 8(3):241-244.

Peters, J.A. and B. Orejas-Miranda. 1970. Catalogue of the neotropical Squamata: Part I. Snakes. United States National Museum Bulletin (297):i-viii + 1-347.

Pregill, G.K. 1981. An appraisal of the vicariance hypothesis of Carribean biogeography and its application to West Indian terrestrial vertebrates. Systematic Zoology 30(2):147-155.

Reagan, D.P. 1984. Ecology of the Puerto Rican boa (*Epicrates inornatus*) in the Luquillo Mountains of Puerto Rico. Caribbean Journal of Science 20(3-4): 119-127.

Rieppel, O. 1980. Green anole in Dominican amber. Nature 286:486-487.

Rodriguez, G.A. and D.P. Reagan. 1984. Bat predation by the Puerto Rican boa, *Epicrates inornatus*. Copeia 1984(1):219-220.

Rosen, D.E. 1975. A vicariance model of Carribean biogeography. Systematic Zoology 24:341-364.

_____. 1978. Vicariant patterns and historical explanations in biogeography. Systematic Zoology 27:159-188.
Ruibal, R., and R. Philibosian. 1974. The population ecology of the lizard Anolis acutus. Ecology 55(3):525-537.
Schmidt, K.P. 1928. Amphibians and reptiles of Puerto Rico, with a list of those reported from the Virgin Islands. Scientific Survey of Puerto Rico and the Virgin Islands. New York Academy of Science 10:1-160.
Schoener, T.W. and A. Schoener. 1980. Densities, sex ratios, and population structure of four species of Bahamian *Anolis* lizards. Journal of Animal Ecology 49:19-53
_____, J.B. Slade, and C.H. Stinson. 1982. Diet and sexual dimorphism in the very catholic lizard genus, *Leiocephalus* of the Bahamas. Oecologia 53:160-169.
Schwartz, A. 1957. A new species of boa (genus *Trophidophis*) from western Cuba. American Museum of Natural History Novitates (1839):1-8.
_____. 1965. A review of the colubrid snake genus *Arrhyton* with a description of a new subspecies from southern Oriente Province, Cuba. Proceedings of the Biological Society of Washington 78:99-114.
_____. 1966. Snakes of the genus *Alsophis* in Puerto Rico and the Virgin Islands. Studies on the Fauna of Curaçao and Caribbean Islands 23:175-227.
_____. 1967. A review of the genus *Dromicus* in Puerto Rico and the Virgin Islands. Stahlia 9:1-14.
_____. 1971. A systematic account of the Hispaniolan snake genus *Hypsirhynchus*. Studies on the Fauna of Curaçao and Caribbean Islands 35:63-94.
_____. 1975. Variation in the Antillean boid snake *Tropidophis haetianus* Cope. Journal of Herpetology 9(3):303-311.
_____. 1978. Some aspects of the herpetogeography of the West Indies. Pp. 31-51 in Gill (ed.). Zoogeography in the West Indies. Academy of Natural Sciences of Philadelphia, Special Publication 13.
_____. 1980. The herpetogeography of Hispaniola, West Indies. Studies on the Fauna of Curaçao and other Caribbean Islands 61:86-127.
_____. 1986. *Darlingtonia, D. haetiana*. Catalogue of American Amphibians and Reptiles (390):1-2.
_____, and O. H. Garrido. 1975. A reconsideration of some Cuban *Tropidophis* (Serpentes, Boidae). Proceedings of the Biological Society of Washington 88(9):77-90.
_____, and O.H. Garridoi. 1981. A review of the Cuban members of the genus *Arrhyton* (Reptilia, Serpentes, Colubridae). Annals of the Carnegie Museum of Natural History 50(7):207-230.
_____, and R.W. Henderson. 1984a. *Uromacer catesbyi*. Catalogue of American Amphibians and Reptiles (356):1-2.
_____, and R.W. Henderson. 1984b. *Uromacer oxyrhynchus*. Catalogue of American Amphibians and Reptiles (358):1-2.
_____, and R.W. Henderson. 1985. A guide to the identification of the amphibians and reptiles of the West Indies exclusive of Hispaniola. Milwaukee Public Museum, 165 pp.

_____, and R.J. Marsh. 1960. A review of the pardalis-maculatus complex of the boid genus *Tropidophis* of the West Indies. Bulletin of the Museum of Comparitive Zoology 123(2):49-84.

_____, and D.A. Rossman. 1976. A review of the Hispaniolan colubrid snake genus *Ialtris*. Studies on the Fauna of Curaçao and Caribbean Islands. 50:76-102.

_____, and R. Thomas. 1960. Four new snakes (*Tropidophis*, *Dromicus*, *Alsophis*) from the Isla de Pinos and Cuba. Herpetologica 16(2):73-90.

Seib, R.L. 1984. Prey use in three syntopic neotropical racers. Journal of Herpetology 18(4):412-420.

Sheplan, B.R., and A. Schwartz. 1974. Hispaniolan boas of the genus *Epicrates* (Serpentes, Boidae) and their Antillean relationships. Annals of the Carnegie Museum of Natural History 45(5):57-143.

Stewart, M.M. 1977. The role of introduced species in a Jamaican frog community. Actas del IV Simposium Internacional de Ecologia Tropical, Tomo 1:110-146.

_____, and G.E. Martin. 1980. Coconut husk-piles: A unique habitat for Jamaican terrestrial frogs. Biotropica 12(2):107-116.

_____, and F.H. Pough. 1982. Population density of tropical forest frogs: relation to retreat sites. Science 221:570-572.

Stuart, L.D. 1941. Studies of neotropical Colubrinae VIII. A revision of the genus *Dryadophis* Stuart, 1939. Miscellaneus Publication of the Museum of Zoology, University of Michigan 49:1-106.

Thomas R. 1963. Cayman Islands *Tropidophis* (Reptilia, Serpentes). Breviora (195):1-8.

_____. 1966. A reassessment of the herpetofauna of Navassa Island. Journal of the Ohio Herpetological Society 5(3):73-89.

_____. 1985. Prey and prey processing in blind snakes of the genus *Typhlops*. American Zoologist 25(4):14A (Abstract).

_____, and O.H. Garrido. 1967. A new subspecies of *Dromicus andreae* (Serpentes, Colubridae). Annals of the Carnegie Museum 39(16): 219-226.

_____, and A. Schwartz. 1965. Hispaniolan snakes of the genus *Dromicus* (Colubridae). Revista Biologia Tropical 13:58-83.

Tolson, P.J. 1982. Phylogenetics of the Boid Snake Genus *Epicrates* and Caribbean Vicariance Theory. Unpublished Ph.D. dissertation, University of Michigan, Ann Arbor, MI 134 pp.

Turner, F.B., and C.S. Gist. 1970. Observations of lizards and tree frogs in an irradiated Puerto Rican forest. Pp. E-25-E-49 in H.T. Odum, and R.F. Pigeon (eds.). A Tropical Rainforest. United States Atomic Energy Commission, Oak Ridge, TN.

Vitt, L.J. 1987. Communities. Pp. 335-365 in R.A. Seigel, J.T. Collins, and S.S. Novak (eds.). Snakes: Ecology and Evolutionary Biology. Macmillan Publishing Co. New York.

Williams, E.E. 1969. The ecology of colonization as seen in the zoogeography of anoline lizards on small islands. Quarterly Review of Biology 44(4):345-389.

_____. 1983. Ecomorphs, faunas, island size, and diverse end points in island radiations of *Anolis*. Pp. 326-370 in R.B. Huey, E.R. Pianka, and T. W. Schoener (eds.). Lizard Ecology. Harvard University Press, Cambridge, MA.

Table 1. - Species of boid, tropidophiid, colubrid and viperid snakes known to occur in the West Indies. Some aspects of their ecology are briefly noted. Literature references provide taxonomic and ecological information. Under "Diet", prey genera or groups are abbreviated. If the record for that prey genus or group came from a source other than our own dissections, it is enclosed in parentheses. Abbreviations used are In = Invertebrates; Fi = Fishes; **Frogs** - El = *Eleutherodactylus*, Hy = *Hyla*, Os = *Osteopilus*; **Lizards** - Sph = *Sphaerodactylus*, He = *Hemidactylus*, An = *Anolis*, Cy = *Cyclura*, Ig = *Iguana*, Le = *Leiocephalus*, Am = *Ameiva*, Ce = *Celestus*, Di = *Diploglossus*, Sa = *Sauresia*, We = *Wetmorena*; **Snakes** - Ty = *Typhlops*, Al = *Alsophis*, Ant = *Antillophis*, Arr = *Arrhyton*; Tu = Turtles; Bi = Birds; Ro = Rodents; Ba = Bats.

Species	West Indian Distribution	Diet	When active; Foraging Mode; Adaptive Zone	Literature
Family BOIDAE				
Boa constrictor	Dominica; St. Lucia	(Ig, Bi, Ba, Ro)	diurnal-nocturnal; sit-and-wait; terrestrial-arboreal	Greene 1983; Lazell 1964
Corallus enydris	St. Vincent; Grenadines; Grenada	(An, Ig, Bi, Ro)	nocturnal; sit-and-wait; arboreal	Pendlebury 1974
Epicrates angulifer	Cuba	(Ro, Ba)	nocturnal; sit-and-wait; terrestrial-arboreal	Sheplan and Schwartz 1974
E. chysogaster	Turks and Caicos Is.; Great Inagua; Acklins I.; Crooked I.	(Ro)	nocturnal; sit-and-wait; terrestrial	Sheplan and Schwartz 1974

Table 1. - Species of boid, tropidophiid, colubrid and viperid snakes known to occur in the West Indies (continued).

Species	West Indian Distribution	Diet	When active; Foraging Mode; Adaptive Zone	Literature
E. exsul	Great Abaco I.	?	nocturnal; sit-and-wait; terrestrial	Sheplan and Schwartz 1974
E. fordi	Hispaniola	An, Ro	nocturnal; active; terrestrial-arboreal	Sheplan and Schwartz 1974; Henderson et al. 1987b
E. gracilis	Hispaniola	An	nocturnal; active; arboreal	Henderson et al. 1987b; Sheplan and Schwartz 1974
E. inornatus	Puerto Rico	(Bi, Ro, Ba)	nocturnal; sit-and-wait; terrestrial	Sheplan and Schwartz 1974
E. monensis	I. Mona; U.S. and British Virgin Is.	(An, Bi, Ro)	nocturnal; active; arboreal	Sheplan and Schwartz 1974

Table 1. - Species of boid, tropidophiid, colubrid and viperid snakes known to occur in the West Indies (continued).

Species	West Indian Distribution	Diet	When active; Foraging Mode; Adaptive Zone	Literature
E. striatus	Hispaniola; Bahamas: Cat I; Bimini Is.; Andros I; Berry I.; Ragged Is.; New Providence I.; Eleuthera I.; Long I; Exuma Cays	(El), An, Bi, Ro	nocturnal; active and sit-and-wait; arboreal	Henderson et al. 1987b; Sheplan and Schwartz 1974
E. subflavus	Jamaica	?	nocturnal; sit-and-wait; terrestrial	Sheplan and Schwartz 1974

Family TROPIDOPHIIDAE

Tropidophis canus	Bahama Is.	(An)	nocturnal; active; terrestrial	Schwartz and Marsh 1960
T. caymanensis	Cayman Is.	(Os, An)	nocturnal; active; terrestrial	Thomas 1963

Table 1. - Species of boid, tropidophiid, colubrid and viperid snakes known to occur in the West Indies (continued).

Species	West Indian Distribution	Diet	When active; Foraging Mode; Adaptive Zone	Literature
T. feicki	Cuba	?	nocturnal; active; terrestrial	Schwartz 1957
T. greenwayi	Caicos Is.	(Sph, An)	nocturnal; active; terrestrial	Iverson 1987
T. haetianus	Hispaniola; Jamaica; Cuba	El,(Hy),Am,An,(Sph,Ce) Ro	nocturnal; active; terrestrial	Schwartz 1975
T. maculatus	Cuba	?	nocturnal; active; terrestrial	Schwartz and Marsh 1960
T. melanurus	Cuba; Navassa Is.	(Os, An, Ce, Bi) (Ro)	nocturnal; active; terrestrial	Schwartz and Thomas 1960
T. nigriventris	Cuba	?	nocturnal; active; terrestrial	Schwartz and Garrido 1975

Table 1. - Species of boid, tropidophiid, colubrid and viperid snakes known to occur in the West Indies (continued).

Species	West Indian Distribution	Diet	When active; Foraging Mode; Adaptive Zone	Literature
T. pardalis	Cuba	?	nocturnal; active; terrestrial	Schwartz and Marsh 1960
T. pilsbryi	Cuba	?	nocturnal; active; terrestrial	Schwartz and Garrido 1975
T. semicinctus	Cuba	?	nocturnal; active; terrestrial	Schwartz 1957
T. wrighti	Cuba	?	nocturnal; active; terrestrial	Schwartz 1957

Family COLUBRIDAE

Alsophis anomalus	Hispaniola	(Ro)	diurnal; active; terrestrial	Cochran 1941

Table 1. - Species of boid, tropidophiid, colubrid and viperid snakes known to occur in the West Indies (continued).

Species	West Indian Distribution	Diet	When active; Foraging Mode; Adaptive Zone	Literature
A. antillensis	Antigua; Montserrat; Guadeloupe; Marie Galante; Iles des Saintes; Dominica	El, An, Ro	diurnal; active; terrestrial	Parker 1933
A. ater	Jamaica	Ce	diurnal; active; terrestrial	Grant 1940
A. cantherigerus	Cuba; Cayman Is.; Little Swan I.	El, Os, Sph, An, Cy, Le, Di Ce, Ant, Arr, Tu, Bi, Ro, Ba	diurnal; active; terrestrial	Henderson 1984; Lando and Williams 1969.
A. melanichnus	Hispaniola	?	diurnal; active; terrestrial	Cochran 1941
A. portoricensis	Puerto Rico; I. Mona; Virgin Is.	El, Sph,(He), An, Ig, Am, (Ty), (Arr)	diurnal; active; terrestrial-arboreal	Schwartz 1966

Table 1. - Species of boid, tropidophiid, colubrid and viperid snakes known to occur in the West Indies (continued).

Species	West Indian Distribution	Diet	When active; Foraging Mode; Adaptive Zone	Literature
A. rijersmai	Anguilla; St. Martin; St. Barthelemy	An	diurnal; active; terrestrial	Brongersma 1959; Henderson, unpubl.
A. rufiventris	Saba; St. Eustatius, St. Christopher; Nevis	El, An, Am	diurnal; active; terrestrial	Brongersma 1959; Henderson, unpubl.
A. sanctaecrucis	St. Croix and Green Cay (U.S. Virgin Is.)	?	diurnal; active; terrestrial	Schwartz 1966
A. vudii	Bahama Is.	El, Os, Sph, An, Am, Al, Ro	diurnal; active; terrestrial	Barbour and Shreve 1935; Henderson, unpubl.
Antillophis andreae	Cuba	El, An, Le	diurnal; active; terrestrial	Henderson, unpubl.; Thomas and Garrido 1967

Table 1. - Species of boid, tropidophiid, colubrid and viperid snakes known to occur in the West Indies (continued).

Species	West Indian Distribution	Diet	When active; Foraging Mode; Adaptive Zone	Literature
A. parvifrons	Hispaniola; Little Inagua I.	El, Os, Sph, An, Am, Sa, We, Ro	diurnal; active; terrestrial	Henderson et al. 1987a; Thomas and Schwartz 1965
Arrhyton ainictum	Cuba	?	diurnal; active; terrestrial	Schwartz and Garrido 1981
A. callilaemus	Jamaica	An	diurnal; active; terrestrial	Buden 1966; Henderson, unpubl.
A. dolichorum	Cuba	?	diurnal; active; terrestrial	Schwartz and Garrido 1981
A. exiguum	Puerto Rico, Virgin Is.	El, An, Sph	diurnal; active; terrestrial	Henderson, unpubl.; Schwartz 1967
A. funereum	Jamaica	El, An	diurnal; active; terrestrial	Buden 1966; Henderson, unpubl.

Table 1. - Species of boid, tropidophiid, colubrid and viperid snakes known to occur in the West Indies (continued).

Species	West Indian Distribution	Diet	When active; Foraging Mode; Adaptive Zone	Literature
A. landoi	Cuba	?	diurnal; active; terrestrial	Schwartz and Garrido 1981 1981
A. polylepis	Jamaica	El, An	diurnal; active; terrestrial	Buden 1966; Henderson, unpubl.
A. taeniatum	Cuba	Ty	diurnal; active; terrestrial	Henderson, unpubl; Schwartz and Garrido 1981
A. tanyplectum	Cuba	?	diurnal; active; terrestrial	Schwartz and Garrido 1981
A. vittatum	Cuba	?	diurnal; active; terrestrial	Schwartz and Garrido 1981
Chironius vincenti	St. Vincent	El	diurnal; active; arboreal	Schwartz and Thomas 1975

Table 1. - Species of boid, tropidophiid, colubrid and viperid snakes known to occur in the West Indies (continued).

Species	West Indian Distribution	Diet	When active; Foraging Mode; Adaptive Zone	Literature
Clelia clelia	Dominica; St. Lucia; Grenada	Ro	nocturnal; active; terrestrial	Henderson, unpubl.; Lancini 1979
Darlingtonia haetiana	Hispaniola	El, An	diurnal; active; terrestrial	Henderson and Schwartz 1986; Schwartz 1986
Hypsirhynchus ferox	Hispaniola	An, Le, Am	diurnal; sit-and-wait; terrestrial	Henderson 1984b; Schwartz 1971
Ialtris agyrtes	Hispaniola	Ty	diurnal; active; terrestrial	Schwartz and Rossman 1976
I. dorsalis	Hispaniola	Os, An, Le, Ce, Ro	diurnal; active; terrestrial	Henderson 1984a; Schwartz and Rossman 1976
I. parishi	Hispaniola	?	diurnal; active; terrestrial	Schwartz and Rossman 1976

Table 1. - Species of boid, tropidophiid, colubrid and viperid snakes known to occur in the West Indies (continued).

Species	West Indian Distribution	Diet	When active; Foraging Mode; Adaptive Zone	Literature
Liophis cursor	Martinique	In, El, An	diurnal; active; terrestrial	Dixon 1981; Henderson, unpubl.
L. juliae	Dominica; Guadeloupe; Marie Galante	In, An	diurnal; active; terrestrial	Dixon 1981; Henderson, unpubl.
L. melanotus	Grenada	?	diurnal; active; terrestrial	Dixon 1981; Henderson, unpubl.
L. ornatus	St. Lucia; Maria Is.	?	diurnal; active; terrestrial	Dixon 1981
L. perfuscus	Barbados	?	diurnal; active; terrestrial	Dixon 1981
Mastigodryas bruesi	St. Vincent; Grenadines; Grenada	El, An, Am	diurnal; active; terrestrial	Henderson, unpubl.; Lancini 1979

Table 1. - Species of boid, tropidophiid, colubrid and viperid snakes known to occur in the West Indies (continued).

Species	West Indian Distribution	Diet	When active; Foraging Mode; Adaptive Zone	Literature
Nerodia fasciata	Cuba	(Fi)	nocturnal; active; aquatic	Garrido and Jaume 1984; Jaume 1965
Pseudoboa neuwiedi	Grenada	?	nocturnal; active; terrestrial	Henderson, unpubl.; Lancini 1979
Tretanorhinus variabilis	Cuba; Cayman Is.	Fi	nocturnal; active; aquatic	Garrido and Jaume 1984; Henderson, unpubl.
Uromacer catesbyi	Hispaniola	Hy, Os, An, Le	diurnal; active and sit-and-wait; arboreal	Schwartz and Henderson 1984a; Henderson et al. 1987c
U. frenatus	Hispaniola	Sph, An, Le, Am	diurnal; sit-and-wait; arboreal	Henderson and Schwartz 1984b; Henderson et al. 1987c

Table 1. - Species of boid, tropidophiid, colubrid and viperid snakes known to occur in the West Indies (concluded).

Species	West Indian Distribution	Diet	When active; Foraging Mode; Adaptive Zone	Literature
U. oxyrhynchus	Hispaniola	An, Le, Am	diurnal; sit-and-wait; arboreal	Schwartz and Henderson 1984b; Henderson et al. 1987c
Family VIPERIDAE				
Bothrops caribbaea	St. Lucia	Ro	nocturnal; sit-and-wait; terrestrial	Henderson, unpubl.; Lazell 1964
B. lanceolata	Martinique	(Bi), Ro	nocturnal; sit-and-wait; arboreal-terrestrial	Henderson, unpubl.; Lazell 1964

In = Invertebrates; Fi = Fishes; Frogs - El = *Eleutherodactylus*, Hy = *Hyla*, Os = *Osteopilus*; Lizards - Sph = *Sphaerodactylus*, He = *Hemidactylus*, An = *Anolis*, Cy = *Cyclura*, Ig = *Iguana*, Le = *Leiocephalus*, Am = *Ameiva*, Ce = *Celestus*, Di = *Diploglossus*, Sa = *Sauresia*, We = *Wetmorena*; Snakes - Ty = *Typhlops*, Al = *Alsophis*, Ant = *Antillophis*, Arr = *Arrhyton*; Tu = Turtles; Bi = Birds; Ro = Rodents; Ba = Bats.

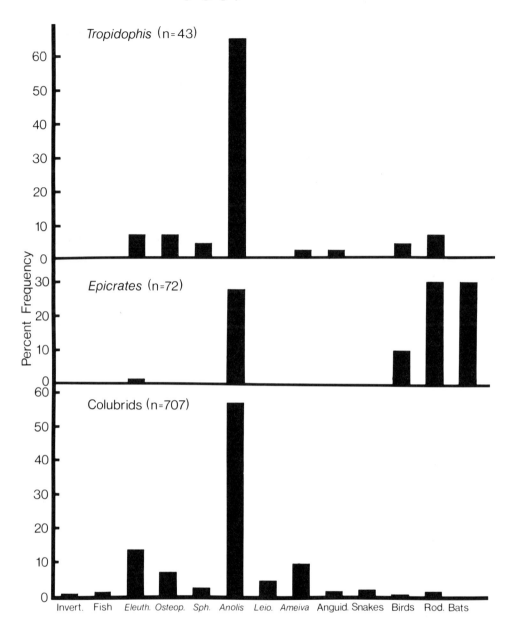

Figure 1. - The percent frequency with which various prey genera/groups appeared in the diets of West Indian *Tropidophis* (Tropidophiidae), *Epicrates* (Boidae), and Colubridae. *Tropidophis* and *Epicrates* data are from dissections and literature records. The colubrid data are from dissections only. Abbreviations used are Invert = invertebrates: Eleuth. = *Eleutherodactylus*; Osteop. = *Osteopilus*; Sph. = *Sphaerodactylus*; Leio. = *Leiocephalus*; Anguid. = Anguidae; Rod. = Rodents.

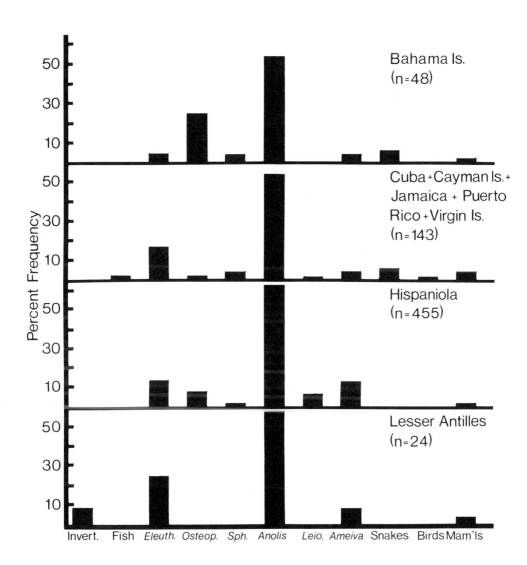

Figure 2. - The percent frequency with which various prey genera/groups appeared in the diets of West Indian colubrid snakes from four geographic areas. See Figure 1 for a key to abbreviations.

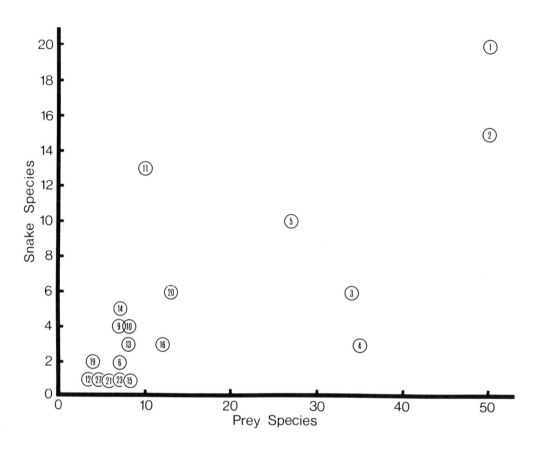

Figure 3. - The relationship between the number of prey species (*Mus* + *Eleutherodactylus* + *Osteopilus* + *Sphaerodactylus* + *Anolis* + *Rattus* + *Leiocephalus* + *Ameiva*) and snake species (Tropidophiidae + Boidae + Colubridae + Viperidae). Numbers refer to individual islands ranked in order of area (no. 1 is the largest, no. 27 is the smallest). If more than one island had the same number of snake and prey species, only one number was plotted. Most island areas were taken from "Baedeker's Caribbean" (1984, Prentice-Hall, Inc.). 1. Cuba (111,463 km^2), 2. Hispaniola (76,484) 3. Jamaica (10,991) 4. Puerto Rico (8,676) 5. Isla de la Juventud (Cuba) (2,200) 6. Guadeloupe (1,510) 6. Grand Bahama (1,372) 6. Martinique (1,106) 9. Dominica (751) 10. St. Lucia (616) 11. Ile de la Gonave (Hispaniola) (660) 12. Barbados (431) 13. St. Vincent (346) 14. Grenada (344) 15. Antigua (280) 16. New Providence (207) 13. Grand Cayman (197) 15. St. Christopher (169) 19. Marie-Galante (158) 20. Isla Saona (Hispaniola) (111) 21. Montserrat (101) 15. Nevis (93) 23. Anguilla (88) 15. St. Martin (88) 21. St. Barthelemy (22) 15. St. Eustatius (21) 27. Saba (13).

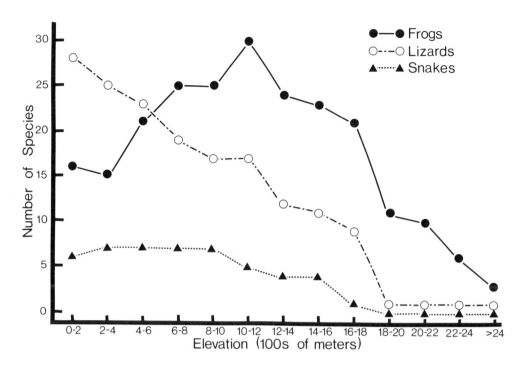

Figure 4. - The number of frog (*Eleutherodactylus* + *Osteopilus* + *Hyla*), lizard (*Anolis* + *Leiocephalus* + *Ameiva*) and colubrid snake (*Antillophis parvifrons* + *Darlingtonia haetiana* + *Hypsirhynchus ferox* + *Ialtris dorsalis* + *Uromacer catesbyi* + *U. frenatus* + *U. oxyrhynchus*) species found at various elevations on Hispaniola. Although the snake species frequently have wide elevational distributions, they may be very rare at the upper limits of the range. Many of the elevation data were taken from Schwartz (1980).

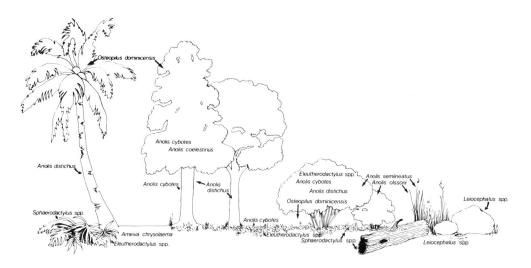

Figure 5. - A generalized habitat profile illustrating the distribution of prey species in an hypothetical habitat on Hispaniola; not all of these species would occur together at the same locality in the same habitat. The prey species would usually be found in places as indicated.

Appendix 1: West Indian snake genera

Family Leptotyphlopidae
Leptotyphlops.– Thread snakes range through the tropics in the Old and New World. They are small and fossorial with vestigial eyes, and feed on ants and termites. Eight species have been described from the West Indies, and four of those occur on Hispaniola.

Family Typhlopidae
Typhlops.– Blind snakes have a pantropical distribution. They are generally small and slender, have reduced eyes, and a toothless lower jaw. They feed on ants and termites (Thomas 1985). Sixty-one described species range through the West Indies, including nine species on Hispaniola.

Family Boidae
Boa.– A monotypic genus that is widespread on the neotropical mainland, ranging from northern Mexico to southern South America. It has limited distribution in the West Indies. The boa constrictor is a large (to about 5 m total length), heavy-bodied constrictor that preys on a wide variety of ectothermic and endothermic prey, on the ground and in trees (Greene 1983b).

Corallus.– A widespread genus of tree boas in Central and South America. Only *C. enydris*, the most widespread of the three known species, has limited distribution in the West Indies. Tree boas are adapted for arboreal existence, having laterally compressed bodies and enlarged anterior teeth that may be used for penetrating the plumage of birds. Diet is fairly catholic (frogs, lizards, birds, bats, rodents) (Dixon and Soini 1986; Duellman 1978).

Epicrates.– Ten species are recognized, nine of which occur in the West Indies. They range in size from 1-4 m total length and may be small and highly adapted for moving on very slender branches (*E. gracilis*), or very long, heavy-bodied, and ground-dwelling (*E. angulifer*) (Shepland and Schwartz 1974). Diets vary from specialization on *Anolis* lizards to fairly catholic. Large species show ontogenetic changes in diet (Henderson et al. 1987b), shifting from anoles to large birds (wild and domestic) and rodents (primarily *Rattus*, but also *Plagiodontia* and *Capromys*). Hispaniola is the only island that has more than one species.

Family Tropidophiidae
Tropidophis.– A genus of small to moderately sized (0.5-1.6 m total length), usually stout, primarily ground-dwelling dwarf boas. Primarily consumers of small frogs and lizards, larger species also eat small birds and rodents. Only Cuba has more than a single species, and it has nine. Three species occur on the South American mainland; none of those species occurs in the West Indies (Bogert 1968, 1969; Greene 1979).

Family Colubridae
Alsophis.– The most geographically widespread genus of snakes in the West Indies, ranging from Dominica to the northern edge of the Bahamas. Ten species occurred throughout the area at one time, but it is likely that at least one species (*A. sanctaecrucis*) is extinct, and two others have not been collected in over half a century (*A. ater* and *A.*

melanichnus). These are largely ground-dwelling, active foraging, trophic generalists that will eat any terrestrial vertebrate of appropriate size. They range in size from 1-2 m total length. Only Hispaniola harbors more than one species (Henderson and Sajdak 1986). More than any other genus of West Indian snakes, *Alsophis* has been especially sensitive to the introduction of the mongoose (Henderson and Sajdak 1986). Representatives occur in the Galapagos Islands and western South America.

Antillophis.— Small (0.8 m total length), ground-dwelling snakes that will eat most or all terrestrial vertebrates of appropriate size. One species occurs on Cuba and another on Hispaniola (and Little Inagua Island). They frequently occur in high relative density in appropriate habitat. Relationship with the widespread neotropical mainland genus *Rhadinaea* has been suggested (Maglio 1970), but affinities are uncertain (Myers 1974).

Arrhyton.— A genus of small (< 50 cm SVL), slender to stocky snakes endemic to the West Indies. At least one species (*A. taeniatum*) is at least semi-fossorial and appears to feed primarily on blind snakes (*Typhlops*). They are active foragers and some species include amphibian and reptile eggs in their diets. Members of the genus occur in Cuba, Puerto Rico, Jamaica, and the Virgin Islands. Although no species occurs on Hispaniola, the closely related (Maglio 1970) *Darlingtonia haetiana* does occur there. More than one genus may be involved here (Maglio 1970; Schwartz and Garrido 1981).

Chironius.— Only a single endemic species of this genus occurs in the West Indies, and it is restricted to St. Vincent. The five known specimens were collected prior to 1894 and it is now likely extinct. If, indeed, it is extinct, the cause of the extinction is unknown. It was probably at least semi-arboreal and therefore less susceptible to mongoose predation than more exclusively ground-dwelling snakes. There is no evidence that the species was ever common since the arrival of European man and perhaps its survival has always been tenuous (A. Schwartz pers. comm.). The genus is widespread on the neotropical mainland, where they are primarily frog predators (Dixon and Soini 1986; Duellman 1978). **Note added in proof:** *C. vincenti* has been rediscovered.

Clelia.— This genus is represented by *Clelia clelia* in the Lesser Antilles. This large (2.0 m) species is widespread on the neotropical mainland. It is catholic in diet, but has a reputation for ophiophagy (Duellman 1978; Lancini 1979).

Darlingtonia haetiana.— This monotypic genus is endemic to Hispaniola. *Darlingtonia haetiana* is a small (< 35 cm SVL) snake that preys almost exclusively on frogs of the leptodactylid genus *Eleutherodactylus*, including their egg clutches. It is related to *Arrhyton* (Maglio 1970).

Hypsirhynchus.— This is a monotypic genus endemic to Hispaniola. *Hypsirhynchus ferox* is a moderately sized snake (up to 75.0 cm SVL) with an attenuated head, vertical pupil, and relatively stout habitus. It is strictly saurophagous, with adults feeding primarily on *Ameiva chrysolaema*, a geographically widespread teiid that frequently occurs at very high densities. It is probably derived from an *Alsophis* (Cadle 1984; Maglio 1970).

Ialtris.— Three species are endemic to Hispaniola. Little is known about the ecology of this genus, but they may be stenophagous to euryphagous (Henderson 1984a; Henderson et al. 1988). They are moderate in size (ca. 0.8 m SVL). Affinities of the genus are uncertain (Schwartz and Rossman 1976), but close relationship to *Alsophis* has been suggested (Dunn 1932; Maglio 1970).

Liophis.— This is a diverse (about 40 described species), geographically widespread genus on the neotropical mainland. Five species occur in the West Indies, four of which are endemic to the area. They are generalized colubrids, generally small in total length (0.6 m), found in a variety of moist habitats, and will exploit a wide variety of invertebrates and vertebrates (Dixon 1980). Species of *Liophis* are the only West Indian colubrids known to prey on invertebrates. Affinities lie with other members of the tribe Xenodontini.

Mastigodryas.— Ten species range from Mexico to Argentina (Peters and Orejas-Miranda 1970), and one species (*M. bruesi*) is endemic to the West Indies. It is a genus of generalized colubrids, racer-like in habitus, moderate in size (90 cm SVL), and probably catholic in diet (Sieb 1984). The closest relative of *M. bruesi* is probably the South American *M. amarali* (Stuart 1941).

Nerodia.— A single species, *N. clarki*, of this largely North American natricine genus occurs on the north coast of Cuba where it is rare (Jaume 1965). It is largely aquatic and feeds on fishes and frogs. It attains a total length of about 80 cm.

Pseudoboa.— A genus (four described species) that is widespread in northern South America, one species, *P. neuwiedi*, reaches the Lesser Antilles. SVL is < 1 m. Diet is probably fairly catholic (frogs, lizards, mammals). *Pseudoboa* shares a common ancestor with *Clelia* (Jenner and Dowling 1985).

Tretanorhinus.— This genus has one representative on Cuba and the Cayman Islands (*T. variabilis*), two in Middle America and one in the Choco of Colombia (Jenner 1981). *T. variabilis* is strictly aquatic; attains a SVL of about 80 cm; and is a piscivore.

Uromacer.— The genus is endemic to Hispaniola and three species are recognized. With the possible exception of *Chironius vincenti*, species of *Uromacer* are the only arboreal colubrids in the West Indies. All species are relatively slender, with two of them being extremely so. The two most morphologically derived species (*U. frenatus* and *U. oxyrhynchus*), are strictly saurophagous. The more generalized species (*U. catesbyi*) preys on tree frogs and lizards (Henderson et al. 1987c). *Uromacer* may be related to *Alsophis* (Maglio 1970). SVL reaches 90 cm.

Family Viperidae

Bothrops.— The genus is widespread on the neotropical mainland and about 60 species are recognized. Two species are endemic to the West Indies. Total length approaches 2 m. These large pit-vipers are noctunal, relatively heavy-bodied, and prey primarily on rodents.

Appendix 2: Island snake faunas (exclusive of *Typhlops* and *Leptotyphlops*)

Bahama Islands.-- This large group of islands at the northern end of the West Indies harbors seven species of snakes. Three species of the boid genus *Epicrates* occur in the Bahamas, but no more than one species occurs on any island. Similarly, two species of *Tropidophis* occur in the island group, but never together. The widespread colubrid *Alsophis vudii* occurs on many islands with an *Epicrates* and/or a *Tropidophis*. The Hispaniolan colubrid *Antillophis parvifrons* is known from but a single record on Little Inagua Island.

Cuba.-- This is the largest island in the West Indies (114,524 km^2). It has a single species of *Epicrates* that attains a total length of at least 4.0 m and is the largest species of snake that is endemic to the West Indies. Cuba has the richest *Tropidophis* fauna with nine species, seven of which are endemic to the island. Several of the *Tropidophis* are morphologically specialized (e.g. *T. feicki* and *T. semicinctus*). Similarly, Cuba has the most species of *Arrhyton* (six), and these snakes show correlation between morphological and trophic specialization (e.g., *A. taeniatum*). Two colubrids that are fairly recent arrivals on Cuba occupy a habitat that is unique among West Indian snakes: *Nerodia clarki* is a natricine that has a fairly wide distribution in the southeastern United States, including the Florida Keys; it is an inhabitant of salt and brackish water, and its distribution in Cuba is restricted to the north coast. Likewise, *Tretanorhinus variabilis* is a xenodontine with its only known congeners occurring on the neotropical mainland from southern Mexico to extreme northern South America. It is aquatic in brackish and freshwater habitats. Two other colubrids on Cuba are *Alsophis cantherigerus* and *Antillophis andreae*. Both are very widespread on the main island and on many satellite islands.

Cayman Islands.-- The largest of these islands (Grand Cayman) harbors three snakes: a *Tropidophis*, a terrestrial colubrid (*Alsophis cantherigerus*) and an aquatic colubrid (*Tretanorhinus variabilis*). The two smaller islands have two of these snake species represented.

Jamaica.-- With an area of 10,991 km^2, Jamaica has a snake fauna of six species. This includes an endemic species of large *Epicrates*, a species of dwarf boa (*Tropidophis haetianus*) that also occurs on Cuba and Hispaniola, a species of racer (*Alsophis ater*) that may be extinct (but see Henderson and Sajdak 1986), and three species of *Arrhyton*.

Hispaniola.-- This is the second largest island in the West Indies (76, 484 km^2) and it has 15 snake species, 11 of which are colubrids. It is the only island with more than one species of the boid genus *Epicrates*: it has one large widespread species (*E. striatus*); one small, largely terrestrial species that occurs in xeric habitats (*E. fordi*); and one small, extremely slender species that is strictly arboreal, occurs in mesic habitats, and feeds predominantly or exclusively on *Anolis* lizards (*E. gracilis*). A single species of *Tropidophis* occurs on the island, and it is geographically widespread. The colubrid fauna is the largest and most interesting in the West Indies. It includes two species of racers (*Alsophis*) (the only island to harbor more than a single species of this widespread genus), but both species are rare on the main island and *A. melanichnus*, which has not been seen or collected for about 70 years, may be extinct. *Alsophis anomalus* has been observed on the main island in recent years, and is apparently not uncommon on two of the satellite islands (Ile de la Tortue and Isla Beata). *Antillophis parvifrons* is widespread on the main island and occurs on many satellite islands; its only congener occurs on Cuba. *Darlingtonia haetiana* is a small, secretive snake that occurs in upland situations on the Hispaniolan "south island." It feeds almost exclusively on *Eleutherodactylus* frogs, including their egg clutches, and has no close relatives on the island. A ground-dwelling lizard specialist, *Hypsirhynchus ferox* belongs to a monotypic genus. It is widespread throughout the island (and several of the larger satellite islands), but is nowhere very abundant. Three species of *Ialtris* occur on the island, but only one has a wide geographic distribution (*I. dorsalis*). There are also three members of the endemic tree snake genus *Uromacer* on the island. The most morphologically generalized of the three species (*U. catesbyi*) has the widest distribution, occurs on most of the satellite islands, and has the most generalized diet (frogs and lizards). The other two species (*U. frenatus* and *U. oxyrhynchus*) have largely allopatric distributions (and where they do occur syntopically, both species are rare), are morphologically and trophically specialized, frequently occur with *U. catesbyi* at high relative densities, and never co-occur on the same satellite island (although each may occur there with *U. catesbyi*).

Puerto Rico.-- Despite its large size (8,676 km^2), Puerto Rico has only three snake species: a large endemic species of *Epicrates*, a moderately sized colubrid (*Alsophis portoricensis*) and a small colubrid (*Arrhyton exiguum*).

Virgin Islands.-- This island group harbors a moderate sized boa (*Epicrates monesis*) and two racers with allopatric distributions: the moderately sized *Alsophis portoricensis* occurs on many islands, and the large (2 m) species *A. sanctaecrucis* has been extirpated on St. Croix, but may still survive on mongoose-free Green Cay (although unlikely).

Anguila, St. Martin, St. Barthelemy.-- These small islands range in size from 22-88 km. Only the racer *Alsophis rijersmai* occurs on these islands, but it has been extirpated on St. Martin (apparently due to mongoose introduction).

Saba, Nevis, St. Eustatius, St. Christopher.-- These Lesser Antillean islands range in area from 13-169 km^2. Only the racer *Alsophis rufiventris* occurs, or occurred, on them; mongoose introduction has eliminated the species from Nevis and St. Christopher.

Antigua, Montserrat.-- Ranging in area from 101-280 km^2, these two islands harbor a single species of racer (*Alsophis antillensis*). The endemic subspecies that occurred on Antigua is now extinct (if the population that occurs on nearby Great Bird Island is taxonomically distinct (Henderson in prep.), due almost certainly to the introduction of the mongoose.

Guadeloupe, Marie-Galante.-- These islands harbored two species of colubrid snakes: *Alsophis antillensis* and *Liophis juliae*. The former has been eliminated from both islands, but the *Liophis* is still known to occur at least on the larger island (1,510 km^2) of Guadeloupe.

Dominica.-- With an area of 751 km^2, Dominica has four snake species. Two are West Indian endemics: a moderate sized colubrid (*Alsophis antillensis*), and a small colubrid (*Liophis juliae*). Also occurring on the island are two species that are widespread on the neotropical mainland: a large boid (*Boa constrictor*) and a large colubrid (*Clelia clelia*).

Martinique.-- With an area of 1,106 km^2, this island has two endemic snake species: a small ground-dwelling colubrid (*Liophis cursor*) (surviving only on an offshore islet), and a large viperid (*Bothrops lanceolata*).

St. Lucia.-- This island has an area of 616 km^2, and four snake species. The one endemic colubrid is *Liophis ornatus*, now known only from offshore islets. The boa constrictor also occurs there, as does *Clelia clelia*. The other snake is the large endemic viperid *Bothrops caribbaea*.

St. Vincent.-- This Lesser Antillean island has an arboreal boid (*Corallus enydris*) that is widespread on the neotropical mainland, an endemic colubrid (*Chironius vincenti*) (the genus is widespread on the neotropical mainland), and a racer (*Mastigodryas bruesi*) that belongs to a genus that is also widespread on the neotropical mainland.

Barbados.-- Only a single endemic colubrid snake species (*Liophis perfuscus*) occurs on this island (431 km^2).

Grenada.-- This island of 344 km^2 contains five snake species, only one of which is endemic to the West Indies (*Mastigodryas bruesi*). The rest of the snake fauna (*Corallus enydris, Clelia clelia, Liophis melanotus,* and *Pseudoboa neuwiedi*) occur on the South

American mainland and are members of genera that are widespread throughout much of the neotropics.

Appendix 3: widespread prey genera/groups

(See Fig. 5 for an illustration depicting the distribution of prey species in a generalized habitat on Hispaniola.)

Frogs

Eleutherodactylus.-- The most speciose (400+ species) vertebrate genus in the world, about 120 species of these leptodactylid frogs are found in a wide variety of habitats from sea level to 2400+ m in the West Indies. Undoubtedly, additional species await discovery, especially on Cuba and Hispaniola. Although they are exploited by a number of snake species in the West Indies, it is, at first glance, surprising that they are not eaten more frequently by more snake species. Only the small Greater Antillean *Arrhyton* and *Darlingtonia* and to a lesser extent *Antillophis*, prey regularly on *Eleutherodactylus*. There is indication that, in the Lesser Antilles, *Eleutherodactylus* may be an important food source for the small ground-dwelling species of *Liophis*. The Hispaniolan colubrid *Darlingtonia haetiana* preys almost exclusively on *Eleutherodactylus* at all stages of the frogs' life cycle (eggs to adults).

Species of *Eleutherodactylus* reach their greatest species diverstiy at elevations where snake species diversity has already declined sharply (Fig. 4). In addition, *Anolis* lizards occupy most of the spatial niches occupied by *Eleuthrodactylus*, and the anoles are diurnal (i.e., active at the same time as most of the colubrid snakes) and frequently occur at very high densities. These factors may explain why anoles, rather than Eleuthrodactylus, comprise a much higher percentage (by frequency and volume) of the diet of most West Indian snakes, even though the two genera have about the same number of species in the area.

Osteopilus.-- Three species of this hylid frog genus occur in the West Indies (the Bahama Islands, Cuba, Cayman Islands, Jamaica, Puerto Rico, Hispaniola, Virgin Islands). The two most important prey species in terms of snake diets are *O. dominicensis* and *O. septentrionalis*. Both are geographically widespread species that occur in a wide variety of habitats. Although the frogs are nocturnal, active foraging snakes (e.g. *Alsophis cantherigerus, Uromacer catesbyi*) seek them out in their daytime retreats. Occasionally, large numbers (up to 26) of recently metamorphosed young are eaten by a snake in a very short period of time.

Lizards

Sphaerodactylus.-- These diminutive (< 1 cm^3) geckos are widespread in the West Indies and frequently occur at high population densities (Cheng 1983). Seemingly, they are not important in the diets of adults of any snake species, and only Hispaniolan *Antillophis parvifrons* preys on them with any regularity (Henderson et al. 1987a). They may, however, be important in the diets of newborn snakes (P.J. Tolson pers. comm.).

Anolis.-- Anoles are small to moderately sized (40-191 mm SVL) lizards found throughout the West Indies (about 125 species have been described to date) and much of the neotropical mainland as well. They are undoubtedly the most ubiquitous and conspicuous vertebrates in the West Indies, and they are eaten by snakes with much greater frequency than any other prey genus that occurs in the area. They are durinal, occur in most habitats from ground level to tree crowns, and often at very high densities. Anoles are the most frequently consumed prey items for most species of West Indian snakes, and it would be surprising if any colubrid snake species under discussion here does not exploit anoles as food to some extent. Guyer and Savage (1986) have recently presented arguments in favor of dividing the genus into five genera.

Leiocephalus.-- Nineteen species of this iguanid lizard genus occur in the Bahama Islands, Cuba, and Hispaniola. Another species, *L. herminieri*, occurred on Martinique, but is now extinct. Species of *Leiocephalus* are primarily ground-dwelling lizards, frequently found in open, rocky situations. Some species, however, do climb trees and live in tree hollows and root systems. Commonly referred to as curly-tailed lizards, they are often very conspicuous members of the West Indian lizard fauna. Frequently occurring at high population densities (Henderson pers. observ.; Schoener et al. 1982), it is surprising that they are not exploited for food by snakes more often.

Ameiva.-- About 18 species of these teiid lizards occur in the West Indies; the genus is well represented on the neotropical mainland as well. They are diurnal, ground-dwelling, active foragers that are conspicuous in a variety of habitats on most islands. Like anoles, they may occur at very high densities (Henderson pers. observ.). Some species of *Ameiva* attain large size (160 mm SVL), and may be too powerful for most West Indian snake species to attack successfully and consume.

Mammals

Rodents.-- Although bats may be of localized geographic importance to some species of West Indian boid and colubrid snakes (Hardy 1957a, b; Rodriguez and Reagan 1984), rodents are the most important mammals in the diets of West Indian snakes. Introduced Mus and Rattus are taken by a wide variety of West Indian snakes, but especially by the larger boids (Boa and *Epicrates*) and viperids (*Bothrops*).

Historically, other species of rodents and insectivores were probably exploited by snakes in the West Indies before the arrival of Europeans and the introduction of *Mus* and *Rattus*. Other species of mammals are known, or believed, to be at least occasionally exploited for food by snakes in the West Indies. This includes the rodent genera *Capromys* (Cuba), *Plagiodontia* (Hispaniola), *Dasyprocta* (Lesser Antilles; Lazell 1964); the viverrid carnivore *Herpestes auropunctatus* (Lesser Antilles; Lazell 1964); and the marsupial *Didelphis* (Lesser Antilles; Lazell 1964).

Note added in proof: Recent work (B.I. Crother, Ph.D. Diss., Univ. of Miami, 1989) on West Indian xenodontine phylogeny and biogeography suggests a phylogenetic history different from that expressed on p. 485, and vicariance is the best explanation for their distribution (except for Jamaica; Crother, in prep.).

THE SHINY COWBIRD *MOLOTHRUS BONARIENSIS* IN THE WEST INDIAN REGION - BIOGEOGRAPHICAL AND ECOLOGICAL IMPLICATIONS

Alexander Cruz, James W. Wiley,
Tammie K. Nakamura, and William Post[1]

Abstract

The Shiny Cowbird (*Molothrus bonariensis*), an avian brood parasite, is endemic to South America, Trinidad, and Tobago, but during the last 100 years the species has spread through the West Indies. The spread of the cowbird through most of the Caribbean has been documented, and the expansion appears to have originated from two separate areas: 1) In the southern Lesser Antilles from where it has moved northward as far as Antigua, and 2) in the Greater Antilles (Puerto Rican Bank) where it has moved westward as far as Cuba. The arrival dates in the Lesser Antilles suggest natural "island hopping" and derivation from Trinidad and Tobago, although introductions cannot be ruled out. The cowbird's presence in the Greater Antilles may represent natural expansions, introductions, or both. The species will likely spread to western Caribbean areas not yet colonized and to the North American mainland.

Successful colonization by the Shiny Cowbird depends on the availability of suitable habitats and host species. The Shiny Cowbird occurs in a wide variety of habitat, but it prefers open areas. In pre-Columbian times, most islands were heavily forested and therefore not suitable for cowbirds. However, with the destruction of forests in post-Columbian times, the conditions necessary for the spread of the cowbird to these islands were created. Man's alteration of natural habitat continues on most West Indian islands. This trend facilitates the continued spread of the Shiny Cowbird through the region.

Many factors are important in the Shiny Cowbird's selection of hosts. A species may be unsuitable because it breeds during different seasons than the cowbird, nests in habitats which are not used by the parasite, or feed the young diets which are physiologically incompatible with the needs of the parasite young. Species which are potentially suitable as hosts meet the following criteria: accessibility of the nest to the egg-laying female cowbird, acceptance of cowbird eggs by the host, and high probability of cowbird

---

[1] Dr. Cruz is Professor, Environmental, Population, and Organismic Biology Department, B-334, University of Colorado, Boulder, CO. 80309. Dr. Wiley is Endangered Species Biologist, United States Fish and Wildlife Service, Patuxent Wildlife Research Center, Laurel, MD. 20708. Tammie K. Nakamura is a graduate student in the Zoology Department, Colorado State University, Fort Collins, CO. 80523. Dr. Post is Curator of Ornithology, Charleston Museum, Charleston, SC. 29403.

fledging success. The Shiny Cowbird has been able to persist in the West Indies because of habitat modification and the suitability of available hosts; i.e., the egg-laying activity of the cowbird coincides with that of hosts, which have shown acceptance of cowbird eggs and chicks.

Given the above criteria for host suitability, we found that a number of suitable hosts are present in the region, including flycatchers (*Myiarchus*), Black-whiskered Vireos (*Vireo altiloquus*), Yellow Warblers (*Dendroica petechia*), and blackbirds (*Icterus and Agelaius*). The availability of these hosts in the region, and the ability of the cowbird to find these resources has enabled the cowbird to successfully colonize many islands.

Before the cowbird's invasion of the region, its potential hosts had no history of exposure to brood parasitism, and are therefore unlikely to have evolved defenses against it. Thus they may be vulnerable to reproductive failure as a result of cowbird parasitism. Our data suggest that nest parasitism is severe enough to threaten some species, notably Yellow-shouldered Blackbirds (*Agelaius xanthomus*). Our data also suggest that some other species (Puerto Rican Flycatchers *Myiarchus antillarum*, Black-whiskered Vireos, Black-cowled Orioles *Icterus dominicensis*, and Troupials *I. icterus*) should be carefully monitored in the future. As forest fragmentation increases in the West Indies, new hosts, particularly forest species, will be parasitized. Forest fragmentation creates small patches of forests surrounded by open habitat and increases the portion of forest habitat available to cowbirds.

Introduction

When the geographic range of a species expands strikingly over a short period of time, the biologist is presented with an unusual opportunity to examine the adaptive processes of the species in the new environment. Such changes in status may bring new pairs of species into contact and provide insight into the role of organisms and the environment in determining how potentially available habitats will be used by a complement of species. When the species involved in the range expansion possesses some unusual adaptations not present in the species found in the new region, such as brood parasitism, the biologist is presented with an even more unusual opportunity to observe community interactions. Brood parasitism involves not only the loss of nesting and parental behavior but also the development of a complex of new patterns of behavior adapting the parasite to the host.

Recent changes in the range of the Shiny Cowbird (*Molothrus bonariensis*) have brought it into contact with bird communities inexperienced with brood parasitism (Fig. 1). The Shiny Cowbird was originally confined to South America, Trinidad, and Tobago, but during the last 100 years the species spread through the West Indies (Bond 1966, 1971, 1973; Ricklefs and Cox 1972; Post and Wiley 1977a; Cruz et al. 1985), and it continues to increase its range and numbers.

The subspecies involved in the Caribbean range expansion is *M. b. minimus*, which was originally confined to northern Brazil, the Guianas, eastern Venezuela, Trinidad, and Tobago. Other subspecies have been expanding in other parts of the range, notably Argentina and Chile (Johnson 1967; Fraga 1985). The Shiny Cowbird's invasion of the Caribbean region is interesting in terms of range extensions in general and of cowbird biology in particular, because: 1) the invasion apparently began within the last 100 years and is continuing; 2) some potential hosts of this parasitic species are patchily distributed in the West Indies and breed in small populations that may be vulnerable to extirpation by the cowbird; and 3) the species is likely to spread from Cuba to Florida and Yucatán, where it would not only come into contact with new host species, but other parasitic species. The Brown-headed Cowbird (*M. ater*) occurs in Florida and the Bronzed Cowbird (*M. aeneus*) occurs in the Yucatán (Friedmann 1929; Friedmann et al. 1977), and from Louisiana west, along the Gulf of Mexico (American Ornithologists' Union 1983).

The purpose of this work is: 1) to review the spread of Shiny Cowbirds in the region, 2) to examine factors that have allowed them to be good colonists, 3) to investigate the biogeography of host suitability, 4) to determine the impact of the cowbird on the native avifauna, and 5) to predict future effects of the Shiny Cowbird's expansion.

Study areas

We conducted field work in Puerto Rico between 1973 and 1983 in mangrove and adjacent coastal habitats at Roosevelt Roads Naval Station, eastern Puerto Rico, and Boquerón Commonwealth Forest, southwestern Puerto Rico. Detailed information on the Puerto Rican study areas is provided by Post and Wiley (1977b), Post (1981), and Wiley (1982, 1985). Investigations in St. Lucia were carried out from 1983 through 1985. In St. Lucia, our studies were concentrated in the arid southern part of the island. This area is a composite of many habitats and vegetation types, most of which have been influenced by man. Natural vegetation included mangrove forest, dry scrub, and dry forest woodlands. Recent clearing of forests has created a mosaic of habitats of pastures and crops, interspersed with acacia and logwood thickets. In the Dominican Republic, we gathered information in the southern coastal plain between 1974 and 1978, and 1981 to 1984. Much of the region is under cultivation (e.g., sugarcane, maize, rice, banana) or grazing.

Methods

We gathered information on the breeding biology of the cowbird and its potential and actual host species, the frequency of use of host species, and the effects of brood parasitism on host breeding success. We located nests by regularly searching the study areas, and causes of nest failures were determined whenever possible. We marked each nest with a coded tag (inconspicuously placed) and plotted it on field maps. At each visit to the study area (usually 2-4 day intervals), we inspected nests to determine number of host and, if present, parasite eggs and chicks. We defined a nest as active when the resident laid at least one egg or if the host incubated cowbird eggs, regardless of whether host eggs were present. We defined egg success as the proportion of eggs hatched/eggs laid, and nest success as the proportion of nests fledging at least one young.

To determine which species reject cowbird eggs, we followed Rothstein's (1971, 1975) technique of artificial parasitism, using both real and artificial cowbird eggs. We obtained real eggs from nests in other parts of the study areas where experiments were not being performed, or from eggs laid by cowbirds in traps. We constructed artificial cowbirds eggs of wood and painted then to simulate the cowbird eggs. Size and weight of the artificial eggs closely approximated real eggs. We experimentally parasitized nests within two hours after dawn to simulate the natural timing of cowbird parasitism (Hoy and Ottow 1964; pers. observ.). After nests were parasitized they were monitored to determine host response. We considered hosts to be rejecters of cowbird eggs if they ejected or damaged the experimental eggs, deserted or rebuilt the nest.

Documentation of the spread into the region

The Shiny Cowbird was first recorded north of Trinidad and Tobago on Vieques Island (Newton 1860), 10 km east of Puerto Rico (Fig. 1). This single bird probably was an escaped captive brought to the island (Bond 1983). Following the collection of the Vieques specimen, no further birds were observed in the Caribbean region until 1891, when the species was recorded on Grenada, Lesser Antilles (Bond 1963, 1983).

The cowbird's spread through most of the Caribbean has been well documented, and this range expansion appears to have originated from two separate areas (Bond 1966; Post and Wiley 1977a; Cruz et al. 1985): In the **Lesser Antilles**, in the region around Grenada, the Grenadines, and Barbados, from where the species has spread northward as far as Antigua (Fig. 1). The arrival dates in the Lesser Antilles suggest natural "island hopping" and derivation from Trinidad and Tobago, although introductions cannot be ruled out. The Lesser Antillean model is supported by the relatively continuous distribution of recent immigrants in the southern tier of the Lesser Antilles. In the **Greater Antilles**, *M. bonariensis* has moved westward as far as Cuba from population centers in the Puerto Rican Bank (St. John and eastern Puerto Rico). The presence of the cowbird in the Greater Antilles may represent natural expansions, introductions, or both. A male collected in 1934 on St. Croix, 120 km east of Puerto Rico, is said to have been one of a pair transported from Barbados (Bond 1956). No additional cowbirds were observed on St. Croix until 1985 (Norton 1986), when a male was noted near mid-island. Over and above the natural dispersal ability and possibility of introduction of the cowbird, much of the West Indian region lies within a region of frequent hurricanes that may have aided dispersal (Ricklefs and Cox 1972).

In the Puerto Rican Bank area, a flock of 10-12 cowbirds appeared on St. John, Virgin Islands, 88 km east of Puerto Rico, in 1955 (Robertson 1962), and in the same year, Grayce (1957) saw a flock of 150-175 cowbirds at Cabezas de San Juan in extreme northeastern Puerto Rico. The species quickly spread through the island (Fig. 1). The large number of cowbirds that were first seen in Puerto Rico, at widely separated points and over a relatively short period, strongly indicate that the cowbird arrived on that island prior to 1955 (Post and Wiley 1977a). At present, the cowbird is common in coastal habitats and in disturbed montane areas.

The cowbird reached Mona Island and Hispaniola in 1971 and 1972, respectively (Bond 1973; Raffaele 1973; Arendt and Vargas 1984; Cruz et al. 1985). In Hispaniola, the Shiny Cowbird was first observed at Hato Nuevo in 1972, 5 km west of Santo Domingo (Dod in Long 1981). Dod reported that cowbirds were well established by 1972, suggesting that they probably had been there for a considerable period. From 1976 to 1978, Arendt and Vargas (1984) observed cowbirds in 13 widely-scattered localities in eastern and central Dominican Republic, including Saona Island (off southeastern Hispaniola). Additionally, they banded 196 cowbirds and documented the spread of the species in Hispaniola by recovering 14 individuals (7.1%) at distances up to 130 km (mean = 25.8 km) from the banding sites. By the 1980's, the Shiny Cowbird had become established in many coastal sites of the Dominican Republic (Cruz et al. 1985).

The cowbird was first recorded from Cuba in 1982, where a bird was captured at Central Progreso, Cardenas (Garrido 1984). In 1983, several more birds were trapped at this locale. The cowbird's occurrence in western Cuba, approximately 900 km west of Hispaniola suggests that the cowbird is probably established in eastern Cuba. More recent reports (Guerra and Alayon 1987) indicate that the cowbird is present in widely separated localities in Cuba. It has been recorded at Holguin (eastern Cuba), Tapaste (north-central Cuba), and San Antonio de los Baños (western Cuba) (Fig. 1).

With the greater diversity of habitats and variety of avian species in Cuba, we expect that the Shiny Cowbird will quickly expand its range through that country. A Shiny Cowbird was found at Lower Matecumbe Key, Florida in 1985 (Kale 1985). In 1986, cowbirds were again found in the keys (Islamorada; Paul 1986). This species will likely spread into the Bahamas, Jamaica, the Yucatán Peninsula, and through Florida via Cuba.

What makes the shiny cowbird a good colonist?

We may now ask what traits allowed the cowbird to colonize the Caribbean region, and distinguish the following possibilities. A good colonist may have intrinsic biological characteristics that allow it to reach islands and suit it to the physical environment one finds there (Simberloff 1981). The ability to persist, following colonization, may derive from the ability to tolerate either the physical environment of the island or find suitable resources (e.g., host species). Traits that make Shiny Cowbirds good colonists are listed below.

Flexibility in habitat selection.-- Graves and Gotelli (1983) note that widespread bird species are disproportionately common on some Caribbean islands. Why should this be so? First, species that are widespread on a mainland may have a higher probability of colonizing an island. Second, species with large geographical ranges might generally inhabit more different habitats than species with restricted ranges and may be more likely to find a suitable habitat on a given island (Moulton and Pimm 1986). Mayr (1965), in his discussion of birds that are good colonizers, similarly emphasized the importance of flexibility in habitat selection and use.

The Shiny Cowbirds appears to have these attributes - it is widespread in South America and it is found in a diversity of habitats. Shiny Cowbirds prefer open country and edge habitats from coastal areas to high elevations depending upon the extent of habitat alteration and animal husbandry practices in such areas (Friedmann 1929; Mason 1986; pers. obs.). In the West Indies, we have seen cowbirds in habitats ranging from mangrove and thorn scrub to the edges of montane forests.

Catholic dietary habits.-- The ability to subsist on a wide variety of foods is often a prerequisite of being successful colonists. Herbivores that feed on a few species with relatively limited distributions will normally be disadvantaged relative to polyphagous species (Ehrlich 1986). A substantial proportion of the birds that have been successful colonizers have been omnivorous, often ones in which seeds make up a substantial portion of the diet (Mayr 1965, Long 1981). The Shiny Cowbird is not an exception to this generalization. It it an omnivore with a diversity in its feeding habits (Long 1981; Wiley 1982; pers. obs.).

Host generalist.-- The Shiny Cowbird is an extreme host generalist, its eggs have been found in the nests of over 201 species (Friedmann and Kiff 1985, Mason 1986). The ability to successfully parasitize different host species has facilitated its spread throughout the West Indies, allowing opportunistic exploitation of species encountered by the parasite in the new regions (Table 3). The abundance and expansion of both the Brown-headed Cowbird (Mayfield 1965) and the Shiny Cowbird (Post and Wiley 1977a) is evidence that a generalist strategy can be successful.

Sociality and flocking behavior.-- The Shiny Cowbird is a gregarious species. Mayr (1965) contends that birds that travel in small flocks would be favored because, once they arrived, reproduction could succeed as a result of the greater number of propagules. He notes that groups of birds that show such a tendency, like the white-eyes (*Zosterops*), starlings (*Sturnus*), sparrows (*Passer*), and certain thrushes (*Turdus*), are among the most successful colonizers of isolated islands, whereas solitary birds like woodpeckers are poor colonizers.

Fecundity.-- Data are currently lacking on the fecundity of the Shiny Cowbird, but it is reasonable to assume that it is similar to the closely related Brown-headed Cowbird. Scott and Ankney (1980, 1983) found that a female Brown-headed Cowbird lays about 40 eggs a year and, because she lives for about 2 breeding seasons, her total fecundity is approximately 80 eggs. Scott and Ankney concluded that the Brown-headed Cowbird has a potential annual production greatly in excess of that known for any other wild passerine bird. Only 1 or 2 days elapse between laying separate consecutive "clutches" or sequences of eggs, whereas nesting passerines require at least 5 days between the loss of a nest and the first egg in the replacement set (Delius 1965; Dixon 1978; Friedmann and Kiff 1985).

In summary, the Shiny Cowbird has several attributes that make a good colonist, including use of a wide array of habitat types and host species, catholic dietary habits, and, probably, high fecundity.

Factors that have allowed the cowbird to persist in the region

The probability of successful colonization (good persistence) of each island depends on the availability of suitable habitat and resources (e.g., host species). These aspects will be discussed next.

Fragmentation of West Indian habitats

In pre-Columbian times, most West Indian islands were densely forested and therefore not suitable for Shiny Cowbirds (Beard 1949; Lugo et al. 1981). However, with the destruction of forests by Afro-Europeans, the conditions necessary for the expansion of the Shiny Cowbird into the West Indies were created (Table 1). The post-Columbian human activity in the Caribbean Islands and its impact on forests can be divided into the following categories (Lugo et al. 1981): 1) The era of early settlement when farming activities were not intense, and forests were not affected significantly by human activities (1493-1630); 2) the era of extensive sugar cane and other monocultures (1630-1880's) - most coastal forests cleared; 3) the era of economic collapse whose beginning is coincident with the abolition of slavery (1880's-1940's) - forests return in abandoned degraded lands, but some montane areas converted to agriculture; 4) the era of increasing energy use, population growth, and urbanization (1940's-present) -continued fragmentation and disturbance of natural habitats, especially on most populated islands throughout the Caribbean.

If Shiny Cowbirds were capable of expanding into the Caribbean region within the last 100 years, then it seems reasonable to postulate that such movements may have been made in the past. But it is only since the turn of the century that the species has expanded extensively in the Caribbean region. Possibly the Shiny Cowbird has been in the West Indies longer than current evidence suggests, although this seems unlikely in view of its conspicuousness and observed rapid expansion through the region. Earlier waves of the cowbirds may have reached the West Indies, but did not become established. The first introduction of the House Sparrow (*Passer domesticus*) in North America, for example, involved the release of eight pairs in New York City in 1851, and it failed. The introduction of about 50 individuals in the same general area in 1853 succeeded (Long 1981). What factor or factors may have permitted the recent expansion of the cowbird into the Caribbean region cannot be determined at this time. Friedmann and Kiff (1985) suggest that the recent expansion of the Shiny Cowbird in the West Indies and in many areas of South America may be the result of increased populations of cowbirds in their original home ranges, although no evidence is available on this point. Fonaroff (1974) and ffrench (1980) have discussed changes in the environment of Trinidad and Tobago, where the spread of the cowbird into the West Indian region may have originated. Vegetation maps indicate that northern Trinidad was heavily forested as late as 1797, but by the 1900's it had been extensively modified (Fonaroff 1974). Major changes have been the reduction of forest and large-scale habitat alterations associated with agriculture, animal husbandry, and urbanization. The net effect of these changes probably has been to increase population levels of the cowbird. At present, the Shiny Cowbird is a common resident on Trinidad and Tobago, and is frequently found in large flocks, especially at roosts in mangroves where the numbers may reach thousands (ffrench 1980, pers. obs.).

Host suitability

Many factors are important in the Shiny Cowbird's selection of hosts from the species present in a given community. A species may be unsuitable because it breeds during different seasons than the cowbird, nests in habitats that are not used by the parasite, or feed the young diets which are physiologically incompatible with the needs of the parasite young (e.g., Wiley 1982). Of the species which are potentially capable of rearing parasite young, additional measures of suitability are important. These measures are generally partitioned into components of temporal and spatial availability of hosts: 1) accessibility of the egg-laying female cowbird to the nest (Rothstein 1975; Robertson and Norman 1977; Wiley 1982; Cruz et al. 1985), 2) cowbird egg acceptance (Rothstein 1975; Friedmann et al. 1977), and 3) high likelihood of cowbird fledging success (op. cit., Manolis 1982; Fraga 1985; Mason 1986). In addition to suitable feeding habitats, suitable hosts are thus important in determining whether the Shiny Cowbird is able to persist in a given region. We will next examine some of the factors involved in host selection by the Shiny Cowbird in the West Indian region.

Breeding chronology.-- Shiny Cowbird breeding activity must approximate that period when hosts are nesting. Although a few species (e.g., Greater Antillean Grackles *Quiscalus niger*) breed throughout the year, nearly all passerine species confine their reproductive activity to the spring and summer. In Puerto Rico, for example, the main breeding season for most species begins in March, which corresponds with the onset of increased precipitation, and continues to September. Most egg-laying activity of the Shiny Cowbird coincides with this period and few parasitized nests were found beyond the main breeding season (Wiley 1982, 1985). Parasite-host breeding overlap may be mediated by a shared environmental response (onset of the rainy season), with the sight of the nestbuilding host possibly bringing about the final stages of cowbird gonadal activity and ovulation (Payne 1977; Wiley 1982).

Accessibility of host nests.-- We found that regularly-parasitized species, when experimentally parasitized, were characteristically accepters, but those species for which a low proportion of nests were parasitized were rejecters (Table 2). In contrast, each of the species that exhibited rejection behavior have experienced low or no parasitism (Table 2) or, parasitism was not detected because of egg rejection. It is possible that a low level of intraspecific nest parasitism has always existed in species exhibiting rejection behavior and that selection may have favored those individuals that reacted to a new or different egg added to their clutches.

In addition to their rejection behavior, a number of other characteristics of these birds reduces the chances of their nests being parasitized. Birds can sometimes protect their nests from being parasitized by driving off the Shiny Cowbird. This defense requires that the nesting individual be present in the nest area at the time the parasite attempts to lay its eggs, and that it responds in an aggressive manner to the presence of the cowbird. For example, some species (Gray Kingbirds *Tyrannus dominicensis*, Pearly-eyed Thrashers *Margarops fuscatus*, Northern Mockingbirds *Mimus polyglottos,* and Antillean Grackles) are aggressive at their nests or have high attendance rates (Wiley 1982). These species were not parasitized or exhibited a low incidence of parasitism (Tables 2 and 3).

Nestling diet.-- Nestling cowbirds require a diet primarily of animal protein. Since most passerines feed their young arthropods (Hamilton and Orians 1965; Skutch 1976; pers. obs.), host selection is little restricted by diet. Some birds do feed their young exclusively on seeds (e.g. *Tiaris* finches), which makes these species unacceptable as hosts (Wiley 1982). Of the 17 species we recorded as being parasitized by the Shiny Cowbird (Table 3), the seed-eating Bronze Mannikin (*Lonchura cucullata*) and the mainly frugivorous Palm Chat (*Dulus dominicensis*) had the narrowest feeding overlap with the Shiny Cowbird. Unfortunately, the cowbird egg did not hatch at the single parasitized Bronze Mannikin nest so we could not not evaluate development of the seed-fed cowbird. Several other seed-eating species (Nutmeg Mannikin *L. punctulata*, Black-faced Grassquit *Tiaris bicolor*, and Yellow-faced Grassquit *T. olivacea*) were not parasitized. None of the several columbids was parasitized on the study areas. Nor were any of the primarily nectarivorous (e.g., Bananaquit *Coereba flaveola*) or frugivorous (Stripe-headed Tanager *Spindalis zena*) species parasitized. The Palm Chat, although primarily frugivorous, feeds the young insects. There was broad feeding overlap between the cowbirds and the remaining 15 species in the study areas (Wiley 1982; pers. obs.). This evidence suggests that cowbirds may choose host species partly on the basis of food habits compatibility.

In summary, the Shiny Cowbird has been able to persist in the West Indies because of continued "improvement" of certain habitats in the region, and the suitability of available hosts; i.e., the cowbird has been able to synchronize its breeding with that of its major hosts, which have shown acceptance of cowbird eggs and chicks.

The use of hosts in the West Indian region

Biogeography of host suitability

Given the above criteria for host suitability, we find that a number of suitable hosts are present in the West Indian region, in particular *Myiarchus* flycatchers, Black-whiskered Vireos (*Vireo altiloquus*), Yellow Warblers (*Dendroica petechia*), and icterines (*Icterus* and *Agelaius*). The wide distribution of these hosts in the region and the ability of the Shiny Cowbird to parasitize them (Table 3) has enabled the cowbird to successfully colonize many islands. Below, we discuss host use by cowbirds in our study areas.

St. Lucia.-- In our study areas, four species were parasitized by the Shiny Cowbird (Table 3); of these species, the Black-whiskered Vireo and the Yellow Warbler were the most heavily used with a parasitism rate of 88% and 47%, respectively. These species occur not only on St. Lucia, but are among the most widely distributed passerines in the region, and occur from the Lesser Antilles to Florida. Hence, the Shiny Cowbird will find these species available on many West Indian Islands. The Caribbean Elaenia (*Elaenia martinica*) was recorded as an infrequent host in St. Lucia, and it has not been recorded as a host in Puerto Rico, where it is a common resident in coastal areas. The Carib Grackle (*Q. lugubris*) was also an infrequent host. This species has been recorded as a host in Trinidad (ffrench 1980) and was the "chief, if not sole, host" in northeastern Venezuela (Friedmann 1963). The St. Lucia Oriole (*I. laudabilis*), an island endemic, was observed feeding a cowbird fledgling.

Puerto Rico.-- Thirteen species were parasitized by the Shiny Cowbird in coastal areas of Puerto Rico (Table 3). Some species were parasitized only occasionally (≤ 17 % of nests examined), and others were regularly parasitized (31-100%) of nests. As in St. Lucia, Yellow Warblers and Black-whiskered Vireos were heavily parasitized. The Puerto Rican Flycatcher (*Myiarchus antillarum*), a member of a superspecies widely distributed in the Caribbean region, was also parasitized. The data also show the importance of blackbirds (icterinae) as host species. All Black-cowled Oriole (*Icterus dominicensis*) and Troupial (*I. icterus*) nests were parasitized and 91% of Yellow-shouldered Blackbird nests were parasitized.

Hispaniola.-- Five species were parasitized by the Shiny Cowbird in the Dominican Republic (Table 4). Since we began our studies, the mean incidence of parasitism for these hosts has increased by an average of 46 percent, ranging from 16 percent for the

Village Weaver (*Ploceus cucullatus*) to 83 percent for the Yellow Warbler (Table 4). In addition to finding eggs and chicks in these hosts nests, we have seen the Palm Chat, Black-cowled Oriole, Black-whiskered Vireo, and Village Weaver feeding cowbird fledglings.

Effects of cowbird parasitism on hosts in the West Indies

Before the cowbird's invasion of the region, potential hosts had no prior experience with brood parasites. Consequently, island bird populations are unlikely to have evolved defensive strategies and thus may be vulnerable to reproductive failure as a result of cowbird parasitism.

Cowbird parasitism reduced the nesting success of Puerto Rican Flycatchers, Black-whiskered Vireos, and Yellow Warblers (Table 5). With the exception of the Yellow Warbler, our samples sizes for unparasitized nests of the other species are still too small to make any definite conclusions about them. Number of host chicks fledged were no different between parasitized and non-parasitized Yellow-shouldered Blackbird and Antillean Grackle nests (Table 5). An earlier study (Post and Wiley 1977b) showed that non-parasitized blackbird nests had higher nest success than parasitized nests.

Although it has been suggested that the cowbird's generalist strategy enables host populations to maintain parasitic loads more easily (Payne 1977), some cowbird populations have had extreme depressive effects on the reproductive success of certain host species (e.g., Brown-headed Cowbird on Kirtland's Warbler *Dendroica kirtlandii*, Mayfield 1977; Shiny Cowbird on the Yellow-shouldered Blackbird, Post and Wiley 1976, 1977a, 1977b, Post 1981; Wiley 1985). Cowbird parasitism has affected the Yellow-shouldered Blackbird more than any other Puerto Rican species. Once abundant and widespread, the endemic Yellow-shouldered Blackbird has declined since at least 1972 (Post and Wiley 1976). In 1975, Post and Wiley (1977a) estimated that about 250 pairs of blackbirds were nesting in the Boqueron Forest area, but only 30 nesting pairs were located in 1985 (Silander, pers. comm.). Many factors have contributed to the blackbird's decline, including disease, loss of feeding and nesting habitats, and introduced mammals (e.g., *Rattus rattus* and *R. norvegicus*). However, an important cause of the blackbird's decline has been extensive parasitism of nests by the Shiny cowbird. From 1975 to 1983, 91% (241/266) of nests examined were parasitized. Cowbird parasitism results in reduced productivity at parasitized nests, mainly puncturing and breaking of host eggs by female cowbirds.

Future trends

Once the Shiny Cowbird has established itself in a given area, its persistence and spread can depend on many factors, but primarily include habitat and host resource availability. On the basic of our studies, we predict the following scenario for the Shiny Cowbirds in the West Indian region.

Continued impact on certain host species and shifts in host use.-- Our long-term data on the rate of parasitism of Yellow-shouldered Blackbirds indicate that once a species is used as a host by the Shiny Cowbird it will continue to be used despite declining availability. It is possible that the continued use of the blackbird as a cowbird host will be an important factor in driving the former to extinction. Populations of House Wrens (*Troglodytes aedon*) on Grenada (Bond 1971) and Yellow Warblers on Barbados (Bond 1966) have also declined to near extinction, and although the evidence is circumstantial, the Shiny Cowbird has been implicated. The cowbird may not be able to maintain high populations numbers if it did not shift its parasitism behavior to include other hosts. Thus, whereas hosts species that occur in low numbers will continue to be used (e.g., blackbird), more common species may be parasitized to a greater degree.

In our studies in Puerto Rico, we initially found Shiny Cowbirds mainly parasitizing Yellow-shouldered Blackbirds, even though other species were nesting in our study areas. We have found that even though the blackbird continues to be heavily parasitized, other species are also being used (Table 6). No Black-whiskered Vireos and Black-cowled Orioles were parasitized in 1975-1976 at Roosevelt Roads, but by 1980-1983 the parasitism rate was 88 and 100 percent, respectively. In 1975-76, the Puerto Rican Flycatcher was not recorded as a host species in Boqueron Forest, but 40 percent of the nests were parasitized by 1980-83. Regional differences in intensity of parasitism and of host use were also observed. Eastern (Roosevelt Roads) Yellow Warbler populations are more heavily parasitized than those in western Puerto Rico (Table 6), and the Puerto Rican Flycatcher is not found in eastern Puerto Rico. Parasitism rate at nests of Yellow Warblers possibly increased over the study period because of dramatic decline in availability of the main host species, Yellow-shouldered Blackbirds.

So far only Yellow-shouldered Blackbirds are known to be seriously affected by cowbirds. Our data suggest that some other species (Puerto Rican Flycatchers, Black-whiskered Vireos, Yellow Warblers, Black-cowled Orioles, and Troupials) ought to receive special monitoring in the future.

Brood parasitism and forest fragmentation.-- Fragmentation creates small patches of forest surrounded by open habitat and increases the portion of forest edge habitat available for cowbirds to use for nest searching. Gates and Gysel (1978) and Brittingham and Temple (1983) found that brood parasitism was more intense along the edges of wooded tracts, where nests are presumably closer to the feeding areas of cowbirds and are therefore more likely to be found. They showed that the distribution of cowbirds and parasitized nests within forest habitat was related to both the proximity and area of open habitat near nest. In the West Indian region, the fragmentation of forest habitats will most likely facilitate the parasitism of forest species (e.g., certain flycatchers, vireos, warblers, and orioles) by Shiny Cowbirds. In the montane region of Puerto Rico, for example, the cowbird is found in forest edge habitats where it is known to parasite the Puerto Rican Vireo (*V. latimeri*) and the Black-cowled Oriole (Peréz-Rivera 1986; pers. obs.). We feel that in the future other forest species will be parasitized, particularly those individuals forced into nesting in close proximity to open habitats created by forest fragmentation.

Predictions of future effects in new regions.-- We believe the Shiny Cowbird will colonize Jamaica, the Bahamas, and the Isle of Pines, and that it will spread quickly through these islands with their diverse habitats and great array of potential host species. Information on the Shiny Cowbird on St. Lucia, Hispaniola, and Puerto Rico may be used to predict the effect of the Shiny Cowbirds on species in other West Indian areas. On some islands, the Shiny Cowbird has parasitized blackbirds (*Quiscalus, Icterus, Agelaius*; Table 3). We do not have any information about host species used by cowbirds on Cuba at this time, but we anticipate that several other icterids will be parasitized by cowbirds on Cuba and on their arrival on other islands, including: the Cuban Blackbird (*Dives atroviolaceus*), Jamaican Oriole (*I. jamaicensis*), Tawny-shouldered Blackbird (*A. humeralis*), Red-winged Blackbird (*A. phoeniceus*), and Common Meadowlark (*Sturnella magna*). Species of the genus *Sturnella* are parasitized by cowbirds in South America (Gochfeld 1979) and North America (Friedmann 1929, 1963; Friedmann et al. 1977). Greater Antillean Grackles and Black-cowled Orioles occur on these islands and will probably be parasitized there, too.

Similarly, *Myiarchus* flycatchers, vireos, and *Dendroica* warblers are parasitized on West Indian islands (Table 3). The Black-whiskered Vireo will probably be an important host in Cuba and other West Indian areas. The Puerto Rican Vireo is used as a host in Puerto Rico and we believe that other vireo species will be parasitized in new areas, including the Cuban Vireo (*V. gundlachi*), Thick-billed Vireo (*V. crassirostris*), and the Jamaican White-eyed Vireo (*V. modestus*). The Yellow Warbler is an important host in other parts of the West Indies and we expect they will be used in Cuba and in other areas. Adelaide's Warbler (*D. adelaidae*) is parasitized in Puerto Rico, and we anticipate that the related Olive-capped Warbler (*D. pityophila*) of Cuba and the Bahamas, as well as other warbler species (e.g., the endemic Yellow-headed Warbler *Teretristis fernandinae* and Oriente Warbler *T. fornsi* of Cuba), will be used as Shiny Cowbird hosts. We also predict that the Zapata Sparrow (*Torreornis inexpectata*), Zapata Wren (*Ferminia cerverai*) and House Wren will be used as hosts in Cuba.

An even greater array of potential host species and habitat types are available to the Shiny Cowbird in North America and the Yucatán Peninsula. The cowbird will probably become established on these mainland areas and make use of population of avian species parasitized in other parts of the cowbird's range as well as species not before encountered in its range expansion. Perhaps the most interesting prospect is how the Shiny Cowbird will fare once it encounters populations of other species of brood parasites (Brown-headed and Bronzed Cowbirds) in North America.

Acknowledgments

We thank the convenor of this symposium, Dr. Charles A. Woods of the Florida State Museum for inviting us to participate. This work was supported by an NSF Grant PRM-8112194 to the University of Colorado, A. Cruz, principal investigator; by the U.S. Fish and Wildlife Service Office of Endangered Species and Patuxent Wildlife Research Center; the New York Zoological Society; University of Colorado Grant-in-Aid; U.S. Forest Service, Institute of Tropical Forestry; Caribbean Islands National Wildlife Refuge; the National Academy of Sciences; and by the Graduate School of the University of Miami.

For their assistance, we thank A. Arendt, W J. Arendt, C. Belitsky, D. Bennett, M. M. Browne, P. Butler, J. Cardona, G. Charles, H. Childress, J. Colon, S. Corbett, C. A. Delannoy, J. DiTomaso, S. Furniss, R. Johnson, E. Litovich, A. Nethery, E. C. Phoebus, C. A. Post, K. W. Post, E. Santana, S. R. Silander, D. Smith, J. Taapken, and T. Vargas Mora. This work was done in cooperation with the Department of Natural Resources of Puerto Rico; the Department of Game and Wildlife and the Natural History Museum of the Dominican Republic; and the Ministries of Agriculture, Lands, Fisheries, Cooperatives and Labour, Castries, St. Lucia. We also thank P. Mackenzie, A. Meier, J. A. Moreno, C. A. Woods, and anonymous reviewers for their useful comments on the manuscript.

Literature cited

American Ornithologists' Union. 1983. Check-list of North American Birds, 6th Edition. American Ornithologists' Union, Washington, D.C. 877 pp.

Arendt, W.J., and T.A. Vargas Mora. 1984. Range expansion of the Shiny Cowbird in the Dominican Republic. Journal of Field Ornithology 55:104-107.

Beard, J.S. 1949. The Natural Vegetation of the Windward and Leeward Islands. Clarendon Press, Oxford, UK 192 pp.

Biaggi, V., Jr. 1963. Record of the White Pelican and additional information on the Glossy Cowbird from Puerto Rico. Auk 80:198.

Bond, J. 1956. Check-list of Birds of the West Indies. 4th ed. Academy of Natural Sciences of Philadelphia, PA. 214 pp.

_____. 1963. Derivation of the Antillean Avifauna. Proceedings of the Academy of Natural Sciences of Philadelphia 115:79-97.

_____. 1966. Eleventh Supplement to the Check-list of the Birds of the West Indies (1956). Academy of Natural Sciences of Philadelphia, 1-13 pp.

_____. 1971. Sixteenth Suppplement to the Check-list of Birds of the West Indies (1956). Academy of Natural Sciences of Philadelphia, 1-15 pp.

_____. 1973. Eighteenth Supplement to the Check-list of the Birds of the West Indies (1956). Academy of Natural Sciences of Philadelphia, 1-12 pp.

_____. 1983. Avian population explosions in the Caribbean. Memorias cuarto simposio sobre la fauna de Puerto Rico y el Caribe. Universidad de Puerto Rico: 87-91.

Brittingham, M.C. and S.A. Temple. 1983. Have cowbirds cause forest songbirds to decline? Bioscience 33:31-35.

Buckley, P.A. and F.G. Buckley. 1970. Notes on the distribution of some Puerto Rican birds and the courtship behavior of White-tailed Tropicbirds. Condor 72:483-486.

Cruz, A., T. Manolis, and J.W. Wiley. 1985. The Shiny Cowbird: a brood parasite expanding its range in the Caribbean region. Pp. 607-620 in P A. Buckley, M. S. Foster, E. S. Morton, R. S. Ridgely, and F. G. Buckley (eds.). Neotropical Ornithology. Ornithological Monographs 36, American Ornithologists' Union.

Danforth, S.T. 1935. The birds of Saint Lucia. Monographs of the University of Puerto Rico, Ser. B., 3:129 pp..

Delius, J.D. 1965. A population of Skylarks (*Alauda arvensis*). Ibis 197:466-492.

Dixon, C.L. 1978. Breeding biology of the Savannah Sparrow on Kent Island. Auk 95:235-246.

Ehrlich, P.R. 1986. Which animals will invade? Pp. 79-95 in H.A. Mooney and J.A. Drake (eds.). Ecology of biological invasions of North America and Hawaii. Ecological Studies 58, Springer Verlag.

ffrench, R. 1980. A Guide to the Birds of Trinidad and Tobago. Harrowood Books, Newtown Square, PA. 470 pp.

Fonaroff, L.S. 1974. Urbanization, birds and ecological change in northwestern Trinidad. Biological Conservation. 6:258-262.
Food and Agriculture Organization of the United Nations. 1975. Production Yearbook. 29.
Fraga, R.M. 1985. Host-parasite interactions between Chalk-browed Mockingbirds and Shiny Cowbirds. Pp. 829-944 in P.A. Buckley, M.S. Foster, E.S. Morton, R.S. Ridgely, and F.G. Buckley (eds.). Neotropical Ornithology. Ornithological Monograph 36, American Ornithologists' Union.
Friedmann, H. 1929. The Cowbirds. A study in the biology of social parasitism. C. C. Thomas, Springfield, IL. 421 pp.
_____. 1963. Host relations of the parasitic cowbirds. Bulletin of the United States National Museum 233: 276 pp.
_____, and L. F. Kiff. 1985. The parasitic cowbirds and their hosts. Proceedings of Western Foundation of Vertebrate Zoology 2:225-302.
_____, L.F. Kiff, and S.I. Rothstein. 1977. A further contribution to knowledge of the host relations of the parasitic cowbirds. Smithsonian Contribution to Zoology 235:1-75.
Garrido, O.H. 1984. *Molothrus bonariensis* (Aves: Icteridae). Nuevo record para Cuba. Miscelanea Zoologica, Academia de Ciencas de Cuba, 19:3
Gates, J.E., and L.W. Gysel. 1978. Avian nest dispersion and fledging success in field-forest ecotones. Ecology 59:871-883.
Gochfeld, M. 1979. Brood parasite and host coevolution: Interaction between Shiny Cowbirds and two species of meadowlarks. American Naturalist 113:855-870.
Graves, G.R., and N.J. Gotelli. 1983. Neotropical landbridge avifaunas: new approaches to null hypotheses in biogeography. Oikos 41:322-323.
Grayce, R.L. 1957. Range extensions in Puerto Rico. Auk 74:106.
Guerra, F., and G. Alayon. 1987. Cowboy bird invades Cuba- Cuban ornithologist Orlando Garrido discusses what do do about the threat. Gramma 8:12 (22 February 1987).
Hamilton, W.J., III, and G.H. Orians. 1965. Evolution of brood parasitism in altricial birds. Condor 67:361-382.
Hoy, G., and J. Ottow. 1964. Biological and oological studies of the molothrine cowbirds (Icteridae) of Argentina. Auk 82:186-203.
Johnson, A.W. 1967. The Birds of Chile. Platt Establecimientos Graficos S.A., Buenos Aires, 398 pp.
Kale H.W. 1985. Rare bird alert news. Florida Audubon Society-Florida Ornithological Society, 11:4.
Kepler, C.B., and A.K. Kepler. 1970. Preliminary comparison of bird species diversity and density in Luquillo and Guanica Forests. Pp. 183-186 in H. Odum (ed.). A Tropical Rain Forest. U. S. Atomic Energy Commission, Oak Ridge, TN.
Long, J. 1981. Introduced Birds of the World. David and Charles, London. 528 pp.
Lugo, A.E., R. Schmidt, and S. Brown. 1981. Tropical Forests in the Caribbean. Ambio 1:318-324.
Manolis, T.D. 1982. Host Relationships and Reproductive Strategies of the Shiny Cowbird in Trinidad and Tobago. Unpublished Ph.D. dissertation, University of Colorado, Boulder, CO. 136 pp.

Mason, P. 1986. Brood parasitism in a host generalist, the Shiny Cowbird: I. The quality of different species as hosts. Auk 103:52-60.

Mayfield, H. 1965. The Brown-headed Cowbird with old and new hosts. Living Bird 4:13-28.

____. 1977. Brown-headed Cowbirds: agent of extermination? American Birds 31:107-113.

Mayr, E. 1965. The nature of colonization in birds. Pp. 29-43 in H.G. Baker and G.L. Stebbins (eds.). The Genetics of Colonizing Species. Academic Press, New York.

Moulton, M. P., and S. L. Pimm. 1986. Species introductions to Hawaii. Pp. 231-249 in H.A. Mooney and J.A. Drake (eds.). Ecology of biological invasions of North America and Hawaii. Ecological Studies 58, Springer Verlag.

Newton, A. 1860. To the editor of "The Ibis" (letter). Ibis 2:307-308.

Norton, R.L. 1986. West Indian Region. American Birds 40:1259-1260.

Paul, R.T. 1986. Florida region. American Birds 40:1193-1197.

Payne, R.E. l977. The ecology of brood parasitism in birds. Annual Review of Ecology and Systematics 8:1-28.

Peréz-Rivera, R.A. 1986. Parasitism by the Shiny Cowbird in the interior parts of Puerto Rico. Journal of Field Ornithology 57:99-104.

Pinchon, R. 1963. Faune des Antilles Francaises. Les Oiseaux. Fort-de-France, Martinique, 285 pp.

Post, W. 1981. Biology of the the Yellow-shouldered Blackbird-*Agelaius* on a tropical island. Bulletin of the Florida State Museum 26:125-202.

____, and J.W. Wiley. 1976. The Yellow-shouldered Blackbird - present and future. American Birds 30:13-20.

____, and J.W. Wiley. 1977a. The Shiny Cowbird in the West Indies. Condor 79:119-121.

____, and J.W. Wiley. 1977b. Reproductive interactions of the Shiny Cowbird and the Yellow-shouldered Blackbird. Condor 79:176-184.

Raffaele, H.A. 1973. Assessment of Mona Island avifauna. Pp. 1-32 in Mona and Monito Islands - an assessment of their natural and historical resources. Junta de Calidad Ambiental, Puerto Rico.

Ricklefs, R.E., and G.W. Cox. 1972. Taxon cycles in the West Indian avifauna. American Naturalist 106:195-219.

Robertson, R.J., and R.F. Norman. 1977. The function and evolution of aggressive host behavior towards the Brown-headed Cowbird (*Molothrus ater*). Canadian Journal of Zoology 55:508-518.

Robertson, W.B., Jr. 1962. Observations on the birds of St. John, Virgin Islands. Auk 79:44-76.

Rothstein, S.I. 1971. Observations and experiments in the analysis of interactions between brood parasites and their hosts. American Naturalist 105:71-74.

____. 1975. An experimental and teleonomic investigation of avian brood parasitism. Condor 77:250-271.

Scott, D.M., and C.D. Ankney. 1980. Fecundity of the Brown-headed Cowbird in southern Ontario. Auk 100:582-592.

_____, and C.D. Ankney. 1983. The laying cycle of Brown-headed Cowbirds: passerine chickens? Auk 97:677-683.

Simberloff, D. 1981. What makes a good island colonists? Pp. 195-201 in R.F. Denno and H. Dingle (eds.). Insect Life History Patterns: habitat and geographic variations, Springer-Verlag.

Skutch, A. 1976. Parent Birds and their Young. University of Texas Press, Austin, TX. 503 pp.

Wiley, J.W. 1982. Ecology of avian brood parasitism an at early interfacing of host and parasite populations. Unpublished Ph.D. dissertation, University of Miami, Coral Gables, FL. 256 pp.

_____. 1985. Shiny Cowbird parasitism in two avian communities in Puerto Rico. Condor 87:165-176.

Table 1. Forested area in countries of the West Indian region based on Food and Agriculture Organization of the United Nations (1975) data.

Country	Percent Forested
Cuba	11.0
Haiti	1.8
Dominican Republic	22.7
Puerto Rico	16.7
Jamaica	44.9
Virgin Islands (UK)	6.7
Virgin Islands (US)	5.9
Antigua	15.9
Martinique	25.5
Saint Lucia	21.0
Saint Vincent	41.2
Grenada	11.8
Barbados	0
Trinidad and Tobago	44.0

Table 2. Dietary suitability of prospective hosts of the Shiny Cowbird and results of experimental parasitism, Roosevelt Roads and Boquerón Forest, Puerto Rico, 1975-1982.

Species	Animal Protein Provided	Total Tests	Rejection No.	Rejection %	Natural Parasitism % nests
Puerto Rican Flycatcher	Yes	10	0	0	52.0
Gray Kingbird	Yes	21	18	85.7	1.1
Red-legged Thrush	Yes	4	4	100	3.7
Pearly-eyed Thrasher	Yes	21	17	81.0	0
Northern Mockingbird	Yes	9	7	77.8	2.0
Black-whiskered Vireo	Yes	3	1	33.3	86.7
Yellow Warbler	Yes	20	0	0	63.9
Bananaquit	Variable	14	9[a]	64.3	0
Stripe-headed Tanager	No	2	2	100	0
Black-faced Grassquit	No	10	0	0	0
Greater Antillean Grackle	Yes	36	32	88.9	9.0
Black-cowled Oriole	Yes	7	0	0	100.0
Yellow-shouldered Blackbird	Yes	11	1	9.1	94.1
Nutmeg Mannikin	No	15	0	0	0

[a] Rejection includes not only ejected or damaged eggs, but also nest desertions and build-overs.

Table 3. Shiny Cowbird parasitism of host species in St. Lucia, Puerto Rico, and Hispaniola.

Host Species	Percent Parasitized (No. nests parasitized)		
	St. Lucia	Puerto Rico	Hispaniola
Caribbean Elaenia	11% (2/18)	0% (000/014)	
Puerto Rican Flycatcher		34% (016/047)	
Gray Kingbird		1% (001/109)	
Red-legged Thrush		4% (001/027)	
Northern Mockingbird		2% (001/052)	
Black-whiskered Vireo	88% (021/024)	89% (017/019)	35% (008/023)
Puerto Rican Vireo[1]		53% (009/017)	
Yellow Warbler	47% (016/034)	63% (118/187)	39% (012/031)
Adelaide's Warbler		31% (004/013)	
Palm Chat			13% (039/305)
Carib Grackle	2% (002/096)		
Antillean Grackle		9% (025/281)	
Black-cowled Oriole		100% (014/014)	43% (013/030)
Troupial		100% (005/005)	
Yellow-shouldered Blackbird		91% (241/266)	
Bronze Mannikin		17% (001/006)	
Village Weaver			3% (033/1074)

[1] Montane areas.

Table 4. Shiny Cowbird parasitism rates at the nests of five host species in the Domincan Republic during two study periods[1].

Host species	1974-1977			1982		
	Total Nests	Parasitized No.	Percent	Total Nests	Parasitized No.	Percent
Palm Chat	243	13	5.3	62	16	25.8
Black-cowled Oriole	24	7	29.2	6	6	100.0
Black-whiskered Vireo	14	2	14.3	9	6	66.7
Yellow Warbler	19	2	10.5	12	10	83.3
Village Weaver	936	12	1.3	134	21	15.7

[1] All data collected in areas where the Shiny Cowbird is known to occur.

Table 5. Comparison of mean number of host chicks fledged from parasitized and non-parasitized nests in Roosevelt Roads and Boquerón Forest, Puerto Rico, 1975-1981.

Species	Mean number of host chicks fledged per nest				Significance[1]
	Parasitized		Non-parasitized		
	n	mean ± SD	n	mean ± SD	
Puerto Rican Flycatcher	11	0.64 ± 1.29	2	3.00 ± 0.0	$P < 0.05$
Black-whiskered Vireo	9	0.33 ± 0.05	2	2.00 ± 0.0	$P = 0.025$
Yellow Warbler	81	0.14 ± 0.47	26	0.58 ± 0.95	$P < 0.05$
Greater Antillean Grackle	22	1.35 ± 1.29	195	1.48 ± 1.29	n.s.
Black-cowled Oriole	12	0.33 ± 0.49	0	-	-
Troupial	5	0.80 ± 0.84	0	-	-
Yellow-shouldered Blackbird	152	0.25 ± 0.65	12	0.3 ± 0.78	n.s.

[1] Mann-Whitney 2-sample test (1-tailed). H_o: number of host chicks from parasitized nests equals number fledged from non-parasitized nests.

Table 6. Changes in host use by Shiny Cowbirds in Roosevelt Roads and Boquerón Forest Puerto Rico between two study periods, 1975-76 and 1980-1983.

	Number of host nests					
	1975-76[1]			1980-83		
Species	Nests	Parasitized	%	Nests	Parasitized	%
BOQUERON						
Puerto Rican Flycatcher	6[2]	0	0	40	16	40
Black-whiskered Vireo	-	-	-	4	3	75
Yellow Warbler	3	0	0	48	11	23
Adelaide's Warbler	-	-	-	13	4	31
Puerto Rican Bullfinch[3]	-	-	-	4	1	25
Greater Antillean Grackle	5	0	0	49	2	4
Yellow-shouldered Blackbird	35	35	100	93	90	97
ROOSEVELT ROADS						
Black-whiskered Vireo	3	0	0	8	7	88
Yellow Warbler	19	2	11	34	31	91
Greater Antillean Grackle	18	1	6	49	5	10
Black-cowled Oriole	3[4]	0	0	9	9	100
Troupial				3	3	100
Yellow-shouldered Blackbird	18	18	100	14	14	100
COMBINED						
Puerto Rican Flycatcher	6	0	0	40	16	40
Black-whiskered Vireo	3	0	0	12	10	83
Yellow Warbler	22	2	9	82	42	51
Adelaide's Warbler	-	-	-	13	4	31
Puerto Rican Bullfinch	-	-	-	4	1	25
Greater Antillean Grackle	23	1	4	98	7	7
Black-cowled Oriole	3	0	0	9	9	100
Troupial	-	-	-	3	3	100
Yellow-shouldered Blackbird	53	53	100	107	104	97

[1] Boquerón data only for 1975.
[2] 1976-77 data.
[3] *Loxigilla portoricensis*, scientific names of other species in text.
[4] 1977 data

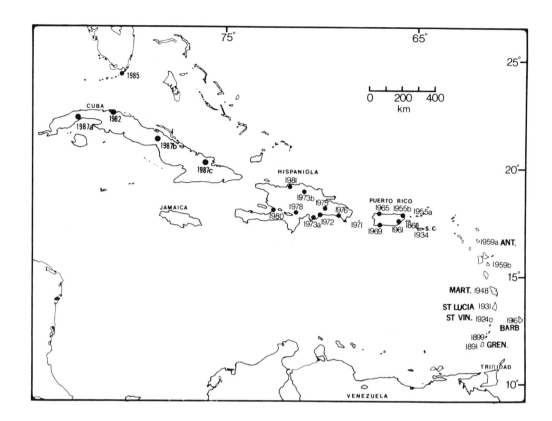

Figure 1.--Range expansion of *Molothrus bonariensis* in the West Indies (dates represent first report of appearance on these localities). 1860: Vieques (Newton 1860). 1891: Grenada (Bond 1963). 1899: Carriacou, Grenadines (Bond 1956). 1916: Barbados (Bond 1956). 1924: St. Vincent (Bond 1956). 1931: St. Lucia (Danforth 1935). 1934: St. Croix (Bond 1956). 1948: Martinique (Pinchon 1963). 1955a: St. John (Robertson 1962). 1955b: Cabezas de San Juan, Puerto Rico (Grayce 1957). 1959a: Antigua (Pinchon 1963). 1959b: Marie-Galante (Pinchon 1963). 1961: Yabucoa, Puerto Rico (Biaggi 1963). 1965: Guajataca Cliffs, Puerto Rico (Buckley and Buckley 1970), 1969: Guanica, Puerto Rico (Kepler and Kepler 1970). 1971: Mona Island (Bond 1973). 1972: Hato Nuevo, Dominican Republic (A. Dod, pers. comm.). 1973a: Najayo Abajo, Dominican Republic (A. Dod, pers. comm.) 1973b: Santiago, Dominican Republic (A. Dod, pers. comm.). 1975: Finca Estrella (pers. observ.). 1976: La Romana, Dominican Republic (pers. observ.). 1978: Neiba, Dominican Republic (Arendt and Vargas 1984). 1980: Port-au-Prince, Haiti (C. Mitchell, pers. comm.). 1981: Monte Cristi, Dominican Republic (Arendt and Vargas 1984). 1982: Cardenas, Cuba (Garrido 1984). 1985: Lower Matecumbe Key (Kale 1985). 1987a: San Antonia de los Baños, Cuba (Garrido in Guerra and Alcyon 1987). 1987b: Tapaste, Cuba (Garrido in Guerra and Alcyon 1987). 1987c: Holguin Cuba (Garrido in Guerra and Alcyon 1987).

THE ECOLOGY OF NATIVE AND INTRODUCED GRANIVOROUS BIRDS IN PUERTO RICO

Herbert A. Raffaele[1]

Abstract

Aspects of the community ecology of two native and four introduced species of small, granivorous birds were studied in Puerto Rico during the summers of 1979 and 1980. Data also are reported for nine other introduced, small, granivores. Food type, food availability, microhabitat, distance to cover, and the effects of human disturbance were investigated.

The native Black-faced Grassquit (*Tiaris bicolor*) was distinctive in some of the habitats it occupied, and it was the most adept species at utilizing areas of low resource abundance. It was followed closely in this regard by its congener, the native Yellow-faced Grassquit (*T. olivacea*). These two species were the most tolerant of human disturbance to their feeding areas, a potentially important factor on an island as heavily populated as Puerto Rico. Other important traits of *T. olivacea* are the food types it eats and the large distances it forages from protective cover.

Introduction

During historic time, Puerto Rico has supported 55 species of native, breeding land birds (pigeons through passerines). Though several are presently endangered, only one, the White-necked Crow (*Corvus leucognaphalus*), has thus far been extirpated. The native avifauna has been gradually augmented by the introduction of exotic species. By 1950 about nine exotics, including four pre-columbids, established feral populations. However, a major burgeoning of breeding exotic land birds occurred during the 1960's and 1970's so that the present number of introduced species breeding or probably breeding stands at 31 (Raffaele 1983a). The total number of extant breeding land bird species now stands at 81, the greatest number of any West Indian island. This is a rather dramatic shift considering that Cuba, nearly 13 times the size of Puerto Rico and more adequately situated to receive founding populations of continental derivation, supports only about 70.

The shift in species composition is most dramatic when one focuses on the small seed

---

[1] Dr. Raffaele is the Western Hemisphere Coordinator of the Office of International Affairs, U.S. Fish and Wildlife Service in Washington, D.C.

eating birds of Puerto Rico's grasslands. Originally consisting of only two native granivores, the Black-faced Grassquit *Tiaris bicolor* and Yellow-faced Grassquit *T. olivacea*, Puerto Rico's avifauna has multiplied seven-fold to presently include no fewer than 15 species that breed or probably do so (Raffaele 1983a; Moreno and Luquis 1985; Raffaele and Kepler unpubl. data).

These data raise a number of interesting questions. What factors have led to this especially sudden increase in species? Has the rapid colonization of the island by numerous exotics disturbed any natural equilibrium that might have existed within the native avifauna? Can Puerto Rico sustain this new level of avian diversity? If not, which species are most likely to be extirpated and by what mechanism?

Two factors are primarily responsible for the sudden increase in avian species in recent years. The first was the clearing of forests by early European settlers, setting the stage for avifaunal shifts by significantly increasing the extent of grasslands at the expense of forests. The second factor was the direct role of humans, an impact unlike that which influenced other lands in previous eras. In the nineteenth century and earlier, organized clubs actively sought out exotic species to embellish local landscapes. Contrarily, the recent rash of avian species invasion in Puerto Rico was unintended, resulting from the common habit of keeping caged birds. Formerly, only local species were typically kept in captivity. However, with the recent, rapid increase in per capita income coupled with the availability of inexpensive exotic birds, a dramatic shift in pet species occurred, particularly in the cities. Whether many of the bird populations now established derived from caged birds that accidentally escaped, or from birds that were intentionally released by disenchanted pet owners, is not know, but probably both mechanisms were important. What is clear is that those species most heavily imported were those that ultimately became established. In fact, ten of the eleven bird species most frequently imported into the U.S. between 1968-1971 (excluding the captively-bred Common Canary *Serinus canaria*) are now wild in Puerto Rico (Raffaele 1983a).

Regarding faunal equilibrium, intensive human pressures (deforestation for pastureland, agriculture, and urban development) have probably thrown local bird populations into a highly fluctuating equilibrium, if not disequilibrium for the past several centuries. In the case of the native granivores, this has been a very positive shift. The exotics do not necessarily create a disequilibrium so much as they modify conditions in an already fluctuating situation. Whether or not all these species can maintain long-term coexistence is difficult to ascertain. The substantial increase in grassland habitat since pre-Columbian times permitted the coexistence of many granivorous birds, but there is little reason to expect that Puerto Rico can sustain the large number of granivores currently established.

How do particular ecological factors influence the community structure of Puerto Rico's small seed eaters and possibly affect the susceptibility of these species to local extinction? Looking first at habitat, Puerto Rico's grasslands can be divided into at least four distinctive habitats: tall grass, short grass, scrubland and city lots. Tall grass principally occurs in sugar cane fields when the cane is young, and along the edges of older cane fields and roads. This habitat accounts for by far the largest proportion of Puerto Rico's grasslands and, likewise, supports the greatest diversity of seed eating bird species (nine). Short grass predominates only in urban areas, where lawns are common. Four species occupy this habitat, two of them being confined to the urban zone. The scrublands used for pasturing cattle are generally overgrazed and consequently low in grass sources. Only two small seed eaters regularly occur there. Finally, city lots bearing patches of sparse, weedy vegetation are relatively scarce and support only two of these small, granivorous birds.

This study concentrates specifically on six of the nine seed eaters in Puerto Rico's tall grass habitat: two native species (*Tiaris bicolor* and *T. olivacea*), two species established over 100 years ago (Bronze Mannikin *Lonchura cucullata* and Orange-cheeked Waxbill *Estrilda melpoda*), and two established only in the past few decades (Black-headed Nun *Lonchura malacca* and Strawberry Finch *Amandava amandava*). Three other seed eaters that occur in tall grass habitat are very rare or extremely local. These, and the six species typical of other grassland habitats, are discussed briefly at the end of the Results and Discussion section.

Although I was unable to document conditions in Puerto Rico prior to any recent colonization, I did arrive very early in the process. Therefore, if the ultimate community structure in grasslands depends upon effects that occur gradually, the data presented here should serve as a useful baseline against which to compare future conditions.

Field research was carried out from 27 May to 5 September 1979, and 27 June to 2 September 1980. The summer appeared to be an excellent season for studying Puerto Rico's seed eaters for two reasons. First, the south coast of Puerto Rico receives substantially less rainfall during the first six months of the year than during the last six months, thus creating the most severe annual drought conditions through the summer months (Morris 1979). Additionally, no fewer than five of the six bird species studied intensively breed during this period, consequently demand for food resources to sustain the young was probably particularly high.

Study sites

All of my study areas, with the exceptions of Isla Grande and those on Vieques, were of the tall grass habitat and contained extensive sugar cane. The Vieques sites had tall grass but no cane, and Isla Grande represented the short grass habitat. My principal study sites were selected along Puerto Rico's south coast for several reasons. First the south coast falls within the rain shadow of the central mountain range and is much more arid than the north coast. Since grass production is related to precipitation (Cable 1975), seed production in the south would likely be lower than in the north, and the lowered resource availability might be expected to foster greater competition among granivores. Second, *E. melpoda*, until recently confined to the south coast, occurs syntopically with *L. cucullata* and the interrelationships between these two long-established species warranted investigation. Additionally, the south coast was more recently invaded by the island's many newly established exotics than was the north coast, and I intended to initiate research on portions of the southeast coast prior to its colonization by the more recently established species. Unfortunately, I arrived a few years too late to achieve this goal.

My primary field site was within the Lajas Agriculture Experimental Station on the outskirts of Lajas in southwestern Puerto Rico. The 4.5 ha Lajas site included the station's four sugar cane study plots and adjacent grassy areas. Each of the four cane plots was sub-divided in checkerboard fashion by two-meter wide grassy swaths, which were the principal foraging areas of the seed eaters. The relative abundances of seeds present at this study site are presented elsewhere (Raffaele 1983b). I selected this site because it had a high seed eater diversity, and because information was available on the past and projected management of the sugar cane plots. A secondary study site at the Experimental Station was at the north gate entrance and is referred to as the Lajas-Gate site. The area consisted of a 300 m road edge evenly bordered by a one m swath of seed-laden Guinea grass (*Panicum maximum*) and Johnson grass (*Sorghum halepense*). One half of this tract had scattered trees and was abutted by areas of moderately dense grass and cane. The other half lacked trees and was bounded by a pasture and cut field. Birds in this section were far from dense cover. The Lajas-Gate site was specifically selected to evaluate the propensity of seed eaters to forage at substantial distances from cover.

A second major study site was approximately 19 km east of the Lajas site and just north of Guanica. It was an area of extensive sugar cane fields bordering Route 332 just north of its intersection with Route 116. Unlike at Lajas, Guanica's sugar cane plots were not dissected by grassy swaths, and the grassy edges bordering its plots were generally narrower (1-5 m). This resulted in seed resources being much less concentrated at the Guanica site. Additionally, almost all of the grassy edges were along cane roads or Route 332. Consequently there was more intensive human disturbance of feeding birds at Guanica. I further divided this study area into subsites based on natural boundaries such as cane roads and highways, as these cane plots and their associated edges were differentially disturbed by human visitors and had varying amounts of food resources. My primary reason for selecting this study area was that it appeared structurally similar to the Lajas site in that patches of cane were fringed by grasses and forbs, yet supported a much smaller proportion of exotics.

Data were collected at three localities on the island of Vieques, which is occupied by only the two native *Tiaris* species and *L. cucullata*. Since sugar cane is no longer grown on the island, study sites similar to those on Puerto Rico proper were not available. One site included open pasture and the adjacent road edge of the main east-west road north of Puerto Mosquito. A second site was an extensive open pasture with scattered giant milkweed (*Calotropis procera*). It was east of the juncture of Route 993 and the central east-west mountain road north of Puerto Real. The third study area incorporated the grassy portions of the U.S. Marine compound in Camp Garcia. Vieques was studied to acquire data from a relatively exotic-free locality, and one which may soon be further colonized.

Limited observations were made at three additional sites on Puerto Rico proper. One was among sugar cane and sparse, open grasslands north of Santa Isabel and west of Route 153 opposite the Libby factory. Another was east of Dorado and Route 165, amongst extensive sugar cane plantations, and the last was in the Isla Grande Naval Reserve in San Juan.

Methods

Data on food type were obtained solely from crop analyses, and three measures of resource overlap were applied to data on five species from the Lajas site (Raffaele 1983b). Discussions of microhabitats usually treat height above ground, perch type, density and height of the vegetation around the bird, and rates of search and feeding of the bird. I have omitted some of these parameters because they are irrelevant to the seed eaters I studied (i.e., the birds usually did not have to search for food, and feeding rates were generally very fast and similar across bird species). Other parameters, such as the height and density of vegetation around the bird, were omitted due to potential bias in the data as discussed elsewhere (Raffaele 1983b). Therefore, my discussion of microhabitat basically treats conspicuous behaviors and readily observable interspecific differences in feeding techniques.

The determination of food availability entailed extensive seed sampling, as described in a previous paper (Raffaele 1983b).

I categorized cover using two different procedures. Closest cover was the nearest form of protection to the bird regardless of its extent and denseness. The second category, dense cover, was a subset of the first and included only thick cover which could potentially provide effective protection. Areas of cover were rated on a scale of 1-9 as follows:

1) A single grass clump
2) A very small tree (1.7 m) or open bush
3) A few small grass clumps
4) Limited vegetation dense to 1 m; a small thorn bush (0.7 m)
5) 1.2-2 m cyclone fence; *Calotropis* (<2 m)
6) A cyclone fence (>2 m); limited vegetation dense to 1.3 m; 1 m thorn bush; moderately dense *Calotropis* (2-3.3 m)
7) A very dense bush (1.1-2 m); *Calotropis* (>3.3 m); 1.3 m thorn bush
8) A single row of tall (2 m) cane; a dense bush (2.1-3 m); a cane field (2-2.5 m)
9) A cane field (>2.5 m); a thorn tree (>2 m); large trees; a dense bush (>3 m); a building.

Cover in the "closest" category could fall anywhere on this scale, while "dense" only included cover rated from 6-9. The distance to each cover type was recorded for every foraging bird and where the rating for closest cover was 6-9 the distance was the same for dense cover.

I divided the subsites of the Guanica study area into four categories to evaluate disturbance due to immediate human activities, such a pedestrian and vehicular traffic. The high disturbance category included grassy borders averaging approximately 3 m in width, along Route 332 (subsites 39-40). Areas of moderate disturbance were a cane road (subsites 35-36) and a grassy patch at its end (subsite 38), which served as a connection between a rural community and Route 332. A cane road (subsites 30-31) was in the slightly-to-moderately disturbed category. The remaining subsites fell into the slightly disturbed category; they were cane roads closed by gates, or were poorly maintained lanes through the cane.

Results and discussion

Food type

Horn's (1966) and Schoener's (1970) measures demonstrated an extremely high overlap between four of five seed eaters adequately sampled (*Tiaris bicolor, Lonchura cucullata, Estrilda melpoda* and *Amandava amandava*) (Raffaele 1983b). The fifth, *L. malacca* was an outlier as a consequence of foraging heavily on the large seeds of *Sorghum halepense* (Fig. 1). Hurlbert's (1978) overlap measure, though derived differently than the previous two, shows a clustering of the same four species and indicates that the intensity with which these species feed on the same foods is approximately 90 percent greater than it would be if they fed randomly with respect to food type (Raffaele 1983b).

The high degree of overlap between *T. bicolor*, *L. cucullata*, *E. melpoda* and *A. amandava* primarily resulted from all of them feeding intensively on barnyard grass (*Echinochloa crusgalli*) seeds (76%, 74%, 69% and 92% respectively) (Fig. 1). This was the most abundant grass at the Lajas site, accounting for approximately 36 percent of the seed.

Tiaris olivacea was insufficiently sampled, but data from two taken at the Lajas plots, and three from similar cane fields 1 km away, are quite suggestive. *Echinochloa crusgalli* accounts for only three percent of these crop contents. On the other hand, *Digitaria* spp, a grass that is generally a minimal dietary component of the other birds (2% in *L. malacca* and *A. amandava* to 11% in *T. bicolor*) represents 43 percent of the food consumed by these five individuals (Fig. 1). Likewise, broadleaf signalgrass (*Brachiaria platyphylla*) is even rarer in the five indexed birds, being absent in the crops of *L. cucullata* and *A. amandava* and occurring maximally in *T. bicolor* where it constitutes seven percent of the diet. In *T. olivacea* this grass represents 21 percent of the food consumed (Fig. 1). Four additional specimens of *T. olivacea* from other localities show the same pattern, suggesting that this species' feeding habits differ from those of the five others.

Data on birds from other sites are noteworthy. Table 1 summarizes *T. bicolor* crop data from Lajas, Guanica and Vieques, and *L. cucullata* data from Lajas and Vieques. Most notable is the reduced proportion of *E. crusgalli* in the diets of both species away from Lajas. While this was apparently due to the near absence of the grass in Guanica, such was not the case in Vieques. *Brachiaria platyphylla* was relatively rare in Lajas, but was common in Guanica and Vieques. It appears that the increased availability of this grass, more than a decline in *E. crusgalli*, is primarily responsible for the shift in the feeding habits of these two bird species among sites.

Most importantly, Table 1 demonstrates how dramatically resource utilization can vary from site to site. To determine whether this affected overlap, I applied Horn's (1966) measure to the Vieques data for these two birds. It is 0.723 as compared to 0.978 for Lajas. However, Schoener's (1970) index is 0.505 and 0.865 for the respective sites, a more substantial difference. Generally, both indices suggest overlap is reduced on Vieques, but they vary as to the amount. These shifts in resource use and overlap can arguably be attributed to differences in the resource spectrum and/or the bird species community at each locality. Three of the Lajas birds are absent from Vieques (*E. melpoda*, *L. malacca* and *A. amandava*). Such shifts are similar to those observed by Wiens and Rotenberry (1979) among grassland birds over several years at a particular site and at different sites during the same year. I believe that though avian species composition may be a contributing factor, the shifts in feeding habits and overlap of *T. bicolor* and *L. cucullata* between Lajas and Vieques are primarily a consequence of differences in food resource availability, but further data are needed to substantiate this hypothesis.

Two of the six bird species differ greatly in food type eaten. *Lonchura malacca*, by far the largest billed species, concentrates heavily on the larger seeds, which are virtually ignored by most other birds, and *T. olivacea* appears to feed extensively on *Digitaria* spp, a grass genus generally bypassed by other seed eaters. *Tiaris olivacea* also seems to avoid *E. crusgalli*, which is a major component of the diets of both larger and smaller billed birds and it is the only species that regularly eats the seeds of the large-awned grass *Andropogon annulatus*.

That *T. bicolor* is among the four species which show extensive overlap in food resource utilization is partially a function of the study site. The Lajas site supported very few and sparsely distributed forb species, and field observations suggest that forbs are an important dietary component of *T. bicolor* (14 % at Guanica, Table 1). Forbs are virtually uneaten by the other bird species, except to a limited extent by *T. olivacea*. *Tiaris bicolor* is almost certainly a much more generalized feeder than the species with which it appears to overlap.

The three remaining seed eaters that show extensive food overlap at Lajas appear to do so at other sites as well. This is particularly true in the case of *L. cucullata* and *E. melpoda*, which displayed similar feeding habits throughout. Presently, there is no evidence to indicate that these species are partitioning food resources. This observation is particularly interesting and unexpected in light of the fact that *L. cucullata* and *E. melpoda* are long-established introduced species that are quite different in bill size, and may even be diverging further in this character (Raffaele 1986). While it appears that under most circumstances these two species overlap in feeding habits, it is possible that in some situations they do not. What these circumstances might be, and whether they occur under resource limited conditions that might have evolutionary consequences, remain to be seen. Food preference studies of caged birds, and crop and resource analyses from other sites, should shed additional light on this intriguing question. *Amandava amandava* overlaps broadly with others in food type but it appears to diverge from them in characteristics to be considered later.

Microhabitat and feeding behavior

Amandava amandava is the most distinctive of the six seed eaters in microhabitat usage. It was conspicuous by its absence at exposed *Panicum* clumps where the five other species regularly forage. Seventy-one percent of all *A. amandava* field observations were of ground-foraging birds and nearly 30 percent of these were in dense cover (Fig. 2). Ground foraging in dense cover is the most difficult type of feeding to observe and utilization of this microhabitat by *A. amandava* probably is much greater than the data indicate.

Of the six seed eaters, *A. amandava* is the most limited in its feeding height. However, it does not solely occupy the ground-foraging microhabitat as Fig. 2 indicates. Crop analyses demonstrated that four of the other species (*L. malacca* excluded) also feed heavily on low grasses and, as indicated below, *T. bicolor* is an excellent ground-forager. Therefore, it would be inappropriate to consider ground foraging as the exclusive domain of *A. amandava*.

Stem thickness of food plants can play an important role with respect to seed accessibility; heavier birds have greater difficulty reaching tall, thin-stemmed grasses. *Panicum maximum* is a tall (2 m) grass with a moderately thick stem averaging 2.1 mm just below the head and 2.7 mm at 25 cm below the head, (n=10), with other representatives of this genus in the study sites being somewhat shorter and thinner. Seeds of these grasses were eaten regularly by all the bird species studied. The two heaviest birds, *T. bicolor* and *L. malacca*, apparently had more difficulty than the others in feeding directly from the grasses on which they perched. Both had significant chi-square values ($P<0.05$) when compared with three of four others (Table 2). *Tiaris bicolor* apparently compensated for its inability to perch on thin stems primarily by feeding on *Panicum* from adjacent, thicker perches. *Lonchura malacca* used this approach as well, but also frequently would land on a *Panicum* stem, sidle up it, and then feed on an adjacent *Panicum* head rather than try to reach the seeds of the plant it was on.

Seed accessibility could play a major role in food resource partitioning among bird species (Cody 1968, 1974), but difficulty of access to *Panicum* does not appear to have deterred *L. malacca* from feeding on it. *Panicum* forms 28 percent of that species' diet, but 18 percent or less in the other seed eaters (Raffaele 1983b). However, comparing relative amounts of each seed type in the diet of *L. malacca* with the stem thicknesses of the respective plant species (Fig. 3) suggests that feeding habits of *L. malacca* are related to grass stem characteristics, and that body size (weight) largely determines the kinds of seeds eaten by *L. malacca*.

Table 3 presents the methods by which *L. malacca* and *T. bicolor* forage on the seeds of three thin-stemmed grasses (*E. crusgalli*, *Digitaria* spp and *Chloris ciliata*). *Lonchura malacca* typically perches on thicker stemmed grasses and reaches over to the seeds of the thinner ones grasping them in its bill (82% of its thin grass foraging). It infrequently (10%) reaches these seeds from the ground. On one occasion a bird reached from the ground to a thin grass, put the seed head under its foot and then fed. However, this behavior apparently is common as seven birds (39%) were seen feeding with thin grasses underfoot and five of these were on adjacent thick-stemmed or medium-stemmed grasses. These five birds almost certainly had obtained the seed heads on which they fed by reaching out with their bills. This is the only alternative method of seed procurement, besides directly landing on a food plant, that I observed in *L. malacca*.

With its limited range of foraging behavior *L. malacca* might have difficulties obtaining seeds of 30 cm to 75 cm high plants other than of those bent close to the ground or near a stable perch. Of course, *L. malacca* could attempt to land on such grasses, carrying them to the ground, and feed. This is rare; only four birds (10%) were observed feeding on the ground, and some of these no doubt on fallen seeds. It appears that *L. malacca* may be deterred from this practice by the instability of the perches. If such is the case, a substantial portion of the thin grass seed heads of an area are relatively inaccessible to this bird as a result of perch characteristics.

If *L. malacca* has difficulty obtaining seeds from thin-stemmed grasses, why is the same not the case for *T. bicolor*? The latter is only slightly lighter than *L. malacca*, but has a diet very similar to many smaller species. It appears that the answer lies in this species' versatile repertoire of feeding behavior.

Unlike *L. malacca*, *T. bicolor* relies only moderately on foraging from thicker stemmed grasses (32%). It feeds primarily on the ground (52%) in diverse ways. Ten (42%) of the ground feeders hopped or fluttered up to grass heads, pulled them down with their bills, and held the heads on the ground or perch with their feet. Three others struck grass stems with their bills to knock off seeds, and two others fluttered in the air as they pecked seeds. In total, fifteen birds (33% of all *T. bicolor* observed foraging), performed behaviors not exhibited by *L. malacca*, whereas there were no cases of the converse. The diverse array of foraging tactics of *T. bicolor* appears to provide it with access to virtually any seed. Therefore, perch characteristics probably have a minimal effect on its diet.

Food availability

The most noticeable difference between the Lajas and Guanica study sites was the substantially greater concentration and abundance of seeding grasses at the former site. This was primarily the result of the numerous grassy swaths crisscrossing the sugar cane plots. I believe it was this difference in resource abundance that was responsible for the dramatic difference in bird species composition between the two study sites. In decreasing order of abundance, based on censuses over two summers, the six granivores studied at Lajas are *L. cucullata*, *L. malacca*, *A. amandava*, *E. melpoda*, *T. bicolor*, and *T. olivacea*. At Guanica the sequence was almost the reverse being *T. bicolor*, *T. olivacea*, *E. melpoda*, *L. malacca*, *A. amandava*, and *L. cucullata* (absent). Most dramatically, *L. cucullata* accounted for 30 percent of the individuals censused at Lajas, whereas the species was not encountered during censuses at Guanica.

To assess whether the bird species were reacting to resource abundance, I divided the subsites of the Guanica study area into three categories of food availability (high, medium, and low) based on the size of each subsite and the concentration of favored seeds in it. The manner in which the seed eaters distributed themselves among these categories is presented in Table 4. I was particularly interested in subsites 33 and 34 bordering a relatively unused cane road. Both subsites were nearly 300 m in length and averaged approximately 2 m and 1 m in width, respectively.

I visited this cane road more than 30 times in the course of two summers and found it occupied almost exclusively by the native *T. bicolor* along with a few *T. olivacea*. The 14 *A. amandava* observations came from a flock of about that size which visited the subsite on two consecutive days. This introduced species did not stop there again, though it was heard calling in flight occasionally.

Although introduced seed eaters did not entirely ignore this cane road, they used it to a very limited extent, and in a manner different than that of the native species. I believe this to be a consequence of the feeding behaviors of these species. The two *Tiaris* species forage individually or in small family groups, whereas all four introduced species forage in flocks of varying sizes. Among the exotics, individuals apparently do not have feeding territories, and, when resources are low, flocks appear to forage nomadically. In fact, this is the manner in which the entire Guanica study area apparently was used by the exotics. Although *L. cucullata* periodically flew over the Guanica site during the observation period, it never stopped. I believe resource scarcity was responsible for its absence, but I have no supportive evidence. As flocks naturally have greater resource requirements than individuals, it is not surprising that they would bypass areas with low seed levels that might prove adequate for one or a few individuals.

Several other aspects of Table 4 require elaboration. The absence of any exotics at subsites 39 and 40, which border Route 332, is apparently a consequence of human disturbance as discussed in the next section. Although disturbance may have also inhibited seed eaters at subsite 38, possibly the second most disturbed area after the borders of Route 332, it should not have completely deterred them; just prior to initiating field observations I observed an *A. amandava* flock foraging here. This subsite was somewhat removed from other feeding areas and may have been overlooked to some degree by nomadic foragers.

Although the native, territorial birds would appear to have an advantage over flocking species in the utilization of low resource areas, they may be at a disadvantage when the road or cane edge vegetation is destroyed by cutting, fire, or herbicide application. In late June 1980 the grasses in two cane plots at the Lajas site were sprayed with herbicide and two others were cut. All bird species fed actively in the sprayed plots, but foraging diminished drastically in the cut area. As discussed earlier, *T. bicolor* is adept at foraging on seeds near the ground and it may well have been overlooked under the cut grass. However, both *Tiaris* species must certainly suffer when grasses are destroyed before they seed.

Distance to cover (two types)

Early in the course of my field research it became apparent that seed eaters differed in how far they would forage from dense cover. I performed an analysis using data from all study sites combined (Fig. 4). Observations of birds feeding in dense cover itself were deleted from the analysis because of the potential for severe bias resulting from their inclusion (Raffaele 1983b). Most notably, *A. amandava* was found to forage regularly at substantial distances from dense cover. Both *A. amandava* and *T. olivacea* were similar to one another ($P>0.50$) and significantly different from all other species ($P<0.05$) in this behavior. On the other hand, *T. bicolor* was intermediate in this regard, foraging significantly closer to dense cover than did *T. olivacea* and *A. amandava* ($P<0.05$), yet significantly further from dense cover than did the remaining three bird species ($P<0.05$).

Another analysis with data for all sites, but using distance to closest cover rather than to dense cover (see distinction in Methods section) showed *A. amandava* is the only seed eater that regularly forages far from the protection of even small grass clumps (Fig. 5). Chi-square values were significant ($P<0.05$) for comparisons of *A. amandava* with all other species. Areas that are too sparsely vegetated to attract other species appear to be acceptable to *A. amandava*. However, *A. amandava* normally feeds under some vegetation, even if it is low and sparse. At the opposite extreme, *E. melpoda* had significantly different ($P<0.05$) chi-square values from all other species; it never foraged more than 8 m from cover. This could be an important limiting factor in its ecology. A list of the species in decreasing order of frequency of individuals foraging at least 8 m from cover reads *A. amandava* (37%), *T. olivacea* (13%), *L. malacca* (10%), *T. bicolor* (6%), *L. cucullata* (6%), and *E. melpoda* (0%).

The distance birds forage from cover can either be advantageous or disadvantageous to their survival depending upon how the potential danger level changes with increased distance. Puerto Rico's introduced seed eaters undoubtedly experienced a very high threat from avian predators in their native lands of southern Asia and Africa. The case may be different for Puerto Rico. Among the eleven species of raptors that occur in Puerto Rico, three are owls, none of which are, or ever were, a serious threat to foraging grassland birds, and six are hawks that are either too large, too rare (in the past as well as the present), or typically do not occur in grasslands. Two falcons remain, of which one is the Merlin (*Falco columbarius*), a winter visitor to Puerto Rico, primarily from October through March. Though its feeding habits are unstudied in Puerto Rico, it regularly feeds on small birds in North America. However, the species is uncommon (and probably always has been) on the island and generally occurs in coastal area and in forests (Raffaele 1983a). For these reasons I believe its impact on grassland seed eaters is negligible.

The other falcon is the American Kestrel (*F. sparverius*), a resident in Puerto Rico and the only hawk to occur regularly in the grasslands. It is generally uncommon on the south coast and is nearly absent from the north. Its diet, based on Wetmore's (1916) analysis of 48 stomachs from Puerto Rico, consists entirely of animal matter with the vast majority of this being insects (50%) and lizards (40%). Mice made up six percent of their diet, and birds, found in two stomachs, accounted for two percent. Miscellaneous items, primarily spiders, composed the remaining two percent. This analysis suggests that although kestrels prey upon small birds, the latter are a very minor component of the diet, or at least were so prior to the recent seed eater introductions. Consequently, raptor predation does not seem to be a serious source of mortality to Puerto Rico's native grassland birds.

Puerto Rico has no native mammalian predators, though introduced rats, cats, dogs and the mongoose (*Herpestes auropunctatus*) no doubt prey on small birds. Additionally, a native snake (*Alsophis portoricensis*), now rare, but formerly common, may also take small birds. However, I do not expect that predation by any of these is greater in open areas than in localities affording dense protective cover.

The low threat of raptor predation suggests that the distance at which seed eaters forage from dense cover is not of major importance for protection. Consequently, it would appear advantageous for birds to utilize resources regardless of their location as this expands the resource base. *Amandava amandava* apparently is most adept at doing this followed by *T. olivacea*. I do not propose that seed resources were limiting during this study, but if they ever are, it appears that a competitive advantage would befall those species that are relatively indiscriminant with respect to cover. Regardless, the distance birds forage from cover at times could be important in the avoidance of competition and in determining community structure.

Human disturbance

Disturbance is treated here as human activities that might alarm a bird, such as automotive and pedestrian traffic. While such a factor is not typically considered among the variables that might contribute to determining the structure of communities, there is good reason to believe it may play an important role in the ecology of these seed eaters. Puerto Rico's human population density is approximately 439/ km^2, and the island's principal grasslands consist of the borders of sugar cane roads and the grassy edges of public roads. Additionally, many people travel via cane roads and others work in the cane itself. Therefore, it seems reasonable to expect that the more wary bird species avoid areas they might otherwise inhabit if the level of human disturbance were lower.

Data on feeding obtained at the Guanica subsites have been divided among four categories of disturbance (Table 5).

To quantify disturbance at the most highly distributed subsites (39-40 along Route 332), I kept traffic records on five different days for a total three hours during the prime morning and evening feeding periods. An average of 78 cars, three bicycles, and five walkers or joggers traversed the road per hour. Route 332 is a typical rural road.

Only the two native seed eaters (*T. bicolor* and *T. olivacea*) occurred in the high disturbance zone along Route 332. This, despite the fact that high, dense stands of protective cane grew along the greater portion of this road during the entire study, and a moderate amount of food was available. On one occasion I observed *T. bicolor* only 6 m from me feeding low in the grass far from any protective cane. Many cars passed within a few meters, but the bird ignored them. At another time I located a *T. olivacea* nest only three m from the edge of the road and 0.2 m above the ground.

The other seed eaters were decidedly more retiring in habits than the two *Tiaris* species (Table 5). On two occasions when I observed *A. amandava* feeding on the ground with *T. bicolor* along a cane road, the former were conspicuously more nervous and were the first to flee. They flushed when I was approximately 35 m away.

Despite the fact that *T. bicolor* and *T. olivacea* were the only species to forage along Route 332, they did not appear as approachable at other study areas. The Route 332 birds may have adjusted their behavior to accommodate the site. The tolerance by both *Tiaris* species to high levels of human disturbance provides them access to substantial amounts of grassland that may be avoided by other seed eaters.

Other small granivores

In addition to the six most abundant granivores of Puerto Rico's tall grass habitat, there are other tall grass species, and six species typical of other habitats. The three inhabiting tall grass are the Spice Finch (*Lonchura punctulata*), the Red Bishop (*Euplectes orix*), and the Red-eared Waxbill (*Estrilda troglodytes*). *Lonchura punctulata* is one of four members of its genus introduced to Puerto Rico and, like *L. malacca* and *L. cucullata*, inhabits the tall grass habitat. Perhaps, then, it is not coincidental that while *L. malacca* colonized the entire coastal plain with extreme rapidity, *L. punctulata* was very successful in San Juan, where *L. malacca* is rare, and it has spread to outlying areas only in small numbers. These two species are native to southeast Asia. The subspecies found in Puerto Rico were derived from Indian stock. Ali and Ripley (1973) report that in India *L. punctulata* is common and widespread in bushclad hillsides, secondary jungle and open country with scrub. It does not occupy similar habitats in Puerto Rico. Ali and Ripley (1973) describe *L. malacca* as being only locally common and principally a marshland bird, but an affinity for marshes is not evident in the Puerto Rico populations. In India the habitats these species utilize seemingly overlap minimally. This is not the case in Puerto Rico. Where *L. punctulata* is found outside of San Juan it typically occurs in grassy edges, as does *L. malacca*. Though I have no quantitative data, my general observations indicate no apparent differences between *L. punctulata* and *L. malacca* in feeding behavior. It appears, therefore, that *L. punctulata* and *L. malacca* could be interacting in some way, *L. malacca* presently being more successful. Quantitative data on the feeding behavior of *L. punctulata* could contribute valuably towards understanding this relationship.

Euplectes orix is locally common within approximately 30 km of San Juan and it appears to be spreading. It is native to much of central and southern Africa where Bannerman (1949) described it as occurring wherever there is long grass, particularly along streams. This vague habitat description is generally applicable to the bird in Puerto Rico. I have recorded the species only once on the dry, south side of the island. This could indicate that the bird is only just beginning to spread into that region, that the area is simply too xeric to inhabit, or that *E. orix* is being excluded from the region. I favor the first hypothesis, but presently have no evidence for discarding the others. *Euplectes orix*, *L. malacca* and *L. punctulata* apparently have very similar feeding habits. But observations of these three species where they occur together are often difficult to make because the birds forage low among the young cane. *Euplectes orix* is larger than *L. malacca* and may be more of a ground forager, though despite intensive searching, I was unable to locate an adequate site for measuring this. A comparison of crop contents may demonstrate greater differences in food preferences between *L. malacca*, *L. punctulata* and *E. orix* not readily apparent otherwise. Presently the successful coexistence of these three species in Puerto Rico remains unexplained.

The Red-eared Waxbill (*Estrilda troglodytes*) is a recently arrived congener of *E. melpoda*. It is native across central Africa, where it is sympatric with *E. melpoda* over about half of its range. In Puerto Rico, it is locally distributed in the lowlands, primarily occurring along edges of cane fields and rarely in arid scrub country. It appeared to be more common in the early 1970's than it is now (Raffaele pers. obs.). The mean bill size of *E. troglodytes* is smaller than that of *E. melpoda* (Raffaele 1986); however, the range of bill sizes of the former is within that of the latter. From casual observations, I noticed no differences in microhabitat utilization or food preferences between the two *Estrilda* species. The scanty data suggest that both species are extremely similar in microhabitat utilization and probably are in competition with each other. It is possible that one species will be competitively excluded.

The six remaining species typically occupy habitats other than tall grass. The Silverbill (*Lonchura malabarica*) is the only common exotic in scrublands, which enables it to flourish despite the presence of three other congeners on the island. Grasses in the scrublands rarely go to seed due to overgrazing by cattle, consequently forbs are a more important part of the seed resource base in this habitat. *Lonchura malabarica* is native to India and adjacent areas where it prefers drier environs than the other sympatric members of its genus (Ali and Ripley 1973). A similar relationship exists in Puerto Rico.

Euplectes afer, like *E. orix*, is native to central and southern Africa. In Puerto Rico *E. afer* is confined to a few swampy areas, a habitat similar to that which it occupies in its native land. Little is known of its habits in Puerto Rico.

Three of the four remaining granivores are very closely associated with civilization. The Saffron Finch (*Sicalis flaveola*), Java Finch (*Padda oryzivora*) and Pin-tailed Whydah (*Vidua macroura*) typically inhabit very short grass, usually that of cultivated lawns. *Sicalis flaveola* and *P. oryzivora* are almost exclusively found on lawns in San Juan, *P. oryzivora* also occurring in small patches of weeds. *Sicalis flaveola* is indigenous to parts of South America, and *P. oryzivora* to Java, Bali and Sumatra. In their native lands both species apparently occupy broader habitat spectra than they do in Puerto Rico. *Sicalis flaveola* is reported from the campos as well as gardens (Meyer de Schauensee 1970), and *P. oryzivora* occurs in mangroves and scrub as well as cities and villages (Long 1981). The heavy bill of *P. oryzivora* enables it to feed on large spiny sandbur (*Cenchrus*) seeds, which go untouched by the smaller billed *S. flaveola*. However, I have noted no other differences between these species in food preferences nor in feeding patterns, and I expect coexistence over a long period will be difficult. *Sicalis flaveola* appears to be slightly less versatile than *P. oryzivora* not only in habitat and food utilization, but also in its nesting requirements as it is very partial to cavities. I therefore expect *P. oryzivora* to be the more successful of the two species. *Vidua macroura* occurs in mowed fields and cut edges in more rural areas in addition to urban lawns. It never occurs in even medium-high grasses and it feeds only on the ground. I have no information on its feeding habits in central and southern Africa where it is native.

One might not have expected *V. macroura* to successfully colonize Puerto Rico.

First, it has the same size bill (culmen length) as does *T. bicolor*, the most versatile granivorous bird on the island (Raffaele 1986). Additionally, *V. macroura* is a near obligate nest parasite on African weaver-finches of the genus *Estrilda*, so its success is contingent upon their's. The astounding success of *V. macroura* on the island highlights several points. First, it suggests that the relative culmen lengths of sympatric species may not be a good indicator of possible competition. Microhabitat utilization may be critical. Second, that such an alien species can be so successful in Puerto Rico demonstrates how much the island has changed. *Vidua macroura* is at home in Puerto Rico with its native African nest hosts present, and with native African grasses. The island's native *Tiaris* may find the grassland habitat more foreign than do the Old World colonizers.

The final granivore to be considered is the House Sparrow (*Passer domesticus*). It typically inhabits weedy, urban lots, which it shares to some extent only with *P. oryzivora* among the other seed eaters. Presently, *P. domesticus* is gradually spreading outward from the city of Ponce in the southwest, whereas *P. oryzivora* is restricted to San Juan and vicinity. *Passer domesticus* has not been particularly successful in the tropics, though it is doing well in Hawaii (Berger 1981). Whether it and *P. oryzivora* will be able to survive together remains to be seen. In Hawaii *P. oryzivora* is presently spreading rapidly on Oahu where *P. domesticus* has been long established. A comparison of developments there with those in Puerto Rico should prove interesting.

Overview

In this paper I have described how seed resources are utilized with respect to seven factors by Puerto Rico's small granivorous bird species, a community that has increased more than seven-fold in species number, primarily since the 1960's.

The methods utilized by the different species are summarized in Table 6. The table immediately suggests that Puerto Rico's two native granivores are unique along many more feeding dimensions than are any of the introduced species. The native species also differ moderately from one another. At the other extreme are the two long-established exotics, *L. cucullata* and *E. melpoda*, which display no unique foraging characteristics. This case is particularly interesting due to the substantial difference between the bill sizes of these birds and their apparently increasing divergence (Raffaele 1986). A possible explanation for different bill sizes in birds with very similar feeding habits is that the birds may feed similarly within a given habitat, but are not distributed similarly among habitats (Pulliam and Enders 1971). Such could have been the case for these two species in the past as *L. cucullata* ranged over Puerto Rico's entire coastal plain, whereas *E. melpoda* was confined to the southwest. However, since at least the early 1970's, *E. melpoda* has been expanding its range on the island to where it is presently syntopic with *L. cucullata*. Continued investigations into the coexistence mechanisms of *L. cucullata* and *E. melpoda* should prove instructive.

To more fully explain the present success of so many seed eaters in Puerto Rico, and whether their differences in food resource utilization will suffice to foster long-term coexistence, requires answers to several critical unknowns. First, are food resources ever limiting to the extent that they affect the granivores? An affirmative answer is most probable because, as Thomson (1980) points out, as long as there are costs involved in obtaining resources there will be relative differences in the efficiency of resource acquisition and this will likely translate into differences in individual fitness. Second, though seed resource availability may affect the granivores, it is not known with what regularity these effects might be important. I believe that pressures which have important effects on granivore populations only occur periodically, minor effects occurring every few years and major ones every few decades. It is also likely that food becomes an important factor on a local basis more regularly than it does on an islandwide scale. For example, burning of a sugar cane plantation before the harvest, or spraying of herbicides to control weeds, might have important effects on local bird populations and, in fact, may create effects counter to islandwide trends. This point, in turn, leads to the problem of which of the dozen or so food resource dimensions might be important at a particular, critical moment. If a food becomes critically scarce as a result of a cane field being burned (yet some seed survives on the field edge), "distance to dense cover" may become extremely important. On the other hand, if a road crew clears several kilometers of road edge vegetation, "human disturbance" assumes greater significance.

The degree of plasticity of Puerto Rico's granivores, with respect to each of the seven resource factors reviewed, particularly under conditions of food stress, is another unknown. Will *E. melpoda* feed in the open when no dense cover is available? Can *L. malacca* obtain seeds from thin-stemmed grasses when there are no adjacent strong perches from which to reach the seed? Precise measurements of foraging plasticity must come from controlled experiments on caged birds.

In addressing the question of what makes coexistence among these granivores possible, several assumptions must be made: 1) food resources periodically occur at levels that affect bird survivorship; 2) most of the food resource dimensions are important, at least periodically, during episodes of food resource limitation; and 3) plasticity of foraging behaviors among the seed eaters is relatively limited.

Given these assumptions, the coexistence of four of the six seed eaters studied can be explained to some extent by the distinctive feeding behaviors summarized in Table 6. The two exceptions are *E. melpoda* and *L. cucullata*, which show no distinctive feeding characteristics or specializations. Until 15 years ago *L. cucullata* had a much broader range in Puerto Rico than did *E. melpoda*. This could reasonably explain the survival of *L. cucullata* on the island, but in the case of *E. melpoda* it would have to be argued that this species could out-compete *L. cucullata* where the two species were syntopic. Since, in recent years, the ranges of *L. cucullata* and *E. melpoda* have come to overlap much more extensively, it is possible that the coexistence of these two granivores could become more tenuous.

Of course, the initial successful coexistence of native and introduced seed eaters may not necessarily be sustained. Even if the three critical assumptions hold, there is no reason to believe the various mechanisms for food resource partitioning are equally important. Also, though limited plasticity is assumed for each bird species, evolutionary shifts in behavior that would affect interspecific relationships are not precluded.

It is particularly important to examine the relationships of Puerto Rico's native seed eaters to the exotics to determine whether the former may be threatened. Fortunately, both indigenous granivores appear to differ from the exotics across several feeding dimensions. In the case of *Tiaris bicolor*, habitat preferences alone are probably enough to ensure its survival. Presently, most of Puerto Rico's scrub thickets and overgrazed pastures are occupied solely by this bird. This broad distribution is probably a consequence of the species' tolerance of areas with sparse food resources and the bird's broad dietary range, which includes forbs. *Tiaris olivacea*, on the other hand, differs from most other species along many food gathering dimensions. This is probably a good sign because even if only a few of the feeding dimensions analyzed are important, the chances are high that *T. olivacea* occupies a unique position in one or more of them.

The relationship between the similar diets, but disparate bill measurements of *L. cucullata* and *E. melpoda* has already been discussed. Another interesting comparison of bill measurements with diet data is between *T. bicolor* and *T. olivacea*. The average culmen lengths of both species are very similar (ratio=1.03), while their diets appear to be very different. This is virtually the converse of the *L. cucullata* and *E. melpoda* comparison and both cases suggest an interesting point. Similarities between species for a commonly studied variable (vogue variable) such as culmen length, or food type, may or may not be important reflections of relationships. Numerous other factors, many of them not in vogue, but none-the-less probably relevant (valuable variables), as suggested here, may be equal to, or more important than, vogue variables in determining how species partition food resources among themselves. Most likely it is the interplay of all of these variables, particularly as food resource availability becomes stressed, that matters.

The present study has not pinpointed all the methods of resource partitioning that enable Puerto Rico's seed eaters to coexist with each other, even among the six species intensively studied. However, this investigation identified factors that may be ecologically important and it forms a base for further studies on the structure and dynamics of this community of grassland granivores. Items that should merit particular attention in subsequent, related research are:

1) Controlled aviary testing of behavioral plasticity in feeding.
2) More intensive research on Vieques, where the two *Tiaris* species and only one exotic occur.
3) Long-term research on study plots over many seasons and covering all granivore species equally.
4) The collection of additional specimens of Puerto Rico's native and introduced seed eaters for comparisons of bill dimensions against old specimens, or as the first samples of newly established species against which to make future comparisons.
5) Morphometric analyses of *E. melpoda*, especially as to bill length and differences between southwestern and northeastern populations.

Acknowledgements

I would like to express my appreciation to Drs. James Thomson, Robert Smolker, Mike Bell, Ronald Pulliam, and an anonymous reviewer for reviewing this manuscript. The Puerto Rico Department of Natural Resources provided me with housing in the Guanica Commonwealth Forest, and with the necessary collecting permits. Housing assistance was also provided by Barbara and Gilberto Cintron and Doug Pool. Access to the Lajas Agriculture Experimental Station cane plots was courteously granted by Gumersindo Ramirez. This research was supported by a grant from Sigma Xi, a SUNY dissertation fellowship, and was completed in partial fulfillment of a Ph.D. degree at the State University of New York-Stony Brook.

Literature cited

Ali, S. and S.D. Ripley. 1973. Handbook of the Birds of India and Pakistan. Vol. 10, Oxford University Press, London, 334 pp.

Bannerman, D.A. 1949. The Birds of Tropical West Africa, vol. 7, Oliver and Boyd, London, 413 pp.

Berger, A.J. 1981. Hawaiian Birdlife. University Press of Hawaii, Honolulu, HI, 260 pp.

Cable, D.K. 1975. Influence of precipitation on perennial grass production in the semi-desert southwest. Ecology 56:981-986.

Cody, M.L. 1968. On the methods of resource division in grassland bird communities. American Naturalist 102:107-147.

_____. 1974. Competition and the Structure of Bird Communities. Princeton University Press, Princeton, NJ, 318 pp.

Horn, H.S. 1966. Measurements of "overlap" in comparative ecological studies. American Naturalist 100:419-424.

Hurlbert, S.H. 1978. The measurement of niche overlap and some relatives. Ecology 59:67-77.

Long, J.L. 1981. Introduced Birds of the World. Universe Books, New York, 528 pp.

Meyer de Schauensee, R. 1970. A Guide to the Birds of South America. Livingston, Wynnewood, PA, 470 pp.

Moreno, J.A. and L. Luquis. 1985. La proliferación de gorriones exóticos en Puerto Rico. Revista Sociedad Ornithología de Puerto Rico 1:8-10.

Morris, G.L. 1979. Regional development and water resource management: implications of a changing agricultural sector on Puerto Rico's south coast. Unpublished Ph.D. dissertation, University of Florida, Gainesville, FL, 250 pp.

Pulliam, H.R. and F. Enders. 1971. The feeding ecology of five sympatric finch species. Ecology 52:557-566.

Raffaele, H.A. 1983a. A Guide to the Birds of Puerto Rico and the Virgin Islands. Fondo Educativo Interamericano, San Juan, PR, 255 pp.

_____. 1983b. Native and introduced seed eating birds: Measurement bias and community structure. Ph.D. dissertation, State University of New York, Stony Brook, NY. 177 pp.

_____. 1986. Bill dimension comparisons of Puerto Rico's native and long established exotic granivores: Adaptation at work? Ornitología Caribeña 2:16-26.

Schoener, T.W. 1970. Non-synchronous spacial overlap of lizards in patchy habitats. Ecology 51:408-418.

Thomson, J.D. 1980. Implications of different sorts of evidence for competition. American Naturalist 116:719-726.

Wiens, J.A. and J.T. Rotenberry. 1979. Diet niche relationship among North American grassland and shrubsteppe birds. Oecologia 42:253-292.

Wetmore, A. 1916. Birds of Porto Rico. Bulletin of the U.S. Department of Agriculture 326:1-140.

Table 1. Comparisons of crop contents and resource availability for two species of seed eating birds from three different sites in 1980 in Puerto Rico and Vieques.

FOOD TYPE	LAJAS Tb Crops (n=11) %	LAJAS Lc Crops (n=18) %	LAJAS Veg (n=93) %	GUANICA (Rd 2) Tb Crops (n=7) %	GUANICA (Rd 2) Veg (n=26) %	VIEQUES Tb Crops (n=5) %	VIEQUES Lc Crops (n=6) %	VIEQUES Veg (estimate)
P	4	18	9	0	0	1	17	Common
Ec	76	74	33	17	1	3	13	Common
Sh	Trace	0	47	5	0	0	0	0
D	11	8	7	0	0	17	6	Common
Bp	7	Trace	2	49	16	69	39	Common
Cc	0	0	0	1	29	0	0	0
RG	Trace	Trace	2	12	30	2	6	Common
F	Trace	0	Trace	14	23	2	0	Common
AM	1	Trace	?	3	?	6	Trace	?
	99	100	100	101	100	100	101	

Tb = *Tiaris bicolor*, Lc = *Lonchura cucullata*, P = *Panicum* spp, Ec = *Echinochloa crusgalli*, Sh = *Sorghum halepense*, D = *Digitaria* spp, Bp = *Brachiaria platyphylla*, Cc = *Chloris ciliata*, RG = Rare grasses, F = Forbs, AM = Animal Matter.

Table 2. Frequency (%) of perch types used by six species of seed eating birds in Puerto Rico when feeding on the seeds of *Panicum* spp.

PERCH	BIRD			SPECIES		
	Tb $n=66$	To $n=91$	Lc $n=334$	Em $n=72$	Lm $n=39$	Aa $n=51$
Panicum, but a different stem	6	9	5	13	26	14
Other perch sp, or ground	38	19	7	7	21	16
The same *Panicum* as fed upon	56	73	88	81	54	71
Total	100	101	100	101	101	101

Tb = *Tiaris bicolor*, To = *T. olivacea*, Lc = *Lonchura cucullata*, Em = *Estrilda melpoda*, Lm = *L. malacca*, Aa = *Amandava amandava*

Table 3. Perches and feeding techniques used by *Tiaris bicolor* and *Lonchura malacca* to feed on three thin-stemmed grasses (*Echinochloa crusgalli*, *Digitaria* spp and *Chloris ciliata*).

Perch Type	T. bicolor Perch n	T. bicolor Perch %	T. bicolor Behavior n	T. bicolor Behavior Type	L. malacca Perch n	L. malacca Perch %	L. malacca Behavior n	L. malacca Behavior Type
Air	3	7	3	E	0	0		
Ground	24	52	1	B	4	10	3	A
			9	C				
			2	D				
E. crusgalli	0	0			2	5		
Digitaria spp	0	0			1	3		
C. ciliata	2	4			0	0		
Medium-stemmed grs.	1	2			6	15	1	B
Thick-stemmed grs.	14	30	1	A	26	67	4	A
			1	B				
Bushes or Trees	2	4	1	C	0	0		
	46	99			39	100		

A = Held plant with foot; B = Reached to plant, put it under foot and fed; C = Hopped or fluttered up, pulled down head with bill, and held it with foot; D = Hopped or flew up and pecked seed off head in air; E = Struck the grass stem with its bill.

Table 4. Feeding observations of seed eating birds at the Guanica study site ranked according to resource availability at each subsite.

BIRD SPECIES	RESOURCE AVAILABILITY									
	High		Moderate						Low	
	27-8	29	30-1	32	35-6	38	39	40	25	33-4
Tb	11	3	23		15	21	11	12	12	26
To	7	2	6		14	22	5	2	3	2
Em	13		3	15	14				1	
Lm		10	43	10	4					
Aa				6	10					14

Tb = *Tiaris bicolor*, To = *T. olivacea*, Em = *Estrilda melpoda*, Lm = *Lonchura malacca*, Aa = *Amandava amandava*

Table 5. Feeding observations of seed eating birds at the Guanica study site sorted by the degree of human distubance at each subsite.

BIRD SPECIES	DEGREE OF HUMAN DISTURBANCE									
	High		Moderate		M-L		Low			
	39	40	35-6	38	30-1	25	27	29	32	33-4
Tb	12	11	15	21	23	12	11	3		26
To	2	5	14	22	6	8	2	2		2
Em			14		3	3	11		15	
Lm			4		43				10	10
Aa					6				10	14

Tb = *Tiaris bicolor*, To = *T. olivacea*, Em = *Estrilda melpoda*, Lm = *Lonchura malacca*, Aa = *Amandava amandava*.

Table 6. Summary of habitat and food resource partitioning by Puerto Rico's 15 species of small granivorous birds.

NICHE DIMENSION	BIRD SPECIES														
	Tb	To	Lc	Em	Lm	Aa	Lp	Eo	Et	Po	Sf	Vm	Ea	Lmb	Pd
Habitat	1	0	0	0	0	0	0	0	0	0	½	½	½	1	1
Food Type	½	1	0	0	1	0					½	½	½	1	
Low Res. Abund.	1	½	0	0	0	0									
Dist. to Dense Cover	0	1	0	0	0	1									
Dist. to Close Cover	0	0	0	0	0	1	UNIVESTIGATED								
Human Disturb.	1	¼	0	0	0	0									
Adept on Thin Perch	0	¼	¼	0	0	0									

1 = The species is unique, or nearly so, for this dimension.
¼ = The species is slightly unique for this dimension.
½ = The species is moderately unique, for this dimension.
0 = The species is not distinctive from many others for this dimension.

Tb = *Tiaris bicolor*, To = *T. olivacea*, Lc = *Lonchura cucullata*, Em = *Estrilda melpoda*, Lm = *L. malacca*, Aa = *Amandava amandava*, Lp = *L. punctulata*, Eo = *Euplectes orix*, Et = *Estrilda troglodytes*, Po = *Padda oryzivora*, Sf = *Sicalis flaveola*, Vm = *Vidua macroura*, Ea = *Euplectes afer*, Lmb = *L. malabarica*, Pd = *Passer domesticus*.

Figure 1. Food resource utilization as determined from crop analyses of six species of seed eating birds taken in the Lajas Experimental Station, P.R. in the summers of 1979 and 1980. Tb = *Tiaris bicolor*, To = *T. olivacea*, Lc = *Lonchura cucullata*, Em = *Estrilda melpoda*, Lm = *Lonchura malacca*, Aa = *Amandava amandava*.

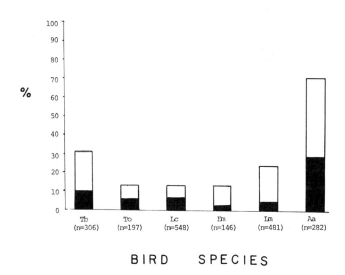

Figure 2. The proportion of ground foraging observed for six species of seed eating birds in Puerto Rico. Tb = *Tiaris bicolor*, To = *T. olivacea*, Lc = *Lonchura cucullata*, Em = *Estrilda melpoda*, Lm = *Lonchura malacca*, Aa = *Amandava amandava*; the darkened portions of the bars represent birds seen foraging under dense cover, the light portions represent birds foraging under moderate or no cover.

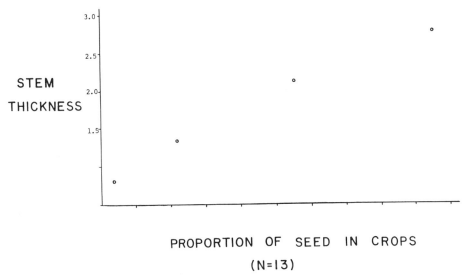

Figure 3. The relationship of stem thickness to the proportion of seed in the crops of *Lonchura malacca*. The grasses, from smallest to largest stemmed, are: *Digitaria* sp., *Echinochloa crusgalli*, *Panicum maximum*, and *Sorghum halepense*.

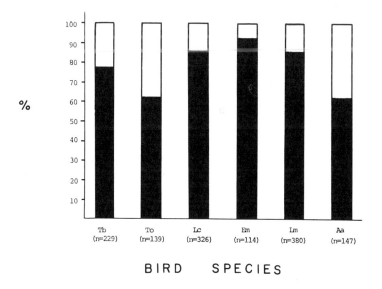

Figure 4. The proportion of individuals of six bird species that foraged greater than or less than 8m from dense cover at all study sites combined. Tb = *Tiaris bicolor*, To = *T. olivacea*, Lc = *Lonchura cucullata*, Em = *Estrilda melpoda*, Lm = *Lonchura malacca*, Aa = *Amandava amandava*; the darkened portions of the bars represent birds seen foraging within 8m or less of dense cover, the light portions represent birds foraging greater than 8m from dense cover.

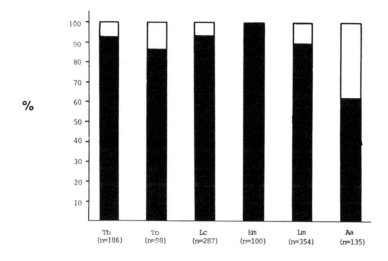

Figure 5. The proportion of individuals of six bird species that foraged greater than or less than 8m from the closest cover at all study sites combined. Tb = *Tiaris bicolor*, To = *T. olivacea*, Lc = *Lonchura cucullata*, Em = *Estrilda melpoda*, Lm = *Lonchura malacca*, Aa = *Amandava amandava*; the darkened portions of the bars represent birds seen foraging within 8m or less of the closest cover, the light portions represent birds foraging greater than 8m from the closest cover.

DISTRIBUTION, STATUS, AND BIOGEOGRAPHY OF THE WEST INDIAN MANATEE

L. W. Lefebvre[1], T. J. O'Shea[1], G. B. Rathbun[2], R. C. Best[3]

Abstract

We review historical and recent information on the distribution, status, and habitat associations of the West Indian manatee, *Trichechus manatus*, summarize threats to its continued survival, and discuss some biogeographical patterns of trichechids. Historical accounts indicate that manatees were once more common and that hunting has been responsible for declining numbers throughout much of their range. Accounts of manatees in the Greater Antilles, however, present little evidence that manatees were ever very abundant there, except possibly in Cuba. Evidence for greatly reduced Central and South American manatee numbers is stronger. A review of recent literature and reports on manatees in Puerto Rico, Jamaica, and Hispaniola indicates that small numbers occur throughout the Greater Antilles, and that opportunistic taking by fishermen is a major source of mortality. Manatees have not been documented to occur in the Lesser Antilles since the eighteenth century, except for rare sightings in the Virgin Islands. Sightings from the Bahamas are also very rare. Manatees are relatively abundant in Belize in comparison with other countries of Central America, and in Guyana in South America. Illegal killing continues to threaten the survival of manatees in many countries, although in some countries manatee hunting appears to be a dying art. Manatees are still reasonably abundant in some areas of Mexico, and on both coasts of Florida in the United States. Despite protective measures to regulate boating activity, collision with boats is still the major cause of man-related manatee mortality in Florida.

Manatees in the Greater Antilles and Central and South America belong to the same subspecies, *T. manatus manatus*. Although recent records of manatees from the Lesser Antilles are lacking, these islands may have been of former importance to manatees migrating between the Greater Antilles and South America. Gene flow may also have been maintained by wanderers between Yucatan and Cuba, and perhaps between the Miskito Bank of Nicaragua and Jamaica. Salinity, temperature, water depth, currents, shelter from wave action, and availability of vegetation are important determining

---

[1] Drs. Lefebvre and O'Shea are biologists with the Sirenia Project, U.S. Fish and Wildlife Service, National Ecology Research Center, 412 N.E. 16th Ave. Rm. 250, Gainesville, FL 32609, U.S.A.

[2] Galen Rathbun is a biologist with the U.S. Fish and Wildlife Service, National Ecology Research Center, P.O. Box 70, San Simeon, CA 93452, U.S.A.

[3] Robin C. Best, formerly with Departamento de Biología de Mamíferos Aquáticos, Instituto Nacional de Pesquisas da Amazônia, Manaus, Brasil, died on 17 December 1986 in Cambridge, England.

factors of manatee distribution. The association of *T. manatus* with freshwater sources is a highly consistent pattern. Throughout most of their range, manatees appear to prefer rivers and estuaries to marine habitats. The Amazonian species may be restricted to the Amazon River because of intolerance of salinity. Physiological studies of metabolic rates as well as seasonal shifts in manatee distribution and their use of available warm water sources (natural springs and heated industrial effluents) in Florida clearly show that energetic requirements influence the range limits of manatees. Cool winters and gaps in suitable habitat on the northern Gulf coast, and the Straits of Florida to the south, serve as geographic barriers that isolate the Florida subspecies, *T. m. latirostris*.

Introduction

There are four living species of the mammalian order Sirenia (sea cows): the dugong (*Dugong dugon*) and three manatees (*Trichechus* spp.). The dugong inhabits tropical and subtropical coastal waters in the Indian and western Pacific Oceans. The Amazonian manatee (*T. inunguis*) occurs only in fresh water in the Amazon river system, and the West African manatee (*T. senegalensis*) occurs in both freshwater and coastal marine habitats of tropical West Africa. The West Indian manatee (*Trichechus manatus*) is the most widely distributed of the three trichechids. The common name 'West Indian manatee' is not fully descriptive of the range of *T. manatus*, which includes the coasts and many of the rivers of Florida, the Greater Antilles, eastern Mexico and Central America, and northern and eastern South America. The purpose of this paper is to review historical and recent information on the distribution, status, and habitat associations of *T. manatus*, to summarize threats to continued survival of this species throughout its range, and to discuss some biogeographical patterns of trichechids.

Because unpublished material is cited throughout this paper, several different types of citations are used: in litt. refers to personal communications by letter; unpubl. reports refers to unpublished reports that do not appear in the LITERATURE CITED section; and pers. comm. refers to oral personal communications. All material that is cited as in litt., unpubl. report, or unpubl. data is in U.S. Fish and Wildlife Service Sirenia Project files, and can be obtained at the address of the first author.

Historical distribution

Historical accounts indicate that manatees were once more common and that hunting has in part been responsible for declining manatee numbers throughout their range (Thornback and Jenkins 1982). McKillop (1985) pointed out that early reports of manatee distribution and exploitation, such as Dampier's (1699), may demonstrate that manatees were previously more common and a reliable meat source, whereas current ethnographic accounts indicate that in many areas manatees are now scarce, difficult to capture, and not a major food source. Nevertheless, many of the areas where manatees were hunted prehistorically and historically still have manatees (Fig. 1). The historical uses of manatee meat, oil, bones, and hide have been described by many authors, for example True (1884), Allen (1942), Rouse (1964), Bertram and Bertram (1973), and Husar (1977). It is probably fair to say that if manatee meat had not been highly esteemed by pre-Columbian inhabitants and early explorers of the West Atlantic region, we would know much less about this species former range and abundance.

With the possible exception of Cuba, little historical evidence exists that indicates manatees were ever very abundant in the Greater Antilles. Columbus (in 1494) reportedly found freshwater springs in the Bahía de Cochinos, Cuba, which attracted "swarms" of manatees (Morison 1942). Oviedo reported manatees on the coast of Cuba in 1520, described the use of tethered remoras by Indians to locate and capture manatees, and noted hunting by Spaniards with cross bows (Cuni 1918). Acosta also saw manatees in Cuba, as well as "St. Dominicke, Portrique and Jamaique" (Hispaniola, Puerto Rico, and Jamaica) in 1588, and ate manatee meat in Hispaniola (Baughman 1946). Dampier (1699) saw manatees in Cuba and had heard of their occurrence on the north coast of Jamaica. According to Cuni (1918), manatees were abundant in Cuba's river mouths and estuaries before 1866, but since then they have been reduced in numbers. Gundlach (1877) also described manatees in earlier times in Cuba as very abundant, but in his day as much reduced although not rare. Barrett (1935) optimistically concluded that manatees were once plentiful in estuaries on the central north coast of Puerto Rico because a town there was named Manatí by the Spanish. Evermann (1902) described the manatee in Puerto Rico as being "of very rare occurrence." Although Fewkes (1907) lists the manatee as one of many species that contributed to the Puerto Rican diet, he reported that fish, crabs, and small mammals were the most important food animals in the West Indies. Duerden (1901) did not mention manatees in his account of fisheries in Jamaica, the Bahamas, or the Leeward Islands (the northern Lesser Antilles), whereas he specifically mentioned them in his coverage of Trinidad and Tobago. Both Duerden (1901) and Neish (1896) reported that eight manatees had been captured in the vicinity of Old Harbour, Jamaica in the previous 7-10 years; Neish described them as uncommon, and Duerden seemed unimpressed with the manatee's potential for greater economic importance, noting that they are "very slow breeders."

A small number of manatee bones or bone fragments have been recovered at Amerindian sites in Jamaica (Wing and Reitz 1982), Haiti (Wing pers. comm.), the Dominican Republic (Miller 1929), Puerto Rico (Wing 1987 pers. comm.), and Vieques Island off Puerto Rico (Narganes-S. 1982). Rouse (1986:133) speculated that flint points crafted by Amerindians in Hispaniola 3,000-5,000 yBP were used as spear heads for hunting manatees, since he assumed that large land mammals were absent. However, Rouse (1986) noted that early inhabitants of Cuba and Hispaniola were oriented primarily towards the land, and he may have overlooked ground sloths as an alternative large prey species (Morgan and Woods 1986). Manatees were commercially exploited in Surinam (de Jong 1961, Husar 1977), Guyana (de Jong 1961, Bertram and Bertram 1973), and Brazil (Domning 1982a) for export to the West Indies during the seventeenth through the nineteenth centuries, which suggests that manatees may not have been very abundant historically around these islands. In the Lesser Antilles, manatee remains have been identified at Amerindian archeological sites on five islands: St. Kitts, Barbuda, Antigua, St. Lucia, and Grenada (Ray 1960; Wing et al. 1968; Wing and Reitz 1982; Watters et al. 1984; Steininger 1986). Vertebrate remains collected at other excavation sites on St. Kitts, Antigua, St. Lucia, and Barbados did not include manatees (Wing 1967; Wing et al. 1968; Wing and Scudder 1980). Except for rare sightings in the Virgin Islands (W.E. Rainey in litt.; D.W. Peterson 1982 in litt.), manatees have not been documented to occur in the

Lesser Antilles since the eighteenth century (Allen 1942; Ray 1960). Several nineteenth century authors included the Islands of Marie-Galante and Martinique in the manatee's range, but gave no new documentation of their presence (Ray 1960).

The evidence for recent large-scale reductions in Central and South American manatee numbers is stronger. Notable declines are thought to have occurred in Honduras (Klein 1979; Rathbun et al. 1983a), Costa Rica and Panama (Husar 1977; O'Donnell 1981), Venezuela (Bertram and Bertram 1973; Mondolfi 1974; O'Shea et al. 1988), and Brazil (Domning 1981a). Dampier (1699) saw manatees along the coasts of Mexico and Central America, and heard of great numbers occurring in the rivers of Surinam. Historical accounts exist for manatees along the coast of Guatemala (Bradley 1983). Barrett (1935) claimed that one of the best known "herds" of manatees on the Caribbean coast inhabited the Indio River and its bayous in Nicaragua, where hunters took many. Von Frantzius reported manatees to be very common in several rivers in Costa Rica in 1868 (Allen 1942). The Bahía Almirante and Bocas del Toro area were documented as a region where buccaneers were familiar with manatees and provisioned themselves with manatee meat during the 1600's (Dampier 1699; O'Donnell 1981). If manatees currently persist at all in this region, their numbers have been severely depressed (Husar 1977; O'Donnell 1981). Maack reported in 1874 that manatees were frequently caught in the Atrato and Cacarica Rivers in Colombia (Allen 1942). True (1884) cited Brandt's 1868 literature review of manatee abundance in South America, which indicated that in many regions, particularly around river mouths or other places where shelter was lacking, manatees were disappearing or extirpated. In the upper reaches of rivers, however, Brandt concluded that hunting pressure may have been less severe (True 1884). By the end of the nineteenth century, West Indian manatees disappeared from much of the east coast of Brazil (Whitehead 1977).

Manatees historically occurred along the entire Mexican coast of the Gulf of Mexico and Caribbean Sea (Campbell and Gicca 1978). Manatee meat contributed to the prehistoric diet in at least two sites in Mexico, near Alvarado and in northern Quintana Roo (Fig. 1) (McKillop 1985). Baughman (1946) cited several sixteenth and seventeenth century authors that reported manatees along the coast of Yucatan. Middens (A.D. 400-700) on Moho Cay near what is now Belize City have yielded the largest number of manatee bones of all greater Caribbean prehistoric sites thus far investigated (McKillop 1985). Manatee remains have been reported from several other coastal and inland archaeological sites in Belize (Bradley 1983), and manatees from Belize were provided to seventeenth century privateers as a meat source (O'Donnell 1981). Prehistoric evidence exists for the occurrence of manatees at the Islas de la Bahía in the offshore Caribbean near present day Honduras (Strong 1935), but they are no longer found in the region (Klein 1979; Rathbun et al. 1983a).

Hartman (1972, 1974) believed that hunting during the seventeenth through the nineteenth centuries reduced the number of manatees in Florida to a few relict groups, but he presented no evidence of their former greater abundance. Harlan (1824) cited Burrows' description of considerable numbers of manatees about the mouths of rivers near "the capes of East Florida" (according to Moore [1956], the south end of Key Biscayne in the Miami area). Allen (1942) concluded that evidence of manatees having once been much more abundant in Florida was subjective and unconvincing, although he stated, without reference to a specific source, that the manatee had become rather scarce in many parts of Florida by about 1890. He may have been overly influenced by information from Bangs (1895), who described the winters of 1894 and 1895 as exceptionally cold, resulting in freeze-related manatee mortality in the Sebastian River. Excessive drainage of wetlands has also been blamed for the manatee's disappearance from some regions of Florida (Trumbull 1949).

Cumbaa (1973) determined that aboriginal use of manatees in Florida was almost exclusively restricted to inland and coastal riverine sites (Fig. 1), and concluded that manatees were not abundant enough to supply a stable food resource to the Paleo-Indians of Florida. In 1773 Bartram (1791) found skeletal remains of a manatee that he thought had been killed by Indians at Manatee Spring on the Suwannee River. There are few other records of manatees using the Gulf coast of Florida from the Suwannee River south to the Chassahowitzka River before the mid-1900's (Powell and Rathbun 1984). Allen (1942) attributed a slow increase in manatee numbers in Florida to an 1893 State law that prohibited their killing.

Present distribution, status, and habitat associations
West Indies

Puerto Rico

Manatees are unevenly distributed around the coast of Puerto Rico (Fig. 2), with greatest concentrations observed during aerial surveys on the south-central and eastern shores and none on the northwestern shore (Powell et al. 1981; Rathbun et al. 1985a). In 10 monthly surveys flown from June 1978 through March 1979, a mean of 22.6 þ 12.6 manatees was sighted, with a range of 11-51 (Powell et al. 1981). In 12 monthly surveys flown from March 1984 through March 1985, a mean of 43.6 þ 13.1 was sighted, with a range of 20-62 (Rathbun et al. 1985a). Both Powell et al. (1981) and Rathbun et al. (1985a) reported that slightly over one-third of their sightings came from around the Roosevelt Roads Naval Station on the eastern end of the island, and the northwestern shore of neighboring Vieques Island (Fig. 2). Manatees were not seen near Mona or Culebra Islands (Powell et al. 1981). Manatees have been seen near Icacos and Diablo Islands, offshore of Cabo San Juan, in the late 1970's (T. Carr pers. comm.), and a manatee was entangled in a fishing net off Culebra in 1982 (Jimenez 1982). A similar proportion of calves was seen in both aerial survey studies: 6.4 $\pm$ 4.9% (Powell et al. 1981) and 7.6% (Rathbun et al. 1985a). Comparison of 1978/79 and 1984/85 survey results may not be entirely valid because of differences in observers, methods, and survey conditions. It appears, however, that the number of manatees in Puerto Rico has at least not declined since 1978.

The greatest difference in the 1978/79 and 1984/85 surveys was the proportion of sightings along the south-central coast: 47% (Powell et al. 1981) versus 19.8% (Rathbun et al. 1985a). Rathbun et al. (1985a) suggested that the observed decline in manatee numbers in this region may be related to industrial development, and noted that three carcasses were recovered from the south coast in the vicinity of the Union Carbide plant at Guayanilla. An increase in commercial boat traffic may be influencing manatee distribution and mortality in this region.

Fifteen manatee carcasses have been recovered or verified in Puerto Rico from 1975 through 1987, (U.S. Fish and Wildlife Service unpubl. data). At least four deaths were human-related (three manatees drowned or were killed following entanglement in fishermen's gill nets and one was struck by a boat) (Rathbun et al. 1985a). Two cases involved dependent calves, and in nine cases cause of death was undetermined (U.S. Fish and Wildlife Service unpubl. data). Five carcasses were verified from the north coast, eight from the south coast, and two from near Roosevelt Roads Naval Station on the east coast. Interviews indicated that unattended gill nets are a major cause of mortality that is not always documented. Manatees are protected by the U.S. Marine Mammal Protection Act and Endangered Species Act, as well as by several Commonwealth of Puerto Rico laws: La Ley de Pesca del Estado Libre Associado (1943), La Ley de Vida Silvestre (1977), and the Regulation to Govern the Management of Threatened and Endangered Species in the Commonwealth of Puerto Rico (1985) (D. Mignogno 1987 in litt.).

Evermann (1902) recognized the association between animal distribution and the physical environment of Puerto Rico. He attributed the rarity of manatees "... to the absence of broad sluggish rivers in which it finds its favorite environment." Barrett (1935) related a decline in manatees along the north coast of Puerto Rico in part to "silting-up" of river mouths, which prevented manatees from grazing on shoreline grasses in river estuaries. Powell et al. (1981) also believed that manatee distribution in Puerto Rico is influenced by the availability of fresh water. Almost all manatees seen during their nearshore surveys were in the ocean; however, 85.8% of the sightings were within 5 km of natural or artificial freshwater sources. Interviews indicated that manatees visit the mouths of the Loiza, Fajardo, and Guanajibo Rivers, and after heavy rains they ascend these rivers for short distances. Powell et al. (1981) concluded that other Puerto Rican rivers are too shallow for manatees to ascend. Rathbun et al. (1985a) did not detect any dramatic seasonal shifts in distribution along the coast, nor was there a significant correlation between total monthly rainfall and the total number of manatees seen per month.

Powell et al. (1981) noted the association of manatee distribution and seagrasses, describing *Thalassia*, *Syringodium*, and *Halodule* as probably the most important manatee foods in Puerto Rico. Seagrass beds are relatively sparse along the northwest coast where no manatees were seen, and are extensive on the eastern and southern coasts of the island (Powell et al. 1981). The 20 m contour is much closer to the shore along the northwest and north-central part of Puerto Rico (Fig. 2), reflecting a steeper drop in the ocean floor. The north shore is also the windward shore, and lacks embayments to buffer the effects of wind and surf. Powell et al. (1981) first documented that manatees use the Cape Hart sewage treatment plant effluents on Roosevelt Roads Naval Station as a source of fresh water for drinking. Rathbun et al. (1985a) further documented the importance of these outfalls by demonstrating that manatees are likely to be present at any time, showing no seasonal, tidal, or diurnal changes in their preference for this site. Rathbun et al. (1985a) also noted the association of manatees and seagrass beds, and suggested that the higher incidence of cow/calf pairs observed at Roosevelt Roads Naval Station and northwestern Vieques Island may have been related to the abundance of seagrass beds at this site, the availability of fresh water at Cape Hart, and the protection from human harassment provided by security procedures at the naval base.

Jamaica

Manatees occur primarily along the southern coast, to the west of Kingston (Fairbairn and Haynes 1982). Between May 1981 and April 1982, 13 aerial surveys were conducted along the entire coastline, and manatees were observed primarily in Portland Bight, off the parishes of St. Catherine and Clarendon, and off St. Elizabeth and Manchester Parishes, between the Black River and the Río Minho (Fig. 3) (Fairbairn and Haynes 1982). Two were observed off St. Thomas Parish, and 1 off St. Ann Parish (Fig. 3). The range in number of manatees seen during the monthly surveys was 1-13, with a mean of 6 (Fairbairn and Haynes 1982). The average number of manatees seen per flight hour in Jamaica (1.27) was higher than the number seen per hour in Haiti (0.64), where only one entire-coast survey was flown, but much lower than the average number per hour seen in Puerto Rico (9.1) (Hurst 1987). Powell (1976, unpubl. report) conducted an aerial survey of the coast from St. Mary Parish clockwise to St. Elizabeth Parish in September 1976, and saw only one manatee, at Alligator Reef off Manchester Parish.

Manatee sightings are occasionally reported on the north coast (Hurst 1987). Powell (1978) reported piscivory from fishermen's nets by manatees off St. Mary and Trelawny Parishes on the north coast, near Port Maria and Río Nuevo. Interviews with local residents during the mid-1970's indicated that the mouth of the Black River, St. Elizabeth Parish, was the only area in Jamaica where manatees were commonly seen, and that fishermen there take manatees opportunistically by net (Crombie 1975 unpubl. report; Powell 1976 unpubl. report). Manatees are now most frequently reported from Treasure Beach (St. Elizabeth Parish), Alligator Pond (Manchester Parish), and Farquhar's Beach (Clarendon Parish), and have not been reported in recent years from the Black River (Haynes-Sutton 1987 in litt.)

Despite public education on the manatee's protected status, Jamaican fishermen continue to take and sell manatees illegally (Hurst 1987). Deliberate capture of manatees is most frequent in regions that are economically depressed (Haynes 1986 in litt.). At least three manatees were captured on the south coast of Jamaica in 1985; two were killed for meat and one was rescued by the Natural Resources Conservation Department and released in the Alligator Hole River (Hurst 1987). In 1987, at least one manatee was killed illegally at Alligator Pond (Haynes-Sutton 1987 in litt.). Hurst (1987) believes that fishermen are responsible for nearly all human-related manatee mortality in Jamaica. Manatees are captured following accidental entanglement in gill nets and beach seines; fishermen claim that manatees frequently drown before their presence is noted (Shaul and Haynes 1986). Hurst (1987) attempted to estimate net entanglement losses from fishing boat statistics, and concluded that St. Elizabeth Parish has the greatest potential for fishing-related mortality. Interviews with local residents indicated that manatee numbers were generally believed to have declined in recent years (Crombie 1975 unpubl. report; Powell 1976 unpubl. report).

The south coast of Jamaica has extensive areas of shallow water, numerous bays and other areas of calm water, freshwater sources, and seagrass beds, which create favorable habitat for manatees (Hurst 1987). The north coast has a deep and rugged shoreline, but also has a number of river mouths. Manatees have been observed feeding on *Ceratophyllum* and the starchy root of *Phragmites* in the rivers of south Clarendon and Manchester Parishes (Shaul and Haynes 1986). Several manatees in the Alligator Hole River were observed feeding on *Ceratophyllum*, *Potamogeton*, and *Phragmites*, and seemed to prefer the lower part of the river, where the undercut banks provide cover (Hurst 1987). These manatees were impounded in the Alligator Hole River as part of a program, Operation Sea Cow, initiated by the Natural Resources Conservation Department in 1980 to publicize manatees and encourage research and conservation. Manatees in Jamaica are protected by the Wild Life Protection Act (1945), which prohibits hunting and possession of protected species (Shaul and Haynes 1986). Law enforcement efforts, however, are inadequate (Haynes-Sutton 1987 in litt.).

Dominican Republic

Husar (1977) reported manatees to be distributed along the southwest and the entire north coasts of the Dominican Republic, with the greatest concentrations occurring in Bahía de Neiba and near Las Terrenas. Belitsky and Belitsky (1980) conducted six entire-coast aerial surveys every other month from February through December 1977. Manatees were sighted along the north coast from Manzanillo to Miches, and along the southwest coast from Isla Beata to Bahía de Ocoa (Fig. 4); the only sighting outside of these areas was northwest of Isla Saona on the southeast coast. Belitsky and Belitsky's (1980) surveys indicated that manatees were more common around Monte Cristi than Las Terrenas on the north coast, and common in Bahía de Ocoa as well as Bahía de Neiba on the southwest coast, but in general their findings conform to the report by Husar (1977). The mean number of sightings per survey on the north coast was 12.3 (range = 2-30), and the mean on the southwest coast was 7.5 (range = 1-11). Seasonal changes in distribution and abundance were not apparent. Interviews indicated that manatees occur in several other areas: on the east coast, from Boca de Yuma south to Isla Saona; near Nizao on the south coast; in coastal waters between Isla Beata and the mainland; and near Pedernales (Belitsky and Belitsky 1980). Local fishermen and residents also reported that manatees once ranged over more extensive areas of the coastline, and Belitsky and Belitsky (1980) attributed apparently reduced numbers to hunting pressure and degradation of habitat by land development.

The reported sources of manatee mortality in the Dominican Republic are poaching and shark predation (the latter may actually be scavenging) (Belitsky and Belitsky 1980). Five manatees were taken in 1976: three near Nizao, one east of Monte Cristi, and one near Pedernales (Belitsky and Belitsky 1980). Manatees were frequently sighted by fishermen in Bahía de Samaná and occasionally caught accidentally in their nets (Mortensen 1975 in litt.). Interviews conducted by Irvine (1975 unpubl. report) indicated that manatee meat is highly prized by fishermen in Bahía de Samaná, and that manatees are rarely seen on the north side of the bay, possibly because of fishing pressure.

Most of the river mouths in the Dominican Republic are periodically blocked by sand bars, and manatee habitat has been described as coastal marine rather than estuarine (Belitsky and Belitsky 1980). Fishermen frequently report that manatees are attracted to springs and river mouths, to drink fresh water (Ottenwalder 1987 in litt.). Interviews conducted in 1975 indicated that manatees frequent river mouths on the north coast during the rainy season, and are regular visitors to Río San Juan and Río Yasica (Campbell and Irvine 1975 unpubl. report). Manatees were seen within 1 km of springs northwest of Isla Saona (Tres Hermanas), in Bahía de Samaná (La Guázuma), and near Barahona (Playa de Saledilla) during at least one aerial survey (Belitsky and Belitsky 1980). Manatees were reported to visit the Massacre, Yaque del Norte, San Juan, Bajabonico Isabela, and Yaque del Sur Rivers; however, none were seen in aerial surveys or site visits to these rivers (Belitsky and Belitsky 1980). Reports of manatees at other river mouths or springs include: Playa Grande, northeast of Río San Juan; Playa de Rincón and Bahía de Rincón; Río Caño del Agua near Barahona; Río Cosón and Arroyo Cañada Salada, near Punta Cosón, Samaná; La Poza, east of Las Terrenas, Samaná; Bahía de Manzanillo; Los Patos y Bahía Regalada, Barahona; Estero Hondo y Punta Rucia, Puerto Plata; and Bahía de Yuma (Ottenwalder 1987 in litt.). Crombie (1975 unpubl. report) conducted a helicopter survey of the coast between El Peñón and Boca de Yuma, including Isla Saona. He found the coast near Peñón (the site of Tres Hermanas springs), Bahía de Palmillas, and the south coast of Isla Saona to have the most suitable manatee habitat in this area (shallow, calm waters, with a few small *Thalassia* beds). He saw no manatees, and interviews indicated that manatees were scarce but most frequently seen during the rainy season when they came closer to the coast, reportedly to drink fresh water in nearshore springs. Manatees are officially protected in the Dominican Republic (Husar 1977). However, law enforcement is not entirely effective (J. Ottenwalder 1980 in litt.).

Haiti

One entire-coast aerial survey was conducted in May 1982, and eight manatees were observed (Rathbun et al. 1985b). Manatee sightings occurred within a very small portion of the western coastline, between Gonaives and Montrouis (Fig. 4). Interviews with coastal residents in 1982 and 1983 indicated that few people under 50 years old had first-hand knowledge of manatees, and the only other area where manatees were reported to have occurred recently was in the Bay of Jacmel in 1977 and 1978.

Manatees are caught opportunistically in beach seines; however, traditional hunting apparently is no longer practiced, probably because of a decline in manatee numbers over the past 50 years (Rathbun et al. 1985b). The status of manatees in Haiti is extremely tenuous because their known range is so restricted, and the areas where they occur are important to fishermen in this densely populated and economically depressed country. Several areas were identified as having shallow and protected waters, extensive submerged vegetation, and rivers (Fig. 4), including the area where manatees were observed at the mouth of the Arbonite River.

Cuba

Manatees occur along both the north and south coasts of Cuba, and in 1978 were reported to be most frequently seen in the Hatiguanico River in the Zapata Swamp and in Ensenada de la Broa (Fig. 5) (Thornback and Jenkins 1982). Fishermen were surveyed to determine manatee distribution and abundance in western Cuba between February 1984 and February 1985 (Estrada and Ferrer 1987). Manatees are reportedly abundant from Mariel to Cabo San Antonio on the north coast and from Cortés to Bahía de Cochinos on the south coast (Fig. 5) (Estrada 1986 in litt.). There are three areas where manatees are particularly abundant: Golfo de Guanahacabibes, between Cortes and La Coloma, and Ensenada de la Broa/Río Hatiguanico (Fig. 5) (Estrada and Ferrer 1987). Sightings were most commonly of solitary manatees, groups of 4 or more, and pairs, in that order.

Cuba has extensive areas of shallow, protected coastal waters (Fig. 5) and many rivers on both the north and south coasts. Estrada and Ferrer (1987) reported that manatees were most frequently sighted along sheltered coasts with extensive shallow areas offshore. Cuni (1918) described the habitat of manatees in Cuba as more riverine than marine, and noted manatees feeding on river bank grasses with part of their bodies out of water. He also noted their use of freshwater springs. Estrada and Ferrer (1987) noted that almost no reports were received of animals in interior waters, and concluded that like manatees in the Dominican Republic and Puerto Rico, Cuba's manatees occupy coastal marine habitat, where they feed in large meadows of aquatic plants. Interviews indicated that *Syringodium filiforme* and *Thalassia testudinum* are preferred by manatees (Estrada and Ferrer 1987).

Varona (1975 in litt.) described the status of the manatee in Cuba as rare and declining alarmingly, because of pollution and pursuit by humans for its flesh, fat, and hide. In contrast, Estrada and Ferrer (1987) reported that more than 50% of those interviewed had seen a manatee within the last year, and believed that manatees were abundant and had increased in number in the last 10 years. Estrada and Ferrer (1987) acknowledged, however, that changes in riverine habitat may have resulted in the manatee's shift to a more marine existence in Cuba. They recommended that the rest of Cuba be surveyed for manatee presence, and that studies be initiated in the areas of greatest manatee abundance to facilitate management and conservation efforts.

Bahamas

The earliest record of a manatee in the Bahamas was in 1904 in the Bimini Islands (Allen 1942). Odell et al. (1978) observed a manatee in a boat basin at West End, Grand Bahama Island in September 1975, and reported several other probable sightings of manatees on Grand Bahama Island between 1965 and 1977. A dead manatee, possibly the same one observed in September, was found at Freeport, Grand Bahama Island in November 1975 (Odell et al. 1978). Odell et al. (1978) described the habitat at West End as "ideal" for manatees because of the presence of extensive *Thalassia* beds, but noted the possibility that limited freshwater sources in the Bahamas could restrict the distribution and size of a resident group of manatees, if one exists there. Andros Island, the largest in the Bahamas, may not have suitable manatee habitat on the east coast because reefs extend out to the Tongue of the Ocean, a deep trough that parallels the island. The west coast of Andros is bordered by extensive shallows, and since there are few people on this side of Andros (Kranz 1981 in litt.), it is possible that manatees occur but are undetected. Alternatively, the rare sightings of manatees in the Bahamas may be of strays from Florida or the Greater Antilles.

North America

United States

Florida is the northern limit of the manatee's year-round range, although they are frequently reported along the southern Georgia coast in warm seasons. Sightings on the Atlantic coast drop off markedly north of Georgia, with the northernmost record from the Potomac River in Virginia (Rathbun et al. 1982). The Suwannee River appears to be the present northern limit of the manatee's usual range on the Gulf coast of Florida

(Powell and Rathbun 1984). Manatees reported in Texas are believed to be strays from Mexico, whereas sightings in Louisiana and further east are believed to represent wanderers from Florida (Powell and Rathbun 1984). The Florida manatee represents a subspecies (*T. m. latirostris*) of the West Indian manatee that is distinct from *T. m. manatus* in the greater Caribbean region and northeastern South America (Domning and Hayek 1986).

Manatees occur along all of the east coast and most of the west coast of Florida from March through October (warm seasons) (Fig. 6a). Foci of abundance on the east coast are the St. Johns River, the Banana and Indian Rivers to Jupiter Inlet, and Biscayne Bay; manatees are most abundant on the west coast at the Suwannee, Crystal, and Homosassa Rivers, the Charlotte Harbor/Matlacha Pass/San Carlos Bay region, and in the creeks, rivers, and bays of the Everglades (Moore 1951; Hartman 1974; Odell 1979; Rose and McCutcheon 1980; Bengtson 1981; Irvine et al. 1982; Shane 1983; Kinnaird 1985).

When ambient water temperatures drop below about 20°C in autumn and winter, manatees migrate to natural or man-made warm water sources (Fig. 6b) (Powell and Waldron 1981; Irvine 1983; Powell and Rathbun 1984). Hartman (1979) believed that manatees began to frequent the headwaters of the Crystal and Homosassa Rivers in the early 1960's. Here their winter numbers have increased markedly over the last 15 years (Powell and Rathbun 1984; Rathbun et al. in press), with a record high count in January 1987 of 204 (U.S. Fish and Wildlife Service, unpubl. data). The heated effluent waters of power plants also attract manatees during the winter. Before the proliferation of coastal electrical generating power plants in the 1950's-1960's, Moore (1951) believed that the northern limit to the normal winter range of manatees was the Sebastian River on the east coast and Charlotte Harbor on the west coast. In January 1985, an aggregation of over 300 individuals was sighted at the Fort Myers power plant, and over 200 individuals were sighted at both the Riviera Beach and Port Everglades power plants (Fig. 6b) (Reynolds and Wilcox 1986). Appreciable numbers also occur at power plants in the upper Indian River and Tampa Bay (Fig. 6b). Manatees often leave warm water refuges to forage during the warmest hours of the day or during warm weather between cold spells. For example, manatees move between the Riviera Beach power plant and Hobe Sound, where they feed on seagrasses (Packard 1981). Manatees frequently return to the same winter ranges each year (Rathbun et al. 1983b; Powell and Rathbun 1984), and some may also return to the same summer ranges (Bengtson 1981; Shane 1983). Individual manatees may use different aggregation sites within winters, and use of sites as far as 850 km apart between winters has been documented (Reid and Rathbun 1986).

The minimum number of Florida manatees is estimated to be 1200 (Florida Department of Natural Resources, unpubl. data), with approximately equal numbers on the east and west coasts. Regional and seasonal fluctuations in manatee abundance do not necessarily indicate changes in the statewide population trend, as manatees on both coasts of Florida migrate seasonally (Hartman 1974) and are capable of moving great distances (Rathbun et al. 1983b). The Florida population shows no regional genetic differentiation (McClenaghan and O'Shea 1988). The manatee's north-south seasonal migrations, promiscuous mating system, and wide-ranging movement patterns of individuals on both coasts contribute to the high rate of gene flow among regions within Florida (McClenaghan and O'Shea 1988). The levels of genic variability observed in Florida manatees suggest that this population has not been subjected to the loss of variability caused by bottlenecks in population size that are characteristic of some endangered species (McClenaghan and O'Shea 1988).

Florida manatees are protected under the U.S. Marine Mammal Protection Act (1972) and are listed as endangered under the Endangered Species Act (1973). From April 1976 through March 1981, 406 dead manatees were recovered in Florida (O'Shea et al. 1985). The major known cause of manatee mortality is collision with boats. Other human-related causes are entrapment in gates of locks and dams, vandalism, drowning in commercial fishing nets, and infection resulting from entanglement in fishing gear (O'Shea et al. 1985). High winter mortality in 1977 and 1981 was related to exceptionally cold weather (O'Shea et al. 1985), and a die-off of manatees in southwest Florida in 1982 was associated with a red tide outbreak (Buergelt et al. 1984). Manatees are thus vulnerable to both human-related and natural threats, and their vulnerability is increased in areas where large aggregations assemble. The U.S. Fish and Wildlife Service and the Florida Department of Natural Resources, under the authority of the Endangered Species Act (1973) and the Manatee Sanctuary Act (1978), have established 21 manatee sanctuaries with restrictions on boat speed. Development of non-legislative programs has helped to raise funds and develop public awareness and support for measures such as slow boat speed zones, sanctuaries, and redirection of waterfront development (Packard and Rose 1985). As the human population in Florida continues to grow, local implementation of regulations affecting manatee habitat will become increasingly important (Packard 1983; Packard and Wetterqvist 1986).

Hartman (1979) listed several factors that influence manatee distribution in Florida: 1) availability of aquatic vegetation, 2) proximity to channels of at least 2m in depth, 3) recourse to warm water during cold weather, and 4) a source of fresh water. He concluded that the Florida manatee's preferred habitats are rivers and estuaries (<25 ppt salt). The rare occurrence of manatees in Florida Bay and in the area between Tampa Bay and the Chassahowitzka River may be related to shallowness and low freshwater input. Packard (1981) noted that manatees using Hobe Sound and Jupiter Sound during the winter were located primarily in areas where there were grassbeds (*Syringodium, Halodule,* and *Thalassia*). Manatees radio-tracked in Charlotte Harbor, Matlacha Pass, and San Carlos Bay were frequently located along the edges of seagrass beds (Lefebvre and Frohlich 1986 unpubl. report), supporting Hartman's (1979) suggestion that manatees tend to stay close to deeper water while in shallow water situations. Radio-tracked manatees in Crystal and Homosassa Rivers typically left warm-water springs during the day and moved downriver at dusk to feed on submerged vegetation *Ruppia maritima* and *Potamogeton pectinatus*), returning to a warm-water source by morning (Rathbun et al. in press). Manatees seen during aerial surveys near the mouths of the Suwannee, Withlacoochee, Crystal, Homosassa, and Chassahowitzka Rivers often feed on beds of *Ruppia maritima* or *Halodule wrightii* growing on sandbars adjacent to channels (Powell and Rathbun 1984). Manatees frequently used those areas of a river that had vegetated shallow shelves or sandbars next to a channel (Powell and Rathbun 1984).

Manatees in Florida also feed on a wide variety of freshwater aquatic plants, particularly *Hydrilla verticillata* in the Crystal River headwaters (Hartman 1979) and *Hydrilla, Eichornia, Vallisneria, Najas, Paspalum,* and floating grasses in the St. Johns River (Bengtson 1981). Hartman (1979) believed that *Eichornia* is the preferred floating vegetation in turbid rivers where submergent vegetation is absent. *Spartina, Vallisneria/Sagi-*

taria, and grasses were the major food items identified in the stomachs of 12 manatees salvaged from Duval County in northeast Florida (Hurst and Rathbun, unpubl. data). *Spartina* is found in fresh and saline marshes of north Florida and southern Georgia, and may be the most important food for manatees in this region. The manatee's capacity for flexibility and opportunism in foraging is well illustrated by their use of live oak (*Quercus virginiana*) acorns in the St. John's River during the winter (O'Shea 1986).

Mexico

Manatees occur along much of the southeastern coast of Mexico from Nautla, Veracruz, to the Belize border, but they are still reasonably abundant in only three principal areas in Mexico (Villa-R. and Colmenero-R. 1981; Colmenero-R. 1984; Colmenero-R. and Zavala 1986). In decreasing order of importance, these include the vast wetland systems in the states of Tabasco and Chiapas, the bays and coastal springs along the northern and eastern coasts of the state of Quintana Roo, and in the rivers near Alvarado in the state of Veracruz (Fig. 7). Historically manatees occasionally were seen during the summer in Texas (Powell and Rathbun 1984). Gunter (1942) proposed that these seasonal sightings were of migrants from Mexico. In recent years, however, there has been a decline in sightings from Texas (Powell and Rathbun 1984; Rathbun et al. in press), which has been attributed to extirpation of manatees from northern Mexico (Powell and Rathbun 1984). Their extirpation from this area was first suggested by Alvarez (1963). Colmenero-R. (1984) found no evidence of manatees north of Veracruz. Overexploitation by fishermen probably is responsible for their decline and possible disappearance in this region.

Colmenero-R. (1986) suggested that ideal manatee habitat in Mexico is characterized by average water temperatures above 22°C and average rainfall above 1000 mm. These criteria coincide with the most important manatee areas in the southern Bahía de Campeche (Fig. 7). Lluch-B. (1965) and Colmenero-R. (1986) also described seasonal movements of manatees in this region and related these to patterns of seasonal rainfall and food availability. During the dry season (February through mid-May) manatees are found in the lower parts of the major rivers and along the coast of Tabasco. When the heavy rains begin in late May the animals move upriver, but they are prevented from reaching the lakes and small tributaries further inland by swift and turbulent water. When the rains moderate in November through January, the animals are able to move up into the lakes and streams and feed on the abundant emergent aquatic vegetation (i.e., *Paspalum* sp., *Chloris* sp., *Panicum* sp., *Eichornia crassipes*) that becomes available during high water. As the water levels drop in these inland areas during the dry months, food becomes limited and the manatees move down towards the coast again. Although fishermen traditionally took advantage of the seasonal migrations and captured manatees while they were in the shallow, restricted inland lakes and creeks (Lluch-B. 1965), Colmenero-R. (1986) reported that most fishermen now respect the law protecting manatees (Decreto de la Sria. de Pesca, Diario Oficial, October 1981), which has resulted in the recovery of manatee numbers in some areas, such as the Bahía de Campeche. Poaching, however, is still a problem in some regions. Manatees are hunted for food in Campeche, and in southern Quintana Roo they are reportedly killed for shark bait (Colmenero-R. and Zavala 1986).

Fishermen and personnel of the Secretary of Fisheries have reported that manatees are abundant in the bays of Ascension and Espíritu Santo on the eastern coast of Quintana Roo (Colmenero-R. and Zavala 1986) (Fig.7). During an aerial survey in September 1977, no manatees were seen in Bahía Espíritu Santo and 17 manatees were seen in the southwestern corner of Bahía de la Ascension (Gicca unpubl. report). Two manatees were seen in southwestern Bahía de la Ascension during an aerial survey in July 1987 by investigators from the Centro de Investigaciónes de Quintana Roo and the U.S. Fish and Wildlife Service (O'Shea and Rathbun 1987 unpubl. report). The eastern coast of the Yucatan Peninsula is very dry, with few rivers, which under normal circumstances would preclude manatees from occupying this area because of their apparent need for fresh water (Lluch-B. 1965). This coast is unusual, however, in that natural artesian springs occur nearshore along the coast, and manatees frequent these sources of fresh water (Lluch-B. 1965). The springs do not attract the large aggregations associated with similar natural springs in Florida (Bengtson 1981; Powell and Rathbun 1984), perhaps because they are used as sources of fresh water (Gallo-R. 1983) rather than as sources of warm water. The Bahía de la Ascension has numerous freshwater springs and seasonally inundated creeks, and manatees were once more abundant in both Bahía de la Ascension and Bahía de Espíritu Santo (O'Shea and Rathbun 1987 unpubl. report).

Colmenero-R. and Zavala (1986) believe that the most pressing conservation problem in Mexico is the enforcement of protective laws to reduce illegal poaching. Manatees continue to be entangled and killed by gill net fishermen in Veracruz and Tabasco (Johnson-M. and Maruri-G.pers. comm.). Colmenero-R. and Zavala (1986) suggested that protection might be most effectively accomplished by creating "protected areas" where manatees would be maintained in semicaptivity. Gallo-R. (1983) recommended that the artesian springs in Quintana Roo be protected from development.

Central America

Belize

Although Belize does not have an extensive Caribbean coastline, manatees are relatively abundant there in comparison with other countries of Central America. In September 1977, Bengtson and Magor (1979) completed five aerial surveys of the shoreline, major rivers, prominent keys, and coastal and inland lagoons of Belize. The resultant 101 sightings reported in 10 hr of search time constitute the greatest rate of manatees seen per unit search effort for any surveys outside of Florida. Charnock-Wilson (1968, 1970) also reported that aerial sightings of manatees were common in Belize and that the outlook for the species was good there in comparison with other areas within their range.

Low human population densities and excellent manatee habitat along much of the coast of Belize are the keys to these optimistic findings. A long offshore barrier reef shields most of the irregular coastline of Belize from strong tidal currents or significant wave action. The coastline itself is marked by numerous sluggish rivers and creeks, both fresh and saltwater lagoons, and extensive mangrove swamps (Charnock-Wilson 1968). Rivers and creeks in southern Belize are faster-flowing and rocky, and are apparently less suitable for manatees. Manatees are seen all along the coast, including the larger keys and intervening waters, but particularly at river mouths. They have been seen between Belize City and St. Georges Cay, up the Belize River to Bermudian Landing, in the Sibun and Sittee Rivers, Deep River, Indian Hill Lagoon, near Corozal and Sarteneja, Turneffe Island, Corker Cay, Drowned Cays, Moho Cays and Ambergris Cay (Charnock-Wilson et al. 1974) (Fig. 8). Bengtson and Magor (1979) made sightings all along the coast except in the southernmost region, where river water clarity was poor. Areas of greatest numbers of aerial sightings were Northern and Southern Lagoons near Belize City ($n=45$) and Four Mile Lagoon at the mouth of the New River ($n=21$) (Fig. 8). Manatees were seen in saltwater areas off keys, and in estuaries and rivers, including as far as 100 km up the New River. The Sarstún and Hondo Rivers along the borders of Guatemala and Mexico were not surveyed. Nine of the 101 sightings were of small calves (Bengtson and Magor 1979).

Manatee killing is illegal in Belize. Charnock-Wilson (1968, 1970) stated that "purposeful hunting by man now seems to be rare" and believed that younger generations of Belizeans had lost the desire to eat or hunt manatees. This lack of recent hunting pressure was thought to have played a role in the relative abundance of manatees in Belize (Charnock-Wilson 1968, 1970). However, illegal take of manatees has not completely ceased. Murie (1935, cited in O'Donnell 1981) noted that earlier in the century the availability of manatee meat in markets in Belize City had declined, presumably because of overhunting. Charnock-Wilson et al. (1974) reported the killing of 12 manatees for sport at Ambergris Cay and for the sale of meat at San Pedro. Barrett (in litt.) noted that manatees are occasionally caught and killed incidentally in fishing nets. Matola (in litt.) observed that poaching attempts continued near Ambergris Cay in 1986 and that manatees are still eaten along the Manatee River. The Belize Zoo Outreach project has recently incorporated a travelling manatee conservation education program targeted at reducing this illegal hunting of manatees.

Guatemala

Guatemala has the smallest Caribbean coastline in Central America and a small number of manatees. Most of the persisting manatees occur in association with freshwater systems (Fig. 8). Janson (1978) noted that manatees may be encountered in the Caribbean along the coast and at minor coastal rivers north of Puerto Barrios. Manatees continue to occur in the inland freshwater Lago Izabal/Río Dulce system, at what are considered to be very depressed levels compared to their former abundance (Janson 1978 in litt.; Godoy 1986 in litt.). They also persist in the Río Sarstún along the Belize border, the Río Motagua system near the Honduras border, and possibly in coastal waters near Punta de Manabique (Godoy 1986 in litt.; Grisko 1986 in litt.) (Fig. 8). No comprehensive aerial surveys for manatees have been conducted in Guatemala. Gicca (unpubl. report) made a single 6 hr survey in 1976. One manatee was seen along the central southern shore of Lago Izabal. Within the Lago Izabal/Río Dulce system, the wetlands at the mouths of the Río Polochic and Río Oscuro at the western end of the lake are considered the best manatee habitat, although numbers there appear to have been greatly reduced by the late 1970's (Janson 1978, 1980).

Manatees are occasionally seen in El Golfete between Lago Izabal and the Río Dulce (Tres 1986 in litt.) (Fig. 8). A manatee reserve, the Biotopo Para La Conservación del Manatí Chocon-Machacas, encompasses the northeastern shore and surrounding habitat of El Golfete. To our knowledge, this is the first manatee reserve designated anywhere in Central or South America, and is indicative of the desire of

Guatemala to protect manatees. Heavy penalties also exist for killing of manatees in Guatemala. However, manatee meat has been sold at market in Livingston as recently as November 1985, and shooting or hunting has been reported from near Cayo Piedra in El Golfete and Laguna Perdida on the Río Motagua during 1986 (Grisko in litt.). Given the reports of severely depressed numbers and continued hunting, the outlook for manatees in Guatemala is not good. Their more substantial numbers in neighboring Belize, however, could provide sources for natural recolonization should conditions become more hospitable in Guatemala through the continued existence of the Biotopo Chocon-Machacas reserve, stronger enforcement of laws, and increased educational efforts.

Honduras

Manatees are found in three areas along the coast of Honduras: rivers and lagoons west of La Ceiba, rivers east of Trujillo, and the lagoons and rivers of La Mosquitia of eastern Honduras (Rathbun et al. 1983a; Klein 1979) (Fig. 8). The latter area provides the most extensive habitat for manatees (Klein 1979), although relatively few sightings were made there during an aerial survey in 1979 (Rathbun et al. 1983a). These three areas are sheltered, include access to fresh water, and are characterized by abundant aquatic vegetation. The remainder of the Honduran coast does not provide suitable habitat, and is characterized by strong surf, steep shorelines, and a lack of broad, slow-moving rivers. Rathbun et al. (1983a) observed manatees 46 times in 16 hours of aerial surveys in 1979 and 1980 at the following locations: the coast near Zambuco, Laguna de Boca Cerada, Laguna de Tansin, and mouths of the Río Lecan, Río Cuero, and Río Salado (Fig. 8). Most of the sightings were not offshore, but at coastal lagoons and rivers. Interviewees in the mid-1970's reported manatees from the following lakes and lagoons in eastern Honduras: L. Brus, L. Ibans, L. Rapa, L. Guarunto, L. Biltamaíra, L. Tilbakan, and L. Caratasca (Klein 1979) (Fig. 8). Reports of manatee sightings and killings during the mid-1970's also exist for the mouth of the Río Chamelecón near Puerto Cortés, and the Río Congrejal near La Ceiba (Klein 1979) (Fig. 8).

Reports based on interviews made during the 1970's are unanimous in the conclusion that manatee abundance has declined markedly in Honduras (Klein 1979; Rathbun et al. 1983a). This decline is the result of hunting by harpoon and widespread gill netting, both incidental and intentional. Manatees have been protected by law in Honduras since 1959 (Article 49 of Fisheries Law, Decree No. 154), but enforcement has not been effective. Manatees may have become nocturnal in response to hunting pressure (Rathbun et al. 1983a); nevertheless, a substantial illegal take has continued into the 1980's (G. Cruz, unpubl. data). Small calves were observed during the 1979-80 aerial surveys, and there is extensive remaining favorable habitat. Effective law enforcement is therefore a major key to the recovery of manatees in Honduras.

Nicaragua

The Caribbean coast of Nicaragua comprises some of the finest and most extensive habitat for manatees in Central America. The Moskito lowlands stretch over nearly the entire eastern coast of the country, and consist of wetlands with numerous long, slow-moving rivers, emptying into many interconnected coastal lagoons. Shallow waters with lush seagrass beds extend far offshore (Phillips et al. 1982) (Fig. 1). Extremely high rainfall creates a corridor of brackish to fresh water parallel to the coast (Murray et al. 1982), which may attract manatees. There have been no aerial surveys for manatees in Nicaragua, and no reports based upon interview surveys are available to us. O'Donnell (1981) noted that manatees may occur in low numbers in the Río San Juan along the Costa Rica border, and that they reportedly occur in Lago de Nicaragua (Fig. 8). Nietschmann (1979) reported a manatee sighting off Set Net, one of the Pearl Cays (Fig. 8). Manatees were described by Nietschmann's (1979) Moskito companion as very scarce compared to their former numbers. Kenworthy (1987 in litt.) was told by Indians along the central coast that manatees were in the area in the mid-1970's.

Human population density in coastal Nicaragua has been relatively low. Manatees have been a traditional food source for indigenous people in this area, but in recent years the take has been low (Nietschmann 1972, Loveland 1976). Manatees have been protected in Nicaragua since 1956 under general hunting laws (Legislative Decree 306).

Costa Rica

Manatees probably continue to occur in reduced numbers in two areas of favorable habitat along the eastern coast of Costa Rica. The most extensive of these areas is the broad coastal plain of northeastern Costa Rica (Llanura de Tortuguero). A few manatees may also occur in southeastern Costa Rica, at the mouth of the Río Sixaola on the Panama border, at the mouth of the Río Estrella, and in Laguna Gandoca (O'Donnell 1981) (Fig. 9). The area between Limón and Panama is in general less favorable habitat, with mountains reaching to the sea and few large rivers or lagoons. D. E. Wilson (1974 unpubl. report) and O'Donnell (1981) conducted interviews and site visits in the 1970's, which suggested that few, if any, manatees occurred in this region. These investigators also made overflights in northeastern Costa Rica between Tortuguero and Barra del Colorado without observing manatees (Fig. 9). Through interviews, however, they concluded that this area contained the last noteworthy but small group of manatees in Costa Rica. Wilson and O'Donnell agreed that much of the habitat along the inland waterways and rivers of the Tortuguero region seemed to provide favorable conditions for manatees. Tortuguero National Park is the most important protected area for the manatee (Vaughan 1983). Sightings were made during the 1970's in Tortuguero National Park, Río Servulo, Laguna Penitencia, Laguna Yaqui, Laguna Coronel, the lower Río San Carlos, Río Sierpe, and around Parismina (O'Donnell 1981) (Fig. 9). Manatees were thought to be absent or very rare in the Río Sarapiqui, Río Puerto Viejo, and the middle Río San Juan, although small numbers may exist in Lago Arenal far up the Río San Carlos (O'Donnell 1981; D. E. Wilson 1974 unpubl. report) (Fig. 9). Evidence of a decline began to be apparent in the 1950's (O'Donnell 1981). The manatee is probably one of Costa Rica's rarest wildlife species because of past and present hunting pressure (Vaughan 1983). Enforcement of recent protective legislation (Law 6919, Ley de Conservación de la Fauna Silvestre of January 1984), establishment of reserves, and a dying interest in hunting linked to manatee scarcity (O'Donnell 1981) provide the best prospects for the species' recovery in Costa Rica.

Panama

Panama has the longest Caribbean coastline of the Central American countries. Information on the present distribution and abundance of manatees in Panama is scant. The range is apparently fragmented by habitat discontinuities and possible depletion.

Most of the current reports suggest that manatees continue to exist in regions of optimal habitat (lower reaches of large, slow-moving rivers and protected lagoons) but are seldom seen elsewhere.

The region from the Costa Rican border to Punta Valiente may constitute the most favorable habitat (Fig. 9). O'Donnell (1981) received reports of manatees in the lower parts of the following rivers in this region: Río Sixaola, Río San San, and Río Changuinola (Fig. 9). MacLaren (1967) reported on the capture of nine manatees from the Río Changuinola in 1963-64 as part of a translocation project. Montgomery (1980 unpubl. report) observed a few manatees in the lower Río San San during a series of brief overflights of this region, but did not observe them elsewhere in very limited surveys of the Changuinola area and the coast to the Río Guarumo, including the Laguna Chiriquí (Fig. 9). Manatees observed feeding on seagrass beds in the Boca del Dragón area by fishermen are thought to move between that area and the Río Changuinola, rather than east to the Bahía Almirante (O'Donnell 1981). No reports of manatees in the Laguna Chiriqui reached O'Donnell, although he was told that manatees persisted in the sparsely settled, heavily vegetated lower reaches of the Río Manatí.

The coastline east of Punta Valiente to Río Coclé del Norte lacks large lagoons, and the rivers are settled at their mouths and swifter than in the Bocas del Toro region (Fig. 9). Manatees are currently only known from the relatively uninhabited Río Veraguas in this region (O'Donnell 1981). Manatees apparently enter the Río Coclé del Norte and Río Miguel de la Borda only in the dry season, when they remain in the lower reaches. None have been reported from the Río Indio. Manatees occurred in the Río Chagres prior to the construction of the Panama Canal (MacLaren 1967). The number occupying the canal system may have been augmented by the translocation of nine manatees from the Río Changuinola area to the Río Chagres in the 1960's, and manatees may now reach as far as the Pacific Ocean (MacLaren 1967; Montgomery et al. 1982; Schad et al. 1981). There is almost no information on the distribution of manatees over the long stretch of Caribbean coastline extending from Colón to the Colombian border. The mountains come close to the shore along this coast, which is marked by swift and shallow rivers and an absence of lagoons or extensive swamplands. This lack of information coupled with uncharacteristic habitat may indicate an absence of manatees. However, Garrett (unpubl. report) has noted that manatees are sometimes seen by Kuna Indians on the San Blas coast, particularly in the area of Playón Chico, Ustupo, and Puerto Obaldia (Fig. 9).

Manatees are still killed by people in Panama. Manatees are not often hunted by harpoon any longer, probably because of protection laws and a lack of interest or skill caused by a decrease in abundance (O'Donnell 1981). They may still be taken incidentally in fishing nets, and have been killed by boats and blasting activities in the Panama Canal (Montgomery et al. 1982). O'Donnell (1981) was not optimistic about the future of manatees in Panama.

South America

Colombia

No systematic survey for manatees has been conducted in Colombia. Manatees persist in isolated pockets of major river systems or estuaries distant from human settlements (Fig. 6) (Powell and Gicca 1975 unpubl. report). The upper reaches of the Río

Magdalena between Baranquilla and Bogotá, and the Río Atrato and its confluence with Bahía Candelaria appear to have the largest remaining groups (Husar 1977; Powell and Gicca 1975 unpubl. report) (Fig. 10). Small numbers occur within the Parque Naciónal Isla de Salamanca (Husar 1977).

Manatees in Colombia are in danger of extirpation because of hunting and the high price of their meat on the black market (T. Shuk 1986 in litt., Powell and Gicca 1975 unpubl. report). They have been protected since 1969 under Resolución No. 574 of INDERENA, Colombia's natural resource agency; however, funding is inadequate for effective law enforcement and research.

Venezuela

Venezuela has the most extensive Caribbean coastline of any country of South and Central America. Manatees are apparently absent from this coastline and its nearshore islands. They continue to occur in reduced numbers in parts of Lago de Maracaibo, and in greater abundance in the Orinoco River system and in drainages of eastern Venezuela along the Golfo de Paria (O'Shea et al. 1988, Mondolfi 1974) (Fig. 10). The absence of manatees along the extensive Caribbean coast of Venezuela appears to be the result of habitat unsuitability, and represents a major discontinuity in the range of this species.

O'Shea et al. (1988) conducted interview and aerial surveys of potential manatee habitat throughout Venezuela in early 1986. Aerial surveys yielded poor results in most areas because of water turbidity, and possibly because of low numbers and dry season inactivity. A small number of manatees exist in Lago de Maracaibo, centered northwest of Maracaibo City in the Río Limón/Laguna Sinamaica region and on the southwestern part of the lake bordering the extensive swamps of the Cienaga Juan Manuel de Aguas Claras y Aguas Negras (Fig. 10). There are no resident manatees along the Caribbean coast between Lago Maracaibo and the Boca del Dragón separating Trinidad and Venezuela, and no recent or historical evidence for their former occurrence on this coast (O'Shea et al. 1988). The northern coast of Venezuela is characterized by long, often desert stretches of rocky, mountainous coastline with high energy beaches, deep water and relatively cool sea surface temperatures; pockets of habitat which are suitable for manatees (sheltered lagoons and bays with freshwater input) have been settled by people for centuries (O'Shea et al. 1988). Eastern Venezuela provides a dramatic contrast to the Caribbean coastline, however, with frequent reports of continued manatee occurrence, numerous broad, slow-moving rivers, estuaries, mangrove swamps, and warmer waters. Manatees continue to occur in most of the waterways of the extensive Orinoco Delta and throughout the middle Orinoco and its tributaries as far as depths and currents allow. Strong seasonal fluctuations in rainfall and water levels may have a dramatic impact on manatee ecology in this region. Manatees have been reported to feed on 18 plant species in Venezuela, including mangrove (*Avicennia* and *Rhizophora*), grasses, sedges, *Montrichardia arborescens*, and *Eichornia*, (O'Shea et al. 1986).

Hunting for local markets in Venezuela was relatively heavy during the middle of the present century and resulted in a reduction in manatee abundance within their current range (Mondolfi 1974; Mondolfi and Müller 1979; O'Shea et al. 1988). Hunting activities appear to have declined substantially in recent years in response to laws protecting the species, education campaigns, and a general lack of interest in manatee hunting by the younger generations (O'Shea et al. 1988). Incidental netting and habitat alteration continue to present problems, particularly in the Llanos (central plains) tributaries, and a few boat kills have been reported. However, the waters of eastern Venezuela and the Orinoco system provide some of the largest continuous manatee habitat anywhere within the species' range. If the current trend to reduced hunting pressure and controlled development continues, Venezuela has the potential to provide a major stronghold for manatees in the future.

Trinidad

Information on manatees in Trinidad was obtained by Bindernagel in the early 1980's (Garrett 1984 unpubl. report, Hislop 1985). Manatees are present in Nariva Swamp, a large, freshwater swamp extending about 20 km north from Guataro Point to Manzanillo Bay (Fig. 10). The swamp is sparsely settled, and local inhabitants claimed that manatees are common. During the dry season much of the swamp becomes inaccessible to manatees, which may become trapped in ponds. Hislop (1985) received unconfirmed reports of manatee sightings and an incidental taking of a manatee entangled in a fishing net in the North Oropuche River. Although manatees are occasionally killed for food in Nariva Swamp, the primary threats to manatees are proposed agricultural or maricultural schemes that would require modification of the wetlands. A 1555-ha wildlife sanctuary on Bush Bush Island in the center of the swamp is apparently not protected from poaching and illegal woodcutting (Garrett 1984 unpubl. report). Manatees are protected by default under the Conservation of Wildlife Act (i.e., they do not appear on a schedule of game animals); however, enforcement of protection for manatees and their habitat is needed.

Guyana

There is little current information on manatees in Guyana, and a systematic survey is needed. Manatees occur all along the coast, but are most frequently reported in river or canal mouths (Bertram and Bertram 1973). The greatest concentrations occur in the eastern region, on both sides of the border with Surinam (Husar 1977). Considerable numbers of manatees were noted in the Canje and Abary Rivers, particularly in wet savannahs in their upper reaches (Bertram and Bertram 1973) (Fig. 11). Manatees were also reported in the Courantyne, Berbice, Demerara, Essequibo, Pomeroon, Arapiako, Akawini, Waini, Barima, and Kaituma Rivers (Fig. 11).

Manatees in Guyana apparently consume a wide variety of marine, estuarine, and freshwater vegetation; *Montrichardia arborescens*, an aquatic herb, was specifically noted by Bertram and Bertram (1964). Allsopp (1969) noted a wide variety of freshwater aquatic plants consumed by manatees in canals. Rivers with large masses of floating grass were described as providing the most suitable environment for manatees (Bertram and Bertram 1964). Bertram and Bertram (1964) estimated that there were some thousands of manatees in Guyana, yet their numbers were generally believed to be much reduced from former times. Manatees have been protected in Guyana since 1956 (Fisheries Ordinance No. 30, Revised No. 13, 1961), and hunting no longer appears to be a serious problem (Bertram and Bertram 1973; Husar 1977). Accidental entanglement in fishermen's nets may result in some deaths, but manatees are not a welcome catch as they can cause considerable damage to nets (Roth 1953). Motorboat strikes may be an increasing problem in Guyana (Roth 1953; Bertram and Bertram 1964).

Surinam

Manatees occur in most of the coastal plain rivers of Surinam, usually not farther than 60 km inland (Bertram and Bertram 1973; Husson 1978) (Fig. 6). Duplaix and Reichart (1978) reported the greatest concentrations in Nanni Creek and the Coesewijne, Tibiti, and Cottica Rivers (Fig. 11). However, they believed that manatees were easier to observe in these smaller creeks and rivers than in broader rivers such as the Corantijn, lower Saramacca, and Commewijne (Fig. 11). Manatees have also been observed or reported in the Maratakka, Nickerie, Wayambo, Coppename, and Suriname Rivers (Husson 1978; Duplaix and Reichart 1978) (Fig. 11).

Mangrove forests provide manatee habitat in the flat coastal regions and river estuaries of Surinam (Duplaix and Reichart 1978). Husson (1978) stated that manatees have never been found in the open ocean off the coast of Surinam. Swamp forests behind the mangroves are also inhabited by manatees, which graze on stands of *Montrichardia arborescens* along the banks at high tide (Duplaix and Reichart 1978). Dekker (1974) believed that manatees in Surinam favored fresh water because of their preference for "Mokko mokko" (*Montrichardia*), which occurs only in fresh water. Savannah swamps, or floating savannahs, are found in upper river reaches, and are characterized by *Cyperus* and *Montrichardia*. Husson (1978) also noted *Montrichardia* as a manatee food in Surinam, as well as *Machaerium lunatum*, *Caladium arborescens*, and *Panicum*. In contrast to Bertram and Bertram's (1973) findings in Guyana, Duplaix and Reichart (1978) reported that manatees are not found in the floating savannahs, but in the small creeks transecting them. Seasonal flooding undoubtedly makes vegetation in the floating savannahs accessible to manatees. Rapids in the upstream portions of Surinam's rivers prevent manatees from travelling further upriver (Duplaix and Reichart 1978).

Duplaix and Reichart (1978) interviewed 89 residents, primarily Amerindians, and found that although some people believed manatees have become more common in recent years because they are no longer hunted, former hunters described the manatee as having disappeared from its usual haunts over the last 30 years. Manatees receive some protection under Surinam's Nature Protection and Game Ordinances. Although no reserve has been created specifically for manatees, they are found in the estuary and mangrove swamps of the Coppename River Nature Reserve, covering 10,000 ha (Duplaix and Reichart 1978). None of the manatees observed bore propeller scars, but commercial river traffic is heavy and the increasing use of outboard motors by fishermen and hunters may eventually become a problem (Duplaix and Reichart 1978). Some incidental taking in nets is also likely to occur. Duplaix and Reichart (1978) suggested that more vigorous enforcement of existing conservation laws and protection of areas with the highest manatee density, such as Nanni Creek, the upper Coeswijne River, and the Perica River, are needed to improve the status of the manatee in Surinam.

French Guiana

The absence of a broad coastal plain in French Guiana led Bertram and Bertram (1964) to conclude that there is little suitable manatee habitat in this country. We know of no aerial survey or interview studies on manatee abundance or distribution in French Guiana. Brazilian hunters in the vicinity of Río Oiapoque on the Brazilian border reported manatees in the rivers Approuague, Mahury, Laughan, and Ouanary (Fig. 11), as well as in some of the smaller rivers in eastern French Guiana (Best and Teixeira 1982).

Brazil

Systematic survey data are generally lacking for Brazil, although interviews of fishermen and coastal residents have been conducted in some areas, and boat surveys were conducted in Paraíba in 1986. *T. manatus* has a disjunct distribution on the northern coast of Brazil, and ranges from Río Grande do Norte to the state of Bahía on the eastern coast (Jackson 1975 unpubl. report; Domning 1981a; Best and Teixeira 1982) (Fig. 12).

On the north coast of Brazil, *T. manatus* occurs along the coast of Amapá north of Cabo Norte, in the Río Mearim in Maranhão (Domning 1981a), and along the coast of Río Grande do Norte (Fig. 12). The species seems to have been exterminated from the coast of Pará, and may be absent from the Marajó region, where the presence of *T. inunguis* has been confirmed (Domning 1981a, 1982a) (Fig. 12). Banks (1985), however, noted a 1983 report by Catuetê and Duarte of the coexistence of *T. manatus* and *T. inunguis* at the mouth of the Río Amazonas. In an interview survey of Amapá, Best and Teixeira (1982) were able to positively identify only one *T. manatus* specimen, in the vicinity of the Río Oiapoque. Hunter interviews indicated that manatees occur in the mouths of the rivers Oiapoque, Uaçá, and Cassiporé (Fig. 12). Interviewees reported that manatees are common along all of the Amapá coastline, whereas Domning (1981a) reported that the results of his interviews with Amapá residents, also conducted in 1978, were inconclusive.

Manatees were observed during 10 of 16 boat surveys conducted on the Río Mamanguape, Paraíba, from January to May 1986 (Mônica Borobia 1987 in litt.). Manatees were found at the mouth of the river and on the seaward side of a reef that protects the river mouth. Single manatees and groups up to three were observed. Interviews with local residents and fishermen indicated that manatees frequent the mouth of the Río Mamanguape from January to March. An adult manatee was captured in nets at Bessa Beach in João Pessoa, Paraíba, in March 1982 (O Norte, 19 March 1982, p.3) (Fig. 12). Jackson (1975 unpubl. report) concluded that manatees still occur from Itamaracá, Pernambuco, to Mangue Seca, Bahía, but are no longer present in Espírito Santo (as Ruschi [1965] reported earlier) or in Baia de Todos os Santos, Bahía (Fig. 12). He did not survey the southern panhandle of Bahía. Jackson believed that Alagoas provided the most favorable manatee habitat on the east coast because of numerous estuaries between Maceió and Penedo, and because inshore waters are protected by reefs running parallel to the coast (Fig. 12). Banks and Albuquerque described beds of *Halodule wrightii* as natural pastures for *T. manatus* in Pernambuco (Banks 1985).

Best and Teixeira (1982) were unable to find any practicing manatee hunters south of Oiapoque, Amapá, and residents of Amapá interviewed by Domning (1981a) claimed that manatees were seldom captured, perhaps no more than two or three a year. Domning (1982a) reported evidence of exploitation of manatees, presumably for meat, in the states of Alagoas (in 1959) and Bahía (in 1964). Occasional taking of manatees for meat may still occur in Paraíba (Mônica Borobia 1987 in litt.).

The seasonally flooded mangrove swamps of the Amapá coast have little or no submergent vegetation, thus manatees appear to depend upon floating and shoreline species such as mangrove (*Avicennia* and *Rhizophora*), aninga (*Montrichardia arborescens*), cai-seca (*Rhabdadenia biflora*), paraturá (*Spartina brasiliensis*), and mururé (*Eichornia*) (Domning 1981a; Best and Teixeira 1982). Domning (1981a) and Best and

Teixeira (1982) recommended that a manatee reserve be established to include the Amapá coast near Cabo Norte and adjacent inland lakes, as this area may be the only place in the world where two sirenian species occur in close proximity (Domning 1981a). Domning (1981a) also recommended that the lower Río Mearim in Maranhão be included in a reserve, because it provides large areas of undisturbed floating meadows in lakes and channels off the main river. Cabo Orange, Amapá, has been proposed as the site of a national park (Best and Teixeira 1982). Both manatee species have been fully protected in Brazil since 1967 (Under Lei No. 5.197, 1967, and Portaria No. 3.481, 1973), however enforcement of these laws is almost impossible because of the lack of enforcement personnel and the large areas involved (Best and Teixeira 1982).

Biogeographical patterns of *Trichechus*

Domning (1982b) pointed out that although manatees make up three of the four living species of the order Sirenia, their representation in the fossil record is minute compared to that of the family Dugongidae. Early sirenians had a "pan-Tethyan" distribution and gave rise to dugongids in the middle Eocene (Domning et al. 1982c). Manatees do not emerge in the fossil record until the Miocene (Domning 1982b). Domning (1982b) speculated that early manatees arose in South America as coastal river and lagoon inhabitants in contrast to the more marine dugongids, their contemporaries in the New World. Manatees developed a unique process of tooth replacement, adapted to a more abrasive diet including true grasses (Gramineae), which may have allowed them to outcompete dugongids (Domning 1982b). The latter disappeared from the western Atlantic during the Pliocene. *T. inunguis* probably evolved in isolation in the Amazon basin following the Miocene Andean orogeny, and exhibits more derived characters than the other two species of manatees (Domning 1982b). Whether competition or some other factor maintains the apparent parapatry of *T. inunguis* and *T. manatus* at the mouth of the Amazon River is unknown. The great similarity between *T. manatus* and *T. senegalensis* led Simpson (1932) and later Domning (1982b) to conclude that manatees dispersed between South America and Africa relatively recently, probably in the Pleistocene or Pliocene. Simpson (1932), however, suggested that manatees dispersed from Africa to South America.

Although recent records of manatees from the Lesser Antilles are lacking, prehistoric and early historic evidence indicates that manatees occurred there. Thus it is possible that linkage between *T. m. manatus* in the Greater Antilles and in South America has been maintained by wanderers that island-hopped across the Lesser Antilles. Dispersal may also occur between the Greater Antilles and Mexico and Central America. The North Equatorial Current, flowing from east to west through the Caribbean Sea and northward offshore of Yucatan, would tend to favor a Yucatan to Cuba crossing more than a Venezuela to Puerto Rico crossing, although manatees may not necessarily depend upon favorable currents to cross open ocean. An extensive area of shallow water (Moskito Bank) between the Honduras-Nicaragua border and Jamaica (Fig. 1) might help to promote ocean-crossing between Central America and the Antilles. Such a crossing is known for green sea turtles (Nietschmann 1972, Carr et al. 1978). Allen (1942) suggested that manatees originally extended their range to the West Indies by way of the Yucatan Peninsula and the intervening shallows. Reynolds and Ferguson (1984) sighted two manatees 61 km northeast of the Dry Tortugas in the Gulf of Mexico, and suggested that they could be wanderers from Florida, Cuba, or Yucatan.

In addition to historical geography, physical and biological environmental factors such as salinity, temperature, water depth, currents, shelter from wave action, and availability of vegetation are important determining factors of manatee distribution. The rareness of offshore sightings of manatees in locations that apparently lack dependable freshwater sources, such as the Bahamas, suggests that their distribution is influenced by the availability of fresh water. Seasonal shifts in manatee distribution and their use of available warm water sources in Florida suggest that energetic requirements influence the range limits of manatees (Irvine 1983). The average lower limit of thermal neutrality is approximately 24°C (Irvine 1983). Whitehead (1977) described the full range of *T. manatus* as falling within the northern and southern limits of the 24°C mean annual isotherm (Figs. 1 and 12). The Florida manatee's low metabolic rate (15-22% of predicted weight-specific values) and high thermal conductance suggest that manatees could not survive in winter water temperatures in much of Florida (Irvine 1983), much less further north. The cold Labrador Current, flowing southwest along the northeast Atlantic coast, meets the warm Florida Current in the area of Cape Hatteras; relatively few manatees are reported north of Cape Hatteras (Rathbun et al. 1982). Natural and artificial warm water refuges ameliorate the effects of winter temperatures on manatees, and may have allowed a recent northward winter range extension in Florida (Hartman 1979, Shane 1983, 1984).

Extensive gaps in suitable habitat, such as the northern coast of the Gulf of Mexico and the Caribbean coast of Venezuela, may also represent geographic barriers. The absence of manatees along the Caribbean coast of Venezuela may be related to cooler water temperatures, and the scarcity of sheltered lagoons and fresh water (O'Shea et al. in press). Deep water and strong currents in the Straits of Florida may be effective barriers to gene flow between *T. m. latirostris* in Florida and *T. m. manatus* in the Antilles (Domning and Hayek 1986). The geo-oceanography of the northern Gulf of Mexico, in addition to cool winters, may contribute to the scarcity of manatees in this region. From the Florida panhandle westward to northern Mexico, the regional coastal type is alluvial: smooth shorelines with sandy beaches interrupted by deltas (Price 1954). The Gulf shoreline in peninsular Florida is drowned karst or biogenous, and is characterized by mangrove swamps and marshes, with few sandy beaches (Price 1954). Along much of the karst shoreline springs are common in streams and offshore, because of artesian groundwaters (Price 1954). Although Louisiana has extensive salt marsh (dominated by *Spartina* and *Juncus*), Alabama and Mississippi have relatively little salt marsh. Discontinuity of suitable habitat along the northern Gulf of Mexico may discourage manatee migration west of Florida even during warm seasons. Domning (1981b) proposed that the lack of diversity among sirenians is a direct result of their coevolution with a relatively undiversified food base, the seagrasses. While the distribution of sirenians is clearly parallel to the distribution of seagrasses in tropical and subtropical regions (Brasier 1975, McCoy and Heck 1976), the origin and evolution of trichechids is theoretically linked to their specializations for feeding on true grasses found in freshwater or estuarine systems (Domning 1982b), and they are known to eat a wide variety of aquatic and semi-aquatic macrophytes (Best 1981). The relative contribution of freshwater and marine macrophytes to the geographic distribution of the West Indian manatee is unknown. The distribution of

T. manatus is probably not influenced by the distribution of particular plant species, as the manatee is highly opportunistic in selecting foods, and many of the freshwater and marine plants consumed by the West Indian manatee have a wide distribution.

Conclusions

Water temperature determines the northern and southern limits to the West Indian manatee's range. Few manatees are reported north of Cape Hatteras on the east coast of the United States, and none are known to occur south of the State of Bahía in Brazil. The winter range of manatees in temperate regions is generally restricted to areas with natural or man-made sources of warm water (20°C).

The association of manatees with fresh water sources is an overwhelmingly consistent pattern, from Columbus' report of their attraction to springs in Bahiá de Cochinos, Cuba, to the recent distribution surveys in various countries throughout their range. Manatees in Florida and Central and South America are frequently found in rivers and estuaries, while those in the Greater Antilles may of necessity occupy more marine environments. The association of manatees with shallow, protected coasts has also been noted by many authors. The occurrence of seagrasses and other submerged vegetation in shallow water accounts for part of this association. Extensive, unprotected coastlines with unvegetated sandy beaches or rocks may act as geographic barriers in some parts of the manatee's range.

Humans have undoubtedly influenced the distribution of manatees by depredation and destruction of habitat. Although hunting may have caused the manatee's disappearance from portions of its former range, the species persists in some regions despite a long history of exploitation by humans. In these regions, for example Guyana and Belize, the abundance of suitable habitat may have supported greater densities of manatees than areas where manatees have apparently been extirpated, such as the Lesser Antilles.

The West Indian manatee has fared better than the extinct Caribbean monk seal (*Monachus tropicalis*) (Kenyon 1981, LeBouef et al. 1986), and the many species of land mammals in the West Indies that have become extinct since the late Pleistocene, largely as a result of human activities (Morgan and Woods 1986). Manatees have adapted to hunting pressure by developing highly secretive behavior, and because they are totally aquatic and relatively far-ranging, they have thus far been less susceptible to complete extirpation than many terrestrial species. Manatees are, however, be increasingly vulnerable to human activities, and new sources of mortality may overshadow the former threat of hunting. The increased and widespread use of nylon and polyester gill nets is of particular concern because manatees may be unable to avoid nets in some locations, e.g. river mouths, and although incidentally caught they are often peremptorily slaughtered. The already high rate of boat strike mortality in Florida may continue to increase with human population growth, and boat-strike deaths are now reported for other countries in the species' range. Only through strengthening of existing conservation efforts and improving enforcement of protective laws will manatees have a chance to endure.

Acknowledgments

Maps were prepared by B.G. Charest, L. Reep, and I.E. Beeler. R.K. Bonde provided computer system assistance for manuscript preparation. Helpful reference material was provided by M. Borobia, R.I. Crombie, C.K. Dodd and W.J. Kenworthy. S.R. Colbert translated several articles and letters from Spanish to English. I.J. Gordon and V.M.F. da Silva assisted with R.C. Best's contribution to the paper. The entire manuscript was reviewed by D.P. Domning, M.E. Ludlow, H. Raffaele, and M.A. Bogan. Selected country accounts were reviewed by M. Borobia, R.I. Crombie, A. Haynes-Sutton, L.A. Hurst, and J.A. Ottenwalder. Discussions with B.J. MacFadden were helpful in writing the biogeography section.

Literature cited

Allen, G.M. 1942. Extinct and Vanishing Mammals of the Western Hemisphere. American Committee for International Wildlife Protection, Special Publication No. 11, The Intelligence Printing Co., Lancaster, PA. 620 pp.

Allsopp, W.H.L. 1969. Aquatic weed control by manatees -- its prospects and problems. Pp. 344-351 in L.E. Obeng (ed.). Man-made Lakes. Ghana University Press, Accra. 398 pp.

Alvarez, T. 1963. The recent mammals of Tamaulipas, Mexico. University of Kansas Publications -- Natural History 14:363-473.

Bangs, O. 1895. The present standing of the Florida manatee, *Trichechus latirostris* (Harlan) in the Indian River waters. American Naturalist 29:783-787.

Banks, N. 1985. Sinopse cronológica dos estudos sobre o peixe-boi, no Brasil. Fifth Northeast Zoology Meeting, Natal, Río Grande do Norte, Brasil. Abstract.

Barrett, O.W. 1935. Notes concerning manatees and dugongs. Journal of Mammalogy 16:216-220.

Bartram, W. 1791. Travels through North and South Carolina, Georgia, East and West Florida. James and Johnson, Philadelphia. (Republished 1955, Dover Publications, Inc., New York), 414 pp.

Baughman, J.L. 1946. Some early notices on American manatees and the mode of their capture. Journal of Mammalogy 27:234-239.

Belitsky, D.W. and C.L. Belitsky. 1980. Distribution and abundance of manatees (*Trichechus manatus*) in the Dominican Republic. Biological Conservation 17:313-319.

Bengtson, J.L. and D. Magor. 1979. A survey of manatees in Belize. Journal of Mammalogy 60:230-232.

_____. 1981. Ecology of manatees (*Trichechus manatus*) in the St. Johns River, Florida. Ph.D. dissertation, University of Minnesota, Minneapolis, MN. 126 pp.

Bertram, G.C.L., and C.K.R. Bertram. 1964. Manatees in the Guianas. Zoologica (New York Zoological Society) 49:115-120.

_____. 1973. The modern Sirenia: their distribution and status. Biological Journal of the Linnean Society 5:297-338.

Best, R.C. 1981. Foods and feeding habits of wild and captive Sirenia. Mammal Review 11:3-29.

_____, and D.M. Teixeira. 1982. Notas sobre a distribuição e "status" aparentes dos peixes-bois (Mammalia: Sirenia) nas costas amapaenses brasileiras. Boletim Fundação Brasileira para a Conservação da Natureza, Río de Janeiro, 17:41-47.

Bradley, R. 1983. The pre-Columbian exploitation of the manatee in Mesoamerica. Papers in Anthropology, University of Oklahoma 24:1-82.

Brasier, M.D. 1975. An outline history of seagrass communities. Palaeontology 18:681-702.

Buergelt, C.D., R.K. Bonde, C.A. Beck, and T.J. O'Shea. 1984. Pathologic findings in manatees in Florida. Journal of the American Veterinary Medical Association 185:1331-1334.

Campbell, H.W. and D. Gicca. 1978. Reseña preliminar del estado actual y distribución del manatí (*Trichechus manatus*) en México. Anales Instituto Biología, Universidad Nacional Autónoma de México, Serie Zoología 49:257-264.

Carr, A., M.H. Carr, and A.B. Meylan. 1978. The ecology and migrations of sea turtles, 7. The West Caribbean green turtle colony. Bulletin of the American Museum of Natural History 162:1-46.

Charnock-Wilson, J. 1968. The manatee in British Honduras. Oryx 9:293-294.

_____. 1970. Manatees and crocodiles. Oryx 10:236-238.

_____, C.K.R. Bertram, and G.C.L. Bertram. 1974. The manatee in Belize. Belize Audubon Society Bulletin 6:1-4.

Colmenero-R., L.C. 1984. Nuevos registros del manatí (*Trichechus manatus*) en el sureste de México. Anales Instituto de Biología, Universidad Nacional Autónoma de México, Serie Zoología 54:243-254.

_____. 1986. Aspectos de la ecología y comportamiento de una colonia de manatíes (*Trichechus manatus*) en el municipio de Emiliano Zapata, Tabasco. Anales Instituto de Biología, Universidad Nacional Autónoma de México, Serie Zoología 56:589-602.

_____, and Ma. E. H. Zavala. 1986. Distribución de los manatíes, situación y su conservación en México. Anales Instituto de Biología, Universidad Nacional Autónoma de Mexico, Serie Zoología 56:955-1020.

Cumbaa, S.L. 1980. Aboriginal use of marine mammals in the southeastern United States. Southeastern Archaeological Conference Bulletin 17:6-10.

Cuní, L.A. 1918. Contribución al estudio de los mamíferos acuaticos observados en las costas de Cuba. Memorias de la Sociedad Cubana de Historia Natural "Felipe Poey" Vol. 3: 83-126.

Dampier, W.A. 1699. A New Voyage Round the World. Vol. I. Fourth edition. James Knapton, London. 550 pp.

Dekker, D. 1974. On the natural history of manatees (*Trichechus manatus*) from Suriname for the Amsterdam Zoo. Aquatic Mammals 2:1-3.

Domning, D.P. 1981a. Distribution and status of manatees *Trichechus* spp. near the mouth of the Amazon River, Brazil. Biological Conservation 19:85-97.

_____. 1981b. Sea cows and sea grasses. Paleobiology 7:417-420.

_____. 1982a. Commercial exploitation of manatees *Trichechus* in Brazil c. 1785-1973. Biological Conservation 22:101-126.

_____. 1982b. Evolution of manatees: A speculative history. Journal of Paleontology 56:599-619.

_____, G.S. Morgan, and C.E. Ray. 1982c. North American Eocene sea cows (Mammalia: Sirenia). Smithsonian Contributions to Paleobiology 52, 69 pp.

_____, and L.C. Hayek. 1986. Interspecific and intraspecific morphological variation in manatees (Sirenia: *Trichechus*). Marine Mammal Science 2:87-144.

Duerden, J.E. 1901. The marine resources of the British West Indies. West Indian Bulletin 2:123-127.

Duplaix, N., and H.A. Reichart. 1978. History, status and protection of the Caribbean manatee *Trichechus m. manatus* in Suriname. Rare Animal Relief Effort and United States Fish and Wildlife Service, unpublished report, 23 pp. + x append.

Estrada, A.R. and L.T. Ferrer. 1987. Distribución del manatí antillano, *Trichechus manatus* (Mammalia: Sirenia), en Cuba. I. Región occidental. Poeyana 354:1-12.

Evermann, B.W. 1902. General report on the investigations in Porto Rico of the United States Fish Commission Steamer Fish Hawk in 1899. Bulletin of the United States Fish Commission 20:24-25.

Fairbairn, P.W., and A.M. Haynes. 1982. Jamaican surveys of the West Indian manatee (*Trichechus manatus*), dolphin (*Tursiops truncatus*), sea turtles (Families Cheloniidae and Dermochelydae) and booby terns (Family Laridae). United Nations Food and Agriculture Organization Fisheries Report 278:289-295.

Fewkes, J.W. 1907. The aborigines of Porto Rico and neighboring islands. Annual Report of the Bureau of American Ethnology 25:48-50 and 192-193.

Gallo-R., J.P. 1983. Notas sobre la distribución del manatí (*Trichechus manatus*) en las costas de Quintana Roo. Anales Instituto de Biología, Universidad Nacional Autónoma de México, Serie Zoología 53:443-448.

Gundlach, J. 1877. Contribución a la Mamalogia Cubana. Montiel y Compañía, Habana, Cuba53 pp.

Gunter, G. 1942. Further miscellaneous notes on American manatees. Journal of Mammalogy 23:89-90.

Harlan, R. 1824. On a species of Lamantin resembling the Manatus Senegalensis (Cuvier) inhabiting the coast of East Florida. Journal of the Academy of Natural Sciences of Philadelphia 3:390-394.

Hartman, D.S. 1972. Manatees. Sierra Club Bulletin 57:20-22.

_____. 1974. Distribution, status and conservation of the manatee in the United States. National Technical Information Service Publication PB81-140725, Springfield, VA. 246 pp.

_____. 1979. Ecology and behavior of the manatee (*Trichechus manatus*) in Florida. American Society of Mammalogists Special Publication 5:1-153.

Hislop, G. 1985. Trinidad and Tabago. Sirenews 3:8. Department of Anatomy, Howard University, Washington, DC.

Hurst, L.A. 1987. Review of the status and distribution of the West Indian manatee (*Trichechus manatus*) in Jamaica, with an evaluation of the aquatic vegetation of the Alligator Hole River. Unpublished M.S. Thesis, University of Florida, Gainesville, FL. 168 pp.

Husar, S.L. 1977. The West Indian manatee (*Trichechus manatus*). United States Fish and Wildlife Service, Wildlife Research Report 7. U. S. Government Printing Office, Washington, DC. 22 pp.

Husson, A. M.1978. The mammals of Suriname. Zoölogische Monographie?n van het Rijksmuseum van Natuurlijke Historie No. 2:334-339.

Irvine, A.B. 1983. Manatee metabolism and its influence on distribution in Florida. Biological Conservation 25:315-334.

____, J.E. Caffin, and H.I. Kochman. 1982. Aerial surveys for manatees and dolphins in western peninsular Florida. Fishery Bulletin 80:621-629.

Janson, T. 1978. Ecology and conservation of the Guatemalan manatee. Marine Mammal Information, Oregon State University: 3. Abstract.

____. 1980. Discovering the mermaids. Oryx 15:373-379.

Jimenez, J.C. 1982. Letter to Editor, San Juan Star Magazine June 6, 1982.

Jong, C. de. 1961. Manatee hunters in Guiana in past centuries. Surinaamse Landbouw 9:93-100.

Kenyon, K.W. 1981. Monk seals-*Monachus*. Pp. 195-220 in S.H. Ridgway and R.J. Harrison (eds.). Handbook of Marine Mammals, Vol. 2 Seals. Academic Press, London. 359pp.

Kinnaird, M.F. 1985. Aerial census of manatees in northeastern Florida. Biological Conservation 32:59-79.

Klein, E.H. 1979. Review of the status of manatee (*Trichechus manatus*) in Honduras, Central America. Ceiba 23:21-28.

LeBouef, B.J., K.W. Kenyon, and B. Villa-Ramirez. 1986. The Caribbean monk seal is extinct. Marine Mammal Science 2: 70-72.

Lefebvre, L.W. and R. K. Frohlich. 1986. Movements of radio-tagged manatees in southwest Florida, January 1985-March 1986. United States Fish and Wildlife Service and Florida Department of Natural Resources, unpublished report, 87 pp.

Lluch-B., D. 1965. Algunas notas sobre la biología del manatí. Anales del Instituto Nacional de Investigaciones Biológico-Pesqueras, México 1:405-419.

Loveland, F.O. 1976. Tapirs and manatees: cosmological categories and social process among Rama Indians of eastern Nicaragua. Pp. 67-82 in M.W. Helms and F.O. Loveland (eds.). Frontier Adaptations in Lower Central America. Philadelphia Institute for the Study of Human Issues. Philadelphia, PA. 178 pp.

MacLaren, J.P. 1967. Manatees as a naturalistic biological mosquito control method. Mosquito News 27:387-393.

McClenaghan, L. R., Jr. and T. J. O'Shea. 1988. Genetic variability in the Florida manatee (*Trichechus manatus*). Journal of Mammalogy 69:481-488.

McCoy, E.D., and K.L. Heck, Jr. 1976. Biogeography of corals, seagrasses, and mangroves: an alternative to the center of origin concept. Systematic Zoology 25:201-210.

McKillop, H.I. 1985. Prehistoric exploitation of the manatee in the Maya and circum-Caribbean areas. World Archaeology 16:337-353.

Miller, G.S. 1929. Mammals eaten by Indians, owls, and Spaniards in the coast region of the Dominican Republic. Smithsonian Miscellaneous Collections 82:1-16.

Mondolfi, E. 1974. Taxonomy, distribution, and status of the manatee in Venezuela. Memoria de la Sociedad de Ciencias Naturales La Salle No. 97. Tomo 34:5-23.

____, and C. Müller. 1979. Investigación y conservación del manatí en Venezuela. Vols. 1 (55 pp.) and 2 (88 pp.), Fundación para la Defensa de la Naturaleza, unpublished report.

Montgomery, G.B., N.B. Gale, and W.P. Murdoch, Jr. 1982. Have manatee entered the eastern Pacific Ocean? Mammalia 46:257-258.

Moore, J.C. 1951. The range of the Florida manatee. Quarterly Journal of the Florida Academy of Sciences 14:1-19.

____. 1956. Observations of manatees in aggregations. American Museum of Natural History Novitates 1811:1-24.

Morgan, G.S. and C.A. Woods. 1986. Extinction and the zoogeography of West Indian land mammals. Biological Journal of the Linnean Society 28: 167-203.

Morison, S.E. 1942. Admiral of the Ocean Sea: A Life of Christopher Colombus. Little, Brown and Co., Boston, MA. 680 pp.

Murie, A. 1935. Mammals from Guatemala and British Honduras. Miscellaneous Publications of the Museum of Zoology University of Michigan 26:1-30.

Murray, S.P., S.A. Hsu, H.H. Roberts, E.H. Owens, and R.L. Crout. 1982. Physical processes and sedimentation on a broad, shallow bank. Estuarine, Coastal and Shelf Science 14:135-137.

Narganes-S., Y.M. 1982. Vertebrate faunal remains from Sorce, Vieques. Unpublished M.S. Thesis, University of Georgia, Athens, GA. 110pp.

Neish, W.D. 1896. The manatee, *Manatus australis*. Journal of the Institute of Jamaica 2:287-288.

Nietschmann, B. 1972. Hunting and fishing focus among the Moskito Indians, eastern Nicaragua. Human Ecology 1:41-67.

____. 1979. Ecological change, inflation, and migration in the far western Caribbean. Geographic Review 69:1-24.

Odell, D.K. 1979. Distribution and abundance of marine mammals in the waters of the Everglades National Park. National Park Service Biological Series 5:673-678.

____, J.E. Reynolds, and G. Waugh. 1978. New records of the West Indian manatee (*Trichechus manatus*) from the Bahama Islands. Biological Conservation 14:289-293.

O'Donnell, D.J. 1981. Manatees and man in Central America. Unpublished Ph.D. Dissertation, University of California, Los Angeles, CA. 110 pp.

O'Shea, T. J. 1986. Mast foraging by West Indian manatees (*Trichechus manatus*). Journal of Mammalogy 67:183-185.

____, C.A. Beck, R.K. Bonde, H.I. Kochman, and D.K. Odell. 1985. An analysis of manatee mortality patterns in Florida, 1976-81. Journal of Wildlife Management 49:1-11.

____, M. Correa-Viana, M.E. Ludlow, and J.G. Robinson. 1986. Distribution and status of the West Indian manatee in Venezuela. International Union for the Conservation of Nature. Contract Report 9132, Gland Switzerland. 101 pp.

____. 1988. Distribution, status, and traditional significance of the West Indian manatee (*Trichechus manatus*) in Venezuela. Biological Conservation 46:281-301.

Packard, J.M. 1981. Abundance, distribution, and feeding habits of manatees (*Trichechus manatus*) wintering between St. Lucie and Palm Beach Inlets, Florida. United States Fish and Wildlife Service Contract Report No. 14-16-004-80-105, 142pp.

____. 1983. Proposed Research/Management Plan for Crystal River manatees. Vol.I. Summary. Technical Report No. 7. Florida Cooperative Fish and Wildlife Research Unit, University of Florida, Gainesville, FL. 31pp.

_____, and P.M. Rose. 1985. Strategies for protection of non-exploited Sirenia populations. Fourth International Theriological Congress. Abstract.

_____, and O.F. Wetterqvist. 1986. Evaluation of manatee habitat systems on the northwestern Florida coast. Coastal Zone Management Journal 14:279-310.

Phillips, R.C., R.L. Vadas, and N. Ogden. 1982. The marine algae and seagrasses of the Moskito Bank, Nicaragua. Aquatic Botany 13:187-195.

Powell, J.A. 1978. Evidence of carnivory in manatees (*Trichechus manatus*). Journal of Mammalogy 59:442.

_____, D.W. Belitsky, and G.B. Rathbun. 1981. Status of the West Indian manatee (*Trichechus manatus*) in Puerto Rico. Journal of Mammalogy 62:642-646.

_____, and J.C. Waldron. 1981. The manatee population in Blue Spring, Volusia County, Florida. Pp. 41-51 in R.L. Brownell and K. Ralls (eds.). The West Indian Manatee in Florida. Florida Department of Natural Resources, Tallahassee, unpublished report. 154 pp.

_____, and G.B. Rathbun. 1984. Distribution and abundance of manatees along the northern coast of the Gulf of Mexico. Northeast Gulf Science 7:1-28.

Price, W.A. 1954. Shorelines and coasts of the Gulf of Mexico. Pp. 39-65 in P.S. Galtsoff (ed.). Gulf of Mexico -- its origin, water, and marine life. U.S. Department of Interior, Fishery Bulletin 89, Vol. 55. 604 pp.

Rathbun, G.B., R.K. Bonde, and D. Clay. 1982. The status of the West Indian manatee on the Atlantic coast north of Florida. In R.R. Odom and J.W. Guthrie (eds.). Proceedings of the Nongame and Endangered Wildlife Symposium. Georgia Department of Natural Resources, Technical Bulletin WL 5:152-165.

_____, J.A. Powell, and G. Cruz. 1983a. Status of the West Indian manatee in Honduras. Biological Conservation 26:301-308.

_____, J.A. Powell, and J.P. Reid. 1983b. Movements of manatees (*Trichechus manatus*) using power plant effluents in southern Florida. Florida Power and Light Co. Contract Report, Cooperative Agreement No. 14-16-0009-82-906, 123 pp.

_____, T. Carr, N.M. Carr, and C.A. Woods. 1985a. The distribution of manatees and sea turtles in Puerto Rico. National Technical Information Service PB86-1518347AS, Springfield, VA. 83pp.

_____, C.A. Woods, and J.A. Ottenwalder. 1985b. The manatee in Haiti. Oryx 19:234-236.

_____, J.P. Reid, and G. Carowan. In press. Manatee (*Trichechus manatus*) distribution and movement patterns in northwestern peninsular Florida. Florida Marine Research Publications 48.

Ray, C.E. 1960. The manatee in the Lesser Antilles. Journal of Mammalogy 41:412-413.

Reid, J.P., and G.B. Rathbun. 1986. 1985 manatee identification catalog update. United States Fish and Wildlife Service and Florida Power and Light Co., unpublished report, 14pp.

Reynolds, J.E. III, and J.C. Ferguson. 1984. Implications of the presence of manatees (*Trichechus manatus*) near the Dry Tortugas Islands. Florida Scientist 47:187-189.

_____, and J.R. Wilcox. 1986. Distribution and abundance of the West Indian manatee *Trichechus manatus* around selected Florida power plants following winter cold fronts: 1984-1985. Biological Conservation 38: 103-113.

Rose, P.M., and S.P. McCutcheon. 1980. Manatees (*Trichechus manatus*): abundance and distribution in and around several Florida power plant effluents. Florida Power and Light Co. Contract No. 31534-86626, unpublished report, 75pp.

Roth, V. 1953. Notes and observations on animal life in British Guiana. Daily Chronicle, Ltd., Guiana Edition No. 3:69-73. Georgetown, British Guiana.

Rouse, I. 1964. Prehistory of the West Indies. Science 144:499-513.

———. 1986. Migrations in Prehistory. Yale University Press, New Haven, CT. 202pp.

Ruschi, A. 1965. Lista dos mamíferos do estado do Espírito Santo. Boletim do Museu de Biología No. 24A:29.

Schad, R.C., G. Montgomery, and D. Chancellor. 1981. La distribución y frequencia del manatí en el Lago Gatun y en el Canal de Panama. ConCiencia 8:1-4.

Shane, S.H. 1983. Abundance, distribution, and movements of manatees (*Trichechus manatus*) in Brevard County, Florida. Bulletin of Marine Science 33:1-9.

———. 1984. Manatee use of power plant effluents in Brevard County, Florida. Florida Scientist 47:180-187.

Shaul, W. and A. Haynes. 1986. Manatees and their struggle to survive. Jamaica Journal 19:29-36.

Simpson, G.G. 1932. Fossil Sirenia of Florida and the evolution of the Sirenia. Bulletin of the American Museum of Natural History 59:419-503.

Steininger, F.F. 1986. Erste Ergebnisse über Untersuchungen zu Ernährungsstrategien des Arawaken-Siedlungsplatzes Pointe der Caille, NNW Vieux Fort, St. Lucia, West Indies. Mitteilungen der Prähistorischen Kommission der Oesterreichischen Akademie Wissenshaften 23:37-50.

Strong, W.D. 1935. Archaeological investigations in the Bay Islands, Spanish Honduras. Smithsonian Miscellaneous Collection 92:1-176.

Thornback, J. and M. Jenkins. 1982. Caribbean manatee. Pp. 429-438 in Red Data Book. Vol. 1. Mammalia, International Union for Conservation of Nature and Natural Resources, Morges, Switzerland.

True, F.W. 1884. The Sirenians or seacows. The Fisheries and Fishery Industries of the United States, Sec. 1, Natural History of Useful Aquatic Animals, Part 1, art. C, pp. 114-136.

Trumbull, S. 1949. Sea cows making comeback. Audubon Magazine 51:337.

Vaughan, C. 1983. A report on dense forest habitat for endangered wildlife species in Costa Rica. United States Department of Interior Contract No. 14-16-0009-79-055 and National University of Costa Rica Research Project No. 782085, unpublished report.

Villa-R., B. and L.C. Colmenero-R. 1981. Presencia y distribución de los manatíes o tlacamichin, *Trichechus manatus* Linneo 1782, en México. Anales Instituto de Biología, Universidad Nacional Autónoma de México, Serie Zoología 51:703-708.

Watters, D.R., E.J. Reitz, D.W. Steadman, and G.K. Pregill. 1984. Vertebrates from archeological sites on Barbuda, West Indies. Annals of the Carnegie Museum 53:383-412.

Whitehead, P.J.P. 1977. The former southern distribution of New World manatees (*Trichechus* spp.). Biological Journal of the Linnean Society 9:165-189.

Wing, E.S. 1967. Aboriginal fishing in the Windward Islands. Proceedings of the Second International Congress of Pre-Columbian Cultures of the Lesser Antilles:103-107.

———, C. Hoffman, and C. Ray. 1968. Vertebrate remains from Indian sites on Antigua, West Indies. Caribbean Journal of Science 8:123-139.

———, and S. Scudder. 1980. Use of animals by the prehistoric inhabitants on St. Kitts, West Indies. Proceedings of the Eighth International Congress of the Pre-Columbian Cultures of the Lesser Antilles. Arizona State University Anthropological Research Papers No. 22:237-245.

———, and E.J. Reitz. 1982. Prehistoric fishing communities of the Caribbean. Journal of New World Archaeology 5:13-32.

Figure 1. Map of the Caribbean Basin showing the distribution of the West Indian manatee (derived from Cumbaa 1988, Mckillop 1985, and others; see text).

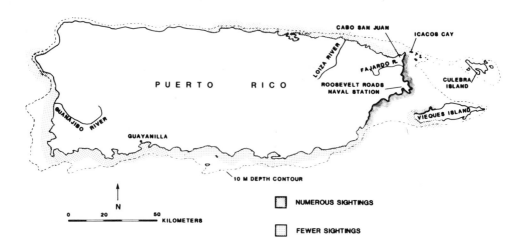

Figure 2. Map of Puerto Rico showing the distribution of the West Indian manatee.

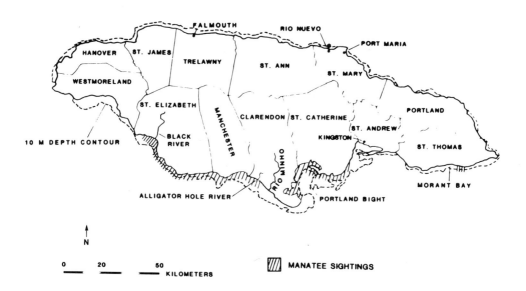

Figure 3. Map of Jamaica showing the distribution of the West Indian manatee.

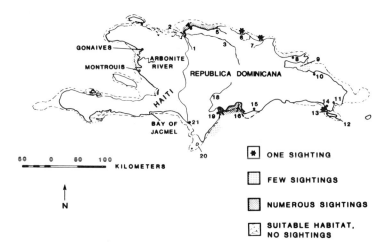

Figure 4. Map of Hispaniola showing the distribution of the West Indian manatee.

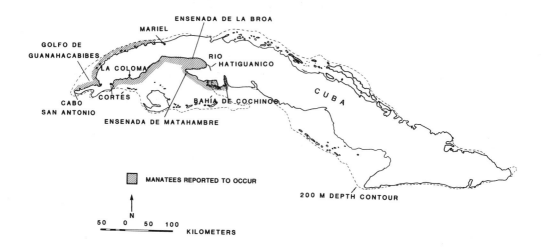

Figure 5. Map of Cuba showing the distribution of the West Indian manatee.

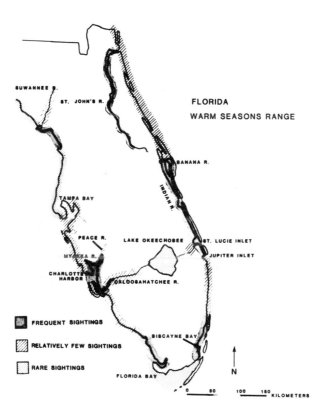

Figure 6a. Map of Florida showing the distribution of the West Indian manatee during the summer.

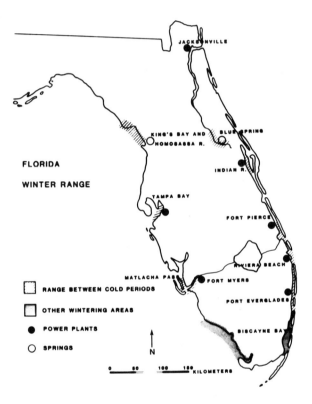

Figure 6b. Map of Florida showing the distribution of the West Indian manatee during the winter.

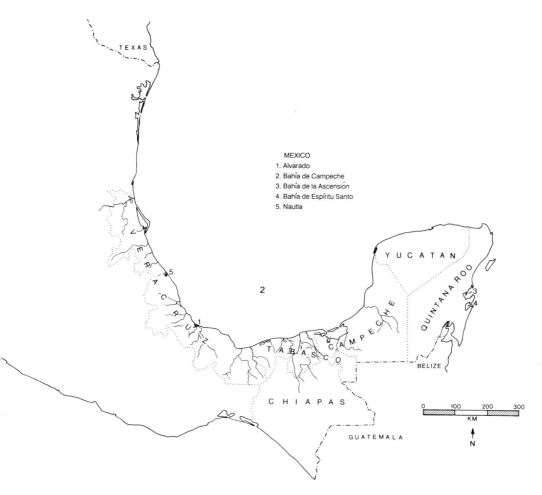

Figure 7. Map of Mexico showing the distribution of the West Indian manatee.

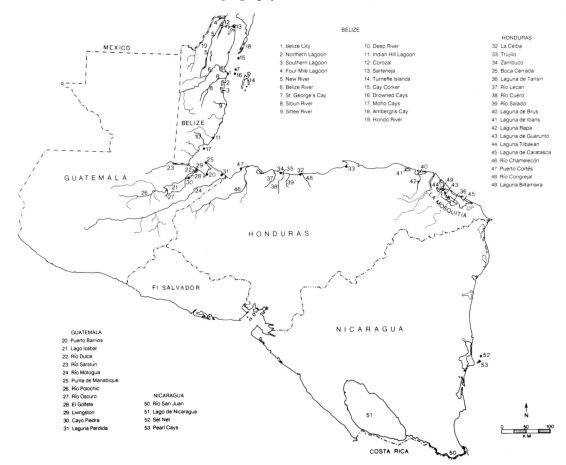

Figure 8. Map of upper Central America showing the distribution of the West Indian manatee.

Figure 9. Map of lower Central America showing the distribution of the West Indian manatee.

Figure 10. Map of northeastern South America showing the distribution of the West Indian manatee.

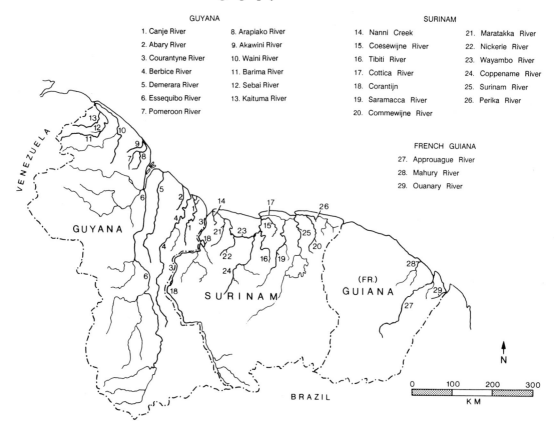

Figure 11. Map of northcentral South America showing the distribution of the West Indian manatee.

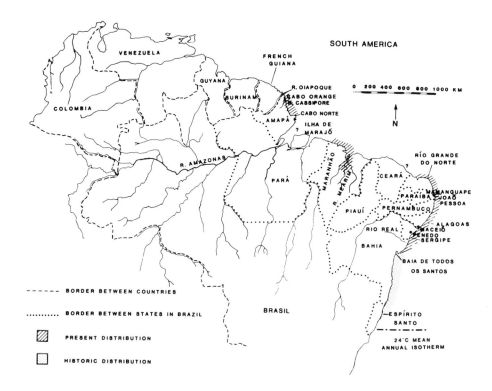

Figure 12. Map of South America showing the change in the distribution of the West Indian manatee in Brazil from historic times.

BIOGEOGRAPHY AND POPULATION BIOLOGY OF THE MONGOOSE IN THE WEST INDIES

Donald B. Hoagland[1], G. Roy Horst[2], and C. William Kilpatrick[3]

Abstract

The small Indian mongoose (*Herpestes auropunctatus*) was introduced into the West Indies from India in the early 1870's and, to date, has been spread to 29 islands. Although most of these subsequent introductions occurred prior to 1900, the source of mongooses and the date of introductions can be established for at least 12 islands. The size of the founding populations, however, can be established for only five West Indian islands.

Mongoose population dynamics were investigated by mark-recapture techniques on the islands of St. Croix (1983-1986) and Jamaica (1986). The sex ratio for each population was not significantly different from the expected 1:1. Individuals of each population were placed in one of four age classes by tooth-wear criteria. Major differences were observed in the age structure between the St. Croix and Jamaican populations. Population densities averaged 6.4 (range 2 to 14) animals per ha on St. Croix and 2.6 (range 1 to 7) animals per ha on Jamaica. Comparisons of these population densities of mongoose with those reported for other islands indicate an indirect correlation between population densities and island size.

Rodent densities in populations sympatric with mongoose were also investigated by mark-recapture techniques. Norway rats (*Rattus norvegicus*) were generally captured in areas of relatively low mongoose densities, whereas black rats (*Rattus rattus*), although somewhat more restricted in habitat, were common in areas with high mongoose densities. The densities of house mice (*Mus musculus*) and of black rats appeared to be directly correlated with mongoose densities, suggesting that the mongoose is not very effective at preying on these taxa. The density of the Norway rat is indirectly correlated with mongoose density, suggesting that it is an effective predator of this taxon.

---

[1] Dr. Hoagland was a graduate student in the Department of Zoology of the University of Vermont, Burlington, VT, 05405. Present address: Department of Physiology and Cell Biology, University of Kansas, Lawrence.
[2] Dr. Horst is a Professor of Biology at the State University of New York at Potsdam, Potsdam, NY, 13676.
[3] Dr. Kilpatrick is an Associate Professor of Zoology at the University of Vermont.

Introduction

The small Indian mongoose, *Herpestes auropunctatus*, is an Old World carnivore that was introduced onto islands of the Caribbean Sea and Pacific Ocean by owners of sugar cane plantations to control the rapidly expanding rat populations. The specific name *auropunctatus* means gold-spotted and accurately describes the pelage of this mongoose. The hair is short and alternately banded gray-brown and yellow giving a speckled appearance.

Although the native range of the small Indian mongoose extends throughout southrn Asia from Iraq through India and Burma to southern China (Honacki et al. 1983), very little is known of the biology of this animal in the Old World. Most published studies concerning the ecology and population biology of *H. auropunctatus* have been conducted on introduced insular populations of the western hemisphere. Nellis and Everard (1983) reviewed the available literature on the food habits of the small Indian mongoose and concluded that this animal is an opportunistic, omnivorous carnivore that prefers to eat small rodents and birds. Most food habit studies, however, indicated that insects are the most common food item in mongoose stomachs (Seaman 1952; Wolcott 1953; Pimentel 1955).

Most population studies of the small Indian mongoose, to date, have concentrated on the mongoose as a nuisance or else on its role as a vector in disease transmission (Nellis and Everard 1983). Few studies have examined the similarities and differences among various insular mongoose populations, other than indicating densities and habitats in which mongooses occur.

This paper presents the results of population biology studies of *H. auropunctatus* on the islands of St. Croix (U.S. Virgin Islands) and Jamaica. This work is part of an ongoing study of the biology and genetics of introduced populations of the small Indian mongoose. The major objectives of this paper are as follows: 1) To review the current distribution of the small Indian mongoose in the West Indies. 2) To reconstruct from the literature the history of the introductions of *H. auropunctatus* into the West Indies. 3) To compare and contrast the population demographies and habitats of mongoose populations on different islands. 4) To determine the role the mongoose plays in controlling rodent populations by examining the relationships between mongoose and rodent population densities.

Biogeography

History of introduction

The first introduction of the mongoose into the West Indies is reported to have occurred in 1870 (Urich 1914; Myers 1931; Husson 1960b) when an unknown number of animals from India (Urich 1914) was introduced to Trinidad. These animals arrived by ship and some escaped in Port-of-Spain before the others were released on a sugar cane estate in the Naparimas district (Urich 1914).

The best documented introduction of the small Indian mongoose is of nine animals, four males and five females (one of which was pregnant), that were imported from Calcutta in 1872 and arrived in Jamaica on the 13th of February aboard the East-Indian ship Merchantman (Espeut 1882). These animals were released on the Spring Garden Estate and within a few months young mongooses were reported about the vicinity (Espeut 1882). This introduction by Espeut in 1872 may represent the only successful New World introduction of mongooses from the Old World. Espeut (1882) indicated that several other Jamaican planters obtained mongooses from India, but the animals were few in number and in some cases were known to have died without leaving progeny. Mongooses were trapped on the Spring Garden Estate and sold to other Jamaican planters. Espeut and other planters sent animals to Cuba, Puerto Rico, Grenada, Barbados, and Santa Cruz (Trinidad) (Espeut 1882).

From 1870 to 1898 little was reported of the mongooses on Trinidad, although Urich (1914) suggested the possibility of additional introduction. According to Urich (1914), the Proceedings of the Agriculture Society of Trinidad records a letter objecting to an introduction of five mongooses to Santa Cruz, Trinidad, in 1898. First, it seems likely that the five Santa Cruz specimens mentioned in the Proceedings of the Agriculture Society of Trinidad are the animals from Jamaica identified by Espeut (1882). The date of introduction of the mongooses to Santa Cruz thus appears to have occured by 1882. Second, the reports of animals being obtained from Jamaica for introduction to Santa Cruz and the lack of reports on the mongoose in Trinidad between 1870 and 1898 indicates that the earlier introductions of the mongoose to Trinidad may not have been successful.

Between 1876 and 1879, seven pairs of mongooses from Jamaica were introduced to Grenada (Allen 1911; Jonkers et al. 1969; Groome 1970). Two to four pairs of mongooses from Jamaica were also introduced to St. Croix between 1882 and 1884 (Heyliger 1884). Additional introduction of mongooses from Jamaica to Barbados, Hispaniola, Puerto Rico, St. Kitts, St. Thomas, and Cuba occurred between 1877 and 1899 (Table 1). Jamaica was also probably the source of mongooses introduced to Guadeloupe, Martinique, St. Vincent, Nevis, and Antigua (Table 1). Mongooses from Jamaica were introduced to Goat Island between 1920 and 1925 according to Lewis (1944).

Guadeloupe was the source of mongooses introduced to St. Martin and probably Desirade and Marie Galante (Table 1). Barbados and Puerto Rico were the probable sources of animals introduced to Sta. Lucia and Vieques, respectively (Palmer 1898; Hinton and Dunn 1967). The mongoose population inhabiting Buck Island was introduced from St. Croix either in 1910 (Nellis and Everard 1983) or in 1952 (Nellis et al. 1978).

Mongooses from Jamaica have also been introduced to Hawaii (Palmer 1898; Bryan 1915; Kramer 1971), Fiji (Gorman 1975), British Guiana (Westerman 1953; Hinton and Dunn 1967), and Colombia (Seaman 1952). The mongooses in Surinam are reported to have been imported from Barbados (Husson 1960a; Husson 1978).

Current distribution

By 1882, ten years after the small Indian mongoose had been introduced onto Jamaica, the population had increased greatly and became a source for other introductions (Espeut 1882). Allen (1911) reported the mongoose was present on eleven islands of the West Indies (Table 2). That report (Allen 1911) extends the known distribution of the mongoose in the West Indies to the islands of Sta. Lucia, St. Vincent, St. Thomas, St. Croix, Vieques, and Hispaniola.

Barbour (1930) not only indicated 17 islands on which the mongoose was known to occur in the West Indies (Table 2) but also listed a number of islands as free of the mongoose, including Water Island, Jost van Dyke, Tobago, and others. The mongoose was reported by Barbour (1930) to occur on seven additional West Indian islands (Tortola, St. Kitts, Nevis, Antigua, Guadaloupe, Maria Galante, and Martinique), to probably occur on Desirad, and possibly to occur on St. Martin. He also reported two unsuccessful attempts to introduce the mongoose to Dominica in the 1880's, once when the animals died in transit and again when ten mongooses were introduced to northern Dominica, but perished.

Varona (1974) listed 21 islands of the West Indies where the mongoose occurs (Table 2). In addition to the 18 islands reported by Allen (1911) and Barbour (1930) as supporting mongoose populations, Varona (1974) reported the distribution of the mongoose to include the islands of St. John, St. Martin, and Desirade. Nellis and Everard (1983) not only list 27 islands on which the mongoose occurs (Table 2), but also list 38 islands from which the mongoose was absent. They reported that the mongoose now occurs on Water Island and Jost van Dyke, although Barbour (1930) had reported these islands were free of the mongoose. Three additional islands (Buck Island, Beef Island, and Louango) were reported by Nellis and Everard (1983) to support mongoose populations in addition to the 21 islands reported by Varona (1974) and Trinidad reported by Espeut (1882). The mongoose is also known to inhabit Goat Island (Jamaica) (Lewis 1944) and Gonave (Hispaniola) (C.A. Woods pers. comm.).

The small Indian mongoose presently occurs on all islands of the Greater Antilles, 20 islands of the Lesser Antilles, and five small islands (near larger islands which support mongoose populations). This carnivore has also been introduced and occurs presently in Surinam (Husson 1960a; Husson 1978), British Guiana (Westerman 1953; Hinton and Dunn 1967), French Guiana (Hinton and Dunn 1967), and Colombia (Seaman 1952). In addition, the mongoose has also been reported to have been introduced into Panama (West 1972) and stray animals have been captured in Florida (Nellis et al. 1978) and Kentucky (Jackson 1921).

Population biology

Methods

Mongoose population dynamics were investigated by recapture techniques on St. Croix, U.S. Virgin Islands, from 1983 to 1986 and Jamaica in 1986. During seven three-week field trips to St. Croix and one five-week field trip to Jamaica, grids containing 30-40 traps were established in representative habitat types at various elevations on the two islands (Table 3 and 4). Twelve grids were trapped during 35 sampling periods on St. Croix for a total of 3,135 trap-days. On Jamaica, eight grids were trapped during 11 sampling periods for a total of 1,070 trap-days. Grids were sampled from one to six times. At Hardwar Gap in the Blue Mountains on Jamaica, trap lines were run instead of grids; a total of 178 trap-days were logged.

Mongoose grids were laid out in five rows of eight traps each, except where physiography prevented that design. Traps were placed in the shade at a 30 m intertrap distance. An estimate of the area sampled by each grid was derived by adding one-half the intertrap distance to all sides of the grids. Estimates of areas sampled by trap lines (Hardwar Gap, Jamaica) were determined by considering the line to be a 50 m wide belt. Tomahawk livetraps (15x15x45 cm) baited with fish were set on grids between 1,000 and 1,300 h each day and checked 24 hours later. During each sampling period, grids or lines were trapped for one to three consecutive days with a minimum of three days between each sampling period.

Each mongoose caught was weighed, sexed, marked, and assigned to an age class by tooth-wear criteria (Pearson and Baldwin 1953), eye color, reproductive status, and overall size. Age class 1 consists of juveniles, which generally are less than one year of age. All teeth are sharp, eyes are blue-grey to green, and body weight is generally below 200 g. Testes are not descended in males and nipples are minute in females. All other age classes have orange eyes, are sexually mature adults, and are differentiated by relative tooth-wear. Animals in age class 2 correspond with Pearson and Baldwin's (1953) age class 2 and are characterized by sharp or slightly rounded teeth and distinct cusps on the first molars. Animals in age class 3 correspond with Pearson and Baldwin's age classes 3-6 and are characterized by worn and rounded teeth, blunt or "flat-topped" canines and an absence of cusps on the first molars. Mongooses in age class 4 correspond with Pearson and Baldwin's age classes 7 and 8 and are characterized by badly worn, broken, and missing teeth. Early in the study, animals were ear-tagged with No. 1 monel fingerling tags (National Band and Tag Co.), but later a series of ear notches was instituted due to ear-tag loss. The processing of each animal was conducted at the capture location and handling time was kept to a minimum, usually less than two minutes.

Population estimates were also obtained for the rodent populations by capture-mark-release-recapture techniques in the same habitats as the mongoose grids (Table 3 and 4). Rodent grids were laid out in five rows of seven traps except where physiography prevented that design. Sherman livetraps were placed at intertrap distances of 10 m in grids adjacent to mongoose grids. Estimates of the areas sampled by each grid were derived by adding one-half the intertrap distance to all sides of the grids. An estimate of the area sampled by trap lines was determined by considering the line to be a 10 m wide belt. Sherman livetraps were baited with scratch grain and placed out at dusk and removed at dawn. During each sampling period, grids or lines were trapped for one to three consecutive days with a minimum of three days between sampling periods. Each rodent captured was weighed, sexed, designated as a juvenile, sub-adult, or adult, then toe-clipped, and released.

Population sizes and their standard errors for each grid were estimated by the Lincoln Index, Fisher-Ford model, Jolly-Seber stochastic model, and Bailey's model (Bailey 1951; Begon 1979; Blower et al. 1981; Seber 1982). Significant deviations from an expected 1:1 sex ratio were tested by Fisher's exact probability or Yate's corrected chi-square.

Sex ratio and age distribution

A total of 572 mongooses were captured on St. Croix, marked, and released. Of that total, 272 were female and 300 were male; the resulting sex ratio did not differ significantly from an expected sex ratio of 1:1. The sex ratios in each of the four age classes likewise did not differ significantly from 1:1 (Fig. 1). Approximately 10 percent of the mongooses captured on St. Croix were juveniles; 80 percent of the adults were assigned equally to age classes 2 and 3 and the remaining 20 percent were adults assigned to age class 4. Age class 1 is under-represented on Fig. 1 because very young mongooses do not leave the nest and many juveniles might escape from live traps undetected.

In Jamaica, 104 mongooses were captured, marked, and released in July and August 1986. Fifty two males and 52 females were recorded and 97 were assigned to age classes (Fig. 1). The Jamaican mongoose population appeared younger than the St. Croix population. Juveniles were 5-7 percent more abundant in the Jamaican population and animals in age class 2 were 18-23 percent more abundant. Animals in age class 3 were 9-16 percent less abundant and those in age class 4 were 13-14 percent less abundant on Jamaica. An August population was separated from the St. Croix data for comparison with the late July and August Jamaican sample. Similar trends to those observed for age class distributions over all of the sampling periods were observed on St. Croix (Fig. 1) as well.

Numerous studies have reported male-biased sex ratios in introduced insular mongoose populations (Baldwin et al. 1952; Pearson and Baldwin 1953; Pimental 1955; Tomich 1969; Nellis and Everard 1983; Coblentz and Coblentz 1985) yet the statistical significance of those biases was reported only for the St. John, U.S. Virgin Islands, population (Coblentz and Coblentz 1985). Generally, the greater male-bias in those studies was produced by removal trapping, whereas mark-recapture trapping yielded sex ratios closer to 1:1. In the St. John study, the authors found the first few months of removal trapping produced significantly more males, but later months yielded significantly more females. The greater male-bias produced by removal trapping might be explained by the greater average distance normally traveled by males than by females (Tomich 1969; Gorman 1979). If all local animals are removed from a trapping area, it is more likely that males will move into the area before females because of their greater mobility. Further evidence is supplied by Coblentz and Coblentz (1985) who reported a significantly greater proportion of females captured later in their study. Trapping efforts in St. Croix and Jamaica indicate local fluctuations in sex ratio; however, the overall sex ratio does not appear to deviate significantly from 1:1. Sex ratios of mongooses in insular populations generally do not appear to deviate from 1:1 and many reported deviations are probably a result of biased trapping technique.

The greater proportion of juvenile mongooses captured on Jamaica than on St. Croix may result from the use of slightly smaller, more escape-proof traps on Jamaica. However, the greater abundance of animals in age class 2 and lower abundance of animals in age classes 3 and 4 on Jamaica compared with St. Croix is not likely to be attributable to trap differences. The age class distribution on St. Croix is similar to that reported for Hawaii (Pearson and Baldwin 1953), whereas the age class distribution on Jamaica is more similar to an earlier study conducted on St. Croix (Nellis and Everard 1983). The discrepancy between the present results and those of Nellis and Everard (1983) for the St. Croix population might be explained by the greater number of juveniles sampled in that study. Recalculating those data after removing the juvenile data, however, confirms a younger St. Croix population in that study. An alternative explanation may be in the smaller number of habitats sampled by Nellis and Everard on St. Croix. In addition, their age determination was based exclusively upon eye lens weights and not on the tooth-wear criteria used in the Hawaiian study and in this study. The St. Croix mongoose population appears to be older than the Jamaican population. About 20 percent of all St. Croix adults have extremel badly worn teeth or no teeth at all. Only one mongoose captured on Jamaica displayed badly worn teeth.

Population densities and habitat use

St. Croix mongoose population estimates derived from the Lincoln Index, Fisher-Ford model, Jolly-Seber stochastic model, and the Bailey model for the August 1983 and January 1984 sampling periods were all similar and had overlapping standard errors. Therefore, only the Lincoln Index was used to estimate population sizes for all other sampling periods. The average capture of unique individuals per grid varied from 7.5 to 22.6 mongooses, whereas estimates of population size from mark-recapture data were from 21 to 130 percent higher (Table 5). The average number of mongooses per ha on St. Croix varied from two in some dry grass-scrub, thorn-scrub, and desert-scrub habitats to approximately 14 in a rain forest habitat. The mean average density on St. Croix was estimated to be 6.4 mongooses per ha (Table 6). The rain forest was the most preferred habitat while the desert-scrub grassland was the least preferred.

The total number of individual mongooses captured in each location on Jamaica varied from three to 33 (Table 7). Population estimates were lower than actual abundance on four grids, equal to the abundance on one grid, and higher on three grids. The lowest estimated density, 0.6 mongoose per ha, was recorded from a cultivated garden-scrub-forest habitat, whereas the greatest density, 6.8 animals per ha, was observed in a mixed plantation of coconuts and bananas. The mean average density on Jamaica was estimated to be 2.6 mongooses per ha (Table 6). The dry scrub-grassland and mixed coconut and banana habitats were the most preferred while the sugar cane field was least preferred.

Mongooses were captured in similar habitats on St. Croix and Jamaica, but density estimates for the St. Croix population were consistently greater than the Jamaican population. The greatest average density on St. Croix was recorded in a rain forest habitat and was approximately four times greater than that of a similar habitat on Jamaica. The grassland-scrub and thorn-scrub habitats on St. Croix yielded densities two or three times greater than the density of similar habitats on Jamaica. Averaging all estimates for all habitats sampled revealed a St. Croix mongoose population density similar to that of the population density of Grenada (Nellis and Everard 1983) and approximately two and one-half times greater than the Jamaica, Trinidad (Nellis and Everard 1983), and Puerto Rico (Pimental 1955) populations densities (Table 6). The St. Croix grassland habitats supported average densities for that island, whereas similar habitats on Jamaica supported greater-than-average densities for that island. A rain forest habitat on Jamaica supported an average Jamaican mongoose density, whereas similar habitats on St. Croix yielded the greatest average densities. On both islands, thorn-scrub habitats supported lower-than-average mongoose population densities for their respective islands. These data indicate mongooses generally prefer drier habitats, as previously reported by Pocock (1941), for their native range in Northern India and also many Caribbean Islands (Nellis and Everard 1983; Pimental 1955) and Pacific Islands (Baldwin et al. 1952; Gorman 1975, 1979; Pearson and Baldwin 1953; Tomich 1969). On the islands of St. Croix and Grenada (Nellis and Everard 1983), however, mongooses have expanded their ranges into, and have flourished in, most habitats.

In an effort to identify causal factors for the trends observed in insular mongoose population demographics, the ecological diversity of each island was estimated by two variables. The first variable is based upon physiographic diversity and is represented by the highest elevation of each; the second variable is a compilation of the terrestrial vertebrates inhabiting each island (Table 6). St. Croix has the least variation in elevation and a high mongoose population density, whereas Jamaica has the highest elevation and a low population density (Table 6). The correlation coefficient between mongoose densities and elevations is -0.59. The diversity of terrestrial vertebrates on St. Croix and Grenada is low, whereas on Jamaica, Trinidad, and Puerto Rico it is high (Table 6). The correlation coefficient between mongoose densities and the number of species of terrestrial vertebrates is -0.90. Although vertebrate faunal diversity does appear to be negatively correlated with mongoose densities, a greater correlation coefficent (-0.98) is observed between mongoose densities and the log of island size. Smaller islands support high mongoose densities and low vertebrate diversity, whereas large islands support low mongoose densities and high vertebrate diversity.

Rodent densities and habitat utilization

Two species of rats (*Rattus norvegicus* and *R. rattus*) were captured in both mongoose and rodent traps, and the house mouse (*Mus musculus*) was captured in rodent traps at various locations on St. Croix. Density estimates of rat populations derived from mongoose trapping grids were much lower than those derived from rodent trapping grids in the same habitats, probably as a result of larger intertrap distances on mongoose grids and greater probability of rats escaping from mongoose traps. Mice were more abundant than both species of rats and the black rat (*R. rattus*) was more abundant, and more frequently encountered, than the Norway rat (*R. norvegicus*) (Table 8). The highest and lowest densities of mice were recorded in a grassland habitat by two different trapping techniques. The difference in density estimates may be due to burning of the habitat between the time when the high density and low density samples were obtained. Black rats were captured more frequently in mesic habitats than in drier grassland-scrub.

The house mouse was more abundant than either species of rat on Jamaica and was captured in all but two areas trapped (Table 9). Black rats were captured only in the Blue Mountains at Hardwar Gap, whereas Norway rats were captured in all locations except sugar cane plantations. The highest density of mice was recorded in a mixed banana and coconut plantation; the lowest density was observed in a moist sugar cane plantation and a dry grassland mixed with thorn-scrub. No mice were captured on the grid supporting the highest density of Norway rats. One of the two grids with the second highest density of Norway rats, Discovery Bay Limestone, also did not support any mice; but, the other grid, Milk River Lower, supported the third highest density of mice (Table 9). The lowest densities of Norway rats were recorded in the areas with the highest densities of mice.

Black rats were captured only in the rain forest on Jamaica and in more mesic habitats on St. Croix. Norway rats were most abundant in lower elevation and drier habitats on Jamaica, and only one Norway rat was captured on St. Croix in a low-elevation grassland habitat. Similar habitat and elevational distributions of these two rat species have been reported in Hawaii (Tomich 1981). Mice on Jamaica and St. Croix were widespread, occuring in all habitats sampled with Sherman traps.

Efficiency of mongoose predation upon three rodent taxa

Although the mongoose was originally introduced into the West Indies to control rodent populations, the effectiveness of mongoose predation in controlling rodent populations is virtually undocumented. Without doubt the mongoose greatly lessened the number of rats in cane fields and coffee plantations (Walker 1945; Lewis 1953) following its introduction, and rodents are the major prey of mongooses inhabiting sugar cane fields or coconut plantations (Pemberton 1925; Kami 1964; Gorman 1975). Other studies of mongoose food habits, however, indicate that rodents constitute a small component of the diet of the mongoose in many habitats (Seaman 1952; Wolcott 1953; Pimentel 1955; Gorman 1975). Unfortunately, most of the food habit studies failed to identify the species of rat(s) found, and none quantitatively considered the relative availability of prey. Thus, little is known of the effectiveness of mongoose predation on Norway rats, black rats, or house mice, although several workers (Walker 1945; Pimentel 1955) have speculated that the mongoose was less effective at preying on the more arboreal black rat than on the ground-dwelling Norway rat. The data available from this study on the densities of mongoose and three species of rodents from sympatric populations from various habitats was examined to gain insight into the efficacy of mongoose predation on each of the three rodent taxa. On St. Croix, house mice and black rats were abundant in areas of high mongoose densities, but the only Norway rat captured was in an area of low mongoose density (Tables 5 and 8). Norway rats were more widely distributed in Jamaica, where their densities appear to be indirectly correlated with mongoose densities ($r = -0.65$) (Tables 6 and 9). Mouse densities (Table 9) and black rat densities (Table 8) demonstrate direct correlations ($r = 0.32$ and 0.33 respectively) with mongoose densities (Tables 5 and 6). These data suggest that the mongoose is an effective predator of the Norway rat. In sugar cane fields, the mongoose may be especially effective since no Norway rats were observed in this habitat. The mongoose, however, does not appear to be an effective predator of either black rats or house mice.

Further studies are needed to determine the predatory efficiency of the mongoose on each of these three taxa of rodents. Comparisons of densities of each taxa of rodent from islands inhabited by the mongoose and islands free of the mongoose would provide some additional insight into the predatory effectiveness of the mongoose.

Summary and conclusions

The small Indian mongoose was introduced to the West Indies a little more than 100 years ago. Although two independent introductions from India seem to have occurred in the 1870's, only the introduction onto Jamaica appears to have been successful. In the 18 years following 1882, the small Indian mongoose was introduced to all islands of the Greater Antilles and at least 17 islands of the Lesser Antilles. At present the small Indian mongoose is known to occur on 29 islands of the West Indies.

The sex ratio determined from the mark-recapture studies reported in this paper do not deviate significantly from a 1:1 ratio. Previous reports of a male-biased sex ratio of introduced, insular populations of the mongoose appear to be a result of the removal-trapping technique. The difference in age structure of the mongoose populations of St. Croix and Jamaica may be attributed, in part, to differences in the size of traps used in the two studies, but the almost total absence of older animals (age class 4) on Jamaica seems to reflect a younger population.

Mongooses were captured in similar habitats on St. Croix and Jamaica, but density estimates for the St. Croix population were consistently higher. The difference in estimates of mongoose densities among islands appears not to be associated with the ecological diversity of the islands, but rather to the size of the islands. Small islands, such as St. Croix and Grenada, support higher population densities of mongooses and more uniform high densities in all habitats.

As a method of rodent control, the mongoose appears to have had an effect only on Norway rats. These rats were generally captured in areas of relatively low mongoose densities. Black rats are more arboreal than are Norway rats, and appear to suffer less predation by mongooses. Black rats, although somewhat more habitat-restricted, were commonly found in areas supporting average or even high mongoose densities. The house mouse is widespread and its population densities appear to be directly correlated with mongoose density.

During the past 100 years the mongoose has been able to colonize and establish populations with relatively high densities in most habitats on small islands. At present, many areas on larger islands either support no mongoose populations, or populations with relatively low density. In the future, it seems likely that the population growth will push the small Indian mongoose into most habitats as population densities increase on these larger islands.

Acknowledgments

This research was supported by grants from the Center for Field Research to Horst and Kilpatrick and by funds from the University of Vermont to Kilpatrick. The authors wish to thank the numerous Earthwatch volunteers who aided in many ways in the collection of population data for mongooses and rodents from St. Croix and Jamaica. The authors also wish to thank Steven J. McKay for his technical assistance in estimating population sizes from our mark-recapture data. We are also indebted to Ms. Ann Haynes, Acting Director of the Natural Resource Conservation Division of Jamaica, and to Dr. J. D. Woodley, Director of the Discovery Bay Marine Laboratory, University of the West Indies, for their cooperation and support.

Literature cited

Allen, G.M. 1911. Mammals of the West Indies. Bulletin Museum Comparative Zoology 54:175-263.

Bailey, N.T.J. 1951. On estimating the size of mobile populations from recapture data. Biometrika 38:293-236.

Baldwin, P.H., C.W. Schwartz, and E.R. Schwartz. 1952. Life history and economic status of the mongoose in Hawaii. Journal of Mammalogy 33:335-356.

Barbour, T. 1930. Some faunistic changes in the Lesser Antilles. Proceedings of New England Zoological Club 11:73-85.

Begon, M. 1979. Investigating Animal Abundance. Edward Arnold Ltd., London, 97 pp.

Blower, J.G., L.M. Cook, and J.A. Bishop. 1981. Estimating the Size of Animal Populations. George Allen & Unwin Ltd., London, 128 pp.

Bond, J. 1980. Birds of the West Indies. Fourth ed. Houghton Mifflin, Boston, MA. 255 pp.

Bowden, M.J. 1968. Water balance of a dry island. Geography Publication Dartmouth 6:11-89.
Bryan, W.A. 1915. Natural history of Hawaii. The Hawaiian Gazette Col. Ltd., Honolulu, HI. 596 pp.
Burdon, K.J. 1920. A handbook of St. Kitts - Nevis. The West India Committee, London, 247 pp.
Coblentz, B.E., and B.A. Coblentz. 1985. Reproduction and the annual fat cycle of the mongoose on St. John, U.S. Virgin Islands. Journal of Mammalogy 66:560-563.
Colon, E.D. 1930. Datos Sobre la Historia de la Agricultura de Puerto Rico Antes 1898. Privately printed, San Juan, PR. 302 pp.
Espeut, W.B. 1882. On the acclimatization of the Indian mongoose in Jamaica. Proceedings of the Zoological Society of London 1882:712-714.
Feilden, H.W. 1890. Notes on the terrestrial mammals of Barbados. Zoologist 14:52-55.
French, R. 1976. A Guide to the Birds of Trinidad and Tobago. Asa Wright Nature Center, Cornell Laboratory of Ornithology, Ithaca, NY. 470 pp.
Goodwin, G.G., and A.M. Greenhall. 1961. A reveiw of the bats of Trinidad and Tobago. Bulletin American Museum Natural History 122:187-301.
Gorman, M.L. 1975. The diet of feral *Herpestes auropunctatus* in the Fijian Islands. Journal of Zoology, London 178:237-246.
_____. 1979 Dispersion and foraging of the small Indian mongoose, *Herpestes auropunctatus*. Journal of Zoology 187:65-73.
Gosse, P.H. 1847. The Birds of Jamaica. Van Vorst, London, 447 pp.
_____. 1849. Illustrations of the Birds of Jamaica. Van Vorst, London, 57 pp.
_____. 1851. A Naturalist's Sojourn in Jamaica. Longmans, London, 508 pp.
Groome, J.R. 1970. A Natural History of the Island of Grenada, W.I. Caribbean Printers, Trinidad, 115 pp.
Hall, E.R. 1981. The Mammals of North America. Second ed. John Wiley and Sons, New York, 2:601-1181.
Heyliger, W.H. 1884. Notice. St. Croix Avis (newspaper), 20 February 1884.
Hill, R.T. 1899. Cuba and Puerto Rico with the Other Islands of the West Indies. The Century Co., New York, 447 p.
Hinton, H.E. and A.M.S. Dunn. 1967. Mongooses: Their Natural History and Behavior. University of California Press, Berkeley and Los Angeles, CA. 144 pp.
Honacki, J.H., K.E. Kinman, and J.W. Koeppl. 1982. Mammal Species of the World. Allen Press and Association of Systematic Collections, Lawrence, KA. 694 pp.
Husson, A.M. 1960a. Het voorkomen van de mungo in Suriname. Lutra 2:12-13.
_____. 1960b. De Zoogdieren van de Nederlandse Antillen. Fauna Nederlandse Antillen, The Hague, 170 pp.
_____. 1978. The Mammals of Suriname. Brill, Leiden, 569 pp.
Jackson, H.H.T. 1921. A mongoose in Kentucky. Journal of Mammalogy 2:234-235.
Jonkers, A.H., F. Alexis, and R. Loregnard. 1969. Mongoose rabies in Grenada. West Indian Medical Journal 18:167-170.
Kami, H.T. 1964. Food of the mongoose in the Hamakua District, Hawaii. Zoonoses Research 3:165-170.
Kenny, J.S. 1969. The Amphibia of Trinidad. Studies on the Fauna of Curaçao and Other Caribbean Islands 29:1-78.

_____. 1977. The Amphibia of Trinidad. addendum. Studies on the Fauna of Curaçao and Other Caribbean Islands 51:92-95.
Kramer, R.J. 1971. Hawaiian Land Mammals. Charles E. Tuttle Co., Tokyo, 347 pp.
Lack, D. 1976. Island Biology Illustrated by the Land Birds of Jamaica. Blackwell Scientific Publishers, Oxford, UK. 445 pp.
Lazell, J.D., Jr. 1966. Studies on *Anolis reconditus*. Bulletin of the Institute of Jamaica, Science Series 18:1-15.
Lewis, C.B. 1944. Notes on *Cyclura*. Herpetologica 6:93-98.
_____. 1953. Rats and the mongoose in Jamaica. Oryx 2:170-172.
Lindblad, J. 1969. Journey to Red Birds, translated by Gwynne Vevers. Hill and Wang, New York, 176 pp.
Lynn, W.G. and C. Grant. 1940. The herpetology of Jamaica. Bulletin of the Institute of Jamaica, Science Series 1:1-148.
Myers, J.G. 1931. A Preliminary Report on an Investigation into the Biological Control of West Indian Insect Pests. Empire Marketing Board, H.M. Stationery Office, London, 172 pp.
Nellis, D.W., N.F. Eichbolz, T.W. Regan, and C. Feinstein. 1978. Mongoose in Florida. Wildlife Society Bulletin 6:249-250.
_____, and C. O. R. Everard. 1983. Biology of the mongoose in the Caribbean. Studies of the Fauna of Curaçao and Other Caribbean Islands 195:1-162.
Palmer, T.S. 1898. The danger of introducing noxious animals and birds. Yearbook United State Department of Agriculture 1898:87-110.
Pearson, O.P., and P.H. Baldwin. 1953. Reproduction and age structure of a mongoose population in Hawaii. Journal of Mammalogy 34:436-447.
Pemberton, C.E. 1923. Investigations pertaining to the field rats and other problems in Hamakua. Hawaiian Planters' Record 27:169-175.
Pimental, D. 1955. Biology of the Indian mongoose in Puerto Rico. Journal of Mammalogy 36:62-68.
Pocock, R.I. 1941. The fauna of British India. Vol. II. Taylor and Francis, London, 503 pp.
Raffaele, H.A. 1983. A Guide to the Birds of Puerto Rico and the Virgin Islands. Fondo Educativo Interamericano, San Juan, PR. 255 pp.
Schwartz, A., and D.C. Fowler. 1973. The anura of Jamaica: a progress report. Studies of the Fauna of Curaçao and Other Caribbean Islands 142:51-142.
_____, and R.T. Thomas. 1975. A checklist of West Indian amphibians and reptiles. Carnegie Museum Natural History Special Publication 1:1-216.
_____, R.T. Thomas, and D.C. Fowler. 1978. First supplement to a checklist of West Indian amphibians and reptiles. Carnegie Museum Natural History Special Publication 5:1-35.
Seaman, G.A. 1952. The mongoose and Caribbean wildlife. Transactions of the North American Wildlife Conference 17:188-197.
Scientific Research Council. 1963. The rainfall of Jamaica. Scientific Research Council, Jamaica, 17 pp.
Seber, G.A.F. 1982. The Estimation of Animal Abundance and Related Parameters. Charles Griffin & Co. Ltd., London, 654 pp.

Tomich, P.Q. 1969. Movement patterns of the mongoose in Hawaii. Journal of Wildlife Management 33:576-584.

_____. 1981. Rodents. Pp. 105-110 in D. Mueller-Dombois, W. Bridges, and H.L. Carson (eds.). Island Ecosystems Biological Organization in Selected Hawaiian Communities. Hutchison Ross, Stroudsberg, PA, 583 pp.

Underwood, G. 1962. Reptiles of the Eastern Caribbean. Unpublished document, Deptartment of Extra Mural Studies, University of the West Indies, St. Augustine, Trinidad, 192 pp.

_____, and E. Williams. 1959. The anoline lizards of Jamaica. Bulletin of the Institute of Jamaica, Science Series 9:1-48.

Urich, F.W. 1914. The mongoose in Trinidad and methods of destroying it. Board of Agriculture Trinidad and Tobago Circular 12:5-12.

Varona, L.S. 1974. Catalogo de los Mamiferos Viventes y Extinguidos de los Antillas. Academia de Ciencias De Cuba, Havana, 139 pp.

Walker, L.W. 1945. The Hawaiian mongoose - friend or foe. Natural History 54:396-400.

Westerman, J.H. 1953. Nature preservation in the Caribbean. Publication of the Foundation for Scientific Research in Surinam and the Netherland Antilles 9:1-106.

Wolcott, G.N. 1953. The food of the mongoose (*Herpestes javanicus auropunctatus* Hodgson) in St. Croix and Puerto Rico. Journal of Agriculture of the University of Puerto Rico 37:241-247.

Table 1. History of introductions of the small Indian mongoose. Information in parenthesis is inferred from references.

Introduced Population	Date of Introduction	Origin	Number	Reference
Trinidad	1870	India	?	1,2,3,4
	by 1882	Jamaica	5	3
Jamaica	1872	India	9	5
St. Croix	1882-1884	Jamaica	4-8	5,6
Buck Island	1910 or 1952	St. Croix	4	7,19
Barbados	1882	Jamaica	(20)	5,8
Hispaniola	by 1895	Jamaica	?	9
Puerto Rico	1877	Jamaica	(20)	9,10,11
St. Martin	1885-1889	Guadeloupe	(20)	12
Grenada	1876-1879	Jamaica	14	5,9,13,14
Guadeloupe	1880-1885	(Jamaica)	?	5,12
Martinique	1889	(Jamaica)	?	15
St. Kitts	1884	Jamaica	?	16
St. Thomas	by 1899	Jamaica	?	10
St. Vincent	by 1900	(Jamaica)	?	9,17
St. Lucia	by 1900	(Barbados)	?	9,17
Cuba	1882	Jamaica	?	18
Vieques	by 1899	(Puerto Rico)	?	10
Tortola	by 1900	?	?	4,17
Carriacou	by 1900	?	?	18
Desirade	by 1900	(Guadeloupe)	?	17,18
Nevis	by 1900	(Jamaica)	?	4,17
Antigua	by 1900	(Jamaica)	?	4,17
Marie Galante	by 1900	(Guadeloupe)	?	4,17
Goat Island	1920-1925	Jamaica	?	20

References: (1) Husson 1960a; (2) Myers 1931; (3) Urich 1914; (4) Westerman 1953; (5) Espeut 1882; (6) Heyliger 1884; (7) Nellis et al. 1978; (8) Feilden 1890; (9) Allen 1911; (10) Palmer 1898; (11) Colon 1930; (12) Husson 1960b; (13) Groome 1970; (14) Jonkers et al. 1969; (15) Hill 1899; (16) Burdon 1920; (17) Hinton and Dunn 1967; (18) Barbour 1930; (19) Nellis and Everard (1983); (20) Lewis (1944).

Table 2. Reported distribution of the mongoose *Herpestes auropunctatus* in the West Indies.

Island	Allen (1911)	Barbour (1930)	Varona (1974)	Nellis & Everard (1983)
Cuba	+	+	+	+
Jamaica	+	+	+	+
Hispaniola	+	+	+	+
Puerto Rico	+	+	+	+
Vieques	+		+	+
St. Thomas	+	+	+	+
Water Island		-		+
St. John			+	+
Jost van Dyke		-		+
St. Croix	+	+	+	+
Buck Island				+
Tortola		+	+	+
St. Martin		?	+	+
St. Kitts		+	+	+
Nevis		+	+	+
Antigua		+	+	+
Guadalupe		+	+	+
La Desirade		?	+**	+
Maria Galante		+	+	+
Dominica		+*	+*	-
Martinique		+	+	+
St. Lucia	+	+	+	+
St. Vincent	+	+	+	+
Grenada	+	+	+	+
Barbados	+	+	+	+
Trinidad				+
Beef Island				+
Lovango				+
Total	11	17	21	27

*Introduction unsuccessful
** Deseada?

Table 3. Habitat types, average annual rainfall, and elevations of grids trapped on St. Croix, 1983-86.

Grid	Habitat Type	Rainfall* (cm)	Elevation (m)
Butler Bay	Hardwood rain forest	125	2-30
Mahogany	Hardwood rain forest, old ruins, human disturbance	125	120-150
Caledonia	Grassland, old pasture, adjacent to rain forest	125	150-180
Contessa	Grassland, abandoned papaya, plantation, some small acacia	125	60-90
Enfield Back-9	Grassland, acacia thorn-scrub	75	3-6
Enfield Green	Grassland, sparse acacia, periodic burning	75	3-6
Cottongarden	Desert-scrub, cacti, grassland	50	12-24
East End	Desert-scrub, cacti, grassland	50	12-24
Enfield Rough	Thorn-scrub, dense acacia	75	3-6
West End	Thorn-scrub, back beach	75	1-3
Concordia	Thorn-scrub, back beach	75	1-3

* From Bowden 1968.

Table 4. Habitat types, average annual rainfall, and elevations for grids trapped on Jamaica July - August 1986.

Grid	Habitat Type	Rainfall* (cm)	Elev. (m)
Milk River Lower	Grassland adjacent to thorn-scrub along a river	<125	2
Milk River Upper	Thorn-scrub, limestone outcrops	<125	10
Milk River Sugar	Sugar cane field	<125	5
Discovery Bay Bauxite	Grassland mixed with thorn-scrub	<125	10
Discovery Bay Garden	Cultivated beans and root crops, scrub-forest, limestone outcrops	<125	30
Discovery Bay Limestone	Scrub-forest, agave, giant bromeliads, limestone outcrops	<125	30
St. Ann's Coconut	Mixed coconut and banana plantation	125-200	5
St. Ann's Sugar	Sugar cane field	125-200	5
Hardwar Gap**	Mountain mist forest	250-500	1300-1500

* Scientific Research Council 1963.
** Trapping lines were run instead of grids.

Table 5. Small Indian mongoose population estimates for grids trapped on St. Croix, 1983-86.

Grid	Number of Sampling Periods	Area Sample (ha)	Mean Number of Unique Animals Captured Per Sampling Period	Average Estimated Lincoln Index Estimate (S.E.)	Relative Density (#/ha)
Butler Bay	1	3.5	21.0	48.0(20.8)	13.7
Mahogany	3	3.5	20.7	27.3(6.8)	7.8
Caledonia	2	3.5	13.0	15.0(3.9)	4.3
Contessa	6	3.0	18.5	27.7(10.9)	9.2
Enfield Back-9	2	3.5	7.5	**	2.1
Enfield Green	5	4.5	11.4	20.4(7.7)	4.5
Madame Carty	5	4.0	22.6	49.7(18.6)	12.4
Cottongarden	3	3.5	11.7	18.9(7.7)	5.4
East End	1	5.3	8.0	10.5(5.1)	2.0
Enfield Rough	2	3.0	20.0	27.6(8.0)	9.2
West End	1	6.0	9.0	12.0(4.0)	2.0
Concordia	4	6.5	16.8	26.0(7.4)	4.0

** Undefined value.

Table 6. Estimated average mongoose population densities on five Caribbean islands.

Island	Area (km^2)	Highest Elevation (m)	Animal Density (#/ha)	Reference	Number of Terrestrial Vertebrate Species*
St. Croix	270	350	6.4	1	197
Grenada	310	840	6.6	2	208
Trinidad	4,800	940	2.5	2	417
Puerto Rico	8,900	1100	2.5	3	338
Jamaica	11,430	2526	2.6	1	307

Reference: 1) this study; 2) Nellis and Everard 1983; 3) Pimental 1955.

* References:

Herptiles - Gosse 1851; Groome 1970; Kenny 1969, 1977; Lazell 1966; Lynn and Grant 1940; Nellis and Everard 1983; Schwartz and Fowler 1973; Schwartz and Thomas 1975; Schwartz et al. 1978; Underwood and Williams 1959; Underwood 1962.

Birds - Bond 1980; French 1976; Gosse 1847, 1849; Lack 1976; Raffaele 1983.

Mammals - Goodwin and Greenhall 1961; Gosse 1851; Hall 1981; Linblad 1969; Varona 1974.

Table 7. Small Indian mongoose population estimates for each area trapped on Jamaica July - August 1986.

Grid	Mean Number of Unique Animals Captured Per Sampling Period	Lincoln Index Estimate (S.E.)	Estimated Relative Density (#/ha)
Milk River Lower	13	10.7(4.2)	3.0
Milk River Upper	8	10.0(3.5)	2.8
Milk River Sugar	3	3.0(1.0)	0.8
Discovery Bay Bauxite	12	13.3(3.2)	3.7
Discovery Bay Gaeden	3	2.0(0.8)	0.6
Discovery Bay Limestone	6	3.0(1.4)	0.8
St. Ann's Coconut	12	24.5(12.0)	6.8
St. Ann's Sugar	10	7.5(2.7)	2.1
Hardwar Gap[L]	33	**	2.6

L - Trap lines were run instead of grids
** - undefined

Table 8. Rodent population estimates for each area trapped on St. Croix, 1983-1986.

Grid	Mean Number of Unique Animals Captured Per Sampling Period	Lincoln Index Estimate (S.E.)	Estimated Relative Density (#/ha)
Mus musculus			
Contessa[R]	17	**	56.7
Madame Carty[R]	2	2.0(0.0)	6.7
Madame Carty[L]	14	**	70.0
Rattus norvegicus			
Enfield Green[M]	1	**	0.2
Rattus rattus			
Butler Bay[M]	4	**	1.1
Mahogany[M]	9	**	2.6
Caledonia[M]	1	**	0.3
Contessa[R]	4	4.0(1.6)	13.3
Contessa[M]	3.3	**	1.1
Enfield Rough[M]	1	**	0.3
Concordia[M]	1	**	0.2

R - rodent trapping grid
L - rodent trapping line
M - Mongoose trapping grid
** - undefined

Table 9. Rodent population estimates for each area trapped on Jamaica July - August 1986.

Grid	Mean Number of Unique Animals Captured Per Sampling Period	Lincoln Index Estimate (S.E.)	Estimated Relative Density (#/ha)
Mus musculus			
Milk River Lower	30	45.5(21.1)	153.3
Milk River Upper	17	35.0(18.7)	116.7
Milk River Sugar	22	52.0(22.5)	173.3
Discovery Bay Bauxite	7	9.0(3.0)	30.0
St. Ann's Coconut	40	66.7(31.9)	223.3
St. Ann's Sugar	16	12.0(4.9)	40.0
Hardwar Gap[L]	78	**	173.3
Rattus norvegicus			
Milk River Lower	7	**	23.3
Milk River Upper	1	**	3.3
Discovery Bay Bauxite[R]	2	2.0(0.0)	6.7
Discovery Bay Bauxite[M]	8	**	2.2
Discovery Bay Garden[R]	12	16.7(7.0)	55.7
Discovery Bay Garden[M]	11	**	3.1
Discovery Bay Limestone[R]	7	**	23.3
St. Ann's Coconut	1	**	3.3
Hardwar Gap[L]	16	**	1.4
Rattus rattus			
Hardwar Gap[L]	36	**	3.2

L - rodent trapping lines
** - undefined value
R - rodent trapping grid
M - mongoose trapping grid

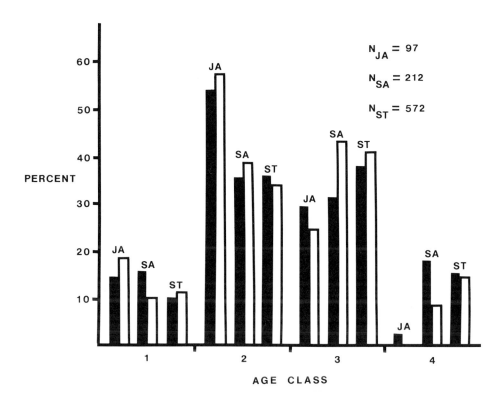

Figure 1. Distribution of mongoose age classes on Jamaica during August (JA), St. Croix during August (SA), and St. Croix during all months sampled (ST). Solid bars represent males, open bars represent females.

Biogeography of the West Indies, 1989:635-644

A REVIEW AND ANALYSIS OF THE BATS OF THE WEST INDIES

Karl F. Koopman[1]

Abstract

A distributional table for West Indian bats is updated and the definition of the West Indies is revised. This involves many taxonomic and distributional changes. It is concluded that probably all endemic supraspecific taxa are of ultimately Greater Antillean origin. Several species endemic to the Lesser Antilles, however, have come from South America. The ultimate source for most Greater Antillean species has been Middle America. A dispersalist explanation for the origin of all West Indian bats is supported.

Updated records

For the purposes of the present paper, the West Indies are redefined to exclude peripheral islands that lack endemic species of mammals. Thus the following islands, which I earlier (Koopman 1959) included, I would now exclude: San Andreas (but not Providencia), Grenada, and the Grenadines. My distributional table is otherwise comparable with that of Baker and Genoways (1978) except for the following particulars. Providencia and the Caymans are added on the western edge of the Greater Antilles. The Virgin Islands (with the exception of St. Croix) are added to Puerto Rico since they are on the same bank, but St. Croix remains as a separate entry. Anguilla, St. Martin, and St. Bartholomew; St. Eustatius, St. Kitts, and Nevis; Barbuda and Antigua; are, in each case, combined as a single entry since I am putting together islands on the same submarine bank (which was of course a single island in the very late Pleistocene). It will be noted, however, that two of the categories (Caymans, Bahamas) include islands on more than one bank.

The table has been updated on the basis of several recent sources (Jones and Carter 1976; Baker et al. 1978; Jones and Baker 1979; Freeman 1981; Hall 1981; Martin and Schmidly 1982; Ottenwalder and Genoways 1982; Buden 1986; Kock and Stephan 1986; Morgan and Woods 1986; Pierson et al. 1986; Handley and Webster 1987). I will therefore only discuss changes in the table and in my own previous ideas, which are in addition to these literature references. As have Baker and Genoways (1978), I have omitted various dubious records of bats from the table (*Lonchorhina, Vampyrum, Carollia, Sturnira* in Jamaica, *Lasiurus cinereus* in Cuba and Hispaniola).

Monophyllus.-- As is generally agreed, this endemic West Indian genus is closely

---

[1] Dr. Koopman is Curator Emeritus of the America Museum of Natural History, New York, NY 10024

related to the largely mainland *Glossophaga*. Since *Monophyllus* occurs in both the Greater and Lesser Antilles, and *Glossophaga* in both Middle and South America (as well as Jamaica and St. Vincent), the problem arose as to whether *Monophyllus* had colonized the West Indies from Middle America through the Greater Antilles, or from South America through the Lesser Antilles. When I first faced this problem in the 1950's, only two mainland species of *Glossophaga* were recognized (the Middle and South American *soricina* and the South American *longirostris*). Since then considerable revisionary work has been done (Webster and Jones 1980) and now five species of *Glossophaga* are recognized, all but *longirostris* being known from Middle America. Whereas early inspection indicated closer resemblance between *Glossophaga longirostris* and *Monophyllus plethodon* than between *Glossophaga soricina* and *Monophyllus redmani*, it is now possible (using characters worked out by Webster and Jones) to compare all five species of *Glossophaga* with both species of *Monophyllus*. After doing a rudimentary character analysis, there seems to be no evidence of special relationship between *Monophyllus* and *G. longirostris*. If anything, the species of *Glossophaga* most like *Monophyllus* is probably *G. commissarisi* of Middle and northwestern South America. I would therefore modify my earlier judgment (Koopman 1968:11) and conclude that in the light of greater diversity in the Greater than in the Lesser Antilles, origin of *Monophyllus* in the former is most likely, its distributional pattern therefore resembling that of *Brachyphylla*.

Artibeus lituratus.-- Prior to 1970, a single species of large *Artibeus* was recognized on St. Vincent. Because it was considerably larger than *A. jamaicensis* of St. Lucia and Barbados and agreed in size with *A. lituratus* of Grenada, it was so identified. Jones and Phillips (1970) and Jones (1978), however, showed that all but one of the specimens from St. Vincent they examined disagreed with *lituratus* in the structure of the orbital region of the skull and were best regarded as a large subspecies of *A. jamaicensis* (*A. j. schwartzi*). Having noted, however, that old specimens of *A. jamaicensis* may become somewhat *lituratus*-like in orbital region morphology, I considered that the single specimen (a skull only) should be reexamined. This I have done (with the kind assistance of Dr. Robert Timm, who loaned me the skull). The specimen is by no means aged and clearly agrees well with *A. lituratus* and not with *A. jamaicensis schwartzi*, thus in complete agreement with Jones and Phillips.

Phyllops falcatus.-- Currently (Hall 1981), two living species of *Phyllops* are recognized; *P. falcatus* of Cuba and *P. haitiensis* of Hispaniola, distinguished on the basis of size, and the shape of the palatal emargination (U-shaped vs. V-shaped). Examination of a number of skulls from both islands shows that neither character will really separate the two alleged species, though there are average differences. I would therefore combine the two, recognizing them only as subspecies (*P. f. falcatus* and *P. f. haitiensis*). The specific status of the extinct *P. vetus* is in no way affected.

Eptesicus serotinus.-- Although most large American *Eptesicus* are almost universally called *E. fuscus*, I am unaware of any paper that presents cogent arguments for this (but see Miller 1897). The last revisionary work that treated both New and Old World representatives of the *serotinus* group (Dobson 1878) included *fuscus*, as a variety, in *serotinus* and gave what still seems good reasons for doing so. To this variety were referred specimens from New York, California, Cuba, and Barbados. Interestingly enough, two specimens from Guatemala (his only representatives of what is now called *E. f. miradorensis*) were not referred to this smaller *fuscus* variety, but to the larger nominate form. I have compared skulls of most of the representatives of the *serotinus-fuscus* complex and concur with Dobson. While Old World representatives tend to be larger than those of the New, each shows considerable variation and there is definite overlap. Until someone presents cogent reasons for separating them on a species level, I will refer to most members of the entire complex as *Eptesicus serotinus*. Two West Indian problems remain, however. The first concerns the Jamaican populations currently treated as an endemic species, *E. lynni* (Hall 1981). It was originally described from western Jamaica and allocated to the otherwise Middle and South American *brasiliensis* group. Previously, Sanborn (1941) had allocated specimens from eastern Jamaica to *fuscus hispaniolae*. Arnold et al. (1980) clearly showed that *lynni* is more closely related to *fuscus* that to the *brasiliensis* group but compared *lynni* only to *f. fuscus* of eastern North America and not with *f. hispaniolae* of Hispaniola, though eastern Jamaican specimens were examined and stated to be *lynni*. No good reasons were given for not treating *lynni* as a subspecies of *fuscus* and its resemblance to *f. hispaniolae* and the possible intermediate position of the eastern Jamaican populations would support this. Actually, I fail to see that *lynni* is any more different from *hispaniolae*, than *bahamensis* of New Providence is from *dutertreus* of Cuba (with southern Bahamian populations somewhat intermediate). These two latter taxa seem at present universally to be regarded as conspecific with *fuscus*. I would therefore call the Jamaican form *E. serotinus lynni*. The other problem involves the Lesser Antillean *Eptesicus*. Prior to 1975, the only specimen representing the *E. serotinus* complex between *s. wetmorei* in Puerto Rico and *s. miradorensis* in Venezuela was an old Barbados record which was generally viewed skeptically. As indicated above, Dobson (1878) allocated the specimen to his small *fuscus* variety of *E. serotinus*. *Eptesicus guadeloupensis*, as described from Guadeloupe (Genoways and Baker 1975), is somewhat larger than *miradorensis* and considerably larger than *wetmorei*. Though derivation from Puerto Rico was favored (obviously on geographical grounds), a derivation from Venezuela seems more reasonable to me, on the basis of size. Very interesting in this regards is the recent record of *Eptesicus* from Dominica (Hill and Evans 1985). Though coming from the next major island south of Guadeloupe, it is quite different from *guadeloupensis* and closely resembles *wetmorei*. The most reasonable solution to me would be double invasion of the Lesser Antilles by *Eptesicus* from the two ends.

Zoogeographical discussion

With the rejection of a Lesser Antillean origin for *Monophyllus*, it now appears that all endemic genera and subgenera arose on the Greater Antilles with the exception of *Ardops*, which is closely related to and forms an endemic suprageneric group with the Greater Antillean genera *Phyllops*, *Ariteus*, and *Stenoderma*. There are, however, five endemic species confined to the Lesser Antilles (*Sturnira thomasi*, *Chiroderma improvisum*, *Myotis dominicensis*, *M. martiniquensis*, *Eptesicus guadeloupensis*), all, I believe coming into the Lesser Antilles from the south. Two other West Indian endemic genera (*Brachyphylla*, *Monophyllus*) occur throughout the Lesser Antilles, both being known along with *Ardops*, as far south as St. Vincent. Another genus (*Phyllonycteris*), though now endemic to the Greater Antilles, is known fossil from Antigua. Morgan and Woods (1986) have emphasized that three species now confined to the Greater Antilles (*Pteronotus parnellii*, *Mormoops blainvillei*, *Macrotus waterhousii*), as well as *Phyllonycteris major*, are known fossil from the northern Lesser Antilles, thus breaking down the faunal distinction between the Greater and Lesser Antilles. However, the amount of actual mixing appears to have been minimal. Of the Lesser Antillean species which I believe to have had a southern origin, only *Myotis dominicensis* (assuming the St. Martin record belongs to this species) occurs north of Montserrat. The others (*Pteronotus davyi, Glossophaga longirostris, Sturnira lilium* and *thomasi, Chiroderma improvisum, Artibeus lituratus, Myotis martiniquensis, Eptesicus guadeloupensis*) do not reach beyond Montserrat or else (*Natalus stramineus, Molossus molossus*) also occur widely in the Greater Antilles. In my opinion, the remaining species in the Lesser Antilles either came from the Greater Antilles (*Artibeus jamaicensis, Tadarida brasiliensis*), or could readily have come either way (*Noctilio leporinus*).

The origin of the Greater Antillean bat fauna is considerably more complex. As mentioned above, I believe two species (*Natalus stramineus, Molossus molossus*) came to the Greater Antilles through the Lesser Antilles. This is based on the much greater resemblance between South American and Lesser Antillean populations than between Central American and Greater Antillean populations. For the remaining species, several means of entry involving crossing of relatively narrow (past or present) oceanic barriers have been proposed: Honduran bank to Jamaica, Yucatan to Cuba, Florida to Cuba, Florida through the Bahamas to Cuba or Hispaniola). The only case in which derivation from Florida (directly to Cuba) is clearly most probable is that of *Nycticeius*, which has an extensive distribution in eastern North America (including southern Florida) but in Middle America does not occur south of northern Veracruz. In the West Indies, it is only known from western Cuba so the northern route is most probable. The northern route is also a reasonable probability for *Mormoops, Eptesicus serotinus, Lasiurus borealis*, and *Tadarida brasiliensis* since each of them occur (or occurred) in Florida and the Florida populations are as similar to those of Cuba as the Central American populations are. However, in each of these cases, a Middle American origin is equally probable. In two other cases (*Lasiurus intermedius* and *Eumops glaucinus*) the same species occurs in Florida, Middle America, and Cuba (also Jamaica in the case of *Eumops*), but in each case, the Middle American population is clearly more like the Greater Antillean than is the Floridian. In these, as in all other cases where the West Indian invaders have not reached a supra-specific level of differentiation (the *Pteronotus* invasions, *Macrotus, Tonatia, Glossophaga soricina*, the *Artibeus* invasions, *Desmodus, Antrozous, Mormopterus*, the *Nyctinomops* invasions, *Eumops auripendulus, E. perotis*), a Middle American derivation is clearly most likely. It should be noted, however, that the closest that mainland *Antrozous* is known is from central Mexico and the closest mainland *Mormopterus* is in Peru. (A similar problem is seen in the Lesser Antilles where the closest relative of *Chiroderma improvisum* is *C. doriae* of southern Brazil.) While there are also no populations of either *Nyctinomops macrotis* or *Eumops perotis* on the Central America mainland, former occurrence is presumed since they are known both to the north and to the south.

Coming down to the endemic genera and subgenera, there is some uncertainty as to whether *Brachyphylla* and the phyllonycterines (*Erophylla* and *Phyllonycteris*) have had a common origin separate from the glossophagines (see Griffiths 1982). In either case, a fairly ancient origin from somewhere in the tropical American mainland seems certain. Since all the species occur in the Greater Antilles and only two have reached the Lesser

Antilles, a Middle American origin through the Greater Antilles is highly probable. I have already discussed the problem of the origin of *Monophyllus*. The four endemic West Indian genera *Phyllops*, *Ardops*, *Ariteus*, and *Stenoderma* clearly form a monophyletic group, almost certainly derived from *Artibeus* (s.l.) or something similar. Again, all this tells us is that the endemic West Indian complex had a tropical American mainland origin. Since *Phyllops*, *Ariteus*, and *Stenoderma* are all Greater Antillean and only *Ardops* (currently regarded as including only a single species) is Lesser Antillean, a Middle American origin through the Greater Antilles is indicated. There remain only the two endemic West Indian subgenera of *Natalus*, both confined to the Greater Antilles. A Middle American origin is very likely but it is not clear whether *Chilonatalus* and *Nyctiellus* differentiated from one another after their common ancestor reached the West Indies. The only species of *Natalus* (or of the Family Natalidae) in Middle America today is *Natalus stramineus* (which has reached San Andreas), but it is doubtful that this species could have been the ancestor of *Chilonatalus* and *Nyctiellus*.

Thus the general picture is of Middle America being the major source of West Indian bats, particularly for the older and more distinct endemics, all of which have probably come from this source. South America has also been an important source, but apparently only for relatively recent elements. Most of these have not extended north of Montserrat, though at least two have reached the Greater Antilles. Invasion of the Lesser Antilles, particularly of the northern islands has been more extensive. In one case (*Eptesicus*), the same species has apparently invaded from both ends.

As should be evident from all the above discussion, I have assumed overwater dispersal of all bats to the West Indies. Though I freely concede the existence of continental drift based on plate tectonics, I do not believe that it has played any role in the colonization of the West Indies by bats. I therefore find myself in full agreement with Baker and Genoways (1978).

Literature cited

Arnold, M.L., R.J. Baker, and H.H. Genoways. 1980. Evolutionary Origin of *Eptesicus lynni*. Journal of Mammalogy 61:319-322.

Baker, R.J. and H.H. Genoways. 1978. Zoogeography of Antillean Bats. Academy Natural Sciences of Philadelphia, Special Publication 13:53-97.

_____, H.H. Genoways and J.C. Patton. 1978. Bats of Guadeloupe. Occasional Papers, The Museum, Texas Tech University 50:1-16

Buden, D.W. 1986. Distribution of Mammals of the Bahamas. Florida Field Naturalist 14:53-63.

Dobson, G.E. 1878. Catalogue of the Chiroptera in the Collection of the British Museum. London, 567 pp.

Freeman, P.W. 1981. A Multivariate Study of the Family Molossidae (Mammalia, Chiroptera): Morphology, Ecology, Evolution. Fieldiana Zoology n. s. 7:1-173.

Genoways, H.H., and R.J. Baker. 1975. A new species of *Eptesicus* from Guadeloupe, Lesser Antilles (Chiroptera: Vespertilionidae). Occasional Papers, The Museum, Texas Tech University 34:1-7.

Griffiths, T.A. 1982. Systematics of the New World Nectar-Feeding Bats (Mammalia, Phyllostomidae), based on the Morphology of the Hyoid and Lingual Regions. American Museum of Natural History Novitates 2742:1-45.

Hall, E.R. 1981. The Mammals of North America, 2nd. ed., vol. 1. John Wiley & Sons, New York, 600 pp.

Handley. C.O., and W.D. Webster. 1987. The supposed occurrence of *Glossophaga longirostris* Miller on Dominica and problems with the type series of *Glossophaga rostrata* Miller. Occasional Papers, The Museum, Texas Tech University 108:1-10.

Hill, J.E., and P.G.H. Evans. 1985. A record of *Eptesicus fuscus* (Chiroptera: Vespertilionidae) from Dominica, West Indies. Mammalia 49:133-136.

Jones, J.K. 1978. A New Bat of the Genus *Artibeus* from the Lesser Antillean Island of St. Vincent. Occasional Papers, The Museum, Texas Tech University 51:1-6.

_____, and R.J. Baker. 1979. Notes on a Collection of Bats from Montserrat, Lesser Antilles. Occasional Papers, The Museum, Texas Tech University 51:1-6.

_____, and D.C. Carter. 1976. Annotated checklist with keys to subfamilies and genera. Pp. 7-38 in R.J. Baker, J.K. Jones and D.C. Carter (eds.). Biology of bats of the New World family Phyllostomatidae, Part I. Special Publications, The Museum, Texas Tech University 10:1-218.

_____, and C.J. Phillips. 1970. Comments on systematics and zoogeography of bats in the Lesser Antilles. Studies on the Fauna of Curaçao and other Caribbean Islands 32:131-145.

Kock, D., and H. Stephan. 1986. Une chauve-souris nouvelle pour la Martinique, Antilles francaises: *Monophyllus plethodon luciae*. Mammalia 50:268.

Koopman, K.F. 1959. The zoogeographical limits of the West Indies. Journal of Mammalogy 40:236-240.

_____. 1968. Taxonomic and distributional notes on Lesser Antillean bats. American Museum of Natural History Novitates 2333:1-13.

Martin, C.O., and D.J. Schmidly. 1982. Taxonomic Review of the Pallid Bat, *Antrozous pallidus* (Le Conte). Special Publications, The Museum, Texas Tech University 18:1-48.

Miller, G.S. 1897. Revision of the North American bats of the family Vespertilionidae. North American Fauna 13:1-185.

Morgan, G.S. and C.A. Woods. 1986. Extinction and zoogeography of West Indian land mammals. Biological Journal of the Linnaean Society, London 28:167-203.

Ottenwalder, J.A., and H.H. Genoways. 1982. Systematic Review of the Antillean Bats of the *Natalus micropus*-complex (Chiroptera, Natalidae). Annals of Carnegie Museum 51:17-38.

Pierson, E.D., W.E. Rainey, R.M. Warner, and C.C. White-Warner. 1986. First record of *Monophyllus* from Montserrat, West Indies. Mammalia 50: 269-271.

Sanborn, C.C. 1941. Descriptions and records of Neotropical bats. Field Museum Natural History, Zoological Series 27:371:387.

Webster, W.D., and J.K. Jones. 1980. Taxonomic and Nomenclatorial Notes on Bats of the genus *Glossophaga* in North America, with Description of a new species. Occasional Papers, The Museum, Texas Tech University 71:1-12.

Table 1 Distribution of West Indian Bats.

	A	B	C	D	E	F	G	H	I	J	K	L	M	N	O	P	Q	R	S	T
Noctilio leporinus	+			+	+	+	+	+	+	+	+	+	+	+		+	+	+	+	+
Pteronotus parnellii	+	+	Æ	+	Æ							Æ								
Pteronotus pristinus (Æ)				Æ																
Pteronotus quadridens	+			+	+	+	+													
Pteronotus macleayi	+			+	+															
Pteronotus davyi														+	+					
Mormoops megalophylla				Æ	Æ	Æ	+													
Mormoops blainvillii	+			+	+	+						Æ								
Mormoops magna (Æ)				Æ	Æ															
Macrous waterhousii	+		+	+	+	+	+					Æ								
Tonatia bidens	Æ																			
Brachyphylla nana	Æ		Æ	+	+	+	+													
Brachyphylla cavernarum								+	+	+	+	+	+	+		+	+	+	+	+
Erophylla sezekorni	+		+	+	+	+	+													
Erophylla bombifrons						+	+													
Phyllonycteris poeyi				+		+														
Phyllonycteris major (Æ)							Æ					Æ								
Phyllonycteris aphylla	+																			
Glossophaga longirostris	+				?															
Glossophaga soricina	+				+															
Monophyllus redmani	+			+	+	+	Æ													
Monophyllus plethodon									+		+	+	+	+		+	+	+	+	+
Sturnira lilium														+		+	+	+	+	
Sturnira thomasi													+	+						
Chiroderma improvisum													+	+						

Table 1 Distribution of West Indian Bats (continued).

	A	B	C	D	E	F	G	H	I	J	K	L	M	N	O	P	Q	R	S	T
Artibeus jamaicensis	+		+	+	+	+	+	+	+	+	+	+	+	+	+	+	+	+		+
Artibeus anthonyi (Æ)				Æ																
Artibeus lituratus															+				+	
Phyllops falcatus				+		+														
Phyllops vetus (Æ)				Æ																
Ardops nichollsi											+		+		+	+	+	+		
Arieus flavescens		+																		
Stenoderma rufum							+													
Desmodus rotundus				Æ																
N. (Natalus) stramineus	+			Æ	Æ	+			+			+	+							
N. (Chilonatalus) micropus	+	+		+		+														
N. (Chilonatalus) tumidifrons					+															
N. (Nyctiellus) lepidus				+	+															
Myotis dominicensis									?							+	+			
Myotis martiniquensis													?	+		+				
Eptesicus serotinus			+	+	+	+	+		+			+	+							
Eptesicus guadeloupensis																+				
Nycticeius humeralis				+																
Lasiurus intermedius				+										+						
Lasiurus borealis	+			+	+		+				+	+	+							
Antrozous pallidus				+																
Mormopterus minutus				+																
Tadarida brasiliensis			+	+	+	+	+		+		+	+	+					+		
Nyctinomops laticaudata				+																
Nyctinomops macrotis				+		+														

Table 1 Distribution of West Indian Bats (concluded).

	A	B	C	D	E	F	G	H	I	J	K	L	M	N	O	P	Q	R	S	T
Eumops auripendulus		+																		
Eumops glaucinus		+		+																
Eumops perotis				+																
Molossus molossus		+	+			+	+	+	+	+		+	+	+		+	+	+	+	+
Total	2	23	8	33	14	19	15	4	7	3	6	11	9	11	1	12	10	8	9	7

A - Providencia; B - Jamaica; C - Caymans; D - Cuba; E - Bahamas; F - Hispaniola; G - Puerto Rican bank; H - St. Croix; I - St. Martin bank; J - Saba; K - St. Kitts bank; L - Antigua bank; M - Montserrat; N - Guadeloupe; O - Marie Galante; P - Dominica; Q - Martinique; R - St. Lucia; S - St. Vincent; T - Barbados;

Æ - fossil only

DISTRIBUTION AND SYSTEMATICS OF BATS IN THE LESSER ANTILLES

J. Knox Jones, Jr.[1]

Abstract

Some 24 species of Chiroptera presently are known to occur on one or more islands in the Lesser Antillean chain, but no more than 12 have been recorded to date from any one island. Lesser Antillean bats are here grouped for discussion under four general headings--widespread, recent invaders from South America, Antillean endemic genera, and enigmatic species (those with a restricted distribution, of questionable affinity, and of uncertain geographic origin). Four additional species are known as Holocene fossils from islands in the northern part of the chain. A review is presented of the systematic status and biogeographic relationships of each modern species occurring in the Lesser Antilles.

Introduction

The Lesser Antilles, which extend some 800 km from Anguilla in the north to Grenada in the south, form an archipelago, along with the Virgin Islands, between the Greater Antilles and Trinidad--Tobago and the South American mainland. Twenty-four species of bats presently are known to occur in the island chain, although no more than 12 have been recorded from any single island. The systematics and zoogeography of Chiroptera in the Antillean region have received considerable attention from mammalogists over the years; bats of the Lesser Antilles, for example, have been the subject of three published summaries in the past two decades (Koopman 1968; Jones and Phillips 1970; Baker and Genoways 1978).

For purposes of discussion, I have divided the 24 bats reported from the Lesser Antilles into four groups: 1) species that are widespread in the Antilles and on the adjacent mainland; 2) species that have invaded the islands, presumably relatively recently, from South America via Trinidad--Tobago; 3) Antillean endemic genera (in one case belonging to an endemic subfamily); and 4) a group of six species I refer to as enigmatic, for reasons explained in that section. Additionally, four species that now occur in the Greater Antilles are known as subfossils from Anguilla or Barbuda. The major islands of the Lesser chain are shown in Fig. 1 and are listed, along with their bat fauna, in Table 1.

---

[1] Dr. Jones is Paul Whitfield Horn Professor of Biological Sciences and Curator at The Museum, Texas Tech University, Lubbock, TX 79409.

The northern islands south to Guadeloupe frequently are referred to as the Leeward Islands, whereas those from Dominica southward, including Barbados, are known as the Windward Islands.

Widespread species

Six species found in the Lesser Antilles are widely distributed on islands in the Caribbean and elsewhere in the mainland Neotropics, and in two instances (*Eptesicus fuscus* and *Tadarida brasiliensis*) also in temperate North America.

Routes by which most of these species, all possibly late Pleistocene or early Holocene colonizers, reached the Antilles are problematic in all but one case--*Artibeus jamaicensis*, which clearly is a double invader from west and south. *Noctilio leporinus* and *Molossus molossus* also could have come from both Middle and South America. *Eptesicus fuscus* occurs on all mainland masses surrounding the Caribbean, but probably reached the islands from the west or north, as did *Tadarida brasiliensis* (although Antillean populations seem more closely allied to those in the southeastern United States than any others). Finally, *Natalus stramineus* most likely invaded the insular region from the west.

Noctilio leporinus

This large, fish-eating bat is widely distributed in the American tropics from western Mexico to northern Argentina. It is known from all four islands in the Greater Antilles, the Virgin Islands, and has been recorded from most Lesser Antillean islands (Table 1).

Davis (1973) studied geographic variation in this bat, and assigned all Antillean material to *Noctilio leporinus mastivus* (Vahl, 1797), with type locality on St. Croix. This race occurs also on the Middle American and northern South American mainlands; thus the origin of Antillean populations cannot be surmised, invasions from west, south, or more probably both, being plausible.

Artibeus jamaicensis

This species occurs throughout the Antilles, and is one of the most common bats found there. Although there is slight variation among them, populations from the Greater Antilles (except Cuba) eastward and southward to St. Lucia and Barbados all are referable to a single subspecies, *Artibeus jamaicensis jamaicensis* Leach, 1821, the precursors of which almost certainly reached the region from the west. A smaller subspecies, *Artibeus jamaicensis trinitatis*, Andersen, 1906, which typically possesses a third upper molar lacking in the nominate race, is found in the Grenadines and on Grenada southward to Tobago and Trinidad. Actually, specimens from Grenada average slightly larger than those from Trinidad and the subspecific name *grenadensis* Andersen, 1906, is available for them, but the differences are slight and recent authors (Koopman 1968; Jones and Phillips 1970) have referred all to *trinitatis*, which obviously invaded the region from the south.

Remaining, then, is St. Vincent where Koopman (1968) believed *A. jamaicensis* did not occur. In point of fact, however, the species does occur there (Jones and Phillips 1970; Jones 1978) and there is no evidence to suggest this represents a recent colonization. Individuals on St. Vincent average significantly larger than those of either *jamaicensis* or *trinitatis* and are suggestive of a classic case of heterosis resulting from hybridization between the two invading groups. In any event, I have recently named the St. Vincent population as a distinct subspecies, *Artibeus jamaicensis schwartzi* Jones, 1978.

Natalus stramineus

This funnel-eared bat has been reported in the Lesser Antilles only from Anguilla, Antigua, Dominica, Guadeloupe, Montserrat, and Saba. Such a known distribution suggests that *N. stramineus* probably will be found on other islands, at least those to the north of Dominica, where suitable habitat prevails. This same species (or a closely related one according to some authorities) occurs also on Hispaniola and Jamaica, and an extinct subspecies is known from Cuba. On the mainland, it ranges throughout most of Middle America and occurs also in northeastern South America. The endemic subspecies of the Lesser Antilles is *Natalus stramineus stramineus* Gray, 1838 (see Goodwin 1959).

Koopman (1968) thought this species "clearly has reached the Lesser Antilles from the south." Jones and Phillips (1970) and Baker and Genoways (1978) agreed but with some reservations. Reexamining this question, it now seems to me more likely that invasion of the Antilles was from Central America to the west because 1) the species occurs or once occurred on three of the four Greater Antillean islands, 2) it is not known in the Lesser Antilles south of Dominica, and 3) it evidently does not occur in Trinidad or the immediately adjacent South American mainland.

Eptesicus fuscus

Dobson (1878) listed a specimen of this big brown bat from Barbados, a record discounted by most subsequent workers as either a result of mislabeling or representing some kind of accidental occurrence. Recently, however, Hill and Evans (1985) reported four specimens from Dominica that were taken in the early 1980's. These, along with the Barbadan specimen, seemed to them to more closely resemble *Eptesicus fuscus wetmorei* Jackson, 1916, of Puerto Rico than any other named subspecies from the region. A distinct but related species of *Eptesicus* is known from Guadeloupe.

The occurrence of *E. fuscus* on Dominica obviously adds credence to the record from Barbados. The specimens were taken (three in bat traps, one in a mist net) within or along the edge of "lower montane primary rain forest" at an altitude of about 700 m. It seems likely that collecting in similar habitats, such as they remain, elsewhere in the Lesser Antilles will reveal a broader distribution for this bat.

This species occurs on the mainland all around the Caribbean rim, making speculation as to possible invasion routes difficult. Baker and Genoways (1978) thought immigration from either the west or north plausible and I am inclined to support their view.

Tadarida brasiliensis

This species is widely distributed in the Caribbean region and occurs in the Lesser Antilles south as far as St. Lucia. It is unknown from Barbados, St. Vincent, the Grenadines, and Grenada. A single specimen has been reported from Tobago (Goodwin and Greenhall 1961), probably representing an accidental occurrence because the species is unknown from Trinidad and the adjacent South American mainland (Koopman 1982). The Lesser Antillean subspecies, which also is found on Puerto Rico, is *Tadarida brasiliensis antillularum* (Miller, 1902), with type locality at Roseau, Dominica. Thus it would seem that this free-tailed bat reached the Antillean islands from the north or west; it occurs in the southern United States and throughout much of Middle America.

The specific relationships of Antillean populations of *Tadarida* remain obscure. The subspecies *antillularum*, for example, is a small, brownish-colored, sedentary bat as contrasted with *T. b. mexicana* of the southwestern United States and Mexico, which is larger, grayish in color, and mostly migratory. *T. b. antillularum* and other races in the Antilles most closely resemble *T. b. cynocephala* of the southeastern United States, also a relatively small, sedentary bat. Jones and Phillips (1970) thought it possible that Caribbean populations represented a distinct species or, alternatively, that they were related to *cynocephala* but not closely to other populations of the *brasiliensis* complex.

Molossus molossus

"This common house bat probably occurs on virtually every Lesser Antillean island" (Koopman 1968). It also ranges throughout the Greater Antilles. Presently, the species is unknown only from St. Bartholemew and Saba in the Lessers, but I agree with Koopman that it no doubt will be found there. There is demonstrable variation among populations, even between samples from the same island (Genoways et al. 1981), and *M. molossus* is in need of critical study, not only in the Caribbean region but on the mainland as well. In recent years, two subspecific names have been applied to these bats in the Lesser Antilles (Hall 1981)--*Molossus molossus molossus* (Pallas, 1776), with type locality on Martinique, from Guadeloupe southward to South America, and *Molossus molossus debilis* Miller, 1913, with type locality on St. Kitts, for northern islands. Different subspecies are thought to inhabit each of the Greater Antillean islands. Because it occurs throughout much of tropical America, invasion of the Antilles by this strong flier could have come wither from the south or the west, or both.

Recent invaders from South America

At least nine kinds of bats have invaded the Lesser Antilles in recent geologic time from South America (Trinidad-Tobago), five of which are believed to be limited to Grenada. Of these, four are of the same subspecies that occupies Trinidad or that island and Tobago--*Micronycteris megalotis megalotis* (Gray, 1842), *Anoura geoffroyi geoffroyi* Gray, 1838, *Carollia perspicillata perspicillata* (Linnaeus, 1758), and *Artibeus cinereus cinereus* (Gervais, 1856). The fifth bat, *Peropteryx macrotis macrotis* (Wagner, 1821), occurs on Tobago and the South American mainland, but another subspecies is endemic to Trinidad.

Three postscripts to the foregoing paragraph are necessary. In restricting *Carollia* to Grenada, I am ignoring Hahn's (1907) report of three specimens from Redonda, an islet between Montserrat and Nevis. Both Koopman (1968) and Jones and Phillips (1970) thought this record to be in error. Secondly, in the first report of *Anoura* from Grenada, Jones and Phillips (1970) noted that specimens averaged slightly larger than those they examined from Trinidad and had a somewhat more inflated braincase. However, they opined that they "probably" represented the same subspecies, a suspicion I here confirm. Lastly, insofar as I know, both *Micronycteris* and *Carollia* are known from Grenada by a single specimen each, and neither has been taken there in recent years. Thus, their status on the island is in question.

The remaining four species treated in this section are distributed northward as far as Marie Galante, off Guadeloupe, and are discussed below.

Pteronotus davyi

This naked-backed species is recorded from Grenada and from three central islands in the chain--Marie Galante, Dominica, and Martinique. Why it is apparently absent from St. Lucia and St. Vincent is a mystery to me. Its absence from Barbados and the Grenadines is considerably more plausible.

Specimens I have examined from the northern islands average slightly larger cranially than do bats from Trinidad, but they do not differ otherwise and the subspecific name *Pteronotus davyi davyi* Gray, 1838, is appropriately applied to all.

Glossophaga longirostris

This species was reported for many years as occurring northward in the Windward Islands to Dominica, but the next northernmost record was from St. Vincent, casting doubt as to the validity of the Dominican record, originally based on two skulls (Miller 1913). Handley and Webster (1987) recently have provided documentation establishing beyond doubt that the skulls in question were mislabeled and actually belong to skins from Grenada. Thus, the true distribution in the Lesser Antilles includes only Grenada, the Grenadines, and St. Vincent.

The Lesser Antillean subspecies, with type locality on Grenada, is *Glossophaga longirostris rostrata* Miller, 1913 (see especially Handley and Webster 1986).

Sturnira lilium

The common yellow-shouldered bat is known from all the major Windward Islands except Barbados. Specimens from Grenada, only recently collected, have not been studied in detail and thus cannot be assigned to subspecies here. Each of the other islands to the north harbors an endemic race as follows, from south to north: *Sturnira lilium paulsoni* de la Torre and Schwartz, 1966 (St. Vincent); *Sturnira lilium luciae* Jones and Phillips, 1976 (St. Lucia); *Sturnira lilium zygomaticus* Jones and Phillips, 1976 (Martinique); *Sturnira lilium angeli* de la Torre, 1966 (Dominica). The genus *Sturnira* also occurs on Guadeloupe, where it is represented by a distinctive species discussed elsewhere in text.

With respect to *S. lilium*, Jones and Phillips (1976) noted that Antillean populations (exclusive of Grenada) generally fall into two groups, those from St.Vincent and St. Lucia bearing a close relationship and, similarly, those from Martinique and Dominica. If forced to synonymize the Windward populations by applying ever increasing minimal conditions for subspecific recognition, those authors first would place *luciae* with *paulsoni*, next *zygomaticus* with *angeli*, and finally recognize a single subspecies for which *angeli* is the oldest available name.

Artibeus lituratus

This large fruit-eating bat is known only from Grenada and St. Vincent in the Lesser Antilles. On both islands, it seems to be less common than the other large member of the genus in the region, *A. jamaicensis*. Because of the large size of the specimens of the latter from St. Vincent, it is quite possible that some earlier authors mistook them for *lituratus*, which does, however, occur there (Jones 1978). The subspecies is *Artibeus litu-*

ratus palmarum Allen and Chapman, 1897, known also from Trinidad, Tobago, and the adjacent South American mainland.

Antillean endemics

The three species treated in this grouping belong to endemic Antillean genera. Each is widely distributed in the island chain here considered and, in addition, each has close living relatives that are broadly distributed on the Greater Antillean islands. These are not the only endemic Antillean bats in the Leeward and Windward islands, of course, but the others have highly restricted ranges, lack close relatives elsewhere in the region, and are not distinct at the generic level; they are discussed as "enigmatic species" in another section of this paper.

Monophyllus plethodon

This glossophagine species is known from 10 islands in the Lesser Antilles--from Barbados and St. Vincent northward to Barbuda and Anguilla. A fossil (or subfossil) subspecies has been named from cave deposits on Puerto Rico, where it was found along with remains of a sibling species (*Monophyllus redmani*), which occurs throughout the Greater Antilles. Were it not for the former sympatry on Puerto Rico, it is possible that Schwartz and Jones (1967), who revised *Monophyllus*, might have recognized but a single species in the genus. *Monophyllus* is clearly related to *Glossophaga*--Varona (1974) regarded them as congeneric--and it is not surprising, therefore, that the two are virtually allopatric in the Lesser Antilles. The only island on which both are known to occur is St. Vincent, where *G. longirostris* is common but where *M. plethodon* is known from but a single specimen (albeit a pregnant female).

Two extant subspecies are recognized, *Monophylus plethodon plethodon* Miller, 1900, from Barbados, and *Monophyllus plethodon luciae* Miller, 1902, from elsewhere in the Lesser Antillean range of the species (holotype from St. Lucia).

Ardops nichollsi

A fruit-eating species, *A. nichollsi* is endemic to the Lesser Antilles, being presently known from St. Eustatius, Montserrat, Guadeloupe, Dominica, Martinique, St. Lucia, and St. Vincent. Probably it will be found on some other islands, most notably St. Kitts and Nevis, encompassed within the currently known range. *Ardops* is related to three other short-faced phyllostomids that are Antillean endemics, *Ariteus* of Jamaica, *Phyllops* of Cuba and Hispaniola, and *Stenoderma* of Puerto Rico and the Virgin Islands. Of these *Stenoderma* is the most distinctive and in my view clearly is deserving of generic rank. The other three resemble each other to a considerable degree and arguably could be regarded as subgenera of a single genus for which *Ariteus* Gray, 1838, is the oldest name. Varona (1974) even went so far as to place all four Antillean genera in *Stenoderma*, with two subgenera. These short-faced bats are related to four other genera of the mainland Neotropics--*Ametrida, Centurio, Pygoderma,* and *Sphaeronycteris*.

Five subspecies of *A. nichollsi* are recognized (Jones and Schwartz 1967) as follows, from north to south: *Ardops nichollsi montserratensis* (Thomas, 1894) on Montserrat and St. Eustatius; *Ardops nichollsi annectens* Miller, 1913, on Guadeloupe; *Ardops nichollsi nichollsi* (Thomas, 1891) on Dominica; *Ardops nichollsi koopmani* Jones and Schwartz,

1967, on Martinique; and *Ardops nichollsi luciae* (Miller, 1902) on St. Lucia and St. Vincent. The one known specimen from the latter island, in poor condition and with fragmentary skull, was referred to *luciae* on geographic grounds. No specimens have been taken on St. Vincent in collecting efforts in recent years.

Like all short-faced phyllostomids from both the Antilles and the mainland, these bats have a prominent whitish shoulder patch and females average noticeably larger than males. Surprisingly in a way, sexual dimorphism is most marked in the smallest subspecies, *A. n. nichollsi* (Jones and Schwartz 1967; Jones and Baker 1979), but it is distinctive in all.

Up until Jones and Schwartz (1967) revised *Ardops*, the four taxa described by that time were regarded as separate species. Those authors reduced all four to subspecific status under *A. nichollsi* (and named a fifth) because they found differences between populations to be relatively slight and quantitative in nature, because the overall variation in size was no greater than that present in some other species of stenodermatines, and because "such a classification best reflects the similarities and obviously close affinities" among these bats. A continuum in size is evident among the five recognized subspecies but it is not clinal in a geographic sense. The smallest race is on Dominica in the middle of the range. Next in size are specimens from St. Lucia followed by those from Guadeloupe, Martinique, and finally specimens from Montserrat and St. Eustatius.

Brachyphylla cavernarum

This is the only representative of the endemic Antillean subfamily Brachyphyllinae that occurs in the Lesser chain. Buden (1977) referred specimens from throughout the broad distribution in the West Indies to the single species *B. cavernarum*. However, Swanepoel and Genoways (1978) recognized two closely related species. Whether one species or two are recognized, only *B. cavernarum* is found in the Lesser Antilles, occurring throughout the islands south to Barbados and St. Vincent (Table 1). The species also is known from Puerto Rico and the Virgin Islands (Swanepoel and Genoways 1978).

Two subspecies occur in the Lesser Antilles, *Brachyphyllum cavernarum minor* Miller, 1913, on Barbados, and *Brachyphyllum cavernarum cavernarum* Gray, 1834, with type locality on St. Vincent, on the other islands and St. Croix. A third subspecies (*B. c. intermedia*) is found on Puerto Rico and in most of the Virgin group.

Enigmatic species

Six bats that occur in the Lesser Antilles have highly restricted ranges--two are found only on Guadeloupe, for example, and one is known only from Dominica--and are of perplexing origin or of uncertain affinity. Five are endemic species of otherwise widespread genera; the sixth is known from too few specimens to properly place them in a named taxon of *Myotis*. It is useful to me to think of these bats in a different vein than relatively broadly distributed endemic genera and I thus treat them separately here.

Sturnira thomasi

Sturnira thomasi de la Torre and Schwartz, 1966, is known only from Guadeloupe and from seven specimens (Owen 1987). It differs from *S. lilium*, which occupies the Windward Islands to the south, in being larger and having a distinctly longer and narrower skull. The relationships of this species are uncertain, but it may be most closely allied to *S. tildae* of Trinidad and the South American mainland. If that is true, it no doubt represents an invader from the south that reached the Lesser Antilles before *S. lilium*. It once may have been more broadly distributed, at least to the south of Guadeloupe, having been subsequently replaced by *lilium*.

Chiroderma improvisum

A unique species, *Chiroderma improvisum* Baker and Genoways, 1976, was described initially on the basis of a single specimen from Guadeloupe. Since that time, two more individuals have been reported, both from Montserrat (Jones and Baker 1979; Pierson et al. 1986). I expect this bat will be found on some other islands in at least the central Lesser Antilles. Variation evident among the three known specimens may signal interisland variability of subspecific worth. It may be worthy of passing note that *C. improvisum* resembles, in a superficial way, *Artibeus jamaicensis*, which is a "trash bat" throughout the Antilles. It is quite possible that an inexperienced collector, in poor light, could confuse the two and not save the *Chiroderma*.

Chiroderma improvisum is the largest species of the genus; its relationship to other known species is uncertain (Baker and Genoways 1976). It differs karotypically from *C. villosum*, *C. trinitatum*, and *C. salvini*, but the karyotype of the next largest species, *C. doriae* of southeastern Brasil, is not known. In any event, it seems highly likely that *improvisum* or its ancestral stock reached the region from the south.

Myotis species

Two species of *Myotis* are endemic to the Lesser Antilles (LaVal 1973), and specimens of another have been reported from Grenada and St. Martin and it has been listed also from Montserrat. Inasmuch as the genus is unknown from the Greater Antilles and the affinities of Antillean taxa seem to lie with other Neotropical species, I assume all *Myotis* reached the Lesser Antilles from the south, probably no earlier than mid-Pleistocene times and quite possibly later than that.

Myotis dominicensis Miller, 1902.-- This small species is a member of the *M. nigricans* species group (LaVal 1973); for many years it was regarded as a subspecies of *nigricans* (Miller and Allen 1928), with which it is closely related. This bat is recorded only from Dominica, where it is relatively common.

Myotis martiniquensis LaVal, 1973.-- This bat, which is medium-sized among Neotropical myotis, "with wooly fur and rather large skull" (LaVal 1973), was initially described as occurring on Martinique and Barbados. Later, the Barbadan population was named as a distinct subspecies, *Myotis martiniquensis nyctor* LaVal and Schwartz, 1975. According to LaVal (1973), this species "may be more closely related to *riparius* than to any other of the mainland species."

Myotis nigricans (Schinz, 1821).-- This bat, which is widespread in the American tropics--from Tamaulipas in Mexico to Paraguay and Argentina--has been reported from Grenada (as *M. n. nigricans* by Miller and Allen 1928) and from St. Martin (as *M. n. nesopolus* by Husson 1960). The latter assignment of a single specimen probably was a matter of convenience inasmuch as *nesopolus* was described from Curacao, the fauna of which was included in the same treatise. Varona (1974) also listed *nigricans* from Montserrat but he did not divulge the basis for this record.

Aside from specimens representing *M. dominicensis* and *M. martiniquensis* (see above), LaVal (1973) did not assign to species those he examined from any other Caribbean island, few of which were represented by more than one specimen. He wrote of these bats as follows: "The available specimens are all small in size . . . suggesting relationships with *dominicensis*." He went on to note that "*Myotis nesopolus* Miller, described from Curacao, is a name which may eventually prove applicable to one or more of the island populations." However, LaVal did not treat *nesopolus* as a recognizable entity nor place it in synonymy.

To further complicate the situation, Carter and Dolan (1978) redescribed the holotype or syntype of *Vespertilio splendidus* Wagner, 1845, from St. Thomas in the Virgin Islands, a name overlooked by LaVal and other recent workers. Carter and Dolan thought *splendidus* possibly represented *nigricans*. In any event, *splendidus* clearly is the earliest name applied to Antillean *Myotis*. As the distribution in the Lesser Antilles and Virgin Islands is clarified on the basis of additional field exploration, the relationships of all members of the genus in the Caribbean region will need to be reevaluated. Baker and Genoways (1978) thought it possible that the two endemics listed above might prove to be "conspecific with mainland *N. nigricans* or that both represent a single endemic species."

Eptesicus guadeloupensis

Only three specimens of *Eptesicus guadeloupensis* Genoways and Baker, 1975, are known, all from Guadeloupe. The describers regarded this interesting bat as a member of the *E. fuscus* group. It is the largest New World member of the genus, characterized by large ears, a long tibia, and a large cranium. Typical *E. fuscus* occurs to the north and west (Greater Antilles and Bahamas) and south (Dominica) of Guadeloupe. *Eptesicus guadeloupensis* possibly represents a descendant of an early invasion of the Antilles by *fuscus*-like bats, which could have come from any mainland direction because at least now *E. fuscus* occurs in North, Central, and South America.

Discussion

No attempt is made here to provide a definitive, critical critique of bat zoogeography in the Lesser Antilles because this is to be the subject of another paper in this symposium. Nonetheless, some closing comments concerning distributional patterns seem appropriate.

Chiroptera is the most diverse and widely distributed of the mammalian orders occurring in the West Indian region, and this is especially true of the Lesser Antilles. Most bats in the latter chain, and elsewhere in the region for that matter, clearly have their roots with Neotropical groups. I agree with Baker and Genoways (1978) that overwater dispersal best explains present chiropteran distribution on Caribbean islands.

Nonetheless, many kinds of bats seem not to cross even modest water barriers as easily as their volant locomotion might suggest is possible. Many species in the Antilles show considerable fidelity to individual islands or island groups. Dominica and Guadeloupe, for example, are separated by only approximately 50 km, with a few islets in between. Guadeloupe has three species (two endemic) that do not occur on Dominica

(Table 1), whereas Dominica has four (one endemic) that do not occur on Guadeloupe, albeit one (*Pteronotus davyi*) has been reported from Marie Galante.

How then do bats disperse to and between Caribbean islands? Several explanations are plausible: 1) strong fliers, such as molossids, would have little difficulty in negotiating, for whatever motivating reason, water gaps such as those in the Lesser Antilles; 2) even weaker flying bats might be able to cross such barriers at times of lowered sea level, such as at the height of glaciation; 3) the exceptionally strong winds that regularly occur in the Caribbean could, from time to time, disperse bats from one island to another; and finally 4) accidental human transport in the past several hundred years cannot be entirely discounted, particularly for species that might roost on boats while at anchor (such introductions are known to have occurred elsewhere but have not been documented for the West Indies).

Of the 24 species of bats currently known to occur in the Lesser Antilles, then, six are widespread taxa. One of these clearly is a double invader and two others probably are so. Two species, *Eptesicus fuscus* (south to Dominica) and *Tadarida brasiliensis* (south to St. Lucia) reached the region either from the north or west, whereas the sixth, *Natalus stramineus* (south to Dominica), seemingly invaded from the west. In the case of the monotypic family Natalidae, however, it is noteworthy that two of the four currently recognized species, including one of the two subgenera, are endemic to the West Indies, and a third species (*stramineus*) occurs there. Is it possible that the family originated in the Antillean region?

Nine kinds of bats clearly reached the Lesser Antilles in relatively recent times from the south. Five have been recorded only from Grenada. One (*Pteronotus davyi*) is known from as far north as Marie Galante, between Dominica and Guadeloupe, whereas *Sturnira lilium* reaches Dominica proper. Two, *Glossophaga longirostris* and *Artibeus lituratus*, are found only as far north as St. Vincent. Northward occurrence possibly is related to earlier time of invasion.

Of the endemic Antillean genera here treated, *Brachyphylla* occurs throughout much of the region, south in the Lessers to Barbados and St. Vincent. *Ardops* is endemic to the Lesser Antilles, from St. Eustatius to St. Vincent. *Monophyllus* is found on all of the Greater Antillean islands and southward to the Windward Island of St. Vincent. Morgan and Woods (1986) commented about the distinctive West Indian groups as follows: "The highly derived endemic phyllostomid subfamily Brachyphyllinae, as well as the four endemic stenodermatine genera, may represent groups of bats that reached the West Indies in the Early or Middle Tertiary." *Monophyllus*, on the other hand, probably is of later origin than other endemic genera, representing a descendent of *Glossophaga*-like stock that could have invaded the Antillean region either from the south or the west.

Of the six Lesser Antillean species I have termed enigmatic, all have restricted, perhaps relictual, distributions. None occurs in the Greater Antilles and only two have been reported from north of Guadeloupe. All save *Eptesicus guadeloupensis* seem to have been derived from bats that initially invaded from the south.

Additionally, four species are known as Holocene fossils from the northern Lesser Antillean islands of Antigua and Barbuda. On Antigua, *Phyllonycteris major* (an endemic brachyphylline genus in the Greater Antilles), *Mormoops blainvillii* (an endemic Greater Antillean species), and *Pteronotus parnellii* (which occurs in the Greater Antilles and on the mainland) were present until at least 2,500 y BP, and probably would still occur there were it not for habitat destruction caused by humans (Steadman et al. 1984). *Macrotus waterhousii*, which is found on Greater Antillean islands west of Puerto Rico (where it is known from fossil cave deposits) and on the Middle American mainland, is known as a fossil from Barbuda (Morgan and Woods 1986). Thus it seems that the Virgin Islands (Koopman 1975) and northern Lesser Antilles have served both as "stepping stones" and a "filter zone" to bat movement between major islands in the Lesser chain and those of the Greater Antilles.

Despite a reasonably good record of bat occurrence in the Lesser Antilles, much remains to be learned. Some of the smaller northern islands have not been well explored zoologically; only one species has been reported from St. Bartholemew and one from Nevis, for example, and only three kinds are known from Saba and four from St. Kitts. In the middle of the chain, 12 species have been recorded from Dominica but only 10 from the larger and equally ecologically diverse island of Martinique, immediately to the south. And two important endemic genera, *Ardops* and *Monophyllus*, are known from but a single specimen each from St. Vincent, where their status thus is problematic. Fossil and subfossil sites are unknown from all but a few islands and discovery and subsequent study of cave deposits could yield invaluable information on past distributions, invasion routes, and possibly evolutionary relationships of species summarized here. Hopefully, this and other papers in this symposium volume will stimulate needed additional work.

Literature cited

Allen, G.M. 1911. Mammals of the West Indies. Bulletin of the Museum of Comparative Zoology 54:175-263.

Baker, R.J., and H.H. Genoways. 1976. A new species of Chiroderma from Guadeloupe, West Indies (Chiroptera: Phyllostomatidae). Occasional Papers of The Museum, Texas Tech University 39:1-9.

_____, and H.H. Genoways. 1978. Zoogeography of Antillean bats. Special Publication of the Academy of Natural Sciences of Philadelphia 13:53-97.

_____, H.H. Genoways, and J.C. Patton. 1978. Bats of Guadeloupe. Occasional Papers of The Museum, Texas Tech University 50:1-16.

Buden, D.W. 1977. First records of bats of the genus *Brachyphylla* from the Caicos Islands, with notes on geographic variation. Journal of Mammalogy 58:221-225.

Carter, D.C., and P.G. Dolan. 1978. Catalogue of type specimens of Neotropical bats in selected European museums. Special Publications of The Museum, Texas Tech University 15:1-136.

Davis, W.B. 1973. Geographic variation in the fishing bat, *Noctilio leporinus*. Journal of Mammalogy 54:862-874.

Dobson, G.E. 1878. Catalogue of the Chiroptera in the British Museum. London, xlii + 567 pp., 30 pls.

Genoways, H.H., R.C. Dowler, and C.H.Carter. 1981. Intraisland and interisland variation in Antillean populations of *Molossus molossus* (Mammalia: Molossidae). Annals of the Carnegie Museum 50:475-492.

Goodwin, G.G. 1959. Bats of the subgenus *Natalus*. American Museum of Natural History Novitates 1977:1-22.

_____, and A.M. Greenhall. 1961. A review of the bats of Trinidad and Tobago. Bulletin of the American Museum of Natural History 122:187-301, pls. 7-46.

Hahn, W.L. 1907. A review of the bats of the genus *Hemiderma*. Proceedings of the United States National Museum 32:103-118.

Hall, E.R. 1981. The Mammals of North America. 2nd ed., John Wiley & Sons, New York, 1:xv + 1-600 + 90.

Handley, C.O., Jr., and W.D. Webster. 1986. Systematics of Miller's long-tongued bat, *Glossophaga longirostris*, with description of two new subspecies. Occasional Papers of The Museum, Texas Tech University 100:1-22.

_____, and W.D. Webster. 1987. The supposed occurrence of Glossophaga longirostris Miller on Dominica and problems with the type series of Glossophaga rostrata Miller. Occasional Papers of The Museum, Texas Tech University 108:1-10.

Hill, J.E., and P.G.H. Evans. 1985. A record of *Eptesicus fuscus* (Chiroptera: Vespertilionidae) from Dominica, West Indies. Mammalia 49:133-136.

Husson, A.M. 1960. Mammals of the Netherlands Antilles. Martinus Nijhoff's Gravenhage, viii + 170 pp.

Jones, J.K., Jr. 1978. A new bat of the genus *Artibeus* from the Lesser Antillean Island of St. Vincent. Occasional Papers of The Museum, Texas Tech University 51:1-6.

_____, and R.J. Baker. 1979. Notes on a collection of bats from Montserrat, Lesser Antilles. Occasional Papers of The Museum, Texas Tech University 60:1-6.

_____, and C.J. Phillips. 1970. Comments on systematics and zoogeography of bats in the Lesser Antilles. Studies on the Fauna of Curaçao and Other Caribbean Islands 32:131-145.

_____, and C.J. Phillips. 1976. Bats of the genus *Sturnira* in the Lesser Antilles. Occasional Papers of The Museum, Texas Tech University 40:1-16.

_____, and A. Schwartz. 1967. Bredin-Archbold-Smithsonian Biological Survey of Dominica. 6. Synopsis of bats of the Antillean genus *Ardops*. Proceedings of the United States National Museum 124(3634):1-13.

Koopman, K.F. 1968. Taxonomic and distributional notes on Lesser Antillean bats. American Museum of Natural History Novitates 2333:1-13.

_____. 1975. Bats of the Virgin Islands in relation to those of the Greater and Lesser Antilles. American Museum of Natural History Novitates 2581:1-7.

_____. 1982. Biogeography of bats of South America. Pp. 273-302 in M.A. Mares and H.H. Genoways (eds.). Mammalian Biology in South America. Special Publication Series of the Pymatuning Laboratory of Ecology, University of Pittsburgh 6:xii + 1-539.

LaVal, R.K. 1973. A revision of the Neotropical bats of the genus *Myotis*. Natural History Museum, Los Angeles County, Science Bulletin 15:1-54.

Miller, G.S., Jr. 1913. Revision of the bats of the genus *Glossophaga*. Proceedings of the United States National Museum 46:413-429.

_____, and G.M. Allen. 1928. The American bats of the genera *Myotis* and *Pizonyx*. Bulletin of the United States National Museum 144:1-218.

Morgan, G.S., and C.A. Woods. 1986. Extinction and the zoogeography of West Indian land mammals. Pp. 167-203 in L.R. Heaney and B.D. Patterson (eds.). Island Biogeography of Mammals. Biological Journal of the Linnean Society 28:1-271.

Owen, R.D. 1987. Phylogenetic analyses of the bat subfamily Stenodermatinae (Mammalia: Chiroptera). Special Publications of The Museum, Texas Tech University 26:1-65.

Pierson, E.D., W.E. Rainey, R.M. Warner, and C.C. White-Warner. 1986. First record of *Monophyllus* from Montserrat, West Indies. Mammalia 50:269-271.

Schwartz, A., and J.K. Jones, Jr. 1967. Bredin-Archbold-Smithsonian Biological Survey of Dominica. 7. Review of bats of the endemic Antillean genus *Monophyllus*. Proceedings of the United States National Museum 124 (3635):1-20.

Steadman, D.W., G.K. Pregill, and S.L. Olson. 1984. Fossil vertebrates from Antigua, Lesser Antilles; evidence for late Holocene human-caused extinctions in the West Indies. Proceedings of the National Academy of Science 81:4448-4451.

Swanepoel, P., and H.H. Genoways. 1978. Revision of the Antillean bats of the genus *Brachyphylla* (Mammalia: Phyllostomatidae). Bulletin of the Carnegie Museum of Natural History 12:1-53.

Varona, L.S. 1974. Catálogo de los mamíferos viventes y extinguidos de las Antillas. Academia de Ciencias de Cuba, Havana, 139 pp.

Table 1. Distribution of bats on major Lesser Antillean islands. Principal sources of data were Allen (1911), Baker and Genoways (1978), Baker et al. (1978), Jones and Baker (1979), Jones and Phillips (1970), Koopman (1968, 1975), and Morgan and Woods (1986). Some records from the Grenadines and Grenada were provided by H.H. Genoways. Puerto Rico, the Virgin Islands, and Trinidad-Tobago are included for reference. A) Puerto Rico, B) Virgin Islands, C) Anguilla, D) St. Martin, E) St. Bartholemew, F) Saba, G) St. Eustatius, H) St. Kitts, I) Nevis, J) Barbuda, K) Antigua, L) Montserrat, M) Guadeloupe, N) Dominica, O) Martinique, P) St. Lucia, Q) St. Vincent, R) Barbados, S) Grenadines, T) Grenada, U) Trinidad-Tobago

SPECIES OF THE LESSER ANTILLES	A	B	C	D	E	F	G	H	I	J	K	L	M	N	O	P	Q	R	S	T	U
Peropteryx macrotis																				X	X
Noctilio leporinus	X	X		X				X		X	X	X	X	X	X	X	X	X	X	X	X
Pteronotus davyi													X^1	X	X					X	X
Micronycteris megalotis																				X^2	X
Glossophaga longirostris																X			X	X	X
Monophyllus plethodon	X^3	X								X	X	X	X	X	X	X	X	X			
Anoura geoffroyi																				X	X
Carollia perspicillata																				X^2	X
Sturnira lilium													X	X	X	X				X	X
Sturnira thomasi														X							
Chiroderma improvisum													X	X							
Artibeus cinereus																				X	X
Artibeus jamaicensis	X	X	X	X			X	X	X	X	X	X	X	X	X	X	X	X	X	X	X
Artibeus lituratus																X				X	X
Ardops nichollsi								X			X	X	X	X	X	X					
Brachyphyllum cavernarum	X	X	X	X			X	X		X	X	X	X	X	X	X	X	X			
Natalus stramineus				X			X			X	X	X	X								
Myotis dominicensis														X							
Myotis martiniquensis															X		X				
Myotis nigricans				X^4									X^4							X^4	X^4
Eptesicus fuscus	X													X		X					
Eptesicus guadeloupensis													X								
Tadarida brasiliensis	X	X		X	X		X	X		X	X	X	X	X	X	X					X^5
Molossus molossus	X	X	X	X						X	X	X	X	X	X	X	X	X	X	X	X

[1] Not recorded from mainland Guadeloupe but from Marie Galante.
[2] Known by a single specimen from Grenada.
[3] Known only from fossil cave deposits.
[4] Not identified to species by LaVal (1973).
[5] Known by a single specimen from Tobago.

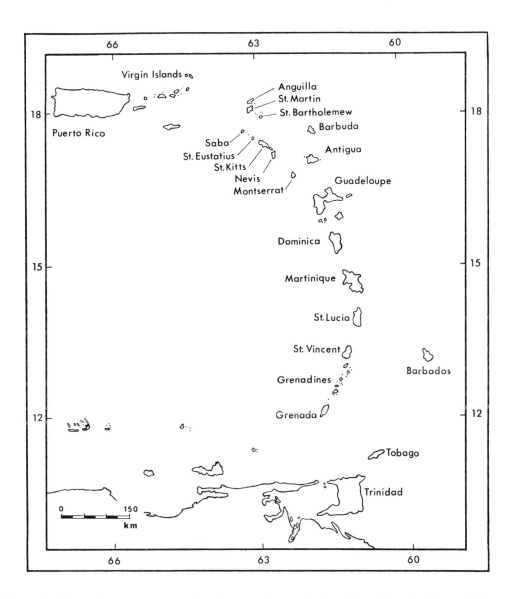

Figure 1. Map of the Lesser Antilles showing their relationship to Puerto Rico and the Virgin Islands to the north and Trinidad, Tobago, and the South American mainland to the south.

Biogeography of the West Indies, 1989:661-684

CARIBBEAN ISLAND ZOOGEOGRAPHY: A NEW APPROACH USING MITOCHONDRIAL DNA TO STUDY NEOTROPICAL BATS

Carleton J. Phillips[1], Dorothy E. Pumo[1], Hugh H. Genoways[2], and Phillip E. Ray[1]

Abstract

Genetic analysis of animal mitochondrial DNA is a new and valuable addition to the battery of techniques available to zoogeographers. This paper describes characteristics of mitochondrial DNA (mtDNA) that make it applicable for the study of island zoogeography.

Some traditional zoogeographic questions are examined using mtDNA from the Neotropical fruit bat, *Artibeus jamaicensis*. The specific questions are: 1) To what extent are island populations isolated (that is, does interbreeding occur between the insular subspecies)? 2) Can a single founding female account for the mitochondrial genomes on specific islands in the Antilles? 3) Is there a correlation between the genomic diversity of an island population and the size of the island or the distance from the mainland?

The mitochondrial genome in *Artibeus jamaicensis* is approximately 16,000-16,500 base pairs. Three major mtDNA groups (designated J, SV, and G), separated by 8 to 17.2 percent divergence in nucleotide sequence, were identified in Antillean *Artibeus jamaicensis*. The J and SV groups each includes two maternal lineages and the G group is represented by three lineages. The sequence divergence between the mtDNA groups is unusually high for conspecific mammals. Either the mtDNA in *Artibeus jamaicensis* can be traced to three relatively old origins or alternatively chiropteran mtDNA evolves at a faster rate than mtDNA in rodents and primates.

Populations on five of eight islands were clearly derived from multiple maternal ancestors. The greatest genetic diversity (judged by numbers of lineages and sequence divergence) was found in *A. j. trinitatus* on Grenada, which probably is a reflection of the proximity of the island to the South American mainland. Genetic diversity did not correlate with island size. One particular mtDNA lineage, J-1, was the most widespread, being found from Jamaica to Barbados and Grenada. These bats trace their ancestry to a common female, underscoring the dispersal and colonization capabilities of *A. jamaicensis*. MtDNA data and paleontological evidence suggest these bats reached the Antilles recently, probably in conjunction with replacement of the xeric Pleistocene environment. The island distribution of mtDNA genotypes somewhat corresponds to the currently recognized subspecies: 77 percent of *A. j. jamaicensis* carry the J mtDNA genotypes;

---

[1] Department of Biology, Hofstra University, Hempstead, NY 11550
[2] The University of Nebraska State Museum, University of Nebraska-Lincoln, Lincoln, NE 68588-0338

93 percent of *A. j. schwartzi* carry the SV mtDNA genotype; and 60 percent of the *A. j. trinitatus* carry the G mtDNA genotypes.

The geographic distribution of specific mtDNA genotypes on different islands documents gene flow between and among islands. For example, the data show that female bats with a common maternal ancestor have moved among the islands of St. Lucia, St. Vincent, Barbados, and Grenada even though they would be required to fly up to 150 km across open ocean. MtDNA and distributional data are used to postulate that the Grenadine islands, between Grenada and St. Vincent have been a partial barrier to gene flow, possibly because of their small size and xeric habitats.

Introduction

The occurrence, zoogeography, and speciation of mammals in the Caribbean have been studied for decades. Overviews of colonization and extinctions have been recently presented by Morgan and Woods (1986), Pregill (1981), Pregill and Olson (1981), and other authors in this volume. Most of what we know about Caribbean mammals has come from conventional data sources such as fossils and museum specimens. With previously available data, it has been difficult to test certain long standing hypotheses such as the following: Do most populations arise from single rather than multiple genetic founding events? Do some island populations receive periodic infusions of new DNA? Can species survival on islands be predicted by island size, ecological diversity, geographic location, or is it the result of chance?

Recent developments in molecular biology make it possible to consider zoogeographic problems from a new perspective--high resolution genetic analysis of individuals. In particular we think that restriction endonuclease analysis of mitochondrial DNA (mtDNA) will improve our understanding of zoogeography of the Caribbean. Our goals with the present paper are to: 1) provide background information about mtDNA analysis; 2) share some practical information about how we are applying the technique to a particular project in the Caribbean; and 3) discuss some recently acquired data in a zoogeographic context.

Mitochondrial DNA has specific characteristics that make it appropriate for certain zoogeographic problems where other methods have been either not particularly helpful or insufficient. Mitochondria are found in virtually all cells throughout an organism. Because each mitochondrion contains at least one copy of the small, circular mitochondrial genome (approximately 15,700-19,500 base pairs in animals; Brown 1983, 1985), and because there are usually many mitochondria per cell (approximately 1,000 in an average liver cell), isolation of the mitochondrial fraction from a tissue supplies the investigator with many identical copies of a small genome. The mitochondrial genome is inherited from the mother (Dawid and Blackler 1972; Hecht et al. 1984), allowing one to determine maternal ancestry and, potentially, to trace the living members of a species to a single female founder (Cann et al. 1987).

The mitochondrial genome evolves at a faster rate than the nuclear genome (Brown et al. 1979; Vawter and Brown 1986). This characteristic, together with its mode of inheritance, make mtDNA valuable for certain kinds of studies, particularly of congeners or conspecifics. The technique has been successfully applied to several kinds of systematic and evolutionary problems, including: relationships among higher primates (Hixson and

Brown 1986); history of the human species (Brown 1980; Wallace et al. 1985; Cann et al. 1987); phylogeny of the genus *Equus* (George and Ryder 1986); pocket gophers, *Geomys* (Avise et al. 1979a; Laerm et al. 1982); ground squirrels, *Spermophilus* (MacNeil and Strobeck 1987); deer mice, *Peromyscus* (Avise et al. 1979b; Lansman et al. 1983; Nelson et al. 1987; Ashley and Wills 1987); several species of the mouse, *Mus* (Boursot et al. 1985; Gyllensten and Wilson 1987); parthenogenetic lizards (Wright et al. 1983); anurans (Spolsky and Uzzell 1986; Lamb and Avise 1986; Carr et al. 1987); crickets (Harrison et al. 1985); and zoogeography of fish (Bermingham and Avise 1986; Avise et al. 1986).

In order to perform the analysis, it is necessary to isolate the mtDNA and cleave it with restriction endonucleases. These enzymes cut all identical mitochondrial genomes into a set of fragments, which can be separated by size using gel electrophoresis (Nathans and Smith 1975). Restriction enzymes can be selected that recognize different palindromic sequences and cut the chromosome in different places. Therefore, each enzyme creates a different pattern of fragments. By comparing fragments obtained with different enzymes, singly and in combination, the relative location of the recognition sites for a particular enzyme can be determined and the genome can be mapped. All individuals from the same maternal lineage will have identical restriction enzyme recognition sites within the chromosome. The longer the time since two individuals shared the same maternal ancestor, the more divergent their restriction fragment patterns are likely to be. The amount of divergence between two different genotypes can be estimated mathematically by comparing the number of shared versus nonshared restriction enzyme recognition sites.

The above approach can be directly applied to zoogeographic problems. For example, in theory one could locate possible geographic sources of island populations by tracing maternal lineages back to a mainland locality. Because the rate at which mitochondrial DNA evolves can sometimes be estimated, it might be possible to determine the minimum time since the invasion and colonization of an island. Several examples of the use of mtDNA to investigate historical zoogeography have been published recently (Boursot et al. 1985; Ashley and Wills 1987; MacNeil and Strobeck 1987) and discussed in detail by Avise et al. (1987).

What group of mammals might serve as a good model for testing the use of mtDNA analysis in studies of island zoogeography? First, the group should be reasonably widespread so that an adequate sample of islands can be compared (see Hastings 1987). Second, there should be some indication that breeding is non-uniform; that is, a geographic pattern of phenotypic differences is desirable. Third, conventional taxonomic data should be available. Fourth, the animals within the study group should be closely related. High-resolution mtDNA analysis is more likely to produce meaningful results with conspecifics or congenerics.

On the basis of these criteria we selected a species of the Neotropical phyllostomid fruit bat, *Artibeus* for our study. Many zoogeographers disregard bats when discussing island populations of mammals (De Beaufort 1951; Darlington 1963; Heaney 1985, 1986) because the flying capabilities of bats appear to lend them easy access to islands. However, systematists have often shown that the distribution of bats within archipelagos is nonuniform. Megachiropteran bats in the Solomon Islands demonstrate striking patterns of morphological variation between adjacent islands and the archipelago is inhabited by endemic genera and species (Phillips 1966, 1968). In Caribbean populations of microchiropteran bats there also are endemics as well as many examples of complex distributional patterns and interisland variation in size, color, and dentition (Koopman 1958, 1968, 1976; Jones and Phillips 1970, 1976; Jones, 1978; Baker and Genoways 1978; Swanepoel and Genoways 1978; Genoways et al. 1981; Ottenwalder and Genoways 1982). These data, obtained by the most commonly used techniques, appear to support the argument that bats can, and do, become reproductively isolated on islands.

Fruit bats of the genus *Artibeus* occur throughout most of Latin America and the Caribbean. The conventional view is that the Antilles are populated by two species (Koopman 1976; Jones 1978). One, *A. jamaicensis*, is very common and geographically widespread. It also varies both morphometrically and morphologically in the Caribbean (Koopman 1968; Jones and Phillips 1970). The other species, *A. lituratus*, is thought to occur only on two islands in the Lesser Antilles. Many specimens have been collected on Grenada and one specimen is reported from St. Vincent (Jones 1978, present volume). Morphologically, behaviorally, and ecologically these two species are very similar; they probably use essentially the same nutrient resources and have similar reproductive cycles (Gardner 1977; Wilson 1979). Their distribution in the Caribbean appears to qualify as an example of competitive exclusion under the criteria outlined by Hastings (1987).

Methodology

Specimen collection

The bats used in this study were collected by mist netting at night in flyways, over water sources, and at cave entrances. Some were sacrificed (with T-61 euthanasia solution) the next morning and their tissues removed and processed. Others were brought alive to our laboratory (with appropriate permits) and processed there. Voucher specimens were deposited in the collections of the Carnegie Museum of Natural History, Pittsburgh, PA, the University State Museum, University of Nebraska-Lincoln, NB, and The Museum, Texas Tech University, Lubbock, TX.

The most appropriate tissue for mtDNA analysis varies somewhat from species to species. Liver is often used, for example, from sheep and goats (Upholt and Dawid 1977) and from mice and gophers (Avise et al. 1979a, 1979b). However, the liver presents problems in some mammals (for example bats, our data and G. McCracken, pers. comm.; and voles, our data and Tegelstr?m [1986]). Generally, animals are homoplasmic with respect to the mtDNA from different tissues (Avise and Lansman 1983), however, heteroplasmy of mtDNA has been described (Solignac et al. 1983; Hauswirth et al. 1984; Densmore et al. 1985; Bermingham et al. 1986). Tissues used successfully include heart, liver, kidneys, and ovaries from lizards (Wright et al. 1983), liver and/or heart from fish (Avise et al. 1986; Bermingham and Avise 1986), placenta from humans (Brown 1980; Cann et al. 1987), liver or brain from dairy cows (Hauswirth et al. 1984), and in the case of small organisms the entire individual (Solignac et al. 1983; Harrison et al. 1985).

We used liver, kidney, heart, and skeletal muscle for our study. Kidney is the most satisfactory, but it must be supplemented with other tissues to provide sufficient DNA for mapping. Laboratory preparation of fresh tissue gives the best yield of mtDNA. Tissue from field-sacrificed specimens was either packed in cryotubes and immediately frozen in liquid nitrogen, or homogenized in eight volumes of buffer (0.25 M sucrose, 10 mM Tris-HCl[pH 7.5], 1 mM ethylenediamine-tetraacetate, 8.3 mM NaCl), centrifuged to remove the nuclei and cellular debris, and incubated for 30 min with Proteinase K (50 μg/ml supernatant) before freezing in dry ice. **Note added in proof:** Southern blot hybridization and the polymerase chain reaction (e.g. Kocher et al. 1989) are now used to study mtDNA. These techniques allow collection of information from partially degraded specimens.

Laboratory methodology

The details of the techniques we used are presented elsewhere (Brown 1980; Wright et al. 1983) and will not be repeated here. Instead, we will outline the techniques and focus on the analysis.

Mitochondria are separated from nuclei and other cellular debris by differential centrifugation and then purified through sucrose. The mitochondria are lysed by addition of sodium dodecyl sulfate (SDS). Many proteins are removed by precipitating the sample with saturated CsCl at 4½C. The intact, circular, mtDNA is separated from contaminating nuclear DNA and nicked or otherwise damaged mtDNA by cesium chloride density gradient centrifugation. The mitochondrial band is removed, dialyzed to remove the salts and dye necessary for the centrifugation, and precipitated with ethanol. The purified DNA is then ready for the analysis.

Small amounts of DNA are mixed with a restriction endonuclease in the appropriate buffer. It is the availability of a variety of these restriction enzymes that is the key to mtDNA analysis. These enzymes recognize specific base pair sequences and cleave the DNA within the recognition sequence (Nathans and Smith 1975). For example, the enzyme *Bam*H I recognizes the six base sequence 5'..GGATCC..3'. The recognition sequences are palindromes, which means that the sequence is the same on the complementary strands of DNA. The enzyme *Hin*d III recognizes 5'..AAGCTT..3', *Mbo* I recognizes 5'..GATC..3' and so on. Because these enzymes only recognize a specific sequence, they will cleave identical pieces of DNA, in our case mitochondria from a single individual, into a specific set of fragments that can be separated and sized by agarose or acrylamide gel electrophoresis (Southern 1979) or electron microscopy (Brown and Vinograd 1974).

We used the following restriction enzymes for the present study: *Bam*H I, *Bgl* II, *Pvu* II, *Hin*d III, *Eco*R I, *Pst* I, *Sal* I, *Xba* I, and *Xho* I. The mitochondrial DNA lineages are defined by the set of restriction fragments obtained after digestion, electrophoretic separation, and autoradiography. Individuals whose mtDNA has the same pattern of restriction endonuclease recognition sites are within the same maternal lineage. An alteration of the restriction enzyme cleavage pattern indicates that one or more mutations occurred in the female ancestors. The more differences between or among the genotypes, the longer the time since the individuals shared a common maternal ancestor.

It is tempting to think that the restriction fragment pattern obtained following gel electrophoresis and autoradiography is sufficient to calculate the amount of divergence. However, not all fragments that migrate the same distance on a gel represent homologous fragments. If one attempts to determine sequence divergence using just the restriction fragment sizes, the result will only be a minimum value. To be certain that two similar size fragments from different genomes are from homologous regions one must map the fragments (Nathans and Smith 1975). This is usually achieved by digesting the same mtDNA with two or more enzymes before electrophoresis. These data are then used to create a map of the genome (Fig. 1). Mapping can be done with the aid of several available computer programs (Fitch et al. 1983; for review, Bishop 1984). In *Artibeus* we found that some enzymes produced several similar sized, seemingly homologous, fragments in different genomes. After mapping, however, it was obvious that the fragments were not homologous, just coincidentally similar in size (Pumo et al. 1988).

Several formulae have been developed to estimate nucleotide sequence divergence based on restriction fragment information. The method of Upholt (1977), which uses unmapped fragment data, gives only a minimum divergence and could be potentially misleading because similar sized fragments are not necessarily homologous. Equation 16 (Nei and Li 1979) is currently widely used to calculate mtDNA sequence divergence from mapped restriction sites. The two disadvantages with this procedure are that no variance can be calculated and one must either know (at present unlikely) or estimate an alpha value. The values obtained with mapped restriction site data and equations 9 and 11 (Nei and Tajima 1983) are more conservative than equation 16 and allow estimation of the variance. The values obtained with equation 9 are the values referred to in the text. Such data can be used for pairwise comparisons of lineages and phylogenetic interpretations. The divergence can also be used to estimate the time of lineage separations (Brown et al. 1979; Cann et al. 1987).

The mitochondrial genome in *Artibeus jamaicensis*

The mitochondrial genome in *Artibeus jamaicensis* is approximately 16,000-16,500 base pairs (bp) in length. The mtDNA genome in *A. jamaicensis* was analyzed with restriction endonucleases that recognize specific 6-bp and 4-bp sequences (Pumo et al. 1988). For the analysis in this paper, we concentrated on 6-bp cutters. The enzymes *Bam*H I, *Bgl* II, and *Pvu* II each recognize one to four restriction sites, *Xba* I recognizes about five, and *Hin*d III recognizes six to ten sites in the *A. jamaicensis* mitochondrial genomes presented here. Three other restriction enzymes, *Eco*R I, *Pst* I, *Xho* I, and *Sal* I each recognize from 0 to 2 sites in the individuals surveyed thus far. The most interesting and complex data have been obtained with Hind III. Each of the major *A. jamaicensis* mtDNA groups known from the Antilles can be recognized with this enzyme because each has its own characteristic restriction fragment pattern that serves as a signature for the genotype (Fig. 1).

The restriction maps for five of the mtDNA lineages are shown in Fig. 1. The maps are linearized at a conserved *Bam*H I restriction site. When this is done numerous similarities and differences among the genotypes are immediately apparent. With this data set the five lineages clearly fall into three distinct groups of mtDNA genotypes based on the number of apparently conserved restriction sites (Fig. 1). Within each group there are less than three site differences, whereas between groups there are at least nine restriction site differences. The differences and similarities among the genotypes appear to be positioned throughout the genome, with no one large region more variable than another. Figure 1 contains representative lineages from each of the three major groupings of mtDNA genotypes, designated J, SV, and G recovered from 164 specimens of *Artibeus jamaicensis* from the islands of Jamaica, Puerto Rico, Anguilla, St. Vincent, St. Lucia, Barbados, Bequia, and Grenada. The numbers of specimens in each mtDNA group and the islands from which they were collected are summarized in Table 1.

All mtDNA groups found in specimens of *Artibeus jamaicensis* are represented in the Antilles by two or more maternal lineages. The J mtDNA group can be divided into a J-1 (84 individuals) and a J-2 (two individuals) lineage; J-2 has only been identified on the island of Jamaica, whereas the J-1 lineage is widespread in the Antilles. The J-2 lineage lacks a *Hin*d III site (Fig. 1) present in the J-1 lineage (Pumo et al. 1988). The SV group also contains two lineages. SV-1 predominates and was isolated from individuals on Barbados, St. Lucia, St. Vincent, Bequia, and Grenada. SV-2 was isolated from only one individual living on Grenada. The restriction site map for SV-2 is identical to SV-1 for all enzymes tested except *Hin*d III. SV-2 lacks two sites found in SV-1, and SV-2 has one site which is not present in SV-1. The G mtDNA group is represented by at least three lineages in the Lesser Antilles. On Grenada, G-1 is represented by six individuals and G-3 by three individuals. G-2 is represented by nine individuals, one from St. Vincent and the remainder from Grenada. The G-2 lineage differs from G-1 in having a *Sal* I restriction site (Fig. 1). The G-3 lineage lacks one *Hin*d III site present in G-1 and G-2 and does not have a *Sal* I site.

Restriction enzyme maps representing individual lineages can be used to estimate the overall sequence divergence between each lineage. An assessment of the difference between genotypes can be gained by using mapped restriction endonuclease cleavage sites (Table 2). Nei and Li's (1979) formula 16 with an alpha=2 is currently widely used for obtaining estimates of sequence divergence (Carr et al. 1986). Our discussions are based on comparisons calculated with Nei and Tajima's (1983) more conservative formulae 9 and 11. On this basis the J-1 and J-2 lineages are estimated to have diverged only 0.4 percent, whereas the J-1 and SV genotypes have 8 percent sequence divergence. The greatest amount of sequence divergence is between the SV and G-2 genotypes, which were estimated to differ by 13.9 percent. Although the 1979 formulae generate larger numbers, the qualitative results are identical with both procedures.

What is the significance of the extensive differences between each of the three major mtDNA groups identified in Antillean *Artibeus jamaicensis*? The best comparative data come from other mammals for which restriction maps are available and where the data have been analyzed by means of Nei and Li's (1979) method. On this basis breeds of horses have been estimated to differ by about 0.55 percent, whereas species of *Equus* differ by up to 7.5 percent (George and Ryder 1986), and California black-tailed deer (*Odocoileus hemionus*) and South Carolina white-tailed deer (*O. virginianus*) differ by 6.9 percent (Carr et al. 1986). Humans have been investigated in considerable detail and the mtDNA differs among lineages by about 0.32 percent (Brown 1980; Cann et al. 1987). Thus, the estimated differences between the J-1 and J-2 lineages and among the G mtDNA lineages are similar to intraspecific mammalian "breed" divergence, whereas the estimated differences among the three mtDNA genotype groups (8.2-13.9 %) exceeds that for most reported intraspecific comparisons among mammals. However, Honeycutt et al. (1987) report intraspecies variation in African mole rats, *Cryptomys hottentotus*, as high as 20 percent. Estimates of sequence divergence in the range of 10-20 percent are similar to those reported for many interspecific comparisons among frogs of the genus *Xenopus* (Carr et al. 1987). The high values for frogs are thought to be the result of divergence accumulated over a great period of time. The previously reported values for mammals are thought to be in keeping with Brown's (1980) calculation that mtDNA sequences in primates evolve at approximately 1 percent per million years.

Does the mitochondrial genome in *Artibeus* evolve at an extremely fast rate or is there some other explanation for the level of sequence divergence estimated from our data? One explanation is that *Artibeus* is a very old genus of phyllostomid bat. Smith (1976) concluded that most of the major microchiropteran families were well-established by at least the middle Oligocene or Miocene and Straney et al. (1979) used allozyme data to estimate that the Phyllostomidae diversified about 40 myBP. The possibility exists, therefore, that the mtDNA genotypes isolated from Antillean *A. jamaicensis* can be traced back to an ancient common ancestry. Another possibility is that the estimates of divergence for these genotypes are greater than expected due to insertions, deletions, or rearrangements in parts of the mitochondrial genome. In making our calculations, restriction sites that differed by more than 400 bp were regarded as non-homologous. Additions or deletions of fairly large (500-700 bp) sections in particular regions of the three genotypes would bring some restriction sites into apparent alignment. Length variation due to additions and deletions in the mtDNA of conspecific animals has been reported in non-mammalian vertebrates (Brown 1985) but its occurrence in the mtDNA of *Artibeus* has not yet been verified.

In summary, three different explanations can be offered for the large estimates of sequence divergence: 1) the mtDNA in *Artibeus* evolves at a significantly faster rate than in other studied mammals; 2) the extant mtDNA lineages are relatively ancient; and 3) evolution of the mitochondrial genome in *Artibeus jamaicensis* has involved additions, deletions, or rearrangements that affect the alignment of restriction maps and exaggerate the degree of sequence divergence. Until more data are available it will be difficult to favor any one of these three hypotheses about the evolution of the mitochondrial genome in *Artibeus*.

Zoogeography

Restriction endonuclease analysis of mtDNA is proving to be valuable to zoogeographic investigations. Some previous applications of the technology have advanced our understanding of human history (Wallace et al. 1985; Cann et al. 1987), have documented the dynamic interaction between two species of deer that occurred when their ranges came into sympatry due to alteration of environment (Carr et al. 1986), have examined the pattern of colonization of the northern Rocky Mountains by Columbian ground squirrels (MacNeil and Strobeck 1987), and colonization of Scandinavia by house mice, *Mus* (Gyllensten and Wilson 1987). The geographic distribution of mtDNA lineages can be interpreted in terms of other, independent, events to gain an insight into historical zoogeography. For example, Gyllensten and Wilson (1987) related the movement of *Mus* into Scandinavia with human agricultural activities and Boursot et al. (1985) correlated North African distribution of mtDNA lineages of *Mus spretus* with the northward expansion of desert and differentiation of isolated floral zones.

Analysis of mtDNA might be useful for island studies because the genome evolves rapidly and it might be possible to trace the dispersal history of individual maternal lineages. Ashley and Wills (1987) applied mtDNA analysis to deer mice (*Peromyscus maniculatus*) on the Channel Islands. Plant and colleagues (1989) used mtDNA analysis to study variation in meadow voles (*Microtus pennsylvanicus*). Both groups compared island populations to the mainland, estimated time of colonization, and compared intrapopulation genetic variation. Ultimately this same type of analysis might also be used to test hypotheses about mainland origins of Caribbean populations of mammals, depending on the availability of specimens and the presence of related maternal lineages on mainland areas.

Colonization and intraisland genetic diversity

Each of the island populations of *Artibeus jamaicensis* consists of descendants of some founding individuals. One could argue that island colonization usually has resulted from small numbers of individuals who reproduced successfully and established a long-term population. The probability of successful colonization from a small propagule is very high if the island can support a large population (K > 1,000) and the species has a per capita birth rate ( ) greater than the death rate (μ) (MacArthur and Wilson 1967). Once an island population reaches, or at least approaches, K, it then would seemingly be difficult for new arrivals to find living space and food. Insofar as fruit bats are concerned, it seems reasonable, *a priori*, to think that each Antillean island offers finite opportunities for success; the islands are limited in size, diversity, total fruit production, suitable roosting sites, and availability of fresh water.

With few exceptions, allozyme analyses have confirmed that island populations of small mammals have less genetic variability than mainland populations (Kilpatrick 1981; Berry 1986). Interestingly, one of the possible exceptions is the population of leaf-nosed bats, *Macrotus waterhousii*, on Jamaica (Greenbaum and Baker 1976). However, this exception might have been exagerated by data from a single locus (Greenbaum and Baker 1976). The more usual finding with allozyme comparisons, reduction in genetic variability, is thought to be the result of a founding effect, which involves intense inbreeding leading to increased frequency of monomorphic alleles and common mainland alleles and the loss of rare, or low frequency, mainland alleles (Kilpatrick 1981). Some authors (e.g., Mayr 1963) have argued that genetic diversity might be gradually restored and others (e.g., Slatkin 1987) have pointed out that occasional genetic input from new arrivals could have a significant "creative" evolutionary effect.

Because it is maternally inherited and not subject to recombination, the mitochondrial genome is not affected by the same factors as the nuclear genome. An island population founded by females carrying highly divergent mtDNA could exhibit a significant decrease in variability in the nuclear genome without any effect on the mitochondrial genotypes. Additionally, an island population might, over time, evolve a number of new mtDNA genotypes, each being derived from the founding mtDNA genotype(s). A long term, isolated, island population thus might carry some number of endemic mtDNA genotypes. However, the survival of new or old mtDNA genotypes is dependent on reproductive success within female lineages and is far from being assured. Indeed, lineage survival might be unlikely over a lengthy period of time (Avise et al. 1984). Consequently, the absence of multiple mtDNA genotypes in an island population could be the result of lineage extinction or, alternatively, a seemingly endemic mtDNA genotype could be the result of its extinction elsewhere. Some mtDNA data thus are difficult to interpret. On the other hand, if two or more geographically wide-spread mtDNA genotypes are also found in an island population, it would be reasonable to think that females in each mtDNA lineage either were part of the founding group, or immigrated separately into the island population, or both.

Another consideration is the amount of sequence divergence in mtDNA within an island population. Large amounts of divergence (e.g., >4%) would require a considerable period of time if mtDNA evolves at the rate of one percent per million years (Brown 1980; Higuchi et al. 1987). Because of stochastic lineage extinction, there is little probability of this much divergence accumulating in a small island population (Avise et al. 1984). For example, in a model population founded by females in two lineages, producing female offspring according to a Poisson distribution with a mean of 0.9, computer-generated data predict mtDNA lineage extinction in about 25 generations (Avise et al. 1984). The data generated by Avise et al. (1984) also predict that once a population reaches carrying capacity (K) of n size (K=n), within 4n generations all of the descendants will trace their ancestry to a founding female. Consequently, unless new females carrying different mtDNA genotypes had immigrated independently into an island, one would expect to detect either a single lineage or, at best, several very closely related lineages (i.e., <1 % sequence divergence) within a long-term insular population (Avise et al. 1987).

The founding of an island population can be viewed in terms of the arrival of a group of individuals, or a single individual, from whom future generations are derived. In terms of mtDNA, the arrival of females carrying divergent mtDNA genotypes might be viewed as separate genetic founding events, at least insofar as mtDNA lineages are concerned. Based on this and the foregoing discussion, we would hypothesize that some Antillean island populations of *Artibeus jamaicensis* resulted from several genetic founding events, probably in the form of periodic or occasional immigrations. For example, on Barbados and St. Vincent we found representatives of two female lineages, SV-1 and J-1. Their mtDNA exhibited so much sequence divergence (8%) that we placed their genotypes in separate mtDNA groups (SV and J; Table 2). These same female lineages are present on many other islands (Table 1). On Grenada, the southernmost island in the Lesser Antilles, we identified at least six female lineages representing all three of the mtDNA groups (SV, J, and G). Within the J mtDNA group, we found a bat in the J-1 lineage living on Grenada. Members of this same lineage also were found on every other island studied (Table 1). Likewise, within the SV group we found bats in the SV-1 lineage, which also constitute part of the *A. jamaicensis* populations on Barbados and St. Lucia and most of the population on St. Vincent (Table 1). Considering the high probability of lineage extinction, it is most likely that the discontinuities that separate the J, SV, and G groups resulted from evolution in geographic isolation (Avise et al. 1987). Thus, we think that the highly divergent mtDNA genotypes within the Grenada, St. Vincent, Barbados, and St. Lucia populations arrived independently from elsewhere and that these genotypes have been maintained by successful interisland movement of females, as documented by the presence of geographically wide-spread maternal lineages.

In contrast to the foregoing, studied populations in the northern Caribbean exhibited relatively little genetic diversity. Additionally, there is little direct evidence of frequent interisland movement although one could argue that broad distribution of bats in the J-1 lineage would require either a recent dispersal or maintainence through continuing interisland movement. Previous analysis of the J-1 mtDNA genotype with restriction enzymes that recognize 4-bp sequences failed to detect differences in the J-1 mtDNA isolated from bats living on Jamaica and St. Vincent, separated by nearly 1,400 km (unpubl. data).

All of the bats examined from the small island of Anguilla were in the same maternal lineage, J-1 (Table 1). The Anguilla population could have been founded exclusively by females in the J-1 lineage but alternatively other J lineages also might have reached the island, or originated there, but became extinct. Lineages from the SV and G mtDNA groups are also absent from Anguilla; in fact the nearest members of lineages in these divergent mtDNA groups live on St. Lucia and St. Vincent, more than 500 km to the south (Fig. 1; Table 1). At least two J lineages are represented on Jamaica. One, J-1, is found throughout the Caribbean but the other, J-2, is known only from Jamaica. The J-2 lineage might have originated on Jamaica, although its absence elsewhere could be due to chance extinction. Again, however, there is no evidence of lineages from either of the other mtDNA groups.

Intrapopulation (=intraisland) genetic heterogeneity ($\hat{h}$) can be estimated by formula [7] from Nei and Tajima (1981). On this basis Grenada is shown to exhibit the most heterogeneity (0.775) among the islands sampled; St. Lucia is second with 0.352 (Tables 1,3). Islands having several equally represented female lineages score much higher than islands on which the majority of the individuals carry one mtDNA genotype and the remaining few individuals represent several other lineages (e.g., St. Vincent versus St. Lucia, Table 3). Overall, the mean genetic heterogeneity for the eight islands sampled was 0.211. By way of comparison, Ashley and Wills (1987) reported that mtDNA heterogeneity in eight island populations of deer mice (*Peromyscus*) ranged from 0 to 0.44, with a mean of 0.20.

To some extent the foregoing data give a potentially misleading impression because Nei and Tajima's (1981) formula is based only on sample size and incidence (x_i) of each lineage in a population. Genetic heterogeneity on Jamaica (h = 0.189), where we found two J mtDNA genotypes with 0.4 percent sequence divergence, should be regarded in a totally different light than St. Vincent where bats trace their ancestries to three separate mtDNA groups with up to 13.9 percent sequence divergence but have a heterogeneity value of only $\hat{h}$ = 0.147 because one maternal lineage dominated the sample (Table 3).

The population of *Artibeus jamaicensis* on Grenada clearly is the most genetically diverse, both in terms of the incidence of different female lineages and the presence of divergent mtDNA groups. The genetic diversity in *A. jamaicensis* on Grenada can not be explained in terms of greater island size because Grenada is less than 50 percent the size of Jamaica or Puerto Rico. The most likely explanation is the geographic location of Grenada, which is only 150 km north of Trinidad and the South American mainland. Taken together, all of the mtDNA data suggest that *A. jamaicensis* on Grenada are not reproductively isolated and that new arrivals help to maintain considerable genetic diversity.

mtDNA and Antillean subspecies of *Artibeus jamaicensis*

In general, the colonization of the Antilles by *Artibeus jamaicensis* has followed a geographic pattern that is similar to the history hypothesized on the basis of analysis of exophenotypic features (Koopman 1968, present volume; Jones and Phillips 1970; Jones 1978, present volume; Baker and Genoways 1978). Four subspecies are thought to occur in the Caribbean. One, *A. j. parvipes*, occurs on Cuba and in the Bahamas but mtDNA has not yet been isolated from this subspecies. The second, *A. j. jamaicensis*, is widespread, occurring from Jamaica to St. Lucia and Barbados. *Artibeus j. schwartzi* occurs on St. Vincent and *A. j. trinitatus* occurs on Grenada south to Trinidad. In the northern

Antilles all 63 of our specimens of *A. j. jamaicensis* had a J mtDNA genotype but on St. Lucia, 79 percent of the *A. j. jamaicensis* carried another genotype, SV-1, as did 12 percent of the specimens collected on Barbados. The SV-1 mtDNA genotype is more typical (90 percent of the specimens) of *A. j. schwartzi* on St. Vincent (Table 1).

Overall, most specimens of *A. j. jamaicensis* (81%) have cytoplasmic genes in the J group. Animals carrying these genes are particularly interesting because of their broad geographic distribution and because of the small amount of sequence divergence discovered thus far in their mtDNA. Indeed, when the J-1 animals from Jamaica were compared to those living on Barbados (approximately 1,400 km apart), no differences were found with eight restriction enzymes (*Hin*d III, *Pvu* II, *Bam*H I, *Bgl* II, *Sal* I, *Eco*R I, *Pst* I, *Xba* I) that recognize different 6-bp sequences (Pumo et al. 1988). The relative uniformity in their mitochondrial genomes documents that these animals are closely related and suggests that this lineage has only recently spread throughout the Caribbean.

The possibility of recent dispersal and colonization by animals carrying the J mtDNA also is supported by paleontological evidence; Williams (1952) and Morgan and Woods (1986) have noted that on Jamaica and in the Cayman Islands, *Artibeus j. jamaicensis* is presently very abundant but is not found in Late Pleistocene or Early Holocene cave deposits. Instead, specimens are only known from the most superficial cave deposits. Likewise, specimens of *A. j. jamaicensis* are lacking from Pleistocene deposits in the Bahamas (Morgan, this volume). Subfossil specimens of *A. j. jamaicensis* from Puerto Rico (Reynolds et al. 1953; Choate and Birney 1968) have been collected from deposits in Cueva Monte Grande, Cueva de Clara, and Cueva del Perro. Although precise stratigraphic or radiocarbon dates are unavailable for these deposits, they are at least Pre-Columbian and possibly sub-Recent (Choate and Birney 1968) but unlikely to be Late Pleistocene or Early Holocene. Based on these data, we conclude that many of the *A. j. jamaicensis* living from Jamaica to Barbados are descendants of a common female ancestry and are quite likely the product of a recent and rapid dispersal over this large geographic area. We hypothesize that this dispersal was triggered by the gradual Holocene replacement of the xeric Pleistocene environment (dry grassland and thorn forest) by more mesic habitats with adequate fruit supply. Physiological and histological studies of *A. jamaicensis* collected in Panama lend support to this hypothesis. These investigations demonstrate that this species is not well-adapted to xeric environments; the kidney in *A. jamaicensis* is not subdivided as it is in xeric-adapted bats and urine concentrating ability is low (mean maximum of 972 mOsm/kg) (Studier et al. 1983a, 1983b). Typically, *A. jamaicensis* are dehydrated when they leave their roosts and depend on fruit juices for rapid rehydration. Finally, it is noteworthy that the recent dispersal of *A. jamaicensis* into the Caribbean also appears to correlate with Late Pleistocene extinctions and range reductions of several xeric-adapted chiropteran species living in the Greater Antilles (Morgan and Woods 1986).

The broad geographic distribution of *Artibeus j. jamaicensis* indicates that these bats have great dispersal power and are excellent colonizers. This is underscored by mtDNA data for the J group of genotypes. Part of their success must be derived from the fact that these bats are capable of flying distances up to 150 km (from St. Lucia to Barbados over open water). Other factors include reproduction, life span, mortality rate, and the carrying capacity of many of the islands. *Artibeus jamaicensis* probably have a bimodal reproductive cycle in the Antilles so females produce two young per year (Wilson 1979).

Per capita reproduction (λ) thus approaches two. For most island populations the value of K is very high, at least to judge from numbers of animals captured per night per mist net. While mortality rate (μ) is unknown, *A. jamaicensis* probably is not subject to predation on the islands. Part of the dispersal and colonization success of *A. jamaicensis*, documented by the mtDNA data for the J mtDNA genotypes, thus can be attributed to a high intrinsic rate of population increase (r), which is = $\lambda - \mu$ (MacArthur and Wilson 1967). It is noteworthy that probability of successful colonizing (r/λ), as predicted by MacArthur and Wilson's (1967) model, is high for *A. jamaicensis* and is borne out not only by its presence on many islands but also by the genetic history of *A. j. jamaicensis* on five islands spread over 1,400 km.

Artibeus j. schwartzi is a large-sized *A. jamaicensis* known from the island of St. Vincent, at the northern end of the Grenadines (Jones and Phillips 1970; Jones 1978). Our specimens from Bequia also can be assigned to this subspecies. Most of the specimens from St. Vincent (93 %) examined carried the SV-1 mtDNA genotype. One specimen had the J-1 mtDNA genotype and another carried the G-2 mtDNA genotype (Table 1). It is perhaps noteworthy that a fossil species, *Artibeus anthonyi*, named by Woloszyn and Silva Taboada (1977) from specimens collected in Cueva del Centenario de Lenin in Cuba, appears to be very similar, if not identical, to *A. j. schwartzi*. Woloszyn and Silva Taboada (1977) described the skull of *anthonyi* as similar in proportion to that of *A. lituratus* and presented cranial and mandibular measurements well within the range for *A. j. schwartzi* (Jones and Phillips 1970; Jones 1978). Is it possible that *A. j. schwartzi* formerly occupied a greater range in the Antilles?

Specimens of *Artibeus jamaicensis* from Grenada were originally named as a separate island subspecies, *A. j. grenadensis*, but later were assigned by Koopman (1968) to *A. j. trinitatus*. As discussed above, the mtDNA data from specimens collected on Grenada have revealed that these animals have an extremely complex genetic history.

Gene flow: are the Grenadines a filter?

Although gene flow generally has been regarded as a constraint to speciation (Mayr 1963), Slatkin (1987) has recently summarized conditions under which limited gene flow between populations might be a "creative" evolutionary force. Theoretical discussions of issues surrounding allopatric, sympatric, and peripatric speciation are common but the actual measurement of gene flow has rarely been achieved. Restriction endonuclease analysis of mtDNA offers one solution because by determining the geographic locations from which particular mtDNA genotypes have been recovered, it is possible to obtain direct documentation of gene flow between islands. Because it is considered unlikely that identical mtDNA lineages would evolve independently on multiple islands, the existence of the same lineage on more than one island is assumed to be due to movement of individuals from one place to another. Gene flow to an island is proportionate to colonization events so an island with four mitochondrial lineages probably has had considerably more divergent genetic input than has an island with but a single lineage. Only instances in which lineage extinction may have occurred would alter this conclusion. But the number of mtDNA lineages represents only a minimum number of founding events anyway.

The broad distribution of the J-1 mtDNA genotype documents recent gene flow among islands from Jamaica to Barbados. Likewise, the distribution of the J and SV mtDNA genotypes on St. Vincent, St. Lucia, and Barbados clearly shows gene flow among these islands. The differential incidence of particular mtDNA genotypes in different island populations might reveal something about gene flow but several other explanations also come to mind. For example, one could argue that incidence of a particular mtDNA genotype in a population could simply be the consequence of stochastic events. Alternatively, unequal distribution of a genotype might indicate recent arrival or might be the result of an asymmetric hybridization as has been discussed with regard to deer in the Southwest United States (Carr et al. 1986) and *Mus* in Scandinavia (Gyllensten and Wilson 1987).

Gene flow in the southern Lesser Antilles is particularly interesting because St. Vincent and Grenada are separated by both 120 km of water and the stepping stone-like Grenadines. A priori one might logically hypothesize that the Grenadines would facilitate gene flow between islands. However, the mtDNA data suggest just the opposite. The J mtDNA group, which has successfully spread throughout most of the Caribbean, is rare on St. Vincent and Grenada (Table 1). Animals carrying the J genotypes have dispersed successfully over a large area but from St. Vincent southward their genes are scarce. Animals carrying mtDNA genotypes in the G group have successfully colonized Grenada, which is about 130 km north of the South American mainland and Trinidad. However, only a single specimen (G-2, which also occurs on Grenada) has been found as far north as St. Vincent (Table 1). Bats carrying the SV mtDNA appear to be the only ones that have been very successful in moving between Grenada and St. Vincent.

An additional indication that the Grenadines have not facilitated gene flow comes from the distributional data for a related species, *Artibeus lituratus*. This large-sized fruit bat is relatively common throughout Middle America and northern South America. It also occurs on Trinidad and Grenada (Jones, this volume). Although it was thought to occur as far north as St. Vincent (Koopman 1968; Jones 1978), it is actually rare on that island. Jones and Phillips (1970) collected and reported on the only known specimen from St. Vincent. Phillips and Pumo (unpubl. data) did not capture any additional specimens during four nights of field work in 1986. This species is also absent from the northern Grenadine islands of Bequia and Mustique (Phillips and Pumo, unpubl. data). The Antillean population of *A. lituratus* thus is mostly confined to Grenada.

In summary, the genetic data as well as species distribution data suggest that the Grenadines have not facilitated movement of *Artibeus* though the Lesser Antilles, even though they provide an intermediate stepping stone connection. During the Pleistocene, Grenada and St. Vincent would have been nearly connected by a land bridge formed by the Grenadines. However, with the exception of bats carrying the SV-1 mtDNA, there is no indication that even this partial land bridge facilitated the movement of *Artibeus*. The Lesser Antilles, like the Greater Antilles, probably were xeric during the Pleistocene. Eshelman and Morgan (1985) have concluded that the late Pleistocene environment of near-by Tobago was similar to the present-day llanos (grassy plains) and dry thorn forest in South America. We postulate that northward colonization of the Lesser Antilles by *Artibeus* was influenced by the post-Pleistocene spread of more mesic habitats from northern South America into the Lesser Antilles (van der Hammen 1974). Current habitat conditions might also explain why the Grenadines have not served as a conduit for gene flow between the northern and southern Lesser Antilles. The Grenadines remain extremely xeric; many lack standing fresh water and fruit trees are nearly non-existent.

Taken collectively, these conditions do not appear suitable for ready colonization by *A. jamaicensis*. Unsuitable habitat probably is accentuated by the small size of the individual Grenadines. As Whitehead and Jones (1969) have pointed out in their discussion of small islands and equilibrium theory, island size is very likely to exert influence over colonization rate. A combination of small size and marginal habitat would result in an extremely small K value (perhaps essentially zero) for a resident population and, based on MacArthur and Wilson's (1967) mathematical analysis of survivorship, permanent populations would easily go extinct without new arrivals. New arrivals would recolonize rather than compete or hybridize with an existing population. The picture that emerges for the Grenadines is one in which these small, dry islands might have been colonized repeatedly but rarely served as a source population for continuing dispersal. The mtDNA data show that on average fruit bats of the genus *Artibeus* have been more successful at crossing expanses of open ocean than moving through the Grenadines. It is possible that bats in the SV-1 mtDNA lineage were slightly better suited to survival in xeric conditions and were able to colonize Grenada and St. Vincent during the Pleistocene. Perhaps these bats are derived from a stock that reached the Antilles earlier than others and this could explain the evolution of the morphologically distinctive *A. j. schwartzi*.

Acknowledgements

Financial support for this investigation came from a variety of sources, which we are pleased to acknowledge: NSF grant BBS-8609231 and NIH grant GM42563 (Pumo and Phillips); Hofstra University HCLAS grants (Pumo, Phillips); Research Corporation grants C-1251 and C-1855 (Phillips); the Mellon Institute of North American Mammal Research, Carnegie Museum of Natural History (Genoways); and The University of Nebraska Research Council (Genoways). We also are pleased to mention the names of a number of individuals who were extremely helpful with collection of specimens, logistics, permits, and technical support: Philipa Newton, Stephen L. Williams, Rafael L. Joglar, Robert J. Baker, Laura Baker, Linda Phillips, Dennis M. Burke, Everett Z. Goldin, Beth Elliot, Albert M. Mennone, and Eugene Kaplan. Finally, the cooperation of the U.S. Fish and Wildlife Service and CDC Atlanta is gratefully acknowledged.

Literature cited

Ashley, M., and C. Wills. 1987. Analysis of mitochondrial DNA polymorphisms among Channel Island deer mice. Evolution 41:854-863.

Avise, J.C., C. Giblin-Davidson, J. Laerm, J.C. Patton, and R.A. Lansman. 1979a. Mitochondrial DNA clones and matriarchal phylogeny within and among geographic populations of the pocket gopher, *Geomys pinetis*. Proceedings of the National Academy of Sciences of the United States of America 76:6694-6698.

_____, G.S. Helfman, N.C. Saunders, and L.S. Hales. 1986. Mitochondrial DNA differentiation in North Atlantic eels: population genetic consequences of an unusual life history pattern. Proceedings of the National Academy of Sciences of the United States of America 83:4350-4354.

_____, R.A. Lansman, and R.O. Shade. 1979b. The use of restriction endonucleases to measure mitochondrial DNA sequence relatedness in natural populations. I. Population structure and evolution in the genus *Peromyscus*. Genetics 92:279-295.

_____, and R.A. Lansman. 1983. Polymorphism of mitochondrial DNA in populations of higher animals. Pp. 147-164 *in* M. Nei and R.K. Koehn (eds.). Evolution of Genes and Proteins. Sinauer, Sunderland, MA.

_____, J.E. Neigel, and J. Arnold. 1984. Demographic influences on mitochondrial DNA lineage survivorship in animal populations. Journal of Molecular Evolution 20:99-105.

_____, J. Arnold, R. M. Ball, E. Bermingham, T. Lamb, J. E. Neigel, C. A. Reeb, and N. C. Saunders. 1987. Intraspecific phylogeography: The mitochondrial DNA bridge between population genetics and systematics. Annual Review of Ecology and Systematics 18:489-522.

Baker, R.J., and H.H. Genoways. 1978. Zoogeography of Antillean bats. Pp. 53-97 in F.B. Gill (ed.). Zoogeography in the Caribbean: The 1975 Leidy Medal Symposium. Special Publications of the Academy of Natural Science, Philadelphia, PA.

Beaufort, L.E. De. 1951. Zoogeography of the Land and Inland Waters. Sidgwick and Jackson, London, 208 pp.

Bermingham, E., and J.C. Avise. 1986. Molecular zoogeography of freshwater fishes in the southeastern United States. Genetics 113:939-965.

_____, T. Lamb, and J.C. Avise. 1986. Size polymorphism and heteroplasmy in the mitochondrial DNA of lower vertebrates. Journal of Heredity 77:249-252.

Berry, R.J. 1986. Genetics of insular populations of mammals, with particular reference to differentiation and founder effects in British small mammals. Biological Journal of the Linnean Society 28:205-230.

Bishop, M.J. 1984. Software for molecular biology. II. Restriction mapping and DNA sequencing programs. BioEssays 1:75-77.

Boursot, P., T. Jacquart, F. Bonhomme, J. Britton-Davidian, and L. Thaler. 1985. Differenciation g?ographique du g?nome mitochondrial chez *Mus spretus* Lataste. Comptes Rendus Academie des Sciences (Paris) 301:161-166.

Brown, W.M. 1980. Polymorphism in mitochondrial DNA of humans as revealed by restriction endonuclease analysis. Proceedings of the National Academy of Sciences of the United States of America 77:3605-3609.

_____. 1983. Evolution of animal mitochondrial DNA. Pp. 62-88 *in* M. Nei and R.K. Koehn (eds.). Evolution of Genes and Proteins. Sinauer, Sunderland, Mass.

_____. 1985. The mitochondrial genome of animals. Pp. 95-130 *in* R. MacIntyre, (ed.). Molecular Evolutionary Genetics. Plenum Press, New York.

_____, M. George, Jr., and A.C. Wilson. 1979. Rapid evolution of animal mitochondrial DNA. Proceedings of the National Academy of Sciences of the United States of America 76:1967-1971.

_____, and J. Vinograd. 1974. Restriction endonuclease cleavage maps of animal mitochondrial DNAs. Proceedings of the National Academy of Sciences of the United States of America 71:4617-4621.

Cann, R.L., M. Stoneking, and A.C. Wilson. 1987. Mitochondrial DNA and human evolution. Nature 325:31-36.

Carr, S.M., S.W. Ballinger, J.N. Derr, L.H. Blankenship, and J.W. Bickham. 1986. Mitochondrial DNA analysis of hybridization between sympatric white-tailed deer and mule deer in west Texas. Proceedings of the National Academy of Sciences of the United States of America 83:9576-9580.

____, A.J. Brothers, and A.C. Wilson. 1987. Evolutionary inferences from restriction maps of mitochondrial DNA from nine taxa of *Xenopus* frogs. Evolution 41:176-188.

Choate, J.R., and E.C. Birney. 1968. Sub-Recent Insectivora and Chiroptera from Puerto Rico, with the description of a new bat of the genus *Stenoderma*. Journal of Mammalogy 49:400-412.

Darlington, P.J., Jr. 1963. Zoogeography: the Geographical Distribution of Animals. John Wiley & Sons, NY, 675 pp.

Dawid, I.B., and A.W. Blackler. 1972. Maternal and cytoplasmic inheritance of mitochondrial DNA in *Xenopus*. Developmental Biology 29:152-161.

Densmore, L.D., J.W. Wright, and W.M. Brown. 1985. Length variation and heteroplasmy are frequent in mitochondrial DNA from parthenogenetic and bisexual lizards (genus *Cnemidophorus*). Genetics 110:689-707.

Eshelman, R.E., and G.S. Morgan. 1985. Tobagan Recent mammals, fossil vertebrates, and their zoogeographical implications. National Geographic Society, Research Reports 21:137-143.

Fitch, W.M., T.F. Smith, and W.W. Ralph. 1983. Mapping the order of DNA restriction fragments. Gene 22:19-29.

Gardner, A.L. 1977. Feeding habits. Pp. 293-350 *in* R.J. Baker, J.K. Jones, Jr., and D.C. Carter (eds.). Biology of Bats of the New World Family Phyllostomatidae, Part II. Texas Tech University Press, Lubbock.

Genoways, H.H., R.C. Dowler, and C.H. Carter. 1981. Intraisland and interisland variation in Antillean populations of *Molossus molossus* (Mammalia: Molossidae). Annals of Carnegie Museum 50:475-492.

George, M., Jr., and O.A. Ryder. 1986. Mitochondrial DNA evolution in the genus *Equus*. Molecular Biology and Evolution 3:535-546.

Greenbaum, I., and R.J. Baker. 1976. Evolutionary relationships in *Macrotus* (Mammalia: Chiroptera): biochemical variation and karyology. Systematic Zoology 25:15-25.

Gyllensten, U., and A.C. Wilson. 1987. Interspecific mitochondrial DNA transfer and the colonization of Scandinavia by mice. Genetic Research, Cambridge 49:25-29.

Harrison, R.G., D.M. Rand, and W.C. Wheeler. 1985. Mitochondrial DNA size variation within individual crickets. Science 228:1446-1448.

Hastings, A. 1987. Can competition be detected using species co-occurrence data? Ecology 68:117-123.

Hauswirth, W.W., M.J. Van De Walle, P.J. Laipis, and P.D. Olivo. 1984. Heterogeneous mitochondrial DNA D-loop sequences in bovine tissue. Cell 37:1001-1007.

Heaney, L.R. 1985. Zoogeographic evidence for middle and late Pleistocene land bridges to the Philippine islands. Modern Quaternary Research in Southeast Asia 9:127-143.

____, 1986. Biogeography of mammals in SE Asia: estimates of rates of colonization, extinction and speciation. Biological Journal of the Linnean Society 28:127-165.

Hecht, N.B., H. Liem, K.C. Kleene, R.J. Distel, and S.-M. Ho. 1984. Maternal inheritance of the mouse mitochondrial genome is not mediated by a loss or gross alteration of the paternal mitochondrial DNA or by methylation of the oocyte mitochondrial DNA. Developmental Biology 102:452-461.

Higuchi, R.G., L.A. Wrischnik, E. Oakes, M. George, B. Tong, and A.C. Wilson. 1987. Mitochondrial DNA of the extinct quagga: relatedness and extent of postmortem change. Journal of Molecular Evolution 25:283-287.

Hixson, J.E., and W.M. Brown. 1986. A comparison of the small ribosomal RNA genes from the mitochondrial DNA of the great apes and humans: sequence, structure, evolution, and phylogenetic implications. Molecular Biology and Evolution 3:1-18.

Honeycutt, R.L., S.V. Edwards, K. Nelson, and E. Nevo. 1987. Mitochondrial DNA variation and the phylogeny of African mole rats (Rodentia: Bathyergidae). Systematic Zoology 36:280-292.

Jones, J.K., Jr. 1978. A new bat of the genus *Artibeus* from the Lesser Antillean island of St. Vincent. Occassional Papers of the Museum of Texas Tech University 51:1-6.

____, and C.J. Phillips. 1970. Comments on systematics and zoogeography of bats in the Lesser Antilles. Studies on the Fauna of Curacao and other Caribbean Islands. The Hague, Netherlands 32:131-145.

____. 1976. Bats of the genus *Sturnira* in the Lesser Antilles. Occasional Papers of The Museum of Texas Tech University 40:1-16.

Kilpatrick, C.W. 1981. Genetic structure of insular populations. Pp. 28-59, 321-378 *in* M.H. Smith and J. Joule (eds.). Mammalian Population Genetics. University of Georgia Press, Athens.

Kocher, T.D., W.K. Thomas, A. Meyer, S.V. Edwards, S. P??bo, F.X. Villablanca, and A.C. Wilson. 1989. Dynamics of mitochondrial DNA evolution in animals: Amplification and sequencing with conserved primers. Proceedings of the National Academy of Sciences of the United State of America 86:6196-6200.

Koopman, K.F. 1958. Land bridges and ecology in bat distribution on islands off the northern coast of South America. Evolution 12:429-439.

____. 1968. Taxonomic and distributional notes on Lesser Antillean bats. American Museum of Natural History Novitates 2333:1-13.

____. 1976. Zoogeography. Pp. 39-47 *in* R.J. Baker, J.K. Jones, Jr., and D.C. Carter (eds.). Biology of Bats of the New World Family Phyllostomatidae, Part I. Texas Tech University Press, Lubbock.

Laerm, J., J.C. Avise, J.C. Patton, and R.A. Lansman. 1982. Genetic determination of the status of an endangered species of pocket gopher in Georgia. Journal of Wildlife Management 46:513-518.

Lamb T., and J.C. Avise. 1986. Directional introgression of mitochondrial DNA in a hybrid population of tree frogs: the influence of mating behavior. Proceedings of the National Academy of Sciences of the United States of America 83:2526-2530.

Lansman, R.A., J.C. Avise, C.F. Aquadro, J.F. Shapira, and S.W. Daniel. 1983. Extensive genetic variation in mitochondrial DNA's among geographic populations of the deer mouse, *Peromyscus maniculatus*. Evolution 37:1-16.

MacArthur, R.H., and E.O. Wilson. 1967. The Theory of Island Biogeography. Princeton University Press, Princeton, NJ, 203 pp.

MacNeil, D., and C. Strobeck. 1987. Evolutionary relationships among colonies of Columbian ground squirrels as shown by mitochondrial DNA. Evolution 41:873-881.

Mayr, E. 1963. Animal Species and Evolution. Belknap Press of Harvard University, Cambridge, MA, 797 pp.

Morgan, G.S., and C.A. Woods. 1986. Extinction and the zoogeography of West Indian land mammals. Biological Journal of the Linnean Society 28:167-203.

Nathans, D. and H.O. Smith. 1975. Restriction endonucleases in the analysis and restructuring of DNA molecules. Annual Review of Biochemistry 44:273-293.

Nei, M., and W.-H. Li. 1979. Mathematical model for studying genetic variation in terms of restriction endonucleases. Proceedings of the National Academy of Sciences of the United States of America 76:5269-5273.

_____, and F. Tajima. 1981. DNA polymorphism detectable by restriction endonucleases. Genetics 97:145-163.

_____. 1983. Maximum likelihood estimation of the number of nucleotide substitutions from restriction sites data. Genetics 105:207-217.

Nelson, K., R.J. Baker, and R.L. Honeycutt. 1987. Mitochondrial DNA and protein differentiation between hybridizing cytotypes of the white-footed mouse, *Peromyscus leucopus*. Evolution 41:864-872.

Ottenwalder, J.A., and H.H. Genoways. 1982. Systematic review of the Antillean bats of the *Natalus micropus*-complex (Chiroptera: Natalidae). Annals of Carnegie Museum 51:17-38.

Phillips, C.J. 1966. A new species of bat of the genus *Melonycteris* from the Solomon Islands. Journal of Mammalogy 47:23-27.

_____. 1968. Systematics of megachiropteran bats of the Solomon Islands. University of Kansas Publications, Museum of Natural History 16:777-837.

Plante, Y., P.T. Boag, and B.N. White. 1989. Macrogeographic variation in mitochondrial DNA of meadow voles (*Microtus pennsylvanicus*). Canadian Journal of Zoology 67:158-167.

Pregill, G.K. 1981. An appraisal of the vicariance hypothesis of Caribbean biogeography and its application to West Indian terrestrial vertebrates. Systematic Zoology 30:147-155.

_____, and S.L. Olsen. 1981. Zoogeography of West Indian vertebrates in relation to Pleistocene climatic cycles. Annual Review of Ecology and Systematics 12:75-98.

Pumo, D.E., E.Z. Goldin, B. Elliot, C.J. Phillips, and H.H. Genoways. 1988. Mitochondrial DNA polymorphism in three Antillean island populations of the fruit bat, *Artibeus jamaicensis*. Molecular Biology and Evolution (in press).

Reynolds, T.E., K.F. Koopman, and E.E. Williams. 1953. A cave faunule from western Puerto Rico with a discussion of the genus *Isolobodon*. Museum of Comparative Zoology, Breviora 12:1-8.

Slatkin, M. 1987. Gene flow and the geographic structure of natural populations. Science 236:787-792.

Solignac, M., M. Monnerot, and J.-C. Mounolou. 1983. Mitochondrial DNA heteroplasmy in *Drosophila mauritiana*. Proceedings of the National Academy of Sciences of the United States of America 80:6942-6946.

Southern, E.M. 1979. Measurement of DNA length by gel electrophoresis. Analytical Biochemistry 100:319-323.
Smith, J.D. 1976. Chiropteran evolution. Pp. 49-69 *in* R.J. Baker, J.K. Jones, Jr., and D.C. Carter (eds.). Biology of Bats of the New World Family Phyllostomatidae, Part I. Texas Tech University Press, Lubbock.
Spolsky, C., and T. Uzzell. 1986. Evolutionary history of the hybridogenetic hybrid frog *Rana esculenta* as deduced from mtDNA analysis. Molecular Biology and Evolution 3:44-56.
Straney, D.O., M. Smith, I.F. Greenbaum, and R.J. Baker. 1979. Biochemical genetics. Pp. 157-176 *in* R.J. Baker, J.K. Jones, Jr., and D.C. Carter (eds.). Biology of Bats of the New World Family Phyllostomatidae, Part III. Texas Tech University Press, Lubbock.
Studier, E.H., B.C. Boyd, A.T. Feldman, R.W. Dapson, and D.E. Wilson. 1983a. Renal function in the Neotropical bat, *Artibeus jamaicensis*. Comparative Biochemistry and Physiology 74A:199-209.
_____, S.J. Wisniewski, A.T. Feldman, R.W. Dapson, B.C. Boyd, and D.E. Wilson. 1983b. Kidney structure in Neotropical bats. Journal of Mammalogy 64:445-452.
Swanepoel, P., and H.H. Genoways. 1978. Revision of the Antillean bats of the genus *Brachyphylla* (Mammalia:Phyllostomatidae). Bulletin of the Carnegie Museum of Natural History 12:1-53.
Tegelstr?m, H. 1986. Mitochondrial DNA in natural populations: an improved routine for the screening of genetic variation based on sensitive silver staining. Electrophoresis 7:226-229.
Upholt, W.B. 1977. Estimation of DNA sequence divergence from comparison of restriction endonuclease digests. Nucleic Acids Research 4:1257-1265.
_____, and I.B. Dawid. 1977. Mapping of mitochondrial DNA of individual sheep and goats: rapid evolution in the D-loop region. Cell 11:571-583.
van der Hammen, T. 1974. The Pleistocene changes of vegetation and climate in tropical South America. Journal of Biogeography 1:3-26.
Vawter, L., and W.M. Brown. 1986. Nuclear and mitochondrial DNA comparisons reveal extreme rate variation in the molecular clock. Science 234:194-196.
Wallace, D.C., K. Garrison, and W.C. Knowler. 1985. Dramatic founder effects in Amerindian mitochondrial DNAs. American Journal of Physical Anthropology 68:149-155.
Williams, E.E. 1952. Additional notes on fossil and subfossil bats from Jamaica. Journal of Mammalogy 33:171-179.
Wilson, D.E. 1979. Reproductive patterns. Pp. 317-378 *in* R.J. Baker, J.K. Jones, Jr., and D.C. Carter (eds.). Biology of Bats of the New World Family Phyllostomatidae, Part III. Texas Tech University Press, Lubbock.
Whitehead, D.R., and C.E. Jones. 1969. Small islands and the equilibrium theory of insular biogeography. Evolution 23:171-179.
Woloszyn, B.W., and G. Silva Toboada. 1977. Nueva especie fÂsil de *Artibeus* (Mammalia: Chiroptera) de Cuba, y tipificaciÂn preliminar de los depÂsitos fosilÀferos Cubanos contentivos de mamÀferos terrestres. Poeyana (Instituto de Zoologia, Academia de Ciencias de Cuba) 161:1-17.
Wright, J.W., C. Spolsky, and W.M. Brown. 1983. The origin of the parthenogenetic lizard *Cnemidophorus laredoensis* inferred from mitochondrial DNA analysis. Herpetologica 39:410-416.

Table 1. Summary of specimens examined, localities, and mtDNA genotypes. The genotypes were assigned to one of three groups, designated J, SV, and G, rather than to specific maternal lineages. See text for details.

	Principal mtDNA Genotypes (% and sample size)			
	J	SV	G	TOTALS
A. j. jamaicensis				
Jamaica	100 (21)	0	0	21
Puerto Rico	100 (33)	0	0	33
Anguilla	100 (9)	0	0	9
Barbados	88 (15)	12 (2)	0	17
St. Lucia	21 (6)	79 (23)	0	29
				109
A. j. schwartzi				
St. Vincent	5 (1)	90 (18)	5 (1)	20
Bequia	0	100 (7)	0	7
				27
A. j. trinitatus				
Grenada	3.6 (1)	36 (10)	60 (17)	28
	Total specimens examined			164

Table 2. Estimates of percent nucleotide sequence divergence in pairwise comparisons of mtDNA lineages isolated from Antillean *Artibeus jamaicensis*. Upper diagonal: percent sequence divergence calculated with Nei and Li's (1979) equation 16 and alpha=2. Lower diagonal: percent nucleotide sequence divergence, standard deviation, and variance (in parentheses) calculated with formulae 9 and 11, Nei and Tajima (1983).

	J-1	J-2	SV-1	G-1	G-2
J-1	---	0.4	9.3	16.7	17.2
J-2	0.4±0.4 (0.001)	---	11.0	15.0	15.5
SV-1	8.2±2.2 (0.05)	9.5±2.6 (0.06)	---	16.1	16.7
G-1	13.5±3.7 (0.12)	12.4±3.3 (0.11)	13.1±3.4 (0.12)	---	0.4
G-2	13.9±3.5 (0.12)	12.8±3.4 (0.11)	13.5±3.5 (0.12)	0.4±0.4 (0.002)	---

Table 3. Intrapopulation genetic variations, $\hat{h}$ (based on mtDNA lineage data; Nei and Tajima, 1981*) in *Artibeus jamaicensis* on eight islands. Islands are listed in order of decreasing area.

Island	Sample size	Minimum No. mtDNA group(s)			Genetic mtDNA lineages	variation ($\hat{h}$)
		J	SV	G		
Jamaica	21	x			2	0.189
Puerto Rico	33	x			1	0
St. Lucia	29	x	x		2	0.352
Grenada	27	x	x	x	6	0.775
St. Vincent	20	x	x	x	3	0.147
Barbados	19	x	x		2	0.221
Anguilla	9	x			1	0
Bequia	7		x		1	0

*
$$\hat{h} = \frac{N}{N-1} (1 - \Sigma_i x_i^2)$$

N is sample size; x_i is frequency of the ith lineage

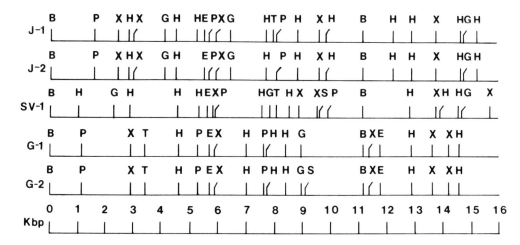

Figure 1.--Restriction maps of five mitochondrial DNA genotypes of *Artibeus jamaicensis* collected in the West Indies. All maps have been linearized at a conserved *Bam*H I site. Enzyme recognition sites are coded as follows: B, *Bam*H I; H, *Hin*d III; E, *Eco*R I; P, *Pvu* II; G, *Bgl* II; T, *Pst* I; S, *Sal* I; X, *Xba* I. J-2 has not been tested with *Pst* I.

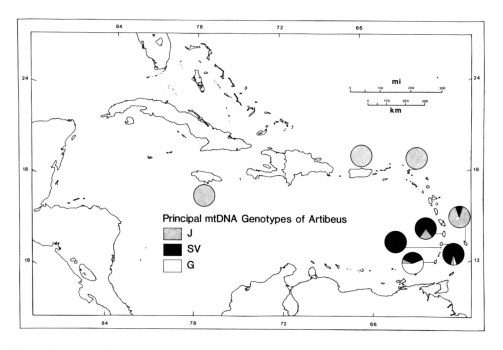

Figure 2.--Map of the Caribbean showing the relative incidence of genotype groups (J, SV, and G) on the islands in this report. Note that only the J group is found throughout the Caribbean.

Biogeography of the West Indies, 1989:685-740

FOSSIL CHIROPTERA AND RODENTIA FROM THE BAHAMAS, AND THE HISTORICAL BIOGEOGRAPHY OF THE BAHAMIAN MAMMAL FAUNA

Gary S. Morgan[1]

Abstract

The zoogeography and systematics of the Recent and fossil mammals from the Bahamas are reviewed, with particular emphasis on fossils from Late Quaternary cave deposits on Andros and New Providence. Fossils are recorded from four deposits on New Providence: Banana Hole, East Cave, Hunts Cave, and Sir Harry Oakes Cave. Banana Hole is the largest and most thoroughly studied fossil vertebrate fauna in the Bahamas, while the other three sites have not been published previously. Fossils are reported from three caves on Andros: Ashton Cave, Coleby Bay Cave, and King Cave. This material constitutes the first evidence of fossil mammals from Andros, the largest island in the Bahamas. Small samples of bones are also reported from Turtle Scratch Cave on Cat Island and an Amerindian archaeological site in Conch Bar Cave on Grand Caicos. The mormoopid bats *Pteronotus macleayii* and *P. quadridens* are recorded from the Bahamas for the first time, both represented solely by fossil remains from Andros and New Providence.

All Recent and fossil species of mammals from the Bahamas were derived from the Greater Antilles. Twelve of the 16 species (75%) of mammals from fossil deposits on islands of the Great Bahama Bank have Cuban affinities, while only one species is of Hispaniolan origin. Six of the 12 Recent Bahamian mammals are Cuban in origin and four species are derived from Hispaniola. The Bahamas are subdivided into two zoogeographic regions, the northern Bahamas and southern Bahamas, separated by the Crooked Island Passage. Ten of the 12 Recent species of Bahamian mammals occur in the southern Bahamas, including five species not presently found in the northern Bahamas: the bats *Noctilio leporinus, Monophyllus redmani, Artibeus jamaicensis,* and *Brachyphylla nana,* and the capromyid rodent *Geocapromys ingrahami.* Only seven species of mammals are found in the northern Bahamas, two of which are absent from the southern Bahamas, the natalid bats *Natalus tumidifrons* and *Nyctiellus lepidus.*

Since the end of the Pleistocene there has been a dramatic decrease in the number of species of mammals inhabiting the Bahamas, most notably on islands of the Great Bahama Bank. The Late Quaternary mammalian fauna of Andros and New

---

[1] Gary Morgan is Senior Biologist in Vertebrate Paleontology, Florida State Museum, Gainesville, FL 32611

Providence consisted of 16 species, 15 bats and *Geocapromys ingrahami*. Eleven of these 16 species (69%) no longer occur on islands of the Great Bahama Bank, including seven bats that are now extinct in the Bahamas: *Mormoops blainvillii, M. megalophylla, Pteronotus macleayii, P. parnellii, P. quadridens, Phyllonycteris poeyi*, and *Natalus major*. Four of the species extinct on Andros and New Providence still survive either in the southern Bahamas (*Monophyllus redmani, Brachyphylla nana*, and *Geocapromys ingrahami*) or on islands in the northern Bahamas peripheral to the Great Bahama Bank (*Natalus tumidifrons*). Only two species of mammals, the bats *Noctilio leporinus* and *Artibeus jamaicensis*, appear to have colonized the Bahamas during the Holocene, both of which have restricted ranges in the southern Bahamas.

The extensive localized extinctions of bats in the Bahamas reflect a similar trend apparent on small islands throughout the West Indies. Twelve of the 15 species of bats from fossil deposits on Andros and New Providence are cave-dwelling forms that generally roost deep within large caves having a hot, humid microenvironment. Seven of these species of specialized cavernicolous bats are now extinct in the Bahamas and three others have disappeared from the Great Bahama Bank. This strongly indicates that there were more extensive cave systems in the Bahamas during the Late Quaternary. Most chiropteran extinctions in the Bahamas probably resulted from either the flooding of these large caves by rising sea levels in the late Pleistocene and early Holocene or by changes in cave microenvironments. The disappearance of *Geocapromys* from most islands in the Bahamas can be attributed to the arrival of humans in the late Holocene.

A species-area curve calculated for Recent Bahamian land mammals demonstrates that there is no correlation ($R = 0.18$) between the size of an island in the Bahamas and the number of species of mammals found on that island. This is in marked contrast to most other islands in the West Indies which show a reasonably strong correlation ($R = 0.83$) between island area and mammalian species richness. Compared to other West Indian islands of similar size, the individual islands in the Bahamas support fewer species of mammals than would be predicted based on their area alone. The disappearance from the Bahamas of nearly half of the species of mammals found there during the Late Quaternary and the range contraction of seven other species are two of the most important factors that have shaped the zoogeography and composition of the Recent Bahamian mammal fauna.

Introduction

The Bahamas archipelago is an extensive group of about 30 large islands and 660 smaller islands or cays located in the western North Atlantic Ocean along the northwestern margin of the West Indies. The Bahamas are oriented in a general northwest to southeast direction and extend for a distance of nearly 1500 km from Florida on the west to Hispaniola on the east. Two countries comprise the Bahamas archipelago, the Bahamas and the Turks and Caicos Islands. The term Bahamas as used in this paper applies to all islands of the Bahamas archipelago. Figure 1 is an outline map of the Bahamas identifying all islands mentioned in the text. The Bahamas can be divided into two major zooogeographic regions (after Buden 1979), the northern and southern Bahamas, separated by the 2,000 m deep Crooked Island Passage located between Long and Crooked islands (see Fig. 1). The northern Bahamas are composed primarily of two large banks, the Great Bahama Bank and Little Bahama Bank. Six large islands are located on the Great Bahama Bank: Andros, New Providence, Eleuthera, Cat, Exuma, and Long. Grand Bahama and Abaco account for the vast majority of the land area on the Little Bahama Bank. The only islands of significant size in the northern Bahamas not located on these two banks are San Salvador and Rum Cay. The Great Bahama Bank is separated from the east coast of Florida by the Florida Straits, 75 km wide and 500-900 m deep, and from the northern coast of Cuba by the Old Bahama Channel, only 20 km wide but 500 m deep. The islands in the southern Bahamas are arranged in small groups including Crooked and Acklins, Great and Little Inagua, and the Caicos Islands, all of which are separated from one another and from Cuba and Hispaniola by deep water passages. There is no geologic evidence that the Bahamas have been connected to other land masses during the Cenozoic. The total land area of the Bahamas is 11,400 km^2, about the same size as Jamaica. However, the area of the shallow submerged banks is vastly greater than that of the present-day islands, on the order of 150,000 km^2.

The Bahamas are low-lying limestone islands with elevations seldom exceeding 20 m. A few ridges and hills reach heights of 30 to 60 m, with the highest point in the Bahamas being 67 m on Cat Island. The surface rocks in the Bahamas consist almost entirely of Pleistocene and Holocene limestone. The Tropic of Cancer passes through the center of the Bahamas, and thus the climate is generally tropical throughout the island group. There is a definite rainfall gradient in the Bahamas with the northernmost islands receiving nearly twice as much rainfall as the southern Bahamas, ranging from 1,300-1,500 mm per year on Grand Bahama, Abaco, and New Providence to about 740 mm on Great Inagua and Grand Turk (from Buden 1979). Based on the Holdridge Life Zone system the northern Bahamas are classified as tropical dry forest, while the southern Bahamas are very dry tropical forest (Buden 1979). Mesic or semi-mesic woodlands are a predominant feature of the northern islands. Pine forests are common on Andros, New Providence, and the two Little Bank islands. The predominant vegetation on the southern islands is xerophytic broadleaf scrub. Hurricanes pass through the Bahamas frequently, often coming from the south facilitating overwater dispersal of organisms from Cuba and Hispaniola.

One of the most important factors influencing the historical biogeography of the terrestrial vertebrate fauna of the Bahamas has been glacio-eustatic sea level fluctuations during the Pleistocene. During the maximum extent of the last or Wisconsinan glaciation approximately 17,000 years Before Present (yBP), it is estimated that sea level was 120 m lower than present (Gascoyne et al. 1979; Bloom 1983). A lowering of sea level of this magnitude would have consolidated most of the Bahamas into five major islands, each exceeding 2,000 km^2 in area. In addition, four large islands would have been created from shallow carbonate banks that are now almost completely submerged: the Cay Sal Bank between Cuba and Florida and the Mouchoir, Silver, and Navidad banks in the extreme southeastern Bahamas north of Hispaniola. Islands not located on extensive submarine banks, such as San Salvador, Rum Cay, Samana Cay, Mayaguana, and Great and Little Inagua would have increased only slightly in area. I have given names to the large paleoislands formed during the late Pleistocene low sea level stand, conforming to the names of the present-day submerged banks (see Fig. 1): Great Bahama Island (consisting of Andros, New Providence, Eleuthera, Cat, Long, and Exuma), Little Bahama Island (including Grand Bahama and Abaco), Crooked-Acklins Island, Caicos

Island (including Providenciales, and North, South, East, West, and Grand or Middle Caicos), and Turk Island. Between 14,000 and 6,000 yBP sea level rose rapidly reaching its current level in the early Holocene, fragmenting the large paleoislands into the smaller islands and innumerable cays that comprise the present topography of the Bahamas. A rise in sea level of as little as 5 m would inundate more than 50 percent of the present land area of the Bahamas. The most recent sea level rise of this magnitude occurred during the last interglacial about 120,000 yBP (Neumann and Moore 1975; Imbrie et al. 1983; Kaufman 1986).

Average sea surface temperatures in the Bahamas during the late Pleistocene glacial maximum were about 4°C cooler than present (Lynts and Judd 1971). During this glacial interval the West Indies and most lowland tropical regions in Central and northern South America had a more arid climate, and grassland savanna and scrub habitats were considerably more widespread (Van der Hammen 1974). Late Quaternary land vertebrate faunas from the Bahamas and elsewhere in the West Indies provide convincing evidence that these changes in climate and vegetation affected biogeographic patterns as well (Pregill and Olson 1981; Morgan and Woods 1986).

Glacio-eustatic changes in sea level dramatically altered the land area of the Bahamas throughout the Pleistocene and must have also had a profound affect on climate, habitats, and animal and plant distributions. Several very large paleoislands in the Bahamas were greatly reduced in area and fragmented into many smaller islands over a geologically short period in the latest Pleistocene and early Holocene. This island fragmentation provides an excellent example of vicariance, in which the ancestral faunas of these large paleoislands were bestowed upon numerous smaller islands. Specifically, in this paper I will compare the Late Quaternary mammalian fauna of Great Bahama Island to the Recent fauna of the six major Great Bahama Bank islands. The relationship between island area and species diversity of the Recent and Late Quaternary mammal faunas of the individual islands in the Bahamas will also be investigated.

Materials and methods

Several geographic names used throughout the paper are defined here to eliminate any confusion regarding their meaning. The term Bahamas applies to all islands of the Bahamas archipelago, including the politically separate Turks and Caicos Islands. Use of the island name Abaco includes both Great Abaco and Little Abaco and likewise Exuma includes both Great Exuma and Little Exuma. The largest and most centrally located of the Caicos Islands is identified as Grand Caicos on nearly every West Indian map I have examined, however, residents of the Caicos Islands and several authors (e.g. Buden) refer to this same island as Middle Caicos. The name Grand Caicos will be used in this paper in keeping with the most widely available maps. A complete list of the Recent and fossil mammals known from the Bahamas and their distribution on the individual Bahamian islands are presented in Table 1. I have not attempted an exhaustive taxonomic review of the Bahamian mammal fauna since the primary emphasis of this paper is on zoogeography. However, in order to assess the zoogeographic affinities of certain species of Bahamian mammals, several taxonomic problems are discussed. The number of taxonomic uncertainties that still exist in a reasonably well-studied group such as the Antillean bats emphasizes the importance of a sound taxonomic basis for biogeographic analysis. All previous literature on fossil mammals from the Bahamas is cited herein, although not all fossils described in these papers have been examined in the preparation of this report. The great majority of the fossils described in this paper are from seven cave deposits on Andros and New Providence and are housed in the Vertebrate Paleontology collections of the Florida State Museum, University of Florida (UF). I have also examined and measured the samples of *Geocapromys* from the Bahamas in the Museum of Comparative Zoology, Harvard University (MCZ) that were described by Lawrence (1934) and G. M. Allen (1937). The Smithsonian Institution has a small collection of *Geocapromys* and bats from Banana Hole on New Providence that was also examined during preparation of this report. Measurements were taken with either dial calipers accurate to 0.1 mm or a Gaertner measuring microscope accurate to 0.01 mm. Recent comparative material of Bahamian and other West Indian mammals is from the Florida State Museum, American Museum of Natural History (AMNH), and the U. S. National Museum of Natural History, Smithsonian Institution (USNM).

Description of fossil deposits

The Bahamas have not been thoroughly explored by vertebrate paleontologists even though these limestone islands have large numbers of caves, which constitute the primary sites for the accumulation of fossil vertebrates on other West Indian islands. Rich, well-sampled Late Quaternary fossil deposits have been discovered on only four islands in the Bahamas, all on the Great Bahama Bank: Andros, New Providence, Great Exuma and Little Exuma. Discussion of the extinct mammalian fauna of the Bahamas will include fossils from these four islands, along with a small sample of bones from Cat, and material from Amerindian archaeological sites on Great Abaco, Eleuthera, Long, Crooked, San Salvador, and Grand Caicos. Olson and Pregill (1982) summarized the available data on most of the published paleontological and archaeological sites from the Bahamas that had produced significant samples of vertebrates. They mentioned only one fossil site on New Providence and none from Andros.

My analysis of the Late Quaternary mammalian fauna of the Bahamas will concentrate primarily on the large and previously unstudied fossil mammal faunas from seven cave deposits on Andros and New Providence. Brief mention will also be made of a small sample of fossils collected in August 1960 from Turtle Scratch Cave on Cat Island by Oris Russell, who at that time was Chief Agricultural and Fisheries Officer for the island. Several new records of bats are reported from a cave deposit associated with an Amerindian archaeological site in Conch Bar Cave, Grand (=Middle) Caicos (Wing and Scudder 1983). Detailed field notes and map data for the fossil sites discussed below are available in the Vertebrate Paleontology locality files, Florida State Museum (FSM).

New Providence
Banana Hole.-- The most important and most thoroughly studied fossil vertebrate locality in the Bahamas is Banana Hole, located on the western end of New Providence at an elevation of about 15 m, immediately west of the main road and 0.5 km north of Clifton (25°01'N, 77°33'W). Banana Hole was discovered in the summer of 1958 by J. C. Dickinson of the FSM and was first excavated during August 1958 by Dickinson and Walter Auffenberg. Additional collections for the FSM were made in 1959 and 1960 by

William Hanon. In August 1978 a Smithsonian field crew consisting of Storrs L. Olson, Helen F. James, Frederick V. Grady, and Charles A. Meister reopened Banana Hole and removed most of the remaining sediments. Olson and Pregill (1982) reviewed the discovery and excavation of the Banana Hole fossil site. The Banana Hole is a limestone sinkhole 6 m deep, with an opening 12 m across. The sink is undercut on the northern and western sides for a distance of about 10 m. It was from this sheltered overhang that the majority of bones were recovered. The bones were deposited in a loose, unstratified, reddish-colored cave earth that formed a layer approximately 20 cm thick. As with all of the fossil sites discussed here, the sediments from Banana Hole were screened through 18 mesh (1.5 mm) window screen to obtain a complete sample of the microvertebrate fossils.

Mandibles, maxillae, isolated teeth, and postcranial elements of the capromyid rodent *Geocapromys* overwhelmingly dominate the fossil vertebrate fauna from Banana Hole. Smaller vertebrates, including frogs, lizards, snakes, birds, and bats are also relatively common in the deposit, along with rare remains of crocodilians and an extinct land tortoise. Brodkorb (1959) described the fossil birds collected during the early excavations at Banana Hole. Etheridge (1966) reviewed the fossil lizard fauna from Banana Hole and Auffenberg (1967) discussed the tortoise remains. In a recent review of the fossil vertebrates of the Bahamas (Olson, Ed. 1982), the amphibians and reptiles (Pregill 1982) and birds (Olson and Hilgartner 1982) from Banana Hole were discussed in detail, whereas the mammals were mentioned only briefly (Olson and Pregill 1982).

A sample of bones of *Geocapromys* from Banana Hole submitted for radiocarbon (^{14}C) analysis yielded a bone collagen date of 7,980 ± 230 yBP (Sample Number-Beta 24481). An early Holocene date for Banana Hole is somewhat younger than the late Pleistocene age previously proposed for this site based on faunal evidence (Brodkorb 1959; Olson and Pregill 1982). Because of the rarity of wood in most Antillean fossil deposits, radiocarbon dates have generally been obtained from either bone collagen or land snail shell carbonate (see review of West Indian radiocarbon dates in Morgan and Woods 1986). Taylor (1980) and Meltzer and Mead (1983) both note that there are analytical reasons to be somewhat skeptical of bone collagen dates. Collagen dates may differ significantly from dates run on other types of organic materials, generally tending to yield ages that are younger than those obtained from charcoal or wood.

The sediments in the Banana Hole deposit appear to be unstratified, a feature observed in many West Indian fossil deposits. Digging in loose cave sediments by land crabs may destroy any original stratification, and more importantly, may cause the mixing of fossils of different ages. Goodfriend and Mitterer (1987) demonstrated that sediments from one stratigraphic level in a cave in Jamaica contained a mixture of land snails of different ages. Furthermore, the ^{14}C date obtained from Banana Hole was not based on a single bone, but on a sample of several hundred hutia bones (long bones, scapulae, innominates, and vertebrae). If the sediments in Banana Hole were disturbed by land crabs the date of 7980 yBP may actually represent an "average" age for the hutia bones. The most accurate method for determining the age of *Geocapromys* from Banana Hole would be to obtain radiocarbon dates on individual bones from different levels using a tandem accelerator mass spectometer (TAMS). Financial constraints do not permit such an analysis at this time. There are two possible explanations for the early Holocene radiocarbon date from Banana Hole: 1) the bone collagen date is either too young or represents a mixture of ages, and should probably be disregarded; or 2) the early Holocene age for the Banana Hole deposit is accurate and the major faunal extinctions observed among mammals and birds from the site must be explained in this context.

Taking into account recent studies on sea levels (Bloom 1983), oxygen isotopes (Mix 1987), and glacial ice volumes (Ruddiman 1987) during the latest Pleistocene and early Holocene in North America and surrounding oceans, it seems reasonable to accept an early Holocene age for Banana Hole. Significant deglaciation apparently did not begin until 14,000 yBP and ended by 6,000 yBP, with the most rapid glacial melting occurring between 13,000 and 9,500 yBP (Ruddiman 1987). These studies indicate that sea levels probably did not attain modern levels until sometime around 6,000 yBP. Extrapolating from the sea level curves of Bloom (1983) and ice volume models of Ruddiman (1987), sea level at 8,000 yBP was probably still at least 10 m below present msl. Since most depths on the Great and Little Bahama Banks are now considerably less than 10 m, these banks were probably still dry land when the Banana Hole fossil deposit formed in the early Holocene. The inundation of these now-submerged banks and the flooding of the extensive cave systems hypothesized to have existed in the Bahamas at this time (see discussion in Zoogeography section), happened sometime after 8,000 yBP. The widespread savanna and grassland habitats that were postulated to have occurred in the Bahamas during the late Pleistocene (Olson and Hilgartner 1982; Pregill 1982) probably also persisted into the early Holocene. If this chronology of events is accurate, the numerous extinctions of mammals and birds on the islands of the Great Bahama Bank (or Great Bahama Island) occurred in the early Holocene, rather than at the end of the Pleistocene as had been previously assumed (Pregill and Olson 1981; Olson and Pregill 1982).

Morgan and Woods (1986) predicted that many West Indian fossil sites generally regarded as late Pleistocene in age may actually be early Holocene instead. The age of most vertebrate fossil sites in the West Indies has been inferred from faunal associations rather than from actual radiocarbon dates (e.g. previous estimates for the age of Banana Hole). Sites that lack evidence of human occupation and have no introduced mammals are usually considered to be late Pleistocene in age, even though these two factors characterize Antillean early Holocene sites as well. The earliest reliable date for the appearance of humans in the Caribbean is about 4,500 yBP (Rouse and Allaire 1978). Therefore, unlike the situation in North America, terminal Pleistocene climatic changes do not coincide with the first arrival of humans into the Caribbean. Most terrestrial mammals in the West Indies appear to have survived the late Pleistocene extinction event that characterizes North American vertebrate faunas (Morgan and Woods 1986). Without radiocarbon dates it is probably not possible to distinguish late Pleistocene and early Holocene vertebrate faunas in the West Indies. For this reason, the undated vertebrate fossil deposits from the Bahamas discussed in this paper, including all sites except Banana Hole, are considered Late Quaternary in age, a broader time designation that includes both the late Pleistocene and early Holocene.

Hunts Cave.-- Hunts Cave is located about 0.5 km south of Harold Pond and 0.8 km north of Carmichael road at an elevation of about 20 m (25°02'N, 77°22'W). The cave was discovered in August 1958 by J. C. Dickinson and Walter Auffenberg. The small entrance to the cave is located in a low cliff about 3 m high. The cave opens into several long narrow passageways. The most productive area for fossils was in a pit in the limestone floor located about 20 m from the entrance. The reddish cave sediments from this small, shallow pit produced mostly remains of small vertebrates such as snakes, lizards, and bats, along with isolated teeth and fragmentary postcranial elememts of *Geocapromys*. The vertebrate fauna from Hunts Cave is particularly rich in cave-dwelling bats.

Sir Harry Oakes Cave.-- This cave is situated near Caves Point about 0.4 km inland from the northern coast of New Providence and 0.2 km north of Interfield Road (25°04'N, 77°27'W) at an elevation of about 15 m. Fossils were first discovered in Sir Harry Oakes Cave by J. C. Dickinson in the summer of 1958. The cave consists essentially of a single small room. The powdery reddish sediments from this cave produced a small sample of *Geocapromys*, lizards, and bats.

East Cave.-- This cave is located near the eastern end of the island just inland from Winton at an elevation of about 20 m (25°03'N, 77°16'W). It was discovered in August 1959 by J. C. Dickinson and W. J. Riemer. Like Sir Harry Oakes Cave, East Cave consists of a single room. Only a small sample of fossils was obtained from East Cave composed almost entirely of microvertebrates, particularly frogs, lizards, and the Bahamian funnel-eared bat, *Natalus tumidifrons*.

Andros

King Cave.-- Three cave deposits have been excavated on Andros by Florida State Museum field crews, King Cave, Ashton Cave, and Coleby Bay Cave, all located within 0.5 km of one another in the northeastern corner of the island about 0.3 km inland from Morgans Bluff (25°10'N, 78°02'W). Wayne King discovered the first of these fossil deposits, here named King Cave, in July 1961. The small entrance to the cave is between 1 and 2 m in height and 2 m across. A small fossiliferous deposit was located about 50 m from the cave entrance and 30 m from the nearest vertical opening to the surface. The bones occurred in a yellowish-orange sediment about 5 cm below the surface. In August 1964 King found a second bone deposit in this same cave nearly 100 m from the main entrance, but less than 10 m from a large vertical chimney that opened to the surface. Bones were found from the surface to a depth of about 20 cm in a small pocket of buff to reddish-colored sediments. Except for a small sample of *Geocapromys* mandibles, maxillae, and limbs, bones of large vertebrates are uncommon in King Cave. Small vertebrates are more abundant, including frogs, snakes, lizards, and a rich sample of bats. Auffenberg (1967) reported a single bone of the extinct land tortoise *Geochelone* from this cave, which he called Cave 1 following King's field notes.

Ashton Cave.-- This cave was originally discovered in August 1980 by Ray Ashton who conducted preliminary excavations there. Ashton's discovery was the impetus for two FSM expeditions to Andros the following year. Margaret Langworthy and Laurie Wilkins excavated Ashton Cave in April 1981 and Richard Franz led a crew who dug the cave in June 1981. The entrance to the cave is between 3 and 4 m across and the cave floor is located 3 m below the surface. The fossiliferous deposits were all located within about 5 m of the entrance. The fossils occurred in cream to buff-colored powdery sediments and were dug to a maximum depth of 30 cm. In one test pit between the surface and 15 cm, fresh-looking bones of *Geocapromys* were found associated with teeth and bones of the domestic pig (*Sus scrofa*) and glass, suggesting that hutias survived on

Andros into historic times. Bones of small vertebrates, including snakes, lizards, birds, and bats are also common in Ashton Cave.

Coleby Bay Cave.-- Coleby Bay Cave was discovered by Margaret Langworthy and Laurie Wilkins in April 1981. The cave has two entrances, the largest of which is 1 m in height and about 5 m across. Bones of hutias were found on the surface throughout Coleby Bay Cave, but the only concentrations occurred in the entrance areas. Two pits dug to a depth of 25 cm in the smaller entrance contained *Geocapromys* and bats, although bones of small vertebrates are uncommon in this cave. A test pit dug to 30 cm in the main entrance produced mostly mandibles and postcrania of *Geocapromys*, all of which were covered with a thick deposit of calcium carbonate.

Systematic paleontology

Order Chiroptera
Family Mormoopidae

Mormoops blainvillii Leach

Material examined.-- NEW PROVIDENCE: Banana Hole- 5 mandibles (UF 27612-27613, 79861-79863), 3 isolated upper molars (UF 79864-79866), 2 proximal and 5 distal ends of humeri (UF 79867-79873), 1 proximal end of radius (UF 79874); Hunts Cave- 24 mandibles (UF 27691, 27714, 79466-79479, 79644, 79813-79819), 3 isolated upper molars (UF 79487-79489), 7 isolated lower molars (UF 79480-79486), 1 proximal and 6 distal ends of humeri (UF 79490-79495, 79820), 2 proximal and 2 distal ends of radii (UF 79496-79499); Sir Harry Oakes Cave- left M^2 (UF 79770), 1 proximal and 2 distal ends of humeri (UF 79771-79772, 79812).

Remarks.-- *Mormoops blainvillii* is endemic to the West Indies where it is now restricted to the four Greater Antilles. This species had a considerably wider distribution in the Late Quaternary (Morgan and Woods 1986), as it has been reported from five islands where it no longer occurs, including Antigua (Steadman et al. 1984) and Barbuda (Morgan and Woods 1986) in the northern Lesser Antilles, La Gonave off the west coast of Haiti Koopman (1955), and two islands in the Bahamas. Koopman (1951) first recorded *M. blainvillii* from the Bahamas based on a single mandible from a cave fossil deposit on Exuma Island. Hecht (1955) clarified that this specimen was collected from Upper Pasture Cave on Little Exuma. Koopman (cited in Olson and Pregill 1982) also identifed *M. blainvillii* from Banana Hole on New Providence. *Mormoops blainvillii* is here reported from three cave fossil deposits on New Providence. This species is uncommon in Banana Hole where only three individuals have been identified. It is considerably more abundant in Hunts Cave with 45 elements representing at least 12 individuals.

Smith (1972) considered all populations of *Mormoops blainvillii* to be monotypic, although he did recognize a size cline within the species, with the smallest bats being from Cuba and the largest from Hispaniola. The fossil mandibles from New Providence are most similar in size to the smaller Cuban individuals of *M. blainvillii* (see measurements, Table 2). The extinct populations of *M. blainvillii* in the Bahamas are both from islands on the Great Bahama Bank, which would also favor a Cuban rather than a Hispaniolan origin. This species is an obligate cave-dwelling bat that is found in both the stable and variable cave microclimates defined by Silva Taboada and Pine (1969) and discussed in more detail beyond.

Mormoops megalophylla Peters

Material examined.-- ANDROS: <u>Coleby Bay Cave</u>- 1 distal end of humerus (UF 79323).

Remarks.-- *Mormoops megalophylla* is unknown in the Recent fauna of the West Indies, its modern distribution being limited to the mainland from southern Arizona and Texas south through Mexico and Central America to northern South America and Trinidad (Smith 1972). Fossils of *M. megalophylla* were first reported from the West Indies by Silva Taboada (1974) who identified this species from two late Pleistocene cave deposits in Cuba. Morgan and Woods (1986) added fossil records of this species from Hispaniola and the Bahamas. A single specimen of *M. megalophylla* from a cave deposit in Jamaica has recently been discovered in the UF vertebrate fossil collection. The occurrence of *M. megalophylla* in the Bahamas is based on a partial humerus from a Late Quaternary cave deposit on Andros. The highly distinctive morphology of the distal end of this humerus readily identifies it as *Mormoops*, particularly the deep furrow between the trochlea and capitulum and the strong ridge on the posterodistal portion of the shaft just proximal to the articular surface. Measurements of the distal end of the humerus of Recent and fossil specimens of *M. megalophylla* and *M. blainvillii* are presented in Table 3. The distal humerus from Coleby Bay Cave and fossil humeri from Cerro de San Francisco in the Dominican Republic (UF 96200) and Swansea Cave in Jamaica (UF 102900) are much larger than any Recent or fossil specimen of *M. blainvillii*, but are very close in size to Recent *M. megalophylla* and to several fossil humeri referred to *M. megalophylla* from Rock Springs, a late Pleistocene cave deposit in central Florida (Ray et al. 1963).

Mormoops megalophylla is one of the rarest bats in West Indian fossil deposits. The records from Andros, Dominican Republic, and Jamaica are based on single specimens, while the larger sample from Cuba is represented by specimens from only two caves among hundreds of Cuban caves that have produced fossils. Two extinct species of mormoopids, *Mormoops magna* and *Pteronotus pristinus*, were described from the same Late Quaternary cave deposits in Cuba from which *M. megalophylla* was initally found (Silva Taboada 1974). The more diverse mormoopid fauna in Cuba and the Bahamas during the late Pleistocene indicates that environmental changes have contributed to the extinction and range contraction of certain of the species. Other fossil occurrences of *M. megalophylla* from outside the modern range of the species include the late Pleistocene of Tobago (Eshelman, and Morgan 1985) and Florida (Ray et al. 1963).

Pteronotus macleayii (Gray)

Material examined.-- NEW PROVIDENCE: <u>Hunts Cave</u>- 10 mandibles (UF 79635-79643, 79833), 1 M^2 (UF 79648), 3 isolated lower molars (UF 79645-79647).

Remarks.-- *Pteronotus macleayii* is currently restricted to Cuba, Isla de Pinos, and Jamaica. The fossils from Hunts Cave on New Providence represent not only the first record of *P. macleayii* from the Bahamas, but also the first extinct population of this species. The morphology of the lower molars, along with the presence of a tiny button-like P_3 wedged between the larger P_2 and P_4, readily identifies these mandibles as *Pteronotus*. The Bahamian fossils are much smaller than *P. parnellii*, comparing most closely in size and morphology with the two small Antillean species, *P. macleayii* and *P. quadridens*. These two species differ primarily in size, with *P. macleayii* being larger. In Table 4, measurements of fossil mandibles from Hunts Cave are compared to those of Recent *P. macleayii* and *P. quadridens* from Cuba. The dental measurements of the Bahamian fossils are within the observed range for *P. macleayii* and are considerably larger than *P. quadridens*. There are no apparent morphological differences between the fossils from New Providence and Recent specimens of *P. macleayii* from Cuba, and the Bahamian specimens are confidently referred to that species. Cuba is the obvious source for the Bahamian population of *P. macleayii* since the species does not occur on Hispaniola. *Pteronotus macleayii* roosts exclusively in caves in Cuba (Silva Taboada 1979) and Jamaica (Goodwin 1970), particularly in chambers deep within extensive cave systems characterized by high temperatures and humidity, (Silva Taboada and Pine 1969).

Pteronotus parnellii (Gray)

Material examined.-- NEW PROVIDENCE: Banana Hole- 40 mandibles (UF 27596, 27614, 27663-27664, 27815-27816, 79875-79908), 6 C^1 (UF 79909), 12 isolated upper cheek teeth (UF 79910-79921), 27 periotics (UF 79922-79948), 12 proximal and 23 distal ends of humeri (UF 79949-79983); Hunts Cave- 114 mandibles (UF 27696-27706, 79501-79595, 79821-79828), 1 maxillary fragment (UF 79500), 14 isolated upper cheek teeth (UF 79596-79606, 79829-79831), 25 periotics (UF 79607-79631), 3 distal ends of humeri (UF 79632-79633, 79832), 1 proximal end of radius (UF 79634); Sir Harry Oakes Cave- 1 periotic (UF 79773).

Remarks.-- *Pteronotus parnellii* occurs widely throughout tropical America from Mexico south through Central America to northern South America, and including the four Greater Antilles. Morgan and Woods (1986, fig. 3) pointed out that *P. parnellii* had a much wider distribution in the West Indies during the late Pleistocene. Extinct populations of this species are known from Isla de Pinos (Silva Taboada 1979), La Gonave (Koopman 1955), Antigua (Steadman et al. 1984), Grand Cayman (Morgan in press a; Morgan and Woods 1986), and New Providence in the Bahamas (Fig. 2). An extinct population of *P. parnellii* is also known from Tobago (Eshelman and Morgan 1985). Koopman first identified *P. parnellii* from the Bahamas (cited in Olson and Pregill 1982) based on several mandibles from Banana Hole. Morgan and Woods (1986) reported *P. parnellii* from New Providence on the basis of the extensive material described here. *Pteronotus parnellii* is the most abundant bat in Hunts Cave, represented by a minimum of 67 individuals. It is also very common in Banana Hole where at least 25 individuals are present. Considering the abundance of *P. parnellii* in Hunts Cave and Banana Hole, its absence from similar cave deposits on Andros is puzzling.

The largest Recent representatives of *Pteronotus parnellii* in the West Indies are the nominate subspecies found on Jamaica and Cuba, while the smallest known living form *P. parnellii pusillus* occurs on Hispaniola (Smith 1972, Table 1). Fossil mandibles of *P. parnellii* from Hunts Cave and Banana Hole are very similar in size to mandibles from Cuba and are considerably larger than Hispaniolan specimens. The extinct Bahamian representatives of this species were clearly derived from Cuba.

Pteronotus parnellii roosts exclusively in caves in the West Indies, preferring the hot, humid microclimate found in the largest chambers of extensive cave systems in Cuba and Jamaica (Goodwin 1970; Silva Taboada 1979). This species is occasionally found in smaller caves in Cuba that exhibit a stable but more temperate microenvironment. The somewhat less specialized roost ecology of *P. parnellii* has probably been a factor in the wider distribution of this species compared to most other mormoopids, specifically *P. macleayii* and *P. quadridens*, which appear to be restricted to the hot phase of the stable microenvironment (Silva Taboada and Pine 1969).

Pteronotus quadridens (Gundlach)

Material examined.-- ANDROS: King Cave- 1 complete and one distal end of humerus (UF 79072, 79207); NEW PROVIDENCE: Banana Hole- 1 proximal and 1 distal end of humerus (UF 79984-79985); Sir Harry Oakes Cave- 3 mandibles (UF 79774-79776), 1 distal end of humerus (UF 79777), 2 proximal ends of radii (UF 79778-79779).

Remarks.-- *Pteronotus quadridens* is a Greater Antillean endemic having a somewhat wider distribution than *P. macleayii*, occurring on Hispaniola and Puerto Rico, as well as Cuba and Jamaica. The extinct populations of *P. quadridens* on Andros and New Providence represent the first records of this species from the Bahamas and the first occurrence of *P. quadridens* outside of the Greater Antilles. *Pteronotus quadridens* is more commonly known as *P. fuliginosus* (e.g. Smith 1972). Silva Taboada (1976) discussed the nomenclatural status of *Lobostoma quadridens* Gundlach, demonstrating that this was the oldest available name for the smallest species of Antillean *Pteronotus* previously called *P. fuliginosus*. Hall (1981) and other recent authors have followed Silva Taboada in recognizing the name *quadridens* for this species.

Measurements of the three small *Pteronotus* mandibles from Sir Harry Oakes Cave, New Providence are presented in Table 4, along with comparative measurements of Recent *P. quadridens* and *P. macleayii* from Cuba. The Bahamian fossils are within the size range of *P. quadridens* and are smaller than both Recent *P. macleayii* and fossils from Hunts Cave referred to the latter species. Smith (1972) recognized two subspecies of *P. fuliginosus* (=*P. quadridens*), the smaller *P. q. quadridens* from Cuba and the larger *P. quadridens fuliginosus* from Jamaica, Hispaniola, and Puerto Rico. The three Bahamian mandibles are very similar in size to Recent mandibles of *P. q. quadridens* from Cuba, suggesting that the extinct populations of this species from Andros and New Providence were probably derived from Cuba, the most probable source on geographic grounds as well. Like *Pteronotus macleayii*, *P. quadridens* prefers to roost in hot, humid, climatically stable microenvironments deep within extensive cave systems in the Greater Antilles (Goodwin 1970; Silva Taboada 1979).

Family Phyllostomidae
Macrotus waterhousii Gray

Material examined.-- ANDROS: Ashton Cave- partial skull (UF 79339), 3 mandibles (UF 79341, 79349-79350), 4 complete and 4 proximal and 3 distal ends of humeri (UF 79340, 79342-79347, 79351-79354), 1 complete and 1 distal end of radius (UF 79337,

79348); Coleby Bay Cave- 3 mandibles (UF 79326-79327, 79329), 1 distal end of humerus (UF 79328); King Cave- 4 mandibles (UF 79073-79075, 79295), 1 distal end of humerus (UF 79076), 2 proximal ends of femora (UF 79209-79210); NEW PROVIDENCE: Banana Hole - 113 mandibles (UF 27597-27603, 27605-27611, 27615-27622, 27636-27643, 27651-27662, 27665-27666, 27679-27686, 27780-27810, 27820-27849); East Cave- 3 mandibles (UF 27727-27729), 1 maxilla (UF 99269), 1 proximal and 3 distal ends of humeri (UF 99273-99276), 6 proximal and 2 distal ends of radii (UF 99277-99284), 6 proximal and 3 distal ends of femora (UF 99285-99293); Hunts Cave- 71 mandibles (UF 27695, 27707-27713, 27715-27729, 79660-79705, 79835-79836), 12 maxillae (UF 79649-79659, 79834), 2 proximal and 10 distal ends of humeri (UF 79706-79715, 79837-79838), 3 proximal and 3 distal ends of radii (UF 79716-79721); Sir Harry Oakes Cave- 10 mandibles (UF 3457, 79780-79788), 1 distal end of humerus (UF 79789), 1 proximal and 1 distal end of radius (UF 79790-79791); GRAND (=MIDDLE) CAICOS)- Conch Bar Cave (FSM zooarchaeology site MC-16)- rostrum (FS 466), 1 maxilla (FS 246), 2 mandibles (FS 238, 248), 1 complete humerus (FS 459).

Remarks.-- *Macrotus waterhousii* is the second most widespread bat in the Bahamas, occurring on 16 islands (Table 1; Buden, 1986). Koopman et al. (1957) reported fossil *M. waterhousii* from Little and Great Exuma. This species is the most abundant and widespread bat in fossil deposits on Andros and New Providence. *M. waterhousii* is here identified from three cave deposits on Andros, four on New Providence, and one on Grand Caicos. According to Buden (1975a, 1985), *M. waterhousii* has not been previously collected on Grand (=Middle) Caicos, although it does occur on three other islands in the Caicos group. Other extralimital fossil records of *M. waterhousii* are from islands east of the species' present range, including Puerto Rico (Choate and Birney 1968) and Barbuda in the northern Lesser Antilles (Morgan and Woods 1986).

In a taxonomic review of Antillean *Macrotus waterhousii*, Buden (1975a) recognized two subspecies in the Bahamas, the smaller *M. waterhousii minor* in the northern Bahamas and the larger nominate race *M. w. waterhousii* in the southern Bahamas. Specimens of *M. waterhousii* from the northern Bahamas, formerly recognized as an endemic Bahamian subspecies *M. waterhousii compressus*, are somewhat larger than typical Cuban *M. waterhousii minor*, but are distinctly smaller than *M. w. waterhousii* from the southern Bahamas and Hispaniola (Buden 1975a). Fossil specimens from Andros and New Providence are indistinguishable from the Recent populations of *M. waterhousii* still surviving on those islands. Fossil mandibles of *M. waterhousii* from Conch Bar Cave on Grand Caicos are larger than those from Andros and New Providence and appear to be referable to the Hispaniolan subspecies. Buden (1975a) postulated a dual origin for *M. waterhousii* in the Bahamas, with the smaller bats in the northern Bahamas derived from Cuban *M. waterhousii minor* and the larger southern Bahamian forms derived from Hispaniolan *M. w. waterhousii*. Fossils of *M. waterhousii* from both the northern and southern Bahamas support Buden's biogeographic scenario.

Based on my personal observations in the Cayman Islands and those of Buden (1975a) in the Bahamas and Silva Taboada (1979) in Cuba, *Macrotus waterhousii* roosts in caves characterized by a variable microclimate, generally in twilight areas near an entrance. This species also roosts in houses and other man-made structures.

Monophyllus redmani Leach

Material examined.-- ANDROS: King Cave- 11 mandibles (UF 79078-79084, 79212-79215), 1 partial rostrum (UF 79077), 1 maxillary fragment (UF 79211), 4 complete and 4 proximal and 8 distal ends of humeri (UF 79085-79092, 79216-79223), 13 proximal ends of radii (UF 79093-79097, 79224-79231), 8 complete and 9 proximal ends of femora (UF 79098-79105, 79232-79240); Ashton Cave- 2 mandibles (UF 79355, 79369), 12 complete and 3 proximal and 4 distal ends of humeri (UF 79356-79368, 79370-79374, 79377), 2 proximal ends of radii (UF 79375-79376); NEW PROVIDENCE: Banana Hole- 4 mandibles (UF 96001-96004), 2 complete and 4 proximal and 19 distal ends of humeri (UF 96005-96029); East Cave- 2 mandibles (UF 27730-27731); Hunts Cave- 1 proximal end of radius (UF 79722); GRAND (=MIDDLE) CAICOS: Conch Bar Cave (FSM zooarchaeology site MC-16)- 1 complete and 1 proximal end of humerus (FS 239, 276), 1 proximal and 1 distal end of radius (FS 247, 459).

Remarks.-- *Monophyllus redmani* was first reported from the Bahamas by Buden (1975b) who collected specimens on five islands in the southern Bahamas, including Crooked, Acklins, North Caicos, Grand Caicos, and Providenciales. With the exception of these Bahamian records, *M. redmani* is otherwise restricted to the four Greater Antilles. *Monophyllus redmani* is one of the more common bats in fossil deposits on Andros and New Providence, where it has been identified from five caves. This material represents the first known extinct populations of *M. redmani*. The Bahamian fossils of *M. redmani* compare very closely to Recent individuals of this species from the southern Bahamas and the Greater Antilles.

Schwartz and Jones (1967) and Buden (1975b) referred the Cuban and Hispaniolan representatives of *Monophyllus redmani* to the subspecies *M. redmani clinedaphus*. Buden noted that the Bahamian specimens of *M. redmani* are somewhat more similar to Cuban individuals than to those from Hispaniola. The differences are slight, however, and although Buden favored a Cuban origin for Bahamian *M. redmani*, he could not completely rule out derivation from Hispaniola. *Monophyllus redmani* is an obligate cave-dwelling bat generally found in the climatically stable microenvironment in caves in the Greater Antilles (Silva Taboada and Pine 1969). This species often roosts in association with *Pteronotus parnellii* (Goodwin 1970; Silva Taboada 1979). Buden (1975b:372) found *M. redmani* in "...relatively large, usually deep, well-aerated caves" in the southern Bahamas.

Brachyphylla nana Miller

Material examined.-- ANDROS: King Cave- 1 mandible (UF 79106), 1 proximal end of radius (UF 79241); Ashton Cave- 1 proximal end of radius (UF 79376); Coleby Bay Cave- 1 mandible (UF 79330); NEW PROVIDENCE.--Banana Hole- 1 mandible (UF 96036), 1 M_3 (UF 96037), 1 C_1 (UF 27817); GRAND (=MIDDLE) CAICOS: Conch Bar Cave (FSM zooarchaeology site MC-16)- 7 mandibles (FS 247, 248, 276, 459, 462), 1 complete and 3 distal ends of humeri (FS 246, 248, 685), 2 proximal ends of radii (FS 459, 462).

Remarks.-- There is little agreement among recent authors on the taxonomic status of the various populations in the endemic Antillean genus *Brachyphylla*. Swanepoel and

Genoways (1978) and Silva Taboada (1976; 1979) recognize the smaller Greater Antillean form *B. nana* as a distinct species from the larger *B. cavernarum* found in Puerto Rico, the Virgin Islands, and Lesser Antilles. Varona (1974), Buden (1977) and Hall (1981) place all populations of *Brachyphylla* in the species *B. cavernarum*. I follow Silva Taboada (1976) who considers *B. cavernarum* and *B. nana* to be distinct species and recognizes two subspecies of *B. nana*, *B. n. nana* from Cuba, Isla de Pinos, and Grand Cayman and *B. nana pumila* from Hispaniola and Grand Caicos. Although Swanepoel and Genoways (1978) synonymized *B. nana pumila* with *B. n. nana*, there appear to be several morphological characters that reliably distinguish them. *Brachyphylla n. nana* has a broader rostrum than *B. nana pumila*, reflected in the greater breadth across the canines and upper molars. The Cuban form has a flat frontal profile, whereas in *B. n. pumila* the maxillae are inflated anterior to the orbit giving a distinct convexity to the cranial profile. The morphology of the upper incisors differs slightly between the two forms. The protoconule is lacking on the M^1 in *B. n. nana*, but is present in *B. n. pumila*. Both the upper and lower molars are essentially smooth in *B. n. nana* while in *B. nana pumila* the enamel is highly crenulated. Although I am tentatively regarding *nana* and *pumila* as subspecies of *B. nana*, a more thorough morphological analysis may show them to be distinct species. Whether these two forms are recognized at the species or subspecies level is not critical to this analysis. The most important point is that *B. n. nana* and *B. nana pumila* can be confidently distinguished at some taxonomic level, thereby allowing a more precise determination of zoogeographic relationships.

Brachyphylla nana is now restricted to Cuba, Isla de Pinos, Grand Cayman, Hispaniola, and Grand Caicos. Extinct populations of this species have been reported from Jamaica (Koopman and Williams 1951), Cayman Brac (Morgan and Woods 1986), and the Bahamas. Koopman (cited in Olson and Pregill 1982) first reported *Brachyphylla* from the northern Bahamas based upon a single canine from Banana Hole. This tooth and several more complete fossils from Andros and New Providence are here referred to *B. n. nana*. Measurements of fossil mandibles from Andros, New Providence, and Grand Caicos are compared to Recent mandibles of *B. n. nana* from Cuba and *B. nana pumila* from Hispaniola and Grand Caicos in Table 5. The fossils from the northern Bahamas are closest in size and dental morphology to Recent specimens of *B. n. nana*. Recent mandibles of *B. nana pumila* average somewhat larger than the Cuban subspecies. The three fossil mandibles from Conch Bar Cave on Grand Caicos are distinctly larger than those of Recent *B. nana pumila* from Grand Caicos and Hispaniola, but do have the crenulated enamel and other dental features of the Hispaniolan form (Table 5). On the basis of these data it appears that the northern and southern Bahamian representatives of *Brachyphylla nana* had a different origin. The small individuals from fossil deposits on Andros and New Providence are derived from Cuban *B. n. nana*, while the larger individuals with crenulated enamel from Grand Caicos are Hispaniolan in origin.

Extinct populations of *Brachyphylla nana* from Jamaica and Cayman Brac also average larger than Recent representatives of *B. nana* (Morgan 1977; in press a). They are thus intermediate in size between Recent *B. nana* and *B. cavernarum*. The large fossil specimens from Grand Caicos, Jamaica, and Cayman Brac are tentatively referred to *B. nana pumila*, but suggest a more complex evolutionary history for *Brachyphylla* than has been previously recognized. *Brachyphylla nana* roosts exclusively in caves, and in Cuba favors the hot phase of the stable microenvironment in large cave systems where it is often found in close association with *Erophylla sezekorni* and *Phyllonycteris poeyi* (Silva Taboada 1979). Buden (1977) found *B. nana* on Grand Caicos inhabiting a large cave that was relatively cool and well aerated.

Erophylla sezekorni (Gundlach *in* Peters)

Material examined.-- ANDROS: King Cave- 1 mandible (UF 79243), 2 maxillary fragments (UF 79107, 79242); Ashton Cave- 9 mandibles (UF 79379, 79381-79382, 79396-79400, 79406), 1 partial rostrum (UF 79395); 5 complete and 2 proximal and 8 distal ends of humeri (UF 79332-79334, 79380, 79383-79390, 79401-79402), 1 complete and 4 proximal ends of radii (UF 79391-70394, 79407), 2 complete and 1 distal end of femur (UF 79403-79405); NEW PROVIDENCE: Banana Hole- 5 mandibles (UF 27604, 27624, 96032-96034), 1 M^1 (UF 96035); East Cave- 2 C^1 (UF 79851); Hunts Cave- 12 mandibles (UF 27692-27693, 27877-27878, 79734-79741), 4 maxillary fragments (UF 79730-79733); Sir Harry Oakes Cave- 1 mandible (UF 79797).

Remarks.-- Buden (1976) synonymized the two previously recognized species of *Erophylla*, *E. sezekorni* and *E. bombifrons*. I agree with Hall (1981) that the broader rostrum, less inflated braincase, longer ears, and paler coloration of *E. sezekorni* readily distinguish this species from *E. bombifrons* of Hispaniola and Puerto Rico. Whether these two forms are considered as distinct species as I believe, or as well-marked subspecies according to Buden (1976), the animals from the Bahamas clearly belong to the Cuban form *E. sezekorni*.

Erophylla sezekorni is the most widespread bat in the Bahamas where it has been recorded from 17 islands, from Great Abaco in the north to East Caicos in the extreme southeastern portion of the Bahamas archipelago (Table 1; Buden 1985). This species is here identified from two fossil deposits on Andros and four on New Providence. It is uncommon in most of these cave deposits except Ashton Cave on Andros. No appreciable differences could be detected between these fossils and Recent individuals of *E. sezekorni* from islands on the Great Bahama Bank. On Great and Little Exuma Koopman et al. (1957:166) found this species "..only in the farthest end of each cave where it was darkest." In the Bahamas, Buden (1976:3) "... found *E. sezekorni* on exposed surfaces and deep within solution cavities on the ceilings of caves, often in areas where much daylight penetrated." In Cuba and Jamaica *E. sezekorni* is generally found in hot, humid caves exhibiting the stable microenvironment (Goodwin 1970; Silva Taboada 1979).

Phyllonycteris poeyi Gundlach

Material examined.-- NEW PROVIDENCE: Banana Hole- 1 mandible (UF 96031); Hunts Cave- 12 mandibles (UF 27879-27880, 79723-79729, 79839-79840, 79849); Sir Harry Oakes Cave- 5 mandibles (UF 79792-79796).

Remarks.--*Phyllonycteris* and the related genus *Erophylla* are readily distinguished by cranial features, such as the incomplete zygomatic arch of *Phyllonycteris*. However, the majority of fossil specimens of these two genera from Bahamian cave deposits consist of fragmentary mandibles. There are several characters that reliably separate mandibles of *Phyllonycteris poeyi* from those of *Erophylla sezekorni*. The mandibular ramus of *P. poeyi* is noticeably deeper and more robust, the lower canine is larger particularly in the antero-posterior dimension, the alveolar length of the molars is noticeably shorter, and the M_2 and M_3 are distinctly curved medially such that the M_3 is located medial to the base of the coronoid process. The lower molars of the two genera are quite different. The M_1 of *Phyllonycteris* is relatively shorter and the M_2 and M_3 are broad with a flat occlusal surface lacking cusps. *Erophylla* has an elongated M_1 and the M_2 and M_3 are narrow, each having a strong anterior and posterior cusp.

Phyllonycteris poeyi is now restricted to Cuba and Isla de Pinos, unless the closely related species *P. obtusa* from Hispaniola is considered conspecific (see Koopman, this volume). Morgan and Woods (1986) first reported *P. poeyi* from the Bahamas based on the fossils from New Providence listed above. No differences could be discerned between the fossil *Phyllonycteris* mandibles from New Providence and Recent Cuban *P. poeyi*. The Bahamian fossils are confidently referred to *P. poeyi*, and undoubtedly reached the Great Bahama Bank from Cuba. The only other fossil record of *P. poeyi* from outside the present range of the species is from Cayman Brac (Morgan in press a; Morgan and Woods 1986). A large extinct species of *Phyllonycteris*, *P. major*, was described from the late Pleistocene of Puerto Rico (Anthony 1917), and was recently reported from a late Holocene deposit on Antigua in the northern Lesser Antilles (Steadman et al. 1984).

Phyllonycteris poeyi is the most specialized cave-dwelling bat in Cuba (Silva Taboada and Pine 1969) where it invariably occupies the hot phase of the stable cave microenvironment ("cuevas calientes" of Silva Taboada 1977, 1979). This species is usually found in blind galleries in the innermost regions of extensive cave systems where the temperature and humidity are extremely high and airflow is imperceptible. Colonies may be huge, often numbering up to a million bats. In these hot caves *P. poeyi* roosts in close association with mormoopids, particularly *Pteronotus quadridens*, and with *Brachyphylla nana* and *Erophylla sezekorni*. According to Silva Taboada (1979), *P. poeyi* is found in much smaller numbers in two comparatively small caves on Isla de Pinos, both of which lack the hot phase of the stable microenvironment. The fossil samples of *P. poeyi* from New Providence are small, seven individuals from Hunts Cave, three from Sir Harry Oakes Cave, and one from Banana Hole. This, along with the small size of these caves, suggests that the roosting ecology of *P. poeyi* in the Bahamas was probably more similar to that of *P. poeyi* on Isla de Pinos.

Family Natalidae
Natalus major Miller

Material examined.-- ANDROS: King Cave- 1 mandible (UF 79110); Coleby Bay Cave- 1 mandible (UF 79324), 1 proximal end of femur (UF 79325); NEW PROVIDENCE: Banana Hole- 5 mandibles (UF 79986-79990), 1 M^1 (UF 79991), 2 distal ends of humeri (UF 79992-79993); Hunts Cave- 15 mandibles (UF 27694, 79751-79759, 79843-79847), 4 isolated upper molars (UF 79761-79764), 2 distal ends of humeri (UF 79765, 79848), 1 distal end of radius (UF 79766); GRAND (=MIDDLE) CAICOS: Conch Bar Cave (FSM Zooarchaeology site MC-16)- 1 mandible and 1 distal end of humerus (FS 246).

Remarks.-- The large Greater Antillean species of *Natalus*, *N. major*, has generally been regarded as a distinct species (e.g. Goodwin 1959). Recently, however, many authors have followed Varona (1974) who relegated the two living and one extinct subspecies of *N. major* to subspecific status under the mainland and Lesser Antillean species, *N. stramineus* (e.g. Silva Taboada 1979; Hall 1981; Koopman, this volume), despite the fact that Varona provided no substantiation for his synonymy.

Goodwin (1959) presented cranial measurements of the three species in the subgenus *Natalus*, *N. major*, *N. stramineus*, and *N. tumidirostris*, including all recognized subspecies. Although closely related to *N. major* and *N. stramineus*, *N. tumidirostris* from northern South America, the Netherlands Antilles, and Trinidad is readily distinguished from these two species by the noticeably inflated maxillary bone dorsal to the toothrow and the deeply emarginated palate. There is virtually no overlap in any measurement between the largest individuals of *N. stramineus* (the nominate race from the Lesser Antilles) and the smallest representatives of *N. major*. Furthermore, morpholological comparisons reveal a number of characters in addition to larger size that reliably distinguish *N. major* from *N. stramineus*. The braincase of *Natalus major* is higher, more inflated, and more sharply elevated above the rostrum than in *N. stramineus*. The anterior edge of the braincase rises from the rostrum at an angle of approximately 70-75° in *N. major*, but only 55-60° in *N. stramineus*. The sagittal crest is better developed in *N. major* extending from the anterior edge of the orbit to the nuchal crest. In *N. stramineus* the sagittal crest extends no farther anteriorly than the interorbital constriction and only rarely extends posteriorly to meet the nuchal crest and then only as a very weak ridge. The interorbital region is more strongly constricted in *N. major*. The interorbital breadth of *N. major jamaicensis* is equal to or less than that of most populations of *N. stramineus*, even though the former is considerably larger in all other cranial measurements (Goodwin 1959). In *N. major*, the rostrum is highest along the midline, slightly concave dorsal to the toothrows, and tapers anteriorly, whereas in *N. stramineus* the rostrum is flatter dorsally, convex above the toothrows, and more parallel-sided. As a result of the more tapered rostrum the nasal opening in the larger species is relatively and absolutely smaller than in *N. stramineus*. Dental characters that distinguish *N. major* from *N. stramineus* include the stronger hypocone on M^1 and M^2, the relatively larger upper incisors, and the larger, higher entoconid and deeper valley connecting the cristid obliqua and postcristid on the lower molars. In summary, the larger size of *N. major*, coupled with the more inflated and upturned braincase, strongly constricted interorbital region, and tapering, less inflated rostrum readily distinguishes this Greater Antillean representative of the subgenus *Natalus* from the closely related species *N. stramineus*.

Natalus major is a Greater Antillean endemic now found only on Jamaica and Hispaniola. Extinct populations are known from Cuba and Isla de Pinos (Anthony 1919; Silva Taboada 1979), Grand Cayman (Morgan in press a), and the Bahamas (Fig. 3). Morgan and Woods (1986) first reported this species from the Bahamas based on the fossils described here from Andros and New Providence. Additional specimens of *N. major* have recently been identified from a cave deposit associated with an Amerindian archaeological site on Grand Caicos. Specimens of *N. major* from Andros and New Providence are much larger than Recent individuals of *N. major jamaicensis* from Jamaica (see measurements in Table 6). Comparison of the most complete fossil mandible from Banana Hole with measurements of three mandibles of the extinct subspecies *N. major primus* from Cuba (Table 6) confirms that the extinct Bahamian form was as large as, or even larger than, the Cuban bat, which is the largest previously known representative of the family Natalidae. The extinct populations of *N. major* from Andros and New Providence were almost certainly derived from the large Cuban subspecies and are tentatively referred to *N. major primus* pending more detailed comparisons with fossils from Cuba. The single mandible of *N. major* from Grand Caicos is much smaller than the fossils from Andros and New Providence, comparing more favorably in size with *N. major jamaicensis* (Table 6). The geographic location of the Caicos Islands would suggest that these smaller individuals of *N. major* were probably derived from the Hispaniolan subspecies, *N. m. major*, which is slightly smaller than the Jamaican subspecies. Thus, like *Macrotus waterhousii* and *Brachyphylla nana*, the populations of *N. major* in the northern and southern Bahamas appear to have had a dual origin, having been derived from both Cuba and Hispaniola, respectively. Although ecological data are scarce for *N. major*, this species is an obligate cave-dweller that is known to inhabit large caves exhibiting a stable microenvironment (Goodwin 1970).

Natalus tumidifrons (Miller)

Material examined.-- ANDROS: King Cave- 28 mandibles (UF 79114-79134, 79244-79250), 3 maxillary fragments (UF 79111-79113), 8 complete and 9 proximal and 24 distal ends of humeri (UF 79135-79165, 79251-79260), 20 proximal ends of radii (UF 79261-79263, 79305-79321), 7 proximal ends of femora (UF 79166-79171, 79264); Ashton Cave- 3 mandibles (UF 79408, 79439-79440); 15 complete and 8 proximal and 16 distal ends of humeri (UF 79409-79417, 79419-79429, 79432-79436, 79441-79453, 79460-79461), 8 proximal ends of radii (UF 79430, 79437-79438, 79454-79458, 79462-79464), 3 proximal ends of femora (UF 79418, 79431, 79459); NEW PROVIDENCE: Banana Hole- 4 mandibles (UF 79994-79997), 1 proximal and 1 distal end of humerus (UF 79998-79999); East Cave- 48 mandibles (UF 27732-27779), 4 maxillary fragments (UF 79852-79855); Hunts Cave- 2 mandibles (UF 79749-79750), 1 distal end of humerus (UF 79842); Sir Harry Oakes Cave- 2 mandibles (UF 79799-79800), 1 distal end of humerus (UF 79801); CAT: Turtle Scratch Cave- 2 mandibles (UF 79809-79810), 1 distal end of humerus (UF 79811).

Remarks.-- *Natalus tumidifrons* is the only endemic species of bat in the Bahamas where it is now restricted to Great Abaco and San Salvador. Several authors have reduced *N. tumidifrons* to a subspecies of *N. micropus,* the Greater Antillean representative of the subgenus *Chilonatalus* (Varona 1974; Hall 1981). I follow Ottenwalder and Genoways (1982) who regard *N. tumidifrons* as a separate species distinguished from the closely related *N. micropus* primarily by its larger size. The large samples of fossils of *N. tumidifrons* from cave deposits on Andros and New Providence are indistinguishable from Recent *N. tumidifrons* from San Salvador.

The fossil record clearly establishes that *Natalus tumidifrons* had a wider distribution in the Bahamas during the Late Quaternary. Koopman et al. (1957) reported a single mandible of *N. tumidifrons* from Max Bowes Cave located near the Forest settlement on Great Exuma. The fossils listed herein extend the former range of this species to include three additional islands on the Great Bahama Bank: Andros, New Providence, and Cat. This species was identified in six of the seven cave deposits from Andros and New Providence and was the most abundant bat in three of these, King Cave and Ashton Cave on Andros and East Cave on New Providence. Three specimens of *N. tumidifrons* were identified from among a small sample of fossils collected in Turtle Scratch Cave on the southwestern end of Cat Island. The disappearance of *N. tumidifrons* from four islands on the Great Bahama Bank is puzzling, since the species still survives on islands to the north (Great Abaco) and east (San Salvador). Most other species of bats that have disappeared from the Great Bahama Bank are now either found only in the Greater Antilles or are restricted to a few islands in the southern Bahamas. Koopman et al. (1957) suggested that competition with *Nyctiellus lepidus*, accelerated by the great decrease in land area at the end of the Pleistocene, may have led to the disappearance of *N. tumidifrons* from much of its former range. No ecological data have been published on *Natalus tumidifrons*. The closely related *N. micropus* has been encountered in Jamaica (Goodwin 1970) and Cuba (Silva Taboada 1979) in both hot caves and in portions of caves having a more temperate but stable microclimate.

Nyctiellus lepidus (Gervais)

Material examined.-- ANDROS: King Cave-2 complete humeri (UF 79108-79109).

Remarks.-- Since Dalquest's (1950) review of the Natalidae, almost all authors have recognized a single genus in the family, namely *Natalus*, including three subgenera: *Natalus*, *Chilonatalus*, and *Nyctiellus*. *Natalus* (*Nyctiellus*) *lepidus*, the smallest species in the family, is quite distinct from all other natalids in a number of cranial, dental, and postcranial characters. These characters will be discussed in some detail here to justify my recognition of *Nyctiellus* as a distinct genus. Besides small size, the most conspicuous features that distinguish *Nyctiellus lepidus* from all other natalids are the lower, less inflated, and less dorsally inflected braincase and the broader, deeper, more inflated rostrum. The zygomatic arches of *Nyctiellus* are relatively and absolutely deeper and more robust than in any species of *Natalus*, especially the anterior portion. *Nyctiellus* has only a tiny cleft in the premaxilla allowing the first pair of upper incisors to nearly meet at the midline. All species of *Natalus* have a deep cleft in the premaxilla that widely separates the first pair of upper incisors. *Nyctiellus* has a single, large, deep, round pit along the midline in the basicranial region immediately posterior to the palate and anteromedial to the periotics. Species of the subgenus *Natalus* lack this pit altogether, while the subgenus *Chilonatalus* has a pair of small, shallow pits in this same position. The tympanic bulla is enlarged and greatly inflated in *Nyctiellus* covering almost the entire periotic. In other natalids the bulla is not inflated and covers barely one-third of the periotic.

There are also a number of dental features that distinguish *Nyctiellus*. The upper incisors of *Nyctiellus* are very different in morphology from other natalids. The I^1 is a short, broad, spatulate-shaped tooth that is strongly inflected medially, while the I^2 is leaf-shaped. Both pairs of upper incisors are simple conical teeth in other natalids. The first upper and lower premolars are greatly reduced in *Nyctiellus* to tiny button-like teeth. The upper canines of *Nyctiellus* are also much reduced in size. The first premolars and canines of *Natalus* are not noticeably reduced. The P_3 and P_4 of *Nyctiellus* are shorter and broader than in other natalids in which all three lower premolars are compressed laterally, but are approximately equal in size. In *Nyctiellus* the talonid is much broader than the trigonid on all lower molars, whereas in other natalids the talonid is only slightly broader than or equal in breadth to the trigonid. Although postcranial characters in the Natalidae have not yet been investigated in detail, the humerus was examined in the West

Indian species in order to confirm the identification of the two fossil humeri from King Cave, Andros referred to *N. lepidus*. Compared to other natalids, the humerus of *Nyctiellus* is much smaller, has a more strongly developed distal spinous process, a reduced medial process, and lacks the medial ridge of the trochlea in the posterior aspect. The two Andros humeri share all of these characters, including small size, with *Nyctiellus*.

The unique, and for the most part highly derived, cranial and dental anatomy of *Nyctiellus* is suggestive of a long and separate evolutionary history from members of the genus *Natalus*. *Nyctiellus* is distantly related to other natalids and has probably been isolated in the West Indies for a long period of time. I disagree with Dalquest's statement (1950:438) that "The differences, however, that separate *Nyctiellus* from the other forms, are all differences in relative size. The differences are all less than those usually considered to separate genera, and are of the magnitude usually used to separate species." Many of the characters of *Nyctiellus* involve striking differences in morphology compared with *Natalus* (e.g. shape of braincase and rostrum, robust zygomatic arches, greatly inflated auditory bullae, presence of deep basicranial pit, form of upper incisors, and reduction of canines and anteriormost premolars). All other species of *Natalus* are much more similar to one another than any is to *Nyctiellus*. The characters that distinguish *Nyctiellus* are comparable to generic-level characters in the Vespertilionidae or Phyllostomidae.

Nyctiellus lepidus is now restricted to Cuba, Isla de Pinos, and Eleuthera, Cat, Long, and Exuma in the Bahamas. Koopman et al. (1957) reported a fossil mandible of this species from Max Bowes Cave near the Forest settlement on Great Exuma. The two fossil humeri from King Cave on Andros represent the first record of *N. lepidus* from that island. The occurrence of this species on Andros is not unexpected considering its presence on four of the six Great Bank islands. Clearly *N. lepidus* has invaded the Bahamas from Cuba as this species is not known from any other island in the Greater Antilles. Like all other species of natalids, *N. lepidus* is an obligate cave-dweller. In Cuba this species roosts in the warm, humid cave microenvironment, preferring caves that are not occupied by other species of bats (Silva Taboada 1979).

Family Vespertilionidae
Eptesicus fuscus Palisot de Beauvois

Material examined.-- ANDROS: Ashton Cave- 1 mandible (UF 79465); King Cave- 1 partial skull (UF 79176), 14 maxillae (UF 79172-79175, 79265-79274), 34 mandibles (UF 79177-79197, 79275-79287), 3 complete and 1 proximal and 3 distal ends of humeri (UF 79198-79200, 79288-79291), 2 radii (UF 79202-79203), 2 complete and 3 proximal and 2 distal ends of femora (UF 79204-79206, 79292-79294); NEW PROVIDENCE: Banana Hole- 33 mandibles (UF 27625-27635, 27644-27648, 27669-27677, 27687-27690, 27811-27814); Hunts Cave- 3 mandibles (UF 79767-79769); Sir Harry Oakes Cave- 1 M_1 (UF 79802), 1 proximal end of radius (UF 79803).

Remarks.-- *Eptesicus fuscus* occurs throughout the Greater Antilles, but has a spotty distribution in the Bahamas where it is known from Grand Bahama, San Salvador, four islands on the Great Bahama Bank including Andros and New Providence, and on Crooked and Acklins in the southern Bahamas. The previously unpublished Grand Bahama record is based on a small sample of bones in the FSM from a Recent barn owl roost. This is the first record of a bat from Grand Bahama and the first occurrence of *E. fuscus* on the Little Bahama Bank, according to distributional data in Buden (1986).

Koopman et al. (1957) recognize two subspecies of *E. fuscus* in the Bahamas. According to them, the small endemic subspecies, *E. fuscus bahamensis*, is restricted to New Providence and San Salvador, while all other Bahamian *E. fuscus* belong to the larger Cuban subspecies, *E. fuscus dutertreus*. Samples of fossil mandibles of *E. fuscus* from Andros and New Providence are very similar in size to one another and to Recent specimens from the respective islands, and all are probably best referred to the subspecies *E. fuscus bahamensis*. *Eptesicus fuscus* occurs in five of the seven fossil deposits analyzed from Andros and New Providence and is particularly abundant in King Cave and Banana Hole. All Bahamian *E. fuscus* are apparently derived from Cuba, as the subspecies of *E. fuscus* from Florida and Hispaniola are larger than either form found in the Bahamas. In the Greater Antilles and Bahamas, *E. fuscus* roosts in caves with a variable microclimate and is also commonly found in man-made structures (Silva Taboada and Pine 1969).

Family Molossidae
Tadarida brasiliensis (I. Geoffroy)

Material examined.-- NEW PROVIDENCE: Banana Hole- 22 mandibles (UF 27623, 27649-27650, 27667-27668, 27818-27819, 96038-96052), 3 isolated upper molars (UF 96053-96055), 10 complete and 42 proximal and 63 distal ends of humeri (UF 96056-96159); GRAND (=MIDDLE) CAICOS: Conch Bar Cave (FSM Zooarchaeology Site MC-16)- 1 proximal end of radius (FS 247).

Remarks.-- *Tadarida brasiliensis* is one of the most widespread of all New World bats occurring from the southern United States south to Chile and Argentina and throughout the Greater and Lesser Antilles. This species has a patchy distribution in the Bahamas where it has been recorded from Abaco on the Little Bahama Bank, Eleuthera, Long, and Exuma on the Great Bahama Bank, and Crooked and Acklins in the southern Bahamas (Table 1; Buden 1986). A single partial radius from a fossil deposit on Grand Caicos represents the first record of *T. brasiliensis* from the Caicos Islands. In the seven caves analyzed from Andros and New Providence, *T. brasiliensis* occurs only in Banana Hole, but it is the third most abundant bat in that deposit represented by a minimum of 40 individuals. Fossil mandibles of *T. brasiliensis* from Banana Hole compare closely in size to Recent specimens from Exuma and Long that have been referred to the endemic Bahamian subspecies *T. brasiliensis bahamensis*. Koopman et al. (1957:172) state that the living Bahamian subspecies, *T. brasiliensis bahamensis* is "...definitely more like the Hispaniolan than the Cuban form." Based on measurements in Shamel (1931), the endemic Bahamian subspecies is the largest Antillean representative of *T. brasiliensis*, and is most similar in size to *T. brasiliensis constanzae* from Hispaniola. In the West Indies, *T. brasiliensis* roosts either in caves having a variable microclimate or in man-made structures (Silva Taboada and Pine 1969).

Order Rodentia
Family Capromyidae
Geocapromys ingrahami (J. A. Allen)

Material examined.-- GREAT ABACO: Imperial Lighthouse Caves- 2 partial skulls (MCZ 2109-2110), 1 mandible (MCZ 2108, type of *G. ingrahami abaconis*); cave near Hole in the Wall- complete skull lacking teeth (UF 96199); ANDROS: Ashton Cave- 16 mandibles (UF 99051-99058, 99062, 99066, 99069-99072, 99075, 99077), 5 maxillae (UF 99059-99061, 99073, 99078); Coleby Bay Cave- 71 mandibles (UF 99082-99152), 1 partial skull (UF 99161), 14 maxillae (UF 99153-99160, 99162-99173); King Cave- 5 mandibles (UF 96186, 96188-96190, 96196), 3 maxillae (UF 96193-96195), 2 partial frontals (UF 96197-96198), 1 jugal (UF 96191); NEW PROVIDENCE: Banana Hole- 223 mandibles (UF 59201-59370, 59471-59523), 110 maxillae (UF 59317-59470, 96160-96169); Sir Harry Oakes Cave- 12 mandibles (UF 99178-99189), 10 maxillae (UF 99190-99199); ELEUTHERA (specific locality unknown): 2 partial skulls, 3 mandibles (MCZ 2678); CAT: Turtle Scratch Cave- 5 mandibles (UF 96170-96174), 8 P_4 (UF 96175-96182); LITTLE EXUMA: Upper Pasture Cave- 1 nearly complete skull, 8 partial skulls, 64 mandibles (all MCZ uncatalogued); CROOKED: Gordon Hill Caves- 1 partial skull (MCZ 2111), 5 skull fragments (MCZ 2665), 5 maxillae (MCZ 2673), 61 mandibles (MCZ 2107-type of *G. ingrahami irrectus*, 2112-2117).

Remarks.-- The only naturally occurring population of *Geocapromys ingrahami* survives on the tiny island of East Plana Cay located about 30 km east of Acklins in the southern Bahamas (Allen 1892; Clough 1972). This species has recently been introduced on Little Wax Cay and Warderick Wells Cay in the Exuma Cays Land and Sea Park (see Jordan, this volume). Bones and teeth of *Geocapromys* are known from fossil cave deposits and Amerindian archaeological sites on nine additional islands in the Bahamas (Table 1, Fig. 4). Lawrence (1934) reviewed the taxonomic status of the extinct populations of *Geocapromys* from Great Abaco, Eleuthera, Long, and Crooked. She named two new extinct subspecies, *G. ingrahami abaconis* from Great Abaco and *G. ingrahami irrectus* from the remaining three islands. Both were described as larger than living *G. i. ingrahami*, differing from one another and from the nominate subspecies in minor cranial and dental features. Almost all of Lawrence's material was from archaeological sites. The great majority of these specimens are very recent in appearance (i.e. unfossilized), except for four mandibles from the Gordon Hill caves on Crooked (MCZ 2114D,E,F,I) that are much darker in color than the other specimens from this same site and appear to have undergone some mineralization. Other archaeological material of *G. ingrahami* has since been reported from San Salvador (Wing 1969). A small sample of incisors and postcranial elements from archaeological sites on Providenciales and Grand Caicos in the Caicos Islands (Wing and Scudder 1983) is tentatively identified as *Geocapromys* pending discovery of more diagnostic material.

Geocapromys fossils were first reported from the Bahamas by G. M. Allen (1937) who referred a large sample of mandibles and partial skulls from Exuma to *G. ingrahami irrectus*. These fossils have since been shown to have come from Upper Pasture Cave on Little Exuma (Hecht 1955). Hecht (1955) and Koopman et al. (1957) reported additional fossil material of *G. ingrahami* from Robertson Cave on Little Exuma and a cave on Pigeon Cay near Mosstown, Great Exuma. Brodkorb (1959), Auffenberg (1967), and

Olson and Pregill (1982) have all commented upon the tremendous abundance of *Geocapromys* from Banana Hole on New Providence, yet before now this remarkable sample has not been studied. New fossil records of *Geocapromys* are here reported from Ashton Cave, Coleby Bay Cave, and King Cave on Andros, East Cave, Hunts Cave, and Sir Harry Oakes Cave on New Providence, and Turtle Scratch Cave on Cat. With the possible exception of the four mandibles from Crooked mentioned above, all other fossils of *Geocapromys* in Bahamas are from islands on the Great Bahama Bank. This undoubtedly reflects the near total lack of paleontological field work conducted on the Little Bahama Bank and in the southern Bahamas, rather than the absence of *Geocapromys* from these regions during the Late Quaternary.

A detailed review of the taxonomic status of the Recent and extinct Bahamian populations of *Geocapromys* is beyond the scope of this paper. However, a brief discussion of the taxonomy of *Geocapromys* is an important prerequisite to understanding the complex zoogeography of this genus in the Bahamas. Morgan (1985a) reviewed the systematics of the three Recent species of *Geocapromys*. Based on a series of cranial characters he suggested that Recent and extinct populations of *Geocapromys ingrahami* from the Bahamas and extinct species of *Geocapromys* from Cuba and the Cayman Islands form a closely related species-group, the *ingrahami*-group, distinct from *G. brownii* of Jamaica and *G. thoracatus* of Little Swan Island, the *brownii*-group. Diagnostic characters of the *ingrahami*-group include the tendency toward anterior convergence of the upper toothrows, shortened and inflated braincase, lack of ventral flexion of the zygomatic arch, narrower jugal, absence of the jugal spine, and inflated auditory bullae. A large extinct member of the *ingrahami*-group, *Geocapromys columbianus*, known from fossil deposits throughout Cuba, is probably closely related to the species that gave rise to the Bahamian populations of *Geocapromys*.

Measurements of Recent *Geocapromys ingrahami* from East Plana Cay and a large series of *Geocapromys* mandibles and maxillae from paleontological and archaeological sites on five islands in the Bahamas are presented in Table 7, primarily to show size trends. Measurements of the largest living species of *Geocapromys*, *G. brownii* from Jamaica, are included in Table 7 for comparative purposes. All of the extinct Bahamian populations average larger than Recent *G. ingrahami*, but are also more variable. The smallest fossil individuals overlap broadly with Recent *G. ingrahami*, whereas some specimens from fossil deposits on Andros, New Providence, and Little Exuma are very large, overlapping in size with *G. brownii*. Habitat diversity and island size, particularly island size during the late Pleistocene, undoubtedly affected body size trends in Bahamian *Geocapromys*. The extinct populations of *Geocapromys* from islands on the Great Bahama Bank represent either a single highly variable species or two closely related sympatric species. The lack of well preserved skulls in most fossil samples from the Bahamas prevents the detailed analysis of cranial characters that is necessary for species-level discrimination in the genus *Geocapromys*. Tentatively, all extinct Bahamian populations of *Geocapromys* are referred to *G. ingrahami*, with the caveat that these may eventually be shown to represent a closely related species complex of two or more species.

Paleontological and archaeological data confirm that the modern distribution of *Geocapromys ingrahami* in the Bahamas is relictual, and that extinction, mostly human-caused, has been the primary factor affecting the zoogeography of this species in the Bahamas. Furthermore, there is some evidence that Amerindian peoples in the West

Indies kept native rodents in captivity and carried them between islands (Olson 1982). Nearly half of the records of *Geocapromys* in the Bahamas are from archaeological sites and therefore may not reflect the natural distribution of the genus prior to the arrival of humans. The presence of the sole surviving naturally occurring population of *Geocapromys ingrahami* on East Plana Cay has no real zoogeographic significance, except to note that on one minuscule, remote, uninhabited island this unique species has managed to escape extinction.

Zoogeography

Recent mammalian fauna

Eleven species of bats and one rodent currently inhabit the Bahamas archipelago (Table 1), however, no single island supports more than six species. Recent mammals have been recorded from only 23 islands in the Bahamas, although most of the smaller cays have not been collected. Three of the 12 Recent mammal species have not been recovered from Bahamian fossil deposits: the bats *Noctilio leporinus*, *Artibeus jamaicensis*, and *Lasiurus borealis*. *Noctilio leporinus* has only recently been reported from the Bahamas (Buden 1985, 1986) where it is restricted to Great Inagua. The population of *N. leporinus* from Great Inagua is of indeterminate origin since all Greater Antillean representatives of this species have been referred to the subspecies, *N. leporinus mastivus* (Davis 1973). *Artibeus jamaicensis* occurs on four islands in the southern Bahamas: Great Inagua, Little Inagua, Mayaguana, and Providenciales. The Bahamian populations belong to the small Cuban subspecies *A. jamaicensis parvipes* (Koopman et al. 1957; Buden 1986). *Lasiurus borealis* has a widespread but spotty distribution in the Bahamas (Table 1). Bahamian red bats have been referred to the small Hispaniolan subspecies, *L. borealis minor* (Koopman et al. 1957). The raccoon, *Procyon lotor*, is often listed as a native Bahamian mammal, but it is clearly a recent human introduction based on its absence from paleontological and Amerindian archaeological sites in the Bahamas and occurrence only on the two most populous islands (Grand Bahama and New Providence).

One of the most abundant and widely distributed bats in the West Indies, *Molossus molossus*, is absent from both the Recent and fossil faunas of the Bahamas. Stenodermatine fruit bats, including both *Artibeus jamaicensis* and endemic Antillean representatives of the *Stenoderma* group, are noticeably absent from the Recent and late Pleistocene fauna of the northern Bahamas. One species of endemic stenodermatine, along with *Artibeus jamaicensis*, occurs on each of the Greater Antilles, as well as in the Virgin Islands and Cayman Islands. Perhaps the more northerly latitude and hence somewhat less tropical climate has prevented the colonization of the Great Bank islands by *Molossus molossus*, *Artibeus jamaicensis*, and *Noctilio leporinus*, which are otherwise three of the most widespread of all Neotropical bats. The limited occurrence of *A. jamaicensis* and *N. leporinus* on a few islands in the southernmost Bahamas further strengthens this hypothesis. Great Inagua, the only island in the Bahamas where both of these species occur, is farther south than much of Cuba. These three ubiquitous Neotropical species are also absent from southern Florida, a large land mass of about the same latitude and distance from Cuba as the Great Bahama Bank islands.

Five species of bats (*Macrotus waterhousii*, *Erophylla sezekorni*, *Eptesicus fuscus*, *Lasiurus borealis*, and *Tadarida brasiliensis*) occur in both the northern and southern

Bahamas, two species (*Natalus tumidifrons* and *Nyctiellus lepidus*) occur only in the northern Bahamas, and four bats (*Noctilio leporinus*, *Monophyllus redmani*, *Brachyphylla nana*, and *Artibeus jamaicensis*) and the capromyid rodent *Geocapromys ingrahami* are restricted to the southern Bahamas. The Recent distribution of at least four of these species (*M. redmani*, *B. nana*, *N. tumidifrons*, and *G. ingrahami*) in the Bahamas is relictual, as they were more widespread in the late Pleistocene.

The mammalian fauna of the Bahamas was derived entirely from the Greater Antilles. All native species of Recent and fossil mammals from the Bahamas are conspecific with or closely related to species from either Cuba or Hispaniola. Although the western edge of the Great and Little Bahama banks would have extended to within 75 km of the Florida peninsula during the late Pleistocene, no indigenous species of Bahamian mammals reached the islands directly from continental North America. Florida and the Bahamas do share three species of bats, *Eptesicus fuscus*, *Lasiurus borealis*, and *Tadarida brasiliensis*, but each of these species occurs in the Greater Antilles as well, and the Bahamian forms are more closely related to the Antillean representative of the species. Buden (1985) recorded a single specimen of *Lasionycteris noctivagans* from Providenciales in the Caicos Islands, a species otherwise restricted to the North American continent. This single Bahamian record, like several specimens of *L. noctivagans* from Bermuda (Van Gelder and Wingate 1961), surely represents an errant migrant rather an established resident population.

The zoogeographic affinities of the Bahamian mammal fauna are summarized in Table 8. Six of the Recent taxa are derived from Cuba, four from Hispaniola, and two could have originated on either island. Among the widespread species in the Bahamas, *Erophylla sezekorni* has a Cuban origin, *Lasiurus borealis* and *Tadarida brasiliensis* were derived from Hispaniola, and *Macrotus waterhousii* has a dual origin with the smaller northern Bahamian form derived from Cuba and the larger southern Bahamian form from Hispaniola. One of the two bats restricted to the northern Bahamas, *Nyctiellus lepidus*, is clearly derived from Cuba, and the other, *Natalus tumidifrons*, is of indeterminate origin, although it too is more likely to be of Cuban than Hispaniolan derivation. Of the species now restricted to the southern Bahamas, *Geocapromys ingrahami* and *Artibeus jamaicensis* are Cuban in origin, *Brachyphylla nana pumila* is derived from Hispaniola, and *Noctilio leporinus* and *Monophyllus redmani* could have originated on either island.

Cuba would appear to be a more likely source for the land mammal fauna of the Bahamas than would Hispaniola on the basis of its longer coastline and closer proximity to most of the islands. During periods of lower sea level in the Pleistocene, Cuba and Great Bahama Island would have been separated by a distance of as little as 20 km. All of the Bahama Islands from Grand Bahama and Abaco at the northwest end of the archipelago to Crooked, Acklins, and Mayaguana are closer to Cuba than to Hispaniola. Great and Little Inagua are equidistant (about 100 km) from these two Greater Antillean islands, while only the Turks and Caicos Islands are closer to Hispaniola. The Little Bahama Bank is actually much closer to Florida than to Cuba, yet it has an Antillean mammal fauna presumably derived from the Great Bank islands to the south.

The mammalian fauna of the Bahamas appears to be rather recently derived, probably during the Pleistocene based on the low overall degree of endemism. There are only two endemic species of mammals in the entire Bahamas archipelago, the Bahamian hutia, *Geocapromys ingrahami*, and the Bahamian funnel-eared bat, *Natalus tumidifrons*, while from three to five endemic subspecies are recognized by various authors. Several of the more distinct endemic forms, such as *Geocapromys ingrahami*, may have reached the Bahamas in the early Pleistocene. However, most Bahamian bats are indistinguishable from Cuban and/or Hispaniolan subspecies, and thus probably arrived in the late Pleistocene. This lack of species-level endemism characterizes the Bahamian land bird fauna as well, which also has but two endemic species, the hummingbird, *Philodyce evelynae*, and the Bahama swallow, *Tachycineta cyaneoviridis*.

Late Quaternary mammalian fauna

The changes that have occurred in the zoogeography and composition of the Bahamian mammal fauna over the last 10,000 years are primarily the result of extinction. Late Quaternary extinctions have significantly decreased the number of species of mammals inhabiting the Bahamas, most notably on the islands of the Great Bahama Bank. It has been known since early in this century that extensive extinctions profoundly altered the native mammalian fauna of the West Indies. The Recent terrestrial mammals of the Antillean islands are but a depauperate remnant of the unique and diverse fauna that inhabited these islands prior to the arrival of humans about 4,500 yBP. Nearly 90 percent of the 76 recognized species of terrestrial or non-volant mammals from the West Indies have gone extinct since the late Pleistocene (Morgan and Woods 1986). Only eight of 59 species (14%) of bats have disappeared from the West Indies during this same time period, however, nearly one-fourth of Antillean bat species have suffered localized extinctions on certain islands, a phenomenon that had a dramatic effect on the Bahamian bat fauna. Two major causes of vertebrate extinctions in the West Indies have been proposed: 1) climatic change and rise in sea level in the late Pleistocene and early Holocene; 2) the arrival of humans in the late Holocene, with subsequent extinctions resulting from predation for food procurement, habitat destruction, and introduction of exotic species.

Only two species of mammals appear to have colonized the Bahamas since the end of the Pleistocene, the bats *Artibeus jamaicensis* and *Noctilio leporinus*, both of which are unknown in Bahamian fossil deposits. Even today these two species have very limited distributions in the southern Bahamas. Although *Lasiurus borealis* has not yet been recorded from fossil deposits in the Bahamas, its widespread Recent distribution in the islands suggests a Pleistocene colonization. Because *L. borealis* roosts in trees it is rare in the fossil record, both in the West Indies and continental North America (Morgan 1985b).

The only well-sampled Late Quaternary vertebrate faunas from the Bahamas are located on islands of the Great Bahama Bank. For this reason the following discussion and comparisons are concerned primarily with the Recent and fossil faunas of these islands, particularly Andros and New Providence. The Recent mammalian fauna of the six largest islands on the Great Bahama Bank totals six species, all bats: *Macrotus waterhousii*, *Erophylla sezekorni*, *Nyctiellus lepidus*, *Eptesicus fuscus*, *Lasiurus borealis*, and *Tadarida brasiliensis*. All six species occur on Long Island at the southeastern corner of the Great Bank, while only four species are known from New Providence and Andros (*N. lepidus* and *T. brasiliensis* are absent). The mammalian fauna of the Great Bahama Bank islands was vastly different in the late Pleistocene and early Holocene. Eleven species of mammals are known from fossil deposits on Andros, 10 species of bats and *Geocapromys* (Table 1). Seven of the bats and the hutia, comprising 73 percent of the fossil fauna, are now extinct on Andros. Similarly, 14 species of mammals have been identifed from fossil deposits on New Providence, 13 species of bats and *Geocapromys* (Table 1). Eleven of these 14 species, constituting 79 percent of the fauna, no longer occur on New Providence. The combined mammalian fauna from fossil deposits on Andros and New Providence numbers 16 species, 15 bats and one rodent. *Geocapromys* and ten of the 15 species of bats have since disappeared from islands of the Great Bahama Bank, constituting more than two-thirds of the mammal species that inhabited Great Bahama Island in the Late Quaternary. Seven species of bats identified from fossil deposits on Andros and New Providence are now extinct in the Bahamas: *Mormoops blainvillii*, *M. megalophylla*, *Pteronotus macleayii*, *P. parnellii*, *P. quadridens*, *Phyllonycteris poeyi*, and *Natalus major*. Five of these species still occur on Cuba, while *M. megalophylla* and *N. major* have been found there as fossils. Four of the species that disappeared from the Great Bahama Bank survive elsewhere in the Bahamas. *Monophyllus redmani*, *Brachyphylla nana*, and *Geocapromys ingrahami* are still found in the southern Bahamas and *Natalus tumidifrons* occurs on two islands in the northern Bahamas peripheral to the Great Bank. The modern distribution of these four species in the Bahamas is obviously relictual as the fossil record establishes that they were once more widely distributed.

The zoogeographic affinities of the fossil mammals from islands on the Great Bahama Bank clarify, and in some instances modify, previous hypotheses on the zoogeography of the Recent fauna. As might be predicted from the proximity of Cuba to Great Bahama Island in the late Pleistocene, the great majority of species on this large paleoisland are of Cuban origin (at least 75 percent of the fauna, see Table 8). Among the 16 species of mammals recorded from fossils deposits on Great Bahama Island, 12 are clearly Cuban in origin, three could have been derived from either Cuba or Hispaniola although these are probably Cuban in origin as well, and only one species, *Tadarida brasiliensis*, appears to have a Hispaniolan origin. Although many of these species are now extinct in the Bahamas, this Cuban influence on the mammalian fauna of the Great Bahama Bank islands is still prevalent in the modern fauna, although to a lesser degree.

Olson and Hilgartner (1982) documented large-scale extinctions in the fossil avifauna from Banana Hole. They identified 32 species of birds from the Banana Hole deposit; 16 (50%) of these birds no longer occur on New Providence of which 12 (38%) are extinct in the Bahamas. These figures compare rather closely to those of the less diverse fossil mammalian faunas from Andros and New Providence in which 11 of 16 species (69%) are no longer found on islands of the Great Bahama Bank, including seven species (44%) that are extinct in the Bahamas. Olson and Hilgartner (1982) identified three extinct species of birds from Banana Hole. None of the mammals that have disappeared from the Bahamas are extinct, as all of the species still survive in the Greater Antilles or in North and South America in the case of *Mormoops megalophylla*. In contrast to the bird and mammal faunas, the herpetofauna of New Providence has suffered minor extinctions during the Late Quaternary. Only four of 17 species of amphibians and reptiles identified from Banana Hole no longer occur on New Providence (Pregill 1982). Moreover, three of these species are large reptiles, including a land tortoise (*Geochelone*), a crocodile (*Crocodylus*), and a rock iguana (*Cyclura*), that may well have been subject to human-caused extinction in the late Holocene.

During the late Pleistocene glacial maximum about 17,000 yBP, when sea level stood approximately 120 m lower than present, Great Bahama Island would have been the second largest island in the West Indies after Cuba with a land area in excess of 100,000 km^2. Great Bahama Island was essentially a fifth member of the Greater Antilles, albeit greatly lacking in topographic, and presumably ecological, diversity. Pregill (1982) hypothesized that during this time period the predominant habitats on the larger islands in the Bahamas probably consisted of vast dry savannas and grasslands. This ecological scenario is supported both by the evidence of a cooler drier climate in the Bahamas during the late Pleistocene (Lynts and Judd 1971; Pregill and Olson 1981) and by the presence in Banana Hole of certain species of birds (Olson and Hilgartner 1982) that favor open grassland or prairie habitats, but which are now extinct in the Bahamas. Most notable among these more arid-adapted birds are the burrowing owl *Athene cunicularia*, a caracara *Polyborus creightoni*, the thick-knee *Burhinus bistriatus*, and the meadowlark *Sturnella magna*. The occurrence in the Banana Hole deposits of several species of arboreal lizards and birds suggests the presence of at least some broadleaf forest on Great Bahama Island (Olson and Hilgartner 1982; Pregill 1982).

The 15 species of bats recorded from Late Quaternary deposits on Great Bahama Island give this island a chiropteran fauna of more or less comparable diversity to that found in fossil deposits in the Greater Antilles. Puerto Rico has 12 species of fossil bats (Anthony 1925; Choate and Birney 1968), Hispaniola has 19 species (Miller 1929; FSM vertebrate paleontology collection), Jamaica has 20 species (Koopman and Williams 1951; Williams 1952; FSM vertebrate paleontology collection), and Cuba has 26 species (Silva Taboada 1979). The composition of the Bahamian Late Quaternary bat fauna suggests that ecological conditions were somewhat different in the Bahamas than in the Greater Antilles. Eleven of the 15 species from Great Bahama Island are insectivorous (all Mormoopidae, all Natalidae, *Macrotus waterhousii*, *Eptesicus fuscus* and *Tadarida brasiliensis*), three species are primarily pollenivorous (*Brachyphylla nana*, *Erophylla sezekorni*, and *Phyllonycteris poeyi*) according to Silva Taboada and Pine (1969), and one species is nectarivorous (*Monophyllus redmani*). No strictly frugivorous species occur in Bahamian fossil deposits even though four species of fruit bats are known from late Pleistocene cave deposits in Cuba: *Artibeus anthonyi*, *A. jamaicensis*, *Phyllops falcatus*, and *P. vetus*. The absence of fruit bats from the Recent and Late Quaternary fauna of the northern Bahamas is probably due to the lack of a year-round fruit source sufficient to maintain a resident population of frugivorous bats, a factor almost surely related to the less tropical climate of this region compared to the Greater Antilles. The cooler, drier climate in the late Pleistocene would have been even less favorable for frugivorous bats than is the present-day climate of the Bahamas.

During the Late Quaternary Great Bahama Island supported an impressive diversity of cave-dwelling bats, including five mormoopids, three natalids, three brachyphylline phyllostomids, and one glossophagine phyllostomid. This fauna is comparable to the Recent fauna of cave-dwelling bats in Jamaica and Hispaniola, and includes four more cavernicolous species than are now found in Puerto Rico. Of the 15 species of bats recorded from fossil deposits on Great Bahama Island, 12 are obligate cave-dwelling forms. Ten of these species are no longer found on the islands of the Great Bahama Bank and seven are totally extinct in the Bahamas. All four of these groups of specialized cave-dwelling bats have tropical affinites, yet they flourished on Great Bahama Island during a cooler and drier period than the present. Morgan and Woods (1986) pointed out that the present mainland distribution of mormoopids suggests that some species may favor more arid habitats. However, the availability of suitable caves for day roosts is probably the single most important factor limiting the distribution of cave-dwelling bats in the West Indies, particularly in the Bahamas.

Localized extinctions of cave-dwelling bats occurred throughout the West Indies in the late Pleistocene and early Holocene, particularly on smaller islands including Isla de Pinos, La Gonave, Grand Cayman, Cayman Brac, Antigua, and Barbuda, as well as many islands in the Bahamas (Morgan and Woods 1986). Three groups of cavernicolous bats were particularly susceptible to these extinctions, the Mormoopidae, Natalidae, and the subfamily Brachyphyllinae in the Phyllostomidae. Most mormoopids, natalids, and brachyphyllines are not only obligate cave-dwellers, but also prefer specialized cave environments. According to Silva Taboada and Pine (1969), there are two general types of microenvironments in West Indian caves. The first type or variable microenvironment is found in caves that are small in size or have many openings, are usually dimly lit, and have variable temperature and humidity. The caves found on most small islands in the West Indies, and specifically most of the caves in the Bahamas, are characterized by a variable microclimate. The second type or stable microenvironment is usually found deep within large caves and is characterized by complete darkness and high temperature and humidity. The Cuban bats that favor the hot climatically stable portions of caves include all four species of mormoopids, *Phyllonycteris poeyi*, *Brachyphylla nana*, and *Erophylla sezekorni*. *Monophyllus redmani* and the two species of natalids are restricted to the more temperate, but stable portions of caves (Silva Taboada and Pine 1969). These are precisely the bats that have suffered the most extensive local extinctions in the Bahamas. All of these specialized cave-dwelling bats, along with extinct species of *Mormoops* and *Pteronotus*, were identified from two Cuban cave fossil deposits described by Silva Taboada (1974). These correspond to "type D" fossil deposits of Woloszyn and Silva Taboada (1977), which are dominated by bats found in the stable cave microenvironment.

Goodwin (1970) described the roost ecology of the chiropteran fauna from St. Clair Cave, Jamaica, an extensive cave system exhibiting the stable microenvironment of Silva Taboada and Pine (1969). Goodwin found eight species of bats roosting in St. Clair Cave about 1500 m from the entrance in a passageway 5-15 m wide and 10-15 m high. Airflow in this portion of the cave was negligible, the temperature was 30°C, relative humidity was near 100%, and the air was permeatated with the smell of ammonia generated by the huge deposits of decaying guano. The bats occupying this hot, humid cave environment (hot phase of the stable cave microenvironment of Silva Taboada and Pine 1979; "cuevas calientes" of Silva Taboada 1979) were: *Pteronotus macleayii*, *P. parnellii*, *P. quadridens*, *Monophyllus redmani*, *Erophylla sezekorni*, *Phyllonycteris aphylla*, *Natalus major*, and *N. micropus*. *Mormoops blainvillii* was also found deep within this same cave. All nine of these species, or closely related congeners in the case of *Natalus tumidifrons* and *Phyllonycteris poeyi*, occur in fossil deposits on New Providence, Bahamas. Eight of the species have been identified in both Hunts Cave and Banana Hole. Remarkably, only one of these nine species, *Erophylla sezekorni*, still survives on New Providence.

There are two probable explanations for the widespread localized extinctions of cave-dwelling bats in the West Indies: 1) there has been a change in the size and distribution of caves, primarily on small islands, resulting from the post-glacial rise in sea level and the flooding of low-lying areas; 2) cave environments on some islands have been altered since the early Holocene, probably reflecting overall climatic changes. Both of these factors probably contributed to the disappearance of most specialized cave-dwelling bats from the remnants of Great Bahama Island. During the late Pleistocene this island would have consisted of a gigantic platform of porous limestone ranging up to nearly 200 m in elevation. Based on these geologic conditions, coupled with a tropical to semi-tropical climate, it is likely that a vast karst terrain would have developed on Great Bahama Island. Extensive cave systems much larger than any known in the Bahamas today almost surely occurred on this large Pleistocene paleoisland, as indicated by the presence of a number of species of bats now restricted to large caves in the Greater Antilles. With the postglacial rise in sea level most of these caves would have been flooded, while other large caves were probably destroyed by erosion. Most of the caves on Andros and New Providence that have produced fossils of specialized cave-dwelling bats are small and have variable microclimates that would not today be suitable for habitation by these same species. Changes in climate undoubtedly affected cave microenvironments as well, although this is difficult to substantiate without more detailed paleoecological data.

The extinction of *Geocapromys* throughout the Bahamas except on East Plana Cay is almost certainly human-caused. The occurrence of *Geocapromys* in Amerindian archaeological sites on many islands in the Bahamas confirms that this large rodent survived the late Pleistocene/early Holocene extinction event. Fresh, unfossilized bones of *Geocapromys* have been found on the surface of caves on Great Abaco, Andros, and New Providence in association with post-Columbian artifacts and *Rattus rattus*, suggesting that this species survived on some larger islands in Bahamas up until a few hundred years ago. *Geocapromys* supposedly inhabited Samana Cay as recently as 1930 (Barbour and Shreve 1935) and was reported from New Providence by Schopf (translated by Morrison 1911) in the late 1700s.

The absence from the Bahamas of terrestrial mammals other than *Geocapromys* is puzzling. During the Late Quaternary, Cuba and Hispaniola each supported more than 25 species of terrestrial mammals, including insectivores, ground sloths, primates, and three families of rodents (Varona 1974; Morgan and Woods 1986). Perhaps the Bahamas supported a richer fauna of terrestrial mammals that disappeared before the late Pleistocene. Only the discovery of pre-late Pleistocene fossil deposits in the Bahamas will resolve this question. A comparison of the Late Quaternary mammal faunas from islands in the Bahamas with fossil faunas from the Cayman Islands (Morgan 1977; in press a) is informative, as the Cayman Islands are topographically and ecologically similar to many of the Bahamas. Grand Cayman and Cayman Brac have a xerophytic vegetation and are even smaller and more remote than most of the Bahamas, yet each island supported three species of terrestrial mammals in the late Pleistocene and Holocene, including one species each of the capromyid rodents *Capromys* and *Geocapromys* and the small nesophontid insectivore *Nesophontes*. Furthermore, the Caymans are surrounded by very deep water and thus would have been only slightly larger during the late Pleistocene. Like the Bahamian *Geocapromys*, these terrestrial mammals appear to have reached the Cayman Islands by overwater dispersal from Cuba during the Pleistocene. The sparse xerophytic vegetation of the Bahamas has been linked to the low diversity of the mammalian fauna (e.g. Koopman et al. 1957), yet arid semi-tropical to temperate regions on the North American continent support an abundant and diverse fauna of terrestrial mammals. On the basis of these comparisons, it would seem that deep water barriers, small size of the islands, and lack of vegetative diversity do not fully explain the rarity of terrestrial mammals in the Bahamas. The more northerly latitude of the Bahamas and thus the somewhat less tropical climate compared to the Cayman Islands or Greater Antilles are probably factors, however, at present there is no satifactory solution to this problem.

Some theoretical considerations

Compared to other West Indian islands of similar size and vegetative diversity, the individual islands in the Bahamas archipelago have fewer species of Recent mammals, specifically bats, than might be predicted on the basis of their area alone. Once again, comparison with the Cayman Islands is instructive. Grand Cayman has an area of 185 km^2, about the same size as Exuma, and has eight species of Recent bats (Morgan, in press b), two more species than are known from any island in the Bahamas regardless of size. Cayman Brac is smaller than 19 of the 20 islands in the Bahamas listed in Table 9, yet it supports 6 species of bats. Long Island is the only Bahamian island that has six species of bats and it is more than ten times larger than Cayman Brac. The primary difference between the bat faunas of these two island groups is the presence of *Molossus molossus* and the stenodermatine fruit bats, *Artibeus jamaicensis* and *Phyllops falcatus*, in the Caymans and their absence in the Bahamas, except for the limited occurrence of *A. jamaicensis* on four islands in the southern Bahamas. Removing these three presumably more tropical bats from the fauna of Grand Cayman and Cayman Brac would give them similar faunas to those on islands of about the same size in the Bahamas.

The species-area plot shown in Fig. 5 demonstrates the relationship between island area and the number of mammal species for most West Indian islands. The letters represent the number of species present in the living fauna on individual islands, while the circled letters are species from fossil deposits. Although only two islands from the Bahamas are plotted here, New Providence (N) and Andros (A"), they accurately reflect the overall depauperate character of the living Bahamian fauna. The Recent faunas of these two islands would have to be nearly doubled for them to fit the predicted species-area curve based on the Recent mammal faunas of islands in the West Indies (solid line marked "recent" on figure). The composite fossil mammal faunas from Andros and New Providence are also plotted on Fig. 5. Using the present land area of the islands, New Providence falls well above the predicted curve generated from fossil data (dashed line marked "fossil" on figure), whereas Andros is slightly below the predicted value. However, as discussed earlier all islands on the Great Bahama Bank would have been consolidated into a huge paleoisland in the late Pleistocene. The fossil mammal fauna and approximate land area of Great Bahama Island are also plotted on Fig. 5 (GB encircled). The number of species of mammals from Great Bahama Island would have to be more than doubled (from the 16 species currently known to about 40 species) for this island to fit the predicted "fossil" curve.

The species-area plot shown in Fig. 6 is based on the present land area and Recent mammal faunas from 19 islands in the Bahamas archipelago (data in Table 9). The predicted species-area curve generated from this data set (solid line on figure) has a correlation coefficient ($r=0.18$) that is not significantly different from 0. The slope or z value of the species-area curve for Recent Bahamian mammals is 0.07, or a nearly horizontal line. In other words, there is no correlation between the size of an island in the Bahamas and the number of species of mammals known to inhabit that island. These values for the

Bahamas do not compare at all closely with the predicted species-area curve ($r = 0.83$, $z = 0.21$) for the Recent mammal faunas of the remainder of the West Indian islands (data from Morgan and Woods 1986), which are similar to theoretical values obtained by MacArthur and Wilson (1967) for vertebrate faunas on oceanic islands, including the West Indies. The reason for the radical departure of Bahamian mammals from most theoretical species-area curves for oceanic islands is unknown, but it is surely related to complex historical and ecological factors, as well as biases such as insufficient field work. For instance, in August and September 1953 Karl Koopman and others collected five species of bats on Great and Little Exuma, two islands from which no mammals had been reported previously. Likewise, extensive field work by Donald Buden and associates in the southern Bahamas over the last 20 years has greatly increased the mammalian fauna known from this region. Similar systematic field studies of bats elsewhere in the Bahamas will undoubtedly add species to the faunas of the individual islands. Until the fauna of a geographic region is completely known (or nearly so), studies of island biogeography based on that fauna, particularly theoretical aspects, may be subject to considerable change.

Concluding remarks

Late Quaternary cave deposits in the Bahamas have produced abundant evidence of a considerably richer mammalian fauna, especially bats, than is currently known from this island group. Seven species of bats identified from fossil sites on Andros and New Providence no longer occur in the Bahamas. Four additional species from these same deposits are now absent from islands on the Great Bahama Bank, although they still occur elsewhere in the Bahamas. Olson and Hilgartner (1982) observed similar large-scale extinctions among the birds from Banana Hole on New Providence. The Late Quaternary herpetofauna of New Providence has survived relatively intact with the exception of the larger species, including a tortoise, a crocodile, and an iguana (Pregill 1982).

Pregill and Olson (1981) attributed the extinction or range contraction of many species of Antillean birds and reptiles to climatic changes and rising sea levels at the end of the Pleistocene. Morgan and Woods (1986) confirmed that many bats in the West Indies underwent localized extinctions at this same time, particularly on smaller islands, whereas most of the extinct terrestrial mammals did not disappear until after the arrival of humans in the late Holocene. Most mammalian extinctions in the Bahamas were probably related to Late Quaternary environmental changes since the Bahamian mammal fauna is composed almost exclusively of bats, a group not strongly affected by human-caused extinction factors. However, the causative factors for bat extinctions in the Bahamas were probably quite different from those that affected the birds. Extinctions in the Bahamian avifauna apparently resulted from changes in vegetation, particularly the transition from widespread arid habitats in the late Pleistocene to more mesic, forested habitats in the Holocene (Olson and Hilgartner 1982). A number of species of birds that favor savannas or grasslands are present in Banana Hole, but are now extinct in the Bahamas. The chiropteran extinctions were not directly caused by vegetational changes, but were more closely tied to the disappearance of suitable cave roosting sites through either the flooding of extensive cave systems by rising sea levels or by changes in cave microenvironments related to overall climatic changes.

Comprehensive field surveys of mammals have not been conducted on most islands in the Bahamas. Many of the larger islands in the southern Bahamas have been adequately collected for mammals (e.g. Buden, 1979, 1985, 1986), as have several islands in the northern Bahamas, including the Exumas and New Providence (Koopman et al. 1957). However, some of the largest islands in the Bahamas (e.g. Abaco and Andros) have been incompletely sampled, while other large islands (e.g. Grand Bahama) have barely been collected at all. There are several hundred potentially habitable islands in the Bahamas, yet only 23 of these islands are presently known to support indigenous species of mammals. Two of the larger islands, Rum Cay and Samana Cay, along with all smaller islands, except Darby Island, Fortune Island, and East Plana Cay, have no recorded mammals. Because of the lack of adequate collections of Recent mammals from many islands in the Bahamas, some of the conclusions reached in this study, particularly those of a more theoretical nature, must be considered tentative.

Fossil mammal faunas from the Bahamas are few in number and are presently known from only five islands, all on the Great Bahama Bank: Andros, New Providence, Cat, Great Exuma, and Little Exuma. Additional evidence of extinct mammals comes from Amerindian archaeological sites on Great Abaco, San Salvador, Eleuthera, Long, Crooked, and Grand Caicos. Considering the abundance of caves, fissures, and sinkholes throughout the Bahamas, most of the larger islands will probably yield some evidence of fossil vertebrates upon thorough field exploration. The diverse Late Quaternary bat faunas from Andros and New Providence strongly indicate that future discoveries on other islands in the Bahamas will add more species to the fauna and expand the known range of other species. The presence of a single species of rodent from fossil deposits and archaeological sites throughout the Bahamas suggests that terrestrial mammals may have always been rare in this island group.

The lack of pre-late Pleistocene land vertebrate fossils from the Bahamas, and for that matter from the entire West Indies, greatly hinders the analysis of the historical biogeography of this fauna. Fluctuating sea levels and the concomitant changes in the size of most of the islands in the Bahamas occurred not only in the late Pleistocene and Holocene, but also periodically throughout the last 3 million years since the advent of Northern Hemisphere glaciation in the late Pliocene. The mammalian fauna of the Bahamas has probably waxed and waned throughout the Plio-Pleistocene depending upon changes in relative sea level. Lowered sea levels led to the formation of several large islands in the Bahamas, and one gigantic island (Great Bahama Island) larger than most of the Greater Antilles, except Cuba. Colonization of the Bahamas by bats probably took place at an accelerated rate during periods of lower sea level as a result of the narrower water gaps separating the islands from potential source areas such as Cuba and Hispaniola, larger island size, and the probable occurrence of extensive cave systems. High sea level stands during the present interglacial and the last interglacial around 120,000 years ago, are (were) periods when the islands in the Bahamas are (were) smaller and more isolated from source areas. Interglacial periods in the Bahamas, most notably the Holocene, are (were) characterized by extinctions, range contractions, reduced colonization, and flooding of large cave systems. Among mammals, only the bats *Artibeus jamaicensis* and *Noctilio leporinus* seem to be good candidates for post-Pleistocene colonizers of the Bahamas.

A number of factors have been suggested that may have contributed to the low species diversity in the Recent Bahamian mammal fauna, including the peripheral location of the Bahamas on the northern margin of the West Indies, the location of many of the major islands outside the tropics (i.e. north of the Tropic of Cancer), small size of most of the islands, lack of topographic and ecological diversity, and inadequate collecting. The extinction of nearly half of the species of mammals present in Late Quaternary fossil deposits in the Bahamas and the range contraction of many other species, must now be regarded as two of the primary factors that have determined the present composition and zoogeography of this fauna. The depauperate Recent mammal fauna of the Bahamas is therefore a net result of the numerous Late Quaternary extinctions and the slow rate of colonization during the Holocene.

Acknowledgements

I would like to extend my deep appreciation to the dedicated field crews from the Florida State Museum who collected the fossil mammals from the Bahamas described in this paper, in particular I thank F. Wayne King, Kay Eoff, Margaret K. Langworthy, Laurie Wilkins, and Richard Franz who excavated caves on Andros and J. C. Dickinson, Jr., Walter Auffenberg, William Hanon, and W. J. Riemer who excavated fossil deposits on New Providence. Ray Ashton and Oris Russell generously donated samples of vertebrate fossils from the Bahamas to the Florida State Museum. Donald W. Buden, Karl F. Koopman, Ann E. Pratt, and Charles A. Woods provided helpful comments on the manuscript. Ann E. Pratt helped draft the figures. This paper is dedicated to the memory of Bonnie McKeen in recognition of her contributions to West Indian vertebrate paleontology. This is University of Florida Contribution to Paleobiology number 322.

Literature cited

Allen, G.M. 1937. *Geocapromys* remains from Exuma Island. Journal of Mammalogy 18:369-370.

Allen, J.A. 1892. Description of a new species of *Capromys*, from the Plana Keys, Bahamas. Bulletin of the American Museum of Natural History 3(2):329-336.

Anthony, H.E. 1917. Two new fossil bats from Porto Rico. Bulletin of the American Museum of Natural History 37:565-568.

_____. 1919. Mammals collected in eastern Cuba in 1917, with descriptions of two new species. Bulletin of the American Museum of Natural History 41:625-643.

_____. 1925. Mammals of Porto Rico, living and extinct- Chiroptera and Insectivora. New York Academy of Sciences, Scientific Survey of Porto Rico and the Virgin Islands Volume 9(1):1-241.

Auffenberg, W. 1967. Notes on West Indian tortoises. Herpetologica 23:34-44.

Barbour, T. and B. Shreve. 1935. Concerning some Bahamian reptiles, with notes on the fauna. Proceedings of the Boston Society of Natural History 40:347-366.

Bloom, A.L. 1983. Sea level and coastal morphology of the United States through the late Wisconsin glacial maximum. Pp. 215-229 in S. C. Porter (ed.). Quaternary Environments of the United States. Volume 1. The late Pleistocene.

Brodkorb, P. 1959. Pleistocene birds from New Providence Island, Bahamas. Bulletin of the Florida State Museum 4(11):349-371.

Buden, D.W. 1975a. A taxonomic and zoogeographic appraisal of the big-eared bat (*Macrotus waterhousii* Gray) in the West Indies. Journal of Mammalogy 56:758-769.

_____. 1975b. *Monophyllus redmani* Leach (Chiroptera) from the Bahamas, with notes on variation in the species. Journal of Mammalogy 56:369-377.

_____. 1976. A review of the bats of the endemic West Indian genus *Erophylla*. Proceedings of the Biological Society of Washington 89:1-16.

_____. 1977. First records of bats of the genus *Brachyphylla* from the Caicos Islands, with notes on geographic variation. Journal of Mammalogy 58:221-225.

_____. 1979. Ornithogeography of the Southern Bahamas. Unpublished Ph.D. dissertation, Louisiana State University, Baton Rouge, LA, 273 pp.

_____. 1985. Additional records of bats from the Bahama Islands. Caribbean Journal of Science 21(1-2):19-25.

_____. 1986. Distribution of mammals of the Bahamas. Florida Field Naturalist 14(3):53-63.

Choate, J.R. and E.C. Birney. 1968. Sub-recent Insectivora and Chiroptera from Puerto Rico, with the description of a new bat of the genus *Stenoderma*. Journal of Mammalogy 49:400-412.

Clough, G.C. 1972. Biology of the Bahaman hutia, *Geocapromys ingrahami*. Journal of Mammalogy 53:807-823.

Dalquest, W.W. 1950. The genera of the chiropteran family Natalidae. Journal of Mammalogy 31:436-443.

Davis, W.B. 1973. Geographic variation in the fishing bat, *Noctilio leporinus*. Journal of Mammalogy 54:862-874.

Eshelman R.E. and G.S. Morgan. 1985. Tobagan Recent mammals, fossil vertebrates, and their zoogeographical implications. National Geographic Society, Research Reports 21:137-143.

Etheridge, R. 1966. Pleistocene lizards from New Providence. Quarterly Journal of the Florida Academy of Sciences 28:349-358.

Gascoyne, M., G.J. Benjamin, H.P. Schwarcz, and D.C. Ford. 1979. Sea-level lowering during the Illinoian glaciation: evidence from a Bahama blue hole. Science 205:806-808.

Goodfriend, G.A. and R.M. Mitterer. 1987. Age of the ceboid femur from Coco Ree, Jamaica. Journal of Vertebrate Paleontology 7:344-345.

Goodwin, G.G. 1959. Bats of the subgenus *Natalus*. American Museum of Natural History Novitates 1977:1-22.

Goodwin, R.E. 1970. The ecology of Jamaican bats. Journal of Mammalogy 51:571-579.

Hall, E.R. 1981. The Mammals of North America, Second Edition. 2 volumes. John Wiley and Sons, New York, 1175 pp.

Hecht, M.K. 1955. The comparison of recent and fossil amphibian, reptilian, and mammalian faunas in the Bahamas. Year Book of the American Philosophical Society 1954:133-135.

Imbrie, J., A. McIntyre, and T.C. Moore, Jr. 1983. The ocean around North America at the last glacial maximum. Pp. 230-236 in S. C. Porter (ed.). Quaternary Environments of the United States. Volume I. The late Pleistocene.

Kaufman, A. 1986. The distribution of ^{230}Th/^{234}U ages in corals and the number of last interglacial high-sea stands. Quaternary Research 25:55-62.

Koopman, K.F. 1951. Fossil bats from the Bahamas. Journal of Mammalogy 32:229.

_____. 1955. A new subspecies of *Chilonycteris* from the West Indies and a discussion of the mammals of La Gonave. Journal of Mammalogy 36:109-113.

_____., M.K. Hecht, and E. Ledecky-Janecek. 1957. Notes on the mammals of the Bahamas, with special reference to the bats. Journal of Mammalogy 38:164-174.

_____. and E.E. Williams. 1951. Fossil Chiroptera collected by H.E. Anthony in Jamaica, 1919-1920. American Museum of Natural History Novitates 1519:1-29.

Lawrence, B. 1934. New *Geocapromys* from the Bahamas. Occasional Papers of the Boston Society of Natural History 8:189-196.

Lynts, G.W. and J.B. Judd. 1971. Paleotemperatures at Tongue of the Ocean, Bahamas. Science 171:1143-1144.

MacArthur, R.H. and E.O. Wilson. 1967. The Theory of Island Biogeography. Princeton University Press, Princeton, NJ, 203 pp.

Meltzer, D.J. and J.I Mead. 1983. The timing of late Pleistocene extinctions in North America. Quaternary Research 19:130-135.

Miller, G.S., Jr. 1929. A second collection of mammals from caves near St. Michel, Haiti. Smithsonian Miscellaneous Collections 81(9):1-30.

Mix, A. C. 1987. The oxygen-isotope record of glaciation. Pp. 111-135 in W.F. Ruddiman and H.E. Wright, Jr. (eds.). North America and adjacent oceans during the last deglaciation. The Geology of North America, Vol. K-3, Geological Society of America, Boulder, CO.

Morgan, G.S. 1977. Late Pleistocene Fossil Vertebrates from the Cayman Islands, British West Indies. Unpublished MS thesis, University of Florida, Gainesville, FL, 273 pp.

_____. 1985a. Taxonomic status and relationships of the Swan Island hutia, *Geocapromys thoracatus* (Mammalia: Rodentia: Capromyidae), and the zoogeography of the Swan Islands vertebrate fauna. Proceedings of the Biological Society of Washington 98:29-46.

_____. 1985b. Fossil bats (Mammalia: Chiroptera) from the late Pleistocene and Holocene Vero fauna, Indian River County, Florida. Brimleyana 11:97-117.

_____. in press a. Late Quaternary fossil vertebrates from the Cayman Islands. In D.R. Stoddart, J.E. Davies, and M. Brunt (eds.). Biogeography and Ecology of the Cayman Islands. Monographiae Biologicae. W. Junk, Dordrecht, Netherlands.

_____. in press b. Mammals of the Cayman Islands. In D. R. Stoddart, J.E. Davies, and M. Brunt (eds.). Biogeography and Ecology of the Cayman Islands. Monographiae Biologicae. W. Junk, Dordrecht, Netherlands.

_____. and C.A. Woods. 1986. Extinction and the zoogeography of West Indian land mammals. Biological Journal of the Linnean Society 28:167-203.

Morrison, A.J. 1911. Travels in the Confederation. Volume 2. W. J. Campbell Co., Philadelphia (Translation of J. D. Schopf, 1788).

Neumann, A.C. and W.S. Moore. 1975. Sea level events and Pleistocene coral ages in the northern Bahamas. Quaternary Research 5:215-224.

Olson, S.L. 1982. Biological archeology in the West Indies. Florida Anthropologist 35:162-168.

_____. (ed.). 1982. Fossil Vertebrates from the Bahamas. Smithsonian Contributions to Paleobiology 48:1-65.

_____. and W.B. Hilgartner. 1982. Fossil and subfossil birds from the Bahamas. Pp. 22-56 in S. L. Olson (ed.). Fossil Vertebrates from the Bahamas. Smithsonian Contributions to Paleobiology 48.

_____. and G.K. Pregill. 1982. Introduction to the paleontology of Bahaman vertebrates. Pp. 1-7 in S. L. Olson (ed.). Fossil Vertebrates from the Bahamas. Smithsonian Contributions to Paleobiology 48.

Ottenwalder, J.A. and H.H. Genoways. 1982. Systematic review of the Antillean bats of the *Natalus micropus*-complex (Chiroptera: Natalidae). Annals of the Carnegie Museum of Natural History 51(2): 17-38.

Pregill, G.K. 1982. Fossil amphibians and reptiles from New Providence Island, Bahamas. Pp. 8-21 in S. L. Olson (ed.). Fossil Vertebrates from the Bahamas. Smithsonian Contributions to Paleobiology 48.

_____. and S.L. Olson. 1981. Zoogeography of West Indian vertebrates in relation to Pleistocene climatic cycles. Annual Review of Ecology and Systematics 12:75-98.

Ray, C.E., S.J. Olsen, and H.J. Gut. 1963. Three mammals new to the Pleistocene fauna of Florida, and a reconsideration of five earlier records. Journal of Mammalogy 44:373-395.

Rouse, I. and L. Allaire. 1978. Caribbean. Pp. 431-481 in R.E. Taylor and C. W. Meighan (eds.). Chronologies in New World Archaeology. Academic Press, New York.

Ruddiman, W.F. 1987. Synthesis; The ocean ice/sheet record. Pp. 463-478 in W. F. Ruddiman and H. E. Wright, Jr. (eds.). North America and adjacent oceans during the last deglaciation. The geology of North America, Vol. K-3, Geological Society of America, Boulder, Colorado.

Schwartz, A. and J.K. Jones, Jr. 1967. Bredin-Archbold-Smithsonian Biological Survey of Dominica. 7. Review of bats of the endemic Antillean genus *Monophyllus*. Proceedings of the United States National Museum 124:1-20.

Shamel, H.H. 1931. Notes on the American bats of the genus *Tadarida*. Proceedings of the United States National Museum 78(19):1-27.

Silva Taboada, G. 1974. Fossil Chiroptera from cave deposits in central Cuba, with description of two new species (genera *Pteronotus* and *Mormoops*) and the first West Indian record of *Mormoops megalophylla*. Acta Zoologica Cracoviensia 19(3):33-73.

_____. 1976. Historia y actualización taxonómica de algunas especies Antillanas de murciélagos de los géneros *Pteronotus*, *Brachyphylla*, *Lasiurus*, y *Antrozous* (Mammalia: Chiroptera). Poeyana (Instituto de Zoología, Academia de Ciencias de Cuba) 153:1-24.

_____. 1977. Algunos aspectos de la selección de hábitat en el murciélago *Phyllonycteris poeyi* Gundlach in Peters, 1861 (Mammalia: Chiroptera). Poeyana (Instituto de Zoologia, Academia de Ciencias de Cuba) 168:1-10.

_____. 1979. Los Murciélagos de Cuba. Editorial Academia, La Habana, Cuba, 423 pp.

_____. and R.H. Pine. 1969. Morphological and behavioral evidence for the relationship between the bat genus *Brachyphylla* and the Phyllonycterinae. Biotropica 1:10-19.

Smith, J.D. 1972. Systematics of the chiropteran family Mormoopidae. University of Kansas Museum of Natural History, Miscellaneous Publications 56:1-132.

Steadman, D.W., G.K. Pregill, and S.L. Olson. 1984. Fossil vertebrates from Antigua, Lesser Antilles: evidence for late Holocene human-caused extinctions in the West Indies. Proceedings of the National Academy of Sciences, USA 81:4448-4451.

Swanepoel, P. and H.H. Genoways. 1978. Revision of the Antillean bats of the genus *Brachyphylla* (Mammalia: Phyllostomatidae). Bulletin of the Carnegie Museum of Natural History 12:1-53.

Taylor, R.E. 1980. Radiocarbon dating of Pleistocene bone: toward criteria for the selection of samples. Radiocarbon 22:969-979.

Van der Hammen, T. 1974. The Pleistocene changes of vegetation and climate in tropical South America. Journal of Biogeography 1:3-26.

Van Gelder, R.G. and D.B. Wingate. 1961. The taxonomy and status of bats in Bermuda. American Museum of Natural History Novitates 2029:1-9.

Varona, L.S. 1974. Catálogo de los Mamíferos Vivientes y Extinguidos de las Antillas. Academia de Ciencias de Cuba, La Habana, Cuba, 139 pp.

Williams, E.E. 1952. Additional notes on fossil and subfossil bats from Jamaica. Journal of Mammalogy 33:171-179.

Wing, E.S. 1969. Vertebrate remains excavated from San Salvador Island, Bahamas. Caribbean Journal of Science 9(1-2):25-29.

_____. and S. J. Scudder. 1983. Animal exploitation by prehistoric people living on a tropical marine edge. Pp 197-210 in C. Grigson and J. Clutton-Brock (eds.). Animals and Archaeology: shell middens, fishes, and birds. BAR International Series, 183, Oxford.

Woloszyn, B.W. and G. Silva Taboada. 1977. Nueva especie fósil de *Artibeus* (Mammalia: Chiroptera) de Cuba, y tipificación preliminar de los depósitos fosiliferos Cubanos contentivos de mamíferos terrestres. Poeyana (Instituto de Zoología, Academia de Ciencias de Cuba) 161:1-17.

Table 1. Distribution by island of the Recent (+) and fossil (*) mammals of the Bahamas. On islands where a species is known in both the Recent and fossil faunas only the Recent record is indicated here. The fossil records of these species are listed under Systematic Paleontology. Distributional data are from Lawrence (1934), G.M. Allen (1937), Koopman (1951), Koopman et al. (1957), Buden (1985, 1986), Morgan and Woods (1986), along with many new fossil records not previously published.

ISLANDS OF THE BAHAMAS ARCHIPELAGO[1]

SPECIES	GB	AB[2]	AN	NP	EL	CA	LO	EX[3]	SS	CR	AC	EP	MA	GI	LI	PR	NC	GC	EC
Noctilio leporinus	-	-	-	-	-	-	-	-	-	-	-	-	-	-	-	-	-	-	-
Mormoops blainvillii	-	*	*	*	-	-	-	*	-	-	-	-	-	-	-	-	-	-	-
Mormoops megalophylla	-	-	-	*	-	-	-	-	-	-	-	-	-	-	-	-	-	-	-
Pteronotus macleayii	-	*	*	*	-	-	-	-	-	-	-	-	-	-	-	-	-	-	-
Pteronotus parnellii	-	-	*	*	-	-	-	-	-	-	-	-	-	-	-	-	-	-	-
Pteronotus quadridens	-	*	*	*	-	-	-	-	-	-	-	-	-	-	-	-	-	-	-
Macrotus waterhousii	-	+	+	+	+	+	+	+	+	+	+	-	-	+	-	+	+	*	-
Monophyllus redmani	-	+	*	*	-	-	-	-	-	+	+	-	-	-	-	+	+	-	-
Artibeus jamaicensis	-	-	-	-	-	-	-	-	-	-	-	-	+	-	+	+	-	-	-
Brachyphylla nana	-	+	+	+	+	+	+	+	+	+	+	-	+	+	-	+	+	-	+
Erophylla sezekorni	-	+	+	+	-	-	-	-	-	+	+	-	+	+	-	+	+	-	+
Phyllonycteris poeyi	-	-	-	*	-	-	-	-	-	+	+	-	-	-	-	-	-	-	-
Natalus major	-	-	*	*	-	*	-	*	+	+	-	-	-	-	-	-	-	*	-
Natalus tumidifrons	+	+	*	*	-	+	-	+	-	+	+	-	-	-	-	-	-	-	-
Nyctiellus lepidus	-	-	*	-	-	-	-	+	+	+	+	-	-	-	-	-	-	-	-
Eptesicus fuscus	-	+	+	+	+	+	+	+	+	+	+	-	+	+	-	-	-	-	-
Lasiurus borealis	+	-	+	-	-	-	-	-	-	-	-	-	+	-	-	+	-	+[4]	-
Tadarida brasiliensis	-	+	*	*	-	-	*	*	-	+	+	-	-	-	-	-	*	*	-
Geocapromys ingrahami	-	*	*	*	*	*	*	*	*	*	-	+	-	-	-	*[5]	-	*[5]	-

1. (continued) Explanation of abbreviations.

viations for islands of the Bahamas Archipelago are as follows: Grand Bahama (GB), Abaco (AB), Andros (AN), New Providence (NP), Eleuthera (EL), Cat (CA), Long (LO), Exuma (EX), San Salvador (SS), Crooked (CR), Acklins (AC), East Plana Cay (EP), Mayaguana (MA), Great Inagua (GI), Little Inagua (LI), Providenciales (PR), North Caicos (NC), Grand (=Middle) Caicos (GC), East Caicos (EC).

includes Great Abaco and Little Abaco.

a includes Great Exuma and Little Exuma.

ding to Koopman et al. (1957) and Buden (1985) this record is based on a specimen which gives "Caicos Islands" as a locality, but does not specify which island. I have arbitrarily placed this record on Grand Caicos as this is the largest and most centrally located of the Caicos Islands.

al incisors and partial postcranial elements from archaeological sites on Grand Caicos and Providenciales in the Caicos Islands are tentatively identified as *Geocapromys* pending discovery of more diagnostic material.

Table 2. Measurements of the mandible and lower molars of fossil Mormoops blainvillii from the Bahamas and Recent M. blainvillii from the Greater Antilles. Mean, observed range (in parentheses), and sample size are provided for each measurment.

Locality	total length of mandible	alveolar length of mandibular toothrow	length m_1-m_3	length m_1	width m_1	length m_2	width m_2	length m_3	width m_3
Mormoops blainvillii New Providence Bahamas (fossil)	11.8 1	7.9 --	4.27 (4.26-4.28) 3	1.57 (1.52-1.63) 11	0.89 (0.85-0.93) 11	1.44 (1.42-1.48) 12	0.86 (0.83-0.88) 12	1.40 (1.36-1.43) 10	0.70 (0.66-0.71) 9
Mormoops blainvillii Cuba (Recent)	11.9 (11.8-11.9) 4	7.9 (7.8-8.0) 4	4.22 (4.19-4.25) 4	1.56 (1.54-1.59) 4	0.86 (0.85-0.87) 4	1.44 (1.42-1.46) 4	0.83 (0.82-0.86) 4	1.38 (1.37-1.40) 4	0.70 (0.67-0.72) 4
Mormoops blainvillii Jamaica (Recent)	12.2 (11.9-12.3) 6	8.0 (8.0-8.1) 6	4.30 (4.22-4.38) 6	1.58 (1.53-1.63) 6	0.88 (0.86-0.90) 6	1.47 (1.43-1.48) 6	0.86 (0.82-0.88) 6	1.42 (1.38-1.46) 6	0.73 (0.71-0.75) 6
Mormoops blainvillii Hispaniola (Recent)	12.4 (12.2-12.5) 2	8.2 (8.1-8.3) 2	4.39 (4.34-4.43) 2	1.66 (1.65-1.66) 2	0.90 (0.88-0.91) 2	1.52 (1.51-1.52) 2	0.85 (0.83-0.87) 2	1.44 (1.41-1.46) 2	0.73 (0.71-0.75) 2

Table 3. Measurements of the distal end of the humerus of *Mormoops blainvillii* and *M. megalophylla*.

SPECIES	width of distal articular surface	distal shaft thickness
Mormoops megalophylla Andros, Bahamas fossil (UF 79323)	3.1	1.7
Mormoops megalophylla Dominican Republic fossil (UF 96200)	3.1	1.8
Mormoops megalophylla Jamaica fossil (UF 102900)	3.2	1.8
Mormoops megalophylla Rock Springs, Florida fossil (N = 4)	x = 3.1 (3.0 - 3.2)	x = 1.7 (1.6 - 1.7)
Mormoops megalophylla Colombia Recent (N = 2)	x = 3.2 (3.2 - 3.3)	x = 1.7 (1.7)
Mormoops blainvillii New Providence, Bahamas fossil (N = 6)	x = 2.6 (2.5 - 2.7)	x = 1.3 (1.2 - 1.4)

Table 4. Measurements of lower molars of fossil *Pteronotus macleayii* and *P. quadridens* from the Bahamas and Recent representatives of these two species from Cuba. Mean, observed range (in parentheses), and sample size are provided for each measurement.

Species	length of m_1	width of m_1	length of m_2	width of m_2	length of m_3	width of m_3
Pteronotus macleayii Hunts Cave, New Providence Bahamas (fossil)	1.49 (1.47-1.51) 3	0.94 (0.94-0.95) 3	1.46 (1.41-1.48) 4	0.89 (0.87-0.91) 4	1.44 (1.42-1.47) 6	0.79 (0.75-0.81) 5
Pteronotus macleayii Cuba (Recent) (N=8)	1.50 (1.47-1.53)	0.93 (0.88-0.95)	1.45 (1.41-1.49)	0.89 (0.85-0.92)	1.46 (1.42-1.48)	0.77 (0.75-0.78)
Pteronotus quadridens Sir Harry Oakes Cave New Providence, Bahamas (fossil)						
UF 79774	1.32	0.86	1.29	0.79	1.30	--
UF 79775	1.33	--	--	--	--	--
UF 79776	--	--	--	--	1.29	0.73
Pteronotus quadridens Cuba (Recent) (N=8)	1.33 (1.31-1.36)	0.84 (0.81-0.88)	1.32 (1.27-1.36)	0.80 (0.78-0.82)	1.29 (1.26-1.32)	0.70 (0.68-0.72)

Table 5. Mandibular measurements of Recent and fossil *Brachyphylla nana* from the Bahamas, Cuba, and Hispaniola and Recent *B. cavernarum* from Puerto Rico. Mean, observed range (in parentheses), and sample size are provided for each measurement.

Species and Locality	length of mandibular toothrow	thickness of ramus at m_3	height of coronoid process
Brachyphylla n. nana Andros and New Providence Bahamas (fossil)	9.7 (9.4 - 10.2) 3	1.3 (1.2 - 1.4) 3	7.6 -- 1
Brachyphylla n. nana Cuba (Recent)	9.8 (9.3 - 10.2) 22	1.2 (1.0 - 1.4) 20	7.1 (6.6 - 7.7) 20
Brachyphylla nana ssp. Grand Caicos (fossil)	10.7 (10.6 - 10.9) 3	1.4 (1.3 - 1.6) 5	--
Brachyphylla nana pumila[1] Grand Caicos (Recent)	10.0 (9.6 - 10.4) 10	1.2 (1.2 - 1.3) 10	7.4 (7.1 - 7.7) 10
Brachyphylla nana pumila Hispaniola (Recent)	10.1 (9.8 - 10.6) 14	1.1 (1.0 - 1.3) 18	7.1 (6.7 - 7.4) 18
Brachyphylla cavernarum[1] Puerto Rico (Recent)	11.0 (10.5 - 11.4) 10	1.4 (1.3 - 1.5) 10	9.0 (8.5 - 9.4) 10

[1]Measurements from Swanepoel and Genoways (1978, Table 9).

Table 6. Measurements of the mandible and lower molars of fossil *Natalus major* from the Bahamas and Cuba and Recent *N. major* from Jamaica. Mean, observed range (in parentheses), and sample size are provided for each measurement.

Taxon and locality	total length of mandible	alveolar length of mandibular toothrow	length of m_1-m_3	length of m_1
Natalus major primus Andros and New Providence, Bahamas (fossil)	15.3 -- 1	9.6 -- 1	5.09 (4.90-5.31) 6	1.77 (1.72-1.83) 10
Natalus major primus[1] Cuba (fossil) (N=3)	14.6 (14.4-14.9)	9.0 (8.9-9.3)	--	--
Natalus m. major Grand Caicos (fossil) (N=1)	--	8.0	4.31	1.49
Natalus major jamaicensis Jamaica (Recent) (N=4)	13.7 (13.6-13.8)	8.3 (8.2-8.3)	4.42 (4.38-4.45)	1.58 (1.57-1.61)

Table 6. Continued

	width of m_1	length of m_2	width of m_2	length of m_3	width of m_3
Natalus major primus Andros and New Providence, Bahamas (fossil)	1.14 (1.07-1.20) 10	1.78 (1.72-1.87) 11	1.14 (1.09-1.24) 11	1.72 (1.67-1.78) 7	0.94 (0.91-0.96) 6
Natalus major primus[1] Cuba (fossil) (N=3)	--	--	--	--	--
Natalus m. major Grand Caicos (fossil) (N=1)	0.93	1.47	0.92	1.46	0.80
Natalus major jamaicensis Jamaica (Recent) (N=4)	1.05 (1.04-1.08)	1.53 (1.50-1.56)	1.01 (1.00-1.02)	1.47 (1.46-1.48)	0.88 (0.86-0.89)

[1] Measurements from Goodwin (1959, Table 1).

Table 7. Cranial, mandibular, and dental measurements of Recent and fossil *Geocapromys ingrahami* from the Bahamas and Recent *G. brownii* from Jamaica. Mean, observed range (in parentheses) and sample size are provided for each measurement.

Species and Locality	alveolar length of maxillary toothrow	breadth of palate anterior to P^4	breadth of palate at posterior edge	length of P^4	width of P^4
Geocapromys ingrahami East Plana Cay (Recent)	15.6 (14.8-16.4) 19	2.4 (1.9-2.8) 19	4.8 (3.8-5.6) 19	4.3 (3.8-4.9) 19	3.1 (2.6-3.4) 19
Geocapromys ingrahami Banana Hole New Providence (fossil)	17.0 (15.5-19.4) 41	2.5 (2.1-3.5) 39	5.5 (4.8-6.5) 23	4.9 (4.4-5.6) 12	3.7 (3.3-4.3) 12
Geocapromys ingrahami Coleby Bay Cave Andros (fossil)	16.4 (14.9-18.5) 6	2.6 (2.4-2.8) 5	5.2 (4.6-5.7) 4	--	--
Geocapromys ingrahami Ashton Cave Andros (fossil)	--	--	--	--	--
Geocapromys ingrahami King Cave Andros (fossil)	17.6 (17.3-18.2) 3	2.5 (2.0-2.9) 3	6.6 -- 1	--	--
Geocapromys ingrahami Upper Pasture Cave Little Exuma (fossil)	16.4 (15.7-17.5) 5	2.5 (2.3-2.8) 8	5.4 (4.7-5.7) 5	--	--
Geocapromys ingrahami Gordon Hill caves Crooked (archaeological)	16.5 (15.6-17.8) 6	2.4 (2.1-2.8) 6	5.3 (4.8-5.8) 5	4.9 (4.6-5.2) 5	3.4 (3.2-3.8) 5
Geocapromys brownii Jamaica (Recent)	19.3 (18.0-20.3) 19	3.4 (2.6-4.2) 18	6.4 (5.5-7.0) 18	5.2 (4.6-6.4) 15	4.1 (3.5-4.6) 15

Table 7 Extended.

total length of mandible (symphysis to condyle)	length of lower diastema	alveolar length of mandibular toothrow	occlusal length of lower cheek teeth	length of P_4	width of P_4
37.7 (35.5-39.5) 16	10.1 (9.2-11.3) 18	15.2 (13.9-16.2) 17	14.6 (13.7-15.6) 17	3.9 (3.7-4.1) 18	2.8 (2.5-3.3) 18
44.3 (42.0-47.1) 11	12.2 (11.1-14.0) 31	17.1 (15.7-20.1) 70	16.6 (15.1-19.2) 48	4.5 (4.0-5.2) 49	3.5 (3.1-4.0) 49
--	11.3 (9.9-13.6) 13	16.5 (14.9-19.3) 33	16.7 (14.7-19.3) 9	4.5 (3.8-5.5) 26	3.3 (2.8-3.8) 26
--	12.4 (11.5-13.2) 2	17.2 (16.3-19.3) 4	17.0 (15.9-18.9) 3	4.8 (4.6-5.2) 3	3.6 (3.4-3.9) 3
--	12.8 -- 1	18.2 (17.9-18.3) 3	18.1 -- 1	4.9 (4.8-5.1) 3	3.8 (3.5-4.0) 3
49.3 -- 1	11.9 (11.0-13.6) 18	17.1 (15.2-19.6) 53	--	4.5 (4.0-5.1) 12	3.3 (3.1-3.6) 12
42.6 (39.9-47.7) 14	10.9 (10.1-11.9) 24	16.7 (15.3-17.7) 51	16.4 (15.9-17.2) 5	4.3 (3.8-4.8) 11	3.7 (2.9-3.7) 11
49.4 (45.5-53.3) 19	13.7 (12.6-16.0) 19	19.1 (17.5-21.0) 19	18.7 (17.3-20.8) 12	4.9 (4.2-5.3) 17	4.0 (3.2-4.5) 17

Table 8. Zoogeographic affinites of the Recent and fossil mammals of the Bahamas.

SPECIES	ZOOGEOGRAPHIC AFFINITIES[1]
Noctilio leporinus	Greater Antilles
Mormoops blainvillii	Cuba
Mormoops megalophylla	Greater Antilles
Pteronotus macleayii	Cuba
Pteronotus parnellii	Cuba
Pteronotus quadridens	Cuba
Macrotus waterhousii minor	Cuba
Macrotus waterhousii waterhousii	Hispaniola
Monophyllus redmani	Greater Antilles
Artibeus jamaicensis	Cuba
Brachyphylla nana nana	Cuba
Brachyphylla nana pumila	Hispaniola
Erophylla sezekorni	Cuba
Phyllonycteris poeyi	Cuba
Natalus major major	Hispaniola
Natalus major primus	Cuba
Natalus tumidifrons	Greater Antilles
Nyctiellus lepidus	Cuba
Eptesicus fuscus	Cuba
Lasiurus borealis	Hispaniola
Tadarida brasiliensis	Hispaniola
Geocapromys ingrahami	Cuba

[1] The island from which the Bahamian population was derived is listed (e.g. Cuba or Hispaniola), except for those species for which the specific zoogeographic origin is indeterminate, in which case the more general affinites are given (e.g. Greater Antilles).

Table 9. Area and number of species of mammals for islands in the Bahamas archipelago. All values are based on Recent land area and fauna except those for Great Bahama Island.

ISLAND	AREA[1] (in km^2)	SPECIES OF MAMMALS
Great Bahama Island (late Pleistocene)	109,400	16
Andros	4,144	4
Abaco[2]	1,680	3
Great Inagua	1,627	5
Grand Bahama	1,114	1
Acklins	524	5
Eleuthera	425	4
Cat	414	4
Long	337	6
Mayaguana	293	3
Grand (=Middle) Caicos	288	4
Crooked	277	5
North Caicos	204	3
Exuma[3]	180	5
East Caicos	178	2
San Salvador	155	4
New Providence	150	4
Little Inagua	127	1
Providenciales	117	5
East Plana Cay	5	2

[1] Land areas for the Bahama Islands are from Webster's Geographical Dictionary (1969) and areas for the Caicos Islands are from Buden (1979).
[2] Abaco includes Great Abaco and Little Abaco.
[3] Exuma includes Great Exuma and Little Exuma.

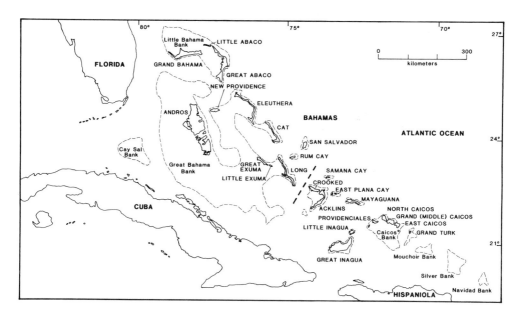

Figure 1. Outline map of the Bahamas archipelago showing all islands mentioned in the text. The thin dashed lines mark the edges of the shallow submarine banks and thus define the approximate size of the islands in the Bahamas during the late Pleistocene glacial maximum. The thick dashed line between Long and Crooked islands marks the Crooked Island Passage, the zoogeographic boundary between the northern Bahamas and southern Bahamas.

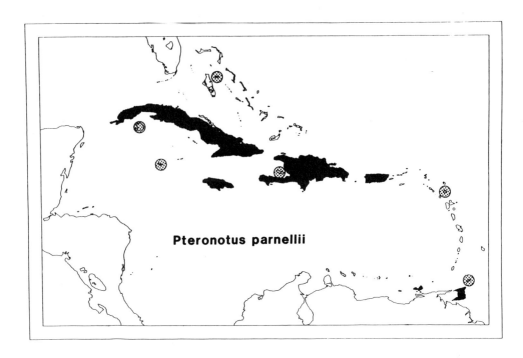

Figure 2. Outline map of the West Indies showing the Recent (Black) and fossil (stippled) distribution of *Pteronotus parnellii*. The mainland distribution of *P. parnellii* is not included.

Fossil Chiroptera and Rodentia from the Bahamas

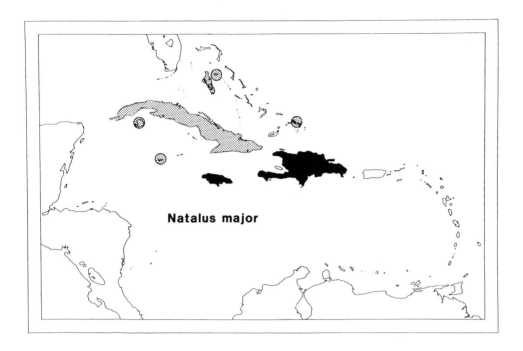

Figure 3. Outline map of the West Indies showing the Recent (black) and fossil (stippled) distribution of *Natalus major*.

738 Biogeography of the West Indies

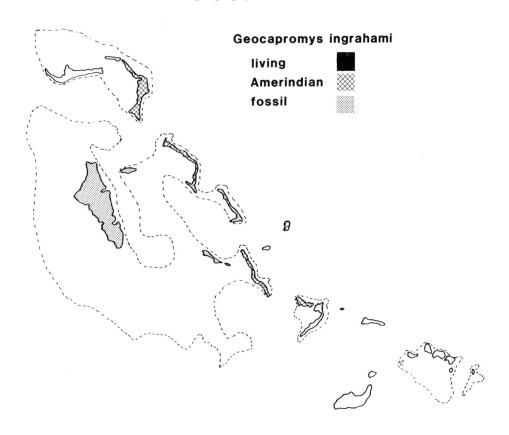

Figure 4. Outline map of the Bahamas showing the Recent (black), archaeological (crosshatched), and fossil (stippled) records of *Geocapromys* in the Bahamas. Archaeological specimens from Providenciales and Grand Caicos are not indicated since they cannot be positively identified as *Geocapromys*.

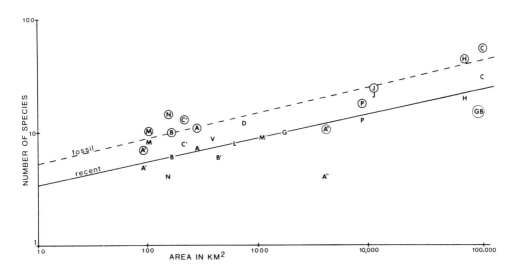

Figure 5. Log-log species-area plot of the Recent and fossil mammalian faunas from selected West Indian islands. Letters represent the number of Recent species recorded for an individual island, while circled letters represent the total number of species recorded from fossil deposits and/or archaeological sites. The continuous line is a log-log species-area plot calculated from the number of Recent species in the modern fauna of each island. The dashed line is a log-log species-area plot based on the species known from late Pleistocene and Holocene fossil deposits and archaeological sites on selected islands that have a reasonably complete record of the pre-human mammalian fauna. Islands on the plot and their abbreviations are as follows: Antigua (A), Anguilla (A'), Andros (A"), Barbuda (B), Barbados (B'), Cuba (C), Cayman Islands (C'), Dominica (D), Guadeloupe (G), Hispaniola (H), Jamaica (J), Montserrat (M), Martinique (M'), New Providence (N), Puerto Rico (P), St. Lucia (L), and St. Vincent (V). Exact land areas and numbers of Recent and fossil species of mammals for all islands in this figure can be found in Morgan and Woods (1986, Table 3, page 192).

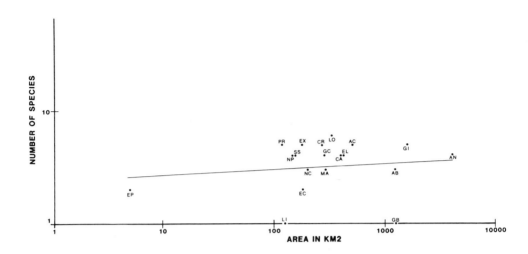

Figure 6. Log-log species-area plot of the Recent mammalian faunas of islands in the Bahamas archipelago. Islands on the plot and their abbreviations are as follows: Abaco (AB), Acklins (AC), Andros (AN), Cat, (CA), Crooked (CR), East Caicos (EC), Eleuthera (EL), East Plana Cay (EP), Exuma (EX), Grand Bahama (GB), Grand Caicos (GC), Great Inagua (GI), Little Inagua (LI), Long (LO), Mayaguana (MA), North Caicos (NC), New Providence (NP), Providenciales (PR), and San Salvador (SS). Exact land areas and the number of species of mammals recorded from each island are given in Table 9.

THE BIOGEOGRAPHY OF WEST INDIAN RODENTS

Charles A. Woods[1]

Abstract

Four families, 18 genera, and 61 species of endemic rodents were widely distributed in the West Indies before the arrival of Amerindians approximately 7,000 yBP. Most forms are now extinct. Except for *Geocapromys ingrahami ingrahami* on East Plana Cay in the Bahamas, at least one species has become extinct on each island where rodents originally occurred. Seventy nine percent of recognized rodent taxa have become extinct since 39,300 yBP. On Hispaniola, 93 percent of the known endemic rodents are extinct, and 100 percent are extinct in the Lesser Antilles. The oldest known endemic rodents of the West Indies are echimyid rodents of the Subfamily Heteropsomyinae, all of which are now extinct. This group appears to have dispersed directly from South America to Puerto Rico or Hispaniola in the late Oligocene or early Miocene, perhaps following the Aves Ridge or Lesser Antilles. Capromyids evolved from echimyids on Hispaniola. Three of the four capromyid subfamilies are confined to Hispaniola. After the differentiation of the capromyid subfamilies on Hispaniola, the Subfamily Capromyinae originated on Cuba, where 87 percent of capromyines still occur. While only two capromyine genera are recognized, 75 percent of all capromyid species are in the Subfamily Capromyinae. The species richness of capromyines on Cuba is likely the result of Cuba's geography. It is a large island with isolated mountain areas surrounded by extensive low lying archipelagos. This geography led to a proliferation of species during Pleistocene climate cycles, and has insulated recent capromyines from the pattern of extinction found on other islands. Seventy seven percent of surviving West Indian capromyids are restricted to the Cuban Archipelago. Giant hutias of Jamaica, Hispaniola, Puerto Rico, and the St. Martin Bank are here split into two subfamilies, but not assigned to a family. One subfamily (Heptaxodontinae) is convergent with the capromyid Subfamily Plagiodontinae, which may explain the strong resemblance between *Quemisia* and *Plagiodontia*. The other subfamily (Clidomyinae) is restricted to Jamaica.

Introduction

The classification of West Indian mammals on which this biogeographical analysis is based differs in many ways from that of Varona (1974), which is the best available summary of the mammals of the region. The present biogeographical analysis is

---

[1] Dr. Woods is Curator of Mammals at the Florida Museum of Natural History, and Professor of Zoology at the University of Florida, Gainesville, FL. 32611.

based mainly on the systematics of rodents in the families Capromyidae, Echimyidae, and giant hutias (family *incertae sedis*). The muroid rodents of the Subfamily Sigmodontinae (Oryzomyinae of some authors) are mentioned, but are not part of the revised classification or cladistic analysis. Sigmodontines were excluded from the cladistic analyses because they are not part of the same suborder (Hystricognathi) as capromyids and echimyids, and their distribution in all but one case is restricted to the Lesser Antilles. The taxonomic diversity of the West Indian sigmodontines is also much more limited than the diversity of capromyids and echimyids (Table 1).

The endemic rodents of the West Indies were diverse in morphology and ranged in mass from approximately 200 gr to over 200 kg. The four known families, 18 genera, and 61 species were widely distributed in the Antilles before the arrival of Amerindians (Tables 1, 5, 7). Endemic rodents are recorded from Barbados and St. Vincent in the south to the western tip of Cuba and Little Swan Island in the west, a distance of over 3,000 kms. The existence of many taxa is well documented by specimens, although some are only known from fragmentary remains. The presence of some taxa in the West Indies cannot be confirmed, however, such as the possible existence of an endemic porcupine, *Sphiggurus pallidus* (see Honacki et al. 1982:571). Others species are still being described, such as a new extinct capromyid from Hispaniola (Woods 1989a) and several extinct rice rats from the Lesser Antilles (Steadman et al. 1984a,b). Other named taxa, such as *Plagiodontia hylaeum*, are now recognized as being conspecific with closely related forms such as *P. aedium* (Anderson 1965). The incomplete record of rodents in the West Indies and the unstable classification has made it difficult to discuss the radiation of West Indian rodents in detail, or to calculate the true rates of extinction. Here I synthesize and evaluate all available data on West Indian endemic rodents, and develop an hypothesis on the origin and evolution of the group based on a revised systematic scheme utilizing data on several newly discovered specimens and a new analysis of the relationships of the various subgroups of Antillean rodents.

Materials and methods

An analysis of 82 morphological features on 18 separate taxa was undertaken using the computer software program PAUP (D. Swofford, Illinois Natural History Survey). The morphological features analyzed are listed in Table 2, and include 33 dental, 16 cranial, four mandibular, three postcranial, and 26 muscle characters. All features were rated as present or absent, or in the case of variable character states, were given a score of 1-5. The data set used in the PAUP analysis is presented in Table 3. The results of the PAUP analysis are presented in Figs. 3-5 as a series of cladograms. The species included in the cladistic analysis are the following: **Hispaniola-** *Plagiodontia aedium, P. araeum, Rhizoplagiodontia lemkei, Isolobodon portoricensis, I. montanus, Hexolobodon phenax, Brotomys voratus*; **Cuba-** *Capromys pilorides, C. nanus, Boromys torrei*; **Bahamas-** *Geocapromys ingrahami*; **Jamaica-** *Geocapromys brownii*; **Puerto Rico-** *Puertoricomys corozalus, Heteropsomys insulans, Homopsomys antillensis, Elasmodontomys obliquus*; **South America-** *Proechimys iheringi, P. semispinosis*. The following taxa were also investigated, but not included in the cladistic analysis: **Hispaniola-** *Plagiodontia velozi, Quemisia gravis*; **Jamaica-** *Clidomys parvus*; **St. Martin-** *Amblyrhiza inundata*; **Martinique-** *Megalomys desmarestii*; **Lesser Antilles-** *Megalomys* sp, *Oryzomys* sp. For all other taxa discussed data were gathered from published accounts. The revised classification of these rodents is presented below, and summarized in Table 1. Some of the data in Table 2 have been previously discussed and analyzed in a series of separate cladograms (Woods and Hermanson 1985) to evaluate the importance of different categories of morphological features, such as dental, cranial, post-cranial, muscular, and biochemical. One conclusion of that study was that dental features are variable and more subject to convergence than are other kinds of data, such as muscle morphology and blood proteins. Most West Indian rodents are extinct and known mainly as dental remains, however, and so by necessity many dental characters have been included in the cladistic analysis that forms the basis of this study. Characters from other anatomical regions have been included in the present analysis (Table 2) to minimize any possible errors resulting from a heavy reliance on dental features.

The analysis of relationships of West Indian echimyids was undertaken by examining a large series of molariform teeth of *Homopsomys antillensis* from Puerto Rico to determine wear patterns. This complete series from unworn to very worn dentitions was then compared with molariform teeth of a series of heteropsomyines from Blackbone 1 cave in Puerto Rico, the type of *Puertoricomys corozalus*, and a series of heteropsomyines from Hispaniola and Cuba. Cranial and mandibular material from these specimens was used in the comparisons when available

Antiquity of the known rodent taxa

The age of most specimens of West Indian rodents is unknown. The fossil record of all but one species is limited to material collected from sediments in caves, sink holes or collapsed cave systems. In Hispaniola fragmentary post-cranial elements that were in close association with known taxa were dated by carbon 14/13 analysis in six caves (Table 4). Dates range from 1,600 yBP 15 cm below the surface to 21,170 yBP 100 cm deep near the rocky base of the sediment profile (Woods 1989a). The top 10 to 15 cms of sediment usually contain taxa mixed with *Rattus*, indicating a post-Columbian date. In Jamaica carbon 14 dates were determined for snail shells located in mammal bearing layers (Goodfriend this volume; Goodfriend and Mitterer 1987) and turtle shell fragments (MacPhee 1984). These data from Jamaica indicate a maximum antiquity 39,300 yBP (see Goodfriend and Mitterer 1987 for a discussion of the difficulties of obtaining accurate dates on shells and bone). Pregill (1981) estimated the age of specimens collected in Blackbone 1 cave in Puerto Rico to be 20,000 yBP. Morgan (this volume) obtained a date of 7,980 yBP for a sample of bones from Banana Hole in the Bahamas. The pattern of dates of specimens collected in diverse cave deposits in Jamaica, Puerto Rico, and Hispaniola indicate that most of the known species from diverse locations on several different islands were roughly contemporaneous in late Pleistocene and Holocene times.

The antiquity of the specimens of *Clidomys* from Jamaica has been a matter of frequent speculation. Anthony (1920) estimated an age of 100,000 yBP. Koopman and Williams (1951) noted that the *Clidomys*-bearing layer of Sheep Pen Cave may have been reworked and much older than the surrounding layers in which remains of the still extant *Geocapromys brownii* are abundant. MacPhee (1984) also noted the possible antiquity of specimens of *Clidomys* he investigated. The most definitive date for an old bone-bearing layer in the West Indies comes from MacPhee et al. (in preparation). Using uranium series and electron spin resonance dating techniques on a series of cave conglomerates and calcite stringers, they demonstrate that the *Clidomys* bearing layers of Wallingford Roadside Cave in Jamaica are in excess of 100,000 yBP. The most certain date in their series is 95,000 þ 4,000 yBP. This is the oldest date for a West Indian cave fauna.

A rodent specimen collected in a limestone crevice near Corozal, Puerto Rico in 1931 may be considerably older than specimens collected in cave deposits. This specimen was named *Proechimys corozalus* by Williams and Koopman (1951). Subsequent attempts to collect additional material from the quarry were not successful (E. Williams pers. comm.). Williams and Koopman noted that the specimen was associated with fragments of a turtle and the femur of a large lizard in addition to other bone fragments too poorly preserved to be identified. The lizard femur is difficult to relate to any extant or known taxon from the West Indies. The rodent specimen, a partial left dentary of a medium sized animal, is dark grey in color and mineralized. It appears to be of considerable age, but no data other than the association of the specimen with an unidentified large lizard and its "old" appearance are available to confirm the antiquity of the fossil.

The oldest confirmed record of a West Indian land mammal is found in amber from the northern region of the Dominican Republic. A piece of amber loaned to the Florida State Museum in 1982 by Dr. Pompilio A. Brouwer contains several 20 mm long hairs from an unidentified mammal. Dominican amber has been dated as early Miocene, 20-23 myBP (Rieppel 1980). Other mammalian hairs found in Dominican amber were recently discussed by Poinar (1988), who concluded that they were from a rodent based on their association with two listrophorid fur mites. The majority of present day listrophorids are restricted to rodents. Other dates for fossil vertebrates in the West Indies also come from amber collected in the northern part of the Dominican Republic (an anole lizard, *Anolis dominicanus* (Rieppel 1980); a gecko, *Sphaerodactylus dommeli* (Schlee 1984); and a hylid frog, *Eleutherodactylus* (Poinar and Canatella 1987). The age of the amber in which the *Eleutherodactylus* was collected is considered to be Eocene in age by Poinar and Canatella. A fossil freshwater fish (*Cichlasoma woodringi*) from the Central Plateau of Haiti has been dated as Miocene (Cockerell 1923). The oldest known fossil vertebrate from the West Indies is a pterosaur (*Nesodactylus hesperius*) from Cuba (Colbert 1969) which is Jurassic (Oxfordian) in age. The fossil pterosaur came from black limestone accretions characteristic of marine facies, so the site may not have been totally terrestrial.

The above data indicate that our knowledge of the history of West Indian rodents is very limited. The presence of terrestrial vertebrates in localities dating back to the early Miocene and perhaps as long ago as the Eocene indicates that non-submerged areas were present in the region for at least the past 23 million years, and that it is possible for rodents to have existed in the West Indies well back into the Tertiary. Poinar (1988) suggests that it is possible that rodents were present on Hispaniola (or at least the Cordillera Septentrional area of the northern Dominican Republic) as far back as the upper Eocene, although most authors consider Dominican amber to be no older than the lower Miocene (Woodruff pers. comm.). The overall number of species and diversity of the mammalian fauna of the Greater Antilles compare closely with the mammalian fauna from Madagascar (Woods and Eisenberg this volume), where the fossil record is much more complete. It appears that there were few major extinctions of terrestrial mammals in the West Indies in the Pleistocene before the arrival of Amerindians approximately

7,000 yBP, and that a good representation of the late Tertiary mammalian diversity of the West Indies exists even though the temporal scope of most collections is limited to the past 40,000 years.

Extinctions

Many West Indian rodents have become extinct since the late Pleistocene (Table 6). Approximately 79 percent of known West Indian rodents have become extinct since 39,300 yBP, the oldest Carbon 14 date for an assemblage of fossil West Indian land mammals. An analysis of the remains of rodents in cave and sinkhole deposits in Hispaniola indicates that all known taxa were present until after the arrival of Amerindians on the island between 4,500 and 7,000 yBP (Rouse and Allaire 1978, Marcio Veloz Maggiolo pers. comm.), and that no known rodent taxon became extinct on Hispaniola before 3,000 BP. The pattern of extinction varies from island to island, and within each major subgroup of West Indian rodents. For example, 93 percent of the 14 species and five subfamilies of Hispaniolan rodents have become extinct. The extinction has been greatest on small islands (less than 3,000 km^2), which include all of the Lesser Antilles, Virgin Islands, Cayman Islands, Turtle Island, Gonave Island, Mona Island, Navassa Island, Beata Island, the Bahamas, and the Swan Islands. On these small islands, with the exceptions of East Plana Cay in the Bahamas and the low lying bank-like archipelagos off Cuba, all endemic rodents have become extinct. The explanation for this high rate of extinction lies in the depauperate nature of small islands (see the species area curves in Morgan and Woods 1986), and the speed at which the effects of disturbances spread throughout small islands. An indication of the latter problem is illustrated at the speed with which the mongoose has reached high densities of six/ha on small islands. In most habitats on large islands, such as Hispaniola and Jamaica, the density of mongoose is only about two/ha (Kilpatrick and Hoagland this volume). Another indication of how fast extinction can occur on a small island is the speed at which *Geocapromys thoracatus* became extinct on Little Swan Island less than ten years after the introduction of cats (Morgan 1985). The *Geocapromys* (þ 1 kg) that survives on East Plana Cay in the Bahamas and the four species of *Capromys* on the small cays of the coast of Cuba probably survive because dogs and cats have not yet been introduced there.

Body size is one of the most important factors associated with extinction of rodents in the West Indies. The largest rodents were the first known forms to become extinct. All giant hutias (usually classified together in the Family Heptaxodontidae but here classified as two subfamilies, Heptaxodontinae and Clidomyinae, in the family *incertae sedis*), are now extinct. Heptaxodontines ranged in size from approximately 20 kg (*Quemisia* on Hispaniola) to an estimated 150 kg (*Amblyrhiza* on St. Martin and Anguilla). Heptaxodontine remains are known only from paleontological sites, so there is no way to confirm that they were present at the time Amerindians arrived. *Quemisia* remains, however, are present in deposits that date after the presumed earliest arrival of Amerindians in Hispaniola. The remains of the other heptaxodontines have been recovered in nearly comparable paleontological sites at approximately the same depths (Table 4), indicating that it is possible that heptaxodontines were still extant 7,000 years ago.

The remains of giant hutias in Jamaica and St. Martin are frequently associated with old breccia deposits, and the assumption has been that giant hutias on Jamaica may have become extinct earlier that many other West Indian rodents. In Jamaica, however, the remains of *Clidomys* are not only found in old breccia deposits that date in excess of 100,000 yBP, but also occur together with the remains of *Geocapromys brownii* in cave fill deposits (MacPhee 1984). Fragmentary remains of *Clidomys* collected on the surface of Wallingford Roadside Cave have a "fresh " appearance according to MacPhee (1984:11). No comparable data are available on the stratigraphic relationships and antiquity of rodent fossils for Puerto Rico or St. Martin and Anguilla. However, fossil remains of *Elasmodontomys* from Puerto Rico and *Amblyrhiza* from the St. Martin bank are abundant in collections as compared with the paucity of fossil remains of the Jamaican and Hispaniolan giant hutias. The abundance of fossils is not directly related to the relative abundance of the animals themselves. However, if *Quemisia* is known to have survived on Hispaniola until after the arrival of Amerindians, it is possible that the remaining giant hutias of the West Indias (all of which are much more abundant in collections than is *Quemisia* in spite of an intensive collecting effort in Hispaniola) also survived until the first arrival of humans on the islands of the Greater Antilles. The hypothesis adopted here, therefore, is that all heptaxodontines, and possibly *Clidomys* as well, were present up until the time of the arrival of Amerindians in the Antilles, and that they were quickly decimated by the first Indians. The larger the island, the longer the taxon appears to have survived. *Quemisia* survived on Hispaniola in the protected and remote Central Plateau of Haiti until about 3,000 yBP.

Among hutias (Capromyidae), large forms also generally became extinct before small to medium sized forms. In Hispaniola the large and apparently terrestrial hutias *Hexolobodon phenax* and *Plagiodontia araeum* (both with an estimated body mass in excess of 5 kg) became extinct 1-3,000 yBP, whereas *P. aedium* (about 1 kg) continues to survive. *Isolobodon montanus*, another large and broad toothed hutia about 2 kg in mass apparently became extinct about 3,000 yBP. The smaller *I. portoricensis* (1 kg) persisted on the mainland of Hispaniola until after the arrival of Europeans, and into this century on Turtle Island off the north coast of Haiti (Woods et al. 1985). Other large capromyids, such as *Plagiodontia velozi*, survived on the higher interior montane plateaus of Hispaniola long after becoming extinct in lowland areas.

West Indian rodents were a favored food item by Amerindians, and are common in midden deposits. The extent of the use of rodents is illustrated by the status of *Isolobodon portoricensis*. This species, a native of Hispaniola, was introduced onto Puerto Rico and the Virgin Islands by Indians where its remains appear in midden deposits. This is the only confirmed case of Amerindians having transported endemic rodents from one island to another in the Antilles, although it is suspected that Amerindians transported sigmodontine rodents around the Lesser Antilles (Steadman et al. 1984a,b), and it is likely that the agouti was transported by Indians to several West Indian islands. *Isolobodon portoricensis* was probably kept in captivity by Amerindians judging by the abundance of remains in midden deposits. Remains of this taxon are more abundant in middens than are remains of any other endemic rodent, including the still extant and widely distributed Hispaniolan Hutia *Plagiodontia aedium*.

In Puerto Rico, *Isolobodon portoricensis* appears to have persisted into the present century, whereas the large heptaxodontine rodent *Elasmodontomys* and two large heteropsomyines (*Homopsomys* and *Heteropsomys*), all of which were larger than *I. portoricensis*, became extinct in pre-Columbian times. The same pattern exists in Jamaica, where both species of *Clidomys* became extinct at about the time Amerindians arrived on the island (see above), while the smaller capromyine *Geocapromys brownii* still survives. The extinction of large taxa occurred in other rodent west Indian mammalian lineages as well. The largest known West Indian insectivore (*Solenodon* sp. from Cuba), which was the size of an American opossum, became extinct in late Pleistocene or Holocene times (Morgan et al. 1980). Smaller insectivores survived until much later on Cuba and Hispaniola. All known West Indian primates and sloths also became extinct before the arrival of Europeans. In all cases the probable cause of extinction of the megafauna was overexploitation by Indians.

Small rodents also are prone to extinction in the West Indies. Rat sized rodents, between 200 and 500 grs in mass, seem to have become vulnerable following the arrival of Europeans. Sigmodontine rodents are now all extinct. Some species were a favored food item of Indians, as judged by the abundance of rice rats in archaeological sites (Steadman et al. 1984a,b). However, other sigmodontines survived in the Lesser Antilles until the last hundred years, as did *Oryzomys antillarum* from Jamaica. These small endemic rodents may have been eliminated by competition from introduced rats in combination with habitat destruction (Steadman et al. 1984a). The genus *Nesophontes* (small insectivores) and the West Indian spiny rats (Subfamily Heteropsomyinae) are also apparently now extinct, even though large areas of suitable habitat still exist in their original range in Cuba, Hispaniola and Puerto Rico. Reports of their presence persisted well into this century (Woods et al. 1985). They were nearly the size of *Rattus norvegicus*, and may have become extinct for the same reason as sigmodontines and the West Indian echimyids. Larger Antillean insectivores (*Solenodon*) continue to survive on Cuba and Hispaniola, and a larger capromyid with similar habits to echimyids (*Capromys nanus*) has survived long after the extinction of echimyids on Cuba. *Capromys nanus* is the most endangered of the surviving capromyids, however, and may have become extinct in the last decade (Woods 1989b). Therefore, it is clear that being about the size of a rat following the introduction of *Rattus* made a species especially vulnerable to extinction.

Rats are now distributed throughout all natural habitats in the West Indies, and inhabit even the most remote forest covered mountain peaks of Hispaniola where *R. rattus* and *R. norvegicus* occur in large numbers (Woods et al. 1985). The introduction of the domestic cat and large dogs by European settlers, also contributed to the extinction of small endemic rodents. The remains of young *Plagiodontia aedium* have been recovered in dog scat at three separate locations in Haiti (pers. obs.). The introduction of the mongoose in the late 19th Century added to the problems faced by small terrestrial rodents. The impact of mongooses on rodents and insectivores in the West Indies has never been documented, although it has been noted by Kilpatrick (this volume) that Norway rats, which are terrestrial, are uncommon or absent in areas of high mongoose densities, but abundant in habitats where mongooses are scarce, such as high mountains. The more arboreal black rat coexists well with mongoose. The combination of competition and predation by introduced forms is the probably reason for the extinction of *Nesophontes, Brotomys, Boromys*, and *Oryzomys antillarum* in the Greater Antilles, and several species of *Oryzomys* and *Megalomys* in the Lesser Antilles.

In summary, 79 percent of rodent species and 84 percent of the genera are now extinct in the West Indies. This high rate of extinction is reflected in all non-volant West Indian mammals, where 100 percent of the Edentates and Primates, and 75 percent of the insectivores have become extinct (Table 6). Only 14 percent of known West Indian bats have become extinct during the same period. Animals on small islands are especially vulnerable, while those on larger islands are most resistant to extinctions. Rodents on small islands where cats, dogs, mongoose, and rats are absent have managed to survive, such as on East Plana Cay and High Cay in the Bahamas, and on small islands of the Cuban Archipelago. Most taxa were still present 3,000 years ago, with the rate of extinction beginning to accelerate about 1,000 yBP. This is the time that Amerindians began to move into remote interior valleys and onto the high mountain plateaus (Rouse pers. comm.) that served as the final refugia for the larger endemic terrestrial mammals. Many taxa were present at the time Columbus discovered the West Indies, and well into recent times (Woods et al. 1985). The West Indian mammals that became extinct following the colonization of the West Indies by Europeans were smaller forms about the size of rats, and the few remaining large forms such as *Plagiodontia velozi*. Intermediate size hutias with generalist food habits such as *P. aedium* from Hispaniola, *G. brownii* from Jamaica, *G. ingrahami* from the Bahamas, and several species of *Capromys* from Cuba, still survive. Most of these species are threatened or endangered (Woods 1989b).

Classification of selected west indian rodents

Family Echimyidae Miller and Gidley 1918
Subfamily Heteropsomyinae Anthony 1918:407

DIAGNOSIS.--Molariform teeth always rooted; lacking cement within reentrant folds or around shafts; rostrum short; paroccipital process moderately long and not closely appressed to auditory bulla; ventral canal in infraorbital foramen present as groove (*Boromys*) or lacking (all other forms).

COMMENTS.--This is the oldest taxonomic assemblage of rodents in the West Indies, and the group with the largest number of characters in common with non-Antillean echimyids. The primitive pattern of the reentrant folds in this group has an oblique fold across the entire occlusal surface, and the derived pattern has three counterfolds opposite the hypoflexus(id). The most primitive heteropsomyines are similar in morphology to the South American subgenus *Trinomys* of *Proechimys*. The previous use of the name Heteropsomyinae by Patterson and Pascual (1968) to include all living echimyids exclusive of the dactylomyines and echimyines is inappropriate and not what I believe Anthony had in mind when he established this subfamily for the Puerto Rican taxa (Woods 1984:436-437).

Genus *Puertoricomys* new gen.

DIAGNOSIS.--Large body size; molar teeth robust; masseteric shelf narrow; masseteric fossa shallow; hystricognathous groove shallow, reentrant folds of molariform teeth with one long fold passing entirely across occlusal surface; incisor very deep with width less than 50 percent of depth (Fig. 1a).

COMMENTS.--This taxon was originally described as *Proechimys* based on the type (and only known) specimen *Proechimys corozalus* (Williams and Koopman 1951). The closest affinity was presumed to be with *Proechimys iheringi* of the subgenus *Trinomys*.

Both forms have in common the pattern of reentrant folds in the molariform teeth in which a single fold passes entirely across the occlusal surface. It differs from *Proechimys iheringi* and all other members of the subgenus *Trinomys* in being larger in size, more robust in morphology, lacking the pronounced masseteric crest and deep masseteric fossa, and in having two anterofossettes instead of one. It is similar in dental morphology to unworn molariform teeth of *Heteropsomys* and *Homopsomys* of Puerto Rico (Fig. 1b). *Puertoricomys corozalus* is part of the same radiation as the Puerto Rican echimyids, and may even be directly ancestral to *Heteropsomys*. It should be classified in its own genus to acknowledge this distinctness from *Proechimys*, and its relationship with the Antillean echimyids. The relationship of this genus with other echimyids is illustrated in Figs. 3-5.

TYPE SPECIMEN.--A partial left mandible with M_1-M_3 (AMNH 17640) collected at Corozal, Puerto Rico, January 19, 1931. Measurements of holotype (in mm); length of alveolar tooth row 13.38; length of M_1 3.20; length of M_2 3.10; length of M_3 2.45; width of M_1 3.30; width of M_2 3.45; width of M_3 3.0; width of incisor 1.45; depth of incisor 3.15.

Genus *Heteropsomys* Anthony 1916:202

DIAGNOSIS.--Mesopterygoid fossa (palatine notch) deep and extending to level of M^1; incisive foramen wide and with lip; post orbital process moderate.

COMMENTS.--Varona (1974) combined *Heteropsomys* and *Homopsomys* in the subgenus *Heteropsomys* and combined *Brotomys* and *Boromys* in the subgenus *Brotomys*. He classified all four forms in the genus *Heteropsomys*. Varona did not present any reasons for these combinations, and the present study identifies many features that clearly categorize each taxon as a distinct genus.

DISTRIBUTION.--Restricted to Puerto Rico.

Genus *Homopsomys* Anthony 1917:187

DIAGNOSIS.--Mesopterygoid fossa shallow and extending to level of $M^{2/3}$; incisive foramen moderate in width with lip; post orbital process large.

COMMENTS.--The largest, most robust, and most hutia-like of the heteropsomyines.

DISTRIBUTION.--Puerto Rico and Vieques Island.

Genus *Brotomys* Miller 1916:6

DIAGNOSIS.--Mesopterygoid fossa shallow and extending to mid-M^3; incisive foramen very narrow and without lip; post orbital process slight; paroccipital process moderate; molariform teeth robust with a 2:1 pattern of counterfolds

COMMENTS.--About same size as *Boromys* from Cuba and much smaller than any of the heteropsomyines from Puerto Rico.

DISTRIBUTION.--Hispaniola and La Gonave Island.

Genus *Boromys* Miller 1916:7

DIAGNOSIS.--Mesopterygoid fossa moderate and extending to level of M^2; incisive foramen moderate and without lip; post orbital process lacking; paroccipital process short; molariform teeth delicate with a 2:1 pattern of counterfolds; ventral canal of infraorbital canal present.

COMMENTS.--Miller (1916:7) described the molariform teeth of *Boromys* as being more robust than in *Brotomys*, while the opposite is true.

DISTRIBUTION.--Cuba and Isle of Pines.

Family Capromyidae Smith 1842

DIAGNOSIS.--Molariform teeth open rooted for at least part of life; cement always present within reentrant folds; paroccipital process long and well separated from auditory bulla (Fig. 2a). Some individuals replace the deciduous molar in old age, especially in *Plagiodontia*, which can live as long as 17 years.

COMMENTS.--This group includes all of the West Indian hutias. A significant morphological feature of the group is the presence of open rooted evergrowing molariform teeth. This character is variable, however, with roots forming soon after birth in *Rhizoplagiodontia*, and at middle age in *Hexolobodon*. Most other dental characters, such as the number and shape of the reentrant folds, the amount and distribution of cement, thickness of enamel, and height and orientation of teeth are also variable. When considered in isolation, the variation seems extreme and led Simpson (1945:98) to place *Isolobodon* in the Echimyidae while retaining all of the other hutias in the Capromyidae. When viewed collectively, however, the relationships of all of the dental features of the group fall into a clade (Woods 1989a). The great morphological diversity, especially within the well defined subfamilies discussed below, suggests strong selective pressures, isolated evolutionary histories, and an initial invasion a long time ago.

Subfamily Capromyinae Thomas 1896:1012 (part)

DIAGNOSIS.--Molariform teeth always open rooted; extensive cement present within reentrant folds which also surrounds the cheek teeth as a uniform layer; reentrant folds nearly perpendicular to body axis (Fig. 2a).

COMMENTS.--Includes *Capromys* (23 species) and *Geocapromys* (seven species). Many taxa are very similar to one another and may be only geographic populations (as in *Geocapromys ingrahami*). All chew or chewed in a mainly anteroposterior pattern.

In spite of the great morphological similarity of the species classified here as *Capromys*, data from blood protein electromorphs (Comache in litt.) and ectoparasites (Jorge de la Cruz pers. comm.) indicate that *Capromys* may be divided into three living genera (*Capromys*, *Mysateles*, and *Mesocapromys*). The distribution of species in these genera are summarized in Table 1 where they are listed as subgenera.

The small dwarf hutia from Cuba, *Capromys* (*Mesocapromys*) *nanus*, was originally named as *Capromys nana* by Allen (1917:54). However, in order for the specific name to be in concordance with the grammatical gender of the genus (masculine), the proper name for the species should be *C. nanus*.

DISTRIBUTION.--Confined to western regions of the Greater Antilles (Cuba, Cayman Islands, Jamaica, Little Swan Island and the Bahamas).

Subfamily Plagiodontinae Ellerman 1940:25

DIAGNOSIS.--Molariform teeth nearly horizontal (i.e. occlusal surfaces on the same plane); moderate cement within reentrant folds, but no cement surrounding teeth; enamel parts of cheek teeth robust; reentrant folds tending towards oblique; metaflexus present as major posterolateral reentrant fold; labial and lingual reentrant folds overlapping significantly in central occlusal area of all upper molariform teeth; deciduous molar replaced in some old individuals Fig. 2a).

COMMENTS.--Includes *Plagiodontia* (five species) and the newly described *Rhizoplagiodontia* (Woods 1988). Dental morphology varies from the condition in *Rhizoplagiodontia* in which the roots of the molariforms are open only briefly (in juveniles) and functional hypsodonty in adults is maintained by massive deposits of cement at the base of the alveolus, to the condition in *Plagiodontia araeum* which has completely open rooted molariforms but has the occlusal surface of the cheek teeth at nearly the same level as the gum line (i.e. low). The most extreme modification in dental morphology in this group is found in *P. velozi* where the completely open rooted molariform teeth have the occlusal surface far above the gum line. *Hyperplagiodontia stenocornoalis* (Rimoli 1976:33) is conspecific with *P. aedium*. *Plagiodontia caletensis* (Rimoli 1976:30) is conspecific with *P. ipnaeum*. This is a species rich group that ranges in mass from 500 gr to an estimated 5 kg.

DISTRIBUTION.--Restricted to Hispaniola and offshore island of La Gonave.

Subfamily Isolobodontinae Woods 1988

DIAGNOSIS.--Molariform teeth very hypsodont and open rooted; cement within reentrant folds and surrounding cheek teeth produced in bands; metaflexus lost in all upper molariform teeth; labial and lingual reentrant folds not overlapping in central occlusal area of upper molariform teeth; opposing reentrant folds with tendency to meet and merge to form lamellar plates; enamel structure of cheek teeth thin and soft (Fig. 2b).

COMMENTS.--The genus *Aphaetrius* (Miller 1922, 1929a) is not separable from *Isolobodon* based on morphological features and is included within this subfamily as *Isolobodon montanus*.

DISTRIBUTION.--Restricted to Hispaniola and its adjacent islands.

Subfamily Hexolobodontinae Woods 1988

DIAGNOSIS.--Molariform teeth very hypsodont in young individuals but roots closing in mid-age so tooth wears to become short and prominently rooted; root of deciduous molar anterior to zygoma, and tooth strongly slanted towards posterior; upper cheek teeth with two lingual reentrants (hypoflexus and posterolingual flexus) to match pattern of two labial reentrants (paraflexus and mesoflexus) to produce a six lobed appearance, hence name hexolobodon; reentrant folds nearly perpendicular to body axis and with moderate amounts of cement; labial and lingual reentrant folds only slightly overlapping (Fig. 2a).

COMMENTS.--Originally included within the Subfamily Capromyinae because of the similar orientation of the reentrant folds perpendicular to the body axis, but distinct and more primitive than capromyines in many features of cranial anatomy.

DISTRIBUTION.--Restricted to Hispaniola and Ile La Gonave.

Family *Incertae Sedis*

COMMENTS.--The giant hutias of Jamaica, Hispaniola, Puerto Rico, and the St. Martin Bank are considered by many authors to be a distinct family with affinities with either the Dinomyidae or Chinchillidae. Patterson and Wood (1982:512) could not find a close relationship of giant hutias with any other known family of hystricognath rodents, however, and assigned the family as *incertae sedis*. Ray (1965) noted many similarities of the Hispaniolan giant hutia *Quemisia gravis* with the capromyid *Plagiodontia aedium*. The main characters separating giant hutias from capromyids (hutias) are: 1) the replacement of the deciduous premolar in giant hutias (capromyids retain this tooth throughout life); and 2) the tendency for giant hutias to have deeply cleft reentrant folds on the molariform teeth that can form laminae (as in dinomyids and chinchillids). There is substantial variation in the dental morphology of giant hutias (MacPhee 1984, fig. 12), and some capromyids are known to form laminae when opposing reentrant folds unite (some individual *Hexolobodon phenax*, *Isolobodon portoricensis*, and *Rhizoplagiodontia lemkei*, and all *I. montanus*). The newly irrupted unworn cheek teeth of all capromyids and most West Indian echimyids are essentially lamellar in configuration. In addition some specimens of old *Plagiodontia aedium* are known in which the dm4 was replaced. *Plagiodontia* has lived to be over 17 years in the Florida State Museum colony. All giant hutias are much larger than any known capromyid. The large body size in giant hutias may have lead to the secondary replacement of the dm4, and to the formation of laminae. This is my working hypothesis. However, to assign giant hutias to the Family Capromyidae would imply that all of the hystricognath rodents of the West Indies are monophyletic and the descendents of an original invasion of echimyids into the central Antilles. There are too few data on cranial anatomy of most giant hutias (especially *Clidomys* from Jamaica and *Quemisia* from Hispaniola) to compare all of the features of the cranial anatomy of giant hutias and capromyids, and dental characters can be very misleading in rodents (Woods 1989a; Woods and Hermanson 1985). Therefore, until more complete fossil material is found, or there is additional data of another kind (such as biochemical or immunological data on the fossil bone), I will continue to follow Patterson and Wood (1982:523) in classifying giant hutias at the family level as *incertae sedis*.

Subfamily Heptaxodontinae Anthony 1917:186

DIAGNOSIS.--The deciduous molar always replaced; reentrant folds of molariform teeth oblique (30° to 55° from perpendicular to the body axis) and not forming complete lamellar bands; cement within reentrant folds and surrounding surfaces of molariform teeth; molariform teeth high crowned (hypsodont) but <u>not</u> open rooted (hypselodont) throughout life.

COMMENTS.--This assemblage includes all giant hutias with obliquely orientated reentrant folds on the molariform teeth (*Quemisia* = 55°, *Elasmodontomys* = 45°, and *Amblyrhiza* = 35°), and which probably chewed plagiognathously (Woods and Howland

1979). The group in analogous (and perhaps convergent) with the capromyid Subfamily Plagiodontinae. Heptaxodontines can be distinguished from the following subfamily by the lack of complete laminae and the presence of closed roots on the molariform teeth. Heptaxodontines are more similar to capromyids than are clidomyines, and are restricted to the eastern Antilles near the proposed origin of capromyids.

DISTRIBUTION.--*Quemisia* on Hispaniola, *Elasmodontomys* on Puerto Rico, *Amblyrhiza* on the islands of the St. Martin Bank (Anguilla and St. Martin).

Subfamily Clidomyinae new

DIAGNOSIS.--Molariform teeth without reentrant folds except as a variant in a few isolated individuals; molariform teeth with three complete laminae in all teeth except in M^3 which have four laminae; each lamina plate separated by wide band of cement which also encases outer surface of whole tooth as a thin sheath; angle of posterior border of laminae semi-oblique (about 10° from perpendicular) in lower molariforms, while essentially transverse (perpendicular) in uppers; molariform teeth hypsodont with open roots (hypselodont) throughout life.

COMMENTS.--This taxon is distinct in dental morphology from the three plagiognathous forms from the eastern Antilles. Clidomyines are analogous to, and perhaps convergent with the capromyid Subfamily Capromyinae. Since Jamaica was largely or totally submerged from the mid-Oligocene to the early Miocene, the invasion of Jamaica by rodents must have occurred later than the origin of plagiodontines on Hispaniola, and heptaxodontines in the eastern Greater Antilles. Vucetich et al. (in press) believe that *Clidomys* is derived from the South American chinchilloids, especially the Neoepiblemidae, and that *Clidomys* and perhaps the other giant hutias may have a distinct evolutionary history from a separate invasion of the Antilles. Chinchilloids during the late Miocene had highly diversified hypsodont teeth with distinct laminae. Considering the differences in dental morphology between *Clidomys* and the remaining giant hutias, and the submergence and isolated history of Jamaica, it is possible that *Clidomys* has an evolutionary history distinct from the heptaxodontines of the eastern Greater Antilles. The trends in dental morphology so closely follow the patterns in capromyids, however, that the more conservative hypothesis would be that clidomyines are fairly recent in their dispersal to Jamaica, and originated from plagiognathous forms already present in the eastern Greater Antilles. Giant hutias and capromyids would share a common ancestor, and originated on Hispaniola or Puerto Rico in the Miocene.

DISTRIBUTION.--Restricted to Jamaica.

Biogeographical discussion

Rodents of the Greater Antilles

Cuban Archipelago.--The 27 known rodents of Cuba and its associated halo of islands belong to two subfamilies: Capromyinae (22 species of *Capromys*, three species of *Geocapromys*); and Heteropsomyinae (two species of *Boromys*). Twenty one rodent species occurred on the 111,463 km^2 main island of Cuba, seven on 3,000 km^2 Isla de la Juventud, and four on the various low lying offshore islands. Of the 25 capromyines and two heteropsomyines, all but ten species (*sensu lato*) are now extinct (63%). On Cuba itself 17 of 21 rodent species (81%) are extinct; on Isla de la Juventud five of seven (71%) are extinct, while on cays off the coast of Cuba where rodents are known all (= four) endemics still survive. Throughout the Greater Antilles, however, 72 percent of all surviving capromyids occur within the Cuban Archipelago. The rodent diversity for the Cuban Archipelago is very high at the species level (species richness) with 27 known forms, the largest cluster of mammalian species in the West Indies. With only two rodent subfamilies on the island, however, higher level taxonomic diversity is very low. There are 33 bats, five insectivores, nine sloths, and two primates for a total of 77 known land mammals (Table 5).

Bahamas Archipelago.--The only known rodent species of the Bahamas, *Geocapromys ingrahami*, belongs to the Subfamily Capromyinae. Three subspecies of *G. ingrahami* are known (Table 1), two of which are now extinct. *Geocapromys i. ingrahami* still survives on East Plana Cay, and was reintroduced to Little Wax Cay (1973) and Warderick Wells Cay (1981) in the Exuma island chain in the original range of *G.i. irrectus*. A previously unknown colony of *Geocapromys* has recently been reported on High Cay two km off the south coast of San Salvador Island (Jane Rose pers. comm.). The species richness for rodents in the Bahamas is low, as is the diversity at higher taxonomic levels. No other terrestrial mammals are known from the islands of the Bahamas. Sixteen bats now occur in the Bahamas, for an overall total of 17 land mammals (Table 5).

Cayman Islands.--The two known rodents of the Cayman Islands are all members of the Subfamily Capromyinae, and are closely related to Cuban capromyines (Morgan 1977). These undescribed species of *Capromys* and *Geocapromys* were found on both Grand Cayman and Cayman Brac. Both forms are now extinct. The species richness and higher level taxonomic diversity of the Cayman Islands is low when compared with Cuba or Hispaniola, but high when considering the small size and remote locations of the individual islands. One extinct species of insectivore (*Nesophontes* sp) also occurred on Grand Cayman and Cayman Brac (Morgan 1977). Therefore, there were almost as many land mammals on the Cayman Islands as Jamaica. Nine bats are known from the islands, for a total number of twelve land mammals (Table 5).

Hispaniolan Archipelago.--Fourteen rodents of five subfamilies: Hexolobodontinae (three species of *Hexolobodon*); Isolobodontinae (two species of *Isolobodon*); Plagiodontinae (five species of *Plagiodontia* and one species of *Rhizoplagiodontia*); Heteropsomyinae (two species of *Brotomys*); Heptaxodontinae (one species of *Quemisia*) are known from Hispaniola, La Gonave, La Tortue, Beata, Saona, and Catalinita islands. Of these 14 species, 13 are now extinct (93%). The species richness of Hispaniola is moderate. The higher level rodent diversity is the highest of any West Indian island. Nineteen bats, five insectivores, six sloths (estimated from available specimens), and two primates occurred on Hispaniola for a total number of 46 known species of land mammals (Table 5). Twenty five species have become extinct (54%), although the extinction rate is much higher for terrestrial mammals where 24 of 27 known species have become extinct (89%).

Part of the reason for the high level of diversity on Hispaniola is because the island is geologically and physiographically more complex than any other West Indian island. The island appears to have been divided into three or four smaller islands that came together to form Hispaniola (Maurrasse et al. 1982; Pindell and Dewey 1982; Sykes et al. 1982; Rosen 1985). The numerous physiographic regions of the island include the highest mountain range in the Antilles, the Cordillera Central of the Dominican Republic with several peaks over 3,000 m. Hispaniola also includes the extensive Massif de La Selle-Sierra de la Baoruco range with many peaks over 2,000 m, several broad, pine covered plateaus, and the remote and isolated Massif de La Hotte with some of the highest, wettest mountains on the island (see Paryski et al. this volume).

Another reason for the high level of taxonomic diversity at upper taxonomic levels on Hispaniola may be because rodents have been on Hispaniola for a longer period of time than other islands of the Antilles. The results of the cladistic analysis of 82 characters indicate that Hispaniolan rodents, along with the rodents of Puerto Rico and the St. Martin Bank, have more plesiomorphic morphological features than do the rodents of Cuba, Jamaica, and the Bahamas. It appears as if rodents in the Greater Antilles evolved in the central Greater Antilles, and that the radiation of capromyids and echimyids on Cuba, Jamaica and the Bahamas occurred later than the radiation on Puerto Rico and Hispaniola.

Jamaica.--Four rodents from three subfamilies are known from Jamaica: Capromyinae (one species of *Geocapromys*); Clidomyinae (two species of *Clidomys*); Sigmodontinae (one species of *Oryzomys*). Of these four rodents, all but *G. brownii* are extinct (75%). Jamaican rodents are more diverse at higher taxonomic levels than are rodents on any other island of the Antilles with the exception of Hispaniola, although there is a low level of species richness. Twenty three bats and one primate are known from Jamaica for a total of 28 known land mammals (Table 5). No sloths or insectivores are known from the island, which is likely the result of Jamaica having been submerged during the late Oligocene and early Miocene (Kashfi 1983). Of the 28 known land mammals on Jamaica, only four have become extinct (4%). The unusually low rate of extinction is largely because most known mammals are bats. Four of the five known species of non-volant mammals of Jamaica have become extinct (80%).

The level of diversity of rodents on Jamaica, with three unrelated subfamilies on an island much smaller than Cuba or Hispaniola, may be a result of three separate invasions of the island: 1) by *Oryzomys* from Central America; 2) by *Geocapromys* from Cuba; and 3) by the clidomyines from an unknown source. Jamaica lacks many of the groups of rodents and other terrestrial mammals found on Cuba and Hispaniola, and is even depauperate in these forms in comparison with the small and remote Cayman Islands. The rodent fauna of Jamaica, therefore, appears to be out of the mainstream of the radiation of rodents in the Greater Antilles.

Puerto Rican Archipelago.--The four endemic rodents of Puerto Rico, Culebra, Vieques, and the Virgin Islands other than St. Croix include members of two subfamilies: Heptaxodontinae (one species of *Elasmodontomys*); and Heteropsomyinae (one species of *Heteropsomys*, one of *Homopsomys*, and one of *Puertoricomys*). The plagiodontine rodent *Isolobodon portoricensis* was introduced by Indians from Hispaniola, and occurs only in midden deposits. Of these five rodents known from Puerto Rico, four endemic and one introduced, all are extinct. The diversity of rodents on Puerto Rico is high at upper taxonomic levels, but there is a lack of species richness. Sixteen bats, one insectivore, and one sloth are also known for a total of 21 endemic land mammals (Table 5). Five have become extinct (24%). All of the endemic non-volant mammals of the Puerto Rican Archipelago are now extinct.

St. Martin Bank.--The islands of St. Martin, Anguilla, and St. Barthelemy are surrounded by shallow water, and would have formed a single island of approximately 5,949 km² at times of low sea levels during the Pleistocene. The two known rodents are from different subfamilies: Heptaxodontinae (*Amblyrhiza inundata* from St. Martin and Anguilla); and Sigmodontinae (an undescribed rice rat from Anguilla). Both species are now extinct. The rodents on the St. Martin Bank are surprisingly diverse at higher taxonomic levels for such an isolated area, but the species richness is low. Seven bats are known from the St. Martin Bank for a total of nine land mammals (Table 5).

St. Kitts Bank.--St. Eustatius, St. Christopher (also called St. Kitts) and Nevis are located on a long narrow bank that would have been approximately 1,546 km² in area at time of low sea levels in the Pleistocene. The only known rodent is in the Subfamily Sigmodontinae (an undescribed rice rat on St. Eustatius and St. Christopher), which is now extinct. Six bats are known from the islands of the St. Kitts Bank, for a total of seven land mammals (Table 5).

Antigua Bank.--The islands of Barbuda and Antigua are located on a bank that would have been approximately 4,274 km² in area during low sea levels during the Pleistocene. Two rodents of the Subfamily Sigmodontinae (*Megalomys audreyae* on Barbuda, and an undescribed rice rat on Barbuda and Antigua) occurred on the bank, but both are now extinct. The undescribed rice rat (species B of Steadman et al. 1984) also occurred on nearby Montserrat, and Guadeloupe and Marie Galante to the south. This distribution of one species on four islands of three separate banks suggests that dispersal was influenced by human transport. Eleven bats are known from the islands, for a total of 13 land mammals (Table 5).

Montserrat.--South of the above islands and banks, but still part of the Leeward Islands of the northern Lesser Antilles, is small but mountainous Montserrat. Two undescribed rice rats of the Suborder Sigmodontinae, both now extinct, occurred on the island. The undescribed rice rats found on Montserrat were widely distributed in the northern Lesser Antilles from Anguilla in the north to Marie Galante off the south shore of Guadeloupe. Montserrat is the only island where both species occur together, suggesting that human transport may have played a role in the dispersal of these two forms. There are nine bats on the island, for a total of eleven land mammals (Table 5).

Windward Islands.--The remaining islands of the Lesser Antilles are south of the Guadeloupe Passage. Unlike the Leeward Islands, which are surrounded by shallow banks, the Windward Islands are each separated by deep water passages. **Guadeloupe** is a large island of 1,510 km² with a high volcanic peak (1,467 m Grande Soufriere). The only known rodent is an undescribed rice rat of the Subfamily Sigmodontinae. There are eleven bats for a total of twelve land mammals. Tiny (153 km²) **Marie Galante** is just south of Guadeloupe. The island has one undescribed rice rat and one bat. South of Marie Galante is **Dominica**, a mountainous 751 km² island capped by 1,447 m Mt. Diablotin. No endemic rodents are known from the island, which is unexpected. There are ten known bats. **Martinique** is 40 km south of Dominica. Mont Pelee is a 1,397 m volcano on Martinique that erupted in 1902 killing all but one of 30,000 residents of the nearby town of St. Pierre, and probably contributed to the extinction of the endemic giant rice rat *Magalomys desmarestii*. This member of the Subfamily Sigmodontinae is the only known non-volant mammal of the island. There are ten known bat species for a total of eleven land mammals. Forty kms south is the island of **St. Lucia**, a mountainous island 616 km² in size. The only known non-volant mammal is an extinct rice rat, *Megalomys luciae*, which probably became extinct in the late 1800s. Eight bat species are known from St. Lucia, for a total of nine land mammals. **St. Vincent** is a mountainous 344 km^2 island with 1,234 m high Mt. Soufriere which also exploded in a volcanic eruption the same year as the eruption of Mont Pelee on Martinique (1902). An endemic rice rat, *Oryzomys victus*, occurred on the island until the late 1800s, and may have persisted into this century. Nine bats are known from St. Vincent for a total of ten land mammals. **Barbados** is located 155 kms east of St. Vincent, and is on a separate underwater ridge. An undescribed extinct rice rat occurred on Barbados, and there are seven known bats for a total of eight land mammals.

Subfamily radiations

Hexolobodontinae.--Capromyids are a diverse rodent assemblage characterized by high crowned molariform teeth with abundant cement. The capromyids with the largest number of plesiomorphic characters belong to the Subfamily Hexolobodontinae, which is known only from Hispaniola. This group was probably confined to northern Hispaniola before the southern parts of the island united with the central and northern to form one large island, as were the non-Puerto Rican heteropsomyines. Hexalobodontines have high crowned teeth with a moderate amount of cement surrounding each molariform tooth. Their molariform teeth are not completely hypselodont, and the pattern of the reentrant folds includes all of the components necessary to derive any of the other capromyid lineages. *Hexolobodon* is too large and too specialized to be directly ancestral to any of the other known genera of capromyids, but it is likely that an early hexolobodontine was close to the ancestor of each capromyid subfamily. The Hexolobodontinae gave rise to two or maybe three radiations (Fig. 2) that appear to be equally diverse and ancient (Plagiodontinae, Isolobodontinae, and perhaps the Heptaxodontinae), and one radiation that appears to have occurred later in time (Capromyinae). The morphological features of *Hexolobodon* suggest that the Subfamily Hexolobodontinae was central in these radiations. If the origin of the capromyid subfamilies occurred during the Oligocene or early Miocene when Hispaniola was separated into at least two, and possibly as many as four smaller islands, then it is reasonable to postulate that plagiodontines, isolobodontines, hexolobodontines, and perhaps heptaxodontines originated and underwent their early evolution in different regions of what is now Hispaniola and Puerto Rico.

Plagiodontinae.--Plagiodontines are medium sized hutias that were mainly distributed in central and southern Hispaniola. One genus (*Rhizoplagiodontia*) is known only from the far south of the southern peninsula. I propose that plagiodontines evolved in southcentral Hispaniola in the area now characterized by the Sierra de Baoruco and Sierra de Neiba ranges of the Dominican Republic and the Massif de la Selle of Haiti. The plagiodontines also underwent a radiation on the far tip of the southern peninsula of Haiti while it was still isolated. The southcentral and southern regions of Hispaniola were isolated from northern Hispaniola and characterized by distinct physiographic and geologic areas that could have served as allopatric nuclei in the generation of new taxa, accounting for the species richness of the genus *Plagiodontia* and the presence of *Rhizoplagiodontia* as an endemic in the Massif de la Hotte of the far southwest of Haiti. The main evolutionary trend in plagiodontines was from a semi-oblique chewer with only a few plagiognathous adaptations, as in *Rhizoplagiodontia lemkei*, towards being an extremely oblique chewer with many plagiognathous adaptations, as in *P. araeum* (Woods and McKeen 1989).

The time of the origin of the Subfamily Plagiodontinae in southern Hispaniola is difficult to resolve because of the lack of old fossils and the controversy over the position and composition of southern Hispaniola during the Oligocene and early Miocene. Sykes et al. (1982) proposed that southern Hispaniola was far to the west of the rest of Hispaniola in the Eocene and Oligocene, and did not collide with the rest of Hispaniola until about 10 myBP. Maurrasse et al. (1982) on the other hand did not believe in a disjunct southern peninsula of Haiti, although they did note that the southern peninsula was divided into two islands. Both Sykes et al. and Maurrasse et al. indicated that the major uplift of southern Haiti occurred about 10 myBP. I propose that plagiodontines evolved from hexolobodontines in southern Hispaniola 10-15 myBP as it neared the rest of Hispaniola, and differentiated into several taxa as its topographical features increased in complexity by massive uplift of the entire southern portion of Hispaniola. The plagiodontines were isolated in southern Hispaniola from other island areas, and did not disperse westward to Jamaica nor northward to Cuba because of significant water barriers. In southern Hispaniola the niche of *Hexolobodon* may have been largely filled by *Plagiodontia velozi*. Both taxa were comparable in body size and many morphological features related to mastication. *Rhizoplagiodontia lemkei* of southern Hispaniola was about the same size as the heteropsomyine *Brotomys*.

Isolobodontinae.--Isolobodontines are another lineage of hutias on Hispaniola similar in size to plagiodontines, further suggesting that the present day island of Hispaniola was separated into more than one island in the past. The Subfamily Isolobodontinae appears to be a specialized side branch of the Subfamily Hexolobodontinae which converged in dental and cranial morphological features with members of the genus *Plagiodontia*. Isolobodontines may have evolved in central or northern Hispaniola in isolation from plagiodontines, and some species of the two groups are ecomorphs. Isolobodontines did not get to Cuba or the Bahamas, although both known species did occur throughout Hispaniola in recent times. Plagiodontine remains are rare or lacking in most areas of northern Hispaniola, while remains of isolobodontines are abundant. *Isolobodon portoricensis* remains occur along with the remains of ground sloths in pre-Amerindian paleontological deposits on Ile de la Tortue, a small (180 km^2) island off the north coast of Haiti, where no plagiodontine remains are known. Isolobodontines are much less species rich than are plagiodontines, which may be the result of the less diverse nature of the physiography of northern Hispaniola, or more likely that plagiodontines more effectively filled the niches of medium to large sized hutias and excluded isolobodontines from these niches after the unification of northern and southern Hispaniola in the Pliocene. Within the Subfamily Isolobodontinae, *I. montanus* is extremely plagiognathous, and has converged with *Plagiodontia aedium* in many features of the masticatory apparatus. The two forms were probably ecomorphs specialized in feeding on roots and bark.

Capromyinae.--Kratochvil et al. (1978:19) and Rodriguez et al. (1979) believed that small species of Cuban *Capromys* (which they separated as the genus *Mesocapromys*) are the most primitive members of the Family Capromyidae and that *Plagiodontia* from Hispaniola is the most derived capromyid. They argued that species of *Mesocapromys* exhibit many characters in common with echimyids (Fig. 1b). The cladistic analysis and faunal diversity studies presented here do not support those conclusions, and suggest that the similarities between the smaller capromyines from Cuba and echimyids may be a consequence of adaptations to similar habitats. The smaller Cuban taxa are lowland forms that inhabit grasslands and swamps. Rodriguez et al. (1979:100) noted that there are "a number of traits in which the life habits of *Mesocapromys* resemble those of Echimyidae" and that there is much less similarity in the habits of the remaining capromyids to echimyids. I believe that the taxa they group in *Mesocapromys* are ecomorphs of typical echimyids, such as *Proechimys*, and that this is the reason for the resemblance rather than a direct phylogenetic link between the small *Capromys* (*Mesocapromys*) and echimyids (Fig. 1b).

The interpretations by Kratochvil et al. (1978) as to which taxa are primitive and derived are not supported by the data available in the cladistic analyses presented here or by the available biochemical data (Sarich and Cronin 1980; C.W. Kilpatrick pers. comm.). The range of morphological diversity in the capromyines of Cuba is not great enough to span the gap between echimyids and capromyids, and is far less than the tremendous diversity exhibited within and among the other capromyid clades. Later in their paper Kratochvil et al. (1978:55) concluded that " the most derived forms occur (and occurred also in the past) on islands in the marginal zone of the Antillean region." They were referring to *Plagiodontia* on Hispaniola. The cladistic analysis, however, indicates that *Plagiodontia* retains more primitive characteristics than do *Capromys* and *Geocapromys*, and as a consequence it is likely that Hispaniola is central, and not peripheral, to the evolution of these rodents.

The Subfamily Capromyinae is restricted to the western Antilles, and appears to be the most recent capromyid clade to have evolved from the Subfamily Hexolobodontinae. The radiation of capromyid rodents, therefore, appears to have spread westward in the Antilles following the origin of these taxa in the central Antilles. In Cuba the various capromyines are not very morphologically diverse, and many of the characters used to differentiate the various forms, especially dental features, are highly variable within species (Kratochvil et al. 1978:41). Indeed, the similarity of the various Cuban capromyines is so great that Varona (1974) placed all but one (*Macrocapromys*) in the same genus (*Capromys*) which he divided into four subgenera (*Capromys*, *Mysateles*, *Mesocapromys*, *Geocapromys*). All of these are considered as separate genera by Kratochvil et al. (1978:18). The level of morphological distinctiveness exhibited by these taxa, however, is much below the levels used to separate other capromyid genera. Only Capromys and Geocapromys among the known capromyines appear to be valid genera. The many species and subspecies of Capromys and Geocapromys may have developed in the Pleistocene when changing sea levels created vast swamps, extensive grasslands, and numerous islands in the area of the Cuban archipelago. Fifty eight percent of all capromyids are of the genus Capromys, and 67 percent of all known capromyids occurred on Cuba or its associated small islands. Kratochvil et al. (1978:20) noted that "on some of these islands the remnants of some populations have been preserved till the present time,... and are of a relict nature, not very numerous and living on small areas, very often only on one island or its part." At the present time 77 percent of the surviving capromyids occur within the Cuban Archipelago.

The genus *Geocapromys* is morphologically the most derived West Indian capromyid. It is likely that *Geocapromys* evolved in Cuba from *Capromys*. Within the Subfamily Capromyinae, 87 percent of the species of *Capromys* and *Geocapromys* occurred on the islands of the Cuban Archipelago. Until recent times *Geocapromys* was widely distributed in the western region of the Greater Antilles from Cuba, the Caymans, the Bahamas, and Jamaica to tiny Little Swan Island. It dispersed to the Bahamas by over water dispersal, probably during the Pleistocene when ocean levels were as much as 120 m lower and the Bahamas were a large bank of low lying islands very close to Cuba. At that time the islands of the Grand Bahama Bank were united into a single large island over 100,000 km^2 in size (Morgan this volume). *Geocapromys brownii* of Jamaica is present in the oldest dated Pleistocene paleontological deposits on that island (Table 4), and so it also dispersed overwater by natural means. It still survives, and is closely related to its sister species *G. thoracatus* of Little Swan Island. Both are part of the same species group (Morgan 1985). Whether the dispersal event of *G. thoracatus* to Little Swan, an isolated, small (2 km^2) island 600 kms southwest of Jamaica was a natural process via over-water dispersal or the result of an early human related introduction is unknown.

It is possible that *Geocapromys* reached Little Swan Island via island hopping along the series of shallow water banks of the Nicaragua Rise extending outward towards Jamaica from Honduras. Many of these mounts would have been exposed as low islands in times of low ocean levels in the Pleistocene. Pedro Bank and the Pedro Cays are only 75 kms from Jamaica. *Geocapromys* sp. from the Cayman islands, the three extinct species of *Geocapromys* from Cuba, and *G. ingrahami* from the Bahamas are sister species in the *G. ingrahami* species group. The relationships of the species groups indicate that *Geocapromys* dispersed to various areas in the western Antilles at different times.

If Amerindians transported *G. brownii* to Little Swan Island 5,000 to 7,000 yBP it is possible that the taxon would have become established and could have differentiated enough as the result of founder effect and the constraints of life on a small and low lying island to be considered distinct at the species or subspecies level. Reintroduction efforts on small islands in the Bahamas have been very successful, with a large population building up on Little Wax Cay in the Exuma Cays fifteen years after the release of twelve individuals trapped on East Plana Cay (Kevin Jordan pers. comm.). I believe that it is more likely that the dispersal to Little Swan Island was by humans rather than natural means, and the observation by Rouse (this volume) that the first Amerindians in the Antilles were from Central America into Jamaica indicates there may have been some human movement backward from Jamaica towards Honduras and Little Swan Island as early as 5,000 to 7,000 yBP.

When the capromyines originated in Cuba is a moot point. Since capromyines are not well differentiated at the subfamily and generic level from one another, it appears likely that it was significantly after the time of the origin of the capromyid subfamiles in the Hispaniolan region. The origin of capromyines was after the early Miocene when Cuba was close to Hispaniola (Sykes et al. 1982) and may have been as late as the Pliocene. It was by overwater dispersal since there is no subfamilial similarity between the capromyid fauna of Cuba and Hispaniola.

Heptaxodontinae and Clidomyinae.--Giant hutias are very large hystricognath rodents found in Jamaica, Hispaniola, Puerto Rico, and the islands of the St. Martin Bank. These forms are usually classified together in the Family Heptaxodontidae. Unlike capromyids and echimyids (Superfamily Octodontoidea), they do not retain the deciduous premolars throughout life, and therefore their superfamily affinities was questioned by Patterson and Wood (1982). *Quemisia* from Hispaniola and *Elasmodontomys* from Puerto Rico have adaptations for oblique chewing, and resemble *Plagiodontia, Isolobo-*

don montanus, and *Myocastor coypus* in many features of the teeth, jaw, and skull. Woods and Howland (1979) and Woods and McKeen (1989) have reviewed these adaptations, which they called plagiognathous adaptations, and determined that they are closely linked with oblique chewing. Convergence in masticatory form and function has occurred in all of these forms (Woods and Mckeen 1989). The giant hutias of the eastern Antilles are large rodents with specializations for oblique chewing similar to the most plagiognathous capromyids, and as such may be ecomorphs of capromyid taxa such a *Plagiodontia araeum*, a very large extinct Hispaniolan hutia. The prevalence of convergence and apparent ecomorphs in different capromyid subfamilies indicate that the similarities between *Plagiodontia aedium* and *Quemisia gravis* discussed by Ray (1965) may not indicate a common phylogenetic heritage for giant hutias and capromyids. In spite of many similarities between giant hutias and capromyids, I am wary at this time of uniting them in the same family, preferring instead to continue classifying giant hutias *incertae sedis* at the family level.

Miller (1929b:22-25) used the historical account of a "quemi" (Oviedo 1959) as the basis for his name *Quemisia gravis*, which he believed was still extant in the 16th Century. However, no specimens have been found that are more recent than 3,000 yBP. *Quemisia* is known from very few specimens, all from northern Hispaniola, whereas the equally large *Plagiodontia velozi* was very abundant in the fossil record and persisted until very recent times (Woods et al. 1985).

The hypothesis here is that giant hutias evolved from a capromyid similar in morphology to the common ancestor of hexolobodontines and plagiodontines. The earliest heptaxodontine may have been similar to *Quemisia*, which would explain the remarkable similarity between *Plagiodontia* and *Quemisia* noted by Ray (1965). If this hypothesis is correct then the origin of heptaxodontines was most likely on Hispaniola, with the direction of the radiation moving eastward to Puerto Rico and the St. Martin bank. The two heptaxodontines that share the greatest number of morphological features in common are *Quemisia* from Hispaniola and *Elasmodontomys* from Puerto Rico. Similarly, the pattern of reentrant folds of *Amblyrhiza* from the St. Martin bank is also easily derived from the pattern in *Quemisia*. It is likely that heptaxodontines evolved from a hexolobodontine ancestor in northern Hispaniola or the combined area of Puerto Rico and Hispaniola, hence the close relationship of the Hispaniolan and Puerto Rican forms. Heptaxodontines would have dispersed overwater eastward to the St. Martin bank, where they became giant in size.

The dental pattern of *Clidomys*, the extinct giant hutia of Jamaica, can be derived from the pattern of a hypothetical common ancestor of *Quemisia*, *Elasmodontomys*, and *Hexolobodon* by following an evolutionary trend towards propalinal chewing and orthognathous adaptations similar to that followed in the evolution of capromyines. However, the lack of any heptaxodontid rodents on Cuba, and the problems of explaining the dispersal of *Clidomys* to Jamaica are formidable problems in uniting *Clidomys* with the remaining genera of giant hutias. *Clidomys* is so distinct in dental morphology from the remaining giant hutias that I have chosen to separate it into its own subfamily. The route that the clidomyine ancestor took to Jamaica could have been by the difficult overwater passage from the southern tip of Haiti to Navassa Island, Formigas Bank and Grappler Bank. A more complete understanding of the endemic fauna of Navassa Island, where we know terrestrial mammals occurred (pers. obs.), might resolve this question. Since

Jamaica was under water from the mid-Oligocene to the early Miocene (Kashfi 1983), the dispersal event would have had to be later than the proposed origin of capromyids and heptaxodontines on Hispaniola. Why is *Clidomys* so distinct in morphology from the remaining giant hutias? Vucetich et al. (in press) propose that *Clidomys* may be the result of a separate and later invasion into the Antilles than the capromyids.

West Indian echimyids (Heteropsomyinae).--The origin of echimyid rodents in the West Indies has long been an enigma because of the lack of an understanding of the relationship between "*Proechimys*" *corozalus* from Corozal, Puerto Rico and the *Heteropsomys*-like echimyids. The fossil specimen from Corozal is so similar to *Proechimys iheringi* of the *Trinomys* group that Williams and Koopman (1951) classified it in the genus *Proechimys*, although not without some hesitation. Moojen (1948) pointed out that the *Trinomys* group became "differentiated" from the more primitive *Proechimys* group when it became isolated southeast of the main range of *Proechimys*. I believe that the same pattern is likely to have occurred in the West Indies, but that there is no direct link between the *Trinomys* group and "*Proechimys*" *corozalus*. I believe that a *Proechimys*-like ancestor of the Heteropsomyinea dispersed to the West Indies from South America. This form secondarily came to resemble the *Trinomys* group by processes similar to the ones that gave rise to *Trinomys*. Moojen (1948:394) believed that "the primitive *Proechimys* probably was large, with a short tail, narrow aristiformes, strongly built skull, and five counterfolds in each molariform tooth." A *Proechimys*-like form with these characteristics would be an appropriate ancestor of *Puertoricomys*, *Heteropsomys*, *Homopsomys*, *Brotomys*, and *Boromys* as well as capromyids. Miller (1916, 1930) and Williams and Koopman (1951:7) pointed out the suitability of *Proechimys* as a potential ancestor to the heteropsomyines. The fossil from Corozal is more closely linked with the heteropsomyine radiation of the Antilles than it is with *Proechimys* and its allies, however, and the cladistic analysis supports a reclassification of the form into a new genus which I have named *Puertoricomys*.

The dental ontogeny of *Homopsomys* shows a close resemblance between the slightly worn occlusal patterns present in *Homopsomys* and the pattern present in *Puertoricomys corozalus* (Fig. 1a). It is likely that a form with the features resembling *Puertoricomys* gave rise to *Homopsomys* and *Heteropsomys*, which in turn gave rise to *Brotomys* and *Boromys*. When did the ancestor of *Puertoricomys* get to the Antilles, and from where? Moojen (1948:309-310) proposed that the isolation of the *Trinomys* group from the primitive *Proechimys* occurred "in (early?) Pleistocene times". He based his assumption on the history of the distribution of tropical forests in the Central Plateau of Brazil, and when drier climatic conditions would have led to an isolation of *Trinomys* in southeastern Brazil. A dispersal event to the West Indies would have been favored at an earlier time when more tropical conditions in the area resulted in larger rivers draining northward into the Caribbean. This pattern would have been present in the lower Miocene (Berry 1942).

The pattern of the radiation of echimyids began in Puerto Rico, and spread westward as a series of radiations including the origin of the Family Capromyidae. The most primitive and diverse assemblage of heteropsomyines is in Puerto Rico with these echimyids diversifying to fill the medium body sized niches occupied on other islands by capromyids. The radiation of heteropsomyines in Hispaniola is less extensive than on Puerto Rico, although the small endemic rodent niche in central and northern Hispaniola is occupied by *Brotomys*. In southern Hispaniola this niche appears to have been occupied by the capromyid *Rhizoplagiodontia*. The genus *Boromys* of Cuba is derived from *Brotomys* of Hispaniola. However, in Cuba the small rodent niche is shared between echimyids and capromyids, with a number of hutias successfully filling the small scansorial rodent niche.

Westward bearing currents pass by the deltas of several major rivers entering the Caribbean from northern South America and could have carried a raft northwest. Puerto Rico may have been closer to South America in the mid-Tertiary, and if so would have been in an excellent position to serve as a target for such rafts. There is a fossil dentary lacking molariform teeth, but resembling *Heteropsomys* and *Homopsomys* in morphology, in the La Venta fauna located along the Magdalena River in Colombia (Fields 1957). The age of this faunal assemblage is estimated to be late Miocene. Rafts from the Magdalena, or any of several rivers that drained the then much higher Guyana highlands, could have carried an overwater disperser directly to the Greater Antilles. This route and chronology would help explain the complete absence of heteropsomyines and capromyids in the Lesser Antilles and the dominance of these forms in the Greater Antilles.

The data on the relationships of heteropsomyines and capromyids indicate that the heteropsomyine echimyids probably invaded the Antilles from South America sometime after 26 myBP. Puerto Rico, or the combined island of Puerto Rico and central Hispaniola, would have been in an appropriate position directly between the mainland of South America at the northwestern tip of present day Colombia in the Oligocene and early Miocene (Pindell and Dewey 1982; Sykes et al. 1982; Rosen 1985). Buskirk (1985) proposed that hystricognath rodents and edentates arrived in the Antilles from northern South America via the Lesser Antilles by the late Miocene. I feel it is more likely that the occurred in the late Oligocene or early Miocene. The dispersal from South America could have been via a direct overwater passage to Puerto Rico, or via a series of shorter island hopping events via the Aves Ridge or the Lesser Antilles.

Rodents of the Lesser Antilles.--The evolution of rodents in the Lesser Antilles followed a separate pattern from the one discussed above for the Greater Antilles. No insectivores or primates are known from these islands. Sloths are known from Curaçao, but have never been reported from Antillean islands outside of the Greater Antilles. Hystricognath rodents were also apparently absent from the Lesser Antilles, although there is a questionable account of the presence of an endemic porcupine, *Sphiggurus pallidus*, in the "West Indies" (Woods 1984:405). If true, this taxon would clearly have come from northern South America by overwater dispersal or human transport.

The radiation of non-hystricognathous West Indian rodents occurred mainly in the Lesser Antilles. This group of muroid rodents of the Subfamily Sigmodontinae (Carleton and Musser 1984) included the genus *Oryzomys* and at least two closely related genera. While at least one of these forms, *Megalomys desmarestii* from Martinique, was 690 mm in total length, most were somewhat smaller. These Lesser Antillean rodents were generally less than 1.5 kg in mass, and quite slender and rat-like in build. Most hystricognath rodents from the Greater Antilles were chunky and frequently larger than 2 kg in mass. The sigmodontine rodents of this Lesser Antillean radiation are now extinct, but until well after the arrival of humans onto the islands included at least three genera and eight species distributed over most of the islands of the archipelago (Steadman et al. 1984a,b). Montserrat and Barbuda had two species, while only one species of sigmodontine occurred on the other islands where they were found. *Oryzomys antillarum* from

Jamaica, which became extinct about 1877, was probably not part of this same Lesser Antillean radiation, but rather represents a separate dispersal (or introduction) directly to the island from Central America. This form is very closely related, and possible conspecific, with *Oryzomys couesi* (Honacki et al. 1982), which is distributed from central Panama to southern Texas.

The Lesser Antilles developed as an island arc after the late Eocene, and the individual islands were available for colonization by overwater dispersing mammals throughout the mid to late Tertiary. It is interesting that no fossil echimyid or even capromyid rodents are known from the area. *Oryzomys*, and its closely related relatives of sigmodontine rodents are descendants of North American forms that dispersed to South America via Central America. When this dispersal event to South America occurred is of great consequence to any hypothesis as to when sigmodontines dispersed to the Lesser Antilles. Patterson and Pascual (1972) suggested that sigmodontines entered South America after the completion of the Panamanian land bridge at the end of the Pliocene. Marshall (1979) established the date of completion of the Panamanian land bridge as 3 myBP. The great diversity of sigmodontines in South American would either be the result of an extensive adaptive radiation in Central America before the completion of the land bridge connection, or the result of an explosive radiation soon after their arrival in South America.

Hershkovitz (1972) and Reig (1980), on the other hand, stated that the invasion of South America by sigmodontines took place by waif dispersal from Central America in the late Miocene or earlier. Marshall (1979) reviewed the late Tertiary history of the flora and fauna of South America, and hypothesized that sigmodontines got to South America by waif dispersal across the Bolivar Trough marine barrier. He proposed that the likely time for this overwater dispersal was between five and seven myBP in the Pliocene, when sea levels dropped by as much as 50 m during the Messinian Low. The earliest known fossil sigmodontines in South America are from near the Atlantic coast in Buenos Aires Province, Argentina, and have been dated as 3.5 my in age (Marshall 1979).

If Marshall is correct in his hypothesis as to the date and ecological requirements of the first South American sigmodontines, then forms similar to *Oryzomys* would have been in areas of South America adjacent to the Lesser Antilles sometime after 3.5 myBP. Since the diversity of sigmodontines in the Lesser Antilles is limited to two genera, I propose that the ancestor of the extant Lesser Antillean *Oryzomys* species dispersed overwater to the southern Lesser Antilles during the Pleistocene. The main trends in evolution of *Oryzomys* and its allies have been an increase in body size, and a change in the relative position of the coronoid process in relation to the condyloid process of the dentary. In *Oryzomys* the coronoid process is low, while in species of *Megalomys* the coronoid process becomes progressively higher and more nearly the same height as the condyloid process. In spite of the broad distribution of the two genera and eight species of endemic Lesser Antillean rats and mice, the diversity of *Oryzomys* and its allies is much less than that found in capromyids. The lack of diversity of sigmodontines in the Lesser Antilles is in contrast to the history of the group in South America, where it has undergone a massive radiation. This lack of diversity supports the hypothesis that sigmodontines are a fairly recent arrival in the Lesser Antilles. The observed distribution of closely related species of *Oryzomys* on several widely separated islands of the Lesser

Antilles may be the result of early humans (Steadman et al. 1984a,b; Morgan and Woods 1986).

Evolutionary patterns and the history of West Indian rodents

Direction of radiations

The results of the cladistic analysis (Figs. 3-5) support the hypothesis that the islands with the greatest taxonomic diversity at higher levels are the center of the oldest part of the radiation of the rodent groups under consideration. The most primitive taxa are in the area of highest taxonomic diversity at the generic and subfamilial levels. Another way of interpreting these data was expressed by Kratochvil et al. (1978:19) and Moojen (1948:394) who noted that the most primitive forms persist in the margins of their range and on isolated islands because these old taxa are less competitive than newer more competitive forms. Kratochvil et al.(1978:55) cited as evidence their hypothesis the distributions of capromyines in Cuba, where the forms they regard as most primitive are restricted to the south coast and some neighboring small islands of Cuba "in the region which is the nearest to its presumed homeland-South America". It could be that the most primitive genera are relictual in peripheral areas of the Antilles, and that the direction of the rodent radiation (or sequence of dispersal events) in the West Indies would move from the center of distribution of the most advanced taxa towards the center of distribution of the most primitive (i.e. from Cuba towards Puerto Rico). I reject this hypothesis for the following reasons.

1) Genera of South American hystricognath rodents tend to survive for long periods in the fossil record of most lineages, and as a result are like living fossils in many cases when compared with the high rate of evolution in groups such as sigmodontine rodents (Woods 1982). Therefore, it is reasonable to expect the older genera of hystricognath rodents also to persist in the West Indies.

2) Evolutionary patterns of mammals on islands, and in taxa with very disjunct distributions, differ from those patterns in taxa occurring on continental areas with more contiguous distributions. Instead of a wave-like replacement of adaptations sweeping towards the periphery of a taxon's distribution, as may be the case in some continental situations, evolution on isolated islands progresses by differentiation rather than replacement. It is difficult for rodents to disperse from one island to another across many kilometers of open ocean. The dispersal event to a remote habitat, such as one of the West Indian islands during the Miocene, would have led to a new radiation or differentiation. This concept was proposed by Moojen (1948:313-314) who postulated that the *Trinomys* group of *Proechimys* was derived from *Proechimys* even though its distribution is peripheral and appears to be relictual to the main groups of South American *Proechimys*. He explained this apparent contradiction by noting that the *Trinomys* group probably became isolated by a great distance from the mainstream radiation of *Proechimys* when an arid belt developed in central Brazil. Once geographically isolated, the species "differentiated" at an accelerated rate from the mainstream of *Proechimys*, and developed a very different pattern of the reentrant folds that make this isolated group recognizable at the subgeneric or even generic level. I believe that the same pattern of isolation by dispersal and subsequent differentiation occurred in the West Indies. The saltatorial

pattern of dispersal led to a situation where the most derived major taxa (genera and subfamilies) are in the most peripheral areas (i.e. the western Antilles in the case of the capromyid radiation) where they most recently arrived.

3) Further evidence against the hypothesis that primitive rodent taxa have been pushed to the periphery of their distributions in the West Indies comes from the theoretical work of MacArthur and Wilson (1967:175) who predicted that "adaptive radiation will increase with distance from the major source region and, after corrections for area and climate, reach a maximum on archipelagos and large islands located in a circular zone close to the outermost dispersal range of the taxon." They referred to this as the "radiation zone".

The presence of the high level of diversity at the subfamily level on Hispaniola in the central Antilles is not a contradiction of the MacArthur and Wilson hypothesis. It is a reflection of the isolated nature of the islands of the Antilles, the history of fragmentation and unification of Hispaniola, and the longevity of hysticognath rodent lineages at the generic and subfamily levels. If adaptive radiation is measured in species richness, as are most of the examples in MacArthur and Wilson, then the "adaptive radiation" of rodents on Cuba with 21 rodent species on the main island and an additional six species on the nearby islands of the Cuban archipelago (total=27 species) greatly exceeds that of Hispaniola with 14 species or Puerto Rico with four species.

How and when did rodents enter the West Indies?

Poinar (1988) hypothesized that the presence of mammalian hairs in amber from the northwestern mountains of Hispaniola, which he concluded were from a rodent, indicated that rodents were likely to have originated in the West Indies by a vicariant event rather than by overwater dispersal. This hypothesis is based on the relative positions of the Greater Antilles in the upper Eocene (see geological reconstructions in Perfit and Williams in this volume). As mentioned earlier, however, it is more likely that the age of the rodent hairs is lower Miocene. I believe that it is more likely that rodents dispersed to the Antilles by over water dispersal rather than a vicariance event. The answer to this question must remain unresolved until better dates and specimens from Dominican amber are available, or until old fossils are found. Here I will examine the various routes that the ancestors of West Indian rodents may have taken in getting to the main islands, as well as the characteristics of the possible ancestors. MacArthur and Wilson (1967:133) point out in a series of figures how important stepping stone islands can be in the dispersal of organisms such as rodents. They determined that "dispersal across gaps of more than a few kilometers is by stepping stones wherever habitable stepping stones of even the smallest size exist." The following is an analysis of some of the ways rodents may have dispersed to the Antilles.

As discussed above, the results of the cladistic analysis indicate that the taxa from Puerto Rico and Hispaniola are central to the radiations of hystricognath rodents in the West Indies. The ancestor of the Heteropsomyinae and Capromyidae is most likely to have initiated the radiations of these groups in Puerto Rico or eastern Hispaniola. This ancestor was an echimyid, and most likely originated in South America near the center of the range of echimyids.

Echimyids in general are medium-sized, semi-arboreal rodents that frequently occur at low elevations and are often found in riparian situations. Animals with these characteristics are good candidates to be associated with mats of floating vegetation originating at the mouths of major river systems. It is reasonable to propose that the founding stock of the West Indian echimyids landed on eastern Hispaniola or Puerto Rico via a form of oversea dispersal. Since the relationships of the morphological features of the taxa examined indicate that West Indian echimyids and capromyids are more derived than echimyids from the Salla and Deseadan formations of southern South America, it is necessary for the dispersal event to have occurred after 24 myBP. The earliest possible date for the dispersal, therefore, was the late Oligocene. During this time the seaway between North and South America was open. The Tethyan Seaway was also open during pre-Miocene times (Holcombe and Moore 1977:50) causing intense westward currents to move through the region of the southern Caribbean.

Northern route.--The question as to whether capromyids, heptaxodontines, and heteropsomyine echimyids could be descendants of an original invasion of hystricognath rodents into South America from North America via the Antilles or proto-Antilles was originally proposed by Landry (1957:91). All capromyid subfamilies appear to be very distinct from one another, which may indicate that the separation from one another was a fairly ancient event. This hypothesis would require the presence of hystricognath rodents in the West Indies sometime long before 26 myBP. Sarich and Cronin (1980) calculated the amount of albumin change in immunological distance units for various hystricognath rodents, and estimated that *Plagiodontia* diverged from *Capromys* and *Geocapromys* 40 units ago. They estimate that the rate of change for rodent albumins is 1.1 units per million years, so that their estimated antiquity for the Family Capromyidae would be 36 my, or early Oligocene. The oldest known fossil rodents in South America are found in the Deseadan and Salla formations of Argentina and Bolivia. These deposits are controversial in age. Marshall et al. (1977) claimed they are earliest Oligocene in age (36 myBP), while MacFadden et al. (1985) proposed that the deposit is 26 myBP. Rodents are not present in the Eocene Mustersan formation (45-48 myBP). If the chronological data of Sarich and Cronin (1980:410) are correct, it would be possible for the ancestors of West Indian hystricognaths to have arrived in South America by the Deseadan, and to have given rise to the radiation of South American hystricognaths.

I reject the hypothesis that capromyids, heptaxodontines, and heteropsomyines are descendants from an Eocene hystricognath invasion of the Antilles from the north via overwater dispersal or a vicariance event from Central America for the following reasons.

1) Capromyids are derived from heteropsomyine echimyids, which include *Puertoricomys*. The heteropsomyines appear to have evolved from a *Proechimys*-like form rather than into any known fossil group in the Deseadan.

2) Heteropsomyines retain the deciduous premolar tooth, as do <u>most</u> capromyids, indicating that the retention of this tooth is a plesiomorphic character in West Indian hystricognaths. The earliest echimyids in the Deseadan replace the deciduous premolar.

3) The most primitive heteropsomyines are in the eastern and southern regions of the Antilles or proto-Antilles in the reconstruction of Sykes et al. (1982), which is the least likely pattern if the original invasion came from the north.

4) The pattern of diversity of capromyid and heteropsomyine rodents on the individual islands of the Antilles indicates it is likely that the sequence of differentiation was from Hispaniola/Puerto Rico northward rather than from Cuba or Jamaica southward. The diversity indices (subfamilies times genera divided by number of species, see Woods

in press) for capromyids and echimyids on the major islands of the Antilles are: Hispaniola (2.1); Puerto Rico (2.0); Jamaica (1.3); Bahamas (1.0); Cayman Islands (1.0); Cuba (0.2). If these indices are compared with the diversity indices for bats (Table 5) the indices for each island are roughly comparable (range 6-7.6). The diversity indices for insectivores, a group that probably entered the Antilles from the north and for which the greatest diversity should be in Cuba (MacFadden 1980), are equal on Cuba and Hispaniola. The evolutionary pattern of West Indian insectivores has followed a very different pattern from that observed in West Indian rodents.

5) While in the work of Sarich and Cronin (1980) the date of the divergence of capromyines from plagiodontines is great, the estimated date of divergence of octodontids from echimyids is even greater (57 myBP). The relative ages in the Sarich and Cronin model, therefore, argue against the West Indian taxa being ancestral to the South American taxa. The West Indian taxa are probably younger than estimated by Sarich and Cronin.

Beata Ridge.--It is tempting to speculate that these strong westerly currents would have carried floating mats of vegetation rapidly westward toward the Beata Ridge, projecting southward like a long peninsula from Hispaniola. The long arm of the Beata Ridge would have gathered in floating islands of vegetation and allowed for an increased chance of a landfall on Hispaniola, helping to explain high level of biological diversity found in many plant and animal groups on Hispaniola. This attractive scenario works well with the distribution of many groups of organisms, and with the geography of the region.

There are two considerations that reduce the possibility that the Beata Ridge/Hispaniola connection was an appropriate route for rafting rodents to take in getting into the central Antillean region. The first shortcoming is that the center of greatest diversity of heteropsomyine rodents is in Puerto Rico rather than Hispaniola. This is not a serious flaw, however, since heteropsomyines may be refugial in Puerto Rico. The more serious complication is that the Beata Ridge was probably submerged during most of the Tertiary, and certainly was completely underwater by the late Oligocene and early Miocene (Holcombe in press).

Aves Ridge.--The available evidence indicates that the Aves Ridge, located between South America and the Saba Bank near Puerto Rico, was an island arc during the Cenozoic. Granodioritic rocks and basalts along this now largely submarine ridge indicate arc magmatism in the past, and seismic studies of the area indicate that the crustal structure along the ridge is similar to that of an island arc (Pindell and Barrett 1988). Most authors believe that the Aves Ridge existed in the late Cretaceous and early Cenozoic, and that the exposed phase of the Aves Ridge terminated as the Lesser Antilles began to emerge in the Eocene about 50 myBP. Holcombe and Moore (1977:50), on the other hand, believed that the Aves Ridge was emergent until the early Miocene. If their conclusions are correct it is possible that the Aves Ridge at that time was between 500 and 1000 m higher than its present relationship with the ocean surface (Holcombe in press). If a conservative 600 m increase in the elevation of the Aves Ridge relative to sea level is assumed, then at least 100 km of continental shelf off Isla Margarita and eight islands between South America and the Saba Bank just east of Puerto Rico would have been exposed (Holcombe and Moore 1977:49). The six most appropriate islands (stepping stones) of the emergent ridge would have been 65, 30, 60, 10, 20, 120, and finally 150 kms apart between South America and the Saba Bank. Some of the crossings from one island to another are formidable, and strong east-to-west wind driven currents would have made an overwater dispersal especially difficult. However, the presence of so many small islands between South America and the central Antilles would have increased the chances of success of an echimyid reaching Puerto Rico from northern South America. If the extreme case of an elevation of the Aves Ridge by 1000 m relative to the oceans surface in the late Oligocene is considered possible, then the potential invader of the Antilles would have started 25 kms farther out on the exposed continental shelf. The waif would have to cross 3 short gaps of 25 km or less of open water between large islands of at least 50 km in diameter before reaching a large and complex island half way between South America and the Saba Bank. The large central island would have been over 500 m above sea level, and would have resisted immersion over a long period of geological time should waifs have landed there. The additional 250 km crossing to the Saba Bank would be over largely open water, but with three intervening small islands. Halfway between the large central island area and the Saba Bank would have been Aves Island, which is even now above sea level but at that time would have been 1000 m in elevation. The final overwater dispersal would have been across 125 km of open sea. However, Puerto Rico lies northwest of this position, so a westerly drifting current would have driven the waif rapidly towards Puerto Rico.

The above scenario of the role that the Aves Ridge may have served in creating a series of stepping stones for rodents dispersing overwater from South America to the central Antilles requires that the Aves Ridge exist as an emergent island arc later than some reliable geologic evidence indicates was the case. The final validation of the availability of the various islands of the Aves Ridge as stepping stones in the Oligocene or Miocene must await further geologic information.

Lesser Antilles.--The Lesser Antilles developed in the late Eocene. Donnelly believes (this volume) that the chain is closer to South America now than at any time in the past, although Coney (1982) believes that the history of the Lesser Antilles is closely tied to South America. The Lesser Antilles could have been more closely associated with the eastern Greater Antilles in the past, and Donnelly (this volume) mentions the possibility of a pre-late Cenozoic terrestrial connection. The Lesser Antilles are clearly a possible factor in assisting rodents in the overwater dispersal to the central Antilles in the region of Puerto Rico. The presence of extensive banks surrounding the Leeward Islands of the northern Lesser Antilles, such as Saba Bank, St. Martin Bank, Antigua Bank, and St. Kitts Bank, would have increased the chance of a successful landfall or eventual arrival of rodents dispersing from South America. Rodents would still have to cross the dangerous water gaps between the Windward Islands of the southern Lesser Antilles. However, of the possible island arcs or ridges that could have assisted rodents in a successful overwater dispersal to Puerto Rico in the late Oligocene and early Miocene, the Lesser Antilles are the least controversial in terms of their location and status. The Lesser Antilles would have been especially important as stepping stones or jumping off points at times of low ocean levels in the mid-Oligocene, and early and late Miocene.

If the Lesser Antilles did facilitate the dispersal of rodents from South America to Puerto Rico, why have no fossil remains of echimyids or capromyids been found, and why do no hystricognaths continue to survive in the region? One possibility might be, as Donnelly points out, that at no time was the chain of islands forming the Lesser Antilles continuously emergent. As sea levels increased in the Miocene and Pliocene (Buskirk 1985, fig. 3) many would have become submerged or very reduced in size. Because of the small size of most of the Lesser Antilles, the chance of extinction over time for echimyids and capromyids is much greater than it is for sigmodontines. Hystricognaths, unlike sigmodontines, have long gestation times and small litters, so they are less able to recover from severe reductions in population sizes. A second factor to consider is that fossil remains of mammals in the West Indies are very uncommon, even on the Greater Antilles. *Puertoricomys corozalus* is known from only one specimen, a fragmentary jaw bone from a fissure in central Puerto Rico. *Quemisia gravis*, the giant hutia from Hispaniola, is known from only two jaw fragments and a few teeth recovered in widely separated regions of interior Hispaniola. Primate remains have been recovered from the entire length of Hispaniola from near sea level to the high mountains, indicating that monkeys were widespread on the island until 3,500 yBP. However, all of the remains grouped together add up to two jaw fragments, one femur, and a few teeth. In a remote section of western Haiti a new genus of rodent has recently been described (Woods 1989a), even though mammalogists and paleontologists have actively searched for fossil remains on the island for over 60 years. The lack of success finding fossils of many mammals in the Greater Antilles, some of which we now know to have been common and widely distributed, should be kept in mind when explaining why no fossil remains are known of hystricognath rodents in the Lesser Antilles. The lack of fossils does not mean that hystricognath rodents did not colonize the region in the Miocene as they dispersed from South America to Puerto Rico as part of the major radiation of rodents that swept westward through the Antilles.

Whichever, if any, of the possible scenarios involving the Beata Ridge, the Aves Ridge, or the Lesser Antilles is correct, the strong currents and substantial overwater crossings would have limited the chances of a successful invasion of an echimyid rodent into the Greater Antilles. As Donnelly mentions (this volume), dispersing overwater from South America to the West Indies was a difficult undertaking that has become progressively easier as geologic events have progressed towards today's arrangement of the islands and South America. In spite of the difficulties involved, the combination of an apparently available island arc coupled with the data on the phylogenetic relationships of heteropsomyine and capromyid rodents, argues strongly for a scenario in which a South American echimyid rodent first arrived in the West Indies on Puerto Rico between the mid-Oligocene and the early Miocene. Sea levels in the late Oligocene and early Miocene were lower than they had previously been in the Tertiary, and as low as they would be again until the late Miocene (Vail et al. 1977). The chance invasion of the central Antilles by rodents in the late Oligocene or early Miocene initiated the explosive radiation of the major subfamilies in the various subregions of Hispaniola and the other Greater Antilles during the Miocene and Pliocene when ocean levels were high and there was limited interchange between the islands of the Greater Antilles. The phenomenal species level explosion of taxa on Cuba and its halo of surrounding islands probably occurred in the Pleistocene when sea level fluctuated rapidly (Vail et al. 1977) and climates varied dramatically (Crowley 1983). Whether it was a single dispersal across 650-700 kms of open sea, or a series of shorter dispersals utilizing the islands of the Aves Ridge or Lesser Antilles cannot be resolved. However, the presence of islands (stepping stones) between northern South America and Puerto Rico as well as the possible close association of the northern Lesser Antilles with Puerto Rico during the late Oligocene and early Miocene increase the chances of a successful invasion of the Antilles by an echimyid, and further points to Puerto Rico and the central Antilles as the area where the radiation of West Indian hystricognath rodents began.

What the central Antilles looked like in the mid-Tertiary is also a moot point. Pindell and Dewey (1982) demonstrated some possible connections between Puerto Rico and eastern Hispaniola in the early Oligocene, but indicated that the two were separate in the early Miocene. Guyer and Savage (1986) incorporated the concept of a united Puerto Rico and central Hispaniola into their vicariance model for the Oligocene. The most extensive central Antillean landmass was proposed by Graham and Jarzen (1969), who analyzed the paleobotany of Puerto Rico based on microfossils (pollen) of 44 plant genera in the Oligocene San Sebastian Formation. They found that plant communities in the Oligocene were similar to those found in the Antilles today, but that there was a temperate element composed of *Fagus*, *Liquidamber*, and *Nyssa*. They explain the presence of this minor temperate flora within a predominantly warm-temperate to subtropical flora by proposing larger geographical areas of more substantial elevation than now exist in the central Antilles. Citing a personal communication from R.P. Briggs of the U.S. Geologic Survey in Puerto Rico, Graham and Jarzen (1969:349) proposed the presence after the middle Eocene of a landmass uniting Puerto Rico, the Virgin Islands, and Hispaniola. This landmass, called "ancestral Puerto Rico" by the authors, reached elevations of 4,600 m and was 330 km in length. This is not dissimilar to the reconstruction by Sykes et al. (1982: fig.11) of the Antilles in the late Eocene. It is not necessary to have a large landmass in the central Antilles in the Oligocene or early Miocene to serve as a platform for the early radiation of West Indian hystricognaths in the Greater Antilles. Such a landmass may have existed in the Eocene and early Oligocene (see fig. 10 in Guyer and Savage 1986:527), but most reconstructions indicate a more fragmentary series of islands in the late Oligocene to early Miocene.

The timing of this origin of rodents in the Greater Antilles relative to the origin of insectivores, primates and edentates must remain a moot point because of the absence of fossil evidence. Because of the possible relationship of West Indian insectivores with apternodontids and possibly geolabidids, which as far as is known became extinct by the late Oligocene in North America, insectivores are likely to have invaded the Antilles in the Eocene or early Oligocene, before the invasion of rodents. Sloths radiated into a variety of niches, and were almost, but not quite, comparable in diversity to rodents. I believe that sloths invaded the Antilles after rodents, perhaps in the late Miocene. Because of the lack of diversity within the two and possibly three lineages of primates, it is assumed that primates entered the Antilles after most primate niches were already occupied by rodents and sloths.

Relationships of West Indian endemics to South American rodents

As discussed above, capromyids appear to be derived from South American echimyids. There is no evidence to indicate that capromyids gave rise to echimyids. The relationship of capromyids to the South American rodent *Myocastor coypus* is more controversial. There are many similarities between *Myocastor* and capromyids, and the two have often been grouped together in the same family by some authors. Morphological and biochemical data, however, do not support the hypothesis of a close phylogenetic relationship between the two groups (Woods and Hermanson 1985; Kilpatrick pers. comm.). The evidence, therefore, does not support the hypothesis that an ancestral myocastorid gave rise to the capromyid radiation, or that capromyids gave rise to *Myocastor* via a secondary dispersal event from the West Indies to South America.

Spencer (1987) proposed that capromyids evolved from an echimyid similar in morphology to *Echimys*. Among capromyids he noted a similarity in dental features along a cline from *Hexolobodon*, *Capromys*, and *Isolobodon* that could have given rise to the pattern of dental morphology found in the South American Chinchilla rat, *Abrocoma*. All of these taxa have extremely high crowned, open rooted teeth. Spencer classified capromyids, abrocomyids, and octodontids together in the Superfamily Octodontoidea based on the common occurrence of high crowned, open rooted teeth. *Myocastor* and the echimyids were excluded from the Octodontoidea because they lack these features. Spencer (1987:107) proposed that capromyids dispersed from the Antilles to South America sometime before the Huayquerian (early Pliocene) to give rise to the ancestor of *Abrocoma*. Further simplification of the dental features led to the condition found in octodontids. The lack of South American descendents of this dispersal of capromyids to South America, other than *Abrocoma* and the fossorial and seimifossorial octodontids, was explained by Spencer as a consequence of competition from sigmodontine rodents. Sigmodontines rapidly radiated after their arrival in South America to fill most available small rodent niches. He noted that there was an explosion of sigmodontine genera from five in the Uquian to 24 genera with 70 species in Argentina alone today. As further evidence for the relationship between capromyids and octodontids he sited the presence of the "octodontid" *Alterodon major* on Jamaica.

No direct fossil evidence in the West Indies, Central or South America helps answer this question. The hypothesis that capromyids and octodontids are closely related is not disproved on the basis of dental morphology. The only known specimen of *Alterodon* is a fragmentary tooth about which nothing conclusive can be stated. MacPhee et al. (1983) examined this tooth fragment, as well as a large series of teeth of other Jamaican rodents, and proposed that the tooth fragment of *Alterodon* is really an incomplete tooth of *Clidomys*. Blood protein electromorph data (Kilpatrick pers. comm.), and immunological comparisons (Sarich and Cronin 1980) both indicate that octodontids have fewer derived characters than do capromyids. Based on these data, octodontids do not appear to have evolved from capromyids.

A cladistic analysis using PAUP on the same data set of 82 morphological features used to analyze the relationships of West Indian hystricognath rodents places *Abrocoma* as the closest ancestor of octodontids. *Ctenomys* has more primitive characters than any of the remaining octodontid taxa, suggesting that ctenomyines may be distinct at the subfamily level. *Octodontomys* is the sister group of the remaining octodontid genera. *Aconomys* is the sister group of *Octodon* and *Spalacopus*. The assemblage of abrocomyids and octodontids cluster between the radiations of capromyids and echimyids when *Echimys* is used as the hypothetical ancestor in the cladogram produced by PAUP. This suggests the possibility that the morphological features examined in octodontids could be derived from the condition of the features found in most capromyids. However, when *Plagiodontia aedium* is used as a hypothetical ancestor (to determine the relative closeness of octodontids and echimyids to capromyids), then octodontids are more distantly grouped with capromyids than are echimyids.

Hypselodont molariform teeth have evolved several times in hystricognath rodents, and are found in the capybara, caviids, and chinchillids among South American forms. This calls into question the usefulness of hypselodonty as a character for uniting capromyids, abrocomyids, and octodontids into a monophyletic clade, especially given how variable this character is in capromyids. The other character used to link capromyids, octodontids, and abrocomyids on the same clade by Spencer was the lack of a mental foramen in the dentary. This character is also variable in capromyids, however, with *Plagiodontia ipnaeum* as well as some specimens of *P. aedium* having a mental foramen. The giant hutia *Elasmodontomys* lacks a mental foramen, while *Quemisia* has one. Therefore, there is no strong evidence to support the hypothesis that capromyids initiated the radiation of abrocomyids and octodontids in South America following a dispersal event from the Antilles. It is more likely that abrocomyids and octodontids evolved from an echimyid, and that the similarities in dental and mandibular features shared by capromyids on the one hand, and abrocomyids, and octodontids on the other are the result of parallelism.

Procapromys geayi was described (as *Capromys geayi*) by Pousargues (1899:150) from a specimen collected in the mountains of northern Venezuela between La Guayra and Caracas. Chapman (1901:322) set the form apart as a new genus based on the unique pattern of the reentrant folds of the M^3, and stated that *Procapromys geayi* "probably represents the ancestral mainland type whence *Capromys* descended" (Chapman 1901:323). The place of origin of this single known specimen of a capromyid in South America has been questioned, however, and the extreme small size and slightly worn molars of the type specimen suggest the possibility that it might be a juvenile. No additional specimens of this species have ever been collected. The area where the specimen was reported from is now near Parque de Avila, which has been the center of numerous studies. Pousargues considered the possibility that the specimen might have been carried to the mainland from the Antilles, but rejected the idea when he became convinced that the specimen was an adult, and therefore much different from *C. pilorides* in size and dental pattern. Anthony (1926:202) argued that the pattern of reentrant folds of the type specimen "seems to be indistinguishable" from an immature *Capromys pilorides* from Cuba. *Procapromys geayi* is here considered to be an immature *Capromys pilorides*.

It does not appear as if there are any known rodent lineages in South America that are descendants of the radiation of capromyids in the West Indies. Capromyids appear to have evolved and radiated only within the boundaries of the West Indies.

Summary
Rodent evolutionary history and biogeography

The heteropsomyine rodents of Puerto Rico, Hispaniola, and Cuba are the remnants of the oldest radiation of rodents in the Antilles. Heteropsomyines are morphologically distinct from capromyids, but resemble South American echimyids enough that they should be included in the same family. The Puerto Rican genus *Puertoricomys* (previously *Proechimys corozalus*) is similar in morphology to the *Trinomys* subgroup of the genus *Proechimys*, which currently has a relictual distribution in southeastern Brazil. In Puerto

Rico a form with dental features similar to *Puertoricomys* gave rise to the much larger *Heteropsomys*, and to *Homopsomys*, a large hutia-like derivative of *Heteropsomys*. Thus in Puerto Rico the niche of most "hutias" was filled by echimyids, indicating it is unlikely that capromyids ever were a major element of the rodent fauna on that island.

A heteropsomyine similar in form to *Heteropsomys* from Puerto Rico gave rise to the Capromyidae by developing high crowned cheek teeth, expanding the deposition of cement from the root area to include the reentrant folds, and continuing the trend begun in heteropsomyines of elongating the paroccipital process. The first capromyid is likely to have resembled a cross between the two Hispaniolan genera *Rhizoplagiodontia* and *Hexolobodon*. The transition from echimyid to capromyid took place in the central part of the Antilles in Puerto Rico or Hispaniola in the Oligocene or early Miocene when Puerto Rico and Hispaniola may have been combined as a single island or were in close proximity to one another (Pindell and Dewey 1982; Graham and Jarzen 1969).

Following the origin of the Capromyidae, and the complete separation of eastcentral Hispaniola from Puerto Rico, the Capromyidae underwent a rapid and extensive radiation in the Hispaniola region. Heteropsomyine echimyids were confined to a minor role as rat-sized rodents on Hispaniola and Cuba by the explosion of larger capromyids with their more adaptive high crowned dentitions. These radiations were associated with various subunits of the West Indies, a fact which is supported by the very distinct morphological characteristics of the genera and subfamilies which are isolated on the main island masses of the Antilles. In the western Antilles the pattern of adaptive radiation is most recent and there is a low level of endemism at the generic and subfamily level.

The cladistic analyses of the Family Capromyidae and Subfamily Heteropsomyinae (Figs. 3-5) support the hypothesis that the center of origin of the group was in the central part of the Antilles rather than the eastern or southern regions. All of the taxa occurring on Puerto Rico and Hispaniola are more primitive than are members of the same clades from Cuba and the western Antilles.

Hispaniola also has the highest level of mammalian "diversity" of any West Indian island (Table 5). In measuring taxonomic diversity on an island an emphasis should be placed on the amount of higher level taxonomic diversity present as opposed to the species richness of an island. Species richness may be the result of island size as well as Pleistocene species multiplier phenomena such as changing ocean levels and climates, both of which created disjunct distributions. Species richness of a particular island may also be a function of the taxonomic attention that the fauna has received, and the amount of field work expended in the search for fossil deposits and cave faunas. The number of species may be over estimated on some islands, such as Cuba (Morgan and Woods 1986; Woods in press), which could bias any conclusions based on species numbers alone.

The high level of endemism on the major islands of the Greater Antilles, as well as the total lack of endemic capromyid and echimyid rodents in the Lesser Antilles points out how infrequent (or unsuccessful) the overwater dispersal events may have been in the past. However, the presence of *Geocapromys*, *Clidomys* and *Oryzomys* on Jamaica, and of *Capromys*, *Geocapromys* and *Nesophontes* in the Cayman Islands demonstrates the possibilities of over water dispersal of these rodents in the Antilles.

The widespread distribution of sigmodontines in the Lesser Antilles: two species of *Oryzomys*, three *Megalomys*, and two widespread undescribed forms (Table 1); indicates that the Lesser Antilles are suitable habitats for mammals, and that mammals can disperse from one island of the Lesser Antilles to another. Only *Oryzomys antillarum* from Jamaica managed to expand the range of simodontines out of the core area of the Lesser Antilles. It appears to be an overwater invader directly from Central America, and may even be conspecific with the mainland form *O. couesi* (Honacki et al. 1982:439). Because capromyids dominate over echimyids in all areas of the Antilles except Puerto Rico, I believe that heteropomyines invaded Hispaniola and Cuba after the origin and initial radiation of capromyids on those islands. The dispersal of heteropsomyines and their subsequent radiation in Hispaniola and then Cuba likely occurred in the Pliocene and Pleistocene.

Acknowledgements

I thank David Klingener and Ernest Williams for their valuable criticism and comments on the manuscript. Bonnie McKeen and Margaret Langworthy helped sort and prepare the fossil material, and I appreciate their assistance. Heidi Norden assisted in the analysis of tooth wear and dental homologies in Puerto Rican echimyids. Dan Cordier, Tia Cordier, Ada Fowler, Dick Franz, Lesley Hay, Margaret Langworthy, Lynette Shirley, John Yarington, Karen Yarnell, and Stott Woods, as well Missy, Patty and Bryan Woods helped collect fossil and recent specimens in the field under difficult working conditions. Their assistance and support is truly appreciated. I thank Drs. Karl Koopman and Guy Musser of the American Museum of Natural History for the loan of several specimens, especially the type of *Puertoricomys corozalus*, and Dr. Richard Thorington of the U.S. National Museum and Dr. Farish Jenkins of the Harvard Museum of Comparative Zoology for the loan of specimens. I am very grateful to Dr. Storrs Olson for loaning me specimens of echimyids he collected in Blackbone cave in Puerto Rico, and to Dr. Luis A. Chanlatte Baik of the University of Puerto Rico for the loan of a large series of *Homopsomys antillensis*. This study was supported in part by grants DEB 7811388, DEB 8216825, and EAR 8212745 from NSF, and by a grant from Earthwatch. I dedicate this paper with respect and affection to the memory of Bonnie McKeen, who died in a bicycle accident near the Florida State Museum in December 1986.

Literature cited

Anderson, S. 1965. Conspecificity of *Plagiodontia aedium* and *P. hylaeum* (Rodentia). Proceedings of the Biological Society of Washington 78:95-98.

Anthony, H.E. 1920. New mammals from Jamaica. Bulletin of the American Museum of Natural History 42:469-475.

_____. 1926. Mammals of Porto Rico, living and Extinct-Rodentia and Edentata. New York Academy of Sciences, Scientific Survey of Porto Rico and the Virgin Islands 9(2):96-238

Berry, E.W. 1942. Mesozoic and Cenozoic plants of South America, Central America and the Antilles. Proceedings of the Eighth American Science Congress 4:365-373.

Buskirk, R.E. 1985. Zoogeographic patterns and tectonic history of Jamaica and the northern Caribbean. Journal of Biogeography 12:445-461.

Carleton, M.D., and G.G. Musser. 1984. Muroid rodents. Pp. 289-379 in S. Anderson, and J.K. Jones (eds.). Orders and Families of Recent Mammals of the World. Wiley, New York.

Chapman, F.M. 1901. A revision of the genus *Capromys*. Bulletin of the American Museum of Natural History 14:313-324.

Cockerell, T.D.A. 1923. A fossil cichlid fish from the Republic of Haiti. Proceedings of the United States National Museum 63(7):1-3.

Colbert, E.H. 1969. A Jurassic pterosaur from Cuba. American Museum of Natural History Novitates 2370:1-26.

Coney, P.J. 1982. Plate tectonic constraints on the biogeography of Middle America and the Caribbean region. Annals of the Missouri Botanical Garden 69(3):432-443.

Crowley, T.J. 1983. The geologic record of climatic change. Review of Geophysics and Space Physics 21:828-877.

Fields, R.W. 1957. Hystricomorph rodents from the late Miocene of Colombia, South America. University of California Publications in Geological Sciences 32(5):273-404.

Goodfriend, G.A. and R.M. Mitterer. 1987. Age of ceboid femur from Coco Ree, Jamaica. Journal of Vertebrate Paleontology 6(3):344-345.

Graham, A. and D.M. Jarzen. 1969. Studies in Neotropical paleobotany 1. The Oligocene communities of Puerto Rico. Annals of the Missouri Botanical Garden 56:308-357.

Guyer, C., and J.M. Savage. 1986. Cladistic relationships among anoles (Sauria: Iguanidae). Systematic Zoology 35(4):509-531.

Hershkovitz, P. 1972. The recent mammals of the Neotropical Region. Pp. 311-431 in A. Keast, F.C. Erk, and B. Glass (eds.). Evolution, Mammals, and Southern Continents. State University of New York Press, Albany, NY.

Holcombe, T.L. and W.S. Moore. 1977. Paleocurrents in the eastern Caribbean: geologic evidence and implications. Marine Geology 23:35-56.

_____. in press. Late Cretaceous and Cenozoic evolution of the Caribbean ridges and rises with special reference to paleogeography. In A. Azzaroli (ed.). Biogeographical Aspects of Insularity. Accademia Nazionale dei Lincei, Rome.

Honacki, J.H., K.E. Kinnman and J.W. Koeppl. 1982. Mammal Species of the World, a taxonomic and geographic reference. Association of Systematic Collections, Lawrence, KS. 694 pp.

Kashfi, M.S. 1983. Geology and hydrocarbon prospects of Jamaica. Bulletin of the American Association of Petroleum Geologists 67:2117-2124.

Koopman, K.F., and E.E. Williams. 1951. Fossil Chiroptera collected by H.E. Anthony in Jamaica, 1919-1920. American Museum of Natural History Novitates 1519:1-29.

Kratochvil, J., L. Rodriguez, and V. Barus. 1978. Capromyidae (Rodentia) of Cuba 1. Acta Scientiarum Naturalium, Brno 12(11):1-60.

Landry, S.O., Jr. 1957. The interrelationships of the New and Old World Hystricomorph Rodents. University of California Publications in Zoology 56(1):1-118.

MacArthur, R.H., and E.O. Wilson. 1967. The Theory of Island Biogeography. Princeton University Press, Princeton, NJ. 203 pp.

MacFadden, B.J. 1980. Rafting mammals or drifting islands?: biogeography of of the Greater Antillean insectivores *Nesophontes* and *Solenodon*. Journal of Biogeography 7:11-22.

____, K.E. Campell, Jr., R.L. Cifelli, O. Siles, N.M. Johnson, C.W. Naeser, and P.K. Zeitler. 1985. Magnetic polarity stratigraphy and mammalian fauna of the Deseadan (late Oligocene-early Miocene) Salla Beds of northern Bolivia. Journal of Geology 93(3):223-250.

MacPhee, R.D.E. 1984. Quaternary mammal localities and heptaxodontid rodents of Jamaica. American Museum of Natural History, Novitates 2803:1-34.

____, C.A. Woods, and G.S. Morgan. 1983. The Pleistocene rodent *Alterodon major* and the mammalian biogeography of Jamaica. Paleontology 26:831-837.

____, D.C. Ford, and D.A. McFarlane. in preparation. Pre-Wisconsinan mammals from Jamaica and models of Late Quarternary extinction in the Greater Antilles.

Marshall, L.G. 1979. A model for paleobiogeography of South American cricetine rodents. Paleobiology 5(2):126-132.

Marshall, L.G., R. Pascual, G.H. Curtiss, and R.E. Drake. 1977. South American Geochronology. Radiometric time scale for middle to late Tertiary mammal-bearing horizons of Patagonia. Science 195:1325-1328.

Maurrasse, F.J-M.R., F. Pierre-Louis, and J.-G. Rigaud. 1982. Cenozoic facies distribution in the southern peninsula of Haiti and the Barahona Peninsula, Dominican Republic, and its relations concerning the tectonic evolution of the La Selle-Baoruco block. Caribbean Geological Conference Contributions 9:1-24.

Miller, G.S. 1916. Bones of mammals from Indian sites in Cuba and Santo Domingo. Smithsonian Miscellaneous Collections 66(12):1-11.

____. 1922. Remains of mammals from caves in the Republic of Haiti. Smithsonian Miscellaneous Collections 74(3):1-8.

____. 1929a. A second collection of mammals from caves near St. Michel, Haiti. Smithsonian Miscellaneous Collections 81(9):1-30.

____. 1929b. Mammals eaten by Indians, owls, and Spaniards in the coast region of the Dominican Republic. Smithsonian Miscellaneous Collections 82(5):1-16.

____. 1930. Three small collections of mammals from Hispaniola. Smithsonian Miscellaneous Collections 82(15):1-10.

Moojen, J. 1948. Speciation in the Brazilian spiny rats (Genus Proechimys, Family Echimyidae). Museum of Natural History, University of Kansas, Publications 1(19):301-406.

Morgan, G.S. 1977. late Pleistocene fossil vertebrates from the Cayman Islands, British West Indies. Unpublished Master of Science Thesis, University of Florida, Gainesville, FL. 282 pp.

____, G.S. 1985. Taxonomic status and relationships of the Swan Island Hutia, *Geocapromys thoracatus* (Mammalia: Rodentia: Capromyidae), and the zoogeography of the Swan Islands vertebrate fauna. Proceedings of the Biological Society of Washington 98:29-46.

____, C.E. Ray and O. Arredondo. 1980. A giant insectivore from Cuba (Mammalia: Insectivora: Solenodontidae). Proceedings of the Biological Society of Washington 93:597-608.

_____, and C.A. Woods. 1986. Extinction and zoogeography of West Indian land mammals. Biological Journal of the Linnean Society 28:167-203.

Oviedo, G.F. 1959. Historia General y Natural de las Indias. Biblioteca de Autores Españoles, Colección Rivadenaira, 5 vols. Madrid.

Patterson, B., and R. Pascual. 1968. New echimyid rodents from the Oligocene of Patagonia, and a synopsis of the family. Museum of Comparative Zoology, Breviora 301:1-14.

_____, and A.E. Wood. 1982. Rodents from the Deseadan Oligocene of Bolivia and the relationships of the Caviomorpha. Bulletin of the Museum of Comparative Zoology 149(7):371-543.

Pindell, J.L., and S.F. Barrett. 1988. Geological evolution of the Caribbean region; A plate-tectonic perspective. In G. Dengo and J.E. Case (eds.). The Geology of North America, Volume H, The Caribbean. The Geological Society of America. Boulder, CO.

_____, J., and J.F. Dewey. 1982. Permo-Triassic reconstruction of western Pangea and the evolution of the Gulf of Mexico/Caribbean region. Tectonics 1:178-211.

Poinar, G.O., Jr. 1988. Hair in Dominican amber: evidence for Tertiary land mammals in the Antilles. Experientia 44:88-89.

_____, and D.C. Cannatella. 1987. An upper Eocene frog from the Dominican Republic and its implications for Caribbean biogeography. Science 237:1215-1216.

Pousargues, E. de. 1899. Sur une nouvelle espèce de Capromys, découverte par M. Geay dans le nord du Vénézuéla. Bulletin de la Musée d'Histoire Naturelle 5:150-154.

Pregill, G. 1981. Late Pleistocene Herpetofaunas from Puerto Rico. Miscellaneous Publication of the University of Kansas Museum of Natural History 71:1-72.

Ray, C.E. 1965. The relationships of *Quemisia gravis* (Rodentia:Heptaxodontidae). Smithsonian Miscellaneous Collections 149(3):1-12.

Reig, O. 1980. A new fossil genus of South American cricetid rodents allied to *Wiedomys*, with an assessment of the Sigmodontinae. Journal of Zoology 192:257-281.

Rieppel, O. 1980. Green anole in Dominican amber. Nature 286:486-487.

Rimoli, R.O. 1976. Roedores Fosiles de La Hispaniola. Universidad Central del Este Serie Cientifica, lll (San Pedro Macoris, Dominican Republic), 95 pp.

Rivero, M. de la Calle, and O. Arredondo. 1988. Craneo de un primate fósil. Juventud Tecnica (Havana, Cuba) 245:35.

Rodriguez, L., V. Barus, and J. Kratochvil. 1979. The genus *Mesocapromys*, a link between the families Echimyidae and Capromyidae. Folia Zoologica 28(2): 97-102.

Rosen, D.E. 1985. Geological hierarchies and biogeographic congruence in the Caribbean. Annals of the Missouri Botanical Garden 72:636-659.

Rouse, I. and L. Allaire. 1978. Caribbean. Pp. 431-481 in R.E. Taylor and C.W. Meighan (eds.). Chronologies in New World Archaeology. Academic Press, New York.

Sarich, V.M. and J.E. Cronin. 1980. South American mammal molecular systematics, molecular clocks and continental drift. Pp. 399-421 in R.L. Ciochon and A.B. Chiarelli (eds.). Evolutionary Biology of of the New World Monkeys and Continental Drift. Plenum Press, New York.

Schlee,D. 1984. Besonderheiten des Dominikanischen Bernsteins. Bernstein-Neuigkeiten, Stuttgarter Beitrage zur Naturkunde, Serie C, Nr. 18.

Simpson, G.G. 1945. The principles of classification and a classification of mammals. Bulletin of the American Museum of Natural History 85:1-350.

Spencer, L.A. 1987. Fossil Abrocomyidae and Octodontidae (Rodentia: Hystricina), a phylogenetic analysis. Unpublished Ph.D. dissertation, Loma Linda University, CA, 126 pp.

Steadman, D.W., G.K. Pregill and S.L. Olson. 1984a. Fossil vertebrates from Antiqua, Lesser Antilles: evidence for late Holocene human-caused extinctions in the West Indies. Proceedings of the National Academy of Sciences, U.S.A 81:4448-4451.

_____, D.R. Watters, E.J. Reitz, and G.K. Pregill. 1984b. Vertebrates from archaeological sites on Montserrat, West Indies. Annals of the Carnegie Museum of Natural History 53:1-29.

Sykes, L.R., W.R. McCann, and A.L. Kafka. 1982. Motion of Caribbean Plate during last seven million years and implications for earlier Cenozoic movements. Journal of Geophysical Research 87(B13):10,656-10,676.

Vail, P.R., R.M. Mitchum and S. Thompson 1977. Seismic stratigraphy and global changes of sea level, Part 4: Global cycles of relative changes in sea level. Memoirs of the American Association of Petroleum Geologists 26:83-97.

Varona, L.S. 1974. Catalogo de los mamiferos vivientes y extinguidos de las Antilles. Instituto de Zoologia, Academia de Ciencias de Cuba. 139 pp.

_____, and O. Arredondo. 1979. Nuevos táxones fósiles de Capromyidae (Rodentia: Caviomorpha). Poeyana 195:1-51.

Vucetich, M.G., R. Pascual and G.J. Scillato-Yane. in press. Extinct and recent South American and Caribbean edentates and rodents: outstanding examples and evolution. In A. Azzaroli (ed.). Biogeographical Aspects of Insularity, Accademia Nazionale dei Lincei, Rome.

Williams, E.E. and K.F. Koopman 1951. A new fossil rodent from Puerto Rico. American Museum of Natural History, Novitates 1515:1-9.

Woods, C.A. 1982. The history and classification of South American hystricognath rodents: reflections on the far away and long ago. Pp. 377-392 in M.A. Mares and H.H. Genoways (eds.). Mammalian Biology in South America. Special Publication, Pymatuning Laboratory of Ecology, University of Pittsburgh, PA.

_____. 1984. Hystricognath Rodents. Pp. 389-446 in S. Anderson, and J.K. Jones (eds.). Orders and Families of Recent Mammals of the World. Wiley, New York.

_____. 1989a. A new capromyid rodent from Haiti: the origin, evolution and extinction of West Indian rodents and their bearing on the origin of New World hystricognaths. Contributions in Science of the Natural History Museum of Los Angeles County, CA..

_____. 1989b. The endemic rodents of the West Indies: the end of a splendid isolation. IUCN/SSC Rodent Specialist Group.

_____. in press. The fossil and recent land mammals of the West Indies: an analysis of the origin, evolution, and extinction of an insular fauna. In A. Azzaroli (ed.). Biogeographical Aspects of Insularity, Accademia Nazionale dei Lincei, Rome.

_____, and J.W. Hermanson 1985. Myology of hystricognath rodents: an analysis of form, function, and phylogeny. Pp. 515-548 in W.P. Luckett and J.L. Hartenberger (eds.). Evolutionary Relationships Among Rodents. Plenum Press, New York.

_____, and E.B. Howland. 1979. Adaptive radiation of capromyid rodents: anatomy of the masticatory apparatus. Journal of Mammalogy 60(1):95-116.

_____, and B. McKeen. 1989. Convergence in New World porcupines and West Indian rodents: an analysis of tooth wear, jaw movement, and diet in rodents. In K. Redford and J. Eisenberg (eds.). Advances in Neotropical Mammalogy, Sandhill Crane Press, Gainesville, FL.

_____, J.A. Ottenwalder and W.L.R. Oliver. 1985. Lost mammals of the Greater Antilles: the summarized findings of a ten weeks field survey in the Dominican Republic, Haiti and Puerto Rico. Dodo, Jersey Wildlife Preservation Trust 22:23-42.

Table 1. List of West Indian Endemic Rodents.

TAXON	DISTRIBUTION	STATUS
RODENTIA		
Capromyidae		
Capromyinae[1]		
Capromys (Macrocapromys) acevedoi	Cuba	Extinct
Capromys (Pygmaeocapromys) angelcabrerai	off Cuba	Living
Capromys (Palaeocapromys) antiquus	Cuba	Extinct
Capromys (Mesocapromys) auritus	off Cuba	Living
Capromys (Mesocapromys) barbouri	Cuba	Extinct
Capromys (Pygmaeocapromys) beatrizae	Cuba	Extinct
Capromys (Pygmaeocapromys) delicatus	Cuba	Extinct
Capromys (Mysateles) garridoi	off Cuba	Living
Capromys (Stenocapromys) gracilis	Cuba	Extinct
Capromys (Mysateles) gundlachi	I. Pines (Juventud)	Living
Capromys (Brachycapromys) jaumei	Cuba	Extinct
Capromys (Mesocapromys) kraglievichi	Cuba	Extinct
Capromys (Palaeocapromys) latus	Cuba	Extinct
Capromys (Mysateles) melanurus[2]	Cuba	Living
Capromys (Mysateles) meridionalis	I. Pines (Juventud)	Living
Capromys (Pygmaeocapromys) minimus	Cuba	Extinct
Capromys (Mesocapromys) nanus	Cuba	Living
Capromys (Capromys) pilorides	Cuba	Living
Capromys (Mysateles) prehensilis	Cuba	Living
Capromys (Palaeocapromys) robustus	Cuba	Extinct
Capromys (Mesocapromys) sanfelipensis	off Cuba	Extinct
Capromys (Pygmaeocapromys) silvai	Cuba	Extinct
Capromys sp	Cayman Islands	Extinct
Geocapromys columbianus	Cuba	Extinct
Geocapromys megas	Cuba	Extinct
Geocapromys pleistocenicus	Cuba	Extinct
Geocapromys brownii	Jamaica	Living
Geocapromys ingrahami	Bahamas	Living
G.i. abaconis	Great Abaco	Extinct
G.i. ingrahami	East Plana Cay	Living
	Little Wax Cay (1973)	Intro.
	Warderick Wells Cay (1981)	Intro.
G.i. irrectus	Andros, Cat,	Extinct
	Eleuthera, Great &	"
	Little Exuma & Long	"
Geocapromys thoracatus	Little Swan	Extinct
Geocapromys sp	Cayman Islands	Extinct

Table 1. List of West Indian Endemic Rodents (continued).

TAXON	DISTRIBUTION	STATUS
Plagiodontinae		
Plagiodontia aedium	Hispaniola	Living
Plagiodontia araeum	Hispaniola	Extinct
Plagiodontia ipnaeum (?)	Hispaniola	Extinct
Plagiodontia spelaeum (?)	Hispaniola	Extinct
Plagiodontia velozi	Hispaniola	Extinct
Rhizoplagiodontia lemkei	Hispaniola	Extinct
Isolobodontinae		
Isolobodon portoricensis	Hispaniola	Extinct
Isolobodon montanus	Hispaniola	Extinct
Hexolobodontinae		
Hexolobodon phenax	Hispaniola	Extinct
Hexolobodon poolei (?)	Hispaniola	Extinct
Hexolobodon sp	S. Hispaniola	Extinct
FAMILY *incertae sedis*		
Heptaxodontinae		
Quemisia gravis	N. Hispaniola	Extinct
Elasmodontomys obliquus	Puerto Rico	Extinct
Amblyrhiza inundata	Anquilla & St. Martin	Extinct
Clidomyinae		
Clidomys osborni	Jamaica	Extinct
Clidomys parvus	Jamaica	Extinct
Echimyidae		
Heteropsomyinae		
Boromys offella	Cuba	Extinct
Boromys torrei	Cuba	Extinct
Brotomys contractus	Hispaniola	Extinct
Brotomys voratus	Hispaniola	Extinct
Homopsomys antillensis	Puerto Rico	Extinct
Heteropsomys insulans	Puerto Rico	Extinct
Puertoricomys corozalus	Puerto Rico	Extinct

Table 1. List of West Indian Endemic Rodents (conclusion).

TAXON	DISTRIBUTION	STATUS
Muroidea		
Sigmodontinae		
Oryzomys antillarum	Jamaica	Extinct
Oryzomys victus	St. Vincent	Extinct
Oryzomys sp	Barbados	Extinct
Megalomys desmarestii	Martinique	Extinct
Megalomys luciae	St. Lucia	Extinct
Megalomys audreyae	Barbuda	Extinct
Undescribed species A[3]	Montserrat & Anguilla	Extinct
	St. Eustatius & St. Kitts	Extinct
Undescribed species B[3]	Montserrat	Extinct
	Antigua	Extinct
	Barbuda	Extinct
	Guadeloupe	Extinct
	Marie Galante	Extinct
<u>Summary</u>		
4 Families		(3 Extinct)
18 Genera[4]		(15 Extinct)
61 Species		(48 Extinct)

1). *Capromys* may be divided into 3 genera (*Capromys, Mysateles, Mesocapromys*) per Kratochvil et al. (1978). The subgeneric names *Brachycapromys, Palaeocapromys, Pygmaeocapromys, Stenocapromys* follow Varona and Arredondo (1979).

2). *Capromys arboricolus* is a young female *C. melanurus* according to Varona (pers. comm.).

3). See Steadman et al. 1984a, 1984b

4). The calculated number of genera includes Steadman's[3] undescribed forms as new genera. If they are distinct only at the species level the number of endemic genera would be 16, and the percent extinction 81%.

Table 2. Characters and character states used in the cladistic analysis. Coded polarities of character states are given in parentheses.

CHARACTER	STATE
1. Deciduous molar (DM) replaced:	Always(0), late(1), sometime(2), never(3)
2. DM position re: zygoma	Rear(0), middle(1), anterior(2)
3. Cement in reentrant folds	No(0), yes(1)
4. Cement surrounding theme	No(0), yes(1)
5. Cement in bands	No(0), yes(1)
6. Amount of cement	Absent(0), thin(1), moderate(2), thick(3)
7. Cheek teeth hypsodont	No(0), yes(1)
8. Cheek teeth hypselodont	No(0), briefly(1), longtime(2), always(3)
9. Upper reentrants P/Ms/Mt:H	No(0), yes(1)
10. Upper reentrants P/Ms/Mt:H/P	No(0), yes(1)
11. Upper reentrants P/Ms:H	No(0), yes(1)
12. Upper reentrants P/Ms:H/P	No(0), yes(1)
13. Upper reentrants Ms:H	No(0), yes(1)
14. Upper reentrants Mt:H	No(0), yes(1)
15. Upper reentrants Ms/Mt:H	No(0), yes(1)
16. Upper DM P/MS/MT:H	No(0), yes(1)
17. Upper DM P/MS:H	No(0), yes(1)
18. Upper DM MS:H	No(0), yes(1)
19. Upper DM P/MT:H	No(0), yes(1)
20. Upper DM P/MS:H/P	No(0), yes(1)
21. Lower DM elongated	No(0), yes(1)
22. Reentrants form from inside	No(0), yes(1)
23. Lower DM differs from molars	No(0), yes(1)
24. Upper DM differs from molars	No(0), yes(1)
25. Incisors procumbent	No(0), slight(1), great(2)
26. Reentrant folds lamellar	No(0), Juv(1), sometime(2), always(3)
27. Upper/lower reentrants differ	No(0), yes(1)
28. Reentrants form from outside	No(0), yes(1)
29. Lower cheek teeth sigmoid	No(0), yes(1)
30. Upper cheek teeth sigmoid	No(0), yes(1)
31. Mandibular foramen medial	No(0), yes(1)
32. Mandibular foramen in fossa	No(0), yes(1)
33. Mandibular foramen height	High(1), middle(2), low(3), bottom(4)
34. Cheek teeth flat	No(0), grading to very(5)
35. Maxillary toothrow converge	No(0), grading to very(5)
36. Reentrant folds oblique	No(0), yes(1)
37. Hook on zygoma dorsomedial	No(0), yes(1)
38. Pterygoid wing-like	No(0), yes(1)
39. Incisive foramen suture	Anterior(1), mid(2), rear(3)

Table 2. Characters and character states used in the cladistic analysis. Coded polarities of character states are given in parentheses (conclusion).

CHARACTER	STATE
40. Incisive foramen distinct	No(0), yes(1)
41. Incissive foramen connection	Broad(0), narrow(1), missing(2)
42. Paroccipital process free	No(0), yes(1)
43. Parocc. proc. longer bullae	No(0), yes(1)
44. Paroccipital process	Short(1), medium(2), long(3)
45. Lateral process	None(0), slight(1), moderate(2), huge(3)
46. Lat. proc. independent bullae	No(0), yes(1)
47. Bulla forms tip of lat. proc.	No(0), yes(1)
48. Pterygoid fossa re: M3	Lateral(1), middle(2), medial(3)
49. Mesopterygoid fossa	Rear M3(1), M3(2), rear M2(3), M2(4)
50. Supraorbital ridge	No(0), yes(1)
51. Post orbital process	No(0), slight(1), large(2)
52. Ventral canal of IO foramen	No(0), slight(1), deep(2), flange(3)
53. Mental foramen of jaw	None(0), slight(1), large(2)

POSTCRANIAL

54. Five sacral vertebrae fused	No(0), yes(1)
55. Acromion process elongated	No(0), yes(1)
56. Cervical vertebrae C2+3 fused	No(0), yes(1)

MUSCULATURE

57. M.cut.max. interdig. thorax	No(0), yes(1)
58. M.cut.max. with humeral head	No(0), yes(1)
59. M.cut.max. p.thoracab. perpend. to p.dorsalis	No(0), yes(1)
60. " " " sweeps over knee	No(0), yes(1)
61. M.cut.max p.femoral. to tail	No(0), yes(1)
62. M.masset. super. p.anter.	No(0), yes(1)
63. " " " massive	No(0), yes(1)
64. M.masset. post lrge w/ jug.f.	No(0), yes(1)
65. M.masset. med. p.post. separt. by masseteric nerv.	No(0), yes(1)
66. M.glossphrg. lrge w/ 2 insrt.	No(0), yes(1)
67. M.cricothyr. long & thin	No(0), yes(1)
68. " " multiparted	No(0), yes(1)
69. M.hyoglossus O. freefloating	No(0), yes(1)
70. M.styloglossus O. pteryg. proc.	No(0), yes(1)
71. M.genioglossus I. freefloating	No(0), yes(1)

Table 2. Characters and character states used in the cladistic analysis. Coded polarities of character states are given in parentheses (concluded).

CHARACTER	STATE
72. M.sternohyoid. I. freefloating	No(0), yes(1)
73. " " O. sep. from M.sternothyr.	No(0), yes(1)
74. M.Omohyoideus lost	No(0), yes(1)
75. M.thyrohyoid. I. freefloating	No(0), yes(1)
76. M.scalenus anticus	No(0), yes(1)
77. M.supraspin.& infraspin. fused	No(0), yes(1)
78. M.pect.major w/ 2 layers	No(0), yes(1)
79. M.lat.dorsi & teres major with separarte I.	No(0), yes(1)
80. Latissimus Achselbogen	No(0), yes(1)
81. M.coracobrach middle & long heads div by median N.	No(0), yes(1)
82. M.brachioradialis	No(0), yes(1)

Table 3. Data set for 82 characters listed in Table 6 for 18 genera of West Indian rodents.

TAXON	CHARACTER STATE[1]
Plagiodontia aedium	2011011300000100001010000110101550100211113210320100101110101111011110100
Plagiodontia araeum	3011011300000100000010000010000110101550100221132103101001109999999999999999
Rhizoplagiodontia lemkei	3211011100000110000000010111000144011020099299922990091099999999999999999
Capromys pilorides	3011031300010000010001000100001043101020111321012110011010101011111111100100
Capromys nanus	3011031300100001000100010000010000143101301011101210001109999999999999999
Geocapromys brownii	3011031300100001000101010001000014320102011122101211101101010101111111100100
Geocapromys ingrahami	3011031300100001000101010001000014320102120102121110101010101011111111100100
Isolobodon portoricensis	3211121300001000100010111011000035501130010222102211109109999999999999999
Isolobodon montanus	3211121300001000100010111311000035511131110221032111091099999999999999999
Hexolobodon phenax	3211021200010000000011000121100030501030110230112310220100910999999999999999
Brotomys voratus	3000000000010000000110101011000100099101010120220100910999999999999999999
Boromys torrei	3000000000010000000110101011000010009301001110240010910999999999999999999
Heteropsomys insulans	3100000000010000100101001110100011001100100099201111210250100919999999999999999
Homopsomys antillensis	3100000000010000100101001110100011001100100099201111210330200910999999999999999
Puertoricomys corozalus	3900000009999999991191090190309999999999999999999999999999999999999999999999
Proechimys iheringi	3000000100010101010010020011100999010100002210241009109999999999999999999
Proechimys semispinosus	3000000100000001000001001001000030100012000022101420290109999999999999999999
Elasamodontomys obliquus	0311011200000000100010003111110545100200113310330219109999999999999999999

[1] Listed in the sequence that they occur in Table 6.

Table 4. Dates for Mammal Bearing Layers in the West Indies.

	Jamaica					Hispaniola				
Depth(cm)	WRC	CRC	LMC	CLB	CLA	CD	CSW	TZ	TG	CSF
15										
25		13,400				1,600				
50			2,145	6,720	9,565 +		6,405	17,405		8,120
65		30,100			8415					
75										
90		39,300								
100					5,270	19,960	3,755	19,800	21,170	
120							7,805			
170				3,860						
200							9,550			
225							9,790			
250							10,320			
300							9,660			
(?)	33,250									
Bottom	95,000[1]									

AVERAGE[2] = 12,929 yBP; RANGE = 1,60C-39,300 yBP (+ 95,000)
JAMAICA: WRC = Cueva la Berna; CRC = Coco Ree Cave (Goodfriend and Mitterer 1987); LMC = Long Mile Cave;DOMINICAN REPUBLIC: CLB = Cueva la Berna; CLA = Cueva las Abejas;HAITI: CD = Cavern Dadier; CSW = Cavern SaWo; TZ = Trou Zombie; TG = Trou Galarie (Isle Tortue); CSF = Cavern San Francisco.

[1] MacPhee et al., see discussion in text.
[2] Average does not include old date by MacPhee et al., in preparation.

Table 5. Data on Selected West Indian Islands and their Native Land Mammals.

ISLAND	SIZE (sq. kms)[1]	ELEVATION (meters)[1]	MAMMALS (# species)				Div. Idx[2]
			Rodnts	Non-volant	Bats	Total	
Cuba	111463	1973	27	44	33	77	7*
Hispaniola	76193	3175	14	28	19	47	13
Jamaica	11424	2256	4	5	23	28	10
Puerto Rico	8865	1338	4	6	16	22	11
Andros	5957	22	1	1	4	5	6
Juventud	3000	310	7	8	18	26	8
Guadeloupe	1510	1467	1	1	11	12	9
Great Abaco	1681	44	1	1	3	4	4
Great Inagua	1544	33	0	0	5	5	5
Grand Bahama	1373	21	0	0	1	1	1
Martinique	1102	1397	1	1	10	11	9
Dominica	751	1447	0	0	10	10	7
La Gonave	680	776	6	8	8	16	11
St. Lucia	616	950	1	1	8	9	7
Eleuthera	518	51	1	1	4	5	5
Long	448	54	1	1	6	7	7
Barbados	431	340	1	1	7	8	7
Cat	389	62	1	1	5	6	6
Acklins	389	43	0	0	5	5	5
St. Vincent	344	1234	1	1	9	10	6
Mayaguana	285	40	0	0	3	3	3
Antigua	280	402	1	1	9	10	11
Crooked	238	47	1	1	5	6	6
St. Croix	207	353	1	1	4	5	5
New Providence	207	37	1	1	13	14	7
Caicos	190	28	1	1	6	7	6
Exuma	187	38	1	1	7	8	7
Grand Cayman	185	42	2	3	8	11	8
La Tortue	180	464	1	3	7	10	8
St. Kitts	180	1155	1	1	6	7	5
Barbuda	160	39	2	2	11	13	8
Marie Galante	153	204	1	1	1	2	2
Nevis	130	985	0	0	6	6	3
Montserrat	102	915	2	2	9	11	6
St. Martin	88	485	1	1	7	8	6
St. Thomas	70	209	1	1	5	6	5
Mona	50	low	1	1	3	4	4
St. John	49	389	1	1	6	7	5
Beata	40	low	2	3	2	5	3

Table 5. Data on Selected West Indian Islands and their Native Land Mammals (concluded).

ISLAND	SIZE (sq. kms)[1]	ELEVATION (meters)[1]	MAMMALS (# species)				Div. Idx[2]
			Rodnts	Non-volant	Bats	Total	
Anegada	34	8	0	0	1	1	1
Cayman Brac	31	138	2	3	6	9	7
Little Cayman	24	23	0	0	2	2	2
Saba	13	870	0	0	3	3	3
E. Plana	5	20	1	1	1	2	2
Swan	2	low	1	1	0	1	1

[1] There are discrepancies among various authoritative sources.
[2] See Woods (in press) for explanation and formula.
* If *Capromys* represents 3 genera, then the DI for Cuba is 8.

Table 6. Extinction Rates (as % of loss) of Endemic Non-volant Mammals of the West Indies. (CUB= Cuba; CI= Cayman Islands; HISP= Hispaniola; JAM= Jamaica; SI= Swan Islands; BAH= Bahamas; PR= Puerto Rico; VI= Virgin Islands; Less Ant= Lesser Antilles.)

	CUB[1]	CI	HISP	JAM	SI	BAH[2]	PR	VI[3]	Less Ant.	TOTAL %
Insectivora										
Fam.	50	100	50	NA	NA	NA	100	NA	NA	50
Gen.	50	100	50	NA	NA	NA	100	NA	NA	50
Spp.	84	100	60	NA	NA	NA	100	NA	NA	75
Edentata										
Fam.	100	NA	100	NA	NA	NA	100	NA	NA	100
Gen.	100	NA	100	NA	NA	NA	100	NA	NA	100
Spp.	100	NA	100	NA	NA	NA	100	NA	NA	100
Primates										
Fam.	100	NA	100	100	NA	NA	NA	NA	NA	100
Gen.	100	NA	100	100	NA	NA	NA	NA	NA	100
Spp.	100	NA	100	100	NA	NA	NA	NA	NA	100
Rodentia										
Fam.	0	100	66	67	100	0	100	100	100	75
Gen.	67	100	83	67	100	0	100	100	100	84
Spp.	63	100	93	75	100	0	100	100	100	79
SUMMARY										
Fam.	66	100	75	75	100	0	100	100	100	78
Gen.	82	100	88	75	100	0	100	100	100	89
Spp.	74	100	89	80	100	0	100	100	100	83

[1] Cuba includes all of its offshore islands in this calculation, which reduces the overall extinction rate for rodents.

[2] While no rodent species are extinct in the Bahamas, two of three subspecies are.

[3] Introduced *Isolobodon portoricensis* included in calculations.

Table 7. Total Numbers of Endemic Non-volant Land Mammals in the West Indies. (CU = Cuba, CI = Cayman Islands, HISP = Hispaniola, JAM = Jamaica, SI = Swan Islands, BAH = Bahamas, PR = Puerto Rico, Less Ant = Lesser Antilles.)

		CU[1]	CI	HISP[1]	JAM	SI	BAH	PR[1]	Less Ant[2]	
										TOTAL
Insectivora										
	Fam.	2	1	2	0	0	0	1	0	2
	Gen.	2	1	2	0	0	0	1	0	2
	Spp.	5	1	5	0	0	0	1	0	12
Edentata										
	Fam.	1	0	1	0	0	0	1	0	1
	Gen.	5	0	6	0	0	0	1	0	12
	Spp.	9	0	6	0	0	0	1	0	16
Primates[3]										
	Fam.	1	0	2	1	0	0	0	0	2
	Gen.	2	0	2	1	0	0	0	0	5
	Spp.	2	0	2	1	0	0	0	0	5
Rodentia										
	Fam.[4]	2	1	3	3	1	1	2	2	4
	SubF	2	1	5	3	1	1	2	2	7
	Gen.	3	2	6	3	1	1	4	5	19
	Spp.	27	2	14	4	1	1	4*	8	61
SUMMARY										
	Fam.	6	2	8	4	1	1	4	2	9
	SubF	6	2	10	4	1	1	4	2	12
	Gen.	12	3	16	4	1	1	6	5	38
	Spp.	43	3	27	5	1	1	6*	8	94

[1] Includes associated smaller islands.
[2] Includes the St. Martin Bank.
[3] Includes the new primate of Rivero and Arredondo (1988).
[4] Includes Muroidea, Capromyidae, Echimyidae, and *Incertae insedis*.
* *Isolobodon portoricensis* was introduced on PR and VI, and is not included in the list of true endemics.

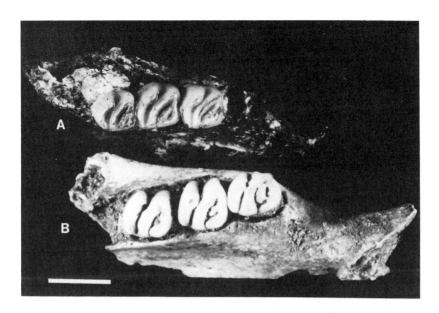

Figure 1a. Views of the left mandibular toothrows of two West Indian echimyid rodents. *Puertoricomys corozalus* (Type from Puerto Rico) on the top (A) with M_1-M_3 present. *Homopsomys* sp. (Blackbone Cave, Puerto Rico) on the bottom (B) with dM1-M_2 present. The white bar is 5 mm.

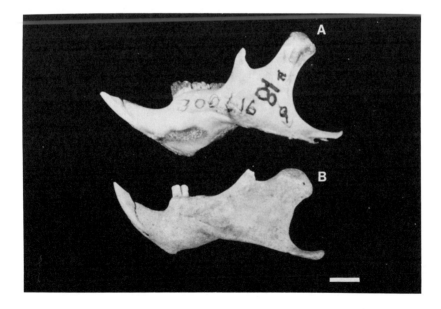

Figure 1b. Lateral views of the left mandible. *Capromys nanus* (Cuba) on the top (A): *Brotomys voratus* (Hispaniola) on the bottom (B). Note the similarity of form, which is likely the result of convergence. The white bar is 5 mm.

794 Biogeography of the West Indies

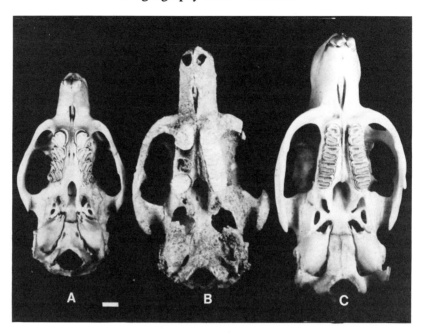

Figure 2a. Ventral view of selected West Indian rodents. *Plagiodontia aedium* (Hispaniola) is on the left (A), *Hexolobodon phenax* (Hispaniola) in the center (B), and *Capromys pilorides* (Cuba) on the right (C). The white bar is 5 mm.

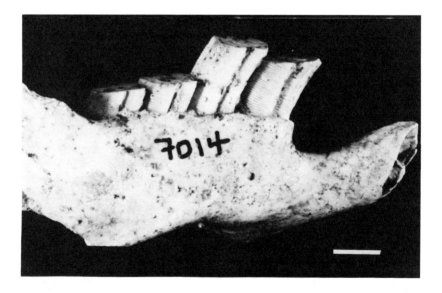

Figure 2b. Ventral view of the right dentary of *Isolobodon portoricensis* (Hispaniola). The dM and M_1 are elevated above their normal positions in toothrow. Note the rings of cement surrounding the teeth, which is diagnostic for this genus. The white bar is 5 mm.

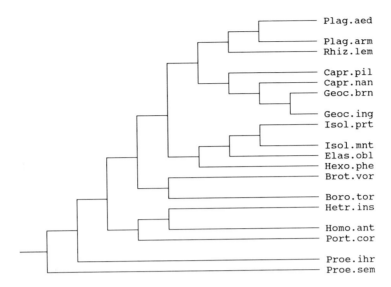

Figure 3a. Concensus tree of rodent relationships based on a PAUP analysis using characters in Tables 2,3.

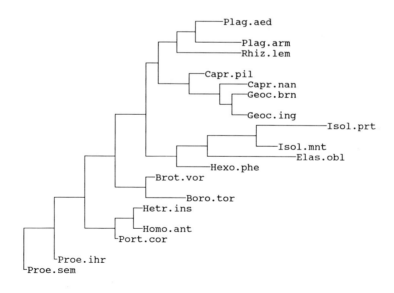

Figure 3b. Tree of rodent relationships showing branch lengths and linkages based on a PAUP analysis using characters in Tables 2,3..

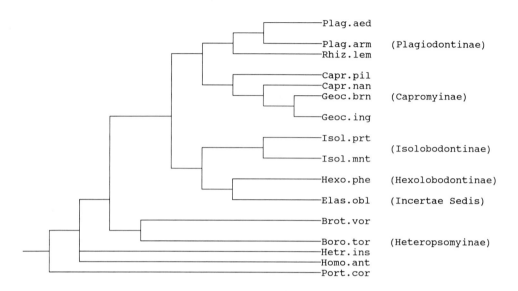

Figure 4. Concensus tree of the relationships of West Indian hystricognath rodent relationships based on a PAUP analysis using characters in Table 2,3.

Biogeography of West Indian rodents 797

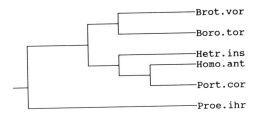

Figure 5A.

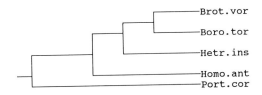

Figure 5B.

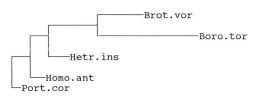

Figure 5C.

Figure 5. Trees of possible relationships of West Indian echimyid rodents. 5a) Concensus tree of West Indian echimyids with *Proechimys ihringi* from South America as ancestor; 5b) Concensus tree of West Indian echimyids with *Puertoricomys corozalus* as ancestor of remaining forms; 5c) Tree of West Indian echinyid relationships showing branch lengths and linkages.

THE LAND MAMMALS OF MADAGASCAR AND THE GREATER ANTILLES: COMPARISON AND ANALYSIS.

Charles A. Woods and John F. Eisenberg[1]

Abstract

In spite of their very different geographical locations, Madagascar and the Greater Antilles share many features in common, and provide for an excellent series of comparisons. The impact of humans on the endemic fauna of Madagascar is not as extreme because humans arrived there only about 2,000 yBP, while there is evidence that Amerindians have been in the Greater Antilles since about 7,000 yBP. While the fossil record from both areas is poor, only representing the last 40,000 years in the Greater Antilles, and the Pleistocene in Madagascar, the mammals that were present at the time of human colonization on each island are well known. Sixteen families, 59 genera, and 133 species of land mammals (including bats) are known from the Greater Antilles, and 18 families, 60 genera, and 105 species are known from Madagascar. Of the non-volant mammals, the forms most sensitive to human exploitation, only 23 percent of the known forms have become extinct on Madagascar, while in the Greater Antilles 78 percent of the families, 88 percent of the genera, and 82 percent of the species of endemic terrestrial mammals are now extinct.

The combination of a longer history of human exploitation and smaller ranges because of insular fragmentation are mainly responsible for the decimation of the endemic mammals of the Greater Antilles. Large mammals were the most vulnerable, and appear to have become extinct first. The second wave of extinctions was in small mammals about the size of rats. The most resistant mammals to extinction are bats, and medium sized (1-3 kg) forms that are able to find shelter in crevices or burrows.

Introduction

This book is primarily concerned with the status and history of West Indian plants and animals. We believe that additional valuable information on the biogeography of the West Indies can be gained by comparing the West Indies with a comparable isolated region. The Phillipine archipelago immediately comes to mind as the most appropriate area for comparison with the West Indies. With over 7,000 individual islands, and an area of about 300,000 km^2, the Philippines is about the same size as the West Indies. The mammalian fauna of the Phillipine islands is large and diverse, with at least 17

---

[1] Charles Woods is Curator of Mammals and John Eisenberg is Ordway Professor at the Florida Museum of Natural History, University of Florida, Gainesville, FL 32611.

endemic genera of rodents, two of insectivores, and four of bats (Heaney 1986). This pattern is similar to that of the West Indies, with 18 endemic genera of rodents, two of insectivores, and four of bats. In the Philippines 101 species of non-volant mammals are known, 85 percent of which are endemic. In the West Indies 94 species of non-volant mammals are known, all of which are endemic. However, when the 14 species of introduced West Indian mammals are included (six rodents, two carnivores, two artiodactyls, two marsupials, one primate, and one lagomorph), there are 109 West Indian mammals, 87 percent of which are endemic. Island size, geological age, and the degree of isolation of each island influence species richness and the degree of endemism in the Philippines, just as in the West Indies (Heaney 1986). Indeed, there are remarkable similarities between the mammalian faunas of the two regions. However, the Pleistocene fauna of the Philippines is less well known, and the Philippines are so similar to the West Indies in their fragmentation and isolation that some contrasts are not possible. Therefore, we have selected Madagascar as the counterpart for this comparison with the West Indies because the fauna of Madagascar is well known and there are many similarities in the composition of the mammalian faunas of both regions. The two regions also share many similarities in climate and geography, and both are adjacent to major continental areas. But most importantly, both may be considered oceanic islands in that within the last 50 million years sealevel subsidence has not connected either area with the mainland. Much of the Phillipine archipelago behaves like a continental shelf island system because of its nearly complete connection with the mainland at times of low sealevels (Heaney 1986). In this comparison, rather than discuss all of the West Indies, which include several distinct subregions, we are concentrating on the Greater Antilles in our comparison with Madagascar. As described below, the geographical scope of the Greater Antilles is remarkably similar to Madagascar when all of the islands and banks of the Greater Antilles are considered together as a single geographical unit.

In addition to the comparisons that are possible because of the geographical similarities between Madagascar and the Greater Antilles, we are also able to contrast the mammalian faunas of the two regions. The Greater Antilles are fragmented into small to medium size units, while Madagascar is a single large island. Humans appear to have entered Madagascar about 2,000 yBP, considerably later than the arrival of Amerindians in the West Indies at 7,000 yBP, and therefore the impact of humans can be effectively contrasted between the two regions. Also carnivores were absent in the Greater Antilles until the introduction of dogs, cats and the mongoose, while on Madagascar carnivores were a part of the endemic fauna.

In the following sections and accompanying tables, data on the land mammals, here defined as all non-marine mammals including bats, are discussed and summarized. The purpose here is to compare trends and patterns between the two regions rather than analyze the land mammals in detail. The land mammals of Madagascar and the Greater Antilles have been discussed in greater detail elsewhere (Eisenberg and Gould 1970; Albignac 1973; Petter 1972; Tattersall 1982; Varona 1974; Morgan and Woods 1986; Woods 1989a, in press).

Geography

Greater Antilles

The Greater Antilles here is broadly defined to include the large islands of Cuba, Hispaniola, Jamaica, and Puerto Rico, as well as the Swan Islands, Cayman Islands, the 600 plus islands of the Bahamas, Turks and Caicos Islands, Virgin Islands, and the nearby islands of Anguilla and St. Martin. At low ocean levels during Pleistocene "climate cycles", many of these islands were much larger, and some were more closely associated with one another or even interconnected. The islands of the Great Bahama Bank, for example, were joined to form one large island, and the Virgin Islands were connected to Puerto Rico 18,000 yBP (Pregill 1981).

The Greater Antilles and associated small islands stretches between 18 and 27° north latitude and 63° to 84.3° west longitude. The land area of this archipelago is 225,676 km^2. The center of the archipelago is at 20° north latitude. The center of Madagascar is 20° south latitude. The land area of the island of Madagascar is 587,041 km^2, which is a close approximation of what the land area of the Greater Antilles would be if all of the islands were connected at low sea levels into a 2,280 km long and irregular shaped single island mass. The Greater Antilles are currently separated from North America by 240 kms between southern Florida and Cuba and 120 kms between Grand Bahama Island and the Florida peninsula.

The Greater Antilles vary in topography from -40 m at Lago Enrinquillo to 3,175 m at Pico Duarte in the Cordillera Central, both areas in the Dominican Republic of Hispaniola. The vegetation of the Greater Antilles is also very diverse. The southern Bahamas and several regions of Hispaniola are characterized by low precipitation and semi desert-like conditions. The high mountains of Hispaniola and Jamaica, and much of Puerto Rico have abundant precipitation. Almost 6,000 mm of rain falls on Pic Formond in SW Haiti. The Greater Antilles, therefore, have a wide range of climates and habitats, including significant areas of high remote mountain plateaus.

Recent evidence from pollen studies indicates that alternate wet and dry periods have allowed expansion and contraction of xeric-adapted vegetation, apparently alternating with north temperate zone glaciation phases. At the present time the climate of the West Indies is in a mesic phase with xeric vegetation being confined to relictual patches, especially in the rain shadows of mountainous areas. Some of these patches are rich in endemic species, such as the Cabo Rojo region of the southwestern corner of the Dominican Republic. During times when the West Indies were much colder and drier, the ocean level was lower and land areas were more extensive. Mesic habitats were largely restricted to relictual patches on the tops of mountains and mountain plateaus during these cool dry episodes, and pine forests were more widespread. These climate cycles had a profound influence on the birds of the West Indies (Woods and Ottenwalder 1983; Woods 1987), and we believe they were also important to the patterns of distribution and speciation of mammals as discussed below.

Madagascar

The standard biological reference for the Island of Madagascar is Battestini and

Richard-Vindard (1972). The island is 1,600 km long and 580 km wide at its broadest point. It is 590,000 km^2 in total area . Madagascar lies south of the equator from 11°57' to 25°32' south latitude. With the exception of the southern tip, most of the area is within the geographically defined tropics. The island stands off the coast of Mozambique, and at its closest lies 350 km from the shores of Africa. There is no doubt that at one time it was joined with the continent of Africa. However, it has been separated for a considerable time, possibly since the late Cretaceous. The late Cretaceous was a period of considerable volcanic activity prior to the separation of Africa and Madagascar The mammalian fauna of the island and its offshore waters is clearly derived from Africa.

Madagascar has a long north and south axis, and the center is dominated by a high plateau, much of it over 1,000 m in elevation. The coastal lowlands on the east are narrow in extent, but in the west are broad. The northern part of Madagascar is quite tropical, especially in the lowlands. The annual average temperature in the north is approximately 3.15°C. In the south the average is 6.85°C. Altitudinal effects are superimposed on this lowland pattern since mean temperature declines at a rate of about 2.7°C per 300 m elevation.

The west coast tends to be quite dry and the southwest has a long history of aridity. The southwestern vegetation is dominated by Euphorbiaceae and Didiereaceae that have converged in form toward the New World Cactaceae. The southwest contains many arid adapted species of vertebrates and the assemblage is obviously a co-evolved ecosystem bearing witness to a long history of aridity on this part of the island. Where man has not cleared the land, the west coast can have multistratal tropical forest in association with major rivers, but otherwise is dominated by dry deciduous forest. At the extreme northwest is a very distinct mesic region known as the sambirano domian (Guillaumet 1984).

The northeast is dominated by multistratal tropical evergreen forest and, indeed, this forest form extends down most of the lowland east coast. Gradually, as one moves west up the escarpment to the plateau the vegetation changes to mixed deciduous tropical forest. The same trend continues as one proceeds further to the west on the plateau itself.

The plateau experiences a cool winter, and in the higher mountains there may even be frost. The vegetation of the plateau has been altered by man. The original colonists from Africa were pastoralists and for centuries have employed fire in conjunction with their grazing operations. As a result, the central plateau is now denuded except for small pockets of vegetation. Many of the larger lemurs were present in the central plateau areas and all are now extinct. The Indonesian colonists arriving around 700-1,100 A.D. are rice cultivators, and their agricultural operations have severely impacted the eastern escarpment.

Because of the activities of man, and the destruction of the high plateau natural vegetation, the modern distribution of the surviving mammals of Madagascar tends to be confined to the lowlands of both coasts and those portions of the eastern escarpment that have not been timbered off. The present distributions of many of the lemurs and tenrecoid insectivores are often extremely limited, and probably do not reflect their original ranges. Given the different vegetation forms between the west and the east coast that reflect climatic trends, species within a genus are often easily distinguishable into east and west coast forms. As indicated previously, the extreme southwest supports a unique fauna adapted to extreme seasonal aridity.

Fossil record

Greater Antilles

The fossil record of land mammals in the Greater Antilles is limited almost completely to the late Pleistocene and Holocene. The oldest dated record of an identifiable land mammal is a primate femur from Coco Ree Cave in Jamaica, which has a corrected carbon date of 39,300 yBP (Goodfriend and Mitterer 1987). In Hispaniola carbon dates of rodent bone fragments range from 1,600 to 21,170 yBP (Woods in 1989a). Mammalian hair is known from pieces of amber from the Dominican Republic, which is at least early Miocene in age. When land mammals first crossed from North America or South America to the Antilles is unknown, although it is likely that the colonizations began in the Eocene and continued into recent times (Woods in press).

Madagascar

Fossil records in East Africa are unknown earlier than the Miocene; the vertebrate fossil history of Madagascar is known only from the Pleistocene to the recent. Therefore, it is impossible to say with certainly when the various stocks of mammals crossed from Africa to Madagascar. The consensus opinion, however, is that colonizations were at irregular intervals since the late Cretaceous or Paleocene and that the Primates, Insectivora and possibly the Tubulidentata arrived first, followed at a later date by the Carnivora, the Rodentia and the Artiodactyla. Sirenians presumably have been in the Indian Ocean for a long period of time.

Summary of land mammals

Greater Antilles

The list of land mammmals (terrestrial mammals plus bats) in the Greater Antilles was compiled from data in Varona (1974, 1986), Varona and Arredondo (1979), Morgan and Woods (1986), Koopman (this volume), and Woods (1989a, in press). Five orders, 16 families, 59 genera, and 133 species of land (=volant and non-volant) mammals are known from the Greater Antilles. In the calculations discussed in this paper all named forms in the literature are discussed with the exception of the supposedly endemic Cuban canids (Morgan and Woods 1986). Nine families, 34 genera and 88 species of non-volant mammals are described in the literature (Table 1). All of these taxa were present in late Pleistocene times, and were encountered by Amerindians when they arrived in the region from South America approximately seven thousand years ago (Rouse and Allaire 1978). As discussed by Morgan and Woods (1986), and Woods et al. (1985) the mammals were not uniformly distributed throughout the Antillean archipelago, and the level of extinction has been very high. Some of the mammals were large, such as most of the megalonychid sloths and a few of the rodents, but many were less than one kg in body size and easily overlooked. Therefore, there is no good record of all of the mammals that were present on the islands at the time of their colonization by Europeans following the arrival of Columbus in 1492. Some species may have persisted undetected as living forms until well into this century (Woods et al. 1985).

Chiroptera.--Bats are numerically the most successful and widespread group of extant land mammals on the islands, with seven of the nine New World families of bats occurring in the region. Seven families, 25 genera, and 45 species of bats are known from the Greater Antilles. The bats of the Greater Antilles show great species richness and comprise 34 percent of the original mammal fauna. Central America, the area of origin for many West Indian bats (Baker and Genoways 1978), is a region with one of the richest bat faunas in the world. Fifty-one percent of the terrestrial mammalian fauna of Panama are bats. There are 59 known bat species in the Greater and Lesser Antilles, 51 of which are still extant. Three bat species are now extinct on the islands but still occur in Central and South America, and five are totally extinct (Morgan and Woods 1986). Griffiths and Klingener (1988) analyze the distributions of Antillean bats and conclude that they follow two patterns. One pattern is associated with highly endemic taxa, which they believe colonized the Antilles in the Oligocene from proto-Central America. The second pattern ("cosmopolitan forms"), occurred during the Pleistocene and centered mainly on dispersals into Cuba and Jamaica.

Insectivora.--Greater Antillean insectivores are divided into two families, two genera, and twelve species, and show extreme morphological conservatism (Table 1). Nine percent of the known endemic land mammals, and 14 percent of the non-volant mammals of the Greater Antilles are insectivores. The Family Solenodontidae has only four known species. *Solenodon* sp from Cuba is extinct and known only from a femur (Morgan et al. 1980). Three species of *Solenodon* survive; two are endangered (*S. cubanus* of Cuba and *S. paradoxus* of Hispaniola), and one previously assumed to be extinct appears to still survive (*S. marcanoi* from Hispaniola, Ottenwalder and Woods in prep.). There appears to be a close relationship between *Solenodon cubanus* and *S. marcanoi*. The features found in *S. paradoxus* from Hispaniola are more derived than either of the previous forms. If these relationships are valid, the resulting sequence of differentiation from Cuba to Hispaniola would support the ideas of Simpson (1956), McDowell (1958), and MacFadden (1980) that the ancestor of the West Indian insectivores reached the Greater Antilles from North America, and that the dispersal of the group in the Antilles moved from west to east. MacFadden (1980:18) hypothesized that during the late Mesozoic and early Cenozoic, soricomorph insectivores (apternodontids and possibly geolabidids) were distributed throughout North America, Nuclear Central America, and the "proto-Antilles", and that a vicariance event isolated the ancestral soricomorph on the Greater Antilles after the "proto-Antilles" drifted eastward relative to the Americas. MacFadden stated that *Solenodon* and *Nesophontes* are descendants of the Greater Antillean part of the ancient soricomorph distribution, and as such are "island relicts" of a once more widespread radiation of soricomorph insectivores which became extinct elsewhere in the late Oligocene.

Insectivores of the genus *Nesophontes* are all extinct. Eight species are known, three from Cuba, three from Hispaniola, one from Puerto Rico, and one from the Cayman Islands. There are accounts of *Nesophontes*-like animals having been seen in the Dominican Republic and Cuba during the last 50 years (Ray in litt.; Varona in litt.). Owl pellets collected in the 1920s in the mountains of the Dominican Republic had tissue and hair attached to bones, suggesting that *Nesophontes* was still extant at that time (Miller 1930).

These observations suggest that it is possible some species of *Nesophontes* may have become extinct in very recent decades.

Morgan (1977:18) determined that the large *Nesophontes edithae* from Puerto Rico may be the most primitive member of the genus, and that the probable sequence of differentiation was from Puerto Rico to Hispaniola, Cuba, and finally the Cayman Islands. If this sequence is correct, then the vicariance hypothesis must be questioned, since the pattern of differentiation in *Nesophontes* would closely resemble the pattern in hystricognath rodents, which appears to have been initiated via an overwater dispersal event to eastern Hispaniola or Puerto Rico directly from South America (Woods this volume). However, Morgan does point out the possibility that the polarity of the characters in *N. edithae* might be reversed. *Solenodon* never occurred on Puerto Rico, and in the absence of a larger sized insectivore, *Nesophontes* became much larger on that island than anywhere else in its range.

The absence of insectivores on Jamaica calls into question the vicariance hypothesis, since Jamaica was possibly closely associated with Nuclear Central America in the Eocene and early Oligocene (Sykes et al. 1982). The most likely explanation for the lack of insectivores on Jamaica is that they became extinct during the mid-Tertiary when the island was largely or even totally submerged from the mid-Eocene to the early Miocene (Kashfi 1983). The absence of insectivores on the Bahamas and the Lesser Antilles, and the lack of diversity of *Nesophontes* on the offshore islands of Hispaniola indicate that insectivores have not been as adept as rodents in the process of overwater dispersal.

Edentata.--The West Indian sloths are all members of the Family Megalonychidae, and probably derived directly from South America. One family, twelve genera, and sixteen species of sloths are known, all of which are restricted to the Greater Antilles (Table 1). Twelve percent of the known endemic land mammals, and 18 percent of the non-volant mammals of the Greater Antilles are sloths. Sloths underwent a remarkable radiation in the Greater Antilles into very diverse forms. Nine species occurred on Cuba, six on Hispaniola, and one on Puerto Rico. Some were quite arboreal, strongly resembling tree sloths in size and morphology, while others were large and terrestrial. Presumably all were herbivorous and occupied niches comparable to some of the larger semi-terrestrial lemurs of Madagascar. The data have not been analyzed in a way that will allow a hypothesis to be tested as to when, where, or how the differentiation of sloths in the West Indies took place. A simple analysis of the number of taxa supports the hypothesis that sloths dispersed from Cuba to Hispaniola and finally to Puerto Rico. No sloths have been found on any other islands of the Greater or Lesser Antilles, although they did occur on Ile de la Tortue off the north coast of Haiti, which is separated from Hispaniola by a deep water channel eight km wide. It is most likely that sloths entered the Antilles via an oversea dispersal directly to Hispaniola from somewhere in northern South America during the mid to late Miocene, and that the history of sloths closely parallels the history of rodents in the Greater Antilles.

Primates.--Primates in the West Indies were restricted to the Greater Antilles where they underwent a limited radiation into two families, five genera, and five species (Table 1). There were two, and possibly three separate lineages of primates which occurred on Jamaica (one form), Hispaniola (two forms), and Cuba (two forms). Each lineage may represent a separate invasion of the Antilles, and no genus has more than one known species. Three percent of the known land mammals of the Greater Antilles are primates. The lack of diversity of primates in the Antilles is probably related to the success of sloths and rodents on the islands. Most primate niches appear to have been occupied by large sloths and medium sized rodents, and at least on Hispaniola there were taxa similar in mass and many morphological features to tree sloths. The difficulty in establishing the phylogenetic affinities of the primate lineages in the Antilles may be the result of an increase in body size of West Indian primates. In the case of *Xenothrix* from Jamaica, for example, its possible relationship to callithricids (small bodied marmosets and tamarins) may be masked by the effects of allometric growth as the lineage increased in body size (Ford and Morgan 1986).

Rodentia.--The rodents of the Greater Antilles have undergone a remarkable radiation into four families, 15 genera, and 55 species, and are morphologically diverse (Table 1). Forty two percent of the known endemic land mammals, and 63 percent of the non-volant mammals of the Greater Antilles are rodents. We are not using the term "adaptive radiation" because when rodents first dispersed to Puerto Rico and Hispaniola there were no competitors, which permitted an unusual opportunity for diversification (Gould et al. 1987). Mammalian taxa evolving on isolated oceanic islands without competition or predators may not be able to adapt to rapidly changing conditions, such as the extensive climatic fluctuations of the Pleistocene or sudden competition and/or predation from introduced forms. When local conditions change dramatically, mammals that have evolved in isolation on remote oceanic islands are especially vulnerable to extinction. At least one rodent species is known to have become extinct on every West Indian island where they occurred with the exception of tiny and remote East Plana Cay in the Bahamas.

Rodents in the Greater Antilles are divided into two separate major radiations. One radiation is related to the old endemic hystricognath rodents of South America, a group that extends back in time at least 26 myBP on that continent (Woods 1982, 1984). These rodents entered the West Indies at a reasonably early date (late Oligocene or early Miocene) and diversified into 14 genera and 54 species, all of which are confined to the boundaries of the Greater Antilles. All but three genera and 13 species are now extinct. Some were highly arboreal and clearly occupied niches similar to primates, squirrels and porcupines, while others were very terrestrial and large (Woods and Mckeen 1989).

The other radiation of at least three genera and eight species of West Indian rodents occurred mainly in the Lesser Antilles. This group included the genus *Oryzomys* and at least two closely related genera, all of which are muroid rodents of the Subfamily Sigmodontinae (Carleton and Musser 1984). This Lesser Antillean radiation is now completely extinct, but until well after the arrival of humans onto the islands included at least three genera and eight species distributed over most of the islands of the archipelago (Steadman et al. 1984).

Madagascar

Insectivora.--The living Insectivora of Madagascar include two families, nine genera, and 22 species (Table 2). One genus and species, *Cryptogale australis*, is extinct. The family

Tenrecidae dominates the fauna and was studied in depth by Eisenberg and Gould (1970). The Soricidae includes a single genus *Suncus* and two species. *Suncus murinus* is clearly a commensal of man, and no doubt was inadvertently introduced during the early human colonization efforts. *Suncus madagascarensis*, however, may be an endemic. It is remarkably similar to *S. etruscus*, a species broadly distributed in the Middle East, Africa, and Asia. It is also possible that this species represents a human introduction. The soricids, therefore, are not included in the list of endemic mammals of Madagascar in Table 3.

Clearly the Tenrecidae are of ancient origin. They are first detectable in East Africa during the Miocene, and with the exception of the aquatic Potomagalidae, were subsequently replaced by the Soricidae that apparently invaded in the late Miocene from Asia (Butler 1985). In Madagascar the Tenrecidae have persisted, radiating into variety of ecological niches including: hedgehog-like forms (*Setifer*); a species that is aquatic (*Limnogale*); species that are adapted for a fossorial or mole-like existence (*Oryzorictes*): and litter foraging forms including some that are semiarboreal (*Microgale longicaudata*; Eisenberg and Gould 1970). Most species are still extant.

Primates.--The five families, 19 genera, and 36 species of primates of Madagascar exemplify a morphologically conservative assemblage that evolved in isolation for a considerable period of time (Table 2). The primates were studied in depth in the years following 1950 by ecologists and physiologists. Recently, the primate studies have been reviewed in the excellent volume by Tattersall (1982). The Madagascan primates are clearly allied to the African galagos and lorises. The exact time at which primates arrived on Madagascar is imperfectly known. It may have been as early as the Eocene. They radiated to fill a variety of semiarboreal and arboreal niches with forms specializing for leaf-eating, frugivory and even in feeding on saps and gums. The 14 extinct species were usually rather large (Fig. 1) and represent several different specializations for locomotion. The largest species presumably locomoted somewhat like the contemporary koala (*Phascolarctos*) and was a semiarboreal folivore.

Chiroptera.--The six families, 14 genera, and 20 species of bats in Madagascar are all derived from African relatives with the exception of *Pteropus*. Presumably colonization by the Chiroptera took place at several different times and being volant they have a more powerful colonizing ability than terrestrial mammals. There is only one endemic bat genus on Madagascar, *Myzopoda*, placed alone in the Family Myzopodidae.

Rodentia.--The rodents were recently reviewed by Petter (1972). There appear to be 13 endemic species belonging to eight genera (Table 2). One genus, *Macrotarsomys* is specialized for bipedal richochetal locomotion and of course evolved in the arid southwest. *Hypogeomys* is a rabbit-like form that was once broadly distributed in the western savannas but now has an extremely restricted range. The other rodent species occupy semiarboreal to terrestrial rodent-like niches and they do not exhibit the diversity of forms exhibited by rodents in continental areas. It is assumed they were relatively late immigrants to the island and many of the rodent-like niches were filled by primates at an earlier date. One genus, the fossorial *Majoria*, is believed to be extinct. They are assigned to their own subfamily, the Nesomyinae.

Carnivora.--The carnivores in Table 2 are listed as a single family. This is rather conservative since Wozencraft (1984, 1986) has made an excellent case for separating the Herpestidae from the Viverridae. *Cryptoprocta* probably should be placed in its own family, thus raising the number of families to three with one endemic family, the Cryptoproctidae. *Viverricula* is clearly an introduction. *Eupleres*, *Galidictis*, and *Fossa* are allied to the viverrids. The remaining species and genera are clearly herpestids. One large, extinct form of *Cryptoprocta* is believed to be specifically distinct from the living form (Lamberton 1939). The carnivores on Madagascar are all rather small with the exception of *Cryptoprocta* which hunts medium-sized birds and mammals. The Madagascar mongooses tend to be diurnal, some are quite terrestrial, others specialize for arboreality. The nocturnal *Fossa* is fox-like in body form. Their natural history has been described by Albignac (1973).

Artiodactyla.--The Artiodactyla include the extinct Madagascan hippopotamus and the African river hog *Potomochoerus* (Table 2). The latter species may have been introduced by man, but the hippopotamus surely was an overwater colonizer.

Tubulidentata.--The Madagascan aardvark reviewed by Patterson (1978) has now definitely been shown to be of an age of 1,500 yBP. This remarkable specimen reflects that the ant-eating niche was occupied on Madagascar prior to man's arrival.

Extinctions

Greater Antilles

Five orders, 16 families, 58 genera, and 132 species of land mammals are known from fossil and recent deposits of the Greater Antilles over the last 40,000 years. Non-volant mammals number four orders, nine families, 33 genera, and 87 species. Fifty percent of the orders, 78 percent of the families, 88 percent of the genera, and 82 percent of the species of non-volant Greater Antillean land mammals have become extinct (Table 3). Many taxa persisted until nearly the time of Columbus, and long after the initial occupation of Hispaniola by Amerindians about 7,000 yBP. In the Greater Antilles all of the ground sloths and primates have become extinct, while approximately 75 percent of the rodents and insectivores became extinct during the same 40,000 year period (Table 3). Body size appears to be the most important factor in the extinction of non-volant mammals in the Greater Antilles. The remains of large terrestrial mammals, such as ground sloths, primates, and giant hutias, are present in cave and sinkhole deposits in Hispaniola that carbon date between 22,000 and 7,000 yBP. All are now extinct, and some species appear to have been lost in the period between 7,000 and 2,000 yBP. This was the time when Amerindians first invaded the Antilles. On remote high plateaus in Hispaniola, such as the Massif de la Selle, the Massif de la Hotte, and the Cordillera Central, elements of the megafauna may have persisted until 2,000-1,000 yBP. Rouse (pers. comm.) believes that Indians did not follow rivers up into these cold and wet interior plateaus until about 1000 yBP.

A second wave of extinction of non-volant mammals began about 500 yBP, when small to medium sized rodents and insectivores began to become extinct. A number of such forms were still extant at the time of colonization of the Greater Antilles following the first voyage of Columbus in 1492. The remains of spiny rats and *Nesophontes* are present in cave and sinkhole deposits along with the remains of rats. We attribute the extinction of these smaller endemic mammals to predation by dogs and cats, and competition from rats (Woods this volume). These extinctions were a long slow process as the ranges of the endemics were reduced to ever smaller areas, and to more and more remote regions. The introduction of the mongoose into the Greater Antilles in the late 1800s (Kilpatrick this volume) was possibly a factor in the extinctions of all of the remaining endemics less than 500 gr in mass (*Nesophontes* and the spiny rats).

Excluding the bats, most genera of surviving Greater Antillean terrestrial mammals have only one, two or at most three species, and many of these are of medium body size. The taxa that have survived the pressures of overhunting, predation, and competition are generally between 1-5,000 gr, and secretive. They have survived longest in areas where forest cover or limestone outcrops provide shelter. Of the 16 surviving endemic terrestrial mammals of the Greater Antilles, the most endangered species is *Capromys nanus* from Cuba, the smallest hutia and the capromyid least likely to use limestone crevices for shelter. Some biologists believe this hutia may now be extinct (Woods 1989b).

Extinction rates of endemic rodents were very high in all taxa and size categories, and we believe that in almost every case the extinction was related to humans in one way or another. However, the high extinction rates of mammals in the Greater Antilles may also indicate that most West Indian mammal groups may be the result of radiations that were not as adaptive as comparable taxa found on the more contiguous land masses. On the isolated islands of the Antilles suboptimality may have characterized evolution in many clades. Organisms in such clades would have been both vulnerable to pressures that lead to extinction, and unresponsive to the evolutionary dynamicism that led to dramatic radiations at the species and genus levels in South American echimyids and sigmodontines. Capromyines are the exception to the rule in the Antilles, which may be the result of their late radiation on the islands of the western Antilles in the presence of increased competition and predation from other vertebrates.

Cuba.--Five insectivores, nine ground sloths, two primates, 27 rodents, and 33 bats are known from Cuba and its associated islands, a total of 76 land mammals (43 non-volant mammals). The mammals of Cuba have fared better than any other island of the Greater Antilles with the exception of the Bahamas. Extinction has occurred in 60 percent of the families, 81 percent of the genera and 76 percent of the species. One important reason that some Cuban mammals have been shielded from extinction may have been the presence of so many uninhabited offshore islands. The sparsely inhabited high mountain areas of eastern Cuba, an ecological insular region surrounded by water or lowland agricultural areas, is also a refugium (Eisenberg and Gonzalez 1985). If the effects of these buffer areas are removed, and if the very rare and possibly extinct small species of *Capromys* (*C. nanus*) is eliminated, then the extinction figure for Cuba becomes 93 percent, which is close to the 89 percent figure for Hispaniola, and far worse than the 80 percent figure for Jamaica. Many of these "islands" of habitat are small, however, and the species area curves presented in Morgan and Woods (1986) suggest that species in these habitats are vulnerable (Woods 1989b). The importance of refugial situations in shielding endemic terrestrial mammals from extinction is apparent in the relative survival .pa of capromyid rodents in the Greater Antilles, where 72 percent of all surviving capromyids occur on Cuba and its associated islands.

Bahamas.--One rodent and 16 bats are known from the Bahamas, for a total of 17 species. The rate of extinction in the Bahamas must have been very high, even for bats. Extinction is known to have been a factor for Andros Island, where five mammalian species survive, but eleven are known from fossil deposits (Morgan this volume). Eleven of the 16 species known from the islands of the Great Bahama Bank are now extinct. It is possible, even probable, that other taxa, such as ground sloths, insectivores, and the various rodent groups which were present on the nearby islands of Cuba, Hispaniola and Puerto Rico, may have colonized the Bahamas at times of low water level in the past but became extinct when the insular fragmentation occurred as ocean levels increased and most islands in the archipelago became very small.

Three subspecies of *Geocapromys ingrahami*, the Bahamian hutia or Coney, are known from ten islands of the Bahamas archipelago. The subspecies probably developed during cycles of high and low water of the Pleistocene that periodically united the large banks of the Bahamas into several mega-islands. The Bahamian Hutia now survives on only one outlying island that was never part of any of these mega-islands, East Plana Cay, although it has been relocated to two other islands within its natural range in the Bahamas in recent years (Kevin Jordan pers. comm.).

Hispaniola.--The rates of extinction have been very high on Hispaniola, where five insectivores, six sloths, two primates, 14 rodents, and 19 bats are known, a total of 46 species. Seventy one percent of the families, 87 percent of the genera and 89 percent of the species have become extinct. The rate of extinction is most remarkable when the case of rodents is examined. Rodents are the group that has persisted longest on most islands of the Antilles, and is most capable of coexisting with humans. On Cuba 61 percent of the rodent fauna became extinct. On Hispaniola 93 percent of the rodents became extinct. Part of the reason that the rate of survival in Cuba has been so high is that there are so many named forms, four of which survive on the small cays off the coast and another on the much larger Isla de la Juventud. No comparable situation exists in Hispaniola.

Jamaica.--One primate, four rodents, and 23 bats are known from Jamaica, for a total of 28 species. Jamaica has few topographical features that have shielded endemic terrestrial mammals on Cuba and Hispaniola from extinction. There are no offshore islands to serve as refugia or to be part of the species generating phenomena during the Pleistocene climate cycles. All but one of its endemic non-volant mammals has become extinct. The percent extinction has been 75 percent at the family and generic level, and 80 percent at the species level.

Puerto Rico.-- One insectivore, one sloth, four rodents, and 16 bats are known from the Puerto Rican archipelago, a total of 22 species. The non-volant mammals of Puerto Rico have been totally extirpated. There are recurring reports of the presence of a large endemic rodent on the island, which is probably *I. portoricensis* rather than either of the species of *Heteropsomys*. However, these reports have not been confirmed (Woods et al. 1985).

Other islands.-- Excluding bats, one insectivore and two rodents occurred on the Caymans. All are now extinct. Anguilla and St. Martin, two small islands in the far east of the Greater Antilles, together with tiny St. Barthelemy are surrounded by shallow water that would unite the three islands into a single land mass of 3,000 km^2 if the sea level were to drop 100 m. Two rodents and seven bats are known from the St. Martin Bank. Both are now extinct. One rodent was a small rice rat of the kind that became extinct throughout the Lesser Antilles in the last 100 years (Woods this volume). The other was a large rodent, described by Cope in 1868 and named appropriately *Amblyrhiza inundata*. This giant rodent had a skull that is estimated to have been at least 400 mm in length. When this species became extinct is still unresolved, but a rodent the size of a large ground sloth would have been very vulnerable to exploitation by the earliest Indians to arrive in the area, and to the consequences of its limited distribution on two islands of less than 100 km^2 in area.

Madagascar

Of the non-volant terrestrial mammals of Madagascar that have suffered extinction, nineteen species have been recorded (Table 3). Of these nineteen, fourteen are from the order primates. Among the extinct primate forms, eight species had a condylar-basal length greater than 80 mm (Table 2; Fig. 1). All extant forms have a condylar basal length less than 80 mm. To date, the history of vertebrate extinctions on Madagascar included three species of giant tortoise that have gone extinct. Madagascar had giant birds resembling ostriches or moas; at least three species have gone extinct.

The island of Madagascar is believed to have been colonized by human beings approximately two thousand years ago. Clearly there were at least two major waves of human colonization prior to the first arrival of Europeans, approximately five hundred years ago. At the time of European arrival, the extinction of the larger vertebrates on the island was almost completed. Although humans are strongly implicated in the extinction of the larger vertebrate fauna, some workers feel that climatic changes may also have been involved (MacPhee 1986). The southeastern portion of Madagascar is exceedingly dry at the present time. However, the plant assemblages and some faunal elements (*Macrotarsomys*) indicate a long history of aridity. One must, however, be cautious because the extent of xeric adapted vegetation may wax and wane or even shift geographically with long-term climatic shifts. The hunting activities of man, and above all the destruction of forests on the high plateau through cutting and burning, no doubt had a very significant impact on the large vertebrates (Battestini and Vein 1967).

Recent research employing radiocarbon dating has indicated that in addition to the larger lemurs and the Madagascan aardvark, large birds and tortoises were all exterminated within the last two thousand years. While climatic effects, including an increase in aridity, may have contributed to the constriction of the ranges of certain of the larger elements of the fauna, certainly there is no doubt that man provided the final push to extinction for the six genera and 14 species of primates, the endemic Madagascan aardvark and the endemic Madagascan hippopotamus (Table 3).

Comparisons and conclusions

In analyzing insular extinctions of higher vertebrates, it is instructive to compare the fauna of Madagascar with that of the Greater Antilles (Table 4). Both land masses or archipelagoes "behave" like oceanic islands. At present the non-volant, terrestrial mammals of Madagascar are still holding on with only 23 percent of the species extinct. The situation in the Antilles is the reverse with 82 percent of the species having become extinct. Both areas demonstrate the vulnerability of large forms to extinction (Fig. 1). Several factors must be considered. First, Madagascar is a large land mass that retains its integrity throughout changes in sea level. Second, Madagascar was colonized by man rather late in history, about 2,000 yBP; whereas the Greater Antilles may have been colonized as early as 7,000 yBP.

The third factor is subtle, but quite real. True carnivores made their way to Madagascar in perhaps two colonization events. Carnivores co-evolved with their prey and the insectivores, rodents, and primates had some experience with mammalian predators. Thus the introduction of cats, dogs, and *Viverricula* by man did not have the same devastating effect as was the case on the Antilles.

Even in the 1960's one could observe the tolerance of some endemic mammals of Madagascar to human intrusion. Figure 2 (adapted from Eisenberg and Gould 1970) portrays a small settlement intruding into the forest. Some species are human commensals (i.e., *Suncus*, *Rattus*). Some endemics are robust and tolerant (i.e., *Hemicentetes*, *Tenrec*, *Setifer*, and *Galidia*, Eisenberg and Gould 1984). Still others are early colonizers of second-growth forest (*Microcebus*, *Cheirogaleus*, and *Lepilemur*). There is hope if primary forest fragments remain. The time for constructive action is all too short.

If the Greater Antilles can be considered as a unit, the single large island would include extensive lowland areas in the north in what are now the Bahamas and in central Cuba. The central part of the island would be composed of an old igneous chain of mountains stretching from eastern Cuba through Puerto Rico. Other ranges formed by uplifted regions of karst and metamorphosed substrate would parallel the high Cordillera Central. In between these ranges are lowland areas, some of which are now, and probably always have been, very xeric. In inhabited regions of moderate topography all of the endemic terrestrial mammals have become extinct. Large mammals in low to middle elevations, and sloths and primates in particular were the most vulnerable creatures in the Greater Antilles. They rapidly became extinct in response to colonization of the islands by humans. In montane regions where there was extensive habitat, especially high plateaus above 1,500-2,000 m elevation, some large mammals, such as ground sloths, primates, and large rodents, persisted until nearly the time of Columbus, and long after the initial occupation of the Greater Antilles by Amerindians. In the Greater Antilles these habitats are largely restricted to Hispaniola. Smaller endemic mammals such as insectivores and rodents instead of becoming extinct between 7,000 and 1,000 yBP vanished after the introduction of rats, cats, swine, and dogs following the discovery and colonization of the Antilles by Europeans Some species such as *Nesophontes*, *Brotomys* and *Boromys*, all about the size of a rat, may have lasted until the introduction of the mongoose in the late 1800s. In regions where montane refugia do not exist, endemic rodents have also persisted in protected insular situations, such as offshore cays in Cuba and the Bahamas where rats, cats, dogs, and the mongoose are absent, and where human activities are minimal. Capromyid rodents are the group that has been most resistant to extinction because of their intermediate body size and broad habitat requirements.

Compared with Madagascar, also an oceanic island long detached from the contiguous continents, there are striking convergences in the mammalian fauna. There are also interesting differences deriving from the stocks that colonized the islands and the sequences during which they arrived. Madagascar is approximately twice as large as the Greater Antilles in terms of the present land area, although as noted above the areas are almost equal if the area of the Greater Antilles at low ocean levels (100 m below present levels) is considered, and the present areas of open water between the islands are connected by hypothetical bridges.

On Madagascar the Insectivora were early colonists, and radiated to fill a variety of terrestrial and scansorial insectivore-omnivore niches. Their radiation was more extensive than that displayed by the Insectivora of the Greater Antilles. Although only two families of Madagascan insectivores are recognized, they radiated into ten genera and 22 species (23 including the introduced *Suncus murinus*).

The 445 known bats of the Greater Antilles show great species richness and comprise approximately 34 percent of the original mammal fauna. This contrasts sharply with the bats of Madagascar that include only 20 species and approximately 19 percent of the total mammalian fauna. However, it is recognized that the species richness of bats in the paleotropics is less than that of the neotropics and the area from which the bat colonization of Madagascar might have taken place, namely East Africa, is even more depauperate with respect to species richness of bats than is the case in West Africa where the number of bat species reaches its maximum. Potential bat colonists from Central America to the West Indies comprises one of the richest bat faunas in the world. Fifty-one percent of the terrestrial mammalian fauna of Panama are bats.

The primates of Madagascar were very early colonists. Before the advent of man they included five families, 19 genera and 36 species. They radiated to fill the scansorial and arboreal frugivore-folivore niches and had adapted to both diurnal and nocturnal activity rhythms. In the West Indies primates clearly came at a much later date and were unable to radiate to the same extent as in Madagascar probably because the typical primate niches were already filled by edentates and rodents.

The edentates of the Antilles represent old South American Endemics. Sixteen species have been identified. Some are quite arboreal, strongly resembling tree sloths in size and morphology. Others are large and terrestrial. Presumably all were herbivorous and occupied niches comparable to some of the larger semi-terrestrial and arboreal lemurs of Madagascar.

Rodents apparently were late arrivals on the island of Madagascar, and may have derived from either one or several colonizations. They occupy more or less typical rodent niches and did not radiate to fill primate-like niches as was the case in the Greater Antilles. Furthermore, the Madagascan rodents did not diversify to occupy the large terrestrial herbivore niches as did the rodents on some of the more distant islands within the eastern margin of the Greater Antilles. Only 13 rodents are known from Madagascar, 23 percent of the number of rodents known to occur in the Greater Antilles. In the Greater Antilles it is clear that rodents entered at a reasonably early date and rapidly diversified. Some of them are highly arboreal and clearly occupied niches similar to primates, squirrels and porcupines.

In total numbers of species of land mammals the Greater Antilles are remarkably similar to Madagascar. There were 132 species of bats and land mammals in the Greater Antilles, and 105 in Madagascar. When bats are eliminated from the list (as well as the soricids from Madagascar, which were probably introduced), 87 endemic terrestrial mammals are known from the Greater Antilles, and 83 from Madagascar. Therefore, in species richness the non-volant mammals of the Greater Antilles and Madagascar are very comparable, as are the two regions in latitude, habitat diversity and degree of endemism. Given the smaller land area of the Greater Antilles, it is remarkable that such species richness was attained. Isolated archipelagoes which characterize the Greater Antilles have resulted in suitable situations for allopatric speciation following dispersal events, and may have increased the species richness of the Greater Antilles. If a geographical analogy can be made between Madagascar and Hispaniola, the island of the Greater Antilles most similar to Madagascar in geomorphology, the large endemic mammals of Madagascar are especially vulnerable to extinction, and the process will be nearly impossible to stop once it is well underway. It is a sobering thought to contemplate the fate of the mammals of Madagascar when on the large and diverse islands of the Greater Antilles like Hispaniola, 71 percent of the families, 87 percent of the genera, and 89 percent of the species of endemic land mammals are gone forever.

Literature cited

Albignac, R. 1973. Mammiferes, Carnivores. Faune de Madagascar 36: 206 pp.
Baker, R.J. and H.H. Genoways. 1978. Zoogeography of Antillean bats. In F.B. Gill (ed.). Zoogeography in the Caribbean. Academy of Natural Sciences of Philadelphia, Special Publication 13:1-128.
Battestini, R. and G. Richard-Vindard (eds.). 1972. Biogeography and Ecology in Madagascar. A. Junk, The Hague, 765 pp.
____, and P. Vein. 1967. Ecological changes in protohistoric Madagascar. Pp. 407-424 in P. S. Martin and H. E. Wright (eds.). Pleistocene Extinctions in Search of a Cause. Yale University Press, New Haven, CT.
Butler, P.M. 1985. The history of African insectivores. Act Zoologica Fennica 173:215-217.
Carleton, M.D., and G.G. Musser. 1984. Muroid rodents. Pp. 289-379 in S. Anderson, and J.K. Jones (eds.). Orders and Families of Recent Mammals of the World. Wiley, New York.
Eisenberg, J.F., and N. Gonzalez-Gotura. 1985. Observations on the Natural History of *Solenodon cubanus*. Acta Zoologica Fennica 173:275-277.
____, and E. Gould. 1970. The Tenrecs: A study in mammalian behavior and evolution. Smithsonian Contributions to Zoology No. 27, 138 pp.
____. 1984. The insectivores. Pp. 155-166 in A. Jolly, P. Oberle, and R. Albignac (eds.). Madagascar. Pergamon Press, New York.
Ford, S.M., and G.S. Morgan. 1986. A new ceboid femur from the late Pleistocene of Jamaica. Journal of Vertebrate Paleontology 6(3):281-289.
Goodfriend, G.A., and R.M. Mitterer. 1987. Age of the ceboid femur from Coco Ree, Jamaica. Journal of Vertebrate Paleontology 7(3):344-345.
Gould, S.J., N.L. Gilinsky, and R.Z. German. 1987. Asymmetry of lineages and direction of evolutionary time. Science 236:1437-1441.

Griffiths, T.A. and D.J. Klingener. 1988. On the distribution of distribution of Greater Antillean bats. Biotropica 20(3):240-251.

Guillaumet, J. L. 1984. The vegetation. Pp. 27-54 in A. Jolly, P. Oberle, and R. Albignac (eds.). Madagascar. Pergamon Press, New York.

Heaney, L.R. 1986. Biogeography of the mammals of SE Asia: estimates of rates of colonization, extinction and speciation. Biological Journal of Linnean Society 28:127-165.

Kashfi, M.S. 1983. Geology and hydrocarbon prospects of Jamaica. Bulletin of the American Association of Petroleum Geologists 67:2117-2124.

Lamberton, C. 1939. Contributions a la Connaissauce de la Fauna Subfossile de Madagascar, Note VIII *Cryptoprocta*. Memoires Academie Malaguche 27:155-93.

MacFadden, B.J. 1980. Rafting mammals or drifting islands?: biogeography of the Greater Antillean insectivores *Nesophontes* and *Solenodon*. Journal of Biogeography 7:11-22.

MacPhee, R. D. E. 1986. Environment, Extinction and Holocene Vertebrate Localities in Southern Madagascar. National Geographic Research 2(4): 441-55.

_____. (in press). The shrew tenrecs of Madagascar: Systematic revision and Holocene distribution of *Microgale* (Tenrecidae, Insectivora). American Museum of Natural History Novitates.

Matthew, W.D. 1918. Affinities and origin of the Antillean mammals. Bulletin of the Geological Society of America 29:657-666.

McDowell, S.B. Jr. 1958. The Greater Antillean insectivores. Bulletin of the American Museum of Natural History 115:113-214.

Miller, G.S. 1930. Three small collections of mammals from Hispaniola. Smithsonian Miscellaneous Collections 82(15):1-10.

Morgan, G.S. 1977. Late Pleistocene fossil vertebrates from the Cayman Islands, British West Indies. Unpublished Masters thesis, University of Florida, Gainesville, FL, 282 pp.

_____, C.E. Ray, and O. Arredondo. 1980. A giant extinct Insectivore from Cuba (Mammalia: Insectivora: Solenodontidae). Proceedings of the Biological Society of Washington 93(3):597-608.

_____, and C.A. Woods. 1986. Extinction and zoogeography of West Indian land mammals. Biological Journal of Linnean Society 28:167-203.

Patterson, B. P. 1978. Pholidota on Tubulidentata. Pp. 268-78 in V. J. Maglie, and H. B. S. Cooke (eds.). Evolution of African Mammals. Harvard University Press, Cambridge, MA.

Petter, F. 1972. Rongeurs. Pp. 661-666 in R. Battestini and G. Richard- Vindard (eds.). Biogeography and Ecology in Madagascar. A. Junk, The Hague.

Pregill, G. 1981. Late Pleistocene herpetofaunas from Puerto Rico. University of Kansas Museum of Natural History, Miscellaneous Publication 71:1-72.

Rivero, M. de la Calle, and O. Arredondo. 1988. Craneo de un primate fósil. Juventud Tecnica (Havana, Cuba) 245:35.

Rouse, I, and L. Allaire. 1978. Caribbean. Pp. 431-481 in R.E. Taylor and C.W. Meighan (eds.). Chronologies in New World Archaeology. Academic Press, New York.

Simpson, G.G. 1956. Zoogeography of West Indian land mammals. American Museum of Natural History Novitates 1759:1-28.

Steadman, D.W., G.K. Pregill and S.L. Olson. 1984. Fossil vertebrates from Antigua, Lesser Antilles: evidence for late Holocene human-caused extinctions in the West Indies. Proceedings of the National Academy of Sciences of the USA 81:4448-4451.

Sykes, L.R., W.R. McCann, and A.L. Kafka. 1982. Motion of Caribbean Plate during last seven million years and implications for earlier Cenozoic movements. Journal of Geophysical Research 87(B13):10,656-10,676.

Tattersall, I. 1982. The Primates of Madagascar. Columbia University Press, NY, 382 pp.

Varona, L.S. 1974. Catálogo de los Mamíferos Vivientes y Extinguidos de las Antilles. Instituto de Zoologia de Academia de Ciencias de Cuba (Havana), 139 pp.

____. 1986. Táxones del subgénero *Mysateles* en Isla de la Juventud, Cuba. Déscription de una nueva especie (Rodentia; Capromyidae; *Capromys*). Poeyana de Instituto de Zoología, Academia de Ciencias de Cuba (Havana) 315:1-11.

____, and O. Arredondo. 1979. Nuevos táxones fósiles de Capromyidae (Rodentia: Caviomorpha). Poeyana, Instituto de Zoologia, Academia de Ciencias de Cuba (Havana) 195:1-51.

Woods, C.A. 1982. The history and classification of South American hystricognath rodents: reflections on the far away and long ago. Pp. 377-392 in M.A. Mares and H.H. Genoways (eds.). Mammalian Biology in South America. Special Publication of the Pymatuning Laboratory of Ecology, University of Pittsburgh, PA.

____. 1984. Hystricognath Rodents. Pp. 389-446 in S. Anderson, and J.K. Jones (eds.). Orders and Families of Recent Mammals of the World. Wiley, New York.

____. 1987. The threatened and endangered birds of Haiti: lost horizons and new hopes. Proceedings of the Jean Delacour/International Foundation for the Conservation of Birds Symposium 2:385-429.

____. 1989a. A new capromyid rodent from Haiti: the origin, evolution and extinction of West Indian rodents and their bearing on the origin of New World hystricognaths. Natural History Museum of Los Angeles County, Contribution in Science Series 33:59-89.

____. 1989b. The endemic rodents of the West Indies: the end of a splendid isolation. Pp. 11-19 in W.Z. Lidicker, Jr. (ed.). Rodents, a World Survey of Species of Conservation Concern. Occasional Papers of the IUCN Species Survival Commission 4.

____. in press. The fossil and recent land mammals of the West Indies: an analysis of the origin, evolution, and extinction of an insular fauna. In A. Azzaroli (ed.). Biogeographical Aspects of Insularity. Accademia Nazionale dei Lincei, Rome.

____, and B. McKeen. 1989. Convergence in New World porcupines and West Indian rodents: an analysis of tooth wear, jaw movement, and diet in rodents. In K. Redford and J. Eisenberg (eds.). Advances in Neotropical Mammalogy. Sandhill Crane Press, Gainesville, FL.

____, and J.A. Ottenwalder. 1983. The montane avifauna of Haiti. Proceedings of the Jean Delacour/ International Foundation for the Conservation of Birds Symposium 1:576-590 + 607-622.

____, J.A. Ottenwalder and W.L.R. Oliver. 1985. Lost mammals of the Greater Antilles: The summarized findings of a ten weeks field survey in the Dominican Republic, Haiti, and Puerto Rico. Dodo 22:23-42.

Wozencraft, W. C. 1984. A Phylogenetic Reappraisal of the Viverridae and its Relationship to Other Carnivora. Unpublished Ph.D. dissertation, University of Kansas, Lawrence, KA.

―――. 1986. A new species of striped mongoose from Madagascar. Journal of Mammalogy 67:561-71.

Table 1. Non-volant Land Mammals of the Greater Antilles (continued).

TAXON	DISTRIBUTION	STATUS
PRIMATES[1]		
Cebidae		
Ateles anthropomorpha	Cuba	Extinct
new gen. et sp	Cuba	Extinct
"*Saimiri*" *bernensis*	Hispaniola	Extinct
Callitrichidae		
"Ceboid M"	Hispaniola	Extinct
Callitrichidae (*incertae sedis*)		
Xenothrix mcgregori	Jamaica	Extinct
RODENTIA		
Capromyidae		
Capromyinae[2]		
Capromys acevedoi	Cuba	Extinct
Capromys angelcabrerai	off Cuba	Living
Capromys antiquus	Cuba	Extinct
Capromys auritus	off Cuba	Living
Capromys barbouri	Cuba	Extinct
Capromys beatrizae	Cuba	Extinct
Capromys delicatus	Cuba	Extinct
Capromys garridoi	off Cuba	Living
Capromys gracilis	Cuba	Extinct
Capromys gundlachi	I.Juventud	Living
Capromys jaumei	Cuba	Extinct
Capromys kraglievichi	Cuba	Extinct
Capromys latus	Cuba	Extinct
Capromys melanurus[3]	Cuba	Living
Capromys meridionalis	I.Juventud	Living
Capromys minimus	Cuba	Extinct
Capromys nanus	Cuba	Living
Capromys pilorides	Cuba	Living
Capromys prehensilis	Cuba	Living
Capromys robustus	Cuba	Extinct
Capromys sanfelipensis	off Cuba	Extinct
Capromys silvai	Cuba	Extinct
Capromys sp	Cayman Islands	Extinct
Geocapromys columbianus	Cuba	Extinct
Geocapromys megas	Cuba	Extinct
Geocapromys pleistocenicus	Cuba	Extinct
Geocapromys brownii	Jamaica	Living

Table 1. Non-volant Land Mammals of the Greater Antilles (continued).

TAXON	DISTRIBUTION	STATUS
Geocapromys ingrahami	Bahamas	Living
G.i. abaconis	Great Abaco	Extinct
G.i. ingrahami	East Plana Cay	Living
	Little Wax Cay	Intro. (1973)
	Warderick Wells Cay	Intro. (1981)
G.i. irrectus	Andros, Cat, Eleuthera, Great & Little Exuma & Long	Extinct " "
Geocapromys thoracatus	Little Swan	Extinct
Geocapromys sp	Cayman Islands	Extinct
Plagiodontinae		
Plagiodontia aedium	Hispaniola	Living
Plagiodontia araeum	Hispaniola	Extinct
Plagiodontia ipnaeum (?)	Hispaniola	Extinct
Plagiodontia spelaeum (?)	Hispaniola	Extinct
Plagiodontia velozi	Hispaniola	Extinct
Rhizoplagiodontia lemkei	Hispaniola	Extinct
Isolobodontinae		
Isolobodon portoricensis	Hispaniola	Extinct
Isolobodon montanus	Hispaniola	Extinct
Hexolobodontinae		
Hexolobodon phenax	Hispaniola	Extinct
Hexolobodon poolei (?)	Hispaniola	Extinct
Hexolobodon sp	S. Hispaniola	Extinct
FAMILY *incertae sedis*		
Heptaxodontinae		
Quemisia gravis	N. Hispaniola	Extinct
Elasmodontomys obliquus	Puerto Rico	Extinct
Amblyrhiza inundata	Anquilla & St. Martin	Extinct
Clidomyinae		
Clidomys osborni	Jamaica	Extinct
Clidomys parvus	Jamaica	Extinct
Echimyidae		
Heteropsomyinae		
Boromys offella	Cuba	Extinct
Boromys torrei	Cuba	Extinct

Table 1. Non-volant Land Mammals of the Greater Antilles (concluded).

TAXON	DISTRIBUTION	STATUS
Brotomys contractus	Hispaniola	Extinct
Brotomys voratus	Hispaniola	Extinct
Homopsomys antillensis	Puerto Rico	Extinct
Heteropsomys insulans	Puerto Rico	Extinct
Puertoricomys corozalus	Puerto Rico	Extinct
Muroidea		
Sigmodontinae		
Oryzomys antillarum	Jamaica	Extinct
Undescribed species A	Anguilla	Extinct

[1] Includes the new primate from Cuba (Rivero and Arredondo 1988).
[2] *Capromys* may be more than one genus (see discussion in Woods this volume).
[3] Includes *Capromys arboricolus* (Varona pers. comm.).
? Questionable validity as separate species.

Table 2. Status and Size of the Non-volant Land Mammals of Madagascar

TAXON	LAST RECORD	SIZE (mm)
Order Insectivora		
Family Tenrecidae		Head and Body
Subfamily Oryzorictinae[1]		
Limnogale mergulus	Living	112-170
Oryzorictes talpoides	Living	115-135
O. hova	Living	115-135
O. tetradactylus	Living	85-130
Geogale aurita	Living	71- 74
Microgale pusilla	Living	40- 50
M. parvula	Living	40- 60
M. longicaudata	Living	40- 60
M. brevicaudata	Living	60- 85
M. cowani	Living	60- 70
M. principula	Living	60- 80
M. dobsoni	Living	90-115
M. gracilis	Living	95-110
M. thomasi	Living	95-100
M. talazaci	Living	115-135
Cryptogale australis	Extinct	--
Subfamily Tenrecinae		
Tenrec ecaudatus	Living	265-390
Hemicentetes nigriceps	Living	160-180
H. semispirosus	Living	160-190
Setifer setosus	Living	150-220
Echinops telfairi	Living	140-180
Family Soricidae		
Suncus murinus	Living	100-110
S. madagasarensis	Living	35- 40
Order Primates		Condylar-Basal
Family Lemuridae		
Subfamily Lemurinae		
Lemur catta	Living	--
L. mongoz	Living	--
L. macaco	Living	--
L. fulvus	Living	78
L. coronatus	Living	--
L. rubriventer	Living	--
Varecia variegata	Living	74
V. insignis	Extinct	--
V. jullyi	Extinct	--

Table 2. Status and Size of the Non-volant Land Mammals of Madagascar (continued).

TAXON	LAST RECORD	SIZE (mm)
Family Lepilemuridae		
Subfamily Lepilemurinae		
Lepilemur mustelinus	Living	46
Subfamily Hapalemurinae		
Hapalemur griseus	Living	66
H. simus	Living	--
H. gallieni	Extinct	--
Subfamily Megaladapinae		
Megaladapis madagascariensis	Extinct	180
M. edwardsi	Extinct	243
M. grandidieri	Extinct	223
Family Indriidae		
Subfamily Indriinae		
Indri indri	Living	87
Avahi laniger	Living	48
Propithecus diadema	Living	--
P. verreauxi	Living	65
Mesopropithecus pithecoides	Extinct	86
M. globiceps	Extinct	88
Subfamily Archaeolemurinae		
Archaeolemur majori	Extinct	116
A. edwardsi	Extinct	131
Hadropithecus stenognathus	Extinct	125
Subfamily Palaeopropithecinae		
Palaeopropithecus ingens	Extinct	150
Archaeoindris fontoynonti	Extinct	175
Family Daubentoniidae		
Daubentonis madagascariensis	Living	59
D. robusta	Extinct	--
Family Cheirogaleidae		
Subfamily Cheirogaleinae		
Cheirogaleus major	Living	51
C. medius	Living	--
Microcebus murinus	Living	28
M. rufus	Living	--
Mirza coquereli	Living	--
Allocebus trichotis	Living	--
Phaner furcifer	Living	45

Table 2. Status and Size of the Non-volant Land Mammals of Madagascar (continued).

TAXON	LAST RECORD	SIZE (mm)
Order Rodentia		
Family Muridae		
Subfamily Nesomyinae		Head and Body
Nesomys rufus	Living	186-230
N. lambertoni	Living	186-230
N. betsileoensis	Living	186-230
Brachytarsomys albicauda	Living	200-250
Brachyuromys ramirohitra	Living	145-180
B. betsileonensis	Living	145-180
Gymnuromys roberti	Living	125-160
Eliurus myoxinus	Living	175
E. minor	Living	80
Macrotarsomys bastardi	Living	80-100
M. ingens	Living	120
Majoria sp.	Extinct	--
Hypogeomys antimena	Living	330-350
Order Carnivora		
Family Viverridae		
Subfamily Galidictinae		Head and Body
Galidia elegans	Living	380
Salanoia concolor	Living	250-300
Galidictis fasciata	Living	320-340
G. grandidieri	Living	420
Mungotictis decemlineata	Living	250-350
Subfamily Euplerinae		
Eupleres goudotil	Living	450-650
E. major	Living	--
Fossa fossa	Living	400-450
Subfamily Cryptoproctinae		
Cryptoprocta ferox	Living	
C. sp	Extinct	610-800
Subfamily Viverrinae		
Viverricula indica	Living	450-630
Order Artiodactyla		
Family Suidae		Head and Body
Potomochoerus larvatus	Living	1000-1200
Family Hippopotamidae		
Hippopotamus lemelii	Extinct	1500-2000

Table 2. Status and Size of the Non-volant Land Mammals of Madagascar (concluded).

TAXON	LAST RECORD	SIZE (mm)
Order Tubulidentata Family Oryctoropidae[2] *Plesiorytoropus madagascarensis*	1,500 BP	<u>Condylar-Basal</u> 100

[1] Modified according to McPhee (in press)
[2] McPhee (pers. comm.) believes this species is a tenrecoid insectivore and not a Tubulidentata.

Table 3. Summary of Non-volant endemic mammals of Madagascar and Greater Antilles.

TAXON	Number Known[1]		Number Extinct		% Extinct	
	MAD	GA	MAD	GA	MAD	GA
Insectivora[2]						
Families	1	2	0	1	0	50
Genera	9	2	1	1	11	50
Species	21	12	1	9	5	75
Edentata						
Families	0	1	NA	1	NA	100
Genera	0	12	NA	12	NA	100
Species	0	16	NA	16	NA	100
Primates						
Families	5	2	0	2	0	100
Genera	19	5	6	4	32	100
Species	36	5	14	4	39	100
Rodentia[3]						
Families	1	4	0	3	0	75
Genera	8	15	1	12	13	80
Species	13	55	1	42	8	76
Carnivora[4]						
Families	1	0	0	NA	0	NA
Genera	7	0	0	NA	0	NA
Species	10	0	1	NA	10	NA
Artiodactyla[5]						
Families	2	0	1	NA	50	NA
Genera	2	0	1	NA	50	NA
Species	2	0	1	NA	50	NA
Tubulidentata						
Families	1	0	1	NA	100	NA
Genera	1	0	1	NA	100	NA
Species	1	0	1	NA	100	NA
SUMMARY						
Families	11	9	2	7	18	78
Genera	46	34	10	30	22	88
Species	83	88	19	72	23	82

[1] Number known includes both living and extinct forms.
[2] Excludes Soricidae.
[3] Heptaxodontidae as separate family; excludes *Mus* & *Rattus*.
[4] Combines all genera within Viveridae; excludes Viverricula on *Madagascar* & *Herpestes* in GA.
[5] Includes *Potomochoerus* on Madagascar.

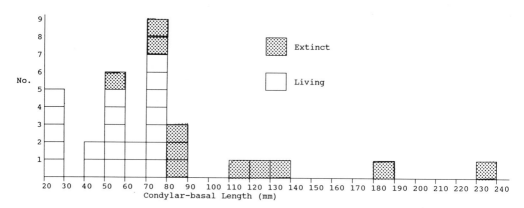

Figure 1. Size and extinction of the Madagascaran Strepsirrhini between 15,000 yBP and present according to Tattersall 1982.

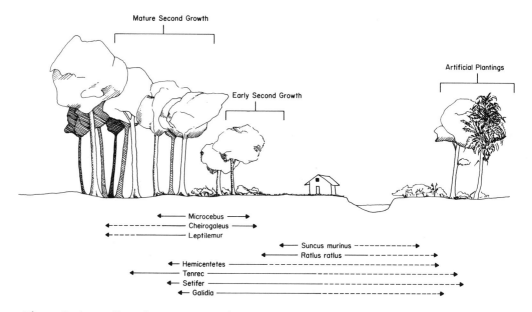

Figure 2. A small settlement on Madagascar showing the ecological relationships of the mammal community.

CONSERVATION TRENDS, AND THE THREATS TO ENDEMIC BIRDS IN JAMAICA

A. M. Haynes, R. L. Sutton[1], and Karen D. Harvey[2]

Abstract

In the last 400 years the natural environment of Jamaica has been radically altered by man. As a result, many species and some ecosystems have become extinct. All remaining areas of natural habitat are severely threatened. At least eight of Jamaica's twenty seven remaining species of endemic birds show evidence of recent population decline. This indicates an overall decline in habitat quality which is affecting all indigenous fauna and flora.

The most serious threats are habitat destruction (for agriculture, forestry, industry, and settlement), and hunting. The development of a system of national parks and protected areas; the improvement of local capacities for public education about the natural environment and conservation; revision and enforcement of the laws relating to the environment; rationalization of the government bodies concerned with the environment and environmental planning; research into the status, distribution, and sustainable use of indigenous species are all urgently needed to ensure the future of Jamaica's indigenous wildlife.

Introduction

Chance alone accounts for the survival of Jamaica's remaining natural areas and endemic species. At least 75 percent of the original forest has been totally cleared, and what remains is almost entirely secondary. Many species of animals and plants have been introduced to the detriment of the original fauna and flora. Hunting for sport and subsistence is a deep rooted tradition. The full extent of what has been lost in the last four hundred years is not known. It includes: 1) the entire lowland forest ecosystem which once covered the fertile coastal plains; 2) the Caribbean Monk Seal (*Monachus tropicalis*), a large marine mammal; 3) the Jamaican Rice Rat (*Oryzomys antillarum*), a small terrestrial mammal; 4) at least three fairly large terrestrial reptiles - the Jamaican Iguana (*Cyclura collei*), the Giant Galliwasp (*Celestus occidus*) and the Black Racer (*Alsophis ater*); and 5) four species of birds - the Jamaican Pauraque (*Siphonorhis americanus*), the Jamaican Macaw (*Ara* sp.), the Jamaican Uniform Crake (*Amaurolimnas concolor*), and the Jamaican Black-capped Petrel (*Pterodroma hasitata caribbaea*). A recent

---

[1] Ann Haynes-Sutton and Robert Sutton reside at Marshall's Pen, P.O. Box 58, Mandeville, Jamaica.
[2] Karen Harvey works for the U.S. Forest Service, P.O. Box 2464, Orofino, ID 83544 USA

survey (Wilkins pers. comm.) indicates that remaining populations of the Coney (*Geocapromys brownii*), the only extant indigenous terrestrial mammal of Jamaica, are also threatened. Losses have occurred in other groups of Jamaican organisms with high levels of endemism, such as flowering plants, insects, and mollusks, but these have not been documented. Very few, if any, areas of virgin habitat are left. Jamaica's natural habitats are being destroyed at an unprecedented rate. It is unlikely that many significant natural areas will survive beyond the next ten years if development continues at the present rate (Haynes pers. obs.). It is, therefore, a matter of great urgency to identify the threats to the indigenous animals and plants, and their habitats, and to implement immediate remedial action while viable populations are present in the wild. Birds have been shown to be useful indicators of environmental quality in Jamaica (Fairbairn and Sutton 1979). an evaluation of threats to the endemic avifauna can help to provide direction for research and conservation.

Jamaica's avifauna consists of 256 species, including 159 breeding species of which 27 are endemic (Gosse Bird Club 1986a). The status and distribution of North American migrants and seabirds in Jamaica have been discussed elsewhere (Sutton et al. 1986; Haynes 1987.). Jamaica ranks fourth among oceanic islands in the number of species of endemic birds (Phillips 1985). Little has been published, however, on the status, distribution, and conservation of Jamaica's endemic birds since Lack (1976). The evaluation of the status of Jamaica's avifauna, and the identification of the factors which threaten its survival are of importance if Jamaica's rich natural heritage is to be conserved.

Methods

The Gosse Bird Club (GBC) in Kingston has records on Jamaican birds dating back to the late 1930's. Since 1963, it has also published the Gosse Bird Club Broadsheet, a bulletin in which members record their observation of Jamaican birds. The present paper is based largely on a preliminary analysis of data collected by the GBC since 1963.

Data on status were extracted from GBC records, supplemented by personal observations by the authors and others and summarized in Tables 1-3. English names and binomials follow the American Ornithologists' Union (1983) with several exceptions. We follow Lack (1976) in treating the Jamaican Parakeet *Aratinga nana* as a species endemic to Jamaica, and we consider Red-billed (western) and Black-billed (eastern) populations of streamertails two separate species, *Trochilus polytmus* and *T. scitulus*, respectively, following Schuchmann (1978). Additionally, we treat the extinct Jamaican populations of the Black-capped Petrel as an endemic subspecies, *Pterodroma hasitata caribbaea* (American Ornithologists' Union 1983).

It is recognized that a more quantitative analysis would provide a superior indication of the status of these species. However, until the GBC's records are computerized and more work is carried out, such data will not be available.

Status of endemic species

Of Jamaica's 27 endemic species, the populations of eight are thought to be decreasing. They are the Ring-tailed Pigeon *Columba caribaea*, Crested Quail-Dove *Geotrygon versicolor*, Yellow-billed Parrot *Amazona collaria*, Black-billed Parrot *Amazona agilis*,

Jamaican Mango *Anthracothorax mango*, Rufous-tailed Flycatcher *Myiarchus validus*, White-eyed Thrush *Turdus jamaicensis*, and Jamaican Blackbird *Nesopar nigerrimus*. Only three species appear to be increasing in number (Jamaican Parakeet *Aratinga nana*, White-chinned Thrush *Turdus aurantius*, and Jamaican Vireo *Vireo modestus*).

Eleven species appear to be maintaining their populations at relatively constant levels. These are the Jamaican Owl *Pseudoscops grammicus*, Jamaican Tody *Todus todus*, Jamaican Woodpecker *Melanerpes radiolatus*, Jamaican Elaenia *Myiopagus cotta*, Sad Flycatcher *Myiarchus barbirostris*, Jamaican Becard *Pachyramphus niger*, Arrow-headed Warbler *Dendroica pharetra*, Orangequit *Euneornis campestris*, and Yellow-shouldered Grassquit *Loxipasser anoxanthus*. The status of the five other species (Jamaican Lizard Cuckoo *Saurothera vetula*, Chestnut-bellied Cuckoo *Hyetornis pluvialis*, Jamaican Crow *Corvus jamaicensis*, Blue Mountain Vireo *Vireo osburni*, and Jamaican Euphonia *Euphonia jamaica*) is not clear.

At least two species and two subspecies have become extinct in the last 150 years. These are the Jamaican Macaw *Ara gossei*, the Jamaican Pauraque *Siphonorhis americanus*, the Jamaican Black-capped Petrel *Pterodroma hasitata caribbaea*, and the Jamaican Uniform Crake *Amaurolimnas concolor concolor*.

It appears that those species whose populations are decreasing are more restricted in their ranges than those that are stable or increasing (Table 1). Fortunately, many of Jamaica's endemic species are widely distributed on the island (Lack 1976) and appear to be sufficiently flexible to adapt to the radical changes which have occurred in Jamaica's natural environment in the last 400 years. These changes include the complete removal of lowland forests, periodical removal of much of the other forest, and the introduction of exotic mammals and birds.

Factors affecting status of endemic birds

Jamaica's endemic bird species are probably more vulnerable to environmental changes than are the migrants (Robbins et al. 1985; Sutton et al. 1986). The relative importance of main factors affecting the endemic bird populations is shown in Table 2. Habitat destruction, especially as a result of forestry and agriculture, is the most serious threat overall, and hunting is affecting the populations of some species. Disease, parasitism, introduced species, and the expansion of the ranges of wild species are among additional factors which appear to be less important at present.

Habitat destruction

The main causes of habitat destruction as described by Sutton et al. (1986) are timber extraction from natural forests, hillside agriculture (including plantings of *Pinus caribaea*, Coffee, *Cannabis sativa*, and garden crops), charcoal burning, urban expansion and human settlement, road construction, and mining and quarrying. The effects of forest destruction on Jamaica's wildlife were discussed by Fairbairn (1986). The problems of land use are exacerbated by the poor enforcement of existing laws that regulate development, the fragmented approach to development control, and by the lack of any formal requirements for environmental impact assessments for agricultural projects.

Hunting

Birds are hunted for five reasons in Jamaica: "pest" eradication; commerce; subsistence; sport; and science. Almost all birds in Jamaica are protected under the Wild Life Protection Act (WLPA). Of the endemic species, only the Jamaican Parakeet and the Jamaican Crow are not fully protected.

Pest eradication.-- Endemic species that are hunted as pests are the two species of parrots (Fairbairn 1981; Cruz and Gruber 1981; Cook 1984), parakeets, Jamaican Crows, and Orangequits.

Commerce.-- There is illegal trade in Yellow-billed and Black-billed Parrots, Crested Quail-Doves, and Ring-tailed Pigeons for pets. Government policy is to use its limited resources to control the most obvious elements of this trade - sales on the streets and in pet shops, and exports. However, the capacity for enforcement is limited and the pet trade is currently affecting the populations of parrots (Fairbairn 1981; Cruz and Gruber 1981; Cook et al. 1984) and probably some columbid species as well.

Sport.--Bird shooting is a traditional upper class sport in Jamaica. The main targets are columbids and ducks. The WLPA provides for hunting seasons for certain species, but no season has been declared for ducks for at least a decade. There is little enforcement of regulations. For example, although the Ring-tailed Pigeon is totally protected at all times, it is a favorite game bird and is hunted illegally throughout the year. Hunters who openly flaunt the law include senior politicians.

In rural communities bird hunting with catapults (sling shots) and snares is a favorite form of recreation for children and young men. As a result, birds are rare around rural settlements.

Subsistence.--The importance of wild bird meat in rural diets has not been assessed in Jamaica, but may be locally important.

Science.--Recent developments in genetic analysis have led to an increase in the pressure on the Jamaica government to allow collection of specimens for reference collections of frozen specimens at foreign museums. If this is not controlled, it may cause a problem for the rarer endemic species.

Other threats

Introduced species.--Jamaica's endemic birds appear to have come into equilibrium with birds which were introduced in the past. The only known conflict is competition between Jamaican Woodpeckers and introduced Starlings (*Sternus vulgaris*) for nest holes. Recently the Veterinary Division appears to have relaxed its policy on importation of pet bird species into Jamaica, and there is a risk that potentially injurious species could be imported and accidentally released (R. Kerr pers. comm.). Jamaica is vulnerable to the same kind of explosion in the numbers of escaped introduced species that has plagued Puerto Rico (Raffaele this volume). The impact of introduced mammals such as rats, cats, and mongooses has not been documented, but it is assumed that ground nesting birds are vulnerable to exploitation by these species.

Range expansion.--The arrival of Cattle Egrets (*Bubulcus ibis*) in Jamaica in the 1950's does not appear to have adversely affected Jamaica's endemic avifauna. However, the spread of the Shiny Cowbird (*Molothrus bonairensis*) into the Greater Antilles is much a more serious threat (Post and Wiley 1977; Cruz et al. this volume). Although is has been reported from both Cuba and Florida (W. Arendt pers. comm.), it has not yet been seen in Jamaica. If it does become established in Jamaica, its effects on endemic species could be severe.

Pollution.--Pesticides including Dieldrin, DDT, and Aldrin are freely available in Jamaica and are often abused. Traces have been found in human foods (A. Mansingh pers. comm.). The effect of the residues of chemicals on Jamaican birds has not been assessed. In 1986, the use of a herbicide ("Glycophos") to eradicate ganja, *Cannabis sativa*, was approved by the Jamaican government. This is sometimes applied by "bombing" areas. Several species of birds feed in the ganja (Gruber 1986). The effects of this policy on Jamaican birds are not known.

Disease and parasitism.--The prevalence of diseases in Jamaican endemic bird populations is not known. Blood parasites are not thought to be a threat (Bennett et al. 1980).

Conservation priorities

Man's "development" of Jamaica has had a high environmental cost. The reasons for this are too many and complicated to be summarized here, and are common to many colonial and post-colonial societies. What is important is to note that the rate of destruction has increased and reached a critical stage. At current rates it is likely that all remaining natural areas will be subject to severe disturbance in the next five to ten years. The costs to Jamaica will go far beyond the loss of natural ecosystems (such as swamp, coastal and riverine forests) and species (such as manatees). Something unique about Jamaica's natural patrimony is lost with each extinction of an endemic species.

Conservation initiatives in Jamaica began in the 1930's. Legislation to control the use of forests, fish, beaches, and wildlife has gradually accreted since then. Currently there are at least 12 laws which relate to the natural environment, 18 government and quasi-government organizations which are involved in protecting and using it, and at least 18 non-government organizations concerned with it. There is an urgent need for coordination between these groups. Habitat destruction is the most important threat to Jamaica's flora and fauna. There are no national parks or reserves managed primarily in the interests of the natural environment. Private landowners who seek to create nature reserves risk having their lands declared "idle" and compulsorily purchased by the government. Existing laws make provision for at least five different types of reserve, and regulate the hunting of birds, protect animals, and restrict the exploitation of marine resources. Very few attempts are made, however, to enforce the regulations, and the existing laws are openly violated.

Most government agencies have declined in strength and have suffered crippling losses of qualified staff over the past seven years. The agencies most closely involved with the environment have been the most severely affected. However, the development of interest in non-government conservation groups is an encouraging sign. Five new groups have formed in the last four years. Existing groups have important roles to play in public education, monitoring, development of protected areas and pressuring the government into action. It is also hoped that the opening of the United Nations Environment Program's Regional Coordinating Unit for the Caribbean Action Plan in Kingston will help to catalyze conservation action in Jamaica. The rate of degradation of Jamaica's natural areas is expected to be reduced by the recent change in attitude towards the environment and environmental projects which has taken place in the international funding agencies including the World Bank and the International Development Bank.

Environmental planning

Stricter control of land development through actions of planning boards and based on sound environmental impact assessments is essential. The planning process needs to be streamlined and made effective. Improvement of coordination between agencies is a necessity. All large scale projects, including agricultural ones, should be subject to environmental impact assessment (EIA) and the proceedings should be subject to public review. A machinery for initiating, implementing and reviewing EIA's should be developed.

Environmental law

Existing laws should be reviewed, consolidated and updated. For example, comprehensive laws to cover national parks are required. The expansion of the Wild Life Protection Act to include plants is also important. Penalties for violations should be increased. Laws covering import and export of wild plants and animals are required. Pesticide and herbicide importation and use should be carefully controlled. The capacity of the environmental agencies to enforce and publicize the laws needs to be upgraded through the expansion of the conservation warden system.

Government agencies

An environmental policy should be developed for Jamaica and selected agencies strengthened. Staffing, funding, facilities and training should be provided at a level appropriate for the responsibilities of such agencies. The use of quasi-government bodies to carry out central government functions should be avoided.

International linkages

Jamaica should treat its commitment to the UNEP Regional Coordinating Unit seriously. The country should also consider joining various conventions, including the Convention on International Trade in Endangered Species (CITES), and Ramsar. International support for wildlife conservation initiatives in Jamaica would be helpful. International agencies have an important role to play in helping the Jamaican conservation NGO's to become more effective.

Public education

There is a great need for increased public awareness of environmental issues and a corresponding lack of materials and resources. NGO's have an important role to play.

Programs which reach young people are specially important because of the age structure of the population.

Research

Locally based, long-term research on the status, distribution, and ecology of endemic species should be encouraged through the University of the West Indies. Much work is needed, particularly studies of the status, ecology and strategies for conservation, and the use of natural ecosystems and indigenous species. Long-term studies are specially important to enable current changes to be put into context. The University of the West Indies has an important function - the encouragement of environmentally feasible development. Assistance is required so that more environmental topics can be included in the curricula and research funds for conservation problems can be make available to post-graduates. Research in the past has tended to concentrate on marine resources (especially coral reefs). Where research has been carried out on terrestrial ecosystems it has tended to be species rather than ecosystem related.

Non-government organizations (NGO's)

Local and foreign NGO's have a crucial role to play in stimulating, assisting, and implementing all the bird conservation initiatives listed above. Non-government participation in conservation should be encouraged. Effective coordination of efforts is essential if Jamaica's natural environment is to be preserved.

National parks and protected areas

There are no national park areas or wildlife reserves in Jamaica. National parks and reserves, including critical habitats for endemic species, must be developed in the next five years. Their creation and maintenance will probably be most easily achieved through a non-government organization.

Literature cited

American Ornithologists' Union. 1983. Check-list of North American birds. 6th Edition. Allen Press, Lawrence, KS, 877 pp.

Bennett, G.F., H. Witt, and E.M. White. 1980. Blood parasites of some Jamaican birds. Journal of Wildlife Diseases 16 (1): 29-38.

Cook, J.M. 1984. Jamaican Amazon parrot expedition. Report to International Council for Bird Preservation, Cambridge, UK.

Cruz, A., and S. Gruber. 1981. The distribution, ecology, and breeding biology of Jamaican Amazon parrots. Pp. 103-127 in R.F. Pasquier (ed.). Conservation of New World Parrots. International Council for Bird Preservation Technical Publication No. 1, Cambridge, UK.

Fairbairn, P.W. 1981. Parrot conservation in Jamaica. Pp. 95-101 in R.F. Pasquier (ed.). Conservation of New World Parrots. International Council for Bird Preservation Technical Publication No. 1, Cambridge, UK.

Fairbairn, P.W. 1986. Conservation of Jamaican Forests with particular reference to wildlife. Pp. 111-119 in P.A. Thompson, P.K. Bretting, and M. Humphreys (eds.). Forests of Jamaica. Jamaica Society of Scientists and Technologists, Kingston, Jamaica.

_____, and R.L. Sutton. 1979. Birds as environmental indicators. Pp. 113-123 in Proceedings of the Symposium on Environmental Studies in Jamaica. University of the West Indies, Kingston, Jamaica.

Gosse Bird Club. 1986a. A Checklist of Birds in Jamaica. Gosse Bird Club, Kingston, Jamaica.

Gruber, S. 1986b. A note on birds and ganja. Jamaica Journal 19(2): 24-26.

Haynes, A.M. 1987. Human exploitation of seabirds in Jamaica. Biological Conservation 41:99-124.

Lack, D. 1976. Island Biology. Studies in Ecology. Vol. 3. Blackwell Scientific Publications, Oxford, UK. 445 pp.

Phillips, B. (comp.). 1985. The I.C.B.P. oceanic islands data base: a preliminary demonstration. International Council for Bird Preservation Study Report No. 5. Cambridge, UK., 45 pp.

Post, W., and J.W. Wiley. 1977. The Shiny Cowbird in the West Indies. Condor 79(1): 119-121.

Robbins, C.S., B.A. Dowell, D.K. Dawson, J. Colon, F. Espinosa, J. Rodriquez, R. Sutton, and T. Vargas. 1986. Comparison of neotropical winter bird populations in isolated patches versus extensive forest. U.S. Fish and Wildlife Service Report, 13 pp.

Sutton, R., A. Downer, A. Haynes, and K. Harvey. in press. Status and conservation of North American migrants in Jamaica. International Council for Bird Preservation, Cambridge, UK.

Schuchmann, K.L. 1980. Die Jamaica-Kolibris *Trochilus polytmus* und *Trochilus scitulus*. Biotropic-Verlag, Frankfurt, West Germany. 55 pp.

Table 1. Habitat use by Jamaican endemic birds.

Endemic Species	A	B	C	D	E	F	G	H	I	J
Ring-tailed Pigeon				X	X			X		3
Crested Quail-Dove			X	X	X	X		X		5
Jamaican Parakeet	X	X	X	X	X	X	X	X	X	9
Yellow-billed Parrot			X	X			X	X	X	5
Black-billed Parrot			X	X			X	X	X	5
Jamaica Lizard Cuckoo			X	X	X	X	X	X		6
Chestnut-bellied Cuckoo			X	X	X	X	X	X		6
Jamaican Owl	X	X	X	X	X	X	X	X	X	9
Jamaican Mango	X	X	X	X	X	X	X	X	X	9
Red-billed Streamertail	X	X	X	X	X	X	X	X	X	9
Black-billed Streamertail	X	X		X	X	X	X	X		7
Jamaican Tody	X	X	X	X	X	X	X	X	X	9
Jamaican Woodpecker	X	X	X	X	X	X	X	X	X	9
Jamaican Elaenia			X	X	X	X	X	X	X	7
Sad Flycatcher	X	X	X	X	X	X	X	X	X	9
Rufous-tailed Flycatcher	X	X	X	X	X	X	X	X	X	9
Jamaican Becard			X	X	X	X		X		5
Jamaican Crow			X	X			X	X		4
White-eyed Thrush			X	X	X	X		X		5
White-chinned Thrush	X	X	X	X	X	X	X	X	X	9
Jamaican Vireo	X	X	X	X	X	X	X	X	X	9
Blue Mountain Vireo				X	X	X				3
Arrow-headed Warbler			X	X	X	X	X	X		6
Orangequit	X	X	X	X	X	X	X	X	X	9
Jamaican Euphonia	X	X	X	X	X	X	X	X	X	9
Yellow-shouldered Grassquit			X	X			X	X		4
Jamaican Blackbird				X	X	X				3
TOTAL SPECIES	13	13	23	27	23	22	21	25	15	

A = Coastal Scrub, B = coastal wetlands, C = dry limestone forest, D = wet limestone forest, E = montane forest, F = el
G = lowland agriculture, H = upland agriculture, I = residential, J = total habitat use.

Table 2. Evaluation of immediate threats to the stability of Jamaica's endemic bird populations. A=Status[1], B=Agriculture[2], C=Forestry[3], D=Industry and Settlement[4], E=Hunting[5], F=Introduced Species, G=Threat Index.

English name	A	B	C	D	E	F	G
Ring-tailed Pigeon	D	4	6	1	5	0	16
Crested Quail Dove	D	7	9	4	4	0	24
Jamaican Parakeet	I	6	2	0	3	0	11
Yellow-billed Parrot	D	6	9	4	7	0	26
Black-billed Parrot	D	4	9	2	7	0	22
Jamaican Lizard Cuckoo	U	6	8	0	0	0	14
Chestnut-bellied Cuckoo	U	4	9	4	0	0	17
Jamaican Owl	S	3	3	9	0	2	19
Jamaican Mango	D	9	10	6	0	0	25
Red-billed Streamertail	S	2	0	0	0	0	2
Black-billed Streamertail	S	1	2	0	0	0	3
Jamaican Tody	S	2	2	1	0	0	5
Jamaican Woodpecker	S	9	10	7	2	3	31
Jamaican Elaenia	S	4	4	6	2	0	16
Sad Flycatcher	S	3	6	3	0	0	12
Rufous-tailed Flycatcher	D	6	10	3	0	0	19
Jamaican Becard	S	2	8	2	0	0	12
Jamaican Crow	U	7	9	1	2	0	19
White-eyed Thrush	D	4	10	2	0	0	16
White-Chinned Thrush	I	0	0	0	0	0	0
Jamaican Vireo	I	0	0	0	0	0	0
Blue Mountain Vireo	U	7	10	1	0	0	18
Arrow-headed Warbler	S	4	8	3	1	0	16
Orangequit	S	4	8	4	0	0	16
Jamaican Euphonia	U	4	9	2	0	0	15
Yellow-shouldered Grassquit	S	6	10	6	0	0	22
Jamaican Blackbird	D	10	10	4	0	0	24
TOTAL		124	180	75	33	5	

[1] Status: D=decreasing, S=stable, I=increasing, U=unknown.
[2] Agriculture: Total effect of coffee, ganja, vegetables.
[3] Forestry: Total effect of pine plantation, timber extraction, charcoal burning.
[4] Industry and Settlement: Total effect of mining and quarrying, human settlement, road construction.
[5] Hunting: Total effect of hunting for sport, commercial, subsistence, scientific, pest control.
Threats: 0=none known, 5=moderate, 10=severe.

Table 3. Relationships between distribution and status of Jamaican endemic birds.

Status of Species		Mean Habitat Index[1]	Mean Threat Index[2]
Decreasing	(8)	5.5	22
Stable	(11)	7.5	14
Increasing	(3)	9.0	4

[1] Data from Table 1.
[2] Data from Table 2.

A SUMMARY OF CONSERVATION TRENDS IN THE BAHAMAS

K. C. Jordan[1]

Introduction

The Bahamas are susceptible to the same two biological hazards that threaten the other nations of the West Indies: the introduction of exotics and the extinction of endemics. Although the Bahamas comprise more than 600 islands, by far the majority of those islands are so small that the impact of a single species can be severe. Current studies by the author (Jordan 1989) indicate a dramatic impact on plant community composition by a transplanted population of the Bahamian hutia (*Geocapromys ingrahami*), an endemic rodent folivore. The hutia was once widespread in the islands, but since the European discovery it has been driven to a single natural refugium in the remote out islands. Hutia extirpation may be almost entirely the result of predation by man and his associated cats and dogs. Two reintroductions have been performed since 1973. Data indicate that some current plant communities may be the result of succession following release from hutia browse. Small island area, low soil organic content, and the paucity or absence of fresh water on many islands all contribute to the sensitivity of Bahamian communities, even by comparison to other West Indian sites, and the alteration of single species places whole communities uniquely at risk.

The isolation of Bahamian islands from each other by water contains such perturbations of the land, and the great number of the islands provides an opportunity for the survival of representative communities of different types. That opportunity requires strong management, however, to protect local sites from disruption, both by agriculture and development, and by the tourism upon which the Bahamian economy relies.

Discussion

The most spectacular example of near extinction in the Bahamas is that of the West Indian flamingo (*Phoenicopterus ruber*). Once abundant on Acklin's, Long, Andros, Abaco, Great Inagua, and on certain sites in the Ragged Islands and Exuma Cays, the species became severely exploited during the nineteenth century for its meat and feathers (Campbell 1978). Populations were so reduced by 1905 that the first annual meeting of the National Audubon Society produced a request to the Bahamian government for flamingo protection. The Wild Birds (Protection) Act soon followed. However, exploitation continued through World War II until the species remained only on Great Inagua.

In 1951 the Society for the Protection of the Flamingo was established. The Society

---

[1] K.C. Jordan is Instructor in Biology, Daytona Beach Community College, Daytona Beach, FL 32015.

placed two wardens on Inagua, Samuel and James Nixon, to enforce sanctions against flamingo harvest. In 1959 the responsibilities of the Society were assumed by the Bahamas National Trust, created by an act of Parliament stating that:

> "The Bahamas National Trust shall be established for the purposes of promoting the permanent preservation for the benefit and enjoyment of the Commonwealth of lands and tenaments (including buildings) and submarine areas of beauty or natural or historic interest and as regards lands and submarine areas for the preservation (so far as practicable) of their natural aspect, features, and animal, plant, and marine life." (Holowesko 1982)

In May of 1963 The Trust leased 74,000 ha of prime flamingo habitat on Great Inagua from the Commonwealth of the Bahamas for 99 years. Today that island is the site of the world's largest breeding colony of *P. ruber*, estimated at close to 50,000 flamingoes. James Nixon and Samuel's son, Henry Nixon, are the wardens.

At present the Trust oversees a total of nine national parks, listed below in order of their establishment.

Date		Area(ha)
1958	The Exuma Cays Land and Sea Park	45,585.0
1963	The Inagua National Park, Great Inagua	74,335.0
1963	Union Creek Reserve, Great Inagua	1,813.0
1968	Peterson Cay, Grand Bahama	0.6
1968	Pelican Cays, Abaco	850.0
1971	Conception Island	809.0
1978	The Retreat, Nassau	4.5
1982	Lucayan National Park, Grand Bahama	16.0
1986	Black Sound Cay, Green Turtle Cay, Abaco	0.5

Located about 60 km southeast of Nassau, The Exuma Cays Land and Sea Park is 35 km long by 13 km wide, comprising dozens of small islands totalling 2,266 ha of land. The remainder of its area is under water, much of which is living reef or grassy shoal. The park thus acts as a sanctuary and showcase, not only of land communities, but also of the vast marine communities for which the Bahamas are uniquely noted.

Within the Bahamian government, the role of the Bahamas National Trust is advisory to the Ministry of Agriculture and Fisheries. The Ministry provides limited funding support, with the greater part deriving from memberships, grants, and private endowments. Though the Trust is charged with the administration of its national park system, the Trust's regulations are subject to Ministry approval. Once ratified by government, those regulations become law and are enforceable by all uniform officers of the government.

The Bahamas National Trust works closely with several North American conservation and research organizations, and encourages the development of biological programs for the Bahamas. Six of the 12 appointed members of the Trust Council are representatives of the following organizations: New York Zoological Society, University of Miami's Rosenteil School of Marine Science, National Audubon Society, U.S. National Parks

Service, American Museum of Natural History, Smithsonian Institution. The remaining six appointed members represent the following: His Excellency the Governor General (two members), Ministry of Agriculture and Fisheries, Ministry of Tourism, Ministry of Health, Ministry of Education and Culture. In addition to the 12 appointed members of Council are nine elected members. Apart from the Council are five Honorary Vice Presidents, an Executive Committee, a Special Advisory Committee, and a Staff (including wardens) to execute the regular duties of administration.

Clearly there is no shortage of scientific advice, nor of provision for its enactment in policy. Several research projects have been conducted recently or are currently in progress. Of particular interest to conservation are those addressing the status and ecology of the following species: the Bahamian hutia and the West Indian flamingo (referred to above), the Bahama parrot (*Amazona leucocephala bahamensis*), the white-crowned pigeon (*Columba leucocephala*), the West-Indian tree duck (*Dendrocygna arborea*), the Bahama pintail (*Anas bahamensis*), Kirtland's warbler (*Dendroica kirtlandii*), the sand tile fish (*Malacanthus palumieri*), the three species and seven subspecies of Bahamian *Cyclura* iguana, four species of marine turtles, the commercially important queen conch (*Strombus gigas*) and crawfish or spiny lobster (*Panulirus argus*), and a newly discovered (Yager 1981) class of marine crustacean (Remipedia: *Speleonectes lucayensis*).

Although the conservation ethic is historically well-established in the Bahamas, it is not unthreatened. The importance of tourism to the national economy places a high premium on the visibility of native assets. But conflicts can arise when the development of tourist facilities, such as those on Nassau's Paradise Island, displace natural communities, or when local agriculture or commerce threatens resource depletion or habitat destruction.

Development is proceeding, although not as rapidly as in other West Indian nations. Part of the reason may be the high cost of material and personnel transport among the islands, which discourages both construction and maintenance.

Agriculture is still an important part of the national economy, and the fertile island of Eleuthera is known for its pineapples, tomatoes, and other produce. Poultry is by far the principal livestock, and operations are found on all of the larger islands. The Ministry of Agriculture takes an active part in advising and regulating all phases of Bahamian agriculture.

The native plants include only a few species of commercial importance. Mahogany (*Swietenia mahagoni*) has been the most valued historically, and consequently has been harvested to low densities and small sizes in the wild. Lignum vitae (*Guaiacum* sp.) has had more limited consumption, and several tree species have been used in bush medicine and the production of commercial dyes (Correll and Correll 1982). The pine forests of the larger northern islands, such as Abaco, have been harvested periodically for pulpwood, but most of the lumber used in the Bahamas is imported from the United States.

Although Byrne (1980) concluded that the flora of the Bahamas is generally both resistant to intrusion by exotics and resilient to moderate levels of human activity, Correll and Correll (1982) cautioned that modern agriculture and real estate development have favored the weedier species, extinguishing the less competitive orchids and ferns from the coppices of many islands. The most successful exotic has been the Australian pine

(*Casuarina litorea*), which tends to form pure stands under which little else grows (Correll and Correll 1982).

The greatest threats to Bahamian patrimony are those imposed by local commerce and cruising yachtsmen on marine resources. The chief problems include poaching, bleach fishing, longlining, the use of "hooka" diving gear, the harvest of spawning grouper and snapper, the illegal export of crawfish and conch to the United States, and the difficulty of law enforcement on open water.

The conch fields of Nassau were exhausted 20 years ago, and every month the power boats and sailing smacks go farther onto the banks in search of adult conch. Many conch fishermen are settling for juveniles, and even the very young "rollers" are appearing on the tables of Nassau.

The Nassau grouper (*Epinephelus striatus*) is still the prime meat fish of the Bahamas, despite reduced populations.

Although the crawfish is protected by a closed season, many foreign yachtsmen take large numbers regardless of season.

As a member of CITES, the Bahamas are obliged to protect all species of marine turtles, but they are still valued in the tourist trade for their meat, shells, and shell products.

Although the Bahamas have no national zoo or natural history museum, one private zoo is located in Nassau. The Ardastra Gardens are a two-hectare botanical and zoological park constructed in the 1950's with emphasis on the endemic West Indian flora. But the greatest attraction to the Gardens is a flock of over 70 West Indian flamingoes. Several semi-captive West Indian tree ducks have raised young within the gardens, and an exhibit of the *Cyclura* iguanas has just been completed.

The diffuse nature of the Bahama Islands, even in a shallow sea, is no doubt responsible for the survival of several endemic species. With over 600 islands scattered over 50,000 square miles of ocean, the Bahamas can lose many individual islands to human influence and still retain hundreds to represent its natural systems. And many of those may be of little use other than as biological exhibits, or as natural laboratories for the evaluation of life history strategies, species interactions, and the causes of extinction. The greater peril is to the Bahamian reefs and shoals, and to their dependent biotic wealth that has so distinguished them throughout history. In 1985, the Bahamian government passed a law prohibiting the removal of any life form by any means from its marine parks. The Exuma Cays Land and Sea Park Warden patrols by boat and maintains radio contact with the Royal Bahamas Defense Force, a local seaplane pilot, and a growing community of sympathetic boaters. Such action and support are good cause for hope.

Acknowledgements

The author wishes to thank Mr. Gary Larson, Executive Director of the Bahamas National Trust, for his help in reviewing the manuscript.

Literature cited

Byrne, R. 1980. Man and the variable vulnerability of island life. Atoll Research Bulletin No. 240 200 pp.

Campbell, D.G. 1978. The ephemeral islands, a natural history of the Bahamas. Macmillan Education Ltd., London, 151 pp.

Correll, D.S., and H.B. Correll. 1982. Flora of the Bahama Archipelago. J. Cramer, Vaduz, 1692 pp.

Holowesko, L.P. 1982. President's report to the annual general meeting of the Bahamas National Trust. Pp. 1-14 in Bahamas National Trust Annual Report (1982), Bahamas National Trust, Nassau.

Jordan, K.C. 1989. An ecology of the Bahamian hutia (*Geocapromys ingrahami*). Ph.D. dissertation, Univ. of Fla., 184pp.

Yager, J. 1981. Remipedia, a new class of Crustacea from a marine cave in the Bahamas. Journal of Crustacean Biology 1(3):328-333.

A SUMMARY OF CONSERVATION TRENDS IN THE DOMINCAN REPUBLIC

José Alberto Ottenwalder[1]

Introduction

The Dominican Republic, the second largest West Indian state (after Cuba), shares the island of Hispaniola with Haiti. Occupying the eastern portion, the country covers approximately two-thirds (48,442 km^2, with 1,575 km of coast line) of Hispaniola (77,914 km^2).

Despite its relatively small area, the Dominican Republic exhibits considerable heterogeneity and variability in local climatic regimes. This allows a variety of life zones and plant communities to exist, ranging from dry thorn scrub to montane pine forest. The highest peaks in the Caribbean archipelago lie within the Dominican Republic, as does the lowest topographic depression in the Caribbean, 40 m below sea level at Lago Enriquillo in the Neiba Valley. As a result Hispaniola is paramount among West Indian islands in the number of indigenous wildlife species, and is considered the center of diversity of a number of plant and animal groups. Its ecological diversity is reflected by its floristic richness, certainly the richest in the Antilles with 5,600 plant species, 36 percent of which are endemic.

Discussion

Protected Areas

At present, the Dominican Republic has nine National Parks and five Scientific Reserves, which represent an approximate surface area of 5,250 km^2 (Table 1). Therefore, about ten percent of the total area of the country is classified as protected natural areas. These areas include a diverse range of lowland, montane and wetland habitats. Protected habitats include: 1) the extremely arid valleys and coastal area of the southwestern portion; 2) pluvial forests in the mountains and the lowlands of the northeast; and 3) fresh, brackish, lacustrine, fluvial, estuarine and coastal marine environments, including two offshore islands.

Recent additions to the list of protected areas include two new marine sites: 1) Parque Marino la Caleta, about 25 km east of Santo Domingo; and 2) The Humpback Whale Sanctuary of Silver Bank, about 80 km north of Puerto Plata. The Silver Bank Sanctuary was created in October of 1986.

---

[1] José Ottenwalder is Director of Research at Parque Zoologico Nacional (ZooDom), P.O. Box 2449, Santo Domingo, Dominican Republic.

Recent inventories of forest resources indicate that only 14 percent of the country remains covered by natural moist broadleaf forests. Pine forests cover about seven percent of the Dominican Republic. Pressure is high on the remaining undisturbed forest resources of the country found inside the boundaries of the nine national parks.

Most native vertebrate species are represented inside the Dominican national parks and reserves. According to an analysis of the recent fossil record, a number of native species of mammals, birds, and reptiles have already become extinct. At present, a number of additional native plant and animal species are either endangered or threatened. A conservative figure for plants indicates that 277 species are endangered or threatened.

Problems and major concerns

One of the major problems hampering the implementation of management plans and conservation programs for native wildlife and protected areas is the lack of knowledge concerning the status of wildlife species, and the ecological processes that regulate ecosystems. Comprehensive faunal and floral inventories have not yet been conducted for most parks. Quantitative data about densities or relative abundance are not available for most species. Nationwide, data on current use of wildlife resources, is inadequate. Faunal resources are exploited mainly for food. Subsistence harvesting of terrestrial and aquatic animals is common, including protected or out-of-season wildlife. The amount of wildlife harvested is unknown, and the use of illegal harvesting methods is a common practice. The enforcement of wildlife regulations is increasing, and some control over exploitation has been established, although it is still deficient in many cases. The most effective enforcement has been achieved only inside the national parks, and even there its success is a function of the manpower available.

Socio-economic development vs conservation

The greatest negative impact upon the native vertebrates of the Dominican Republic is through environmental degradation. Human exploitation (for food, trade, malicious killing), and the effect of exotic species are also significant negative factors.

The Dominican Republic is predominantly an agricultural country, and agricultural produce in some years accounts for up to 79 percent of its exports. Sugar is the most important crop (40-50% of export by value), followed by coffee, cocoa, bananas, and tobacco. The agriculture-livestock-forestry-fishing sectors accounted for 17.3 percent ($529 million) of the total productivity of the country in 1982.

Dominant land use patterns in the Dominican Republic reflect the influence of agricultural activities in the economic infrastructure of the country. Exploitation of the natural habitats for shifting agriculture, wood and charcoal have caused considerable deforestation, erosion, and the degradation of water resources. Approximately 45 percent of Dominican houses depend on firewood and charcoal for energy. Charcoal consumption in 1985 was estimated to have been 582 million bags (45 kg/bag). Deforestation rates were recently estimated as 100,000 ha/year, of which only 20,000 ha are left for regeneration.

The small farmer is blamed for causing most environmental problems, both inside some national parks, and nationwide. The causes are: 1) rapidly expanding human population (therefore increased demands for food and energy to be produced locally); 2) lack of access to productive land; and 3) failure in the service infrastructure.

Evidently, economic development is an important component that must be taken into serious consideration for any local conservation strategy to have some success. At present, the Dominican Republic is implementing aggressive (though not always rational) development programs. The relationship between natural resources and population is always a controversial issue. This relationship is further complicated by socio-political factors (eg. frequently the government deliberately avoids facing the delicate situation of peasant invasions of lands inside national parks). Increasing and high population density, inequitable distribution of land services, highly variable land capability (and the resultant poverty and low productivity of small farmers), and ultimately environmental degradation are in turn fed back into a system where natural resources are lost forever. The population/resource issue (carrying capacity) is the basic component of appropriate management and conservation strategies.

The annual rate of population increase in the Dominican Republic is 2.8 percent. In spite of this, the average human population density for the whole country in still relatively low ($135/km^2$ calculated from population statistics projected for 1986). However a recent analysis of the Dominican macroeconomy by the World Bank indicates that the country is confronting the worst economic crisis in its recent history. The report concludes that, unless major policy initiatives are taken, growth in the second half of the 1980's will not recover to the high rate of growth experienced in the early 1970's. The present government has taken this challenge very seriously; that is, to increase production, particularly agricultural productivity.

In spite of the difficult economic crisis it inherited, the current administration has demonstrated itself to be aware of the long term importance of the balance between agricultural productivity and the maintenance of natural habitats. In fact, a number of significant developments in conservation of natural resources have been recently achieved in the Dominican Republic with support and encouragement of the President.

1) Environmental and wildlife regulations have been strengthened. The best example is the launching of Operación Selva Negra in August 1986 to enforce the already existing forest regulations. Tree cutting, or any other form of forest exploitation in the Dominican Republic, is forbidden. Permits to cut trees are required even for private landowners. Sawmills were declared illegal and were closed down. Operaciónón Selva Negra is carried out daily with reconnaissance planes, helicopters, and military personnel of the Dominican Air Force. The units land wherever clearing of the forest (fires, clear cutting) is detected.

2) Ten energy farms have been created throughout the country. These farms were selected on the basis of geographic location, local energy demand, forest community type, life zone etc, and are designed to help provide for the energy needs of the country.

3) Effective in January 1987, a five-year ban was passed on the capture, killing, mutilation, or any other form of exploitation of eggs, nest, feathers, parts, etc. of all native wild vertebrates in the territory of the Dominican Republic. Excluded from this protection are pest species and feral carnivores.

4) The importation and introduction of exotic freshwater species is now regulated.

5) The following regulations concerning fisheries resources were legislated during 1986:

a) the capture, possession and commercialization of the heavily exploited marine mollusks *Strombus*, *Cassis*, and *Cittarium* are regulated, and a minimum size has been established;

b) additional regulations for sea turtles, and minimum sizes have been established;

c) a seasonal ban and minimum size has been established for the exploitation of native freshwater turtles;

d) a new seasonal ban has been established for the exploitation of all commercial lobsters and crabs species, and a minimum size has been established.

e) Exploitation and commercialization of coral reefs is prohibited.

Environmental education

There are a number of government and non-government agencies (universities, conservation groups, etc.) carrying out environmental education programs, some of which receive international support and cooperation from organizations such as the World Wildlife Fund. Some of these programs are particularly oriented for the training of primary and secondary level school teachers, and on methods of environmental education.

Courses (and related activities) on environmental education and the conservation of natural resources for both urban and rural schools are being proposed for inclusion into the official national education curriculum.

Proposed conservation strategies and management plans

Conservation planning should be integrated with developmental planning as provided by the World Conservation Strategy. Priorities should be identified to insure the greatest sustainable benefit to present generations while maintaining the potential for future generations. The emphasis should be on the management of the human use of resources rather than on management of the resource itself. In this context, the approach should be on the preservation, maintenance, sustainable utilization, restoration and enhancement of the resource. Government planners and administrators should be able to predict the consequences of development, and establish priorities for conservation.

Management plans must take into account not only the distribution and abundance of wild animals, but also the interactions among them in the ecosystem, and the role that each species plays within the system. Therefore, inside of national parks and protected areas there should be a minimum amount of manipulation, and natural ecological process should be allowed to continue. On the other hand, outside of protected areas systems should be managed. Effective watershed, soil, water, and forest management can facilitate the preservation of endemic species, the maintenance of natural biotic diversity, and the protection of natural ecological processes. Adequate "management" of critical habitats will result in the positive management of native plant and animal communities characterizing these habitats.

Table 1. Protected natural areas in the Dominican Republic

Category	Area[1] (km^2)	Year Created
National Parks		
Jaraqua	900	1983
Bahoruco	700	1983
J. Armando Bermudez	766	1956
J. del Carmen Ramirez	764	1958
Monticristi	550	1983
del Este	434	1975
Los Haitises	530	1976
Isla Cabritos	26	1974
Cabo Frances Viejo[2]	1.25	1974
Scientific Reserves		
Valle Nuevo	409	1983
Lagunas Limon y Redonda	101	1983
Laguna de Rincon	47	1983
Isabel de Torres	22	1983
Villa Elisa (La Cacatica)	0.14	1975
Total Area	5,250.4	

[1] Marine areas excluded in size estimates.
[2] Promulgated as a National Park in the legislation, but actually should be in the category of Scientific Reserve.

A SUMMARY OF CONSERVATION TRENDS IN PUERTO RICO

Peter R. Ortiz[1]

Introduction

Puerto Rico is the smallest and the easternmost of the Greater Antilles. It has a surface area of only 8,897 km², and is situated between 18°31' and 17°55' N latitude, and 65°37' and 67°16' W longitude. By the turn of the century the island had lost over 80 percent of its forest cover. Agricultural activities in the early 1900s were high and the pressure on the usable land of Puerto Rico was at an all time high. The only remaining forested areas were located on mountain peaks, and several coastal forests composed mainly of mangroves.

Under the Treaty of Paris, signed on the tenth of December 1898, all land belonging to the Spanish Crown was surrendered to the United States. In 1902, the Congress authorized the President to reserve from such lands those considered necessary for public or national purposes. Thus, the first forest reserve originated with the Act of July 1, 1902 whereby the Luquillo Forest Reserve, presently known as the Caribbean National Forest, was established by President Roosevelt. A year later all lands not previously reserved by the President were transferred automatically to the use of the people of Puerto Rico.

On November 22, 1917, the First Section of the Ninth Legislature Act authorized the creation and development of a forest system for the protection of watersheds and wood production, to be owned and managed for the benefit of the people of Puerto Rico. The "Forest Law" and the Forest Service within the Department of Agriculture were hereby created. The Forest Law of 1917 provided for the creation of "insular forests" by decrees issued by the Governor of Puerto Rico. By 1920 seven forest units constituted the Insular Forest System, including five coastal areas, an island, and a highland area. In 1921 the law was improved making it possible to enlarge forest units by purchase, donation, legacy, and (if necessary) by dispossession.

The President's Executive Order No. 7057 created the Administration for the Reconstruction of Puerto Rico in 1935. Among its functions was land acquisition for wood production and watershed protection. This resulted in the creation of a forest division and a land acquisition program. These programs, and through land ownership transfer, 8,892 ha of additional forest land were acquired. These lands were divided into five new insular forest units.

In 1951 and 1952 the Puerto Rico Land Authority transferred 1,011 additional ha, and created two other forest units. In 1962 another forest reserve was created when 2,797 ha were acquired from the Federal Government through permutation, donation, and transfer.

---

[1] Puerto Rico Natural Heritage Program, Department of Natural Resources, Box 5887, Puerta de Tierra, PR 00906.

The aforementioned events explain how the Forest System of the Commonwealth of Puerto Rico was initiated. The system is presently composed of fourteen forest reserve units, with a total area of roughly 22,628 ha. Additional public lands in forest bring the total of Puerto Rico in forests to 33,988 hectares.

All the land set aside as reserves remained practically unattended for years, and no surveillance, management plans, or real protection was initiated on the land because of the lack of a proper governmental structure to deal with the conservation of natural resources. The beginning of real conservation efforts in Puerto Rico dates from the initiation of measures for the creation of governmental structures designed to administer and evaluate conservation programs and lands on the island.

Discussion

Most of the events of conservation importance in Puerto Rico occurred during the 1970's. It should be noted, however, that these were the result of actions that were initiated previously

1970.-- The House of Representatives passed Law No. 9 on June 18, 1970 establishing the Commonwealth Environmental Quality Board. This agency was created to establish and administer a permit system and formulate regulations concerned with the preservation of the environment and the conservation of natural resources.

1972.-- The House of Representatives Law No. 23 on June 20, 1972 created the Department of Natural Resources. This agency was established to create and administrate programs for the sustained use and conservation of the natural resources of Puerto Rico.

1976.-- As a product of the forest planning process initiated by a resolution of the House of Representatives in 1973, the Master Plan for the Commonwealth Forests of Puerto Rico was prepared in October of 1976. The absence of management plans had previously resulted in the segmentation of contiguous forest areas, inappropriate land use within forest boundaries, and the subsequent endangerment of wildlife species and recreational opportunities.

1977.-- The Puerto Rico House of Representatives by virtue of Law No. 1 created the Ranger Corps of the Department of Natural Resources on June 29, 1977. The objective was to give the Department of Natural Resources the mechanism to enforce the laws it administers.

The adequate protection of the natural resources heritage is also promoted by the enforcement of the following laws: P.R. Wildlife Act (Law No. 70); Puerto Rico Forest Act (Law No. 133); Puerto Rico Fishing Act (Law No. 83); Puerto Rico Water Conservation, Development and Use Act (Law No. 136); P. R. Flood Preservation, Beach and River Conservation Act (Law No. 6); Bylaws to regulate the extraction of materials from earths crust (Law No. 144); and others.

1978.-- The Puerto Rico Coastal Management Program submitted in October of 1978. Principal elements of the program are: the guidance of development on public and private property; the active management of coastal resources involving the improvement of resource management (i.e. building up field services and facilities, organizational changes, establishing a system of natural reserves, other measures to protect coastal resources, etc.); and the promotion of coastal development and research.

1983.-- In May of this year an agreement between the Puerto Rico Department of Natural Resources, the Conservation Trust of Puerto Rico, and the Nature Conservancy Program created the Puerto Rico Natural Heritage Program (Conservation Data Center). The objectives of this new program were the establishment of a permanent process of biological inventory, and the creation of a process to monitor the status of ecosystems and species in danger of extinction. Its immediate responsibilities were: to act as a tool for environmental assessment; to direct development away from ecologically sensitive areas; and to identify areas of priority throughout the island in need of conservation action.

Other institutional support

Additional conservation support and efforts through the years have come from federal agencies with jurisdiction on environmental matters in Puerto Rico . Their support has basically come through research, resources management, and law enforcement. The agencies involved have been the U.S. Forest Service (i.e. Institute of Tropical Forestry), U.S. Fish and Wildlife Service (i.e. Wildlife Refuge System), the National Park Service, and the U.S. Corps of Engineers.

Other support has come from non-governmental organizations. The Conservation Trust of Puerto Rico has made noteworthy contributions through land acquisition and protection. Although other private conservation organizations (i.e. Natural History Society, Puerto Rico Ornithological Society, Fondo de Mejoramiento) exist and have supported conservation issues through the years, their development has been very slow.

Present conservation trends

In 1987 52,861 ha of the land area of Puerto Rico is receiving some form of protection. This acreage includes the fourteen units of the Commonwealth's forest reserve system; the eight units of the Puerto Rico Conservation Trust reserve system; the fifteen natural reserves designated by the Commonwealth Planning Board; the wildlife refuges; a federal forest reserve; and a marine sanctuary. Of these forty-two units, only seven are actively managed due to limited financial and human resources. All areas are administered respectively by the institutions in custody: The Puerto Rico Department of Natural Resources, the Conservation Trust, the U.S. Forest Service, and the U.S. Fish and Wildlife Service.

Additional areas are still under consideration. Eleven proposed natural reserves are awaiting designation. A group of additional areas of ecological importance have been identified through the methodology of the Puerto Rico Natural Heritage Program based on the presence of critical floristic and faunistic contents. These areas are currently being described, their biotic resources characterized, and their conservation priority established by the Natural Heritage Program.

Most conservation efforts in the past were carried out through governmental attempts to protect the natural resources of the island. Although some public awareness has developed, support of conservation issues by most residents of Puerto Rico is not strong. Progress in conservation cannot be only measured in terms of how much land is protected. In Puerto Rico it must also be measured in terms of the availability of government created mechanisms to assess and direct development activities, and the enforcement of "unequivocal" policies assuring proper land use. Although the mechanisms exist, public policies regarding adequate use of natural resources are not clear. This has resulted in unnecessary loss of land to "short-sighted" development activities, which are much too common on the island. Such activities have found some public support because of their attempts "to promote economic growth". Unfortunately, more compatible alternatives of development have not been studied.

Conscientious conservation programs will entail much more than the protection of natural resources through "reserves". The conflict between high economic interests, and resource protection and planning must be carefully assessed and resolved. The recognition that in protecting biotic resources (i.e. diversity) today, the future of mankind is improved, must not be forgotten.

Conservation is not an obstacle to progress. It is the guarantee that men will have what is necessary to meet the demands of the future. It is the discovery of strategies that integrate the satisfaction of human needs with development activities that guarantee the permanence of our natural heritage, for this and future generations.

Let this, be our challenge!

Literature cited

Department of Natural Resources. 1976. The Master plan for the Commonwealth forests of Puerto Rico. 259 pp.
Anonymous. 1978a. The public forests of Puerto Rico. Office of the Forest Service, Area of Resources Administration. 22 pp.
Anonymous. 1978b. Puerto Rico Coastal Management Program. National Oceanic and Atmospheric Administration: Office of Coastal Zone Management; Puerto Rico, Department of Natural Resources and the Puerto Rico Planning Board, 176 pp.
Law No. 1. 1977. The Natural Resources Rangers of the Department of Natural Resources Act.
Law No. 9. 1970. The Environmental Public Policy Act.
Law No. 23. 1972. The Organic Act of the Department of Natural Resources.

Commonwealth Owned Forests of Puerto Rico: 1) **Ceiba** (143 ha); 2) **Pinones** (613 ha); 3) **Carite** (2,617 ha); 4) **Aguirre** (968 ha); 5) **Vega** (448 ha); 6) **Toro Negro** (2,730 ha); 7) **Cambalache** (373); 8) **Rio Abajo** (2,272 ha); 9) **Monte Guilarte** (1,417 ha); 10) **Guanica** (3,882 ha); 11) **Susua** (1,313 ha); 12) **Maricao** (4,154 ha); 13) **Guajataca** (926 ha); **Boqueron** (802 ha). Federal Owned Forests of Puerto Rico: 15) **Caribbean National Forest** (11,330 ha).

Figure 1. Forests of Puerto Rico.

CONSERVATION STRATEGIES AND THE PRESERVATION OF BIOLOGICAL DIVERSITY IN HAITI

Paul Paryski, Charles A. Woods, and Florence Sergile[1]

Abstract

Haiti is one of the most environmentally degraded countries in the world, and faces serious economic and social problems. Haiti is also one of the most biologically significant countries of the West Indies. Many endemic plants and animals, some of which are confined to Haiti, occur on the island of Hispaniola. Developing conservation strategies in Haiti, therefore, is both challenging and important. The most effective conservation strategy in a country with so many socio-economic problems, and with so much habitat already seriously degraded, is to link conservation with issues of high national priority and sustainable economic development.

In Haiti conservation has advanced in the past decade in spite of many other problems by associating conservation with major environmental issues such as watershed management and soil conservation. Two significant national parks have been established with the continued support of the government partly because increasing numbers of people in Haiti have become aware of the relationship of forested montane regions to water resources in surrounding areas. National parks and conservation have also been justified in Haiti because of the tourist value of its flora and fauna, and an effort is now underway to link historical patrimony with natural patrimony. The proof that the association of conservation with watershed management, sustainable economic development, and historical patrimony is effective is demonstrated by the establishment of new national parks at a time when economic and political problems have dominated the country. As environmental degradation and species extinction continue at alarming rates throughout the world, Haiti can serve as an example of the possibility for the development of effective conservation strategies in the face of difficult socio-economic hardships.

---

[1] Paul Paryski is Field Manager for the Florida State Museum National Parks Project in Haiti and former National Parks Project Director for ISPAN in Haiti. Charles Woods is Curator of Mammals at the Florida State Museum, and Project Director of the University of Florida National Parks of Haiti Project. Florence Sergile is a graduate student in Conservation Studies at the University of Florida, and an employee of MARNE in Haiti. The address of all three authors is Florida Museum of Natural History, University of Florida, Gainesville, FL. 32611.

Introduction

The study of the biogeography depends on the presence of a representative sample of the natural flora and fauna of a region. Since many modern systematic analyses rely on biochemical or even behavioral data, living examples of island endemics are especially important in seeking to reconstruct the phylogenetic history of West Indian plants and animals. In certain groups, however, especially in mammals, the rate of extinction has been so high that few species still exist, and entire families, subfamilies, genera and even orders have become extinct. Many surviving endemic species are threatened or endangered. Without active conservation efforts on all of the islands of the West Indies many more species will become extinct.

The struggle to preserve biological diversity is underway throughout the West Indies. Many of the best programs are located in the Lesser Antilles where the Eastern Caribbean Natural Areas Management Program is active, and population pressures are often less severe than in the islands of the Greater Antilles. In this volume conservation trends within the Greater Antilles are summarized for the Bahamas (Jordan), Dominican Republic (Ottenwalder), Puerto Rico (Ortiz), and Jamaica (Haynes and Harvey). The struggle for conservation is more difficult, and the extent of the ecological damage is more widespread in Haiti than in any other country or island of the West Indies. The contrast between the status of the flora and fauna of Haiti and the Dominican Republic, both of which share the island of Hispaniola, provides distressing evidence of the damaging effects of long and uncontrolled over-exploitation of natural resources in Haiti. While the Dominican Republic struggles to accommodate both growth and conservation, and supports a large national park program (Ottenwalder this volume), Haiti is only now beginning to face the important ecological and environmental issues that have been ignored far too long, and which have now reached crisis levels.

The status of the flora and fauna of Haiti is an environmental bellwether for the entire West Indian region. The situation now common in Haiti will be found everywhere in the region in future decades without active conservation efforts on the part of each country. Therefore, we believe that it is worthwhile to analyze the efforts to preserve what is left of Haiti's natural patrimony in order to understand what can and should be done in the region's most damaged ecosystem. If conservation can succeed in Haiti, it can succeed anywhere in the West Indies. Indeed, if conservation is going to succeed in the West Indies, it must succeed in Haiti since Hispaniola, and especially the southern peninsula of Haiti, harbors so many of the remaining endemic species of the West Indies.

Haiti is one of the most unique countries in the world. Due to the geologic history of Hispaniola (having been formed of several paleoislands), and the long isolation of the island in a region where sea levels and climates have changed markedly, Haiti is rich in endemic plants and animals. Haiti has been often called the "Perle des Antilles". It is populated by the descendants of African slaves who fought for and gained their independence from France in 1804 after a long and bloody struggle. The army was paid in part by the distribution of land, thus beginning a process by which peasant farmers spread throughout remote regions of the country. As a result of this process, and the accompanying destruction of the country's natural resources, Haiti is now classified by many as an environmental disaster area.

"Few countries in the world face a more serious threat to their own survival from environmental catastrophe than Haiti. Overpopulated, its resources over-exploited and trends towards further environmental deterioration everywhere, the country should brace itself for widespread famine, social upheaval and the potential breakdown of its socio-economic and political structures." (USAID 1987)

Due to the political unrest and economic decline that followed the departure of the Duvalier government in 1986, the provisional military governments were not able to effectively control the continuing environmental degradation. In fact, degradation and deforestation rates have significantly increased since Duvalier's departure. A decline in the national economy, and a lack of effective governmental control during the administration of Haiti by transitional governments, and a decrease in international financial support for the Government of Haiti (especially from the United States) following the "election" of President Leslie Manigat have resulted in an increase in the rate of deforestation and environmental degradation.

Although time is running out for conservation efforts in Haiti, and most of the country's once lush mountain cloud forests have long since disappeared, it may still be possible to conserve what remains of Haiti's unique natural heritage. Two new national parks have been established in the remote and still forested areas of Haiti's highest mountain ranges in order to protect the country's natural heritage and fragile watersheds. USAID has funded studies that document the extent of environmental degradation and suggest positive steps to solve the problem, including a multi-million dollar project to protect the entire Massif de la Hotte watershed, an area that contains one of the new parks and many of Haiti's rarest endemic plants and animals. Other donor organizations, both private and governmental, are developing projects to save and protect the environment. A new National Conservation Strategy has been drawn up by the Haitian Government. And many of the newly formed political parties are including environmental concerns in their programs. The main reason for these conservation efforts is not to preserve biological diversity, but rather the increasing recognition of the relationship between forest cover and water conservation in Haiti's high mountains. The conservation efforts in the Massif de la Hotte, for example, which started as an effort to set aside Parc National Pic Macaya (the site of origin of four major rivers in southwestern Haiti), has now been expanded to include the establishment of a biosphere reserve in the whole area.

Discussion

Physiogeography of Haiti

Haiti occupies the western third of Hispaniola, covering 27,700 km^2 of the island's 76,193 km^2 (Fig. 1). The prevailing winds are from the northeast, and much of Haiti is in the rain shadow of the high mountains of the Dominican Republic. Pico Duarte (3,175 m), the highest point in the Antilles, is one of three peaks over 3,000 m and 22 over 2,000 m in the Dominican Republic's Cordillera Central which shield much of Haiti from the moist Trade Winds blowing from the east.

Mountains cover much of the land area of Haiti. Over eighty percent of the terrain has slopes in excess of twenty-five percent, and a number of peaks in Haiti are over 2,000 m. The highest point in Haiti is Pic la Selle (2,674 m) in the Massif de la Selle. A series of low ranges, some rather isolated from one another, make up the southern peninsula of

Haiti. The towering Massif de la Hotte mountain range of southwestern Haiti arises 165 km west of the termination of the Massif de la Selle. The steep sided slopes of the Massif de la Hotte include the second tallest peak in Haiti, Pic Macaya (2,347 m).

The physiographic complexity of Haiti creates a heterogeneous terrain and causes much variation in local climatic regimes. While abundant rainfall (2,000 mm/yr) occurs in the Massif du Nord west of Cap Haitien, the Montagnes Noires of central Haiti, and the Massif de la Selle in the south, these areas are surrounded by arid zones of less than 750 mm/yr rainfall. The most abundant rainfall in Haiti occurs in the Massif de la Hotte where over 3,000 mm of rain falls in a broad belt extending 35 km east and west from Pic Macaya. This abundant precipitation supplies water for the important agricultural regions surrounding the cities of Les Cayes and Jérémie, and nurtures a mesic montane forest rich in endemic plants and animals.

Forest cover in Haiti

The patterns of rainfall and soils in Haiti are capable of supporting a mesophytic mixture of plant life. This is well documented in the work of L.R. Holdridge, who developed his ideas of plant zones while working on a doctoral dissertation investigating the pine forests of Haiti (Holdridge 1947). Holdridge described the plant associations based on humidity, precipitation, potential evapotranspiration, altitude, and temperature. Taking into account all of these factors, forests should occupy 55 percent of the land area of Haiti, and all of the mountainous regions of the country should be well forested. However, the exploitation of these forests began soon after Amerindians arrived on the island approximately 7,000 yBP. At the time Columbus discovered Hispaniola in 1492, it is believed that over one million Indians were present on the island. Most of these early inhabitants lived in lowland areas. As the search for gold in Hispaniola declined, the Spanish lost interest in the island and in 1697 ceded the part that is now Haiti to the French by the Treaty of Ryswick.

In the early 1700's the exploitation of the forest resources of the new French colony of "Saint Domingue" accelerated as whole mountain ranges were deforested for the valuable forest hardwoods, and land was cleared for plantations and other forms of agriculture. In the 1790's the bloody and destructive battle for independence raged throughout the country. In 1804 the victorious slaves declared their independence from France and created the new Republic of Haiti. Small farms were created in many areas of the country as vast public land holdings were distributed to the people. In the search for new farm sites, peasants pushed farther and farther into the mountains, and the exploitation of forest resources of the country continued. Part of the legacy of the revolution was a large debt to France, and the cutting and export of valuable hardwoods helped pay the newly independent country's debt to France. In 1845 alone, 18,600 cubic meters of mahogany were exported from Haiti. As the cities grew, the demand for fuel wood, charcoal and construction materials increased. A law requiring land holdings to be divided among the surviving sons led to increased fragmentation of the land, and over-exploitation of the soil.

As soil productivity decreased, peasants and larger land holders turned to the exploitation of forest trees, mostly in dry areas, for the production of charcoal as a resource. People pushed ever farther into the mountains in search of forest resources and available land. Public lands in the mountains became occupied and exploited by generations of peasants who now act as though the land belongs to them, and who have no alternate place to live or work. When the peasants cleared off hillside montane wet forest, two or three good harvests were obtained but this was followed by severe erosion and, eventually, irrevocable degradation of the land. The result of many generations of land abuse in Haiti is an almost completely deforested countryside and a tradition of land stewardship by which most peasants expect the land to be exploited. We have interviewed some peasants who are proud of having cleared all of the trees off of their land, and who do not see the economic value of many trees, since most do not produce an annual crop. Even those peasants who are aware of the value of maintaining forest cover are forced to cut trees by the economic pressures of daily survival; they cannot afford to think of tomorrow. The only trees that exist in many regions are secondary growth scrub used to make charcoal, and fruit trees such orange, avocado, and breadfruit. By the spring of 1988 conditions in Haiti had become so desperate that stacks of newly cut mango trees were beginning to line the road between Port-au-Prince and Les Cayes, and recently cut giant mango trees could be seen next to houses near the highway. A tradition of land abuse is now firmly established. Peasant life is almost totally dependant on cutting trees and other inappropriate exploitation of the natural resource base. The very existence of the peasants depends heavily on cutting trees not only for construction, for fuelwood and charcoal, but for clearing new agricultural land to replace that rendered unproductive by erosion and overuse. In fact, over 6,000 ha of arable land is lost each year because of erosion. Where perennial streams and rivers once flowed, seasonal torrents ravage farms and towns during rainstorms, and become dry ravines during much of the year. In Anse-à-Galet on Ile de la Gonâve west of Port-au-Prince, for example, extreme deforestation of the nearby hills and ravines has exposed the town to the hazards of flash floods. Heavy rains in November of 1987 caused such severe flooding that the town was isolated into three separate parts by floodwaters that carved 7-10 meter deep gullies, and washed people and houses out to sea.

No reliable data exist on the extent of the original forest cover of Haiti. The "beautiful forests of pine mingled with areas of rain forest jungle" described by Wetmore and Swales (1931) and documented by Ekman (1926, 1928, field notes) are now mostly gone. The estimated forest cover in the country as a whole was down to seven percent in the 1950s, with much of this described as a mixture of degraded hardwoods and a few pines (Burns 1954). The only analysis to quantify the extent of forest cover and habitat loss is by Cohen (1984), who documented the extent of forest cover at three sites in two regions of the country by analyzing aerial photographs taken in 1956, 1978, and 1984. This study produced striking results (Fig. 2). In southwestern Haiti Cohen analyzed data from a transect across the eastern foothills of the Massif de la Hotte (Camp Perrin to Roseaux), as well as a 5,000 ha block surrounding Pic Macaya in the center of the la Hotte mountains. The natural vegetation in this region of complex topography and abundant rainfall should include six types of mesic forest based on the Holdridge (1947) system. Species profiles for montane plant communities of southern Haiti are listed in Holdridge (1947) and Judd (1986, 1987). Cohen (Fig. 2) found that only 26.2 percent of the Camp Perrin to Roseaux study site was virgin forest in 1956, and 44.2 percent of the land was covered by forests that were not seriously degraded. By 1978 the amount of virgin forest cover had declined to 2.4 percent and only 17.6 percent of the area was in non-degraded forests. In the nearby high mountain wilderness of Pic Macaya, 100 percent of the land area was covered by virgin forest in 1956. By 1978 the area in virgin forest had declined to 14.8 percent and in 1984 only 3.6 percent of the "mountain wilderness" was in virgin forest. In the Anse Rouge to Jean Rabel study site in northwestern Haiti, where there is a long history of exploitation of the land, the region is much more arid. In 1956 no virgin forest cover remained; however, 45.4 percent of the area was still covered by some form of forest, 99.8 percent of which was severely degraded. Bare soil was exposed in 16.6 percent of the area. By 1978 only 23.6 percent of the region was forested, all of that severely degraded, and bare soil was exposed in 41.3 percent of the area. If these trends continue (Fig. 2), the study area in northwestern Haiti will be completely deforested in eight years, the large zone of southern Haiti in 14 years, and the "wilderness" of Pic Macaya in less than five years (Cohen 1984).

Forest cover in Haiti is now less than 1.5 percent, and the country is one of the most densely populated nations in the world with 720 people/ha of arable land. With 80% of Haiti's six million inhabitants living in rural areas, many of which are remote, the impact of the expanding population numbers on the countryside is especially significant. Each year the population increases by about two percent, meaning it will double every 35 years. Most of this expanding population will live in rural areas where agricultural production has decreased during the past ten years. As impossible as it is to imagine, these data suggest that Haiti may be completely deforested by the year 2000, with the exception of isolated fruit bearing trees and special areas where reforestation has succeeded. When the remaining areas of the countryside are deforested, even these environments will be threatened.

Flora and fauna of Haiti

Haiti is a country of exceptional ecological diversity. Nine different life zones, ranging from desert to mountain pine forests, are found in its borders, giving Haiti more ecological diversity than the eastern part of the United States (OAS 1972; Holdridge 1947). In these life zones grow 5,000 known species of plants, two thirds of which are woody, 600 species of ferns, and an estimated 300 species of orchids (Dod 1984). About 36 percent of the plants of Hispaniola are endemic (Liogier 1981).

Haiti also once had a rich and diverse fauna, but habitat destruction and the introduction of exotic species (cats, dogs, rats, mice, cattle, goats, sheep, pigs, and mongooses) have contributed to the extinction of a number of species. Of the 28 known species of terrestrial mammals once present in Haiti, for example, only two survive (*Plagiodontia aedium* and *Solenodon paradoxus*), and both are highly threatened. There still are, however, significant numbers of endemic plants, invertebrates, fish, amphibians, reptiles, and birds (see USAID reports summarized in Woods and Harris 1986). For example, there are 220 birds in Haiti, 75 of which are resident and 20 (28%) endemic to Hispaniola. One species, the Grey-crowned Palm Tanager (*Phaenicophilus poliocephalus*), is endemic to Haiti. Most of the surviving species of endemic birds now reside in what remains of Haiti's montane forests, including the largest remaining breeding colonies of the Black-capped Petrel (Woods and Ottenwalder 1983, 1986; Woods 1987). Coastal zones are the only ecosystems that have remained relatively untouched and include spectacular coral reefs, extensive mangrove wetlands (about 180 km^2), and large estuaries.

Past conservation efforts

Haiti's remaining wildlife is seriously threatened by: 1) habitat destruction by human activities (many life zones have lost most of their endemics); 2) the introduction of exotic species that successfully compete with endemics for the same ecological niches, or that prey on endemics; 3) structural and institutional weakness in the government of Haiti; 4) commercial export of plants and animals, particularly of coral, turtles, parrots, and boas.

The Haitian government organization responsible for the protection of forests, watersheds, the environment, coastal resources and natural resources is the Division of Natural Resources (DRN) of the Ministry of Agriculture (MARNE). Until recently, however, DRN has restricted its conservation efforts to regulating hunting and fishing, to small hillside terracing projects, and to very limited reforestation projects. Serious conservation actions have been limited by low budgets, overlapping institutional responsibilities, a lack of trained and motivated personnel, the lack of an agency fully responsible for conservation, and changing and contradictory government priorities and policies. The overwhelming poverty of the largely uneducated rural populations, and generally inefficient and sometimes corrupt government bureaucracies in the countryside, have effectively curtailed most conservation initiatives. Over the past five decades successive Haitian governments have passed legislation to protect the environment (Woods and Harris 1986). These laws forbid agriculture on lands with steep slopes, protect certain species of plants and animals, regulate hunting and fishing, create reserves and even small parks. Unfortunately, these laws have been generally neither observed nor enforced.

Alexander Wetmore, the American ornithologist who visited many of the most remote regions of Haiti in the 1920s, argued persuasively for the protection of the flora and fauna of Haiti, and the creation of national parks at Morne la Visite and Pic Macaya (Wetmore and Swales 1931). Erik Ekman, the famous Swedish botanist, also wrote and lectured in Haiti about the special importance of the Morne la Visite and Pic Macaya (1926,1928). Both of these biologists were active in Haiti during the occupation and administration of the country by American Marines (1915-1934), when agricultural programs modeled after the American system were established. It is a special tragedy that national parks or national forests were not established in Haiti at the same time, for, if they had, it would have been one of the most lasting legacies of the occupation and the landscape of the country today might be considerably different.

In 1979, the Institut de Sauvegarde du Patrimoine National (ISPAN) was created to protect and conserve Haiti's natural and cultural heritage. The observations of Wetmore and Swales (1931), and Ekman (1926, 1928) were evaluated, as well as a series of recommendations for national parks by Woods and Rosen (1977), and funding was obtained from the USAID Mission in Haiti for the establishment of two parks in the highest mountains in Haiti. To convince the Haitian government to create the parks and to convince USAID to fund the project, the vital role of the parks in protecting important watersheds and catchment areas was stressed. Included in this project was a biogeophysical survey of the potential national parks sites. The results of these surveys were presented as a series of reports to USAID in Haiti (Judd 1986; Franz and Cordier 1986; Gali and Schwartz 1986; MacFadden 1986; Thompson 1986; Woods 1986; Woods and Harris 1986; Woods and Ottenwalder 1986).

In 1983 two national parks were created by Presidential decree, one at Morne la

Visite and the other at Pic Macaya. The creation of the parks was also dependant on a consistent effort on the part of ISPAN, without which the parks would never have been established or continued. Many conservation measures were discussed, and Haiti has some very good conservation legislation. However, in Haiti there is a tendency to assume that once a measure has been enacted no further effort is necessary. Success depends, in large part, on dogged persistence, even in the face of apparent failure, over a number of years.

The purpose of establishing the new parks was (in order of priority) the: 1) protection of two of Haiti's most important watercatchment/watershed areas and maintenance of natural ecological processes, critical for sustainable agricultural and economic development; 2) protection of Haiti's flora and fauna, particularly the endemics; 3) protection of two of the most beautiful and spectacular sites of natural beauty on the island, an invaluable and unique national heritage; 4) creation of living laboratories for scientific research; and 5) encouragement of national and international tourism. A brief description of the major national parks follows.

National parks in Haiti

Parc National la Visite (Figs. 1,3,4)
AREA: Approximately 3,000 ha. ESTABLISHED: April 1983 by Presidential decree. ADMINISTRATION: Jointly by ISPAN and MARNE. VEGETATION: Pine forest, savannah and montane wet cloud forest.
COMMENTS: Located high in the western section of the Massif de la Selle, Haiti's highest mountain range, Parc National La Visite includes three peaks: Morne la Visite (2,170 m), Morne Cabaio (2,282 m), and Morne Tete Opaque (2,268 m), and a high, saucer-shaped plateau mostly covered by a thick top soil underlaid by limestone and conglomerate rock. The northern part of the park is characterized by 1,000 m limestone cliffs, many of which are overgrown with virgin montane cloud forest. On these steep northern cliffs (Fig. 3) is one of three known breeding colonies of Black-capped Petrels, which nest in burrows, as well as Hispaniolan Hutias (*Plagiodontia aedium*), called the Zagouti in Haiti.

Within the park's 3,000 ha are pine forests (*Pinus occidentalis*), savannahs, montane cloud forests, and the remnants of an endemic juniper forest (*Juniperus ekmanii*). The Ekman juniper, a tall and beautiful conifer, was once abundant within the park. A number of stumps can still be found, and within the past five years several large specimen trees were still present. Judd (1987) lists the species as "rare/nearly extinct". A second rare juniper (*J. urbanii*), more shrubby in growth form, is also found in the Massif la Selle region. The presence of these rare endemic junipers gives special importance to Parc National La Visite as a biological refuge. There are many other endemic plants and orchids, numerous springs, cave systems and sinkholes, the headwaters of two major rivers (Rivière Blanche, the highest river in the Caribbean, and Rivière Grise), as well as many waterfalls and small canyons, making it an area of unique and varied natural beauty. The development of this park has just been started, and an adequate infrastructure and management program have yet to be established. Some peasant squatters live within park boundaries and contribute to the destruction of habitat by cutting trees and using inappropriate and destructive agricultural practices. Recently, a new pattern of deforestation has evolved: the destruction of pine forest by fire following by the introduction of gardens. Figure 4 illustrates the rate of deforestation in the central region of the park over the past decade.

Parc National Pic Macaya (Figs. 1,5)
AREA: Approximately 5,500 ha. ESTABLISHED: April 1983 by Presidential decree. ADMINISTRATION: Jointly by ISPAN and MARNE. VEGETATION: Montane cloud and wet forest, pine forest, lower montane wet forest.
COMMENTS: Parc National Pic Macaya, located near the western tip of the Haiti's southern peninsula, provides the water supply for the Plaine des Cayes, Haiti's most productive agricultural region. Two major east-west and extremely steep mountain ridges, Morne Macaya (Pic Macaya, 2,347 m) and Morne Formond (Pic Formond, 2,250 m), form the center of the park (Fig. 5a). Due to its remoteness and inaccessibility, the park site has remained undisturbed until recently, and still has large tracts of impenetrable montane wet and cloud forest. The central area of the park is surrounded by a highly disturbed region, and the steep slopes of the highest mountains are now being deforested for gardens (Fig. 5b).

This park contains a rich and varied fauna and flora with a high degree of endemism. Six new species of orchids have recently been discovered in the park. Pines 45 m high and nearly two m in diameter loom over the peaks and trap moisture from the almost ever present cloud cover, giving some parts of the park an estimated annual precipitation rate of six meters. Seven major rivers begin within park boundaries, rivers that are essential to the agricultural production of the surrounding areas, particularly the Plaine des Cayes. The park contains a recently discovered breeding colony of Black-capped Petrels, and provides habitat for numerous other birds. Endemic mammals such as *Plagiodontia aedium*, and *Solenodon paradoxus*, as well as numerous endemic orchids, plants, reptiles and mollusks are found within the park.

The first steps towards developing the park have been taken. A boundary survey has been completed and ministerial decrees have made cutting trees in the area illegal, but the deforestation of the area has continued. Only a handful of peasants actually live in the central portion of the park. The deforestation of the Macaya area and its disastrous effects have become widely known throughout Haiti due to a number of conferences, television interviews, and newspaper articles. It is hoped that public pressure will push the government to act more effectively to protect this vital zone.

Recently, USAID has agreed to provide funding for the continued development and protection of the park and surrounding Macaya Biosphere Reserve. These funds are in addition to USAID funding for a larger project designed to protect the entire Massif de la Hotte watershed zone, in particular the Plaine des Cayes. The USAID funds for Parc National Pic Macaya were originally designated to be used by a contract non-profit organization (the Florida Museum of Natural History of the University of Florida) to build up the infrastructure of the Haitian government organizations responsible for the park, and to assist in the development of the park. As discussed in the following section, this has not worked according to plan.

Parc National Historique la Citadelle, Sans Souci, Ramiers (Figs. 1,6)
AREA: 2,200 ha. ESTABLISHED: 1968 by Presidential decree. ADMINISTRATION:

ISPAN and National Tourist Office. VEGETATION: Subtropical wet forest. OBJECTIVES: 1) Protection of national historic monuments. 2) Promotion of tourism. 3) Protection of endemic flora and fauna.

COMMENTS: Although this park was established primarily to assure the restoration and conservation of Haiti's most important historic monuments, the Citadelle (Fig. 6a), the Palace of Sans Souci, and the fortifications of Ramiers, all built by King Henry Christophe in the early 1800s, it does have an important role as a natural park. The park contains subtropical wet forest and rugged karst outcroppings that serve as a refuge for many upland and lowland birds (Woods and Ottenwalder 1983). The ecology of the park, however, has not been protected and is deteriorating. The park is adjacent to the Plaine du Nord, one of the most important agricultural areas in Haiti (Fig. 6b), and should become part of a biosphere reserve if it is going to be properly protected (Sergile in prep.).

Problems facing Haitian parks

Unfortunately, progress in making these parks functional has been slow. Some of the problems include: a) the decree creating the parks failed to assign the final responsibility for the administration of the parks to a single government agency resulting in a conflict between MARNE and ISPAN as to which group has the primary responsibility for national parks; b) very few people in high positions in the government are concerned about protecting Haiti's natural heritage; c) the parks project was taken over from ISPAN by INAHCA, a highly political organization created and supervised by ex-President Jean-Claude Duvalier's wife (after the departure of the Duvaliers, ISPAN regained control of the parks project); d) the deforestation of the park sites was carried out not only by the peasants living in and around the parks, but by politically connected businessmen; e) ISPAN and MARNE have neither adequate and sufficiently trained staff, nor sufficient budgets to effectively develop and protect the parks; f) in spite of this already inadequate funding, the United States government has withdrawn all of its USAID financial support of MARNE and ISPAN (after December 1987) because of political irregularities surrounding the presidential election in Haiti; g) and finally, a fire destroyed a significant part of the main MARNE building at Damien, with the loss of some important papers, maps, and documents that relate to the parks project.

Perhaps the most difficult aspect of establishing parks in Haiti is the complex problem of displacing extremely poor peasants from the park sites and limiting their activities in and around the parks. This is particularly true in a period of social and political unrest such as has been experienced following the departure of the Duvalier family. In order to undertake conservation programs in remote but populated regions of Haiti, reasonable and worthwhile alternatives must be given to people who depend upon working the land to scrape out a living. The problems faced by the peasants suddenly became much more severe in the last decade following the planned destruction of the Creole pigs of the country, which carried a disease which threatened North American pigs. This took away one of the peasants most dependable sources of food and income, with disastrous consequences. A peasant previously was able to sell a pig when cash was needed for seeds or a financial emergency. The increased pressure on peasant to survive has placed increased pressure on the surrounding environment, and is one reason mango trees are now being cut in a desperate last ditch effort to generate cash. To compensate for this problem, the planting of communal forests next to the park sites to provide lumber and fuelwood is being considered, and it is hoped that Christmas trees (endemic conifers) can be planted so that the peasants can profit from this lucrative seasonal market at the same time that stocks of junipers and pines are replenished. Some peasants will work for the parks, while others will be given training in more profitable and appropriate agricultural methods. Another solution might be the formation of a rural conservation corps, perhaps modeled after the American Civilian Conservation Corps of the 1930's. This corps would employ peasants, providing them with an alternate source of income, to build up rural infrastructure such as schools, clinics and roads, and undertake major conservation projects such as reforestation and communal forests.

Experience in Haiti has shown that involving local peasant leaders in the planning and execution of such projects, although involving some risks, helps to assure success. Experience has also shown that informing the local population of the goals and methods of projects and soliciting their opinions is very helpful. Severe measures will have to be used to stop those peasants who refuse to stop destroying forests and valuable habitats.

Recently, DRN has initiated a number of new projects to protect the environment. Among these are: 1) a national forestry project designed to protect and develop forest resources; 2) a watershed protection project; 3) a meteorological project designed to obtain base data about weather patterns; 4) a freshwater resources inventory project. DRN is also fully participating in the National Parks Project by providing personnel, technical assistance, and on site services.

The national conservation strategy

The publication of the World Conservation Strategy in 1980 by the International Union for the Conservation of Nature (IUCN) was followed in 1984 by the publication of a special document outlining the methodology for the preparation of national conservation strategies. In May 1985 the United Nations Environment Program (UNEP) invited the Haitian government to prepare a national conservation strategy. An inter-agency government team, which included representatives of DRN, ISPAN, and the Ministries of Public Health, Mines and Public Works, was formed for this purpose under the direction of the Ministry of Plan.

In April of 1987 a draft document was finished. Among the recommendations formulated were the following. 1) The creation of an independent government agency to be responsible for the planning and execution of conservation policies and programs. 2) The promulgation of more appropriate and adequate conservation legislation. 3) The preservation and protection of natural ecosystems. 4) The protection of the endemic gene pool. 5) An increase in scientific research. 6) The establishment of a conservation education program. 7) The integration of the national conservation strategy into the national development program to assure its effective execution and to assure that development projects are environmentally sound. 8) The protection of watersheds. 9) The increase of forest reserves. 10) The development of national parks.

Unfortunately, just after the publication of the National Conservation Strategy, the Ministry of Plan was abolished to be replaced by a Commissariat of Planning. Changes in personnel, priorities and the physical location of the planning organization stopped any further development or execution of the National Conservation Strategy, which although incomplete, was a very positive step towards a functional and integrated conservation policy.

Other measures

Recently, the Government has removed the tariff on imported wood in an effort to lower its price and thus reduce pressure on local forests.

The Division of Natural Resources (DRN) of MARNE has introduced measures to control the trade in animals and plants by making it illegal to export native species of plants and animals.

In cooperation with ISPAN, DRN has drawn up lists of endangered species of plants and animals and of natural sites in need of protection. It is hoped that at least some of these sites can be made into national parks when adequate funding and personnel become available.

USAID has developed a $15 million project to assure the protection of the Massif de la Hotte watershed. The project also provides for the protection and appropriate agricultural development of the buffer zone surrounding the Park by working with selected non-governmental organizations such as ORE and UNICOR. It is hoped that the Macaya Park and surrounding areas will be managed as a biosphere reserve.

USAID is continuing its agro-forestry outreach project which finances the planting of about ten million trees each year. Some of these trees are fruit trees and others are trees planted for lumber and charcoal which should relieve the pressure on Haiti's remaining forest. Other donor countries are also becoming interested in forestry projects.

Recently, some of the newly formed political parties have included conservation activities in their political programs. The largest labor union, CATH, has called out for emergency measures to start reforestation projects and to protect what remains of Haiti's forests.

The general public is now aware of the disastrous consequences of continuing and progressive environmental degradation. A new conservation lobby group called the "Fédération des Amis de la Nature" (federation of the friends of nature) has formed and is planning to fight to reforest Haiti. But, in general, this consciousness has yet to be translated into positive actions either by private groups or the government.

Current status of national parks in Haiti

The discussion above makes it abundantly clear that many serious problems face Haiti's environment and expanding human population. The effort to set aside national parks in the country is floundering as well. As of April 1988 the governmental agency responsible for the national parks in Haiti is ISPAN. In an effort to improve the quality of the administration of the parks, a special high administrative post has been established by ISPAN with sole responsibility for the national parks program. It appears as if a highly qualified Haitian with international experience in conservation and the management of natural resources will accept this new position. However, all USAID funding to ISPAN for the national parks has ended. In addition, USAID has decided to change the emphasis of its funding. Beginning in April 1988 it committed its funding to non-governmental organizations working on rural development projects in the areas surrounding Parc National Pic Macaya in the hope of luring peasant activities away from the park itself. This innovative (but perhaps ecologically dangerous) step is clearly aimed at the core of the problem, the poor peasant farmers living in the Macaya region. However, it further isolates the government of Haiti from the problem, and does not support ISPAN nor train the future generation of personnel with the tools needed to take charge of their own environmental destiny. The new hope for conservation and the preservation of the unique natural patrimony of Haiti is increasingly getting displaced by the consequences of the unfortunate political events of the past two years, and the resulting international political countercurrents. The political paralysis is resulting in an accelerating ecological decline.

Conclusions

If urgent measures are not undertaken in the immediate future, the present environmental crisis in Haiti will attain tragic proportions, resulting not only in the loss of great numbers of species of plants and animals, but in the loss of human life and the breakdown of social and economic systems. The fragility of Haiti has been vividly demonstrated by the events that have crippled the country as the scheduled national election of November 29, 1987 approached. Bloodshed, human suffering, accelerated environmental degradation, and finally the withdrawal of international aid funds all have occurred in response to a lack of responsible government, and the breakdown in the electoral process. The ecosystem of the country is very disturbed, perhaps irrevocably so. With all of these problems, Haiti provides a useful case study of how to develop and maintain conservation strategies in the face of such serious socio-economic problems. Few countries face the same problems to the extent that Haiti does, but all of the islands of the Antilles are vulnerable, and the example of Haiti is an important lesson for all who are concerned with the conservation of natural resources on islands.

It is not, however, a hopeless cycle. There may be hope that if a strong environmental and conservation program is implemented these trends might be reversed, and Haiti's precious natural heritage partially saved. The key to preserving the greatest amount of biological diversity remaining in Haiti is protecting large blocks of Mature Broadleaved Forest and Mature Pine Forest from deforestation. These habitats form the heart of the Massif de La Hotte (Pic Macaya), and Massif de La Selle (Morne la Visite). In both of these areas (Figs.1,3,4,5) the broadleaved forest protects the soil and helps hold the water from torrential cloudbursts, as well as the dripping of dew drops off of the pine needles, to be slowly released into the rivers. The pine forest along the peaks and ridges helps comb additional moisture from the ever present clouds, increasing the overall amount of precipitation and quality of the watershed (Vogelmann et al. 1968). It is clear, therefore, that finding ways to meet the future growth and development needs of Haiti does not have to be at the expense of much of the remaining endemic flora and fauna of the country. It is in the best economic interest of the future of Haiti that broadleaved and pine forests flourish in carefully selected and protected mountain habitats.

The lost horizons of Haiti are the vast areas of forest that once made Haiti the Perle des Antilles and the Magic Isle. These forests are gone forever, and with them countless scores of endemic plants and animals. The new efforts to develop a meaningful watershed conservation program and national parks service in Haiti are strong indications that the Government of Haiti is serious about the need to preserve habitats in conjunction with watershed management and reforestation. Another good sign is the availability of factual documentation of the scope of the developing disaster in the natural systems of

Haiti. For the first time data on the extent of the pattern of deforestation are available, and the specific ways necessary to improve the status of the endangered and threatened flora and fauna are understood. With these tools, and an improved awareness in Haiti of the link between the conservation of natural resources and the quality of life for all, comes a glimmer of hope. The remaining natural resources of Haiti may yet survive to be part of a new century, and be available for future generations of biogeographers to study.

Haiti is still a country of great biological diversity and a unique natural heritage. To preserve this heritage for not only Haitians but for the rest of the world, a number of specific conservation measures are suggested as follows.

Developmental measures

1). Provide alternate sources of employment and income for the peasant class, such as the creation of a rural conservation corps modeled after the U.S. Civilian Conservation Corps of the 1930's.

2). Establish communal forests to provide a renewable source of energy and animal habitat.

3). Enforce existing environmental and conservation laws strictly, so that agricultural activities in certain extremely degraded areas can benefit from the presence of nearby conservation zones without further degrading these zones.

4). Provide alternate sources of energy to replace charcoal and fuelwood. For example, subsidized kerosene or natural gas could replace wood in dry cleaners and bakeries.

5). Control or subsidize the price of imported wood and charcoal.

6). Establish rural environmental education programs, especially in areas near conservation zones.

7). Increase the salaries of government technicians working on environmental projects to insure decent salaries and working conditions.

8). Undertake cadastral studies of the most ecologically critical areas to determine land ownership and to provide secure land titles for peasants.

9). Precede all development projects with an environmental impact study that includes a long term environmental component.

10). Base development projects on an integrated agriculture/conservation approach reflecting the realities of watersheds and ecosystems.

11). Involve the local community in the planning and execution of development projects and conservation measures.

12). Emphasize planting native species in reforestation projects in mountain zones.

Special conservation measures

1). The I.U.C.N. World Conservation Strategy should be adopted as a guideline for conservation activities.

2). An independent National Park Service (such as the new National Parks Division of ISPAN) should be created with direct responsibility for the planning, creation and management of all national parks. Under this service the national park system should be expanded to include representatives of all of Haiti's life zones. Etang Saumâtre, a large freshwater lake providing habitat to significant populations of water fowl, crocodiles and iguanas, and les Arcadins, a spectacularly beautiful coral reef system, would be two good choices for the next national parks.

3). Scientific research on Haiti's flora and fauna should be significantly increased to determine the minimal habitat needs of animal species, particularly those of the threatened endemics, and to determine what constitutes the natural ecosystems in Haiti, especially what species are part of "virgin plant associations." Other areas of research might include determining the minimal species survival populations for endemics and the exact role of introduced species in the extinction of endemics. In addition, coastal ecosystems, which are the least disturbed in Haiti, should be analyzed and inventoried.

4). The survival of certain threatened species of plants and animals should be assured by the establishment of a program of controlled artificial reproduction.

5). A national program of conservation education should be established for adults and children in both rural and urban areas using the services of public and private schools and the mass media.

6). All fragile areas in Haiti, and especially the regions surrounding the national parks, should be managed as "biosphere reserves" with careful planning resulting in well supervised zoning of the complex topography of the entire region.

In very poor, highly populated and environmentally degraded countries such as Haiti, it is difficult to undertake conservation measures, especially when such measures must limit or alter the basic economic activities of the impoverished peasant class and change the daily habits of much of the rest of the population. It is virtually impossible to ask a peasant farmer not to exploit certain areas in order to assure the survival of a rare or endangered species of plant or animal when the very existence of the peasant's family is threatened by hunger and poverty. It also is unreasonable to expect any government of such a country to undertake conservation measures to save endangered species, especially if such measures involve direct action against the poorest segments of the population. It is for these reasons that we feel that effective conservation strategies, and especially the long term stability of "national parks" in Haiti, will come to fruition only if the efforts are coupled with activities that improve the status of peasants living adjacent to the parks. The "national parks" should be "Core Areas" of well planned biosphere reserves where the emphasis is on sustainable agriculture as well as the conservation of natural resources and endemic species.

We feel that the most productive approach in such situations is first to evaluate and, if possible, quantify the economic cost of environmental degradation so that all concerned, both in the government and private sectors, understand that the protection of the environment is an economic necessity, and a prerequisite for the very survival of the country. Conservation must be directly linked to economic issues and development. It was only after convincing the Haitian government and USAID that protection of mountain water catchment zones was necessary for the continued agricultural development of three of Haiti's most productive plains, that the La Visite and Macaya parks were created and funding obtained for their establishment. The study by Lowenstein (1984) on the economic effects of deforestation in the region surrounding Pic Macaya on the Plaine des Cayes estimated the impact of deforestation to be nine million dollars each year, and was a very effective argument for the continued support of Parc Macaya.

We suggest that further studies be undertaken to evaluate the cost of environmental degradation, and the economic cost benefits of environmental rehabilitation and conservation measures. These studies would be invaluable for the cause of conservation in Haiti, and many other third world countries, where the relationships between the conservation of natural resources and economic and development programs of high national priority are poorly understood.

Acknowledgements

We thank Lance Jepson, Catherine McIntyre, and Robert Wilson of the US Agency for International Development in Haiti, and the International Foundation for the Conservation of Birds for their support of conservation efforts in Haiti. We thank ISPAN, particularly Albert Mangones, and Robert Cassagnol, Edmond Magny, and Raoul Pierre-Louis of MARNE. We also thank the following people who contributed to this project. Dan and Tia Cordier, Donald Dod, Tom Greathouse, John Hermanson, Kevin Jordan, James Keith, Jose Ottenwalder, William Oliver (of the Jersey Wildlife Preservation Trust), and the students of the University of Vermont Field Naturalist Program. We are also appreciative of the support of the Florida Museum of Natural History and University of Florida. This paper is dedicated with appreciation to Dr. Leonce Bonnefil for his dedication as a teacher and researcher, and as someone who saw the need for conservation in Haiti before it was too late.

Literature cited

Burns, L.V. 1954. Report to the government of Haiti on forest policy and its implementation. Report 346, United Nations FAO, Rome.

Cohen, W.B. 1984. Environmental degradation in Haiti: an analysis of aerial photography. Report, USAID/Haiti, Port-au-Prince, Haiti. 35 pp.

Dod, D. 1984. Massif de la Hotte Isla Peculiar: Orquídeas nuevas iluminan su historia. Moscosoa Contribuciones Cientificas del Jardin Botanico Nacional (Moscosoa), Santo Domingo 3:91-100.

Direction d'Aménagement du Territoire et Protection de l'Environement(Ministère du Plan). 1982. Cartographie Thématique d'Haiti/Notice explicative. Port-au-Prince, Haiti.

Ekman, E.L. 1926. Botanizing in Haiti. U.S. Naval Medical Bulletin 24(1):483-497.

_____. 1928. A botanical excursion in La Hotte, Haiti. Svensk Botanisk Tidskrift 22(1-2):200-229.

_____. Unpublished field notes. Catalogue of Hispaniolan plants. Department of Phanerogamic Botany, Swedish Museum of Natural History, Stockholm.

Gali, F., and A. Schwartz. 1986. The butterflies (Lepidoptera: Rhopalocera) of Morne la Visite and Pic Macaya, Haiti. USAID/Haiti, Port au Prince, Haiti. 16 pp.

Holdridge, L.R. 1947. The pine forest and adjacent mountain vegetation of Haiti considered from the standpoint of a new climatic classification of plant formations. Unpublished PhD Dissertation. University of Michigan, Ann Arbor, MI. 186 pp.

Judd, W.S. 1986. Floristic study of La Visite and Macaya National Parks, Haiti. USAID/Haiti, Port-au-Prince, Haiti. 98 pp.

_____. 1987. Floristic study of Morne la Visite and Pic Macaya National Parks, Haiti. Bulletin of the Florida State Museum 32(1):1-136.

Liogier, A.H. 1981. Flora of Hispaniola: Part 1. Phytologia Memoirs 3:1-218.
Lowenstein, F. 1984. Le déboisement du périmètre "Pic Macaya" et son impact sur la Plaine des Cayes. Report, MARNE, Port-au-Prince, Haiti. 20 pp.
MacFadden, B.J. 1986. Geological setting of Macaya and la Visite National Parks, southern peninsula of Haiti. USAID/Haiti, Port au Prince, Haiti. 33 pp.
Ministère du Plan. 1986. Stratégie National de Conservation. Port-au-Prince, Haiti.
OAS. 1972. Haiti: Mission d'Assistance Technique Intégrée. Secretary General, Organization of American States, Washington, D.C., 656 pp.
Thompson, F.G. 1986. Land mollusks of the proposed national parks of Haiti. USAID/Haiti, Port au Prince, Haiti. 19 pp.
USAID. 1987. Country Environmental Profile of Haiti. USAID, Port au Prince, Haiti. 120 pp.
Vogelmann, H.W., T. Siccama, D. Leedy, and D.C. Ovitt. 1968. Precipitation from fog moisture in the Green Mountains of Vermont. Ecology 49:1205-1207.
Wetmore, A. and B. Swales. 1931. The birds of Haiti and the Dominican Republic. Bulletin of the U.S. National Museum 155:483 pp.
Woods, C.A. 1986. The mammals of Parc National La Visite and Parc National Pic Macaya, Haiti. USAID/Haiti, Port-au-Prince, Haiti. 80 pp.
_____. 1987. The threatened and endangered birds of Haiti: lost horizons and new hopes. Proceedings of the Second Delacour/IFCB Symposium. pp. 385-430.
_____, and L. Harris. 1986. Stewardship plan for the national parks of Haiti. USAID/Haiti, Port-au-Prince, Haiti. 272 pp.
_____, and J.A. Ottenwalder. 1983. The montane avifauna of Haiti. Proceedings of the Jean Delacour/IFCB Symposium. pp. 576-590 + 607-622.
_____, and J.A. Ottenwalder. 1986. The birds of Parc National La Visite and Parc National Pic Macaya, Haiti. USAID/Haiti, Port-au-Prince, Haiti. 241 pp.
_____, and R. Rosen. 1977. Evaluation Biologique d'Haiti: Statut du *Plagiodontia aedium* et *Solenodon paradoxus*- Recommendations en ce qui concerne les reserves naturelles et les parcs nationaux. MARNE, Port-au-Prince, Haiti. 32 pp.

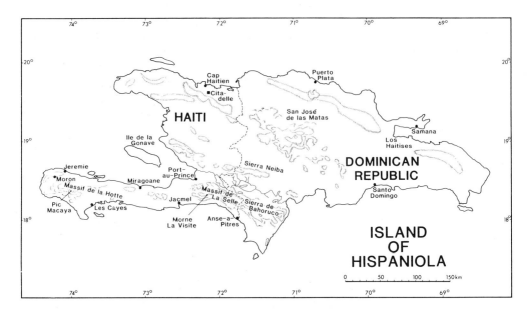

Figure 1. Map of Hispaniola showing the locations of Parc National Pic Macaya surrounding Pic Macaya, Parc National La Visite surrounding Morne la Visite in southern Haiti, and Parc National Historique la Citadelle surrounding the Citadelle in northern Haiti.

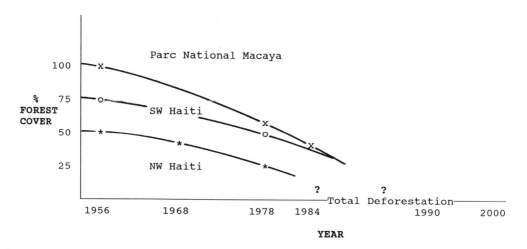

Figure 2. Rate of deforestation in three selected regions of Haiti (Modified from Cohen 1984).

Figure 3a. Aerial view of the Massif de la Selle. Morne la Visite (2,170 m) is near the cloud in the upper right corner (west). Morne Cabaio (2,282 m) is the dark mountain in the center-right. Pic la Selle (2,674 m), the highest mountain in Haiti, is in the upper left. In the distance is the Sierra Baoruco of the Dominican Republic. (Woods, January 1977)

Figure 3b. Aerial view of Parc National la Visite. Morne la Visite is in the foreground, Morne Cabaio in the left-center, and Tête Opaque (2,268 m) is near the top, with the pine covered plateau to the right (south). Black-capped Petrels nest on the steep cliffs to the left (north). (Woods, May 1983)

Figure 4a. View of the interior of Parc National la Visite in 1977 to show the extent of the pine forest on the plateau. (Woods, October 1977)

Figure 4b. Same view of the interior of Parc National la Visite in 1983 to show the extent of deforestation within the park boundaries. (Woods, October 1983)

Figure 5a. View of Pic Formond (2,250 m) at the top left, and Pic Macaya (2,347 m) at the top right in the center of Parc National Pic Macaya. (Woods, January 1984)

Figure 5b. View of Morne Cavalier (1,500 m) and the ridge of Formond above the Plaine Formond on the south edge of Parc National Pic Macaya. The forest in the foreground persists because it is in a region of karst topography. (Woods, January 1984)

Figure 6a. View of La Citadelle Laferrière from the steep slopes of the karst mountains north of the monument. Note how much forest cover remains in spite of the long history of occupation of the area, and the high population density. (Woods, June 1982)

Figure 6b. View from the Citadelle towards Cap Haitian (at the left corner of the bay) and the Plaine du Nord. (Woods, March 1982)